Nuclear Analytical Methods in the Life Sciences

Nuclear Analytical Methods in the Life Sciences

Edited by

Rolf Zeisler

National Institute of Standards and Technology,
Gaithersburg, MD

and

Vincent P. Guinn

University of California,
Irvine, CA

Humana Press • Clifton, New Jersey

This work also appears as volumes 26 and 27 of the Humana Press journal,
Biological Trace Element Research Editor-in-Chief: Gerhard N. Schrauzer

ISBN: 0-89603-202-7

Crescent Manor
PO Box 2148
Clifton, NJ 07015 USA

Printed in the United States of America.

This publication is printed on acid-free paper. ∞

Preface

The 1989 International Conference on Nuclear Analytical Methods in the Life Sciences was a continuation of a series of conferences held by the International Atomic Energy Agency. The first took place in Amsterdam in 1967, the second in Bled in 1972, and the third in Vienna in 1978. The aim of these conferences has been to stimulate discussions between scientists who are working as biologists, environmentalists, and physicians, and those who are working on the advancement of nuclear analytical techniques.

The 1989 Conference was held at the National Institute of Standards and Technology (NIST, formerly the National Bureau of Standards) in Gaithersburg, Maryland. The focal point of the Conference was the presentation of results from multidisciplinary research involving nuclear analytical techniques and their applications to the life sciences. We have obtained contributions from life science fields that relate the nuclear analytical methods to a broad scope of biological, medical, and environmental applications. Deliberately, our definition of nuclear analytical techniques was made flexible. Methods and applications were treated in a more comprehensive way than those at established meetings. Particular consideration was given to contributions from developing countries.

We are indebted to many people and organizations for their assistance in making this symposium possible. The Conference was organized by the US Department of Commerce's NIST in cooperation with the International Atomic Energy Agency, by supplying both financial support and scientific expertise. The meeting was cosponsored by the American Nuclear Society, the US Department of Energy, and the Food and Drug Administration, who provided both organizational and financial support.

We hope that the results of this Conference, presented here, will stimulate the development of new collaborative research efforts between the life sciences and analytical fields. A continuation of this series of conferences will be a measure of the success of this interdisciplinary collaboration.

Rolf Zeisler and Vincent P. Guinn

CONTENTS

Neutron Activation and Related Methods

A Short History of Nuclear Activation Analysis

VINCENT P. GUINN

Department of Chemistry, University of California, Irvine, CA 92717

Received April 17, 1989; Accepted November 16, 1989

ABSTRACT

The major developments in the field of nuclear activation analysis, from 1936 to 1989, are discussed. The developments are grouped into five consecutive time periods. The impact of various scientists on the development of the field in the first 35 years is also discussed.

Index Entries: Nuclear Activation Analysis; Nuclear Activation Analysis, history of; Nuclear Activation Analysis method, origination of; the nuclear reactor, impact on NAA; the NaI(T1) detector, impact on NAA; the germanium detector, impact on NAA; other Nuclear activation analysis developments; application areas of NAA; major early contributors to NAA.

INTRODUCTION

It all started some 53 years ago in Copenhagen—with the 1936 publication of the first paper on the method of nuclear activation analysis by George Hevesy and Hilde Levi (*1*). Many activation analysts had the great pleasure of meeting Hilde Levi and hearing her fascinating account of their discovery of the method—activating rare earth elements with thermal neutrons from a radon-beryllium source, followed by β-counting with a Geiger counter, with radionuclide identification by halflife—at the Seventh International Conference on Modern Trends in Activation Analysis. To celebrate the 50th anniversary of this important discovery, the MTAA-7 conference was held in Copenhagen in 1986.

MAJOR ADVANCES IN THE METHOD, 1936 TO DATE

In the following, the NAA abbreviation will be used for the more general topic of Nuclear Activation Analysis although it is frequently used just for activation analysis using neutrons as the bombarding nuclear particle. Thermal neutrons have been, and continue to be, the most generally useful bombarding particle, but considerable useful work has been, and is today, accomplished with fast neutrons, energetic charged particles, and energetic photons as the bombarding particles. Most NAA work utilizes the detection and measuring of induced radioactive species, but considerable valuable work is also based on prompt-gamma emissions, rather than on radioactive-decay emissions.

It is convenient to break the history of NAA down to approximate time periods that featured major advances brought about by developments of new kinds of equipment. These periods are each briefly summarized below.

1936–1944

In these early years, the development of the great potential of the NAA method was limited and relatively slow, owing to lack of high-flux neutron sources and lack of gamma-ray spectrometry equipment. With the advent of charged-particle accelerators (cyclotrons, Cockcroft-Walton accelerators, and Van de Graaff accelerators) in the 1930s, charged-particle activation analysis (CPAA) became possible, but the absence of gamma-ray spectrometry equipment was still a severe limitation. In this early period, the NAA of multielement samples had to be conducted via tedious, though elegant, postirradiation radiochemical separations with carriers.

1944–1950

A major step forward was made in 1944 with the availability of high fluxes of thermal neutrons in the first research-type nuclear reactor—the X-10 reactor at the Oak Ridge National Laboratory. Thermal-neutron fluxes of $5 \times 10^{11} n/cm^2/s^1$ became available (as opposed to maximum fluxes of only about $5 \times 10^5 n/cm^2/s$ available with isotopic neutron sources)—providing a million-fold increase in the potential measurement sensitivity of the NAA method. Early in this period, the advent of accelerator sources of high-energy photons (via the use of electron linear accelerators) is also of note. However, in both of these cases, practical work was still restricted to tedious radiochemical separations with carriers, followed by counting with Geiger or proportional counters. In addition, full utilization of the high thermal-neutron fluxes of the nuclear reactor was limited by World War II security restrictions and the presence of only one reactor (the ORNL X-10 reactor) in the world until near the end of this period.

1950–1960

This period was marked by the worldwide proliferation of research reactors and by the advent of the thallium-activated sodium iodide scintillation detector (small crystals first becoming commercially available about 1950). These developments were of great importance to the advance of the NAA method. Coupled with the newly-developed pulse-height analyzers (PHAs)—first, vacuum-tube circuits and then transistor circuits, gamma-ray spectrometry became readily available and with a detector of much higher efficiency for the counting of gamma rays than was previously possible.

Progress in the field of gamma-ray spectrometry was very rapid during this period, soon leading to its use for the instrumental NAA (INAA) of multielement samples. The author still remembers the fun and thrill of producing his first gamma-ray pulse-height spectra in the early 1950's (at his Shell Development Company laboratory)—using only a tiny NaI(T1) canned crystal, coupled by black tape to a small photomultiplier tube (PMT) and using only a single-channel PHA. Scanning a complete pulse-height spectrum one channel at a time, manually, was very tedious—so the author attached a synchronous motor to the low-energy side of the analyzer window setting. This made it possible to scan a complete pulse-height spectrum, with reasonable energy resolution, in about 20 min. The analyzer output was to a rate meter and a strip-chart recorder. In another few years during this period, the first multichannel PHAs became available—first a 20-channel instrument,with 20 outputs, that had to be recorded manually. Soon, PHAs of 50, 100, 128, 256, 400, and 512 channels became available, with digital outputs to a printer. On a personal note, in the mid-1950s, the author began using the ORNL X-10 reactor for Shell's INAA studies, followed by gamma-ray spectrometry with their 20-channel PHA. Then, he began using the 250 kw TRIGA Mark I reactor at General Atomic for Shell's INAA studies, using their 256-channel PHA.

By 1956, the author was able to do his own Shell Development Company INAA work with his own neutron source (with a thermal-neutron flux of 10^8 n/cm^2/s^1 from his newly-acquired 3 MeV Van de Graaff electron accelerator, with a water-cooled gold converter to bremsstrahlung, a 5-in (12.7 cm) cube of metallic beryllium, and a large drum of *p*-terphenyl moderator). Using 30 cm^3 samples in a ring around the beryllium, detection limits in the range of 1 ppm could be attained for quite a few elements—even at this rather low thermal-neutron flux. For gamma-ray spectrometry, he used a 3-in $\times$ 3-in NaI(Tl) well detector (7.62 cm $\times$ 7.62 cm) and a newly-acquired 100-channel PHA. Soon, this INAA system was put into routine operation (using the Van de Graaff and the PHA daily in the evenings), analyzing a great variety of samples from Shell's research projects for a large number of major, minor, and trace elements. By 1960, the annual publication rate of NAA papers had climbed to about 100—compared with only about 10 in 1950, and only a few in 1940 (2).

1960–1970

The INAA method, based on NaI(Tl) gamma-ray spectrometry, although a powerful technique, was limited by the rather poor energy resolution of the NaI(Tl) scintillation detector—even with the help of least-squares computer fitting of the pulse-height spectra. A dramatic advance took place about 1960 when the lithium-drifted germanium semiconductor detector (Ge(Li)) was invented. This new kind of detector for gamma radiation provided an energy resolution some 20–30 times better than was possible with a NaI(Tl) scintillation detector (i.e., with total absorption peaks having full-width-at-half-maximum (FWHM) values only 1/20 to 1/30 as large as corresponding NaI(Tl) peaks).

Although the Ge(Li) detector is about as efficient as the NaI(Tl) detector for gamma radiation—on an equal-volume basis—large Ge(Li) detectors are much more difficult and, hence, much more expensive to make than NaI(Tl) detectors. For example, the largest germanium detectors available today only have sensitive volumes of about 150 cm^3, whereas even the common 3-in × 3-in (7.62 cm × 7.62 cm) NaI(Tl) crystal has a volume of 347 cm^3. The author remembers well his first Ge(Li) detector, purchased early in the 1960s while he was with General Atomic: it only had a sensitive volume of 0.9 cm^3! However, its energy resolution was marvelous! By 1970, Ge(Li) detectors with sensitive volumes approaching 100 cm^3 had become commercially available, multichannel PHAs of 4096 channels had become commonplace; the INAA method was now the dominant form of the NAA method, applications in a great variety of fields were mushrooming, and the annual publication rate had risen to about 1000 (2).

1970 to Date

During this period, no spectacular new advances in the capabilities of the NAA method have occurred, but the use of the method has continued to grow and then to level off somewhat—at an annual publication rate of perhaps 2000. Part of this leveling off may be attributed to competition from nonnuclear methods of trace element determinations that have advanced rapidly during this period: particularly methods such as atomic absorption spectrophotometry (AAS) and inductively-coupled plasma (ICP) atomic emission spectroscopy. The NAA method (utilizing reactor fluxes of thermal neutrons) was essentially the only highly-sensitive, quantitative, multielement analytical method available up to the mid-1960s, but it now must compete increasingly with methods such as AAS and ICP. The NAA method, and particularly its INAA form, still has advantages over these other methods for many specific analytical problems, but its need for use of an expensive device—a nuclear reactor—is a barrier to many analytical chemists who prefer to be able to do their analyses in their own laboratory, using less expensive equipment. Many NAA groups, of course, make effective use of nearby nuclear reactors that they do not own.

MAJOR CONTRIBUTIONS MADE BY THE NAA METHOD

Through the years, but particularly during the past 30 years, the NAA method has contributed immensely to our knowledge in many fields. Fields to be especially noted are those of biology and medicine, industry, pure materials, environmental pollution, art and archaeology, geochemistry, and forensic chemistry. A great deal of our knowledge about essential trace elements (and toxic trace elements, such as arsenic and mercury) has come from NAA studies. The semiconductor industry owes much to important early NAA studies of high-purity silicon. Major environmental pollution studies have been, and still are, based on the NAA method. The multielement instrumental form of the method has become a major contributor to studies in the field of archaeology. Applications of the NAA method in the field of forensic chemistry—such as for the detection of primer gunshot residue (detecting barium and antimony) and the analysis of evidence specimens of bullet lead and shotshell pellets (for Sb, As, Ag, Cu, and Sn)—are special to the author and are used on a large scale in the investigation of gunshot homicide criminal cases, especially by the FBI Laboratory. In 1977, as part of the reinvestigation of the President Kennedy assassination, the author's reanalysis of all the bullet-lead evidence specimens, by INAA, produced decisive results.

OTHER IMPORTANT DEVELOPMENTS IN THE FIELD OF NAA

Although most of this presentation has concentrated on the developments of the radiochemical-separation and purely-instrumental forms of the NAA method, using nuclear-reactor fluxes of thermal neutrons, many other important developments should be mentioned. In this short historical summary, limitations of space here allow only a brief listing of these other developments:

Charged-particle NAA, Photon NAA, Epithermal-neutron NAA, NAA with ^{252}Cf neutrons, Delayed-neutron counting, Derivative NAA, Prompt-gamma NAA, Cyclic NAA, Reactor-pulsing NAA, NAA with 14 MeV neutrons, monostandard methods, certified reference materials, and Advance Prediction Computer Program.

MAJOR EARLY CONTRIBUTORS (THROUGH 1971)

Thus far, this presentation has discussed major developments in the field of NAA from 1936 to date, but with essentially no mention of the scientists who particularly contributed to the advancement of this method. It is difficult to summarize who have been the principal contributors to this field—except for the first 35 years (1936–1971). Fortunately, an excellent, detailed, and well cross-indexed bibliography of the field covering these first 35 years exists: the National Bureau of Standards publication "Activation Analysis: A Bibliography Through 1971" (2).

Review of this very useful bibliography reveals that, although many hundreds of scientists contributed a total of some 9000 publications in the field during these first 35 years, some 17—in terms of number of publications—can be classed as major early contributors to the field of NAA. The following 17 scientists each published 40 or more papers or reports during this period (two of them even more than 100): Philippe Albert, Saadia Amiel, William Ehmann, Francesco Girardi, Vincent Guinn, Julien Hoste, Robert Jervis, George Leddicotte, E.M. Lobanov, Richard Lukens, Wayne Meinke, M. Okada, M. Radovic, Albert Smales, Hamilton Smith, Eiliv Steinnes, and Richard Wainerdi.

Of this group of 17 major pioneers, more than half are still active, a few are now deceased, and several have retired. These 17 are from 10 different countries, the largest number, 6, are from the US. Nine of the 17 have received the Hevesy Medal in Radioanalytical Chemistry. The above list of major researchers, of course, is only based upon their number of publications in the field during 1936–1971, so it should be remarked that many others not listed also made a significant number of important contributions during this period—many of them just commencing their careers. Thus, if the list were extended to 1989, additional prominent names could be added. Unfortunately, no such highly useful bibliography of the field since 1971 is available.

ACKNOWLEDGMENTS

The author is grateful to the hundreds of major workers in the field of NAA with whom he is well acquainted and who are close friends. Discussions with many of them, in his laboratories and theirs and at major national and international conferences, over the past 35 years have been helpful, informative, and enjoyable. It is a closely-knit group of inventive, productive, and friendly scientists. It is an honor and a privilege to be a member of this fine group.

REFERENCES

1. G. Hevesy and H. Levi, *Det. Kgl. Danske Videnskabernes Selskab. Mathematisk-Fysiske Meddelelser.* **14,(5),** 3 (1936).
2. G. J. Lutz, R. J. Boreni, R. S. Maddock, and J. Wing (eds.), *Activation Analysis: A Bibliography Through 1971*, NBS Technical Note 467, National Bureau of Standards, Washington, DC, 1972.

Instrumental Neutron Activation Analysis of Biological Samples

VINCENT P. GUINN*,[1] AND MARIA GAVRILAS[2]

[1]*Department of Chemistry, University of California, Irvine, CA 92717; and [2]Department of Chemical and Nuclear Engineering, University of Maryland, College Park, MD 20742*

Received April 17, 1989; Accepted September 14, 1989

ABSTRACT

The elemental compositions of 18 biological reference materials have been processed, for 14 stepped combinations of irradiation/decay/counting times, by the INAA Advance Prediction Computer Program. The 18 materials studied include 11 plant materials, 5 animal materials, and 2 other biological materials. Of these 18 materials, 14 are NBS Standard Reference Materials and four are IAEA reference materials. Overall, the results show that a mean of 52% of the input elements can be determined to a relative standard deviation of ±10% or better by reactor flux (thermal plus epithermal) INAA.

Index Entries: Instrumental neutron activation analysis; biological samples, elemental analysis of; biological reference materials; Advance Prediction Computer Program.

INTRODUCTION

The method of neutron activation analysis (NAA), originated by G. Hevesy and H. Levi in 1936, did not begin to approach its detection sensitivity potential until 1944—when the first nuclear reactor became available for NAA. This reactor, the X-10 reactor at the Oak Ridge National Laboratory, provided thermal-neutron fluxes of about 5 × $10^{11} n/cm^2/s^1$ for the activation of samples, providing lower limits of detec-

*Author to whom all correspondence and reprint requests should be addressed.

tion (LOD's) of the order of 10^{-9} g for many elements. However, to achieve such detection sensitivities, tedious post-irradiation radiochemical separations with carriers were necessary, followed by beta counting.

In 1950, the purely instrumental form of the method (INAA) began to be developed and used, owing to the advent of commercially available NaI(T1) scintillation detectors and accompanying multichannel pulse-height analyzers (PHA's). Now the NAA method, used with thermal-neutron fluxes of 10^{12}–$10^{14} n/cm^2/s^1$, available in the increasing number of research-type nuclear reactors around the world, could not only attain great detection sensitivities for a large number of elements, but it could also attain them nondestructively, simultaneously for many elements (rather than one at a time), and with much less effort. During the period of 1950–1960, many multielement INAA studies were published, especially for biologically/medically interesting trace elements.

The major shortcoming of the INAA method utilizing NaI(T1) gamma-ray spectrometry is the relatively poor energy resolution of the NaI(T1) detector. Numerous computer programs were developed to resolve such pulse-height spectra, involving many overlapping photopeaks, into their radionuclide components, but the energy resolution problem was still significant. This problem was greatly decreased in magnitude by the advent, in 1960, of the germanium semiconductor detector—with its 20–30 times narrower total absorption gamma ray peaks. Increasingly, from that time on, the use of Ge(Li) or intrinsic Ge detectors for the INAA of all types of activated samples, using reactor neutron fluxes, has become the method of choice for many quantitative and sensitive multielement analyses.

It is quite simple to calculate NAA LOD's for various elements in the absence of other activities, such as NAA coupled with radiochemical separations—for any selected irradiation and counting conditions, and various "interference-free" LOD tables exist in the literature, e.g., one by Guinn and Hoste (*1*). However, no such tables exist for INAA, because (1) each element microgram LOD depends on the other elements present in the sample matrix and may range anywhere from about the interference-free LOD to even 100 times higher for a given element in different matrices, and (2) counting rate restrictions may limit usable sample sizes to as low as milligrams, instead of to about a gram—or else larger samples at a lower neutron flux—thus greatly increasing the ppm LOD for an element in that matrix in such cases.

In an effort to put the INAA method on a consistent and useful basis—taking into account all of the induced activities present in a sample—one of the authors originated the concept of an INAA Advance Prediction Computer Program (APCP) in 1973 (*2*), developed it into an operating and tested program largely during the period of 1977–1984 (*3–5*), and has been refining and extending it further to the present time (*6,7,8,9*). The second author has recently revised the APCP and written it for use with a personal computer.

Using a simple, but realistic model of Ge gamma ray pulse-height spectra and best literature values for isotopic abundances, (n, γ) thermal-neutron cross-sections, epithermal-neutron resonance integrals, half-lives, and decay schemes, the APCP calculates all significant features of the resulting pulse-height spectra, for any desired set of neutron fluxes, irradiation time (t_i), decay time (t_d), and counting time (t_c) for any specified detector, and the maximum allowable sample weight, for any input elemental composition (either approximate or, in the case of standard reference materials, quite exact). It has been successfully tested by comparison of its predictions with experimental results on a number of reference materials.

The purpose of the APCP is to ascertain, in advance, which input elements will be detectable to which precisions under various sets of analysis conditions. The counting statistics standard deviation of the net photopeak counts (NPPC) in each significant gamma ray peak is calculated and printed—taking into account the cumulative Compton-continuum level beneath each peak. For any given thermal-neutron flux (ϕ_{th}) and/or epithermal-neutron flux (ϕ_{epi}), the program calculates the pulse-height spectrum—for the calculated maximum allowable sample weight—for each selected combination of t_i, t_d, and t_c. In order to scan all the possible (n, γ) products with half-lives ranging from fractions of a second up to years, the program is usually run for 14 stepped combinations. For the first nine combinations, $t_i = t_d = t_c$, each increasing by a factor of about 3 from one combination to the next. Set 1 uses 0.01 min (i.e., 0.6 s) for each; Set 9 uses 100 min for each. From Sets 10–14, the t_i is held constant at 300 min and the t_c held constant at 100 min (both simply for practical reasons), but t_d continues to increase, stepwise, from 300 min up to 30,000 min (3 wk) in Set 14.

METHODS

In the present study, the purpose was to utilize the APCP to ascertain how many and which elements could be precisely measurable (to a relative standard deviation of $\pm 10\%$ or better) in various kinds of biological materials of plant, animal, or other nature. For a rather typical nuclear reactor-neutron flux of 10^{13} thermal, plus 10^{12} epithermal n/cm^2/s^1, the 14 stepped condition sets described above were applied (for a "15%" Ge(Li) detector and 2 cm distance) to 18 biological reference materials. The APCP conditions used are summarized in Table 1.

RESULTS

The 18 biological materials treated by the APCP are shown in Tables 2–4, along with a summary of the results for each material. Table 2 lists the eleven plant reference materials studied. As shown in Table 2, the

Table 1
INAA APCP Conditions Calculated

- 14 stepped t_i, t_d, t_c sets of conditions (t_d increasing from 0.01 minute up to 30,000 minutes)
- φ_{th} of 10^{13} n cm^{-2} s^{-1}, plus φ_{epi} of 10^{12} n cm^{-2} s^{-1}
- Maximum sample size of one gram, or weight giving 5000 total gamma cps at SOC (whichever is the smaller)
- Counting 2 cm above a "15%" Ge(Li) detector

number of input elements averaged 41 (range of 12 to 73); the number of precisely-measurable input elements averaged 21 (range 9 to 32); and the percent of the input elements that are precisely measurable averaged 55% (range of 40 to 77%).

Table 3 lists the five animal reference materials studied. The number of input elements averaged 40 (range 18 to 60); the number of precisely-measurable input elements averaged 18 (range of 9 to 35); and the percent of the input elements that are precisely measurable averaged 45% (range of 32 to 66%).

Table 4 lists two other biological materials studied. The number of input elements averaged 28 (range of 16 to 40); the number of precisely-measurable input elements averaged 14 (range of 11 to 17); and the percent of the input elements that are precisely measurable averaged 56% (range of 42 to 69%).

Of the 18 biological reference materials studied, 14 are National Bureau of Standards (NBS) Standard Reference Materials (SRM's) and 4 are International Atomic Energy Agency (IAEA) reference materials. Table 5 summarizes the findings for the 14 NBS SRM's regarding which elements are frequently precisely measurable by INAA under the specified, rather typical conditions. The elements included are only those in which the element was an input element in at least half of the 14 SRM's and in which it was precisely measurable in at least half of the materials where it was an input element. For these 14 SRM's, Table 5 shows that 15

Table 2
Plant Reference Materials Studied and Results

REFERENCE MATERIAL		NUMBER OF INPUT ELEMENTS	NUMBER OF PRECISELY MEASURABLE ELEMENTS	PERCENT OF INPUT ELEMENTS PRECISELY MEASURABLE
Citrus Leaves	NBS–1572	52	30	58
Corn Kernel	NBS–8413	12	9	75
Corn Stalk	NBS–8412	13	10	77
Cotton	IAEA V–9	28	16	57
Orchard Leaves	NBS–1571	73	31	42
Pine Needles	NBS–1575	56	28	50
Rice Flour	NBS–1568	37	18	49
Rye Flour	IAEA V–8	21	13	62
Spinach	NBS–1570	62	25	60
Tomato Leaves	NBS–1573	61	32	52
Wheat Flour	NBS–1567	41	18	44
Means :		41	21	55
Minimum :		12	9	40
Maximum :		73	32	77

Table 3
Animal Reference Materials Studied and Results

REFERENCE MATERIAL		NUMBER OF INPUT ELEMENTS	NUMBER OF PRECISELY MEASURABLE ELEMENTS	PERCENT OF INPUT ELEMENTS PRECISELY MEASURABLE
Animal Blood (Dried)	IAEA A-13	18	9	50
Bovine Liver	NBS-1577	60	19	32
Milk Powder	IAEA A-11	38	16	42
Milk Powder (Non-Fat)	NBS-1549	32	11	34
Oyster Tissue	NBS-1566	53	35	66
Means :		40	18	45
Minimum :		18	9	32
Maximum :		60	35	66

Table 4
Other Biological Reference Materials Studied and Results

REFERENCE MATERIAL		NUMBER OF INPUT ELEMENTS	NUMBER OF PRECISELY MEASURABLE ELEMENTS	PERCENT OF INPUT ELEMENTS PRECISELY MEASURABLE
Brewer's Yeast	NBS-1569	40	17	42
Mixed Diet	NBS-RM8431	16	11	69
Means :		28	14	56
Minimum :		16	11	42
Maximum :		40	17	69

Table 5
Elements Readily Measurable by INAA
in 14 NBS Biological Reference Materials

PERCENT OF MATERIALS	ELEMENTS
90-100	Na, Mg, Al, Cl, K, Ca, Se, Fe, Co, Zn, Br, Rb, Cs, Ce, Ta
80-90	Cr, Mn, Sb, Eu, Tb, Th
70-80	Ag, La, Sm
60-70	Se, Dy
50-60	Cu, As, Sr, Ba, Hf

elements are precisely measurable in 90–100% of these materials where they were input elements, 6 more elements in 80–90% of the materials, 3 more elements in 70–80% of the materials, 2 more elements in 60–70% of the materials, and 5 more elements in 50–60% of the materials, for a total of 31 elements frequently precisely measurable by INAA in these various biological materials.

Tables 2–5 only include elements that are measurable to a percent σ_{rel} of ± 10% or less. However, the APCP also provides output for various other input elements that are detectable, but only to poorer measurement precisions (10–50%). For the NBS SRM's, input elemental compositions were taken from either NBS Special Publication 260-88 (*10*) or 260-111 (*11*). For the IAEA reference materials, input elemental compositions were taken from IAEA reports.

DISCUSSION

Whereas NAA under the calculated irradiation/counting conditions used in these APCP calculations—but followed by radiochemical separations with carriers—could precisely determine an even higher percentage of the input elements in these 18 biological reference materials than possible by INAA; it is evident that the INAA method can precisely measure quite a high percentage of them: a mean of 52% of the input elements. In terms of an analyst's time, the INAA method is, of course, far more rapid. In addition, if several $t_i/t_d/t_c$ sets of conditions are employed, additional noninput elements (i.e., elements not known in advance to be present) may be detected and measured.

Examination of Table 5 reveals, of course, that some trace elements of particular biomedical interest are not precisely measurable in many of these biological materials by INAA. In such cases, if there is a radioactive (n, γ) product, NAA with radiochemical separations (RNAA) can be used. Overall, the APCP results show that it is feasible, under quite practical conditions, to precisely determine a large number of elements of interest in biological materials via the INAA method.

SUMMARY

Processing of the elemental compositions of 18 biological reference materials by the INAA Advance Prediction Computer Program shows that a large number of input elements of interest can be precisely determined by reactor flux INAA.

REFERENCES

1. V. P. Guinn and J. Hoste, *Elemental Analysis of Biological Materials,* IAEA Tech. Report No. 197, Intl. Atomic Energy Agency, Vienna, 1980, pp. 105–140.

2. V. P. Guinn, J. Radioanal. Chem. **15,** 473 (1973).
3. V. P. Guinn, E. Garzanov, and E. Cortes, *J. Radioanal. Chem.* **43,** 599 (1978).
4. V. P. Guinn, J. Leslie, and L. Nakazawa, *J. Radioanal. Chem.* **70,** 513 (1982).
5. V. P. Guinn, L. Nakazawa, and J. Leslie, *J. Radioanal. Nucl. Chem. Articles* **84,** 103 (1984).
6. H. S. Hsia and V. P. Guinn, *J. Radioanal. Nucl. Chem. Articles* **112,** 223 (1987).
7. V. P. Guinn, *J. Radioanal. & Nucl. Chem. Articles* **110,** 5 (1987).
8. N. L. Truglio and V. P. Guinn, *J. Radioanal. Nucl. Chem. Articles* **110,** 41 (1987).
9. V. P. Guinn, H. S. Hsia, and N. L. Truglio, *J. Radioanal. Nucl. Chem. Articles* **123,** 249 (1988).
10. E. S. Gladney, C. E. Burns, D. R. Perrin, I. Roelandts, and T. E. Gills, *1982 Compilation of Elemental Concentration Data for NBS Biological, Geological, and Environmental Standard Reference Materials,* NBS Special Publication 260-88, National Bureau of Standards, Gaithersburg, MD (1984).
11. E. S. Gladney, B. T. O'Malley, I. Roelandts, and T. E. Gills, *Compilation of Elemental Concentration Data for NBS Clinical, Biological, Geological, and Environmental Standard Reference Materials,* NBS Special Publication 260-111, National Bureau of Standards, Gaithersburg, MD (1987).

Neutron Activation Analysis of Biological Samples with a Preirradiation Chemical Separation

ROBERT R. GREENBERG,*[,1] H. M. KINGSTON,[1] ROLF ZEISLER,[1] AND JOOST WOITTIEZ[2]

[1]*Center for Analytical Chemistry, National Institute of Standards and Technology, Gaithersburg, MD 20899; and* [2]*Netherlands Energy Research Foundation (ECN), 1755-ZG Petten, The Netherlands*

Received May 5, 1989; Accepted September 14, 1989

ABSTRACT

A preirradiation separation procedure has been developed to separate Al, Cu, Mn, and V from biological materials. Chelex-100 resin is used as the separation medium, and the resin is irradiated directly. Three NIST biological Standard Reference Materials and five samples of human blood serum, obtained under carefully controlled conditions, have been analyzed by NAA following this separation.

Index Entries: Blood serum; Chelex-100; human; neutron activation analysis (NAA); preconcentration; preirradiation separation; Standard Reference Materials (SRMs); trace element analysis.

INTRODUCTION

Traditionally, most chemical separations employed for neutron activation analysis (NAA) have been performed after the irradiation step to preserve the nearly "blank-free" nature of the technique. There are certain types of analyses, however, where it has been advantageous to perform the chemical separations before irradiation. For example, preconcentrating elements of interest from a freshwater sample could signif-

*Author to whom all correspondence and reprint requests should be addressed.

icantly improve the detection limits for most elements, owing to limitations on the amount of water that could be irradiated. In addition, preconcentrating the elements of interest (i.e., transition metals) from a saltwater sample, and at the same time separating these elements from the alkali metals, alkaline earths, and halogens, greatly reduces the background level of radiation under the gamma-ray peaks of interest, and thus significantly improves the detection limits for most elements. In addition, removal of the salt matrix effectively reduces the radiation dose received by the analyst.

A preirradiation chemical separation may also be advantageous when determining elements in solid samples that have short-lived neutron activation products. It is often difficult, and sometimes impossible, to dissolve a complex biological material totally and then separate elements like Al or V in the limited time available before they completely decay. In addition, separating Al from P in a biological sample before irradiation eliminates the interference produced via the fast neutron reaction P-31 (n, alpha) Al-28. In many cases, this P interference limits the determination of Al in biological materials. This separation also eliminates the potential fast-neutron reaction on Si that also interferes with Al determinations. Finally, as with salt water samples, removal of Na and Cl from the sample can greatly reduce the radiation dose received by the analyst.

A previously developed, preirradiation separation procedure used for the determination of 15 elements in saline and freshwater samples (*1*) has been modified to determine Al, Cu, Mn, and V in biological samples. This procedure employs Chelex-100 chelating resin to separate the elements of interest from the biological matrix. A number of different biological Standard Reference Materials (SRMs) and five samples of human blood serum have been analyzed following this preirradiation separation procedure.

EXPERIMENTAL SECTION

Materials and Methods

High purity water, nitric acid, perchloric acid, and glacial acetic acid were prepared at NIST by subboiling distillation (*2*). Ammonium hydroxide was prepared by bubbling filtered ammonia gas through high purity water until room temperature saturation was achieved. Ammonium acetate solutions were prepared by carefully mixing acetic acid with ammonium hydroxide. Samples of human blood were collected through a Teflon catheter (*3*) and allowed to flow directly into cleaned Teflon TFE containers. The blood was allowed to clot, the vials centrifuged, and the serum decanted. All sample dissolution and separation steps prior to irradiation were performed under class-100 clean air conditions. The basic separation procedure used for this study is as follows: the samples were first digested in nitric, hydrofluoric, and perchloric acid, adjusted to a pH between 5.2 and 5.7, passed through Chelex-100 resin that had been precleaned with dilute, high purity nitric acid and ammonium hydroxide, and washed with ammonium acetate to remove residual Ca and Mg. The separated samples on the Chelex-100 were then washed with water, air dried (or sometimes only partially dried), and transferred to acid-cleaned bags for irradiation. Materials used to prepare bags for this study included linear polyethylene (LPE), Parafilm, and Teflon FEP. The separation procedure has been described in more detail in (*1,4*). The samples were irradiated for 1–2 min at a thermal-neutron flux of either 5×10^{17} or 1×10^{18} n/m^2-s. After irradiation, the resin was removed (washed with a stream of water) from the irradiation container to reduce the blank from the container material and counted on a high-resolution gamma spectrometry system. Alternatively, the samples could have been counted directly in the bags if the levels of the elements of interest were high enough.

RESULTS AND DISCUSSION

Retention on Chelex-100

The retention of Cu, Mn, and V, along with 11 other elements, has been studied previously (*1*) with radioactive tracers. Retention of most elements is high and very reproducible, i.e., typical variations observed for most elements were only a few tenths of a percent. Recoveries for Cu and Mn were greater than 99.9%, whereas V recovery was 97.6 ± 0.3%. The recovery for Al could not be checked directly with a radioactive tracer; however, a comparison of Al results obtained for SRM 1643a (Trace Elements in Water), using the Chelex-100 procedure and INAA (*1*), indicated that Al retention was essentially 100%.

Assessment of Blank Values

For any preirradiation separation, blank levels, as well as consistency, are critical. It is vital to prepare a large number of blank samples in a manner as close as possible to that used for the samples under investigation. In this work, we have analyzed similar numbers of blanks and samples for those materials containing relatively high levels of the elements of interest and, typically, twice as many blanks as samples for low-level materials. Despite the large quantities of reagents used in the dissolution and separation procedures, the major fraction of the blank usually originated from the irradiation containers and not the reagents, lab equipment, or environment. One exception is discussed below. Even though the samples were transferred after irradiation, radioactive nuclides near the surface of the container were transferred to the sample as a result of nuclear recoil following neutron capture. Typical blank values

Table 1
Typical Blank Values (ng) Observed for Chelex-100 Separation Procedure Using Different Irradiation Containers

	Al	Cu	Mn	V
LPE, not transferred	1850 ± 100	<2	38 ± 20	<0.5
LPE, transferred	34 ± 10	<2	<0.1	<0.05
Parafilm, transferred	29 ± 9	<2	<0.08	<0.05
FEP, transferred	<1	<1	<0.1	<0.05

for the different irradiation container materials examined are given in Table 1. A comparison of these results indicates that the LPE polyethylene and Parafilm produced similar blank values, whereas those for the Teflon FEP were considerably lower. However, the high degree of radioactivity from the 11.0 second ^{20}F produced from the Teflon required an additional decay interval of 1–2 min prior to handling the samples. This additional decay interval was significant for the determination of Al, V, and, to a lesser extent, Cu via the Cu-66 isotope. Alternatively, Cu can be determined via the longer-lived Cu-64 isotope, approximately 24 h after irradiation, when the Chelex-100 separation procedure is used (*5*).

One blank-related problem encountered with this preirradiation separation procedure resulted from the Cl remaining on the Chelex-100 after separation. Typically, approximately 0.05–0.2 mg of Cl was present on the resin, and this often elevated background levels of radiation, resulting in degraded signal-to-noise ratios. This Cl did not originate from the sample, because the samples and blanks had identical levels of this element even when no Cl-containing reagent (i.e., perchloric acid) was used. In addition, tracer studies indicated that chloride ion was quantitatively eluted from the Chelex-100. This Cl can originate from two potential sources: Cl strongly bound to the polymeric backbone of the resin itself and/or usage of HCl in other experiments conducted at the same time in the clean room. Gaseous HCl could easily dissolve on the moist surface of the resin and remain there during the drying stage. Methods of reducing these Cl levels are under investigation.

SRM Results

Three biological SRMs have been analyzed by this procedure, and the results obtained are given in Table 2. The NIST certified values, and values obtained by additional analytical methods, are given for comparison. It is apparent from this table that the concentrations determined via the Chelex-100 procedure are in good agreement with the other values. Unfortunately, none of these materials have certified values for V, and only one is certified for Al.

Table 2
Elemental Concentrations (ng/g) and 95% Confidence Intervals Determined in Bovine Serum, SRM 1598, Urine (Elevated) SRM 2670, and Powdered Milk SRM 1549 After Chelex-100 Preirradiation Separation

Element	Bovine Serum SRM 1598	Urine (Elevated) SRM 2670	Powdered Milk SRM 1549
Al, this work	4.1 ± 0.7	90 – 120[a]	240 – 2300[a]
Al, certified	3.7 ± 0.9		
Cu, this work	705 ± 19	360 ± 20	604 ± 46
Cu, certified	720 ± 40	370 ± 30	700 ± 100
Cu, RNAA	722 ± 22		606 ± 10
Mn, this work	3.72 ± 0.15	333 ± 5	241 ± 4
Mn, certified	3.78 ± 0.32		260 ± 60
Mn, INAA			233 ± 13
V, this work	0.055 ± 0.035	112 ± 3	9.7 ± 0.7
V, IDMS		118 ± 6	

[a]Inhomogeneous behavior noted.

Human Serum

In view of the dearth of information on the concentrations of Al and V in uncontaminated human blood serum, and because of the lack of recent values for Mn in blood serum for people living in North America, we have tested the applicability of this method to the analysis of human blood serum collected via noncontaminating methods. Vanadium was of particular interest in that very few data were available worldwide, and no agreement between different laboratories could be observed. Mean values for V in human serum of 0.024–0.047 ng/mL have been reported by Cornelis et al. (*6,7*) and 0.67 ± 0.33 ng/mL by Simonoff et al. (*8*). Higher values have also been reported in the literature (*9*).

For this experiment, serum was separated from blood collected from four individuals, one in duplicate (one sample collected from each arm). The five samples of serum, approximately 5 mL each, were dissolved, separated, and analyzed along with eight blanks. The Chelex-100 resin was encapsulated for irradiation in Teflon FEP bags. In view of our experience with the Bovine Serum SRM (20 mL sample size analyzed), it was expected that the blank levels (in FEP) would be sufficiently low to not affect the accuracy significantly. Unfortunately, the blank levels for Al and Mn were unexpectedly high, leading to significantly degraded results for Al. The cause for these elevated blank levels is unknown and had not been observed previously. The Cl blank was also on the high side of the range typically observed, despite efforts to reduce Cl levels by processing the samples and blanks in the clean room during a period of little HCl usage. The Cl levels on the resin produced a significant elevation of the background, especially under the V peak. The combined effects of the background elevation and the additional decay time needed, since Teflon bags were used for sample encapsulation, resulted in poor V determinations. All values had counting statistics worse than 50%.

The results of the analysis of human serum are given in Table 3. Agreement between the Cu concentrations, determined in the duplicate samples from individual R1, agreed within 4%. Some of this difference may be owing to biological variability since the samples were collected at slightly different times and from different arms. The range of values observed lies very close to the median literature value for Cu of 1100 ng/mL (*10*) in human serum.

The blank values for Mn were approximately equal to the values for the samples. Since eight blanks were analyzed, a reasonable estimate of both the mean value and the variability of the blanks could be determined, thus allowing an accurate subtraction of the blank. Uncertainties for Mn and Al, listed in Table 3, were calculated in quadrature by combining the counting statistics and the standard deviation of the mean for the blanks. The range of Mn values observed was 0.26–0.41 ng/g, which corresponds to the low end of the previously reported concentrations for this element. The mean value was approximately half of the currently accepted reference value for this element, i.e., median = 0.63 ng/mL (*10*). However, in reviewing recent literature, it is apparent that all of the Mn values at this level appear to be from people living in Europe (*9,11*). Perhaps average Mn levels may be different for people living in the Washington, DC area, owing to differences in diet or the environment. Concentrations of Mn in human milk have been correlated with geographical location (*10*).

The blank for Al was approximately an order of magnitude greater than the Al concentrations observed in the serum. The individual values for Al in the serum are quite imprecise owing to the variability of the blank; some values are negative. However, 95% confidence intervals, calculated from the listed uncertainties (combined counting statistics and blank variability), extend into the positive range for all samples. Very little quantitative information can be obtained from these Al data. The most likely value for the average Al concentration in the serum samples analyzed appears to be about 1–2 ng/g; however, the 95% confidence interval overlaps zero. An upper limit value for the mean at the 95% confidence level was calculated, and a value of $\leq$3.8 ng/g was obtained.

The V determinations in the human serum samples suffered from poor counting statistics owing to the relatively high level of Cl present on the resin, as well as the relatively small sample size used (approximately 5 g). The observed counting statistics were worse than 50% for all samples, and only upper limit values are meaningful for the individual samples. These upper limit values ranged from less than 0.05 to less than 0.12 ng/g at the 95% confidence level. However, more information can be

Table 3
Concentrations (ng/g) of Al, Cu, Mn, and V Observed in 5 mL Samples of Human Blood Serum and Blanks

	Al	Cu	Mn	V
Blank	9.9 ± 1.2[a]	<0.4	0.25 ± 0.03	−0.001 ± 0.011
Individuals (Blank Corrected)				
J1 (male)	3.0 ± 1.6	942 ± 10	0.32 ± 0.05	0.058 ± 0.040
T2 (female)	−1.0 ± 1.4	826 ± 9	0.29 ± 0.05	−0.006 ± 0.032
B1 (male)	4.7 ± 1.8	994 ± 11	0.41 ± 0.04	0.044 ± 0.045
R1 (male)	1.2 ± 1.3	1055 ± 7	0.26 ± 0.04	0.041 ± 0.026
R1 (duplicate)	−0.9 ± 1.3	1017 ± 8	0.37 ± 0.04	0.004 ± 0.026
Mean	1.4	967	0.33	0.028
Observed SD	2.5	89	0.06	0.02
SD of Mean	1.1	40	0.03	0.012
Weighted Mean	1.0	979	0.34	0.023
A priori SD of mean	0.7	4	0.02	0.014
Red. Chi. Sq.	2.6	111	2.4	0.7
Probability[b]	4%	<0.0001%	5%	59%
Upper limit[c]	3.8			0.047

[a]Uncertainties for the blank values are 1 SD of the mean for eight blanks. Uncertainties for individual serum samples are the observed counting statistics and the standard deviation of the mean for the blanks, combined in quadrature.

[b]This is a comparison of the observed and *a priori* standard deviations. A probability of 4% indicates that if the *a priori* standard deviation truly accounts for all variability in the samples, replication of the measurements would yield an even greater observed standard deviation 4% of the time.

[c]Upper limit of the mean at the 95% confidence level.

obtained for the mean than for the individual samples if care is taken in peak integration. The V peaks for all samples and blanks were integrated via a summation process where peak and background regions were first determined from the standards. A channel-by-channel summation of the background regions were then subtracted from similar summations of the peak regions. Identical regions were used for samples and blanks. Concentrations (possibly negative) and uncertainties (counting statistics) were next calculated from the summation results. Means, both weighted and unweighted, and standard deviations were than calculated for the samples and blanks. The magnitude of the mean for the blanks provides some indication of whether a bias is introduced via this method (or if a true blank is present), and a comparison of the observed standard deviation with the counting statistics (*a priori* uncertainty) for the blanks indicates whether any additional source of variance is present. The weighted mean for the blank concentration was −0.001 ng/g with an *a priori* standard deviation of the mean of 0.011 ng/g, indicating little, if any, bias. The reduced chi squared value for this measurement was 0.59, indicating that the observed variation in the blank values was consistent with the counting statistics. This value of chi squared (for 7 degrees of freedom) corresponds to a "probability of exceeding" value of 76%. This indicates that if sample variability was solely owing to counting statistics, 76% of the time even greater sample-to-sample variation should be observed. The weighted mean value for V in the serum samples was 0.023 ng/g with an *a priori* standard deviation of the mean of 0.014 ng/g. The reduced chi squared value was 0.70 and the "probability of exceeding" value was 58%, again indicating all variations could be explained by counting statistics alone. Thus,the mean V concentration in these human serum samples is on the order of a few hundredths of a nanogram/gram, and the upper limit for the mean, at the 95% confidence level, is 0.047 ng/g. These results are in good agreement with the lowest literature values reported by Cornelis et al. (*6,7*).

Areas for Future Research

In view of the problems uncovered in the analysis of human serum samples, additional research is needed to further improve the method. We need a better understanding of the sources of the blank and need to find ways to minimize the run dependency. We also need to find ways to reduce the levels of Cl present in the Chelex-100 after separation. Various cleanup methods need to be examined. In addition, reducing the amount of resin used should reduce the total amount of Cl present. We also need to try to locate batches of Chelex-100 with lower Cl levels, if possible. Finally, this separation/analysis procedure should be performed again on uncontaminated human serum samples using the above improvements, and also increasing the amount of serum processed, to further improve the signal-to-noise and sample-to-blank ratios.

ACKNOWLEDGMENTS

The authors wish to acknowledge Saul Malozowski of the National Institute of Child Health and Human Development, National Institutes of Health, for his help in the collection of human blood samples.

Certain commercial equipment, instruments, or materials are identified in this paper in order to describe the experimental procedure adequately. Such identification does not imply recommendation or endorsement by the National Institute of Standards and Technology, nor does it imply that these materials or equipment identified are necessarily the best available for the purpose.

REFERENCES

1. R. R. Greenberg and H. M. Kingston, *Anal. Chem.* **55,** 1160–1165 (1983).
2. E. C. Kuehner, R. Alvarez, P. J. Paulsen, and T. J. Murphy, *Anal. Chem.* **44,** 2050 (1972).
3. R. R. Greenberg, R. Zeisler, and S. Malozowski, *Trace Element Analytical Chemistry in Biology and Medicine,* vol. 5, P. Bratter and P. Schramel, eds., de Gruyter, Berlin, pp. 220–228, 1988.
4. R. R. Greenberg, R. Zeisler, H. M. Kingston, and T. M. Sullivan, *Fresenius Z. Anal. Chem.* **332,** 652–656 (1988).
5. R. R. Greenberg and H. M. Kingston, *J. Radioanal. Chem.* **71,** 147–167 (1982).
6. R. Cornelis, L. Mees, J. Hoste, J. Ryckebusch, J. Versieck, and F. Barbier, *Nuclear Activation Techniques in the Life Sciences,* IAEA, Vienna, p. 165 (1979).
7. R. Cornelis, J. Versieck, L. Mees, J. Hoste, and F. Barbier, *J. Radioanal. Chem.* **55,** 35 (1980).
8. M. Simonoff, Y. Llabador, C. Hamon, B. Berdeu, G. Simonoff, C. Conri, B. Fleury, P. Couzigou, and A. Lucena, *J. Radioanal. Nucl. Chem.* **113,** 107–117 (1987).
9. J. Versieck and R. Cornelis, *Anal. Chim. Acta* **116,** 217–254 (1980).
10. V. Iyengar and J. Woittiez, *Clin. Chem.* **34/3,** 474–481 (1988).
11. G. V. Iyengar, *Concentrations of 15 Trace Elements in Some Selected Adult Human Tissues and Body Fluids of Clinical Interest From Several Countries: Results from a Pilot Study for the Establishment of Reference Values,* KFA Julich (FRG) Report No. Jul-1974 (1985).

Nondestructive Determination of Arsenic in Urine by Epithermal Neutron Activation Analysis and Compton Suppression

SHELDON LANDSBERGER,* GARY SWIFT, AND JON NEUHOFF

Department of Nuclear Engineering, University of Illinois, 214 Nuclear Engineering Laboratory, 103 South Goodwin Ave., Urbana, IL 61801

Received April 17, 1989; Accepted September 14, 1989

ABSTRACT

Epithermal neutron activation analysis, in conjunction with Compton suppression, has been employed to determine arsenic levels in artificially doped urine samples. Typical detection limits were of the order of 10 ng/g. Replicate determinations gave precision values between 2 and 12%, whereas accuracy measurements were between ±1 and ±20%. Biological and geological reference materials from the National Institute of Standards and Technology (NIST) were also analyzed for arsenic content. Typically, the precision achieved again was between 2 and 12%, whereas the accuracy measurements were in excellent agreement with the certified values.

Index Entries: Arsenic analysis; urine analysis; biological trace elements; epithermal neutron activation analysis; Compton suppression.

INTRODUCTION

The determination of heavy metals in biological fluids continues to be a very important part in environmental, nutritional, and toxicological programs. In particular, for urine analysis, reliable methodologies are

*Author to whom all correspondence and reprint requests should be addressed.

needed for concentrations of metals at the ppb level. This is important because the metal may be toxic at low levels, or baseline studies are needed. Quite often, very serious matrix effects may arise in traditional chemical methods for the analysis of biological fluids. Instrumental neutron activation analysis (INAA) is not particularly well-suited for the determination of short- or medium-lived nuclides in many biological samples, primarily owing to the presence of sodium. In urine samples, the background continuum, owing to sodium, easily masks the signals of many metals of interest, including antimony, arsenic, barium, molybdenum, strontium, and vanadium (usually detected in short- and medium-lived activation procedures). Some previous work in arsenic determination in urine samples have appeared (*1–3*), usually accompanied by chemical separations. Iyengar (*4*) has shown that very few studies have been performed either for baseline or toxicological determinations of arsenic in urine samples. Heydorn (*5*) has pointed out that few investigators have employed nuclear methods in urine samples although other organs have been analyzed. Cornelius et al. (*6*) have employed various radiochemical techniques in conjunction with NAA to determine trace elements in urine. Epithermal NAA can be used advantageously to increase the sensitivity of certain elements in biological samples, and this has been previously demonstrated (*7–8*). Landsberger and Simsons (*9*) have previously employed epithermal NAA in the determination of arsenic in urine samples, and detection limits below 50 ng/g were not achievable.

As part of a previous interlaboratory investigation (*9*) to determine arsenic in artificially doped urine samples, we have reanalyzed a set of samples using epithermal NAA in conjunction with a Compton suppression system. To ensure accuracy, we also investigated several NIST biological reference materials.

METHODS

To perform epithermal NAA, a cadmium-lined rotating tube was specifically constructed in the 1.5 mW University of Illinois TRIGA Reactor. The epithermal flux was extremely well characterized (*10*) as 2.1×10^{12} n/cm^2/s at 1.5 mW and a cadmium to gold ratio of 1.83. The horizontal flux profile varied less than 2%, whereas the vertical profile varied $\pm 30\%$ depending on the placement of the samples in the tube, which can be 0–40 cm from the bottom.

Calibration was performed by preparing a 10 ppm solution of arsenic from 1000 ppm commercially available atomic absorption standard solution. All samples, weighing 0.7–0.8 g, and duplicate standards were heat-sealed, packaged, and irradiated from 30 min at 1.5 mW. The samples and standards were transferred into new inert vials and reweighed. In addition to the determination of arsenic in the samples, we also evaluated the optimal decay time for counting. Furthermore, we also counted the samples both in the normal and suppressed modes to see if any counting losses were being observed when using the anti-Compton system. Counting times were between 0.5–2 h for real-time and corrected for dead-time and pulse pileup using a pulser. All dead-times were kept at 5% or less.

For the Compton suppression, a 12-in by 12-in NaI annulus and a 3-in by 3-in NaI cylindrical plug surround an intrinsic 18% efficient germanium detector that has a resolution of 2.0 keV at the 1332 keV ^{60}Co peak. The system includes five photomultiplier tubes in parallel and associated electronic modules. To make the system more versatile, the germanium detector could easily be removed by having it on an elevator/cart mechanism. At present, the detectors are not enclosed by lead shielding. In the suppressed mode, an event in the germanium detector is prevented from contributing to a spectrum if it coincides with an event detected by the NaI detectors. This greatly reduces the Compton background normally seen in the gamma-ray spectra. Compton background resulting from the ^{24}Na peaks, which can interfere with arsenic detection, is further reduced owing to coincidences in its decay scheme. System performance was characterized by the ^{137}Cs peak to Compton edge ratio, which measured to be about 450 to 1.

RESULTS AND DISCUSSION

The results for five sets of urine samples and four NIST certified reference materials as shown in Tables 1 and 2, respectively. One problem that occurred with the urine samples is that some particulate matter was always present in the liquid. A reasonable effort was made not to include any of this solid matter by letting the samples settle for several days before removing an aliquot. However, it is not clear what portion of the arsenic may have adhered to the particulate part, if any. The high value for the replicate precision error may be an indication of this type of problem in general trace metal analysis of urine samples. Typical detection limits, based on Currie's criterion (*11*), were of the order of 10 ng/g. These values represent some of the best nondestructive NAA sensitivities for arsenic determination in biological samples. Zeisler et al. (*12*) reported a detection limit of 5 ng/g using a Compton suppressed system on the analysis of human livers. However, a strict comparison is not possible since the ^{24}Na activity may be substantially different in human liver and urine.

To give an indication of the dramatic effect seen with the anti-Compton system, we present Fig. 1 and 2, which represent the normal and suppressed modes, respectively, for the urine sample irradiated and counted under identical conditions. As can be seen, a 10-fold decrease in the background is observed with the added advantage that the bromine

Table 1
Interlaboratory Comparison for Arsenic Determination in Urine

Sample	Doped amount, ng/g	All methods, ng/g			NAA work, ng/g
		Mean	Median	Range	
1	155	163	160	99–262	135 ± 14
2	95	79	91	11–175	96 ± 11
3	225	216	223	63–325	278 ± 16
4	220	206	210	71–277	265 ± 4
5	45	42	42	20–105	56 ± 3

Table 2
Arsenic Determination in NIST Certified Reference Materials

Sample	Certified value, μg/g	NAA Work, μg/g
NIST 1577 Bovine Liver	0.047 ± 0.006	0.048 ± 0.006
NIST 1575 Pine Needles	0.21 ± 0.04	0.190 ± 0.009
NIST 1572 Citrus Leaves	3.1 ± 0.3	2.97 ± 0.09
NIST 1632b Coal	3.72 ± 0.09	3.71 ± 0.08

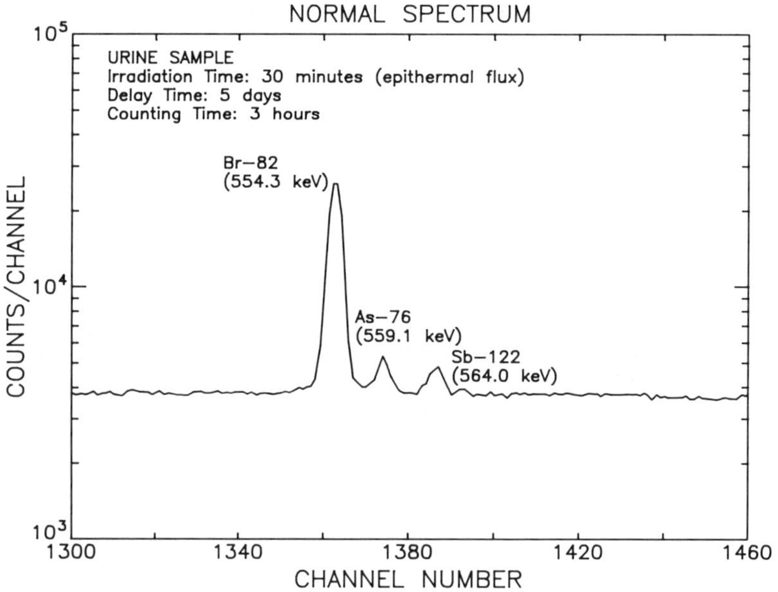

Fig. 1. Normal epithermal-neutron-activated urine sample.

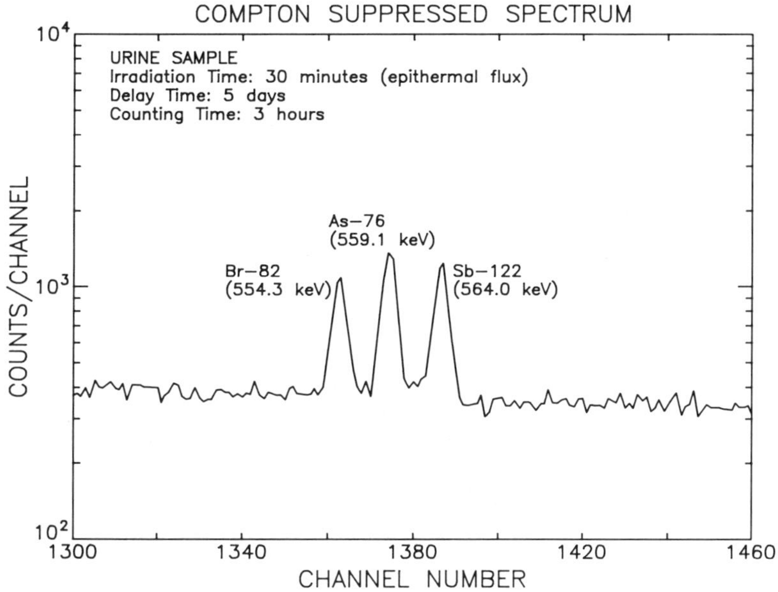

Fig. 2. Compton-suppressed epithermal-neutron-activated urine sample.

peaks are also reduced owing to random coincidences. Although not determined in this present study, it appears that the sensitivity for antimony detection is also greatly improved.

Our experiments indicated that the ideal time for counting the samples was between 4–6 d after the end of the irradiation. The detection limits decreased from 40 ng/g after a 2-d decay to 10 ng/g after 5-d decay. No statistical difference in the results was seen between the normal and anti-Compton mode. Two series of experiments on one of the urine samples and two NIST reference materials confirmed this observation. The results presented in Table 1 are in good agreement with the target values and results from other laboratories. Typically the precision of an individual sample was between 2 and 6%. However, the replicate precision varied from 2 to 13%.

The results for several NIST standards (Table 2) are in excellent agreement with either NIST values or information values from the Gladney et al. (*13*) publication on NIST reference materials. Typically, the individual precision varied from 3% for the NIST SRM 1632b Coal standard to 8% for the NIST SRM 1577 Bovine Liver standard. The precision replicate determinations were between 2 and 14%. In general, the accuracies were in better agreement with the NIST values than the doped arsenic samples of urine.

SUMMARY

We have demonstrated that the use of epithermal neutron activation analysis, in conjunction with a Compton suppression system, is ideally suited to determine arsenic in difficult matrices, such as urine samples. The validity of these techniques for biological samples is born out by the very good accuracy and precision for low amounts of arsenic detection. Furthermore, these techniques appear to be useful for any certification program or interlaboratory exercise.

ACKNOWLEDGMENTS

We are thankful for the cooperation of the University of Illinois Nuclear Reactor staff and operators for the construction of the epithermal facility and irradiation of the samples. We also appreciate the efforts of J. Williams and R.O. Peach for the calculation of neutron flux and profile measurements.

REFERENCES

1. E. L. Baker, C. G. Hayes, P. J. Landrigan, J. L. Hanke, R. T. Leger, W. J. Houseworth, and J. M. Harrington, *Am. J. Epidemiol.* **106,** 261 (1977).
2. F. Peter, G. Growcock, and G. Strung, *Anal. Chim. Acta* **104,** 177 (1979).
3. J. P. Buchet, R. Lauwerys, and H. Roels, *Int. Arch. Occup. Environ. Health* **46,** 11 (1980).
4. G. V. Iyengar, *Biol. Trace Element Res.* **12,** 263 (1987).
5. K. Heydorn, *Neutron Activation Analysis for Clinical Trace Element Research,* vol. 2, CRC Press, Boca Raton, FL, 1984, pp. 79–80.
6. R. Cornelius, A. Speecke, and J. Hoste, *Anal. Chim. Acta* **78,** 317 (1975).
7. N. V. Bagdavadze, and L. M. Mosulishvili, *J. Radioanal. Chem.* **24,** 65 (1974).
8. L. E. Wangen, and E. S. Gladney, *Anal. Chim. Acta* **96,** 271 (1977).
9. S. Landsberger and A. Simsons, *Biol. Trace Element Res.* **13,** 357 (1987).
10. R. O. Peach, *Neutron Spectrum Unfolding of the University of Illinois Advanced TRIGA,* M. S. Thesis, Dept. Nuclear Eng., University of Illinois, Urbana, IL, 1988.
11. L. A. Currie, *Anal. Chem.* **40,** 540 (1968).
12. R. Zeisler, R. R. Greenberg, and S. F. Stone, *J. Nucl. Radioal. Chem.* **124,** 47 (1988).
13. E. S. Gladney, B. T. O'Malley, I. Roelandts, and T. E. Gills, *Compilation of Elemental Concentration Data of NBS Clinical, Biological, Geological, and Environmental Standard Reference Material,* NIST Special Report 260-111 US Department of Commerce, Washington, DC, 1987.

CADEIA and REJECT

Codes to Compute Relevant γ–γ and γ–X True-Coincidence Lines in Absolute Counting of Gamma Rays with a LEPD

M. CARMO FREITAS,*[,1] M. CARMO LANÇA,[1]
ALEXANDRA M. CARVALHO,[1] AND F. DE CORTE[2]

[1]LNETI/ICEN, Departamento de Energia e Engenharia Nucleares, 2686 Sacavém Codex, Portugal; and [2]Institute for Nuclear Sciences, Ghent University, Proeftuinstr. 86, 9000 Ghent, Belgium

Received April 17, 1989; Accepted September 14, 1989

ABSTRACT

CADEIA and REJECT are computer codes for generating γ–γ and γ–X true-coincidence lines for calculation of correction factors for absolute counting of gamma rays in a low energy photon detector. They are written in FORTRAN 77. True-coincidence lines are generated by CADEIA; they are submitted to selection by REJECT according to the rejection criteria defined by L. Moens et al. and F. De Corte. Final relevant lines for true-coincidence correction factors are given as output of REJECT.

Index Entries: Activation analysis; coincidences; correction factors; gamma rays; low energy photon detector; neutron activation analysis; spectra analysis.

INTRODUCTION

True-coincidence effects occur when two or more cascading radiations—emitted in the decay of a radionuclide with negligible time delay—give rise to a total or partial energy deposition in the germanium

*Author to whom all correspondence and reprint requests should be addressed.

crystal. For a few interesting gamma rays in neutron activation analysis (NAA), one should take into account the finite lifetime for the intermediate energy levels when calculating the true-coincidence correction factors. Figure 1 gives a survey of the analytically relevant true-coincidence effects in the case of gamma-ray measurements performed on a low energy photon detector (LEPD). Counting with a LEPD becomes very interesting for gamma rays below 150 keV. In this energy range, the detection efficiency is high, and the resolution is much better than a high-energy coaxial germanium detector (Ge(Li) or HPGe). Many nuclides emit the most intense lines, or even have a single line, in the energy range below 150 keV; this is the case, for example, for most of the lanthanide isotopes. Table 1 shows some nuclides for which counting with a LEPD improves their detection limits in NAA.

The use of the k_0-standardization method in NAA was developed up to now for counting with coaxial Ge(Li) or HPGe detectors (ideal working range: 150–2000 keV). Extension to counting in a LEPD is now being considered. This requires, in the 3–400 keV energy range, procedures for absolute peak efficiency calibration, determination of the peak-to-total ratio, determination of the total detection efficiency, and calculation of correction factors for true-coincidences caused by γ–γ, γ–X (internal conversion, IC), and γ-X (electron capture, EC). Furthermore, k_0-factors and related nuclear data will be experimentally determined with high accuracy for the analytically interesting low-energy gamma rays of some isotopes formed upon (n, γ) activation.

In this work, the computer codes CADEIA and REJECT are presented. Their aim is to compute relevant γ–γ, γ–X (IC), and γ–X (EC) true-coincidence lines for calculation of correction factors for absolute counting of gamma rays with a LEPD. The basic theory and rejection criteria follow procedures previously described (*1–3*). CADEIA gives the true-coincidence lines for any gamma ray of any nuclide. The output of CADEIA includes lines of 2–5 gamma rays for loss coincidences and 2 or 3 gammas for sum coincidences. The output of CADEIA, together with the necessary nuclear data, is the input of REJECT, a code that classifies γ- and X-ray as relevant or not relevant in each coincidence line. The γ–γ and γ-X true-coincidence lines, given as output of REJECT, will enter the code COINLEPD (in elaboration) for calculating the true-coincidence correction factors, which will be checked experimentally.

METHODS

Figure 2 shows the flow chart of CADEIA. For each nuclide, a triangular matrix representing the decay scheme is built: in this matrix, a column and a line stand for the initial and final levels of a gamma transition. For each gamma, CADEIA calls two subroutines: PERDA and SOMA. There are lines of 2, 3, 4, or 5 gammas. According to the position

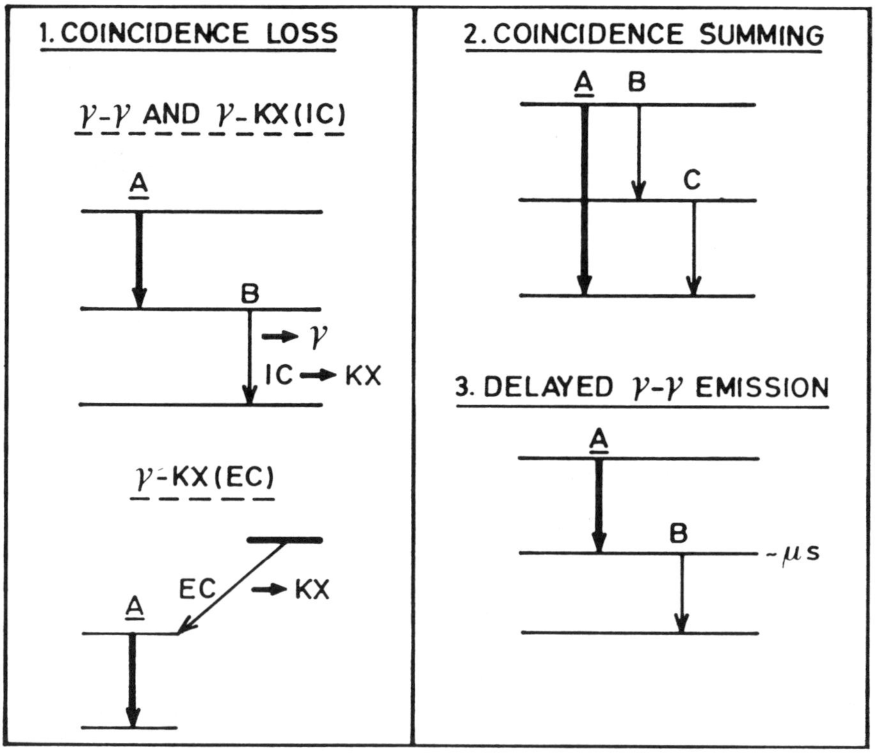

Fig. 1. Survey of analytically relevant true-coincidence effects. The full-energy peak area of emitted gamma ray A ($E_\gamma \leq 150$ keV) is measured with a low energy photon detector (LEPD).

of the measured gamma in the line, PERDA builds the coincidence loss lines. SOMA builds the coincidence sum lines of 2 or 3 gammas. The CADEIA output is processed manually to subtract repeated lines. Using the correct output of CADEIA, the necessary input for REJECT is prepared.

Figures 3 and 4 show the flow chart of REJECT code. The input of REJECT consists of three different matrices containing pertinent gamma lines given by CADEIA and the necessary X-rays (EC or IC). For gamma transitions, the gamma energy (keV), Eγ, the branching ratio, a, $c = (1 + \alpha_t)^{-1}$, where α_t is the total internal conversion coefficient, the gamma intensity, Iγ, and the K-internal conversion coefficient, α_k. For X-rays, the X-energy (keV), E_x, the probability that an electron that fills the hole in the K-shell belongs to the L-shell, K_α, and the fluorescence yield, w_k, are included, and true-coincidence sum- or loss-lines for each gamma to be measured in a LEPD and for which k_0-factors are or will be available.

For each loss true-coincidence line, REJECT calls the subroutine GAMA, and also XIC (or XEC) if there is internal conversion (or electron

Table 1
Some Nuclides for which Counting with a LEPD Improves Their Detection Limit in NAA

Nuclide	Energy (abs. intensity)[4], keV (%)	k_0[3] ● recommended ○ tentative x not available
^{46m}Sc	142.5 (62.0)	x
^{104m}Rh	51.4 (48.3)	x
$^{109}Pd/^{109}Ag$	88.03 (3.6)	x
^{141}Ce	145.4 (48.4)	●
^{143}Ce	57.88 (11.8)	x
^{147}Nd	91.1 (28.0)	●
^{149}Nd	97.0 (1.53); 114.3 (18.8); 155.9 (6.1)	x; x; x
^{151}Nd	116.71 (46.8); 138.9 (7.8); 85.3 (2.72)	x; x; x
^{151}Pm	99.99 (2.53); 64.85 (1.93); 65.82 (1.14); 101.9 (1.29); 104.8 (3.52)	x; x; x; x; x
^{152m}Eu	121.8 (7.2)	x
^{152}Eu	121.8 (28.4)	x
^{153}Sm	69.7 (5.32); 103.2 (28.3)	○; ●
^{154}Eu	123.1 (40.5)	x
^{155}Sm	104.3 (75); 141.4 (2.03)	x; ●
^{159}Gd	57.94 (2.27)	x
^{160}Tb	86.8 (13.2)	●
^{161}Gd	56.29 (3.8); 102.5 (14.0)	x; x
^{161}Tb	25.7 (21.0); 48.9 (16); 74.6 (9.8)	x; x; x
^{165m}Dy	108.2 (3.01)	○
^{165}Dy	94.7 (3.58)	●
^{166}Ho	80.6 (12.8)	●
^{169}Yb	63.12 (47.3); 109.8 (17.4); 130.5 (11.1)	x; x; x
^{170}Tm	84.3 (3.26)	○
^{171}Er	111.6 (20.6); 116.7 (2.30); 124.0 (9.1)	●; ●; ●
^{175}Yb	113.8 (1.91)	●
^{176m}Lu	88.4 (8.9)	●
^{177}Lu	112.9 (21.5)	x
^{177}Yb	121.6 (3.4); 138.6 (1.33); 150.4 (20.0)	○; ○; ○
^{182}Ta	67.75 (41.2); 100.1 (14.0)	x; x
^{187}W	71.98 (10.77); 134.2 (8.56)	x; ●
^{197}Pt	77.3 (17.1)	x
^{197}Hg	77.3 (18.0)	x
^{233}Th	29.36 (2.6); 86.53 (2.6)	x; x
^{239}U	43.54 (4.44); 74.67 (52.2)	x; x
^{239}Np	106.1 (22.7)	x

capture). The flow chart of GAMA and XIC are shown in Fig. 4. GAMA searches for nonrelevant gammas in each line; XIC searches for nonrelevant X-rays originated by internal conversion in each line, and XEC will be written in the near future. For each sum true-coincidence line, REJECT calls the subroutine SGAMA; it searches for nonrelevant gammas in each line. The output of REJECT, after a few necessary changes, will feed

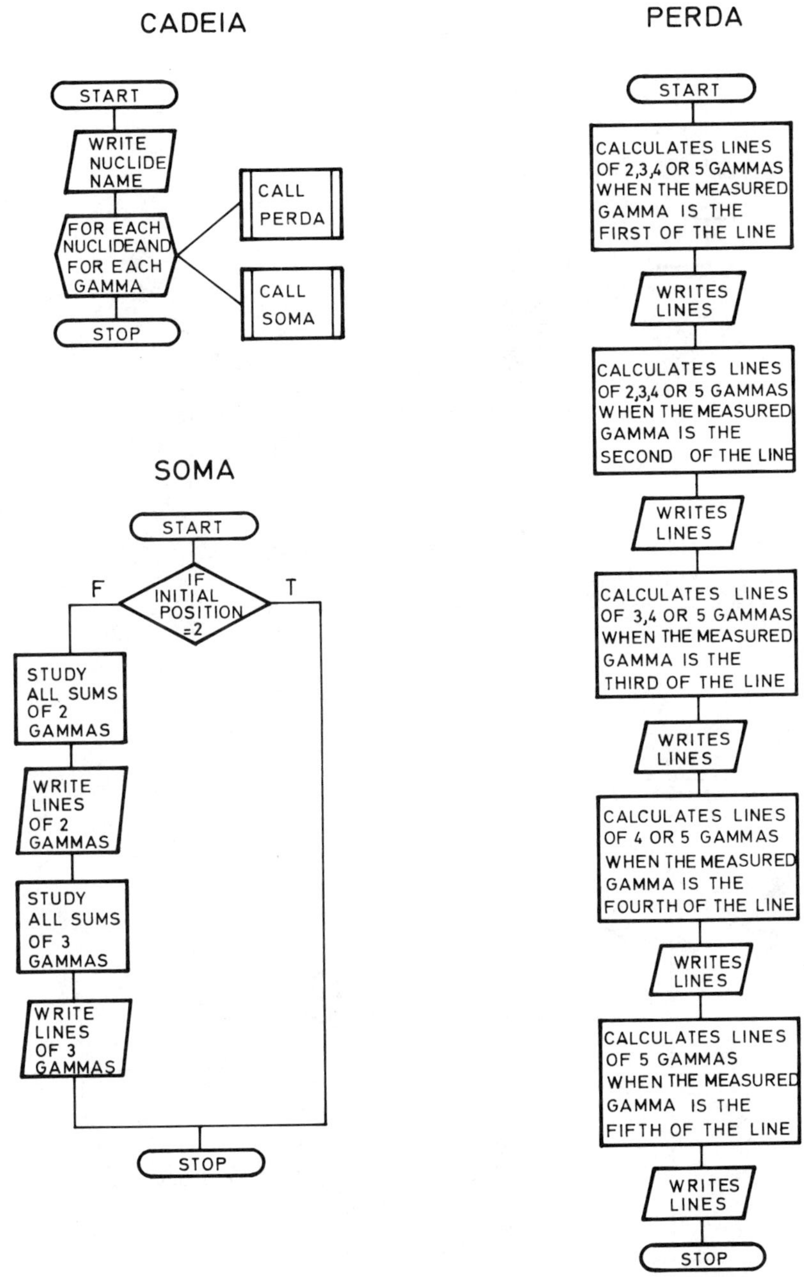

Fig. 2. Flow chart of CADEIA computer code and its subroutines.

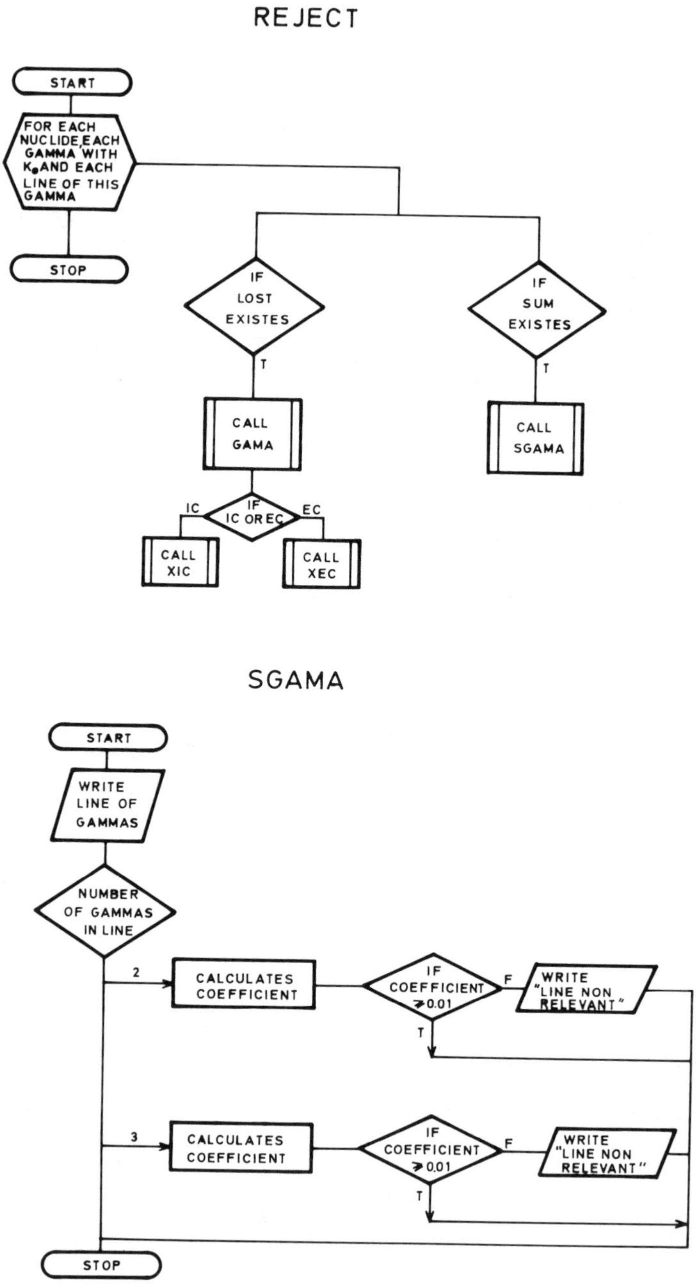

Fig. 3. Flow chart of REJECT computer codes and its subroutine SGAMA.

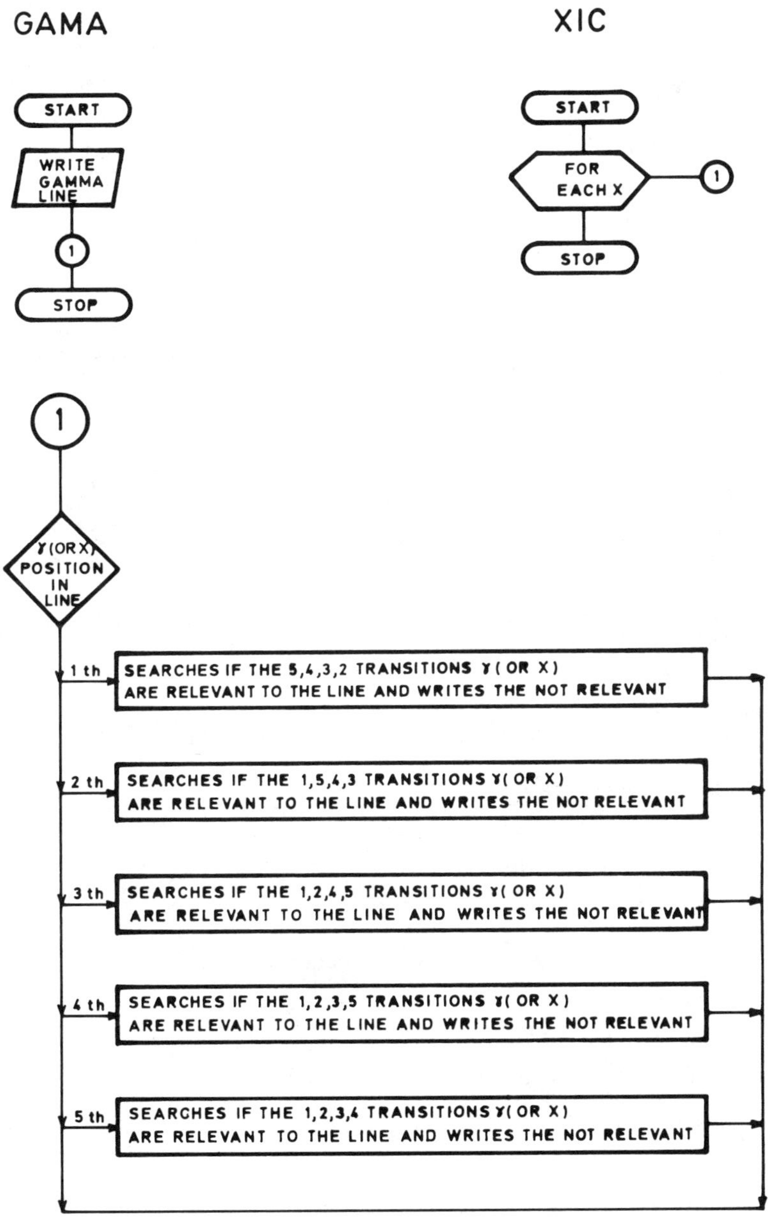

Fig. 4. Flow chart of REJECT subroutines GAMA and XIC.

INPUT OF CADEIA: EXAMPLE OF A DECAY SCHEME MATRIX Yb177

FINAL LEVEL	INITIAL LEVEL 1	2	3	4	5	6	7	8	9	10	11	12
1	0.0	121.62	150.4	268.79	0.0	0.0	0.0	1063.0	1497.0	1230.7	1241.8	1336.4
F 2	0.0	0.0	0.0	147.17	0.0	0.0	927.8	0.0	1028.0	1109.0	1119.7	1214.8
I 3	0.0	0.0	0.0	0.0	138.6	0.0	899.0	912.8	0.0	1080.2	0.0	0.0
N 4	0.0	0.0	0.0	0.0	0.0	0.0	0.0	0.0	0.0	961.8	0.0	0.0
A 5	0.0	0.0	0.0	0.0	0.0	162.5	760.4	0.0	0.0	941.6	0.0	0.0
L 6	0.0	0.0	0.0	0.0	0.0	0.0	0.0	0.0	0.0	779.1	0.0	0.0
7	0.0	0.0	0.0	0.0	0.0	0.0	0.0	0.0	0.0	0.0	0.0	0.0
L 8	0.0	0.0	0.0	0.0	0.0	0.0	0.0	0.0	0.0	0.0	0.0	0.0
E 9	0.0	0.0	0.0	0.0	0.0	0.0	0.0	0.0	0.0	0.0	0.0	0.0
V 10	0.0	0.0	0.0	0.0	0.0	0.0	0.0	0.0	0.0	0.0	0.0	0.0
E 11	0.0	0.0	0.0	0.0	0.0	0.0	0.0	0.0	0.0	0.0	0.0	0.0
L 12	0.0	0.0	0.0	0.0	0.0	0.0	0.0	0.0	0.0	0.0	0.0	0.0

INPUT OF REJECT (necessary data)
EXAMPLE OF A NUCLEAR CONSTANT MATRIX Yb177
Matrix 1

E_γ	a	c	I_γ	α_κ
1214.80	0.680	0.996	0.028	0.004
1119.69	0.140	0.994	0.540	0.005
1109.02	0.024	0.994	0.176	0.005
1080.25	0.760	0.998	5.500	0.001
961.85	0.003	0.992	0.019	0.007
941.64	0.140	0.998	1.010	0.002
779.14	0.017	0.998	0.121	0.002
1028.04	0.500	0.993	0.630	0.005
912.80	0.660	0.992	0.033	0.008
927.82	0.050	0.992	0.039	0.008
899.05	0.870	0.991	0.640	0.008
760.44	0.070	0.976	0.055	0.001
162.50	1.000	0.513	0.060	0.800
138.61	1.000	0.416	1.330	1.170
147.17	0.520	0.500	0.181	0.900
150.40	1.000	0.643	20.00	0.416
121.62	1.000	0.366	3.400	1.270
IC: E	k_α	W_κ		
52.960	0.273	0.953	0.000	0.000
54.070	0.473	0.953	0.000	0.000
61.220	0.101	0.953	0.000	0.000

EXAMPLE OF A LOSS COINCIDENCE LINES MATRIX
Yb177 Interesting : 150.40
Matrix 2

ORIG. X	POS. γ	NUM. γ	NUM. X	COINCIDENCE LINE				
1	2	2	3	11	16	0	0	0
1	2	2	3	9	16	0	0	0
1	2	2	3	4	16	0	0	0
1	3	3	3	12	14	16	0	0
1	3	3	3	6	14	16	0	0
1	4	4	3	7	13	14	16	0

EXAMPLE OF A SUM COINCIDENCE LINES MATRIX :
Sm155 Interesting : 104.35
Matrix 3

NUM.	INT.	COINCIDENCE LINE		
2	34	35	36	0

OUTPUT OF CADEIA (OUTPUT OF REJECT) Yb177

```
          121.62
  LOSS
 927.82   121.62 (LINE NOT RELEVANT)
1028.04   121.62 (X RAYS NOT RELEVANT)
1109.02   121.62 (X RAYS NOT RELEVANT)
1119.69   121.62 (X RAYS NOT RELEVANT)
1214.80   121.62 (LINE NOT RELEVANT)
 981.85   147.17   121.62 (GAMMA 961.85 AND X RAYS NOT RELEVANT)
  SUM
          150.40
  LOSS
 899.05   150.40 (X-RAYS NOT RELEVANT)
 912.80   150.40 (LINE NOT RELEVANT)
1080.25   150.40 (X-RAYS NOT RELEVANT)
 760.44   138.61  150.40 (GAMMA AND X-RAYS NOT RELEVANT FOR 760.44)
 941.70   138.61  150.40
 779.14   162.50  138.61  150.40 (LINE NOT RELEVANT)
  SUM
          138.61
  LOSS
 760.44   138.61  150.40 (X-RAYS NOT RELEVANT FOR 760.44)
 941.70   138.61  150.40 (X-RAYS NOT RELEVANT FOR 941.70)
 779.14   162.50  138.61  150.40 (X-RAYS NOT RELEVANT FOR 779.14 AND 162.50)
  SUM
```

Fig. 5. Input-output example of CADEIA and REJECT for ^{155}Sm and ^{177}Yb (γ-energy is in keV).

COINLEPD, which will calculate the coefficients for coincidence corrections.

RESULTS AND DISCUSSION

Figure 5 shows a typical input-output of CADEIA and REJECT for ^{155}Sm and ^{177}Yb. The input matrix of CADEIA contains the decay scheme of ^{177}Yb. The three input matrices of REJECT contains: matrix 1—necessary nuclear data of ^{177}Yb; matrix 2— for an interesting gamma of ^{177}Yb (150.40 keV), the loss lines and the necessary data are given: for example in the first line of the matrix, the internal conversion (IC = 1), the measured gamma (second in the line), the line's number of gammas (2), the number of X-rays (3), and the line composed by the 11th (899.05 keV) and the 16th (150.40 keV) gammas (cf. 1st column of matrix 1); matrix 3—for an interesting gamma of ^{155}Sm (104.35 keV), the single sum line and the necessary data are given: the line's number of gammas (2), the measured gamma (34th in the 1st column of the nuclear constants matrix of ^{155}Sm), and the line composed by the 35th (25.72 keV) and the 36th (78.63 keV) gammas of the 1st column of the nuclear constants matrix of ^{155}Sm.

In the outputs for ^{177}Yb, the figures and words in brackets correspond to the CADEIA and REJECT outputs, respectively. According to the rejection made by REJECT, the CADEIA output becomes simplified.

This is a small part of a joint project, the goal of which is to extend the k_0-standardization method to counting in a LEPD, involving both the LNETI/ICEN (Portugal) and the Ghent University/Institute for Nuclear Sciences (Belgium). Later communications will complement this work.

REFERENCES

1. L. Moens, *Ph. D. Thesis,* University Ghent, Ghent, Belgium, 1981.
2. L. Moens, F. De Corte, A. Simonits, L. Xilei, A. De Wispelaere, J. De Donder, and J. Hoste, *J. Radioanal. Chem.* **70,** 539 (1982).
3. F. De Corte, *Agrégé-Thesis,* Ghent University, Ghent, Belgium, 1987.
4. E. Browne and R. B. Firestone, *Table of Radioactive Isotopes,* Wiley, New York, 1986.
5. C. M. Lederer and V. S. Shirley, *Table of Isotopes,* 7th ed., Wiley, New York, 1978.
6. F. Lagoutine, N. Coursol, and J. Legrand, *Tables de Radionucléides,* vol. 3, LMRI, CEA, Grif-pur-Yvette, France, 1985.
7. J. Lyttkens, K. Nilson, and L. P. Ekstrom, *Nuclear Data Sheets,* vol. 16, pp. 157–159, 1975.
8. J. Lyttkens, K. Nilson, and L. P. Ekstrom, Nuclear Data Sheets, vol. 50, pp. 590–600, 1987.

Determination of Trace Elements in Standard Reference Materials by the k_0-Standardization Method

B. SMODIŠ,*[,1] R. JAĆIMOVIĆ,[1] S. JOVANOVIĆ,[2] AND P. STEGNAR[1]

[1]*Jožef Stefan Institute, E. Kardelj University, Jamova 39, 61000 Ljubljana, Yugoslavia; and* [2]*Institute for Mathematics and Physics, V. Vlahović University, Cetinjski put bb., 81000 Titograd, Yugoslavia*

Received April 17, 1989; Accepted September 14, 1989

ABSTRACT

The k_0-standardization method is suitable for routine multielement determinations by reactor neutron activation analysis (NAA). Investigation of NIST standard reference materials SRM 1571 Orchard Leaves, SRM 1572 Citrus Leaves, and SRM 1573 Tomato Leaves showed the systematic error of 12 certified elements determined to be less than 8%. Thirty-four elements were determined in NIST proposed SRM 1515 Apple Leaves.

Index Entries: Neutron activation analysis (NAA); K_0-standardization; standard reference materials; trace elements; multielement determination.

INTRODUCTION

In recent years, an increasing interest has been manifested in the determination of trace elements in biological materials in relation to biological and environmental studies. In this respect, NAA plays an

*Author to whom all correspondence and reprint requests should be addressed.

important role owing to its accuracy, sensitivity, and specificity. However, in most cases, a relative method is used in which the sample and a standard containing known amounts of the elements of interest are irradiated simultaneously, followed by measurement of their activities under identical conditions.

This approach works well when only one or a few elements have to be determined. When a great number of elements in one sample have to be determined, the treatment of the standards regarding preparation, dissolution, and counting becomes cumbersome and time-consuming, besides the difficulty of irradiating a large number of standards and sample together in one capsule.

The demand for multielement analyses enhanced the need for an absolute method. Unfortunately, it requires an arbitrarily selected nuclear data set, thus introducing corresponding uncertainties in the results. To overcome this difficulty, the "single-comparator" method was introduced in the mid-60s (*1*), using a primarily assessed constant, *k* that was summarized from all nuclear data involved in activation and counting and determined experimentally for each element, with respect to a chosen standard element. In the period 1973–1987, a new approach, named the k_0-method, was developed at the Institute for Nuclear Sciences, Ghent, Belgium (*2,3*). Essentially, it is an absolute technique where the uncertain nuclear constants are replaced by compound nuclear constants, the k_0-factors. These factors, which characterize the nuclides, were experimentally determined with high accuracy by the authors of the method. During 1987–1988, the k_0-standardization method was implemented at the Jožef Stefan Institute in Ljubljana, Yugoslavia (*4–7*).

The aim of the present work was to study the possibilities of applying this method to the analysis of biological materials. Three National Institute for Standards and Technology (NIST, formerly NBS) biological standard reference materials (SRM), Orchard Leaves, Citrus Leaves, and Tomato Leaves, were analyzed for about 30 elements. In addition, the NIST-proposed SRM 1515 Apple Leaves was characterized for 34 elements.

EXPERIMENTAL

Two-hundred milligram aliquots were packed in polyethylene ampules and irradiated for 20 h in the carousel facility of the TRIGA Mark II reactor of the Jožef Stefan Institute, Ljubljana (thermal-neutron fluence rate 1.1×10^{12} n/cm^2/s^1; alpha = -0.014). An aliquot of each material was separately dried in accordance with the recommendation given by the distributor in order to obtain the moisture content. An Al-0.1% Au alloy wire (Central Bureau for Nuclear Measurements, Geel, Belgium) of 1.0 mm diameter and 125 μm Zr foil (Goodfellow, Cambridge, England)

Table 1
Relevant Nuclear Data for the Nuclides Used for the Analysis

Element	Nuclide	Eγ,keV	$k_{0,Au}$
Ag	^{110m}Ag	657.8; 884.7	3.44E−02; 2.65E−02
As	^{76}As	559.1; 657.1	4.83E−02; 6.61E−03
Au	^{198}Au	411.8	1.0
Ba	^{131}Ba	216.1; 373.2; 496.3	2.91E−05; 2.03E−05; 6.84E−05
Br	^{82}Br	698.4; 776.5;1044.0	9.38E−03; 2.76E−02; 9.14E−03
Ca	^{47}Ca	1297.1	9.54E−07
	^{47}Sc	159.4	8.59E−07
Ce	^{141}Ce	145.4	3.66E−03
Co	^{60}Co	1173.2;1332.5	1.32E+00; 1.32E+00
Cr	^{51}Cr	320.1	2.62E−03
Cs	^{134}Cs	569.3; 604.7; 795.8	7.34E−02; 4.76E−01; 4.15E−01
Eu	^{152}Eu	964.0;1112.1;1408.1	1.08E+01; 1.00E+01; 1.53E+01
	^{154}Eu	1274.4	7.77E−01
Fe	^{59}Fe	1099.2;1291.6	7.77E−05; 5.93E−05
Gd	^{159}Gd	363.6	8.28E−04
Hf	^{181}Hf	133.0; 345.9; 482.2	2.37E−02; 7.93E−03; 4.56E−02
Ho	^{166}Ho	1379.4	6.91E−03
K	^{42}K	1524.7	9.46E−04
La	^{140}La	328.8; 815.8;1596.5	2.87E−02; 3.32E−02; 1.34E−01
Lu	^{177}Lu	112.9; 208.4	3.63E−02; 6.05E−02
Na	^{24}Na	1368.6	4.68E−02
Nd	^{147}Nd	91.1; 531.0	1.02E−03; 4.56E−04
Pr	^{142}Pr	1575.6	6.12E−03
Rb	^{86}Rb	1076.6	7.65E−04
Re	^{186}Re	137.2	4.33E−02
Sb	^{124}Sb	602.7; 722.8;1691.0	2.96E−02; 3.19E−03; 1.41E−02
Sc	^{46}Sc	889.3;1120.5	1.22E+00; 1.22E+00
Se	^{75}Se	136.0; 264.7	6.89E−03; 7.25E−03
Sm	^{153}Sm	103.2	2.31E−01
Sr	^{85}Sr	514.0	9.15E−05
Tb	^{160}Tb	879.4; 966.2;1178.0	9.42E−02; 7.84E−02; 4.71E−02
Th	^{233}Pa	312.0	2.52E−02
Tm	^{170}Tm	84.3	4.30E−02
U	^{239}Np	228.1; 277.6	2.77E−03; 3.40E−03
Yb	^{175}Yb	282.5; 396.3	1.46E−02; 3.12E−02
Zn	^{65}Zn	1115.5	5.72E−03

were coirradiated together with the samples as comparator and fluence rate monitors.

After a cooling time of 2 d, the samples were counted on a calibrated 106 cm^3 HP Ge detector (FWHM = 1.8 keV) (*4*) connected to a Canberra Series 90 multichannel analyzer. The measurements were repeated after 8 d cooling time. The spectra were evaluated by a PC version of the SEQAL program (*8*). The effective solid angles of measured samples and

Table 2
Comparison of Elemental Concentrations in SRM 1571 Orchard Leaves

Elem.	Units	Concn. Found, Mean +SD	(*n*)	NBS, Mean +SD	Consensus (12), Mean +SD
As	μg/g	10.6 + 0.5	(5)	10 + 2	10.7 + 1.3
Au	ng/g	2.6 + 0.4	(3)	–	1.4 + 0.4
Ba	μg/g	38 + 3	(5)	44	43 + 4
Br	μg/g	9.49 + 0.12	(5)	10	9.5 + 1.1
Ca	%	1.89 + 0.07	(5)	2.09 + 0.03	2.04 + 0.12
Ce	ng/g	927 + 48	(5)	–	990 + 120
Co	ng/g	136 + 3	(5)	200	160 + 37
Cr	μg/g	2.55 + 0.11	(5)	2.6 + 0.3	2.6 + 0.3
Cs	ng/g	42 + 14	(4)	40	38 + 9
Eu	ng/g	16 + 8	(3)	–	24 + 3
Fe	μg/g	280 + 7	(5)	300 + 20	286 + 28
Hf	ng/g	28 + 12	(5)	–	30 + 5
Hg	ng/g	100	(2)	155 ± 15	155 ± 14
K	%	1.45 ± 0.02	(5)	1.47 ± 0.03	1.44 ± 0.07
La	μg/g	1.10 ± 0.02	(5)	–	1.17 ± 0.11
Lu	ng/g	7.3 ± 4.0	(5)	–	5.1 ± 2.5
Na	μg/g	94 ± 6	(5)	82 ± 6	89 ± 15
Nd	ng/g	847 ± 341	(4)	–	510 ± 130
Pt	μg/g	1.3 ± 0.6	(4)	–	0.43 ± 0.67
Rb	μg/g	11.2 ± 0.6	(5)	12 ± 1	11.4 ± 1.2
Re	ng/g	3.0 ± 0.9	(5)	–	–
Sb	μg/g	3.0 ± 0.3	(5)	2.9 ± 0.3	2.9 ± 0.3
Sc	ng/g	65.6 ± 1.4	(5)	–	63 ± 14
Sm	ng/g	110 ± 4	(5)	–	114 ± 20
Sr	μg/g	47 ± 8	(5)	37 ± 1	36 ± 3
Tb	ng/g	12 ± 3	(4)	–	13 ± 3
Th	ng/g	78 ± 14	(4)	64 ± 6	58 ± 12
U	ng/g	30 ± 8	(5)	29 ± 5	29 ± 3
Yb	ng/g	24 ± 4	(5)	–	25 ± 5
Zn	μg/g	24 ± 2	(5)	25 ± 3	25 ± 2

monitors were calculated by the SOLANG program (*9*) and the concentrations of the elements by the SINGCOM program (*10*), both running on a VAX 11/750 computer.

RESULTS AND DISCUSSION

The gamma lines considered, and the relevant k_0 factors taken from ref. (*11*) for the particular nuclides used for the analyses of biological materials, are presented in Table 1. The analytical results for NIST SRM 1571 Orchard Leaves, SRM 1572 Citrus Leaves, and SRM 1573 Tomato Leaves are presented in Tables 2–4, respectively. The results are pre-

Table 3
Comparison of Elemental Concentrations in SRM 1572 Citrus Leaves

Elem.	Units	Concn. Found, Mean +SD	(*n*)	NBS, Mean +SD	Consensus (12), Mean +SD
As	µg/g	3.55 ± 0.14	(5)	3.1 ± 0.3	3.0 ± 0.3
Au	ng/g	0.13 ± 0.03	(5)	–	0.11
Ba	µg/g	17.3 ± 0.7	(5)	21 ± 3	23.5
Br	µg/g	7.89 ± 0.13	(5)	8.2	8.36
Ca	%	2.98 ± 0.07	(5)	3.15 ± 0.10	3.13 ± 0.04
Ce	ng/g	334 ± 30	(5)	280	453
Co	ng/g	15.3 ± 1.7	(4)	20	16
Cr	µg/g	0.65 ± 0.08	(5)	0.8 ± 0.2	1
Cs	ng/g	85 ± 5	(4)	98	93 ± 16
Eu	ng/g	5.3 ± 0.7	(4)	10	13.5
Fe	µg/g	77 ± 7	(5)	90 ± 10	101 ± 6
K	%	1.80 ± 0.04	(5)	1.82 ± 0.06	1.83 ± 0.04
La	ng/g	165 ± 21	(5)	190	198
Lu	ng/g	2.9	(2)	–	1.55
Na	µg/g	160 ± 4	(5)	160 ± 20	163 ± 1
Nd	ng/g	320	(1)	–	317
Rb	µg/g	4.63 ± 0.12	(5)	4.84 ± 0.06	–
Re	ng/g	5.2 ± 0.7	(5)	–	–
Sb	ng/g	40 ± 4	(4)	40	34
Sc	ng/g	10.7 ± 1.0	(5)	10	10.4 ± 0.5
Sm	ng/g	46 ± 2	(5)	52	50
Sr	µg/g	116 ± 10	(5)	100 ± 2	98 ± 3
Tb	ng/g	10 ± 7	(3)	–	9
Th	ng/g	22 ± 5	(3)	–	–
U	ng/g	38 ± 3	(5)	<150	40 ± 2
Yb	ng/g	8.0 ± 1.1	(5)	–	11.5
Zn	µg/g	27.8 ± 0.9	(5)	29 ± 2	29.9 ± 1.4

sented as the arithmetic means of *n* determinations with associated standard deviations. The NBS and CONSENSUS values, in the case where no certified values are available; were taken from Ref. (*12*).

It should be noted that, in this work, only relatively long-lived nuclides ($T_{1/2} > 10$ h) have been used for the analysis. The short-lived ones will be investigated in our future work. Relatively good agreement was found between the results obtained and the certified values (better than 8%, on average) or consensus values for the noncertified elements. Systematically higher results were found for Sr, owing to the fact that the computer program could not resolve the 514 keV line of ^{85}Sr well enough from the 511 keV annihilation peak.

In addition, encouraged by the results obtained, we analyzed the NIST proposed SRM 1515 Apple Leaves for 35 elements. In Fig. 1, the

Table 4
Comparison of Elemental Concentrations in SRM 1573 Tomato Leaves

Elem.	Units	Concn. Found, Mean +SD	(*n*)	NBS, Mean +SD	Consensus (12), Mean +SD
As	ng/g	264 ± 83	(5)	270 ± 50	253 ± 36
Au	ng/g	0.53	(2)	–	0.8
Ba	μg/g	53 ± 2	(5)	–	57 ± 9
Br	μg/g	24.2 ± 0.3	(5)	26	21 ± 2
Ca	%	2.85 ± 0.09	(5)	3.00 ± 0.03	2.83 ± 0.23
Ce	μg/g	1.2 ± 0.1	(5)	1.6	1.3 ± 0.2
Co	ng/g	470 ± 14	(5)	600	525 ± 46
Cr	μg/g	4.5 ± 0.3	(5)	4.5 ± 0.5	4.0 ± 0.5
Cs	ng/g	59.9 ± 3.4	(5)	–	57 ± 8
Eu	ng/g	20 ± 7	(5)	40	22 ± 6
Fe	μg/g	636 ± 35	(5)	690 ± 25	580 ± 110
Hf	ng/g	223 ± 50	(5)	–	250
K	%	4.52 ± 0.11	(5)	4.46 ± 0.03	4.44 ± 0.24
La	ng/g	657 ± 89	(5)	900	710 ± 70
Lu	ng/g	20	(2)	–	9.3 ± 2.5
Na	μg/g	628 ± 15	(5)	–	470 ± 110
Rb	μg/g	15.7 ± 0.5	(5)	16.5 ± 0.1	17.3 ± 2.5
Re	ng/g	6.1	(2)	–	–
Sb	ng/g	52.3	(1)	–	36 ± 7
Sc	ng/g	160 ± 8	(5)	130	173 ± 26
Sm	ng/g	86 ± 9	(5)	–	92 ± 16
Sr	μg/g	57.7 ± 5.5	(5)	44.9 ± 0.3	42 ± 5
Tb	ng/g	13.6 ± 6.1	(3)	–	9 ± 5
Th	ng/g	205 ± 74	(5)	170 ± 30	205
U	ng/g	61.8 ± 4.3	(5)	61 ± 3	59 ± 6
Yb	μg/g	53 ± 7	(5)	–	63 ± 16
Zn	μg/g	60.5 ± 3.4	(5)	62 ± 6	61 ± 4

spectra of Apple Leaves are presented, taken 2 and 8 d after the 20 h irradiation. It can be clearly seen that the main activities in the spectra arise from rare earth elements. This is even more evident from Table 5, where the elemental concentrations are summarized. In comparison with SRM 1571 Orchard Leaves, values 3.5–35 times higher were found for particular rare earth elements, considering the consensus values (*12*) as reliable ones. The second difference found in comparison to the other SRMs analyzed are the values found for As and Sb, which are much lower in Apple Leaves.

From the results of this work, it appears that k_0-standardization NAA can be as precise and accurate as the relative method when all the relevant parameters are accurately measured and controlled. The method is particularly powerful as a multielement technique used on a routine basis for the analysis of a larger number of samples.

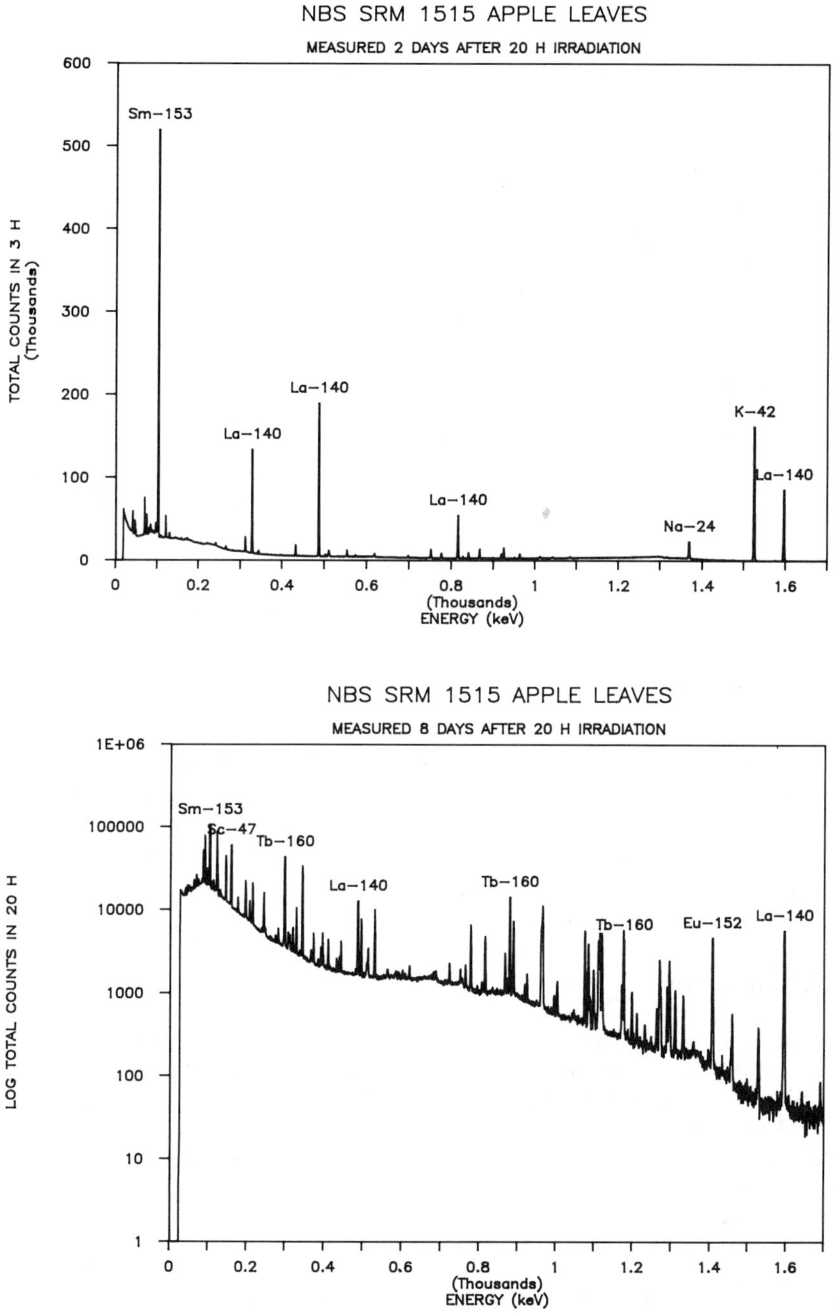

Fig. 1. Gamma spectra of proposed NIST SRM 1515 Apple Leaves obtained after different decay times.

Table 5
Concentration of Elements in NIST Proposed SRM 1515 Apple Leaves

Elem.	Units	Concn. Found, Mean +SD	(*n*)	Elem.	Units	Concn. Found, Mean +SD	(*n*)
Ag	ng/g	193	(2)	Lu	ng/g	45.8 ± 3.5	(5)
As	ng/g	40.2 ± 4.3	(4)	Na	μg/g	25.3 ± 0.5	(5)
Au	ng/g	1.44 ± 0.15	(3)	Nd	μg/g	17.6 ± 0.7	(5)
Ba	μg/g	40.8 ± 2.8	(5)	Pr	μg/g	3.94 ± 0.27	(4)
Br	μg/g	1.61 ± 0.09	(5)	Rb	μg/g	9.65 ± 0.10	(5)
Ca	%	1.52 ± 0.03	(5)	Re	ng/g	1.6	(1)
Ce	μg/g	3.49 ± 0.18	(5)	Sb	ng/g	17.5 ± 2.0	(3)
Co	ng/g	96.3 ± 1.7	(5)	Sc	ng/g	31.3 ± 0.4	(5)
Cr	ng/g	503 ± 176	(5)	Se	ng/g	280	(1)
Cs	ng/g	13 ± 5	(3)	Sm	μg/g	2.84 ± 0.01	(5)
Eu	ng/g	220 ± 7	(3)	Sr	μg/g	34 ± 6	(5)
Fe	μg/g	84 ± 10	(5)	Tb	ng/g	391 ± 23	(5)
Gd	μg/g	3.0 ± 0.8	(4)	Th	ng/g	30 ± 22	(5)
Hf	ng/g	37.9	(1)	Tm	ng/g	64.3	(2)
Ho	ng/g	473	(1)	U	ng/g	31.4	(1)
K	%	1.64 ± 0.04	(5)	Yb	ng/g	193 ± 11	(5)
La	μg/g	21.5 ± 0.3	(5)	Zn	μg/g	12.4 ± 0.6	(5)

ACKNOWLEDGMENTS

The financial support of the Research Communities of Slovenia and Montenegro, Yugoslavia and the National Institute for Standards and Technology Washington, DC, is highly appreciated. The authors wish to thank Frans De Corte (Institute for Nuclear Sciences, Ghent, Belgium) for his valuable help in various ways concerning this work.

REFERENCES

1. F. Girardi, G. Guzzi, and J. Pauly, *Anal. Chem.* **37,** 1085 (1965).
2. A. Simonits, L. Moens, F. De Corte, A. De Wispelaere, A. Elek, and J. Hoste, *J. Radioanal. Chem.* **60,** 461 (1980).
3. L. Moens, F. De Corte, A. De Wispelaere, J. Hoste, A. Simonits, A. Elek, and E. Szabo, *J. Radioanal. Chem. Articles* **82,** 385 (1984).
4. B. Smodiš, R. Jaćimović, S. Jovanović, P. Stegnar, and P. Vukotić, *Vestn. Slov. Kem. Drus.* **35,** 397 (1988).
5. S. Jovanović, B. Smodiš, R. Jaćimović, P. Vukotić and P. Stegnar, *Vestn. Slov. Kem. Drus.* **35,** 409 (1988).
6. S. Jovanović, P. Vukotić B. Smodiš, R. Jaćimović, N. Mihaljević, and P. Stegnar, *J. Radioanal. Nucl. Chem. Articles* **129,** 343 (1989).
7. S. Jovanović, B. Smodiš, R. Jaćimović, P. Vukotić and P. Stegnar, *J. Radioanal. Nucl. Chem. Lett.* **135,** 59 (1989).
8. J. Op De Beeck, *Internal report*, INW, Ghent, Belgium (1976).

9. L. Moens, J. De Donder, Lin Xilei, F. De Corte, A. De Wispelaere, A. Simonits, and J. Hoste, *Nucl. Instr. Methods* **187,** 451 (1981).
10. Lin Xilei, F. De Corte, L. Moens, A. Simonits, and J. Hoste, *J. Radioanal. Nucl. Chem. Articles* **81,** 333 (1984).
11. F. De Corte, A. Simonits, A. De Wispelaere, and J. Hoste, *INW/KFKI Interim Report*, Ghent, Belgium (1986).
12. E. S. Gladney, B. T. O'Malley, I. Roelandts, and T. E. Gills, *NBS Special Publication 260-111*, US Department of Commerce, Washington DC (1987).

A New Microcomputer-Controlled Neutron Activation and Analysis System

Description and Applications for the Analysis of Bioenvironmental Specimens

J. B. EDWARD,[1,3] P. A. BEELEY,[2,3] L. G. I. BENNETT,*[,1,3] AND J. S. POLAND[2]

[1]Department of Chemistry and Chemical Engineering, Royal Military College, Kingston, Canada; [2]Department of Chemistry, Queen's University, Kingston, Canada; and [3]SLOWPOKE-2 Facility at the Royal Military College of Canada, Kingston, Canada

Received April 17, 1989; Accepted September 14, 1989

ABSTRACT

A microcomputer-controlled irradiation and measurement system and a microprocessor-controlled sample changer have been installed at the SLOWPOKE-2 Facility at the Royal Military College of Canada (RMC). These systems can provide the gamut of instrumental neutron activation analysis (INAA) techniques for the analyst. Custom software has been created for system control, data acquisition, and off-line spectral analysis using programs that incorporate Gaussian peak-fitting methods of analysis. The design and use of the equipment is discussed, and the performance is illustrated with results obtained from the analysis of marine sediment and biological reference materials.

Index Entries: Instrumental neutron activation analysis (INAA); automated systems; SLOWPOKE-2 reactor; activation, thermal and epithermal; cyclic activation; biological samples; reference materials.

*Author to whom all correspondence and reprint requests should be addressed.

INTRODUCTION

The SLOWPOKE-2 research reactor at RMC was commissioned in the fall of 1985 and is the first of this type to use low-enriched (20% ^{235}U) uranium dioxide fuel. One benefit has been that the constant and reproducible neutron flux expected of the SLOWPOKE-2 is now available for longer operating periods (*1*). The facility serves a large and diverse research community within the Department of National Defence, as well as other government agencies, and Queen's University at Kingston. In order to provide efficient neutron activation analysis (NAA) services for this community, an Apple IIe microcomputer-controlled irradiation and measurement system and a separate microprocessor-controlled sample changer have been installed in collaboration with Atomic Energy of Canada Limited-Radiochemical Company (now NORDION International).

As a first use of these systems and a means to test their capabilities, various bioenvironmental reference materials were studied. A great deal of data exists in the literature for the INAA of these materials, issued by the US National Bureau of Standards (NBS). In contrast, published data on standards issued by the Marine Analytical Chemistry Standards Program of the National Research Council of Canada (NRC) are not as exhaustive. To augment these data, a systematic study of elemental concentrations in six marine reference materials, namely, MESS-1, BCSS-1, PACS-1 (marine sediments), TORT-1 (lobster hepatopancreas), DORM-1, and DOLT-1 (dogfish muscle and liver), was therefore undertaken. In fact, for the NRC reference materials, data have been obtained for elements whose concentrations have not been previously reported.

In addition, an extensive study is underway on possible metal contamination of submerged weedbeds and marsh biota that may contribute to the low density and biomass of macrophyte beds and Typha marshes found in the Bay of Quinte, Lake Ontario (*2,3*). The large number of samples involved has necessitated the use of the newly installed automated systems.

METHODS

Besides conventional INAA, additional sensitivity for the measurement of short-lived activation products was obtained with the microcomputer-controlled irradiation system by routing samples (pneumatically) back to the reactor for cyclic instrumental neutron activation analysis (CINAA) (*4,5*). This system was also used to irradiate samples automatically for extended periods of time and, at a later date, to act as a sample changer for the measurement of long-lived activation products of such irradiations. A block diagram of the irradiation and measurement system is shown in Fig. 1. This system can also be used for pseudocyclic

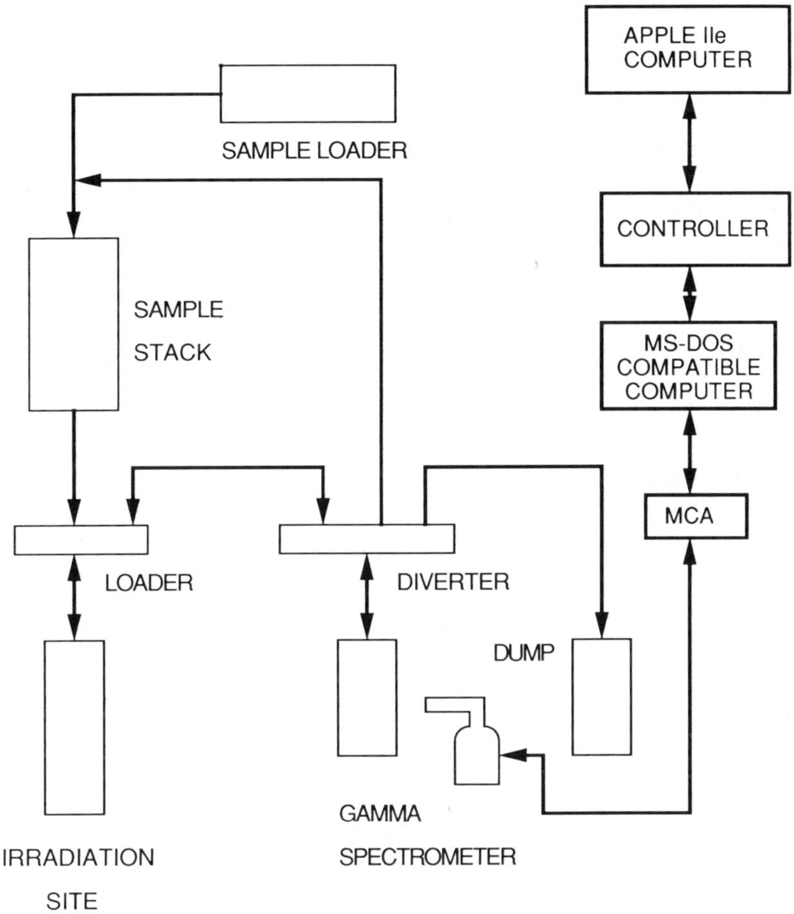

Fig. 1. Block diagram of microcomputer-controlled irradiation and measurement system.

neutron activation analysis (PCINAA) (*4,6*) by routing the irradiated samples back to the top of the sample stack. A second microprocessor-controlled sample changer also has been installed and was used exclusively for the measurement of long-lived activation products. This system is similar to that shown in Fig. 1, with the omission of the irradiation site components and the Apple IIe microcomputer.

Standards were prepared by depositing aliquots of monoelement atomic absorption standard solutions, multielement "plasma" standards (from Spex Industries, Edison, NJ), and electroanalytical standards on Whatman #1 or #41 paper filters, which were evaporated to dryness under a heat lamp and sealed in 1.5 mL polyethylene sample vials. High-purity solid reagents were used for silicon, potassium, calcium, and iron.

Samples of 300 to 500 mg of the NBS and NRC reference materials were pretreated according to the recommended procedures and weighed into 1.5 mL sample vials, which were then heat sealed. These vials were then placed in 7 mL vials that served as transfer capsules. Samples from the Bay of Quinte consisted of plant material (*Myriophyllum spicatum*), snails (*Planorbella trivolvis*), and core and surface sediments. These samples were prepared for analysis using standard analytical procedures that have been previously described (*2,3*). The NBS reference materials, Oyster Tissue (SRM1566), Citrus Leaves (SRM1572), Bovine Liver (SRM1577a), River Sediment (SRM1645), and Estuarine Sediment (SRM1646), were used primarily as control samples. However, for the determination of calcium, chloride, and sodium, standardization based on the NBS reference materials gave more consistent results. Throughout this work, triplicates of each reference material were analyzed.

The analytical methodology used in this work was based on our advance prediction program, from which three main activation and analysis schemes were chosen. The reactor was operated at half power, and samples were irradiated at a nominal thermal flux of 5×10^{11} n/cm^{2}s or at the epithermal flux of 4.6×10^9 n/cm^{2}s, available in the cadmium-lined site. The nuclear properties of the activation products and the analytical conditions used for this study are presented in Table 1.

The gamma-ray spectrometer used with the automatic irradiation system consisted of an ORTEC GMX detector (relative efficiency of 18% FWHM of 1.9 keV for the 1332.5 keV peak of ^{60}Co), coupled to a Nuclear Data Micro-MCA, together with a 7 μs fixed conversion time ADC (Nuclear Data 583) and a Westphal loss-free-counting module (Nuclear Data 599). The spectrometer used with the sample changer consisted of a CANBERRA coaxial Ge detector (relative efficiency of 18%, FWHM of 1.9 keV for the 1332.5 keV peak of ^{60}Co) coupled to a CANBERRA 35+ MCA. For both systems, spectra were transferred to MS-DOS compatible microcomputers for data storage and analysis. Gamma-ray spectra were analyzed using a Gaussian peak fitting method, GAMANAL-PC (*8*), and NAA calculations were carried out using a semiabsolute algorithm (*9*).

RESULTS AND DISCUSSION

The elemental composition of six reference materials, issued by the NRC's Marine Analytical Program, have been determined using INAA methods. The results for the sediment reference materials BCSS-1, MESS-1, and PACS-1 are given in Table 2, and the results for the biological reference materials DOLT-1, DORM-1, and TORT-1 are given in Table 3. The concentrations reported for this study represent the weighted means and standard deviations of triplicate measurements. Where an element was not observed, a detection limit (preceded by <) is reported. Where applicable, nuclear interference corrections for (n,p) and (n,α)

Table 1
Table of Nuclear Properties (7)

Element	Nuclide	Half-life	Energy, keV	Protocol
Ag	Ag-110	24.42 s	657.8	la,b
Al	Al-28	2.24 min	1779.0	la,2
As	As-76	26.32 h	559.1	3
Ba	Ba-139	82.9 min	165.8	2
Br	Br-82	35.34 h	776.5	3
Ca	Ca-49	8.72 min	3084.4	la,2
Ce	Ce-141	32.55 d	145.4	3
Cl	Cl-38	37.28 min	1642.2	la,2
Co	Co-60	5.2719 y	1332.5	3
Cr	Cr-51	27.701 d	320.1	3
Cs	Cs-134	2.062 y	604.7	3
Cu	Cu-66	5.10 min	1039.2	la,2
Eu	Eu-152	13.2 y	344.3	3
F	F-20	10.996 s	1632.6	la,b
Fe	Fe-59	44.56 d	1099.2	3
Hf	Hf-181	42.45 d	133.0	3
I	I-128	24.99 min	442.9	la,2
In	In-116m1	54.12 min	416.9	la,2
K	K-42	12.36 h	1524.6	la,2
La	La-140	40.27 h	1596.5	3
Lu	Lu-177	6.71 d	208.3	3
Mg	Mg-27	9.45 min	843.8	la,2
Mn	Mn-56	2.568 h	846.7	la,2
Na	Na-24	15.03 h	1368.5	la,2
Rb	Rb-86m	1.02 min	556.1	la,b
Sb	Sb-122	2.681 d	564.1	3
Sc	Sc-46	83.80 d	1120.5	3
Se	Se-77m	17.38 s	161.9	la,b
Si	Al-29	6.56 min	1273.2	la,b
Sm	Sm-153	46.8 h	103.2	3
Sr	Sr-85	64.85 d	514.0	3
Tb	Tb-160	72.1 d	879.4	3
Th	Pa-233	26.95 d	311.9	3
Ti	Ti-51	5.80 min	319.7	la,2
U	Np-239	2.346 d	277.6	3
V	V-52	3.746 min	1434.1	la,b
Yb	Yb-175	4.19 d	396.3	3
Zn	Zn-65	244.0 d	1115.5	3

1. Cyclic irradiations (LFC on)
$t_i = 15$ s; $t_d = 1$ s $t_c = 15$ s $t_r = 1.6$ s 20 cycles
(a) $\phi_{th} = 5 \times 10^{11}$n/cm^{2}s; (b) $\phi_{epi} = 4.6 \times 10^9$n/cm^{2}s.
2. Short irradiations (LFC on)
$t_i = 60$–180 s; $t_d = 60$–900 s; $t_c = 600$–900 s
$\phi_{th} = 5 \times 10^{11}$n/cm^{2}s.
3. Long irradiations (LFC off)
$t_i = 2$ h; $t_d = 100$–670 h; $t_c = 1$–4 h
$\phi_{th} = 5 \times 10^{11}$n/cm^{2}s.

Table 2
Composition of NRC Sediment Reference Materials

	Concentration (μg/g) except where noted, BCSS-1			MESS-1			PACS-1	
Element	This work	Cert.[a]	Lit.[b]	This work	Cert.[a]	Lit.[b]	This work	Cert.[a]
Ag	<0.60	–	–	<0.57	–	–	1.57 ± 0.28	–
Al%	6.30 ± 0.21	6.26 ± 0.22	6.19 ± 0.26	5.64 ± 0.11	5.84 ± 0.20	5.98 ± 0.11	6.17 ± 0.06	6.47 ± 0.12
As	10.97 ± 0.54	11.1 ± 1.4	11.7 ± 1.9	12.8 ± 0.7	10.6 ± 1.2	11.0 ± 0.3	234 ± 15	211 ± 11
Ba	297 ± 8	–	392 ± 52	263 ± 10	–	263 ± 22	623 ± 38	–
Br	75.70 ± 0.51	–	77.1 ± 7.6	69.06 ± 0.86	–	150 ± 15	189.7 ± 4.5	–
Ca%	0.578 ± 0.011	0.543 ± 0.053	0.564 ± 0.014	0.456 ± 0.030	0.482 ± 0.046	0.429 ± 0.050	2.010 ± 0.066	2.09 ± 0.09
Ce	46.6 ± 1.0	–	24.0 ± 1.6	75.6 ± 6.4	–	80.2 ± 4.8	29.3 ± 2.5	–
Cl%	1.168 ± .050	1.12 ± 0.05	1.10 ± 0.02	0.862 ± 0.009	0.82 ± 0.07	0.84 ± 0.03	2.375 ± 0.037	2.39 ± 0.09
Co	10.38 ± 0.21	11.4 ± 2.1	11.3 ± 0.77	12.30 ± 0.65	10.8 ± 1.9	11.6 ± 0.71	18.98 ± 1.27	17.5 ± 1.1
Cr	113.5 ± 3.9	123 ± 14	124 ± 14	77.7 ± 3.2	71 ± 11	70.9 ± 5.3	116.3 ± 4.2	113 ± 8
Cs	3.71 ± 0.08	–	4.73 ± 0.70	4.98 ± 0.52	–	4.15 ± 0.16	1.89 ± 0.53	–
Cu	<69	18.5 ± 2.7	–	<67	25.1 ± 3.8	–	324 ± 16	452 ± 16
Eu	0.865 ± 0.019	–	1.53 ± 0.25	1.151 ± 0.054	–	1.46 ± 0.02	0.784 ± 0.042	–
F	807 ± 161	–	–	431 ± 187	–	–	<677	–
Fe%	2.85 ± 0.05	3.29 ± 0.10	3.22 ± 0.19	3.16 ± 0.09	3.05 ± 0.17	3.07 ± 0.13	4.66 ± 0.25	4.87 ± 0.08
Hf	5.57 ± 0.17	–	7.90 ± 1.02	16.5 ± 1.1	–	18.5 ± 1.5	3.57 ± 0.28	–
I	46.12 ± 1.64	–	52.7 ± 1.7	16.22 ± 0.41	–	68.0 ± 2.7	57.62 ± 1.23	–
In	<0.029	–	50.9 ± 16.6	0.050 ± 0.004	–	57.2 ± 11.7	2.150 ± 0.025	–
K%	1.444 ± 0.090	1.80 ± 0.03	1.89 ± 0.04	1.588 ± 0.034	1.86 ± 0.03	2.05 ± 0.59	0.996 ± 0.050	1.26 ± 0.07
La	25.96 ± 0.25	–	40.1 ± 3.0	38.2 ± 1.1	–	45.9 ± 4.94	14.95 ± 0.44	–
Lu	0.461 ± 0.011	–	0.84 ± 0.02	0.764 ± 0.035	–	1.74 ± 1.14	0.336 ± 0.014	–
Mg%	1.408 ± 0.067	1.47 ± 0.14	1.47 ± 0.07	0.859 ± 0.019	0.868 ± 0.054	0.832 ± 0.102	1.325 ± 0.020	1.45 ± 0.05
Mn	232.4 ± 8.7	229 ± 15	237 ± 8	487.8 ± 6.5	513 ± 25	514 ± 21	441.4 ± 1.7	470 ± 12
Na%	1.978 ± 0.076	2.02 ± 0.16	1.97 ± 0.02	1.805 ± 0.076	1.85 ± 0.11	1.89 ± 0.06	3.215 ± 0.030	3.26 ± 0.08
Rb	89.1 ± 18.6	–	9.03 ± 4.13	99.1 ± 1.4	–	78.0 ± 5.8	49.8 ± 23.4	–
Sb	0.633 ± 0.025	0.59 ± 0.06	–	0.927 ± 0.051	0.73 ± 0.08	–	200.8 ± 8.5	171 ± 14
Sc	9.46 ± 0.18	–	13.5 ± 1.44	10.48 ± 0.29	–	11.7 ± 1.1	12.15 ± 0.47	–
Se	<20	0.43 ± 0.06	–	<10	0.34 ± 0.06	–	<61	1.09 ± 0.11
Si%	28.9 ± 1.9	30.9 ± 0.5	31.5 ± 1.4	31.1 ± 2.4	31.55 ± 0.89	30.2 ± 0.6	27.2 ± 2.1	26.0 ± 0.2
Sm	2.361 ± 0.069	–	6.61 ± 1.71	4.019 ± 0.022	–	6.42 ± 0.81	1.693 ± 0.035	–
Sr	<90	–	107 ± 17	<103	–	100 ± 5.3	369 ± 131	277 ± 11
Tb	0.528 ± 0.018	–	1.35 ± 0.235	0.894 ± 0.095	–	2.63 ± 0.72	0.422 ± 0.056	–
Th	6.25 ± 0.11	–	12.2 ± 1.45	12.8 ± 1.7	–	14.7 ± 1.2	3.73 ± 0.23	–
Ti%	0.419 ± 0.014	0.440 ± 0.014	0.442 ± 0.025	0.496 ± 0.016	0.542 ± 0.017	0.539 ± 0.037	0.392 ± 0.006	0.421 ± 0.007
U	2.23 ± 0.13	–	2.42 ± 0.15	4.70 ± 0.10	–	3.54 ± 0.54	3.24 ± 0.28	–
V	89.3 ± 3.7	93.4 ± 4.9	91.3 ± 1.5	74.0 ± 4.0	72.4 ± 5.3	68.9 ± 2.1	114.0 ± 5.3	127 ± 5
Yb	2.02 ± 0.06	–	–	3.73 ± 0.28	–	3.93 ± 0.31	1.59 ± 0.25	–
Zn	100 ± 4	119 ± 12	–	201 ± 16	191 ± 77	–	858 ± 59	824 ± 22

[a]Certified values from (*10*).
[b]Literature values from (*11*).

Table 3
Composition of NRC Biological Reference Materials

	Concentration (μg/g) except where noted,								
	DOLT-1			DORM-1			TORT-1		
Element	This work	Cert.[a]	Lit.[b]	This work	Cert.[a]	Lit.[b]	This work	Cert.[c]	Lit.[d]
Ag	0.967 ± 0.033	–	–	0.205 ± 0.013	–	–	10.17 ± 0.17	–	13.3 ± 0.3
Al	36.6 ± 2.1	–	–	57.3 ± 1.8	–	–	79.9 ± 8.7	–	–
As	10.1 ± 0.9	10.1 ± 1.4	8.83 ± 0.32	16.9 ± 1.0	17.7 ± 2.1	16.4 ± 0.5	24.7 ± 3.4	24.6 ± 2.2	24.8 ± 0.7
Ba	<21	–	–	<24	–	–	<35	–	–
Ca	616 ± 25	–	–	1910 ± 170	–	–	8810 ± 290	8950 ± 580	9050 ± 30
Ce	<0.63	–	–	<0.65	–	–	3.339 ± 0.081	–	4.86 ± 0.21
Cl%	0.840 ± 0.018	0.688 ± 0.022	–	1.401 ± 0.034	1.13 ± 0.03	–	5.943 ± 0.142	5.58 ± 0.10	5.54 ± 0.25
Co	0.314 ± 0.076	0.157 ± 0.037	0.157 ± 0.034	<0.14	0.049 ± 0.014	0.057 ± 0.004	0.571 ± 0.039	0.42 ± 0.05	0.48 ± 0.02
Cr	<1.3	0.40 ± 0.07	0.68 ± 0.04	5.13 ± 0.99	–	3.85 ± 0.09	2.67 ± 0.49	2.4 ± 0.6	2.5 ± 0.2
Cs	<0.078	–	–	0.227 ± 0.100	–	–	<0.092	–	–
Cu	19.0 ± 1.8	20.8 ± 1.2	20.4 ± 1.4	<8.5	5.22 ± 0.33	5.24 ± 0.18	334 ± 8	439 ± 22	325 ± 9
Eu	0.056 ± 0.014	–	–	<0.023	–	–	0.106 ± 0.005	–	0.079 ± 0.003
F	8.9 ± 2.8	–	–	22.7 ± 5.8	–	–	<87	–	–
Fe	612 ± 64	712 ± 48	677 ± 7	<85	63.6 ± 5.3	59.2 ± 4.4	206 ± 53	186 ± 11	184 ± 7
Hf	<0.13	–	–	<0.14	–	–	<0.14	–	0.025 ± 0.004
I	1.989 ± 0.165	–	–	2.371 ± 0.025	–	–	24.8 ± 1.4	–	40 ± 3
In	<0.021	–	–	<0.025	–	–	<0.050	–	–
K%	0.979 ± 0.104	1.01 ± 0.10	–	1.582 ± 0.089	1.59 ± 0.10	–	1.909 ± 0.126	1.041 ± 0.040	1.09 ± 0.03
La	<0.10	–	–	<0.12	–	–	3.89 ± 0.26	–	4.9 ± 0.2
Lu	<0.02	–	–	<0.02	–	–	<0.02	–	–
Mg%	0.114 ± 0.003	0.110 ± 0.015	0.106 ± 0.003	0.139 ± 0.004	0.121 ± 0.013	0.116 ± 0.006	0.258 ± 0.008	0.255 ± 0.025	0.20 ± 0.02
Mn	9.58 ± 0.24	8.72 ± 0.53	9.12 ± 0.13	2.16 ± 0.11	1.32 ± 0.26	1.29 ± 0.06	23.72 ± 0.71	23.4 ± 4.0	23.1 ± 0.7
Na%	0.777 ± 0.014	0.726 ± 0.073	0.797 ± 0.030	0.907 ± 0.010	0.800 ± 0.060	0.850 ± 0.081	3.947 ± 0.046	3.67 ± 0.20	3.75 ± 0.11
Rb	3.00 ± 0.31	–	–	6.05 ± 0.12	–	–	<9	–	3.3 ± 0.2
Sb	<0.06	–	–	<0.07	–	–	<0.15	–	0.13 ± 0.01
Sc	<0.008	–	–	0.018 ± 0.005	–	–	0.019 ± 0.007	–	0.011 ± 0.001
Se	6.13 ± 0.14	7.34 ± 0.42	–	1.436 ± 0.038	1.62 ± 0.12	–	5.21 ± 0.24	6.88 ± 0.47	6.5 ± 0.2
Si%	<0.50	–	–	<0.63	–	–	<1.2	–	–
Sm	<0.06	–	–	<0.06	–	–	0.133 ± 0.006	–	0.40 ± 0.02
Sr	<24	–	–	<33	–	–	111.6 ± 1.5	113 ± 5	123 ± 5
Tb	<0.056	–	–	<0.056	–	–	<0.070	–	–
Th	<0.066	–	–	<0.061	–	–	<0.077	–	–
Ti	<12	–	–	<30	–	–	<52	–	–
U	<0.40	–	–	<0.24	–	–	<0.50	–	0.114 ± 0.005
V	0.240 ± 0.042	–	–	0.104 ± 0.045	–	–	0.94 ± 0.24	1.4 ± 0.3	1.4 ± 0.1
Yb	<0.12	–	–	<0.13	–	–	<0.14	–	–
Zn	76.56 ± 6.58	92.5 ± 2.3	94.4 ± 2.9	22.68 ± 0.62	21.3 ± 1.0	20.6 ± 0.4	145.5 ± 2.0	177 ± 10	177 ± 5

[a]Certified values for DOLT-1 and DORM-1 from *(12)*.
[b]Literature values for DOLT-1 and DORM-1 from *(13)*.
[c]Certified values for TORT-1 from *(14)*.
[d]Literature values for TORT-1 from *(15)*.

reactions have been applied, except in the case of $^{31}P(n,\alpha)^{28}Al$ for DOLT-1 and DORM-1, for which no phosphorus data were available. In general, good agreement between these data and certified or published values have been obtained. However, some discrepancies were observed, such as the In values for BCSS-1 and MESS-1. This discrepancy is probably owing to an incorrect reporting of units by Fong and Chatt (*11*). It is also interesting to note that for some elements, such as chlorine and zinc, our data are more consistent with certified and published data for the sediment reference materials than for the biological reference materials. Furthermore, the major discrepancies found in this work occur where no certified values are available for the reference materials. These findings possibly reflect some of the analytical problems outlined by Becker (*16*). Nevertheless, it is felt that sufficient consistent data have been obtained to demonstrate the reliability of the new automated systems and supplement the available data for the NRC reference materials.

Both of the automated systems are presently being used in support of a metal contamination study of the Bay of Quinte, where drainage from the Moira River has carried sediments from mines into the Bay for 70 years (*2*). To date, over 1000 samples of animal, plant, and sediment have been analyzed. It was expected that higher levels of metal contamination would be present closer to the source (Belleville) than in remote locations (Hay Bay). This hypothesis appears to be correct for arsenic, cobalt, manganese, and sodium for the three sample types, although not for copper and iron. The objectives of this extensive study, which is ongoing, are to determine the contamination pattern in the area and ascertain the usefulness of animal and plant species as bioassays.

CONCLUSIONS

Both the microcomputer-controlled irradiation and measurement system and the microprocessor-controlled sample changer have provided a valuable capability to the SLOWPOKE-2 Facility at RMC. As a trial study, additions to the data for reference materials, issued by the NRC's Marine Analytical Standards Program, have been achieved. An extensive bioenvironmental study, such as that in the Bay of Quinte, could not have been supported without these automated systems.

ACKNOWLEDGMENTS

This work has been supported, in part, from the Department of National Defence, Ottawa, the Ontario Ministry of the Environment, the World Wildlife Fund, and Queen's University at Kingston. The authors gratefully acknowledge G. Burbidge and A. Anderson of Nordion International for their assistance in constructing both systems and

A. Crowder, Department of Biology, Queen's University, for making the Bay of Quinte data available.

REFERENCES

1. P. A. Beeley, L. G. I. Bennett and H. W. Bonin, L. S. Fisher, and L. S. Wright, *Proceedings of the Eighth Annual Conference*, Canadian Nuclear Society, St. John, New Brunswick, 1987.
2. A. Crowder, W. T. Dushenko, J. Greig, and J. S. Poland, *Hydrobiologica*, in press.
3. A. Crowder, W. T. Dushenko, and J. Greig, *Science and the Total Environment*, in press.
4. A. Egan, *J. Radioanal. Nucl. Chem. Art.* **110**, 47 (1987).
5. A. Chatt, H. S. Dang, B. B. Fong, C.K. Jayawickreme, L. S. McDowell, and D. L. Pegg, *J. Radioanal. Nucl. Chem. Art.* **124,** 65 (1988).
6. K. N. DeSilva and A. Chatt, *J. Trace & Microprobe Techniques* **1**, 307 (1983).
7. C. M. Lederer and V. S. Shirley (eds), *Table of Isotopes*, 7th edition, Wiley-Interscience, New York, 1978.
8. M. S. Heimlich, P. A. Beeley and J. A. Page, *J. Radioanal. Nucl. Chem. Art.* **132,** 281 (1989).
9. M. S. Heimlich, *M.Sc. thesis*, Queen's University, Kingston, Ontario, 1988.
10. National Research Council of Canada, *Marine Sediment Reference Materials* (MESS-1, BCSS-1, PACS-1), Ottawa, 1987.
11. B. B. Fong and A. Chatt, *J. Radioanal. Nucl. Chem. Art.* **110,** 135 (1987).
12. National Research Council of Canada, *Dogfish Muscle and Liver Reference Materials* (DORM-1, DOLT-1), Ottawa, 1986.
13. D. Beauchemin, J. W. McLaren, S. N. Willie, and S. S. Berman, *Anal. Chem.* **60**, 687 (1988).
14. National Research Council of Canada, *Lobster Hepatopancreas Reference Material* (TORT-1), Ottawa, 1987.
15. A. Simsons and S. Landsberger, *J. Radioanal. Nucl. Chem. Art.* **110**, 555 (1987).
16. D. A. Becker, *J. Radioanal. Nucl. Chem. Art.* **113**, 5 (1987).

The Use of Compton Suppression Spectrometers for Trace Element Studies in Biological Materials

M. ROSSBACH,*[,1] R. ZEISLER,[2] AND J. R. W. WOITTIEZ[3]

[1]Institute of Applied Physical Chemistry, ICH-4, Nuclear Research Center, KFA-Jülich, PO Box 1913, 5170 Jülich, FRG; [2]Center for Analytical Chemistry, National Institute of Standards and Technology, Gaithersburg, MD 20899; and [3]Netherlands Energy Research Foundation ECN, PO Box 1, 1755 ZG Petten, The Netherlands

Received April 17, 1989; Accepted September 14, 1989

ABSTRACT

A straightforward method for demonstrating the powerful background reduction of Compton suppression spectrometers for neutron activation purposes is presented. The shorter acquisition time needed in Anti-Compton mode (A/C on) for peaks of appropriate counting statistics, compared to normal gamma counting (A/C off), allows a much higher sample throughput, thus compensating for the higher cost of the instrument. Two examples of artificial mixtures of radionuclides demonstrate the drastic time saving for measurement of monoenergetic decaying isotopes. The comparison of results from three different instruments proves the general usefulness of Compton suppression spectrometers for Neutron Activation Analysis of biological samples.

Index Entries: Compton suppression spectrometers; time gain factor; biological samples; Neutron Activation Analysis; background reduction; intercomparison.

*Author to whom all correspondence and reprint requests should be addressed.

INTRODUCTION

Compton suppression recording of gamma spectra for the reduction of background from Compton scattered events and cascade emitting radionuclides has been successfully used to increase the sensitivity in decay scheme analysis, in-beam experiments (including prompt gamma neutron activation analysis, PGNAA), and environmental radio analysis. The application of the HPGe-NaI guard detector devices to Neutron Activation Analysis (NAA) purposes, as originally described by Cooper and Perkins in 1972 (*1*), nevertheless have not yet received the full acceptance of nuclear chemists. Either certain users have been disappointed by severe problems in the electronic setting of the early assemblies, or others had been repelled by the high costs of such an instrument, compared to a plain gamma counter. They could not be convinced by the obvious advantages of Compton suppression counting for the detection of trace elements in biological samples. Part of the skepticism toward such equipment is owing to poor and inconsistent determinations of a figure of merit for anti-Compton (A/C) spectrometers. Attempts have been made to demonstrate the usefulness of such instruments by giving the peak-to-Compton ratio of single nuclides (e.g., ^{137}Cs) (*2*) by ^{88}Y gain factors (*3,4*) or a relatively complicated semiempirical advantage, factor F (*5*). All of these approaches seem to be unsatisfactory for real laboratory needs, where multinuclide spectra of complex biological matrices are analyzed.

A new approach of comparing suppressed and unsuppressed spectra, in terms of collection time necessary for equal peak statistics and independent of the sample type and internal quality of the instrument (resolution and efficiency of the detector), is presented and applied to the performance of three single instruments. Thus, time savings up to a factor of 11 have been determined for single gamma ray emitting nuclides by Compton suppression counting.

METHODS

The methods and instrumentation for suppression of the Compton effect in gamma ray spectroscopy has been extensively discussed in the literature and is reviewed, e.g., by op de Beek et al. (*2*). The instruments used in this trial consist of commercially available HPGe and NaI(Tl)-guard detectors and associated electronics, the heart of which is a "Harshaw Anticompton Analyzer" NC 26 R that provides the pulses to trigger the gate for rejection of scattered events. This instrument replaces a multitude of delays and coincidence units of the past with complicated settings and makes the A/C spectrometer into a stable and easy to set up device.

Table 1
Some Selected Peak Areas from Natural Background Measured with the Same HPGe-Detector (25%, 1.9 keV resolution) normalized for a 60,000 s counting interval[a]

Peak	Ge out of lead cave	Ge inside lead cave	Ge and NaI, A/C off	Ge and NaI, A/C on
integral count rate	65.	1.0	0.9	0.2
185 keV (U-235)	4082.	257.	338.	273.
338 keV (Ac-228)	5674.	–	–	49.
511 keV (Annihil. rad.)	8020.	895.	803.	63.
608 keV (Bi-214)	10,570.	168.	231.	105.
661 keV (Cs-137)	7679.	–	–	12.
910 keV (Ac-228)	9612.	–	102.	79.
1460 keV (K-40)	50,993.	84.	294.	312.

[a]Passive shielding consists of 1 mm Cu, 150 mm of "pre-World War I" armor steel plates and 50 mm normal lead bricks (inner dimension: 60 × 90 × 90 cm).

An important feature, especially if low activity measurements with long counting intervals are considered, is passive shielding of the detector system. The inner dimensions should be large enough to minimize Bremsstrahlung and backscatter from the shielding material, which should be selected carefully (cleaned lead, old steel) so as not to introduce natural decay products from uranium and thorium or ^{60}Co from the recent processing of steel. The count rates for normalized counting intervals of 60,000 s of some selected background peaks are listed in Table 1.

To characterize the suppression capabilities in the measurement of activated samples by the A/C spectrometer, we take advantage of the fact that the printout of all current gamma peak evaluating software includes an error on each evaluated peak in a spectrum. The counting error is a function of the net peak area and the integrated right and left background. Both values are more or less affected by the suppression mode. One can use the peak statistics as a measure for the improvement of the sensitivity or the equivalent time gain for obtaining equally well-resolved peaks in suppressed and unsuppressed measurements. Thus, one can determine the usefulness of a Compton suppression spectrometer for every sample and nuclide by comparing the respective collection time needed to achieve equivalent peak statistics.

This counting error of a given peak is a function of the inverse square root of the collect time t_c

$$f = c \cdot t_c^{-1/2} \quad (1)$$

f = peak error in %; and c = constant.

This applies to all measurements where the counting interval $t_c < t_{1/2}$ of the nuclide. Using eq. (1), it is possible to calculate the appropriate time to reach a desired error level for a peak from any measured spectrum. Thus, one can calculate the realistic figure of merit for Compton suppression spectrometers individually for every nuclide by simply comparing the necessary collect times for equal peak statistics in unsuppressed and Compton suppressed counting.

EXPERIMENTAL

In order to approximate real activated samples, a number of pure radionuclides were mixed in quantities and combinations critical for instrumental determination with cellulose powder and dried in a desiccator. Pellets of equal weight and size (*diam.* (= Ø) = 13 mm) were pressed and sealed in polyethylene foil. By distribution among A/C users, along with an adequate record, the measurements of these samples led to a basic intercomparison of three different instruments of the same kind. Block diagrams of the spectrometers involved are shown in Figs. 1–3.

RESULTS AND DISCUSSION

In a general discussion on Compton suppression spectrometers, Cooper remarked in 1970 (3) that the "count rate of the ACS-in system" (A/C-spectrometers with the sample inside the NaI-well) "is limited by the count rate capabilities of its shield and coincidence circuitry." Thus, he claims that these kinds of detection systems are restricted to source intensities of less than about 37 kBq (= 1 μCi). The samples used in this study were produced by mixing 50 kBq ^{46}Sc, 20 kBq ^{65}Zn, and 0.5 kBq ^{85}Sr with 1 g pure cellulose powder. After drying and thorough mixing, pellets of 100 and 200 mg were pressed and sealed into polyethylene. By counting both samples separately, and in combination, samples of 6, 12, and 18 kBq were processed.

The combination of ^{65}Zn at 1115 keV with unfavorable quantities of ^{46}Sc at 1120 keV is sometimes particularly difficult to resolve by instrumental NAA. Hence, Zn determination is largely jeopardized by ex-

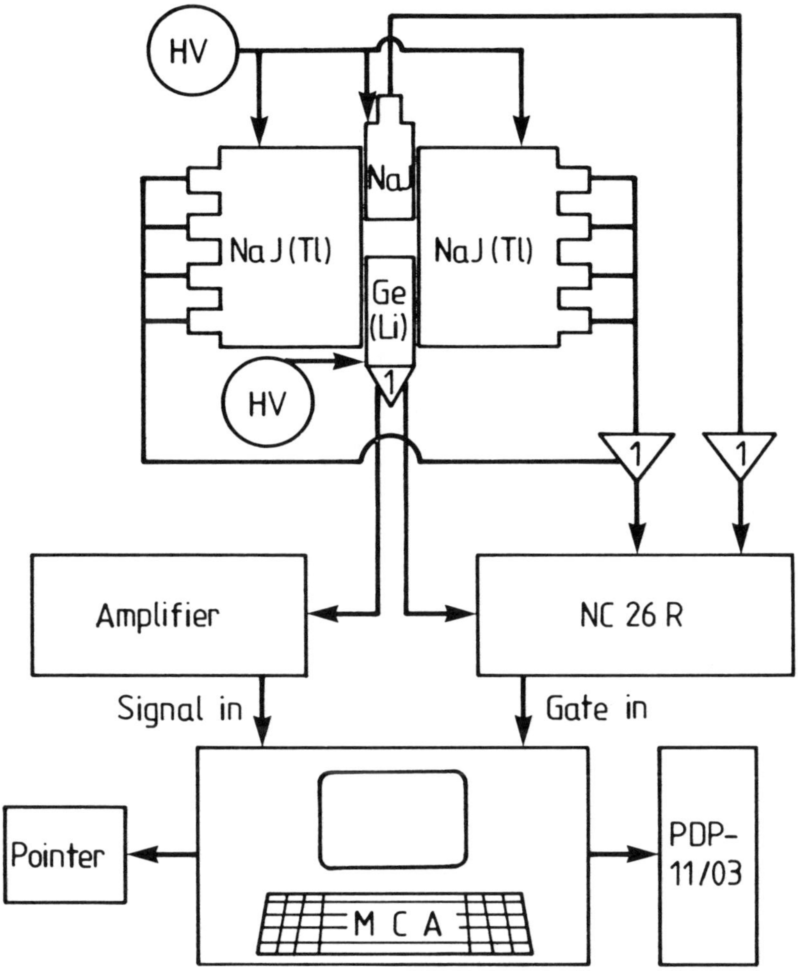

Fig. 1. Block diagram of A/C-spectrometer #1 : HV = high voltage; 1 = preamplifier.

tremely high Sc concentrations in NBS coal (SRM 1632a) (*6*) or in NBS fly ash (SRM 1633) (*7*). The same difficulties arise in the determination of ^{85}Sr (514 keV) close to the annihilation peak at 511 keV.

As can be seen from Fig. 4 and 5, interfering peaks (511 and 1120 keV) in Compton suppression counting are reduced by more than a factor of 4, whereas the peaks of interest (514 and 1115 keV) stay virtually the same. The calculated time gain factors for 1% ^{65}Zn peak statistics for all the participating analysts are shown in Table 2.

A second case where Compton suppression shows particular improvements is whenever low activities of monoenergetic radionuclides appear in the presence of a relatively intense rationuclide with high

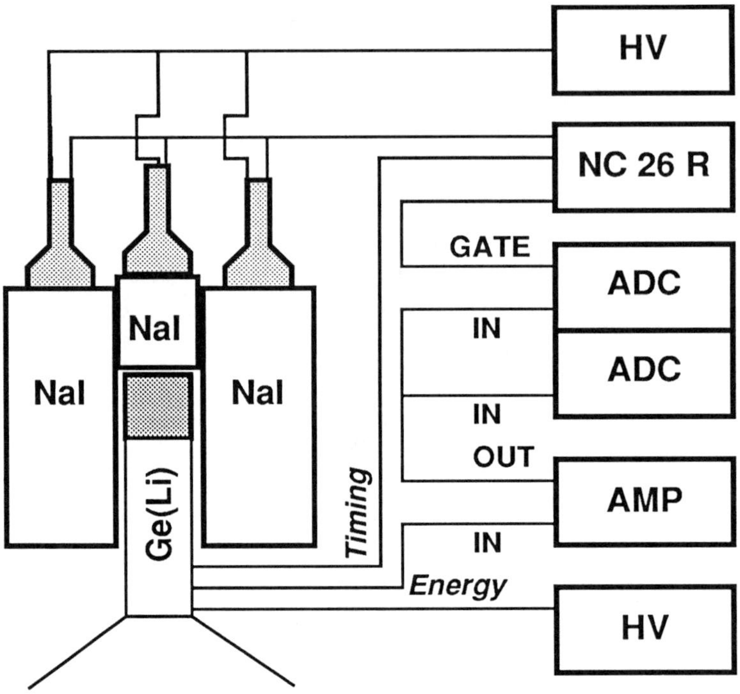

Fig. 2. Block diagram of A/C-spectrometer #2.

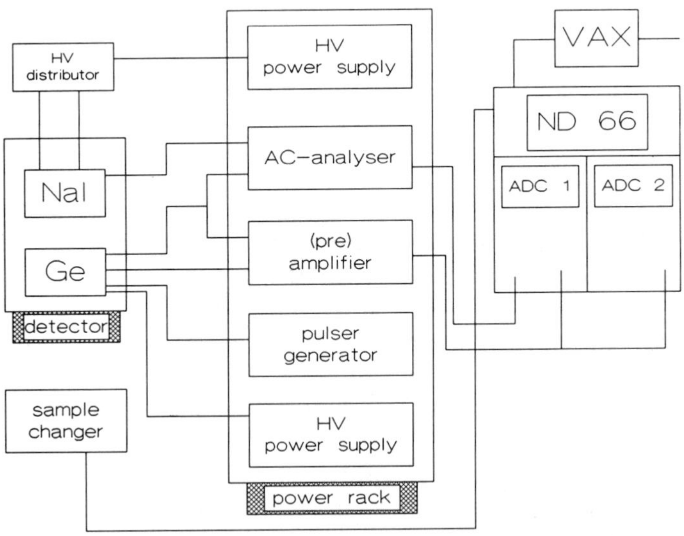

Fig. 3. Block diagram of A/C-spectrometer #3.

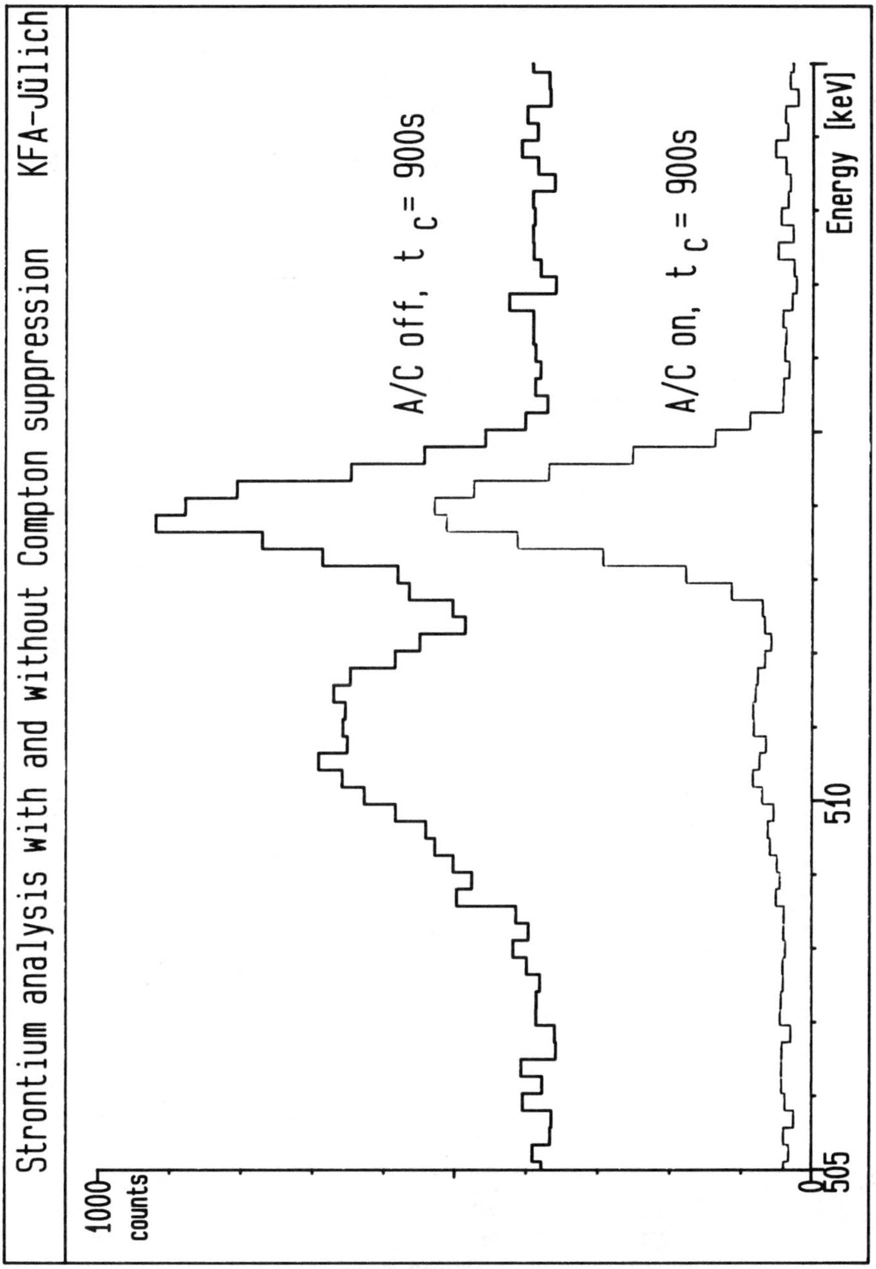

Fig. 4. Strontium-85 analysis at 514 keV is strongly facilitated by Compton suppression of 511 keV annihilation radiation.

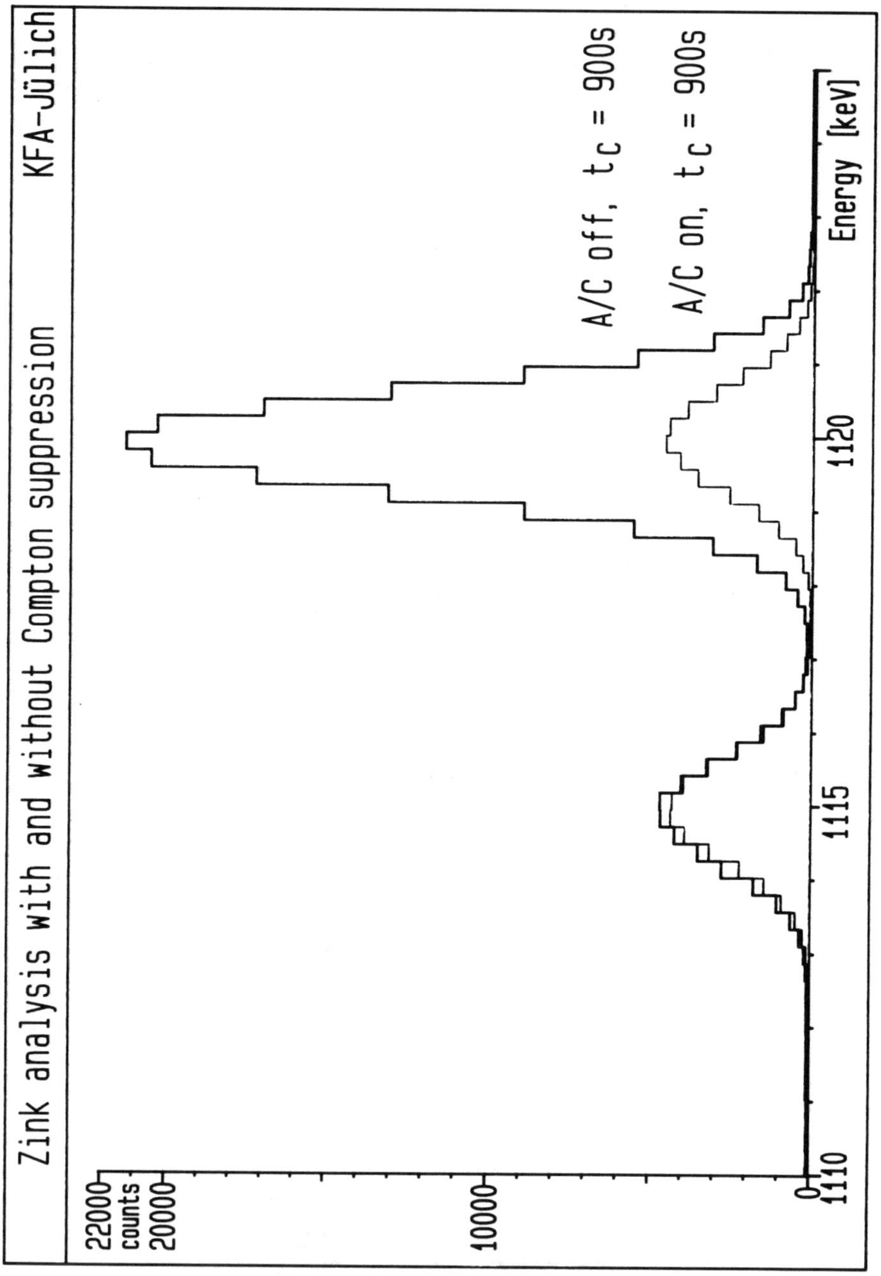

Fig. 5. Zinc-65, in the presence of Sc-46, is determined with higher precision by the suppression of Sc-46 at 1120 keV.

Table 2
Time Gain Factors of Three Separate A/C Spectrometers for Mixed Radionuclide Samples (Sc-46, Zn-65, and Sr-85) of Different Weight

	Instrument 1			Instrument 2			Instrument 3		
Element	100	200	300 mg	100	200	300 mg	100	200	300 mg
Zn	7.65	4.0	3.26	1.75	1.52	1.56	4.0	1.0	4.0
Sr	6.94	5.01	3.38	8.63	3.55	10.91	5.76	4.0	7.56

energy gamma rays, producing a large Compton continuum (^{24}Na, ^{59}Fe, ^{60}Co, and so on). This situation was produced by mixing the low activity of five monoenergetic gamma ray emitters (^{203}Hg, ^{113}Sn, ^{85}Sr, ^{137}Cs, and ^{65}Zn) with 15 kBq/g ^{60}Co activity. 500 milligram samples were distributed for measurements. One can recognize from Fig. 6 that all of the peak areas, except ^{60}Co, in the source are calculated with improved peak error in A/C on mode although the spectrum was taken for only 1000 s, resulting in time gain factors from 4 to 11 (*See* Table 3).

CONCLUSION

The data represented suggest that a straightforward characterization of Compton suppression spectrometers by calculating nuclide specific time gain factors for equal peak statistics in A/C-on and -off is possible. All of the participating spectrometers function in a comparable way so that the results can be generalized. Improvements are still possible regarding the geometry of the detectors (size and shape of the guard detectors, BGO-guard ring below the Ge surrounding the cryostat, and so on) and electronic devices (high throughput ADC's). Some general aspects of A/C counting are summarized below

1. Natural background must be efficiently shielded.
2. A/C spectrometers up to now do not properly work below 100 keV (because of 180° scattered gamma rays escaping undetected toward the cryostat of the Ge-detector in the vertical sample/detector arrangement).
3. A list of 36 elements offers favorable monenergetic (or nearly) radionuclides for improved detection by Compton suppression counting.
4. If sample throughput can be increased by a factor of 3 to 4, owing to better peak statistics, the doubling of costs of the Compton suppression spectrometer, compared with an ordinary HPGe detector system, can easily be recovered within a year or two.

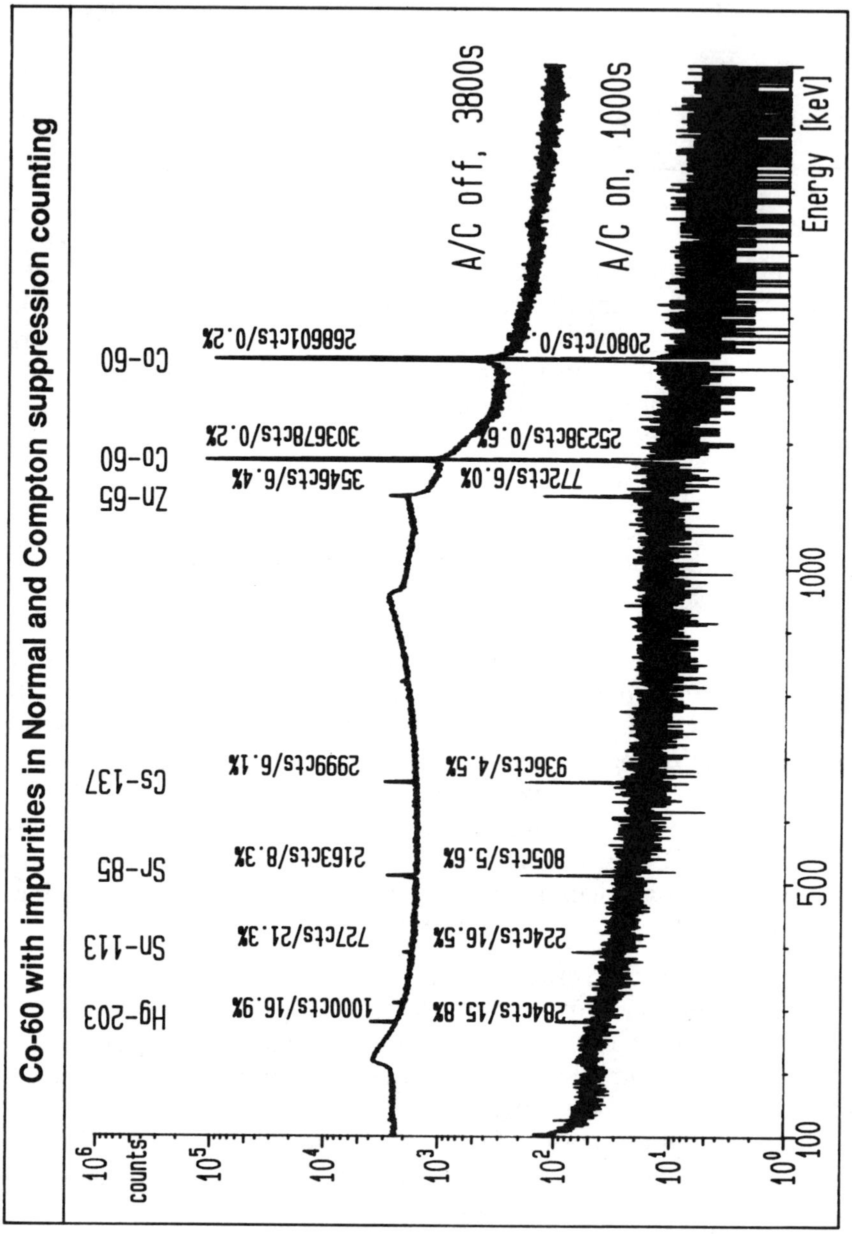

Fig. 6. Co-60 spectrum with impurities of Hg-203, Sn-113, Sr-85, Cs-137, and Zn-65. Precision of the peak area calculation is better for Compton suppressed counting.

Table 3
Time Gain Factors of Three Separate A/C Spectrometers for Traces of Nuclides (Hg-203, Sn-113, Sr-85, Cs-137, and Zn-65) Mixed to a 7.5 kBq Co-60 Activity

Element	Instrument 1	Instrument 2	Instrument 3
Hg-203 (279 keV)	9.92	6.2	?
Sn-113 (391 keV)	4.64	6.3	14.5
Sr-85 (514 keV)	11.03	3.0	5.67
Cs-137 (661 keV)	5.84	8.4	10.6
Zn-65 (1115 keV)	10.16	4.0	9.0

ACKNOWLEDGEMENTS

Many thanks for valuable discussions belong to Prof. Chatt from Dalhousie University, Canada and M. Rossbach acknowledges the financial support by the Bundesministerium für Umwelt, Naturschutz und Reaktorsicherheit (BMU), Bonn within the environmental specimen bank program coordinated by the Umweltbundesamt, Berlin. J. R. W. Woittiez wishes to acknowledge the great technical assistance of Juul Ritkema and Jan Zonderhuis.

REFERENCES

1. J. A. Cooper and R. W. Perkins, *Nucl. Inst. Meth.* **99,** 125–146 (1972).
2. J. op de Beeck, A. Isebaert, and J. Hoste, *J. Radioanal. Chem.* **76,** 225–239 (1983).
3. J. A. Cooper, *Nucl. Inst. Meth.* **82,** 273 (1970).
4. U. Rösick and P. Brätter, *Z. Anal. Chem.* **286,** 336–342 (1977).
5. H. S. Das, *J. Radioanal. Nucl. Chem.* **115,** 159–173 (1987).
6. M. Rossbach and M. Stoeppler, *J. Radioanal. Nucl. Chem.* **113,** 217–223 (1987).
7. H. R. Vogt and R. Zeisler, *Internation Conference on Nuclear Analytical Methods in the Life Sciences,* Gaithersburg, MD, 1989, p. 154.

Tomography and Elemental Analysis of Biological Systems

NICHOLAS M. SPYROU

Department of Physics, University of Surrey, Guildford, Surrey GU2 5XH, UK

Received April 17, 1989; Accepted September 14, 1989

ABSTRACT

The various forms of tomography, expressed as transmission and emission modes using gamma rays and neutrons, are generally discussed in terms of providing information nondestructively about the distribution of elemental composition in a plane through an object. The combination of the principles of tomography with neutron activation analysis is the basis for neutron-induced gamma-ray emission tomography. The concept of detection limit in induced gamma-ray emission tomography, as proposed, incorporates a further factor that is a measure of the quality of the image produced. A specific example is given for the elemental analysis and imaging of a bone specimen.

Index Entries: Gamma-ray transmission and emission tomography; neutron tomography; neutron activation analysis and tomography; elemental detection limits and tomography.

INTRODUCTION

In many biomedical studies, determination of the distribution of an elemental concentration in a biological system, whether to investigate structure or function, is important. To achieve this, the system could be examined either by using a suitable probe that will produce a characteristic and measurable signal from the element of interest, or by introducing into the system a compound appropriately labeled by the element, e.g., in the form of a tracer emitting radiation, to target onto the region of interest. If accurate quantities of the element in the region of interest are to be determined, attenuation of the signal by the intervening medium must be taken into consideration. Then, application of the principles of reconstructive tomography will provide a nondestructive method by which the distribution of the quantity of interest, in a plane through the object, can be obtained without contributions from underlying and overlying planes (*1*).

Over the last twenty years, the growth of medical imaging, in its different modalities and, in particular, tomography in its various forms, applied to the diagnosis of disease, as well as planning and monitoring treatment, has been invaluable. Many of these techniques have been adapted for use in industrial nondestructive testing and other nonmedical appllications (2). However, little use has been made of tomographic imaging as a tool for the determination of elemental composition, either directly or in combination with other complementary radioanalytical methods, for bioenvironmental studies in vivo or in vitro. For example, photon transmission tomography is being used to determine "bone mineral content" or "bone mineral density" in osteoporosis. Also, positron emission tomography is being applied to the study of brain function, e.g., in Parkinson's disease and senile dementia, by imaging the distribution of neurotransmitters, glucose metabolism, and blood flow (*3*). This information, in combination with the determination of trace element levels in appropriate tissues of the same subjects, may prove useful in testing hypotheses currently proposed.

METHODS

Tomographic techniques can be categorized, in general, into transmission and emission. The former relates to the detection of the radiation transmitted through the slice of interest, and the latter requires measurement of the radiation emitted from within it, in all cases maintaining narrow beam geometry. The forms of tomography can be further expanded since the probing beam may also be scattered (elastically and inelastically), and the probe on absorption in the medium may induce the emission of another type of radiation. When gamma rays and neutrons are used as probes, to which we will be confined, the induced radiation may be X-ray fluorescence and neutron capture gamma ray, respectively. These are emitted promptly. The interaction may also result in the production of a radionuclide that emits delayed gamma rays. The tomographic technique employed, therefore, is usually described by the probe and the type of radiation resulting from the interaction with the object, e.g., neutron induced gamma-ray emission tomography (delayed mode). Table 1 (a and b) summarizes the various forms of tomography that have been attempted using gamma rays and neutrons as probes and lists the physical quantity of interest for which the distribution in the tomographic plane is obtained. Positron emission tomography is also included

Table 1
Forms of Tomography in Gamma-Ray and Neutron Tomography

Type	Radiation measured	Distribution represented
(a) Gamma-Ray Tomography		
(1) Transmission	transmitted gamma rays	linear attenuation coefficient
(2) Scattering		
Compton	inelastically scattered photons	density × atomic number
Rayleigh	elastically scattered photons	density × (atomic number)^n
(3) Emission		
radionuclide tracer	emitted gamma rays	concentration of activity
positron emitter	annihiliation radiation	
(4) Induced emission		
X-ray fluorescence	characteristic X-rays	elemental composition
photon activation	prompt gamma rays	elemental (isotopic) composition
	delayed gamma rays	elemental (isotopic) composition
(b) Neutron Tomography		
(1) Transmission	transmitted neutrons	macroscopic neutron cross-section
(2) Scattering	neutrons	scattering centers, e.g., H
(3) Induced emission	prompt gamma rays	elemental (isotopic) composition
	delayed gamma rays	elemental (isotopic) composition

under gamma-ray tomography, for convenience. It is the annihilation radiation that is measured outside the object.

Gamma-Ray Tomography

The first practical realization of the technique of X-ray transmission tomography by Hounsfield in 1972 led to the well-documented eruption of clinical interest in the years that followed. However, it should be noted that because of the bremsstrahlung continuum in the spectrum from the X-ray generators used in medical scanners, it was difficult, until recently, to obtain sufficiently accurate linear attenuation coefficient measurements to characterize the tissue over the broad energy spectrum and thus determine its elemental composition. By using isotopic gamma-ray sources, discrete energy lines allowed accurate measurement of the linear attenuation coefficients to be made, but at the expense of a considerably reduced photon source intensity and, hence, very long scanning times. Gamma-ray transmission tomography was proposed about 10 years ago as a method of analysis based on the detection of differences in photon mass attentuation coefficients, as applied to biological matrices (4,5). Although detection limits for elements in these matrices are poor compared to neutron activation analysis, nevertheless the method provides a means of examining the homogeneity of composition of a sample in terms of the distribution of its density and major elements in the slice of interest, which can be very useful. For example, in specimen banking, where whole organs are maintained for many years for future reference, transmission tomography can be used as a means of monitoring the state of the organ, as well as detecting any changes between fresh and frozen conditions. Present day X-ray CT-scanners take less than 4 s to image a slice. There is also an associated minimum detectable volume, when imaging the distribution of these variations, in the tomographic plane.

Compton scattering tomography and Rayleigh scattering tomography are being explored for application to industrial and biomedical problems. It is possible to acquire a gamma-ray spectrum, at an angle to the incident photon beam, that contains information about the number of Compton and Rayleigh scattered photons. The ratio of the two counts will give a quantity independent of density and dependent on a power function of the atomic number. Thus, variations in density and atomic number can be imaged separately.

Gamma-ray emission tomography is widely recognized as a method for obtaining information about physiological function and has been used in nondestructive applications in industry (*6*). Single photon emission tomography (SPET), as it is more commonly known in medical imaging, where the radionuclides used normally emit monoenergetic gamma rays, has also been applied to the diagnosis and investigation of dementing illnesses. Regional cerebral blood flow patterns in patients with Alzheimer's Disease, multi-infarct dementia, and controls have been sufficiently distinctive to allow differentiation between them (*7*).

The possible combination of gamma-ray emission tomography with perturbed angular correlation experiments, where the angular distribution between two gamma rays emitted in cascade is measured using detectors in coincidence, is an exciting prospect since both the location and molecular environment of a radionuclide may be determined (*8*). The onset of positron emission tomography (PET) systems with multiple rings, which acquire photons from several slices simultaneously and comprise several thousand detectors, give an extra impetus to this. Recent technological advances will allow the setting of two energy windows in the detected photon spectrum.

Neutron Tomography

The advantages of using neutrons as probes to examine the structure and composition of materials, compared to gamma rays or X-rays, are well known in neutron activation analysis and neutron radiography, where the differences in the nuclear cross-sections between neighboring elements or isotopes can be considerable and the characteristic gamma rays emitted following neutron capture by deexcitation of the compound nucleus and decay of the unstable product nucleus, carry information about elemental composition.

Therefore, if a neutron beam is used as a probe and the principles of reconstructive tomography are applied, it is possible to obtain the distribution of the macroscopic cross-section in a selected plane in the object of interest, as well as the distribution of elemental concentrations in that plane. Three tomographic methods have been developed and are being employed at Surrey, using thermal neutrons (*9–11*). They can be briefly summarized as follows: (1) Neutron transmission tomography, NTT, where the neutrons transmitted through the object are detected and images representing the neutron macroscopic cross-section distribution are reconstructed; (2) neutron induced gamma-ray emission tomography, NIGET (prompt mode), where the prompt gamma rays, emitted as the scanning neutron beam traverses the object, are detected and from the gamma-ray spectra, obtained tomographic images of elemental concentrations are produced for planes of interest; and (3) neutron induced gamma-ray emission tomography NIGET (delayed mode), where after irradiation of the object in a neutron flux, the delayed gamma rays from the resulting unstable nuclides are detected. This then is reduced to a problem of multienergetic gamma-ray emission tomography. In all three cases, the object is translated and rotated before stationary detectors in order to obtain the required number of projections to reconstruct an image. The problems that arise in methods (2) and (3) are: attenuation of the probing neutron beam as it traverses the object, resulting in a reduced reaction rate with depth (and therefore a decreased yield of gamma rays emitted) and attenuation of the characteristic gamma rays emitted from the point of production in the object to the detector. Ideally, therefore, the distribution of the macroscopic neutron cross-sections in the object plane is required so that neutron attenuation can be calculated. This can be obtained from a neutron transmission tomography experiment for correction to NIGET (prompt mode) results, but is usually not necessary when the object is uniformly irradiated in the reactor core in the delayed mode case (*10,12*). In addition, the distribution of the photon linear attenuation coefficient is required for gamma-ray attenuation correction, and this can be obtained by performing a gamma-ray emission tomography experiment. The latter problem also occurs in gamma-ray emission tomography, where best results for attenuation correction are obtained when transmission measurements are used, rather than analytical functions, except in simple cases where photon attenuation in the material can be considered uniform.

DETECTION LIMITS AND IMAGE QUALITY

The formulation of detection limit in gamma-ray emission tomography is difficult to describe. The reconstructed values of the function of interest, i.e., the distribution of the concentration of activity, deviate from the true values owing to noise introduced during data acquisition and processing. The noise processes that should be considered are the statistical fluctuations in the projection data, the use of insufficient projection angles, and the use of large spatial sampling intervals, as well as approximations in the reconstruction algorithm and deficiencies in the detecting system (e.g., scattering and septal penetration of the collimator). Therefore, the detection limit in NIGET (delayed mode) depends on the same factors as neutron activation analysis, plus those arising from gamma-ray emission tomography.

The number of photons detected along a line L (e,θ) in the plane of interest (x,y) of finite thickness, a, determined by the collimator aperture, is given by the basic formula

$$D(\ell, \theta) = \epsilon a t \int_{L(\ell,\theta)} ds.\ A(x,y) \exp\left[-\int_{L(\ell,\theta)} \mu(x',y')ds\right] \quad (1)$$

where $A(x,y)$ is the activity concentration (Bq/vol) at (x,y) and $L(e,\theta,x,y)$ is the line segment from (x,y) to the detector, with $\mu(x',y')$ the linear attenuation coefficient at point (x',y') along this line segment; ϵ is the absolute detector efficiency for the gamma ray of interest and t is the time of counting, assuming no decay of activity (otherwise an appropriate factor can be introduced).

For uniform concentration and assuming no attenuation of gamma rays in the medium, Eq. (1) becomes:

$$D = \epsilon a L t A \quad (2)$$

where A is the radionuclide concentration and L is the length of the radionuclide activity along the collimator direction (aL is the volume). The minimum detectable concentration, therefore, can be found by setting the detector response, D, equal to the standard deviation (multiples thereof) of the background underlying the full energy photopeak. Hence, the detection limit (the minimum mass of the element of interest per unit mass of the sample) can be calculated using the neutron activation equation. However, this does not take into account the reconstruction process.

If these minimum values are used, will a clear image be produced, which represents the distribution of radionuclide concentration in the plane of interest? It is clear that the sampling intervals per projection in linearly scanning the object, the number of projections obtained at equally spaced angles over 360 degrees and the reconstruction algorithm, which will provide the image as a square matrix of (n × n) pixels of a given size (area), must be taken into account.

Image quality can be expressed quantitatively by comparing the image obtained with the expected or ideal image, and noise is represented as the deviation between these two images. The fidelity factor is defined as (*13*)

$$Q = 1 - \sum_{i,j}^{n,n} \frac{[A(i,j) - A^*(i,j)]^2}{[A^*(i,j)]^2} \qquad (3)$$

where $A(i,j)$ is the value obtained representing the concentration of activity in pixel (i,j) and $A^*(i,j)$ is the expected value for the same pixel. We have found experimentally (*14*), using a uniformly distributed source of Se-75 of various sizes and different detectors and collimators and a filtered back projection algorithm, that the fidelity factor is a function of the ratio of the number of counts in the line integral in the uniformly distributed source (signal) to the number of counts in the line integral outside the region of the distributed source (background). The latter may be owing to scattering, environmental background, and so on. Figure 1 shows the variation of the fidelity factor vs background-to-signal ratio for a specific case. A signal-to-background ratio of 3 produced an image with a Q value of 0.26. The concept of detection limit, therefore, should incorporate a measure of image quality in terms of the fidelity factor.

By way of illustration, the detection limits for a number of elements have been calculated for a bone sample of 20 mm diameter, scanned by a collimated Ge(Li) detector at 250 mm from the center axis of rotation of the sample, with a 2 mm sampling interval and 10 projections taken (i.e., 100 measurements) to produce a tomographic image of a 2 mm thickness, representing elemental distribution on a matrix of 10 by 10 (i.e., each pixel of 2 mm $\times$ 2 mm area). It was assumed that the bone sample was irradiated in a thermal neutron flux of $2 \times 10^{12}/n/cm^2/s$. for 8 h, and the time of counting for each sampling interval was 540 s. A uniform distribution of the elements in the section of interest was assumed, an average background underlying the full energy photopeaks of interest of 0.5 counts/s was accepted with a photon limit of detection set at twice the standard deviation of this background, and no photon attenuation correction. The detection limits are in (μg/g) except where otherwise indicated: Na-24 (114), Ca-47 (10), Mn-56 (6), Cu-64 (7.6 mg/g) Zn-65, (455), Br-82 (126), Sb-124 (187), and Au-198 (6.3). The calculations suggest that, under these conditions, the distribution of Na, Ca, and Mn may be imaged in bone. Similar calculations for an irradiation time of 1 h at the same irradiation position and a counting time of 20 s, but a background of 5 counts/s, indicate that Mg, C*l*, P, through the (n, alpha) reaction, and possibly V may be imaged in bone.

It should be noted that these detection limits can be significantly improved if NaI(T*l*) or BGO detectors are used, provided the gamma lines of interest are well separated. Furthermore, the use of more than

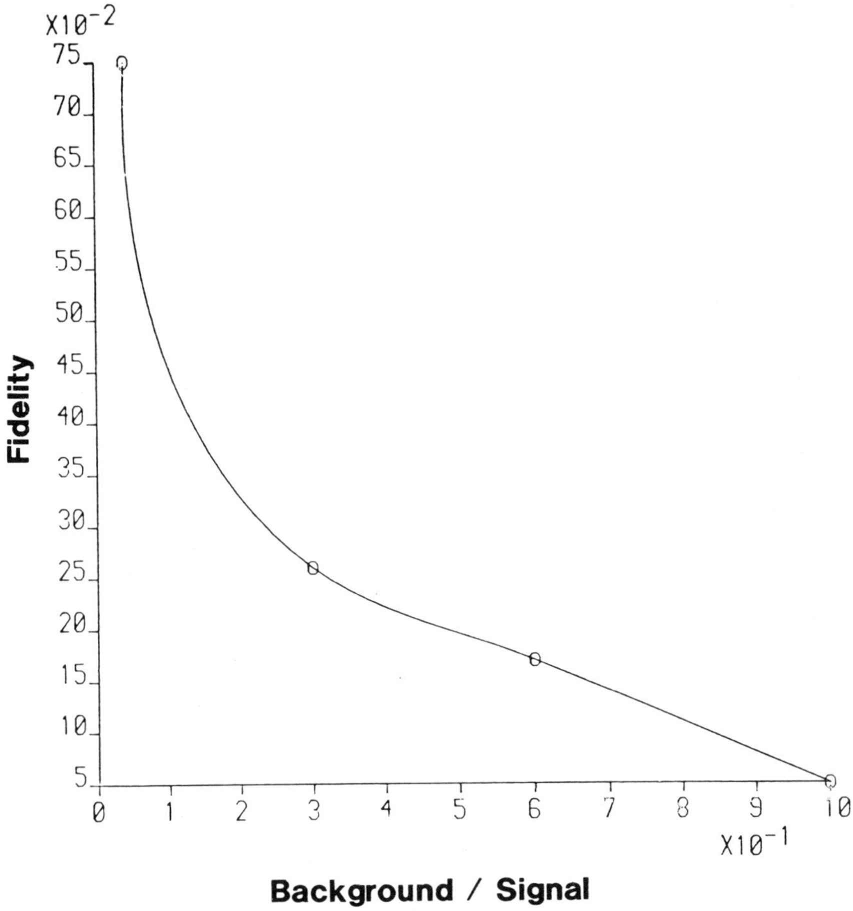

Fig. 1. The fidelity factor as a function of background-to-signal ratio using a filtered backprojection algorithm. The case is for a uniform distribution of ^{75}Se activity of different concentrations in a cylindrical vial using a NaI(Tl) detector with an energy window set at (265 + 280) keV. The sampling interval was 1.5 mm over a length of 45 mm and 24 projections obtained.

one detector in the scanning rig would make better use of the time available for data acquisition and further improve detection limits.

CONCLUSION

A number of tomographic techniques have been briefly described that can be useful in determining the distribution of the elemental composition of biological materials in sections of interest. The detection limits for NIGET (delayed mode) have been calculated for bone as an example for a specific case. The onset of systems with multiple detector arrays of high spatial resolution and spectrometric capabilities should considerably improve these values.

ACKNOWLEDGMENTS

Thanks to a number of research students for experiments and calculations, but in particular, Anwar Dhani and Asad Farooqi.

REFERENCES

1. K. Kouris, N. M. Spyrou, and D. F. Jackson, *Imaging with Ionising Radiations, Progress in Medical and Environmental Physics*, vol. 1, Surrey University Press/Blackie, Glasgow and London, 1982.
2. P. Reimers, W. B. Gilboy, and J. Goebbels, *NDT International* **17**, 197 (1984).
3. E. S. Garnett, G. Firnau, and C. Nahmias, *Nature* **305,** 137 (1983).
4. K. Kouris and N. M. Spyrou, *Nucl. Instr. Meth.* **153**, 477 (1978).
5. N. M Spyrou, V. Neofotistou, and K. Kouris, *Proc. 3rd Int. Conf. on Nuclear Techniques in Environmental and Energy Research*, Technical Information Centre, USDE, Oak Ridge, Tennessee, CONF-771071, 482 (1977).
6. G. Davies, N. M. Spyrou, I. G. Hutchinson, and J. Huddleston, *Nucl. Instr. Meth. Phys Res.* **A242**, 615 (1986).
7. W. J. Jagust, T. F. Budinger and B. R. Reed, *Arch. Neurol.* **44**, 258 (1987).
8. N. M. Spyrou, O. A. Akanle and A. Dhani, *TANSAO* **56**, 149 (1988).
9. N. M. Spyrou, *J. Radioanal. Nucl. Chem.* **110**, 641 (1987).
10. N. M. Spyrou, Kusminarto, and G. E. Nicolaou, *J. Radioanal. Nucl. Chem.* **112**, 57 (1987).
11. Kusminarto and N. M. Spyrou, *Neutron Radiography*, J. P. Barton, G. Farny, J. Person, and H. Rottger, eds., Reidel, Dordrecht, Holland, 1987, p. 719.
12. N. M. Spyrou, F. A. Balogun, and G. Davies, *J. Radioanal. Nucl. Chem.* **113**, 417 (1987).
13. A. Evans, *The Evaluation of Medical Images*, Adam Hilger, Ltd., Bristol, 1981.
14. A. Dhani, *Ph.D. Thesis*, University of Surrey, 1989.

Application of Polyacrylamide Gel Electrophoresis/Neutron Activation Analysis for Protein Quantification

SUSAN F. STONE,*[1,2] ROLF ZEISLER,[1]
AND GLEN E. GORDON[2]

[1]*Center for Analytical Chemistry, National Institute of Standards and Technology, Gaithersburg, MD 20899; and* [2]*Department of Chemistry and Biochemistry, University of Maryland, College Park, MD 20742*

Received June 8, 1989; Accepted September 14, 1989

ABSTRACT

A combination of two methods, polyacrylamide gel electrophoresis (PAGE) and neutron activation analysis (NAA), has been applied to solutions containing phosphoproteins for the purpose of protein quantification. The proteins were separated by molecular weight using PAGE, and then the whole gel was activated by neutron bombardment. Densitometric measurements of the developed bands from ^{32}P, taken from autoradiographs of the activated gels, resulted in quantification of the phosphorus, and then the related protein. This PAGE/NAA method was applied to several phosphoprotein-containing materials, including commercial milk products and reference materials, i.e., IAEA A-11, milk powder, and SRM 1845, Cholesterol in Egg Powder.

Index Entries: Casein; milk; milk products; neutron activation analysis (NAA); phosphoprotein; phosvitin; polyacrylamide gel electrophoresis (PAGE); protein; protein quantitation.

INTRODUCTION

The study of trace elements, associated with macromolecules, has rapidly grown in importance in recent years. Such speciation studies are

*Author to whom all correspondence and reprint requests should be addressed.

necessary to help in understanding the role of trace elements in the structures and functions of biological macromolecules. A combination of methods is required to accomplish this work since a single analytical method is not sufficient. Several combinations have been used successfully in past speciation studies, both for qualitative and quantitative analysis of trace elements associated with macromolecules. These combinations include gel filtration/neutron activation analysis (NAA), ultracentrifugation/NAA, ion-exchange chromatography/NAA (*1*), thin layer chromatography/NAA (*2*), polyacrylamide gel electrophoresis (PAGE)/ proton-induced X-ray emission (PIXE) (*3*), and PAGE/NAA (*4*).

Until now, two different approaches have been taken in studying trace element–macromolecule associations. In one approach, the emphasis is on the detection of the trace elements that indicate the presence of a separated protein. In the other approach, the methods have used trace element information for quantification of the protein. So far, these methods have used isolated proteins or nucleic acids, for the most part, and not yet more complex samples. In this work, we have taken the second approach.

We have previously demonstrated the use of PAGE/NAA for the quantitative determination of proteins through the measurement of an associated element (*5*), using isolated phosphoproteins. The proteins are separated by molecular weight, using PAGE, and a NAA procedure is then performed to quantify the amounts of phosphorus. The proteins can then be quantified if trace element–protein associations are well-characterized, as is the case with many phosphoproteins.

In this work, the application of the PAGE/NAA method to quantify phosphoproteins in natural matrices is presented. This includes reference materials, such as milk powders, and some commercial dairy products.

MATERIALS AND METHODS

Samples were separated by gradient polyacrylamide gel electrophoresis in a discontinuous buffer system. Slab gels of 1.5 mm thickness were prepared with a stacking gel of 5% polyacrylamide and a resolving gel with 10–17% linear polyacrylamide gradient. Sodium dodecyl sulfate (SDS) was included in both gels. The running buffer contained tris(hydroxymethyl)-aminomethane (Tris), glycine, and SDS, and the buffer for sample dissolution contained Tris, SDS, and 2-mercaptoethanol. The final molar concentrations of reagents in the gels and the pH of the Tris buffers were identical to those described by Laemmli (*6*), except for the acrylamide concentrations. For the composites and dairy samples, ethylene diamine tetraacetic acid (EDTA) was also included in the sample buffer (*7*). Two duplicate gels were run simultaneously on a vertical electrophoresis system (Hoefer 600), with cooling. Following electrophoresis, the gels were fixed in 50% ethanol/0.1% formaldehyde and then either put through a detergent removal process for SDS (*8*) or, for the duplicate gel, stained with Coomassie Brilliant Blue.

The gels were irradiated at the University of Virginia, Charlottesville reactor, at a fluence rate of $1.7 - 10^{17} n/m^2/s^1$. The irradiation facility and postirradiation treatment, including autoradiography and densitometry measurements, have been described previously (*5*). Briefly, the irradiated gels are used to expose X-ray films to make autoradiographs of the activated nuclides on the gels, in this case, primarily the ^{32}P in the separated phosphoproteins or the applied standards. The absorbance of the imaged areas on the autoradiographs are then measured using a laser densitometer (LKB Ultroscan 2202 LKB-Produckter, Bromma, Sweden), which is interfaced to an Apple IIE computer for area density integration.

The proteins and protein mixtures or composites used in this work included α-casein and sodium caseinate (Sigma Chemical Co., St. Louis, MO), NIST Standard Reference Material (SRM) 1549, Milk Powder and SRM 1845, Cholesterol in Egg Powder, and the International Atomic Energy Agency's (IAEA) Certified Reference Material A-11, Milk Powder. Also analyzed were some commercial dairy products: low-fat (2%) milk and half-and-half (cream and milk). Molecular weight standards were used to monitor the separation and determine molecular weights. These were purchased as mixtures and were either prestained (Diversified Biotech, Newton Centre, MA) or plain (Sigma Chemical Company). Ammonium dihydrogen phosphate (Baker, Phillipsburg, NJ) solutions were prepared for use as phosphorus standards, ranging in concentration from 0.3 to 45 ng/mL. The composite solutions, made from lyophilized powders, contained 2 mg/mL, whereas the commercial milk products were sampled directly as liquids, diluted 1:1 with buffer. The isolated protein α-casein was prepared as a 2 mg/mL solution.

RESULTS AND DISCUSSION

The separated phosphoproteins were quantified by simultaneous irradiation and evaluation by densitometric measurements of phosphorus standards applied to each gel. The resulting calibration curves for each autoradiograph were necessary because of variable exposure time and developing conditions. The amount of phosphorus in a separated protein was then quantified, and this amount was then related to the mass of the protein by the protein stoichiometry.

An example of absorbance measurements of protein bands (as phosvitin) in the autoradiograph of SRM 1845, Cholesterol in Egg Powder, vs sample volume is shown in Fig. 1. The values for phosvitin obtained from the calibration curve were 10–54 μg protein. Error bars represent uncertainties in repetitive measurements of a single band, and in the fluence, corrections made to the measurements. The measured areas corresponding to phosvitin solution volumes of 10, 20, and 30 μL show a

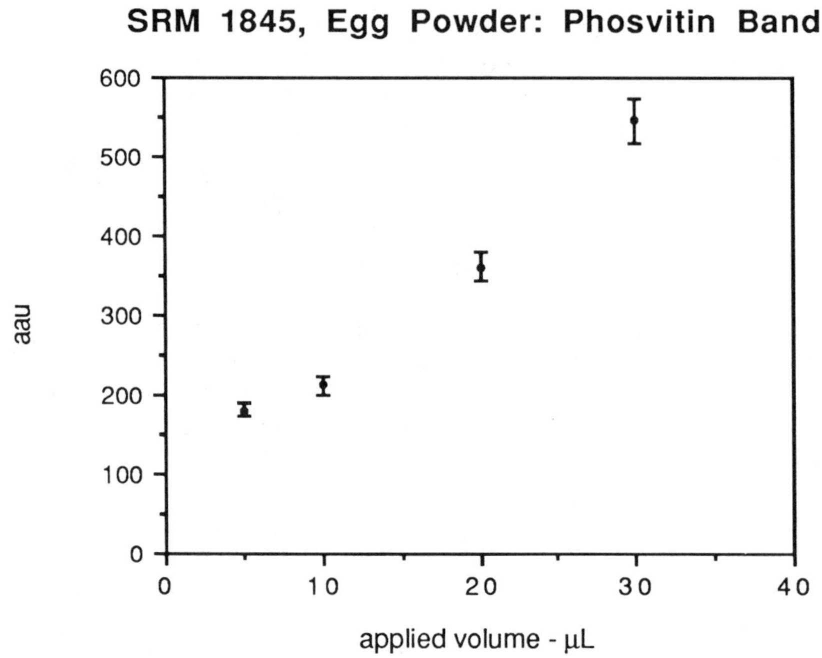

Fig. 1. Absorbance measurements (aau) of protein band, phosvitin, vs sample volume of SRM 1845, Cholesterol in Egg Powder. Uncertainties were estimated to be 5% relative owing to the uncertainties in the determination of fluence corrections and the reproducibility of the absorbance measurements.

linear response, but the area corresponding to the 5 µL volume does not. The imaged band on the autoradiograph, corresponding to this sample volume, was barely visible over the background and did not have enough contrast to be measured accurately. The uncertainties for absorption measurements of low density areas like this one are probably higher than the estimated uncertainties, and in this case, the calculated area is overestimated.

The autoradiograph of separated SRM 1845 (Cholesterol in Egg Powder) samples is given in Fig. 2. Comparing the autoradiograph to its duplicate gel stained with Coomassie Blue reveals that not all the proteins are visible; only the phosvitin band can be seen. The broad, dark band in track 4 is a phosvitin standard that was run along with the Egg Powder samples. The exposure time necessary to detect the phosvitin in the Egg Powder caused the band, resulting from the high level of isolated phosvitin, to be overexposed.

The method was also used to determine α-casein in several milk products and reference materials. The accurate determination of casein is still an analytical challenge. In most attempts to determine milk proteins, only ratios of the caseins have been determined, relative to each other,

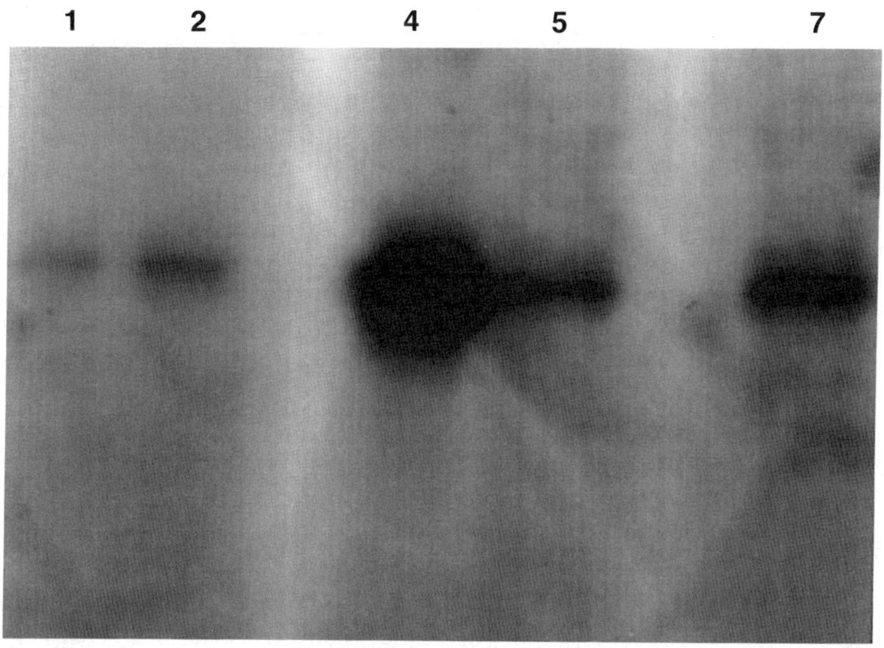

Fig. 2. Autoradiograph of separated SRM 1845. Cholesterol in Egg Powder at different sample volumes. The protein bands visible in tracks 1, 2, 4, 5, and 7 represent phosvitin. Track 4 is purified phosvitin at a level 10 times greater than the highest phosvitin amount in the egg powder.

rather than a true quantitative determination (*7,9*). Alternately, amino acid analysis/sequencing can be used to quantify the caseins after precipitation and isolation from the other milk proteins (*10*). Results for this procedure depend on the completeness and recovery of precipitation. In more recent work (*11*), caseins were determined after PAGE separation by quantifying the Coomassie Blue stained protein bands by densitometry, relative to an internal standard. This procedure yielded reproducible results in the determination of the amount of adulteration of sheep's milk with cow milk, but is still a measure of the degree of staining, which may vary from run to run. A more accurate phosphoprotein measure is obtained by quantifying the phosphorus, which is covalently bound in the protein, and does not depend on the reproducibility of staining. The PAGE/NAA method would separate the characteristic phosphoproteins and quantify by the determination of phosphorus.

A major difficulty with the determination of α-casein in the reference materials was that, in the process of preparing the original reference material, the proteins were altered and probably damaged. This was seen after separation of these samples since the expected bands of the separated proteins were not clearly defined as in other protein samples, but

Table 1
Concentrations of Phosphoproteins in Several Materials[a]

Sample	α-Casein	Phosvitin
Sodium caseinate	1.4 μg/μL (±18%)	
IAEA A-11, Milk Powder	0.08 μg/μL (±10%)	
Low-fat (2%) milk	25.7 μg/μL (±20%)	
"half and half" (50% milk, 50% cream)	2.0 μg/μL (±8%)	
SRM 1845, Egg Powder		2.9 μg/μL (±11%)

[a]Uncertainties represent observed standard deviation relative to the mean, except for IAEA A-11, where the uncertainty is that of a single measurement.

were nondistinct and broader. This effect had been observed previously (*7*) in the separation of non-fat dry milk and whey protein concentrate, which had been spray dried. All of the reference materials used in this work had this characteristic to some degree, with SRM 1549, Milk Powder, being the extreme case. Still, quantitation could be accomplished on the IAEA A-11 material although with an increased uncertainty because of the band broadening.

A list of values for the phosphoproteins quantified in this work is given in Table 1. The α-casein values are for low-fat (2%) milk, "half and half," IAEA A-11, Milk Powder, and sodium caseinate, and a phosvitin value for SRM 1845, Cholesterol in Egg Powder. The uncertainties in parentheses are standard deviations of the mean. Values for sodium caseinate and egg powder represent an average of five and four values, respectively, referring to the different levels of protein sample applied and separated. The reference material milk powder could be quantified only at the highest sample loading owing to the diffuseness of the protein bands, as described above. The value for this sample represents only one measurement. No result was obtained for the α-casein in SRM 1549, Milk Powder, in which the protein denaturation was the most extreme of the reference material samples studied here. The values for the two dairy products are averages of two values each. The value of α-casein calculated for low-fat milk compared favorably to a previous result of 17.7 ± 1.3 μg/μL found in skim milk from various breeds of cows (*12*).

The method of PAGE/NAA has been successfully applied to protein mixtures or composites to determine the level of several phosphoproteins. The detection limits for other phosphoproteins may be lowered by improvements in the method, especially in the areas of separation and detection after irradiation. Significant improvements in instruments to perform the densitometric measurements can also be used to improve the uncertainties in quantification.

ACKNOWLEDGMENTS

The authors acknowledge the assistance of the personnel at the University of Virginia Nuclear Reactor Facility, especially B. Hosticka. Certain commercial equipment, instruments, or materials are identified in this paper in order to adequately specify the experimental procedures. Such identification does not imply recommendation or endorsement by the National Institute of Standards and Technology, nor does it imply that the materials or equipment identified are necessarily the best available for the purpose. This work will be included in the dissertation of S. F. Stone, to be submitted as a requirement for a Ph.D. degree from the University of Maryland.

REFERENCES

1. C. K. Jayawickreme and A. Chatt, *J. Radioanal. Nucl. Chem.* **124,** 257 (1988).
2. K. S. Kwok, R. M. Snapka, J. A. Bernard, O. K. Harling, and A. Varshavsky, *Trans. Am. Nucl. Soc.* **49,** 157 (1985).
3. Z. Szökefalvi-Nagy, I. Demeter, C. Bagyink, and K. L. Kovács, *Nucl. Instr. Meth. Phys. Res.* **B22,** 156 (1987).
4. S. F. Stone, R. Zeisler, and D. K. Hancock, *J. Radioanal. Nucl. Chem.* **112,** 95 (1987)
5. S. F. Stone, R. Zeisler, and G. E. Gordon, *Trace Element Analytical Chemistry in Medicine and Biology,* vol.5, P. Brätter and P. Schramel, eds., Walter de Guyter, Berlin-New York, 1988, p. 157.
6. U. K. Laemmli, *Nature* **227,** 680 (1970).
7. J. Basch, F. W. Douglas, L. G. Procino, V. H. Holsinger, and H. M. Farrell, *J. Dairy Sci,* **68,** 23 (1985).
8. K. Olden and K. M. Yamada, *Anal. Biochem.* **78,** 483 (1977).
9. K. F. NG Kwai-Lang and E. M. Kroeker, *J. Dairy Sci* **67,** 3052 (1984).
10. B. Ridadeau Dumas, *Biochim. Biophys. Acta* **168,** 274 (1968).
11. M. Ramos, R. M. Sanchez, and A. Olano, *Chem. Milkrobiol. Technol. Lebensm.* **9,** 24 (1985).
12. G. D. Rolleri, B. L. Larson, and R. W. Touchberry, *J. Dairy Sci.* **39,** 1683 (1956).

Location and Quantification of Metal Ions in Enzymes Combining Polyacrylamide Gel Electrophoresis and Particle-Induced X-Ray Emission

ZOLTÁN SZÖKEFALVI-NAGY,*[,1] CSABA BAGYINKA,[2] ISTVÁN DEMETER,[1] KORNÉL L. KOVÁCS,[2] AND LE HUEN QUYNH[2]

[1]*Central Research Institute for Physics, Hungarian Academy of Sciences, PO Box 49, Budapest 114, H-1525 Hungary; and* [2]*Institute of Biophysics, Biological Research Center, Hungarian Academy of Sciences, PO Box 521, Szeged, H-6701 Hungary*

Received April 17, 1989; Accepted September 14, 1989

ABSTRACT

A method is presented to identify and determine the relative amounts of protein-bound metal ions *in situ*. Proteins or their subunits are separated by SDS-PAGE, the appropriately dried gel sections are directly scanned by a collimated proton beam of 3 MeV energy, and the characteristic X-rays produced are detected. The determination of Fe content of an iron–sulfur protein (HiPiP), as well as the Fe and Ni analysis of the hydrogenase from *Thiocapsa roseopersicina*, have shown the feasibility of this technique.

Index Entries: PIXE; particle-induced X-ray emission; trace element analysis; polyacrylamide gel electrophoresis (PAGE); metalloprotein; metalloenzyme; bacterial hydrogenase.

INTRODUCTION

A certain group of proteins contains metal ions, in addition to the light major constituents. The importance of metal ions in the active centers of these metalloproteins is widely recognized in well-known

*Author to whom all correspondence and reprint requests should be addressed.

examples, such as the oxygen carrier hemoglobin and myoglobin. A number of other key metabolic processes also involve enzymes containing metal ions, among these are electron transport chains, generation of membrane potential, energy production, photosynthesis, and nitrogen fixation. Both the number of ions per molecule and the location of binding places of these essential metals are very important to understand the functional mechanisms of the enzymes. Keeping track of metal ions throughout biochemical separation processes could be a useful strategy for this purpose.

Polyacrylamide gel electrophoresis (PAGE) is a very effective and sensitive method to separate proteins (*1*). Most biological macromolecules are electrically charged and, therefore, will move in an electric field. The transport of molecules through a solvent is called electrophoresis. The use of polyacrylamide gels as supporting media enhances resolution, particularly for proteins and nucleic acids. Gel electrophoresis can also be appled to whole native proteins, but its most effective application involves the use of denaturing agents, such as sodium dodecyl sulfate (SDS). SDS qualitatively binds to all proteins the same, at about 1.4 g/g protein. There is one negative charge on each SDS molecule, and the protein–SDS complexes assume a random-coil configuration. Therefore, proteins treated in this way behave as though they have a uniform shape and an identical charge-to-mass ratio. In an electric field, the lower molecular weight proteins migrate more rapidly through the gel, and fractions of different molecular weights form well-isolated narrow bands. Proteins that are composed of more than one polypeptide chain in vivo break down under SDS-PAGE conditions, and the unfolded polypeptide subunits are separated during electrophoresis. The most common approaches to the detection of these bands involve the use of stains, autoradiography of radiolabeled analytes, or immunoreaction with specially prepared antisera (*1*). The quantitative measurement of the metal ions is rather difficult since the absolute amounts of these metals are very small and hardly accessible to standard chemical methods. Particle-induced X-ray emission (PIXE) spectroscopy (*2*), however, could be sensitive enough to determine the metal content even in these minute gel volumes.

In the PIXE method, a collimated beam of energetic charged particles, usually a proton beam of 2–4 MeV energy, bombards the sample and ionizes its atoms. During the electron-rearrangement processes, characteristic X-rays can be observed, usually by a Si (Li) semiconductor detector of high energy resolution. By measuring the energy and the intensity of the the emitted X-rays, the atomic number and quantity of emitting atoms in the bombarded volume can be determined. Under usual experimental conditions, all elements from Al to U can be measured simultaneously, with sensitivities in the ppm range.

This paper is devoted to describing the combination of the PAGE and PIXE methods and also demonstrating that PIXE analysis of the subunits, separated and immobilized in the PAGE matrix, can provide information, both on the metal content of a particular protein and the distribution of metals among the subunits. (Similar information can be gathered only by X-ray diffraction; that method, however, is much more difficult and requires much more of a precious protein sample, which is cumbersome to crystallize). For demonstration of the method, the so-called high-potential iron–sulfur protein (HiPIP) was used, whereas new enzymological results were obtained from the PIXE-PAGE analysis of a hydrogenase enzyme. Both proteins were prepared from the bacteria, *Thiocapsa roseopersicina*.

MATERIALS AND METHODS

The purple sulfur photosynthetic bacterium, *Thiocapsa roseopersicina* strain BBS, was a kind gift from E. N. Kondratieva (Moscow State University, USSR). The bacterium was cultivated as described earlier (*3*). HiPIP and hydrogenase were purified from the frozen cell paste as previously reported (*4,5*). The purity of the final preparation was checked by SDS-PAGE, and the functional integrity of the purified hydorgenase was monitored by direct measurement of its enzyme activity (*6*).

For PAGE, the discontinuous buffer system developed by Neville (*7*) was used. Acrylamide concentration was 10%, crosslinking 1%, 0.1 × 5 × 5 cm slab gels were cast. Samples were incubated with an equal volume of a sampler buffer containing 2% SDS and 5% 2-mercaptoethanol at room temperature or were inactivated at 100°C for 5 min. One-hundred micrograms of pure protein were applied in a single slot along the gel width. Proteins are migrating in an electric field of about 100 V. After separation, a portion of the gel was stained for enzyme activity (*5*) and/or protein with Coomassie Brilliant Blue R-250. The remaining part of the gel was covered with a 1 mg/cm^2 thick Nuclepore filter of 12 μm pore size and dried on a GSD-4 (Pharmacia Fine Chemicals, Uppsala, Sweden) gel slab drier. The Nuclepore filter was easily removed from the surface of the dried gel, leaving the gel surface directly accessible for proton irradiation. By aligning this part of the gel with the portion of the same gel showing the stained protein bands, it was possible to indicate the regions of interest for irradiation. (The staining process regularly washes out the metal ions, therefore a stained gel can not be analyzed.)

A detailed description of the PIXE technique has been given elsewhere (*2,8*). The experimental arrangement is shown in Fig. 1. Sections of the dried gel were placed on the Al holder frame and placed into the vacuum chamber of our 5 MV Van de Graaf accelerator. The samples were sprayed with a conducting film aerosol spray (Polaron Equipment Ltd., Watford, UK) to prevent disturbing charge-up owing to the proton

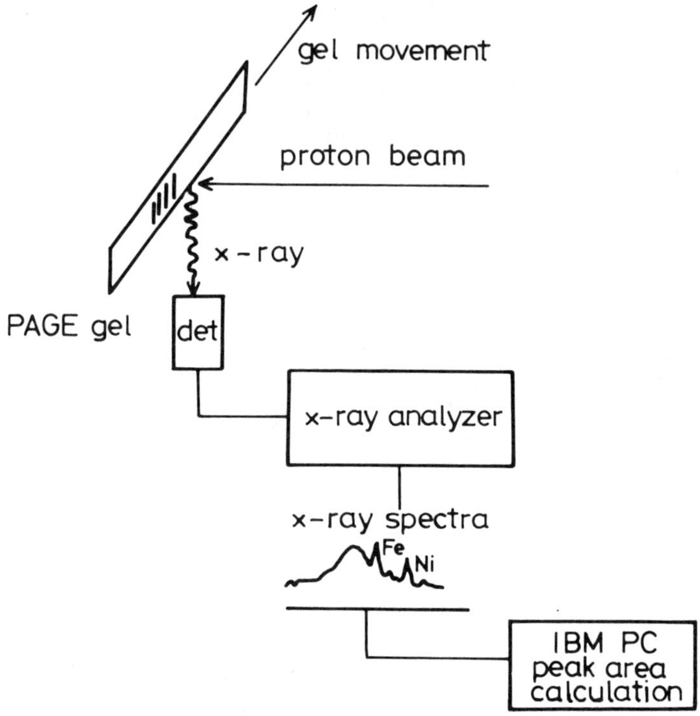

Fig. 1. Schematic diagram of the PIXE-PAGE measurement.

bombardment. The X-ray spectra, induced by irradiation of the sample with a proton beam of 3 MeV energy, were analyzed with a Canberra X-ray spectrometer. The proton beam was focused on a $0.6 \times 7\ mm^2$ area at the gel surface. The target sample was moved in 1 mm steps perpendicularly to the proton beam in order to scan the entire gel length. At each position, an X-ray spectrum was recorded, and 1 μC of incoming proton charge was collected.

After proton irradiation, the sample was rehydrated and stained with Coomassie Brilliant Blue R-250, as above. The areas hit by the proton beam were very clearly deteriorated, and the impact marks of the bombardment were clearly visible as sharp traces superimposed on the protein bands. This phenomenon made the accurate assignment of the X-ray spectra to the protein bands possible.

RESULTS AND DISCUSSION

HiPIP is a relatively small protein of 10 kD molecular mass. It is fairly stable, easy to purify in large quantities, and the Fe ions are bound in the form of 4Fe-4S clusters. The results of a PIXE scan are displayed in Fig. 2.

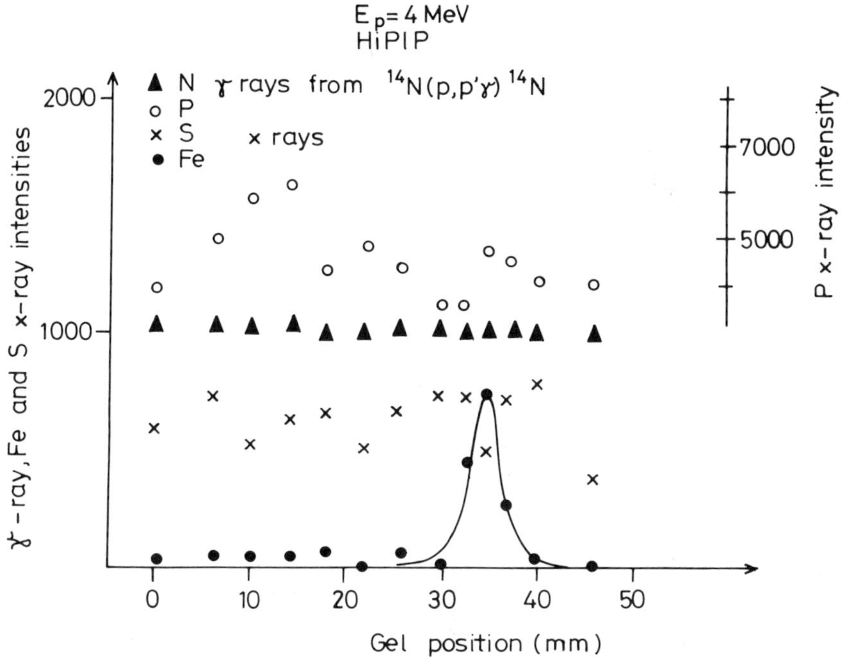

Fig. 2. P, S, and Fe X-ray intensities and the intensity of the γ-rays from the $^{14}N(p,p'\gamma)^{14}$ reaction along a polyacrylamide gel containing the electrophoretic band of HiPIP.

The sharp Fe peak at the gel position of about 35 mm (filled circles), corresponding to the HiPIP band, clearly shows that a metalloprotein band can be unambiguously identified on the basis of the X-ray spectra of a dried PAGE sample. The comparison to the stained half of the gel also showed that the resolution of metal detection by PIXE is about as good as the resolution of the protein staining process in PAGE (*8*). Figure 3 demonstrates the quantitative capabilities of the PIXE-PAGE combination, where the total Fe contents of bands containing different amounts of HiPIP are plotted. The integrated Fe intensities are proportional to the amount of protein supplied for electrophoresis. A simultaneous and independent quantitative measurement of both the total amount of protein and metal content would make the method more powerful. The measurement of sulfur could be a choice for this purpose, but, as the crosses in Fig. 3 show, the gel itself contains so much S originating from the buffer that no peak can be detected at the protein band. The rather irregular P distribution is also attributed to buffer. Nitrogen cannot be measured with conventional PIXE, but the sharp resonance in the $^{14}N(p,p'\gamma)^{14}N$ nuclear reaction at about 4 MeV proton energy offers another possibility (*11*). The very uniform nitrogen distribution obtained shows that the expected small peak of the protein-bound nitrogen cannot be detected on the large background coming from the polyacrylamide gel. In the case of nitrogen-free supporting media, however, this ap-

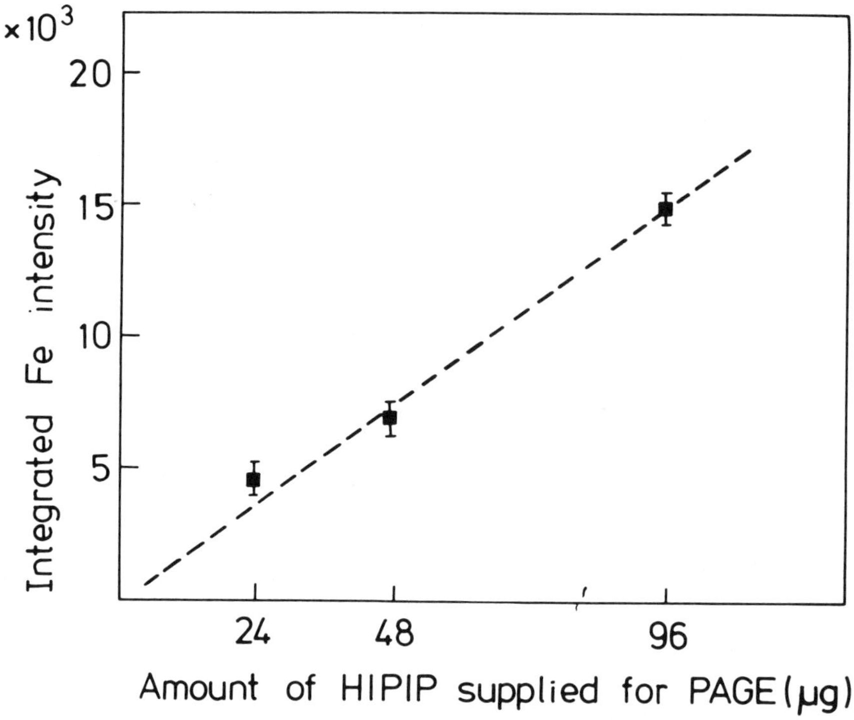

Fig. 3. The total Fe content of a HiPIP band as a function of the amount of protein supplied for electrophoresis.

proach could be successful. (The intention to test this nuclear reaction explains the use of 4 MeV protons in this particular case instead of the regularly used energy of 3 MeV.)

Hydrogenase is a more complicated metalloprotein, and it represents a challenge to the PAGE-PIXE technique. Hydrogenase of *Thiocapsa roseopersicina* is composed of two nonidentical subunits of 64 and 34 kD molecular mass. Thus, the native enzyme has a 98 kD molecular mass and contains at least one 4Fe-4S cluster and one Ni ion in one molecule. Hydrogenases are unstable in general; however, the membrane-bound enzyme isolated from *T. roseopersicina* displays remarkable stability against inactivation by oxygen and heat. The enzyme retains its activity even under SDS-PAGE conditions, unless the sample is boiled for several minutes before the electrophoresis (*5*). The metals play a crucial role in the enzymatic function of hydrogenases (*9*). Hydrogenases from different organisms show striking structural similarities (*10*). Although several of them have been thoroughly studied, the intramolecular location of the metal atoms is unknown. Scanning the protein components, separated with SDS-PAGE by the proton beam along the gel, gives direct information about the distribution of elements in the gel. In the active hydrogenase sample, the Fe and Ni peaks are at the same position, which

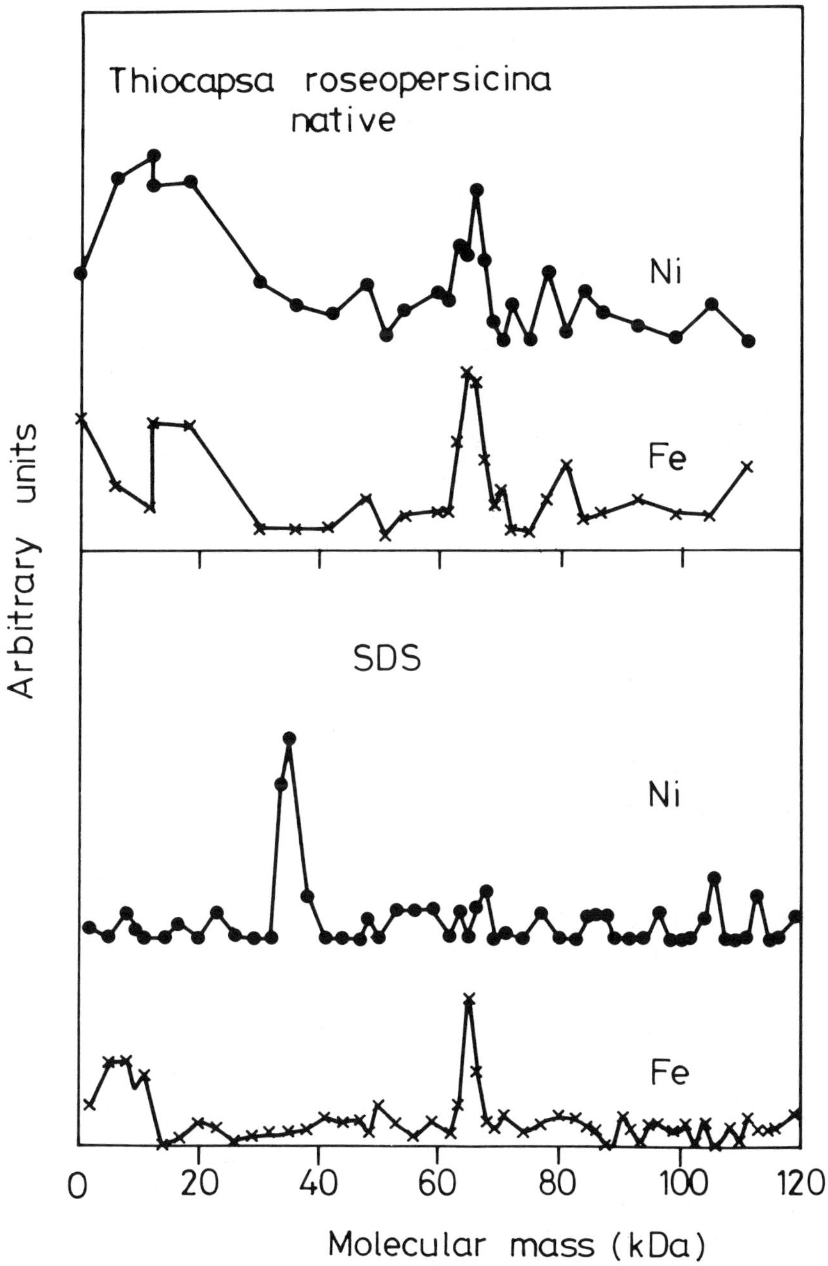

Fig. 4. The net Fe and Ni peak areas, measured by PIXE scans, on native and SDS gels.

corresponds to the migration of the native, unfolded hydrogenase (upper part of Fig. 4). This band can also be stained for activity so the nature of the protein in the protein band can be accurately determined. The fact that the hydrogenase is active, however, means that the polypeptide chains are in the native three dimensional structure, and therefore, the mobility of the native hydrogenase band is not proportional to its molecular mass since the molecule does not assume a rod shape. At any rate, one can conclude that the native enzyme contains Fe and Ni, as predicted from chemical analysis and electron spin resonance measurements, but no other metals. By help of the calibrated sensitivity curve of the PIXE system, the ratio of iron to nickel was also calculated from the integrated X-ray counts along the gel regions of interest, and the value of 7 ± 1 was obtained. This value is far from the previously reported value of 4 (*12*). The subunit polypeptides are separated according to their molecular mass in SDS-PAGE if the hydrogenase is heat denatured prior to the electrophoretic migration. The results of PAGE-PIXE clearly demonstrate that Fe atoms migrate together with the large (64 kD) subunit, whereas Ni is apparently bound to the small (34 kD) subunit (*see* the lower part of Fig. 4).

These findings provide direct evidence that metals of different kinds can be studied in the same biological sample. It is remarkable that the Fe and Ni metal centers are close to each other, according to a number of observations with electron spin resonance and magnetic circular dichroism spectroscopy. Since they are bound to separate polypeptides, both metal centers should be at or close to the surface of their respective subunits in the area where the subunits are held together.

ACKNOWLEDGMENT

This work was supported in part by the Hungarian National Science Foundation (OTKA II.) under Research Contract No. 1017.

REFERENCES

1. J. W. Jorgenson, *Anal. Chem.* **58**, 743 A (1986).
2. L. Keszthelyi, L. Varga, I. Demeter, K. Hollós-Nagy, and Z. Szökefalvi-Nagy, *Anal. Biochem.* **139**, 418 (1984).
3. Cs. Bagyinka, K. L. Kovács, and E. Rak, *Biochem. J.* **202**, 255 (1982).
4. G. Ciszewska, Cs. Bagyinka, and K. L. Kovács, *Proc. Int. Symp. Biol. Solar Energy Bioconv.*, Ed. I. V. Berezin, Soviet Acad. Press, Pushchino, USSR, 1984, p. 330.
5. Cs. Bagyinka, N. A. Zorin, and K. L. Kovács, *Anal. Biochem.* **142**, 7 (1984).
6. K. L. Kovács, G. Tigyi, and H. Alfonz, *Prep. Biochem.* **15**, 321 (1985).
7. D. M. Neville, Jr., *J. Biol. Chem.* **246**, 6328 (1971).
8. Z. Szökefalvi-Nagy, I. Demeter, Cs, Bagyinka, and K. L. Kovács, *Nucl. Instr. Meth. Phys. Res.* **B22**, 156 (1987).

9. M. W. W. Adams, L. E. Mortenson, and J. S. Chen, *Biochem. Biophys. Acta* **594**, 105 (1981).
10. K. L. Kovács, L. Seefeldt, G. Tigyi, C. M. Doyle, L. E. Mortenson, and D. J. Arp, *J. Bacteriol.* **171**, 430 (1989).
11. R. Hänninen and J. Räisänen, *Nucl. Instr. Meth.* **B5**, 34 (1984).
12. I. N. Gogotov, N. A. Zorin, and E. N. Kondratieva, *Biokhimiya* **41**, 836 (1976).

Analysis of Biological Samples Using Prompt Gamma Radiations Induced by 14.7-MeV Neutrons

C. L. HOLLAS,* L. E. USSERY, K. B. BUTTERFIELD, AND R. E. MORGADO

Los Alamos National Laboratory, Los Alamos, NM 87545

Received April 17, 1989; Accepted September 14, 1989

ABSTRACT

We investigate a method for determining the elemental composition of biological samples that uses prompt gamma rays induced by 14.7-MeV neutrons. Alpha particles are produced simultaneously with the neutrons, which exit opposite the alpha detector through the vacuum chamber wall. The sample under investigation is irradiated and emits gamma radiations in a spectral energy distribution characteristic of the material. Barium-fluoride (BaF_2) and high-purity germanium (HPGe) gamma detectors view the sample and record the spectrum of gamma radiation.

Index Entries: Associated-particle technique; prompt gamma radiation; time-coincidence measurement; barium fluoride; high-purity germanium; coincidence measurement; two-dimensional matrices; fluorine; multielement analysis.

INTRODUCTION

At the present time, there are a number of nuclear analytical techniques available for material analysis that are often complementary in the information they provide. We have been investigating another method, the associated-particle technique, which may add to the arsenal already

*Author to whom all correspondence and reprint requests should be addressed.

available. Based on the d-t reaction, it is used to produce 14.7-MeV neutrons and (approximately) 3.5-MeV alpha particles.

Figure 1 illustrates a schematic of the technique. A deuteron beam with 100- to 150-keV incident energy strikes a tritiated titanium target, producing neutrons and alpha particles that are emitted 180 degrees from each other. An alpha-particle detector detects alpha particles from the reaction and defines a cone of correlated neutrons that can be used to analyze the material. The neutrons interact inelastically with the material under investigation and produce excited nuclei that decay, emitting prompt gamma radiation. This radiation, characteristic of the material that was excited by the neutrons, can then be used to identify the material.

METHOD

The technique consists of performing a time-coincidence measurement between the detected alpha particle and the gamma ray resulting from the neutron inelastic reaction. The coincidence requirement results in a substantial reduction in the background events present in the spectrum and allows real-time analysis to be performed.

The alpha-particle detector, consisting of 50 μm of a fast plastic scintillator (BC-400) deposited on an acrylic disk, detects alpha particles at a distance of 10 cm from the production target. The detector's thickness was chosen because it has ~100% efficiency for detecting alpha particles produced in the reaction and very little efficiency for detecting neutrons, X-rays or gamma-rays. The diameter of the detector is 1.9 cm; this can be modified with collimators. A thin nickel foil (0.00013 cm thick), located on the front surface of the detector, eliminates low-energy deuterons that are elastically scattered from the tritium target.

A number of gamma-ray detectors have been used, including barium fluoride (BaF_2) and high-purity germanium (HPGe). The BaF_2 provides moderate energy resolution (with an efficiency comparable to a sodium iodide) and the best time resolution for the coincidence measurement, resulting in the best volume localization for the material under investigation. The HPGe provides very good energy resolution with moderate time resolution. However, for material analysis, the latter is the best candidate because it allows for the interpretation of complex spectra that often result from composite materials.

The data acquisition system comprises NIM and CAMAC electronics interfaced to a VMEbus computer. This enables us to acquire and store data in two-dimensional matrices of time vs energy and to project, in real time, the energy spectra associated with the object under investigation. Spectra for background gamma radiation sources are also available from the same measurement, allowing for immediate elimination of gamma

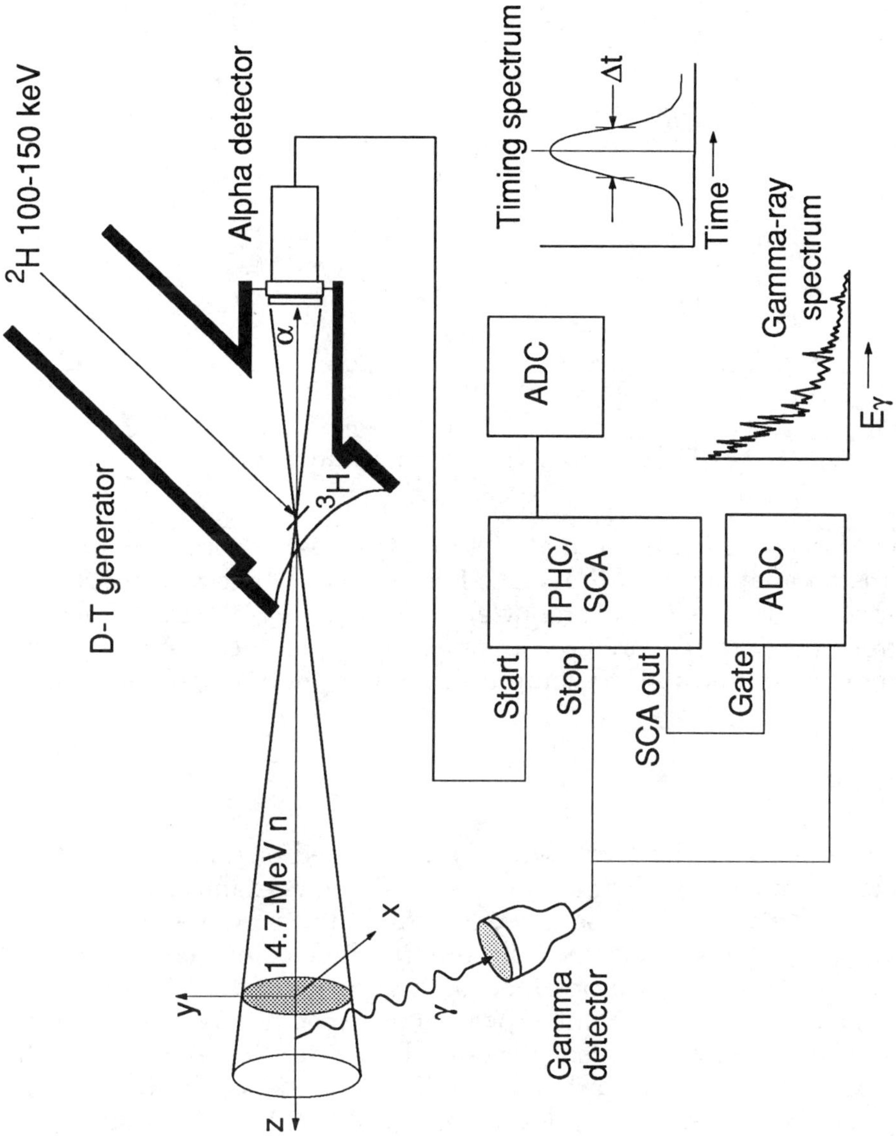

Fig. 1. Schematic of the associated-particle technique.

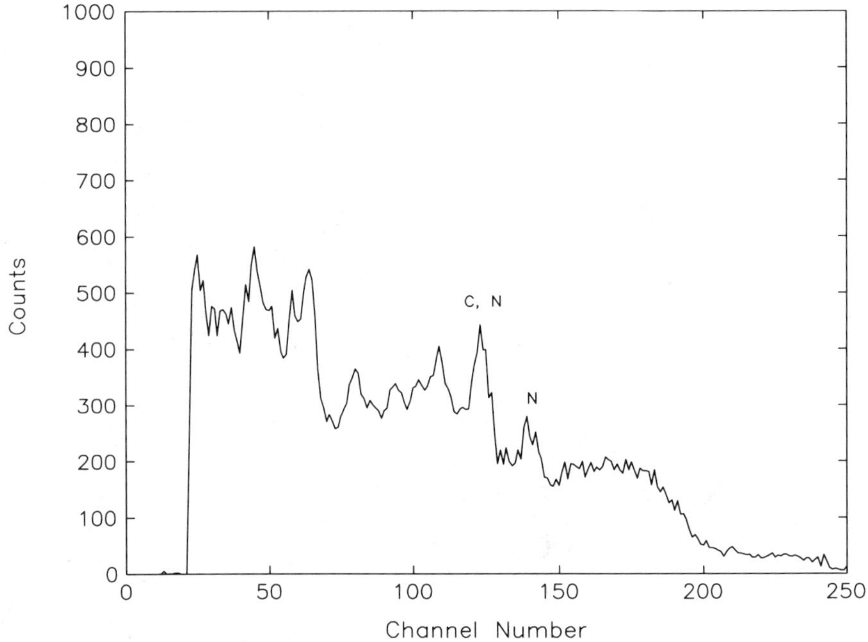

Fig. 2. BaF_2 spectrum for nitrogen.

rays not associated with the material under investigation. Our present system is an adaptation of the VALET Plus system that was developed and supported by the European nuclear laboratory, CERN. Data from the measurements may be stored on streamer tape or on the hard disk of the peripheral computer used to communicate with the VMEbus processor.

RESULTS

To demonstrate the technique, we performed qualitative measurements on a limited set of targets. Figures 2–7 show examples of spectra obtained with BaF_2 and HPGe detectors. All of these figures correspond to energy projections associated with the true-time-coincidence window. Figures 2 and 3 are spectra obtained with the BaF_2 and HPGe detectors for nitrogen. The most prominent peak in this spectrum is at 5.105 MeV. The single-escape peak is also present, along with the 4.439-MeV full-energy peak from carbon. Carbon is produced via the (n,n′d) reaction on the nitrogen target. Another photopeak at 2.313 MeV is also a signature for nitrogen. Figures 4 and 5 are the spectra for oxygen. The major features in the HPGe spectrum are the 6.13-Mev peak, the single-escape 5.619-MeV peak, and the double-escape 5.108-MeV peak. When both elements are present in a sample, the double-escape peak from oxygen

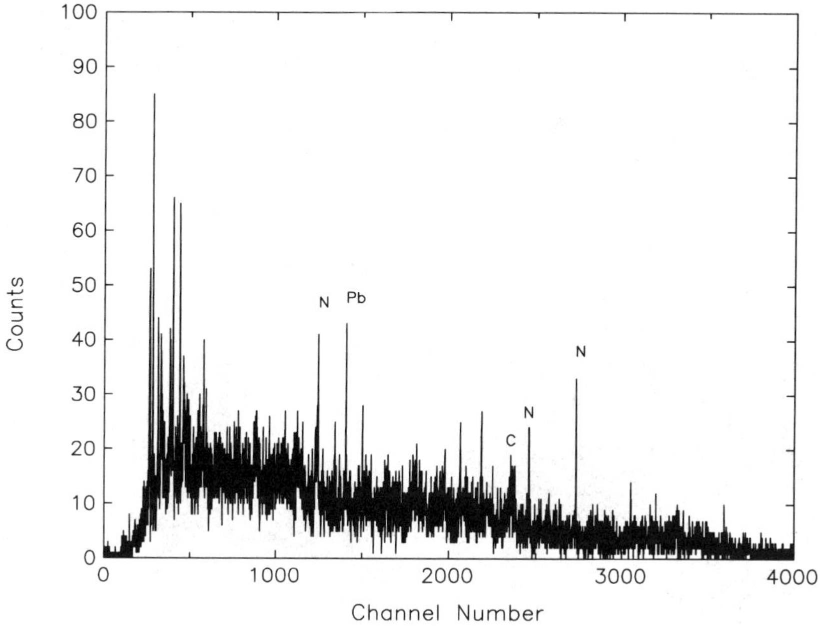

Fig. 3. HPGe spectrum for nitrogen.

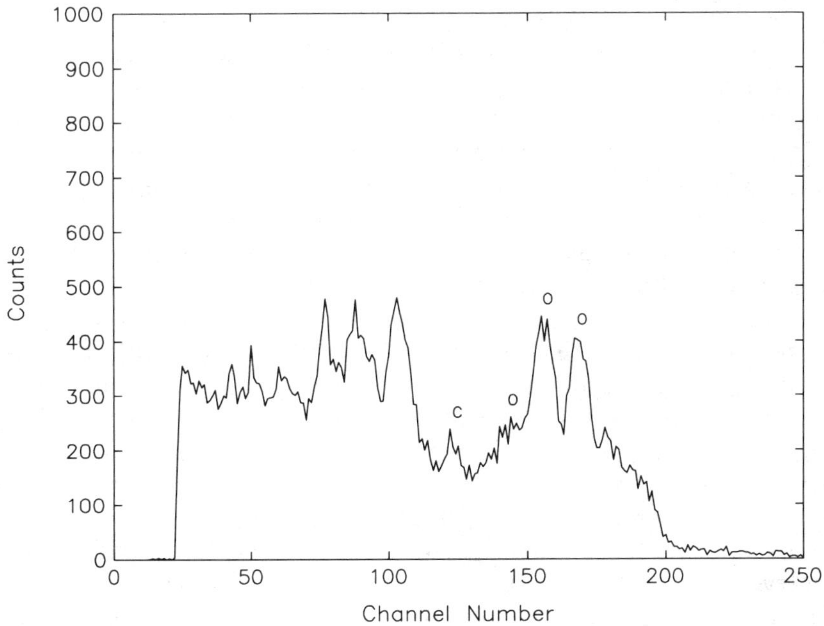

Fig. 4. BaF_2 spectrum for oxygen.

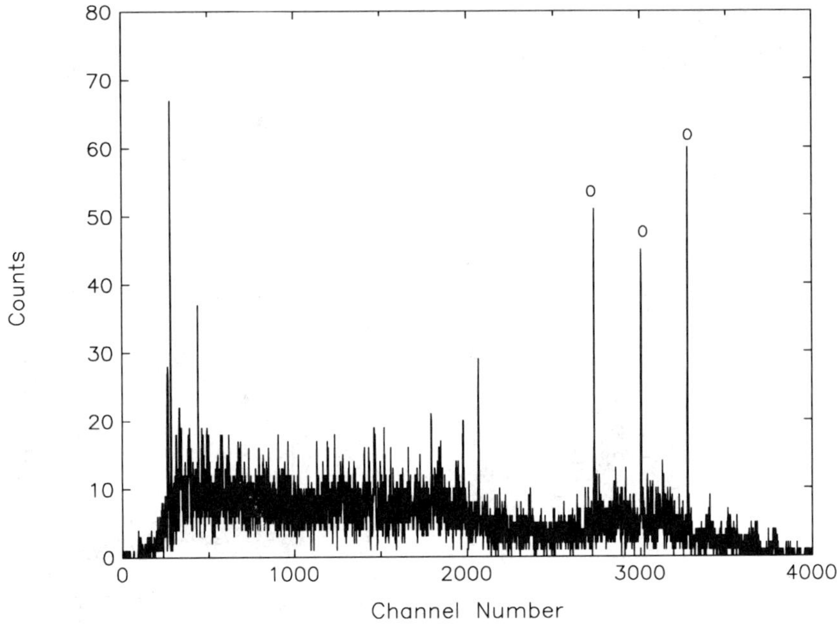

Fig. 5. HPGe spectrum for oxygen.

interferes with the 5.105-MeV full-energy peak from nitrogen. The peaks discussed in Figs. 2–5 are the fingerprints that are used to identify nitrogen and oxygen. Carbon also has a simple spectrum with a full-energy peak at 4.439 MeV and a single-escape peak at 3.928 MeV.

Figures 6 and 7 show examples of spectra from a composite material containing carbon, nitrogen, hydrogen, and oxygen for the BaF_2 and HPGe detectors, respectively. Clearly, the primary signature gamma-ray peaks are present for all of the major constituents of the material. Also present in the HPGe spectrum is a gamma-ray peak for lead, which is used as part of the gamma-ray shield surrounding the detector.

One element that is difficult to observe with standard activation techniques, especially in the presence of oxygen, is fluorine. To demonstrate that the associated particle technique encounters no problems in distinguishing fluorine from oxygen, we have made a measurement using a fluorine target. Figure 8 shows the fluorine spectrum measured with the HPGe detector; the most prominent gamma ray is at 1.98 MeV. This corresponds to the decay of the first excited state in ^{18}O, which is formed via the (n,d) reaction on fluorine. There are also at least three gamma rays present that result from the de-excitation of levels in ^{19}F. A comparison of Figs. 5 and 8 shows that the spectra for oxygen and fluorine are quite distinct.

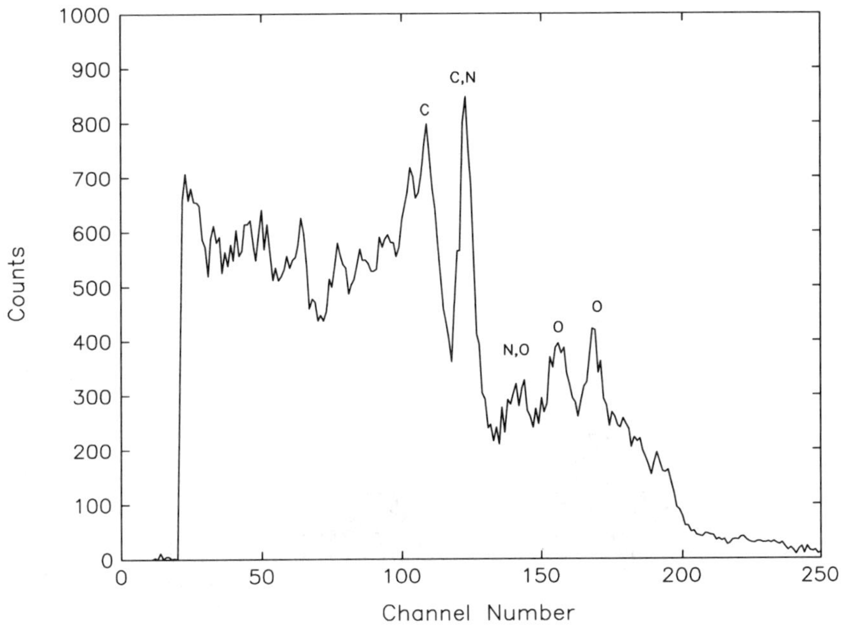

Fig. 6. BaF_2 spectrum for composite material (C, N, H, and O).

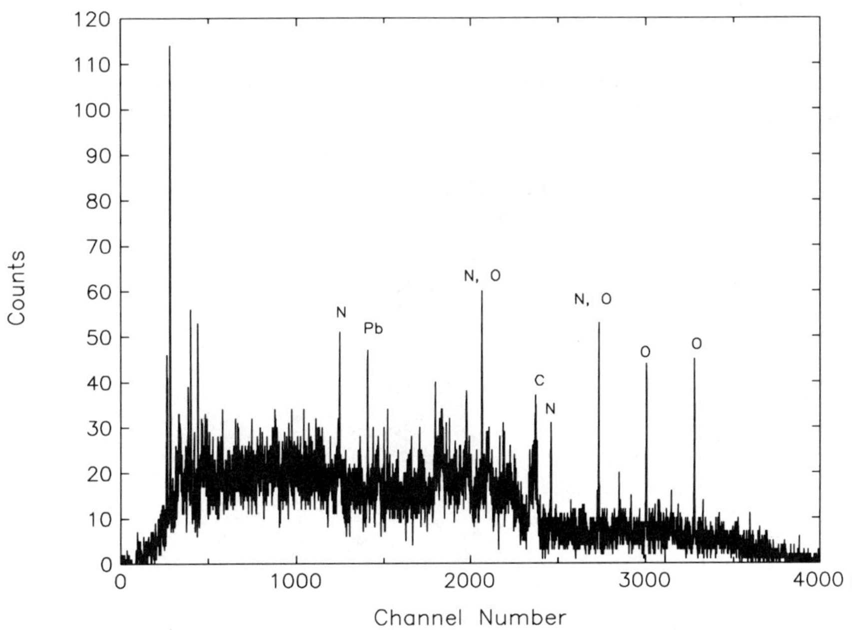

Fig. 7. HPGe spectrum for composite material (C, N, H, and O).

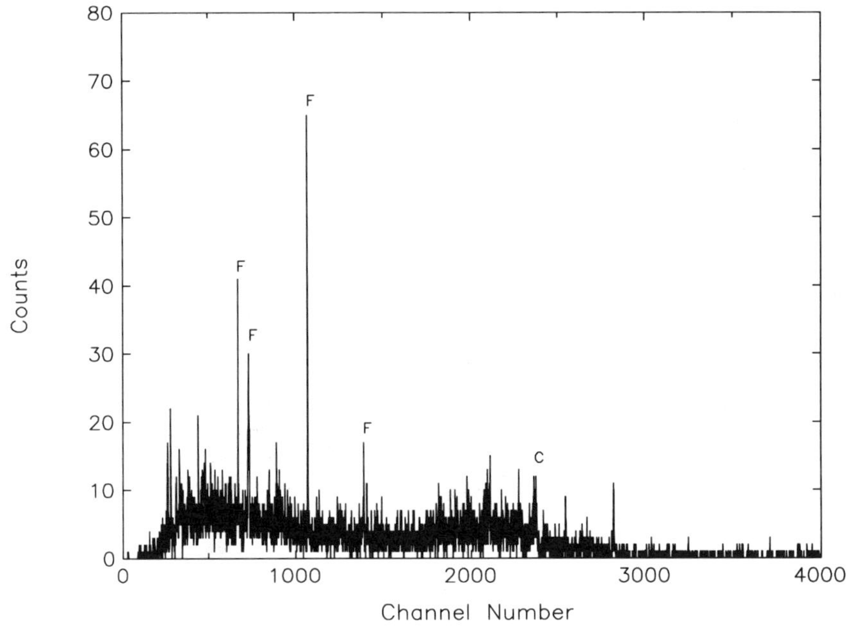

Fig. 8. HPGe spectrum for fluorine.

SUMMARY

We have demonstrated that the associated-particle technique has excellent potential for qualitative and, probably, quantitative analysis of materials. The technique has several strong points, including the ability to observe essentially all elements, except hydrogen and helium, and to analyze many elements in a single measurement. Furthermore, this multielement analysis can be done in real time so that the analysis is finished when irradiation is completed.

A New Technique for the Study of Erythrocyte Survival by Double Labeling and Use of a Well-Type Ge Detector

PETER BODE,[1] MARCEL DE BRUIN,*[,1,2] SIEWART OLDENBURG,[2] ALBERT VAN DER WIEL,[3] AND BERT WOLTERBEEK[1]

[1]Interfaculty Reactor Institute, Delft University of Technology, The Netherlands; [2]Center for Radionuclide Applications, State University Utrecht, The Netherlands; and [3]Hospital "De Lichtenberg," Amersfoort, The Netherlands

Received April 17, 1989; Accepted September 14, 1989

ABSTRACT

A double label procedure with ^{57}Co and ^{58}Co has been developed for detailed in vivo studies of erythrocyte survival. A well-type Ge detector is used in the measurements. The activities necessary for these experiments are very low, and the associated dose received by the test persons can be neglected.

Index Entries: Labeling; double-label; erythrocyte survival.

INTRODUCTION

Over the last 25 years, instrumentation for use in the radiochemical laboratory has developed very rapidly. Through the introduction of semiconductor detectors and laboratory computers, high resolution γ-ray spectroscopy has become available on a routine scale. The new instrumental possibilities have been applied mainly in activation analysis, studies of environmental radioactivity—natural as well as artificial—and in basic radiochemical research. However, γ-ray spectrometry using Ge

*Author to whom all correspondence and reprint requests should be addressed.

detectors can also be very attractive for certain radiotracer applications. Initial restrictions owing to the low efficiency of Ge detectors have disappeared with increasing size of these detectors. The absolute efficiency of presently available well-type detectors is very high and comparable to the commonly used well-type NaI(TI) detectors (Fig. 1). The very good energy resolution allows for precise multitracer experiments, and the combination of high efficiency and excellent energy resolution results in excellent properties for detecting ultra low activities (*1*).

A well-type Ge detector γ-ray spectrometer played an essential role in a series of multitracer experiments at very low activity levels, performed as part of a study of the in vivo properties of components of blood transfusions. A double-label technique was used for simultaneously observing the behavior of the components of the transfused blood and the test persons' own blood. Thereby, the disturbing effects of biological variations in time and between individuals are largely reduced, and relatively small differences, otherwise obscured by biological variations, can be established. When using 5 mL blood samples, an administered dose of maximally 10^4 Bq already yields sufficiently precise results. The associated radiation dose received by the test person is only a few percent of the annual dose from natural sources, and the experiments fall within the lowest risk category, as defined by the WHO.

Double-label procedures have been developed for erythrocytes and thrombocytes; a procedure for leukocytes is under study. Application so far has only concerned erythrocytes, with as main points of interest

1. comparison of erythrocytes survival in autologous and homologous blood transfusion;
2. the occurrence of immunological reactions after multiple transfusions;
3. influence of storage conditions on the survival of erythrocytes; and
4. quality of erythrocytes recycled during operation after passing through a cell-saver.

A first series of experiments was focused on validating the experimental method, comparison of autologous and homologous blood transfusions, and possible adverse effects of multiple transfusions. For these experiments, hospital patients have volunteered as test persons. Some of the results obtained in these experiments are discussed in the present paper.

EXPERIMENTAL DETAILS

Selection of Radionuclides

Most relevant criteria to be met by the radionuclides to be applied in the double-label study of the in vivo properties of erythrocytes are: isotopes of the same element, half-lives in agreement with the time-scale

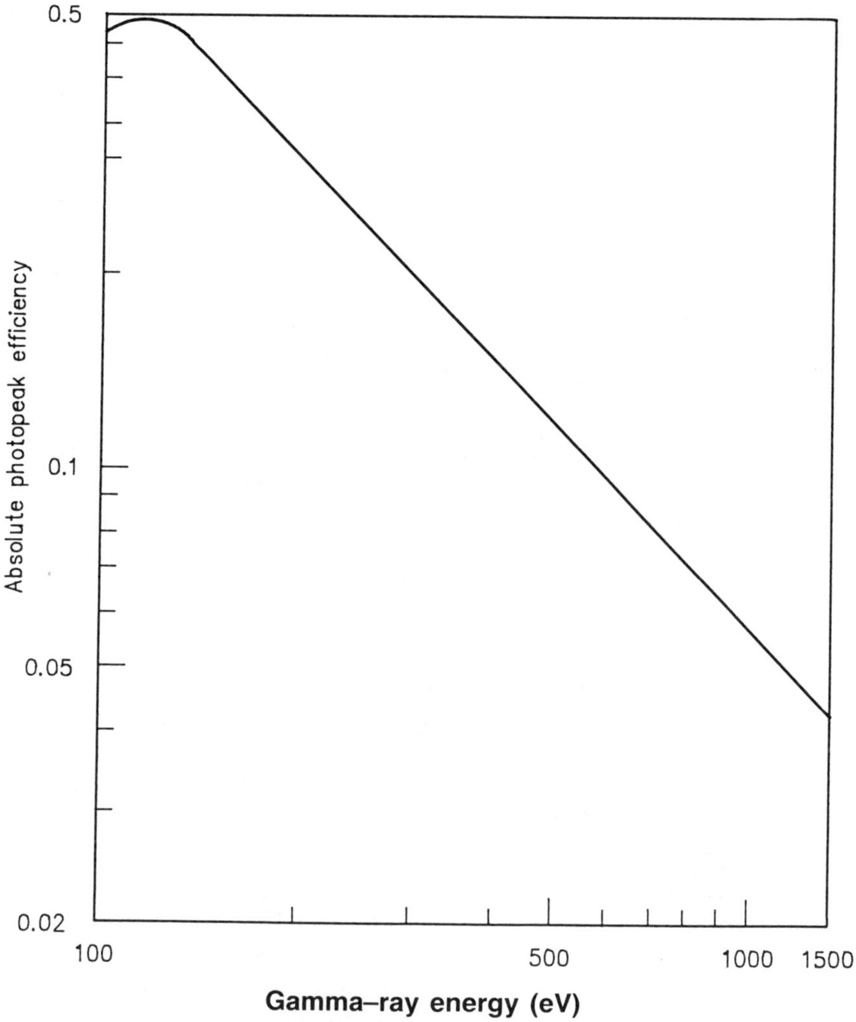

Fig. 1. Photopeak efficiency of a 125 cm^3 well-type Ge detector as a function of γ-ray energy.

of about 100 d of the processes under study, emission of γ-rays with high yield, low effective dose equivalent per unit activity, availability with high specific activity, stable incorporation, and no influence on cell vitality. On the basis of these criteria, isotopes of the element cobalt were selected for the double labeling of erythrocytes. The relevant properties of the five major radioactive Co isotopes are listed in Table 1 (*2,3*). As far as half-lives are concerned, four Co isotopes are applicable in a study of erythrocytes, so that, theoretically, four different erythrocyte populations can be followed simultaneously in one test person. Since only two isotopes were needed in our study, ^{57}Co and ^{58}Co were selected because of the low radiation doses associated with the application of these nuclides.

Table 1
Major Radio Isotopes of Cobalt

Isotope	^{55}Co	^{56}Co	^{57}Co	^{58}Co	^{60}Co
Half-life	17.5 h	78.8 d	271 d	70.8 d	5.27 y
Main γ-rays					
Energy (keV)	931	511, 847	122	511, 811	1173, 1332
Abundance (%)	41	38, 100	86	30, 100	100, 100
H_{50} (10^{-9} Sv/Bq)	3	8	0.8	3	20

Labeling Procedure

Labeling of the erythrocytes with Co isotopes was carried out using Tropolon (2-hydroxy-2,4,6-cycloheptatriene-1-on) as a lipophilic complexing agent. A 10 mL blood sample was centrifuged and the erythrocytes washed two times and resuspended in physiological salt solution. "No carrier added" ^{57}Co or ^{58}Co were added to the suspension in a $10^{-2}M$ Tropolon solution. After incubation for 1 h at 37°C, the erythrocytes were washed to remove free Tropolon and cobalt and injected as a suspension in a physiological salt solution. The efficiency of the labeling procedure was approximately 80%. In each test, 9 kBq of ^{57}Co and 5 kBq of ^{58}Co were applied, leading to corresponding effective dose equivalents of 7 and 15 μSv, respectively.

Sampling and Measurements

Five-milliliter blood samples were taken from 1 h until 70 d after administration, with intervals increasing from 6 h to 7 d. The blood samples were centrifuged and the erythrocyte fractions stored at 0–4°C. In some of the initial tests, both erythrocytes and plasma were counted. The samples were measured using a 125 cm^3 germanium well-type detector with a well of 18 × 55 mm. The detector was equipped with a thin Zr lining to prevent spectrum distortions owing to low-energy X-rays. The absolute photopeak efficiency was 41% at 122 keV (^{57}Co) and 7.5% at 810 keV (^{58}Co). For most samples, a counting period of 1 h was sufficient to obtain adequate precision; for the plasma samples, longer counting periods were required.

Patient Experiments

In a series of six experiments, two samples of autologous blood were labeled with ^{57}Co and ^{58}Co, respectively, to verify the identical behavior of the two differently labeled fractions. In these experiments, activities were measured both in erythrocytes and plasma. In three experiments, two samples of autologous blood were labeled with ^{57}Co and ^{51}Cr to compare the results obtained from Co-labeled blood with the results from the commonly used procedure with ^{51}Cr. Eight experiments were aimed at detecting possibly existing differences between autologous and homologous blood. In these experiments, autologous blood was labeled with ^{57}Co, whereas ^{58}Co was used as label for the homologous donor blood. Part of the experiments are still in progress.

RESULTS

Autologous Erythrocytes Labeled with ^{57}Co and ^{58}Co

In all experiments, the activities of erythrocytes and plasma as a function of time showed a more or less similar behavior. A typical example of the results obtained in these experiments is given in Fig. 2. An initial sharp drop of the erythrocyte activity during the first day is followed by a slow exponential decrease during the further 70-d observation period. The plasma activity starts at a relatively high level, shows a rapid exponential decay, and disappears within a week. In all experiments, no difference could be observed between the course in time of the ^{57}Co and ^{58}Co activity, except for the drop during the first days after injection.

Autologous Erythrocytes Labeled with ^{57}Co and ^{51}Cr

Figure 3 shows the results obtained when comparing the properties of ^{57}Co- and ^{51}Cr-labeled erythrocytes. The ^{57} Co activity again shows the initial drop observed for all experiments with Co-labeled erythrocytes. But except for this first part of the curve, the behavior of ^{57}Co and ^{51}Cr activity is identical.

Autologous and Homologous Erythrocytes Labeled with ^{57}Co and ^{58}Co

Data available so far show no or only very small differences between the course in time of the two types of activity. Figure 4 shows the results of a completed experiment, in which the largest differences were observed. Fitting of an exponential function to the slowly decaying parts of the two curves yields half-lives of 22 and 20 d for the ^{57}Co-labeled autologous and the ^{58}Co-labeled homologous donor erythrocytes, respectively. The difference between these two values is just on the border of a statistically significant difference.

CONCLUSIONS

The present study demonstrates that, when using sophisticated equipment, in vivo experiments in humans can be carried out with very low amounts of radioactivity and a negligible associated radiation risk. The additional effective dose equivalent received by the test persons is limited to 20 μSv or 2 mrem, which is about 1% of the annual natural

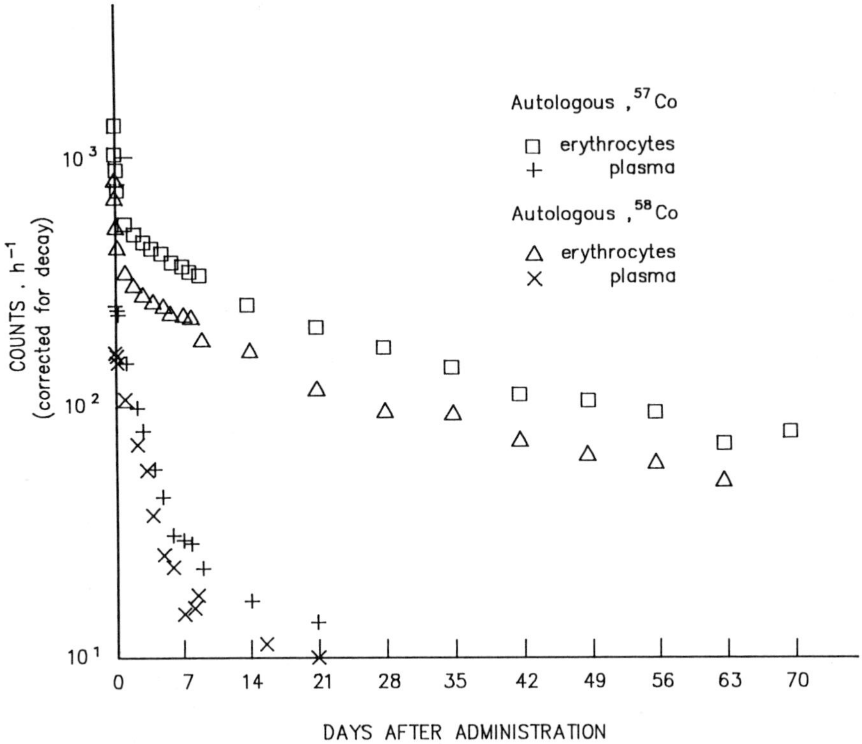

Fig. 2. Results obtained with autologous erythrocytes labeled with ^{57}Co and ^{58}Co.

dose. It is well below the limit of 100 μSv set by WHO (5) for the lowest risk category of applications of radioactivity to humans for research purposes.

The results of the experiments with autologous erythrocytes labeled with ^{57}Co and ^{58}Co show an identical behavior of the two labeled fractions over a period of approximately 2–70 d after injection. This indicates that the labeling procedure is sufficiently reproducible to yield a real double label for studying long-term effects.

The sharp drop of the erythrocyte activity shortly after injection needs further investigation. It obscures possibly existing, interesting short-term effects, and the radiation dose associated with the rapidly released activity is unnecessary and does not yield useful information. The experiments with erythrocytes labeled with ^{57}Co and ^{51}Cr have shown that, after the initial loss of Co activity, both labeled fractions behave identically. This means that results obtained with Co-labeled erythrocytes can be compared directly with the vast amount of information obtained with the regularly applied ^{51}Cr test.

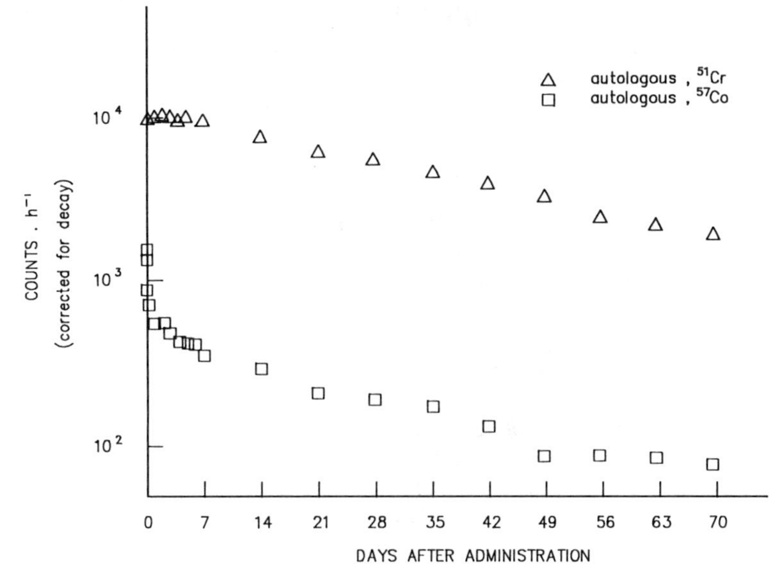

Fig. 3. Results obtained with autologous erythrocytes labeled with ^{57}Co and ^{51}Cr.

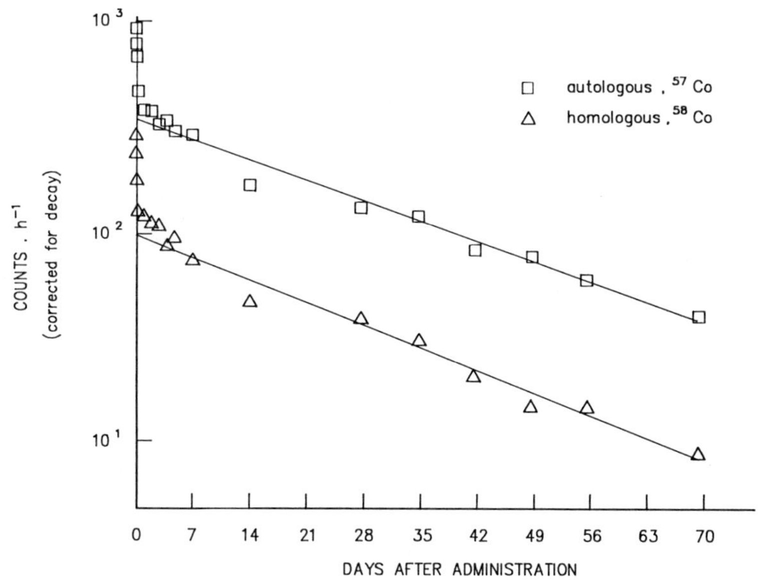

Fig. 4. Results obtained with ^{57}Co-labeled autologous and ^{5} ^{8}Co-labeled donor erythrocytes.

The tests aimed at comparing autologous and homologous erythrocytes are still in progress. Available results are not yet sufficient to draw firm conclusions.

REFERENCES

1. M. de Bruin and P. J. M. Korthoven, *Radiochem. Radioanal. Lett.* **44,** 139 (1980).
2. C. M. Lederer and V. S. Shirley, *Table of Isotopes*, 7th ed., Wiley, New York, 1978.
3. *Limits for intakes of radionuclides by workers* (and supplements), Publication 30, International Commission on Radiological Protection, Pergamon Press, Oxford.
4. H. I. Glass, *Br. J. Haematol.* **21,** 21 (1971).
5. *Use of ionizing radiation and radionuclides on human beings for medical research, training and nonmedical purposes*, WHO Technical Report Series **611,** Geneva, 1977.

Charged-Particle Activation Analysis of Biological Material

J. HOSTE* AND C. VANDECASTEELE

Laboratory for Analytical Chemistry, Rijksuniversiteit Gent, Institute for Nuclear Sciences, Proeftuinstratt 86, 9000 Gent, Belgium

Received April 17, 1989; Accepted October 31, 1989

ABSTRACT

Charged-particle activation analysis offers a number of interesting possibilities for the determination of trace elements in biological material. It allows the determination of those elements that are difficult or impossible to determine by neutron activation, such as Li, B, Al, Si, V, Cr, Ni, Cd, Sn, Tl, and Pb. Up to now, protons have been successfully applied to samples of both vegetale and human origin. A number of difficulties have to be overcome, one of which is excessive heating of the samples owing to the limited range of the charged particles, thus giving rise to a high energy deposition in a small volume. Moreover, the sample composition has to be known to allow the calculation of the range of the particles. An interesting alternative has been proposed using an internal standard together with a standard additions procedure. Proton activation analysis was tested on a wide variety of reference materials, giving evidence that accurate results can be obtained for many trace elements, even when applying a purely instrumental method. Thus, the method can also be applied in the certification of reference materials, since nuclear methods are independent of chemical properties of the sample.

Index Entries: Charged-particle activation analysis (CPAA); trace elements; biological materials analysis.

INTRODUCTION

For the determination of trace and ultratrace elements in biological materials and, more particularly, in human tissues and fluids, neutron activation analysis (NAA) has been widely and succesfully applied. In-

*Author to whom all correspondence and reprint requests should be addressed.

deed, this technique offers a number of advantages owing to its high sensitivity for a large number of elements, specificity, and avoidance of contamination, at least if avoided before irradiation.

After freeze-drying of a serum for instance, a sample can be irradiated in a nuclear reactor. The elements Ag, As, Br, Co, Cr, Cs, Cu, Fe, Hg, Mn, Mo, Rb, Sb, Sc, Se, Sn, V, and Zn can be determined, usually after postirradiation chemical separation of the radionuclides of interest. The determination of a number of elements such as Li, B, Al, Si, Ti, Cd, Tl, Pb, and Bi, is either difficult or impossible by NAA, so charged-particle activation analysis (CPAA) may offer a possible alternative nuclear method of analysis.

PRINCIPLES

In CPAA, simultaneous irradiation of a sample and a standard is generally impossible, so beam monitoring is required, for instance, by means of a Faraday cup or by placing a thin foil in front of the sample and the standard and measuring an induced activity.

It can be shown that if thick targets are irradiated, i.e., thicker than the range of the charged particles, the concentration of the analyte in the sample will be given by

$$c_x = c_s \frac{\left(\frac{N_p}{t_m \cdot D \cdot C}\right)_x}{\left(\frac{N_p}{t_m \cdot D \cdot C}\right)_s} \cdot \frac{i_s\left(1-e^{-\lambda t_{irr,s}}\right)}{i_x\left(1-e^{-\lambda t_{irr,x}}\right)} \cdot F(E_i) \qquad (1)$$

and

$$F(E_I) = \frac{\int_{E_T}^{E_I} \frac{\sigma(E)dE}{S_s(E)}}{\int_{E_T}^{E_I} \frac{\sigma(E)dE}{S_x(E)}} \qquad (2)$$

where: c_x, c_s: analyte concentrations in the sample and the standard; i_x, i_s: beam intensities on sample and standard; E_T: threshold energy of the considered reaction; E_I: incident energy of the charged particle; $\sigma(E)$: cross-section of the considered reaction as a function of energy; and S_x, S_s: stopping power of sample and standard.

It is obvious that Eq. (1) requires the knowledge of the excitation function σ(E) and the stopping power S(E) and the use of thick targets, i.e., thicker than the range for pure elements.

The stopping power data for protons, deuterons, helions-3, and helions-4 can be obtained from the compilations of Andersen and Ziegler (*1*) and Ziegler (*2*). The above authors give equations that allow the calculation of the stopping powers. In general, they conclude that the agreement between calculated and experimental results is better than 1% for p, d, ^{3}He, and ^{4}He in the energy range of 1 to 25 MeV, i.e., for energies usually applied in CPAA. Stopping powers (MeV/cm^2/g) and ranges g/cm^2 and μm) are illustrated in Table 1 for 10 and 20 MeV protons for a number of elements. The ranges were calculated by

$$R(E_I) = \int_0^{E_I} \frac{dE}{S(E)} \tag{3}$$

where *R* is the range of a particle, with energy E_I, i.e., the mean path length in the target. From Table 1, it is clear that a target thicker than 1 mm is required at 20 MeV.

When irradiating targets of biological origin, for instance after incineration to remove the organic components, a mixture of several elements is obtained. The approximate composition of an incinerated human serum is given in Table 2.

The stopping power in this mixture is calculated by means of the additivity rule of Bragg and Kleeman (*3*) given by Eq. (4):

$$S(E) = \sum_{i=1}^{n} \omega_i S_i(E) \tag{4}$$

S(E): stopping power with *n* components at energy E; $S_i(E)$: stopping power of element i at energy E; ω_i: mass fraction of component i.

There is sufficient experimental evidence that the Bragg and Kleeman rule does not introduce an appreciable error in the energy range usually applied in CPAA. Using Eq. (3), the range can be calculated. The composition of Table 2 gives 0.2294 g/cm^2.

The correction factor $F(E_I)$ not only necessitates knowledge of the stopping power, but also the excitation function $\sigma(E)$. If the excitation function is known with sufficient accuracy, the $F(E_I)$ can be calculated by numerical integration. However, even when the available excitation functions show discrepancies, the induced systematic error is often quite small. This problem is discussed in detail for several reactions by Vandecasteele (*4*). In general, it appears that the systematic error is below 1%. In cases where the excitation function is not known, approximations can be used, as given in Table 3.

Numerical integration is preferred if the excitation function is known, but the above approximations can be applied with small system-

Table 1
Stopping Power and Range of Protons in the Elements Indicated

	10 MeV			20 MeV		
Element, Z_A	S, $MeV/cm^2/g$	R, g/cm^2	R, μm	S, $MeV/cm^2/g$	R, g/cm^2	R, μm
C (6)	40.8	0.138	613	23.3	0.477	2120
Al (13)	33.9	0.170	630	19.7	0.573	2122
Ni (28)	28.6	0.207	233	17.1	0.677	761
Cd (48)	22.5	0.270	313	13.7	0.860	995
Pb (82)	17.8	0.353	311	11.1	1.089	960

Table 2
Approximate Composition
of Incinerated Human Serum

Element	Component	% elemental occurrence
Ca	CaO	0.965
Cl	(Na)Cl	38.15
Cu	CuO	0.011
Fe	Fe_2O_3	0.011
Mg	MgO	0.225
K	K_2O	1.946
Na	Na(Cl)	32.31
P	P_2O_5	1.36
Zn	Zn	0.014
S	SO_3	10.8
O		14.2

atic errors if the threshold of the reaction is high and the atomic number of sample and standard are similar. To avoid the problem of beam intensity monitoring and correction for the differences in range between sample and standard, Cantone et al. (*8*) applied the following standard addition method referred to earlier. The sample is doped with a known quantity of a reference element. As a standard, the same sampled doped with the same amount of reference element and with a known quantity of the element under study, large with respect to its natural concentration, is used.

CHOICE OF ACTIVATION PARTICLES AND NUCLEAR REACTIONS

Up to now, only a limited number of papers has been published on CPAA of biological materials, e.g., by Zikovsky and Schweikert (*9*), Bonardi et al. (*10*), Cantone et al. (*11,12*), Yagi and Masumoto (*13–15*),

Table 3
Approximate Standardization Methods

Author	Method	Ref.
Ricci and Hahn	$F(E_I) = \frac{\Delta R_s}{\Delta R_x}$ $\Delta R = R(E_I) - R(E_T)$	(5)
Ricci and Hahn	$F(E_I) = \frac{R_s(E_I)}{R_x(E_I)}$	(6)
Chaudri et al.	$F(E_I) = \frac{S_s(E_M)}{S_x(E_M)}$ $E_M = \frac{1}{2}(E_I + E_T)$	(7)

and Masumoto and Yagi (*16*). All these articles use protons on different matrices, namely Orchard Leaves (SRM 1571), Bovine Liver (SRM 1577), Bowen's Kale, and human serum, and apply either nondestructive techniques or chemical separations. The applied proton energy varied from 12 to 31 MeV, with beam intensities up to 500 nA. The considered reactions were usually of the (p,xn) type. It is also interesting to mention systematic studies of proton activation carried out by Debrun et al. (*17*) and Barrandon et al. (*18*), with 10 MeV protons, for nuclides with a half-life longer than 1 min. Debrun et al. (*19*) also described instrumental proton activation analysis vis short-lived radionuclides, i.e., with a half-life between 1 s and 1 min. Similar studies were carried out by Barrandon et al. (*20*), Kormali and Schweikert (*21*), and Krivan and Krivan (*22*).

Some detection limits are given in Table 4, defined as a photon peak equal to three times the square root of the background measured directly after irradiation of 1.8 $t_{1/2}$ or 60 h, whichever is shorter, with a Ge detector of 22% efficiency for an irradiation time of 1 h at 1 μA intensity.

It is apparent that good detection limits are obtained for a large number of elements. Moreover, one should consider that 10 MeV protons are by no means always at optimal energy especially for high-Z elements. Chemical separations can improve the background by eliminating interfering activities although time-consuming separations will, of course, reduce the sensitivity. Such chemical separations to isolate ^{111}In, ^{203}Pb, and ^{206}Bi, produced from Cd, Tl, and Pb in geological and environmental samples, were developed by Wauters et al. (*23*). These could be applied in a slightly modified form for biological samples. It is also obvious that, for biological samples, mineralization of the matrix can increase the sensitivity by a factor of about 10 and, in the case of serum, even by a factor of about 100 if referred to the liquid specimen. As a general conclusion, proton activation analysis can give a satisfactory analysis for the measurement of low-concentration elements that are difficult or impossible to determine by NAA, such as Li, B, Al, Si, V, Cr, Ni, Cd, Sn, Tl, and Pb.

Boron, which is of biomedical importance, can also be determined by CPAA, namely by the $^{10}B(d,n)^{11}C$ reaction with a detection limit of ca. 3.5 ng/g for a deuteron energy of 5.6 MeV and a beam intensity of 2 μA. A chemical separation is required. The measurements are made by coincidence counting with NaI(Tl) scintillation detectors.

Other particles, such as ^{3}He and ^{4}He, can also be used. A number of possibilities, together with detection limits and interferences, were described by Sastri et al. (*24*). A survey of the elements of interest is given in Table 5. From Table 5, it appears that a number of elements can be determined with excellent sensitivities. Many interfering reactions are to be considered, but are sometimes negligible or can be avoided or minimized by choosing an adequate ^{3}He energy. ^{3}He particles have a much shorter range than protons of the same energy. Therefore, problems relative to destruction of the sample during the irradiation are even more severe.

SAMPLE TREATMENT AND STANDARDIZATION

Obviously, the major components when analyzing samples of biological origin are organic ones, e.g., polysaccharides, proteins, lipids, and so on. These are highly sensitive to decomposition by a possible high temperature during irradiation. This problem is also present for NAA at high neutron fluxes and often gives rise to explosion of quartz containers. It is even more severe in CPAA since the energy is dissipated in a quite small volume, the range of 10 MeV protons being usually only a few hundred micrometers. Zikovsky et al. (*9*) mixed SRM 1571 (Orchard Leaves) and SRM 1577 (Bovine Liver) with nuclear-grade graphite, pressed it into a pellet, and wrapped it in aluminum-foil. This can stand a 12 MeV proton irradiation for 1 h at 0.6 $\mu A/cm^2$. Others (*11*) used dried serum pressed into a pellet of 0.7 mm thickness, sandwiched in a Mylar foil (6 μm), and an aluminized Mylar foil of 30 μm. These samples, cooled at −5°C with freon, can stand an irradiation of 23 MeV protons for 3 h at 300 nA. Cantone et al. (*25*) measured the temperature of serum samples (1 mL dried at 35°C and pressed into a pellet of 12 mm diameter and 0.7 mm thickness). This was inserted in a frame and cooled at −5°C. It was irradiated in air at 300 nA for 5 h with 13.8 MeV protons. It appeared that a steady temperature of 40°C was reached rapidly. Bonardi et al. (*10*) analyzed SRM 1577 (Bovine Liver), SRM 1575 (Pine Needles), and Bowen's Kale for lead using 31 MeV protons at a beam current of 0.5 to 0.6 μA. They used a high-purity, freon-cooled graphite container in which 300–500 mg amounts of the material was pressed. Xenoulis et al. (*26*) have examined the stability of biological specimens during proton irradiation (2.4 and 4.3 MeV) and measurement of the prompt gamma

Table 4
Detection Limits for a Number of Elements of Interest in Biological Material

Element	Radionuclide	Half-life, h	γ-Energy keV	Detection limit, ng/g
Li	^{7}Be	1279.2	447.6	24
Ca	^{44}Sc	3.95	1157	8.8
Ti	^{48}V	386.4	1311.8	9
V	^{51}Cr	667.7	320	43
Cr	^{52m}Mn	0.35	1434.4	0.4
	^{52}Mn	134.4	744.1	28
Fe	^{56}Co	1855.2	846.7	61
Ni	^{60}Cu	0.39	1792	4.5
Cu	^{63}Zn	0.64	669.8	3
Zn	^{66}Ga	9.4	1039.3	6
As	^{75}Se	2880	136.1	40
Se	^{76}Br	15.9	559.3	10
Rb	^{87m}Sr	2.83	388.5	2
Sr	^{86}Y	14.6	1077.4	13
Mo	^{95}Tc	20	765.8	5
Ag	^{107}Cd	6.5	93	17.2
Cd	^{111}In	67.44	171.4	12.5
Sn	^{117}Sb	2.8	158.7	7.5
Tl	^{203}Pb	52.1	279.2	103
Pb	^{201}Bi	149.8	802.8	520

Table 5
Nuclear Reactions and Detection Limits for 14 MeV ^{3}He Activation

Element	Nuclear reaction	Threshold, MeV	Half-life	Detection limit, ng/g[a]	Interf. Reaction
Al	$^{27}Al(^{3}He, 2p)^{28}Al$	Q > 0	2.24 min	17	$^{26}Mg(^{3}He, p)^{28}Al$
Ti	$^{48}Ti(^{3}He, 2n)^{49}Cr$	4.6	41.9 min	3	$^{50}Cr(^{3}He, \alpha)^{49}Cr$
V	$^{51}V(^{3}He, 2n)^{52m}Mn$	2.9	21.3 min	22	$^{52}Cr(^{3}He, p2n)^{52m}Mn$
Cr	$^{52}Cr(^{3}He, 2n)^{53}Fe$	6.0	8.51 min	120	$^{54}Fe(^{3}He, \alpha)^{53}Fe$
Mn	$^{55}Mn(^{3}He, 2p)^{56}Mn$	0.4	2.576 h	40	$^{54}Cr(^{3}He, p)^{56}Mn$
Fe	$^{56}Fe(^{3}He, p)^{58}Co$	Q > 0	71.3 d	270	$^{59}Co(^{3}He, \alpha)^{58}Co$
Ni	$^{60}Ni(^{3}He, pn)^{61}Cu$	3.0	3.41 h	40	$^{59}Co(^{3}He, n)^{61}Cu$
Zn	$^{64}Zn(^{3}He, pn)^{65}Ga$	2.9	15.2 min	10	$^{63}Cu(^{3}He, n)^{65}Ga$
Mo	$^{95}Mo(^{3}He, n)^{99m}Tc$	1.2	6.02 h	70	–

[a]For 1 h irradiation or 1 half-life at 2 μA.

rays. Beam intensity was limited to 10 nA, with a beam charge varying from 1 to 60 μC on a surface of 1–6 mm^2. They find that the prompt gamma yields are not constant. For instance, in the case of leaves of *Medicago sativa*, the intensity is constant for ^{13}C, decreases for ^{16}O, and increases for ^{24}Mg, ^{23}Na, and ^{14}N, as it appears from Fig. 1. The differences observed depend on the nature of the samples and their pretreatment (drying at 60°C or 100°C, freeze-drying). These results are, in part, explained by losses of bound water and hydrocarbons.

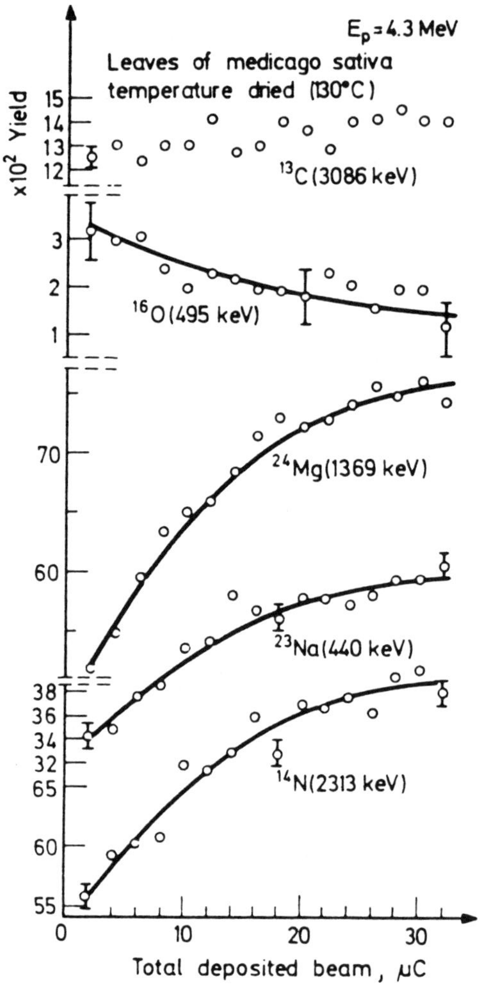

Fig. 1. Intensity of prompt gamma rays from leaves of *medicago sativa*, irradiated with 4.3 MeV protons, as a function of the deposited beam charge (*28*).

Wauters et al. (*27*) examined the temperature of low heat-conducting samples, e.g., SRM 1633a (Fly Ash) irradiated with 23 MeV protons (beam diameter 8 mm) at different intensities under vacuum or under helium (Fig. 2). Although irradiation under helium gives an important decrease in temperature, the maximum temperature can still be quite high, as it appears from Fig. 3. Effects of sample damage can be avoided, of course, by mixing with a heat-conducting material, such as graphite (*9*), or by the use of an internal standard (*11*).

In general, however, it seems more advantageous to mineralize the sample either by LTA or incineration at a relatively low temperature,

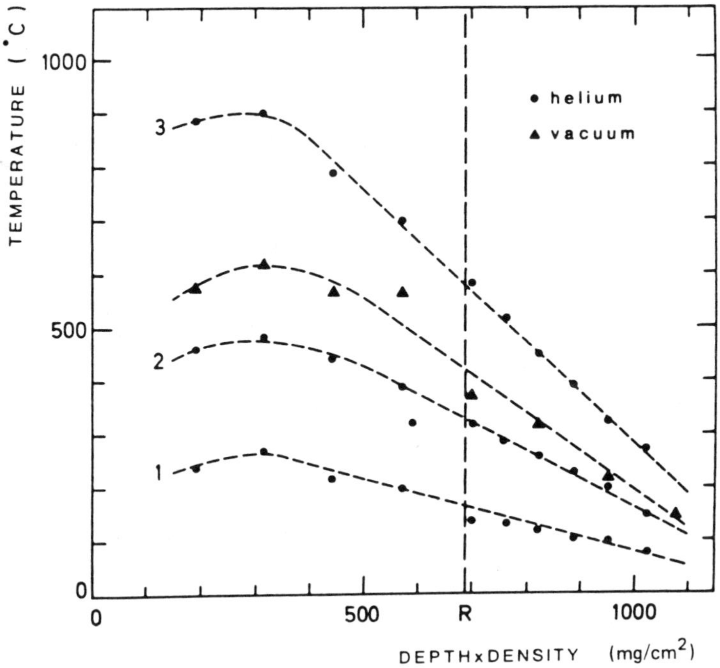

Fig. 2. Temperature curves for SRM 1633a irradiated with 23 MeV protons: ▲ vacuum, 200 nA; ● $3 \cdot 10^5$ Pa helium, 200 nA (curve 1), 500 nA (curve 2), 1000 nA (curve 3); R is the charged particle range (*29*).

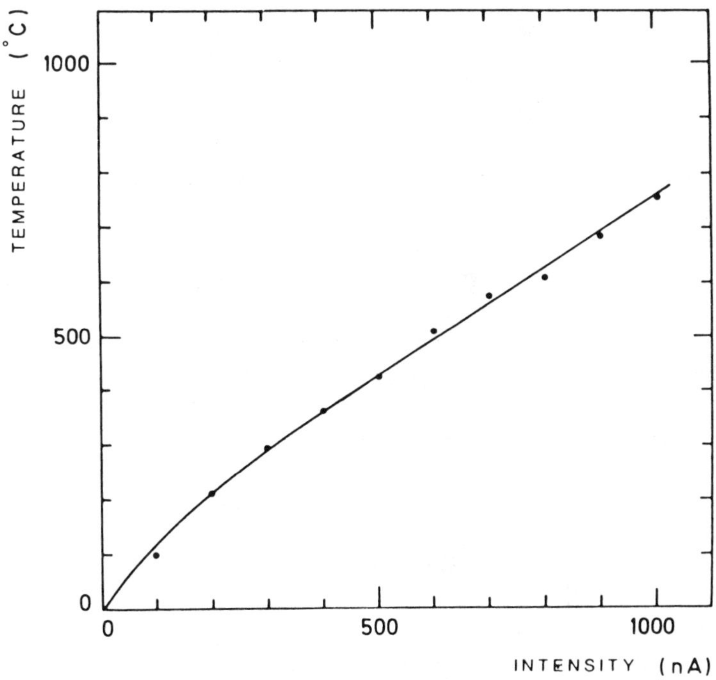

Fig. 3. Maximum temperature in SRM 1633a as a function of the beam intensity (23 MeV protons, $9 \cdot 10^5$ Pa helium) (*29*).

e.g., 400–450°C. Although losses will occur for some volatile elements, such as Br, Se, Hg, and so on, a considerable increase of sensitivity is also obtained. As already mentioned in serum, for instance, the residual ash is only ca. 1% referred to the original liquid and ca. 10% referred to the lyophilized sample.

Calculation of the range from the constituents of the ash is not a serious problem, and one of the major constituents can be used as an internal standard for beam monitoring, e.g., ^{48}Sc from ^{48}Ca(p,n)^{48}Sc. Instead of using pure elements or compounds for standardization, it also seems advisable to add known amounts of the elements of interest in a standard addition procedure mentioned above.

Yagi and Masumoto (*13–15*) and Masumoto and Yagi (*16*) have suggested an internal standard method for proton activation analysis of NBS SRM 1566 (Oyster Tissue), SRM 1569 (Brewer's yeast), and NIES CRM-No6 (Mussel). They spike the samples with 100 μg Y and 500 μg La and add 1–1000 μg of the elements to be determined in the reference sample. The former gives rise to the reaction ^{89}Y(p,n)^{89}Zr ($t_{1/2}$ = 78.5 h) and the latter to ^{139}La(p,n)^{139}Ce ($t_{1/2}$ = 137.2 d). The method can be summarized as follows. When a sample contains W_a grams of a trace element (A) to be determined, together with another element (B) that can be used as an internal standard and when the comparative standard is prepared by adding a known amount of element A, W_a^* grams to a duplicate sample, then

$$W_a = W_a^* \, [(A_R^*/A_R) - 1] \tag{5}$$

where A_R and A_R^* is the ratio of the count rates produced from the element A and B in the sample and comparative standard, respectively. About 1 g of the sample is mineralized with HNO_3, tetra-ethyl-silicate is added, and the mixture dried and incinerated at 400°C. The resulting powder is pressed into a pellet thicker than the range. The pellets are wrapped in 10 μm aluminum foil and placed in a sample changer. The 12 samples and references are irradiated simultaneously in an external proton beam at 13 MeV, with an intensity of 2–4 μA for 2 h. The samples are measured by Ge gamma-ray spectrometry.

RESULTS

A number of results obtained by proton activation are given in Table 6.

It appears that the sensitivity for a number of elements is excellent, as could be expected, and analysis in the submicrogram per gram range is possible. Moreover, the results found are often in good agreement with the data available from reference materials. The results obtained for

Table 6
Results of Proton Activation Analysis of Biological Samples[a]

	Reaction	Found	Present
Orchard Leaves (SRM 1571) *(9)*			
As	$^{75}As(p, n)^{75}Se$	13 ± 3	14 ± 2
Cu	$^{65}Cu(p, n)^{65}Zn$	14 ± 2	12 ± 1
Mo	$^{96}Mo(p, n)^{96}Tc$	4 ± 2	3.2
Pb	$^{206}Pb(p, n)^{206}Bi$	57 ± 17	45 ± 3
Sr	$^{88}Sr(p, n)^{88}Y$	45 ± 15	37
Ti	$^{48}Ti(p, n)^{48}V$	2.4 ± 0.4	–
Zn	$^{66}Zn(p, n)^{66}Ga$	20 ± 6	25 ± 3
Zr	$^{92}Zr(p, n)^{92m}Nb$	3.8	–
Bovine Liver (SRM 1577) *(9)*			
Cu	$^{65}Cu(p, n)^{65}Zn$	278 ± 14	270 ± 20
Ti	$^{48}Ti(p, n)^{48}V$	2 ± 1	–
Zn	$^{66}Zn(p, n)^{66}Ga$	128 ± 14	130 ± 10
Zr	$^{92}Zr(p, n)^{92m}Nb$	4 ± 3	–
Bovine Liver (SRM 1577) *(10)*			
Pb	$Pb(p, xn)^{206}Bi$	0.36	0.34 ± 0.08
Bowen's Kale *(10)*			
Pb	$Pb(p, xn)^{206}Bi$	2.65	2.645
Pine Needles (SRM 1575) *(10)*			
Pb	$Pb(p, xn)^{206}Bi$	10.8	10.8 ± 0.5
Human Serum (Mixture of several individuals)[c] *(28)*			
Cd	$^{111}Cd(p, n)^{111}In$	0.025 ± 0.005	0.00018[b]
Ti	$^{48}Ti(p, n)^{48}V$	0.090 ± 0.009	–
Cu	$^{63}Cu(p, 2n)^{62}Zn$	0.57 ± 0.06	1.0
Fe	$^{56}Fe(p, 2n)^{55}Co$	1.69 ± 0.31	2.35
Se	$^{78}Se(p, 2n)^{77}Br$	0.19 ± 0.06	0.095
Sr	$^{88}Sr(p, 2n)^{87}Y$	0.076 ± 0.008	–
Zn	$^{68}Zn(p, 2n)^{67}Ga$	1.45 ± 0.14	0.87
Rb	$^{85}Rb(p, 3n)^{83}Sr$	0.44 ± 0.04	0.17
Cr	$^{52}Cr(p, n)^{52}Mn$	0.0170 ± 0.0019	0.000069
Tomato Leaves (NBS 1573) *(28)*			
Cu	$^{63}Cu(p, 2n)^{62}Zn$	11.0 ± 2.4	11 ± 1
Fe	$^{56}Fe(p, 2n)^{55}Co$	675.8 ± 18.9	690 ± 25
Pb	$^{206}Pb(p, n)^{206}Bi$	6.23 ± 0.97	6.3 ± 0.3
Sr	$^{88}Sr(p, 2n)^{87}Y$	42.9 ± 0.5	44.9 ± 0.3
Zn	$^{68}Zn(p, 2n)^{76}Ga$	62.5 ± 1.2	62 ± 6
Oyster Tissue (SRM 1566) *(15)*			
Ca	$^{44}Ca(p, n)^{44m}Sc$	1390 ± 40	1500 ± 200
Ti	$^{48}Ti(p, n)^{48}V$	20.7 ± 2.0	–
Fe	$^{56}Fe(p, n)^{56}Co$	234 ± 7	195 ± 34
Zn	$^{67}Zn(p, n)^{67}Ga$	826 ± 8	852 ± 14
As	$^{75}As(p, n)^{75}Se$	13.1 ± 0.2	13.4 ± 1.9
Sr	$^{88}Sr(p, n)^{88}Y$	11.0 ± 0.7	10.36 ± 0.56
Zr	$^{90}Zr(p, n)^{90}Nb$	3.4 ± 0.3	–

(continued)

Table 6
Results of Proton Activation Analysis of Biological Samples[a] (*Cont.*)

	Reaction	Found	Present
Orchard Leaves (SRM 1571) (*9*)			
Brewer's Yeast (SRM 1569) (*15*)			
Ca		2410 ± 80	2400 ± 100
Ti		36.3 ± 2	38
V	$^{51}V(p, n)^{51}Cr$	3.7 ± 0.2	3.3 ± 1.6
Fe		664 ± 10	630 ± 80
Zn		60.1 ± 2.5	65 ± 3
As		0.70 ± 0.11	0.59 ± 0.07
Sr		8.2 ± 0.5	–
Zr		3.0 ± 0.2	–
Mo		3.8 ± 0.3	3.6 ± 0.3
Mussel (NIES CRM-No6) (*15*)			
Ca		1380 ± 20	1300 ± 100
Ti		9.2 ± 0.3	–
V		0.6 ± 0.1	–
Fe		160 ± 2	158 ± 8
Zn		106 ± 3	106 ± 6
As		9.9 ± 0.6	9.2 ± 0.5
Sr		17.4 ± 0.9	17
Zr		0.4 ± 0.1	–
Mo	$^{96}Mo(p, n)^{96}Tc$	1.0 ± 0.1	–

[a]All results are in μg/g; except for human serum.
[b]By comparison: Certification of a second generation biological standard reference material (Freeze-dried Human Serum) (*29*).
[c]Results in μg/l of original serum.

human serum (*28*), however, seem to deviate significantly from generally accepted values, even for such elements as Cu, Fe, and Zn.

CONCLUSIONS

It is evident that CPAA of biological materials is not a simple and inexpensive analytical technique. Cyclotron irradiation is expensive, as is the counting equipment. Instrumental analysis is possible in a number of cases although radiochemical separations are often necessary to avoid interferences or enhance the sensitivity. Excessive heating of the samples during the irradiation can be avoided by mixing the samples with graphite or irradiation in air, together with cooling of the sample holder. As an alternative, ashing of the samples can be recommended, together with an irradiation under helium instead of vacuum. Uncertainties concerning the range can be avoided by using an internal standard and standard addition.

Up to now, about 15 elements have been determined in different biological materials, mainly by proton activation and instrumental analysis by gamma-ray spectrometry with a Ge detector. CPAA is certainly useful as an independent nuclear analytical method, especially where NAA is not feasible or insufficiently sensitive. This could be particularly important for the certification of reference materials.

REFERENCES

1. H. H. Andersen and J. F. Ziegler, *Hydrogen Stopping Powers and Ranges in all Elements*, Pergamon Press, New York, 1977.
2. J. F. Ziegler, *Helium Stopping Powers and Ranges in all Elemental Matter*, Pergamon Press, New York, 1977.
3. W. Bragg and R. Kleeman, *Phil. Mag.* **10,** 318 (1905).
4. C. Vandecasteele, *Activation Analysis with Charged Particles*, Ellis Horwood Ltd., Chichester, England, 1988.
5. E. Ricci and R. L. Hahn, *Anal. Chem.* **37,** 742 (1965).
6. E. Ricci and R. L. Hahn, *Anal. Chem.* **39,** 794 (1967).
7. M. A. Chaudri, G. Burns, E. Reen, J. E. Rouse, and B. M. Spicer, *Proc. Int. Conf. Mod. Trends in Activation Analysis*, Münich 1976.
8. M. C. Cantone, N. Molho, and L. Pirola, *Clin. Phys. Physiol. Meas.* **3/1,** 67–77 (1982).
9. L. Zikovsky and E. A. Schweikert, *J. Radioanal. Chem.* **37,** 550 (1977).
10. M. Bonardi, C. Birattari, H. C. Gilardi, R. Pietra, and E. Sabbioni, *J. Radioanal. Chem.* **70,** 337 (1982).
11. M. C. Cantone, N. Molho, and L. Pirola, *Clin. Phys. Physiol. Meas.* **3,** 67 (1982).
12. M. C. Cantone, N. Molho, and L. Pirola, *J. Radioanal. Chem.* **91,** 197 (1985).
13. M. Yagi and K. Masumoto, *J. Radioanal. Nucl. Chem.* **83,** 319 (1984).
14. M. Yagi and K. Masumoto, *J. Radioanal. Nucl. Chem.* **91,** 379 (1985).
15. M. Yagi and K. Masumoto, *J. Radioanal. Nucl. Chem.* **111,** 359 (1987).
16. K. Masumoto and M. Yagi, *J. Radioanal. Nucl. Chem.* **109,** 591 (1987).
17. J. L. Debrun, J. N. Barrandon, and P. Benaben, *Anal. Chem.* **48,** 167 (1976).
18. J. N. Barrandon, P. Benaben, and J. L. Bebrun, *Anal. Chim. Acta* **83,** 157 (1976).
19. J. L. Debrun, D. C. Riddle, and E. A. Schweikert, *Anal. Chem.* **48,** 1386 (1972).
20. J. N. Barrandon, J. L. Debrun, and A. Kohn, *J. Radioanal. Chem.* **16,** 617 (1973).
21. S. M. Kormali and E. A. Schweikert, *J. Radioanal. Chem.* **31,** 413 (1976).
22. K. Krivan and V. Krivan, *Z. Anal. Chem.* **295,** 413 (1979).
23. G. Wauters, C. Vandecasteele, K. Strijckmans, and J. Hoste, *J. Radioanal. Nucl. Chem. Articles* **112,** 23 (1987).
24. C. S. Sastri, H. Petri, and G. Erdtmann, *Anal. Chem.* **49,** 1510 (1977).
25. M. C. Cantone, G. Gambarini, N. Molho, and L. Pirola, *J. Radioanal. Nucl. Chem.* **111,** 351 (1987).
26. A. C. Xenoulis, A. E. Arabantinos, and C. E. Douka, *J. Radioanal. Chem.* **77,** 207 (1983).
27. G. Wauters, C. Vandecasteele, and J. Hoste, *J. Radioanal. Nucl. Chem.* **98/2,** 345 (1986).
28. M. C. Cantone, N. Molho, and L. Pirola, *Fisica in Medicina* **4,** 57 (1983).
29. J. Versieck, L. Vanballenberghe, A. De Kesel, J. Hoste, B. Wallaeys, J. Vandenhaute, N. Baeck, H. Steyaert, A. R. Byrne, and F. W. Sunderman, Jr., *Anal. Chim. Acta* **204,** 63 (1988).

An X-Ray Microprobe Facility Using Synchrotron Radiation

B. M. GORDON,*[,1] K. W. JONES,[1] A. L. HANSON,[1] J. G. POUNDS,[1] M. L. RIVERS,[2] P. SPANNE,[3] AND S. R. SUTTON[2]

[1]Department of Applied Science, Brookhaven National Laboratory, Upton, NY 11973; [2]Department of Geophysical Sciences, University of Chicago, Chicago, IL 60637; and [3]Department of Radiation Physics, Linköping University, Linköping, Sweden

Received April 17, 1989; Accepted October 10, 1989

ABSTRACT

An X-ray microprobe for trace elemental analysis at micrometer spatial resolutions, using synchrotron radiation (SR), is under development. The facility consists of two beamlines, one including a 1:1 focusing mirror and the other an 8:1 ellipsoidal mirror. At present, "white light" is used for excitation of the characteristic X-ray fluorescence lines. Sensitivities in thin biological samples are in the range of 2–20 fg in 100 μm^2 areas in 5 min irradiation times. Scanning techniques, as well as microtomography and chemical speciation, are discussed. Application to a specific biomedical study is included.

Index Entries: Trace element; X-ray microprobe; X-ray fluorescence analysis; synchrotron radiation; chemical speciation; biological reference materials; microanalysis; nuclear analytical techniques.

INTRODUCTION

X-ray fluorescence analysis has been an important technique for elemental analysis for many years. The adoption of excitation and detection techniques from the nuclear sciences, such as PIXE and the use of

*Author to whom all correspondence and reprint requests should be addressed.

solid state detectors, has extended XRF sensitivities to trace element concentration ranges at spatial resolutions of the order of a micrometer or in the case of very thin targets on the electron microprobe, fractions of a micrometer. Recently, there has been considerable interest in the development of dedicated synchrotron radiation (SR) sources that provide a continuous energy spectrum of photons extending into the hard X-ray region. This paper is concerned with description of a SR-induced X-ray microprobe at the 2.5 GeV electron storage ring of the National Synchrotron Light Source (NSLS) at Brookhaven National Laboratory. Presently, it is the only dedicated SR microprobe for fluorescence analysis in the US and is the focus of this paper, because it is available to the general user community. Other probes do exist in the world, with varying degrees of dedication (*1*).

METHODS

Synchrotron radiation is electromagnetic radiation emitted by charged particles when accelerated in a circular orbit. Electrons of relativistic energies emit a radiation pattern that transforms in the laboratory frame into a strongly forward-peak distribution with a vertical divergence angle of $\approx$0.2 mrad at energies equal to K-absorption edges. The most important properties of SR for application to an X-ray microprobe are

1. a wide and continuous spectral range of very high intensity extending at the NSLS to 30 keV at a bending magnet and to over 100 keV at a wiggler magnet source;
2. a high degree of polarization, thereby reducing background scattering into a detector positioned at 90° to the beam in the plane of the ring;
3. a natural collimation that permits the focusing of X-rays with mirrors positioned at small grazing angles;
4. an ability to work in air or helium atmosphere and providing the opportunity to analyze living cells in a natural environment;
5. a relatively high efficiency for X-ray fluorescence production, minimizing the radiation damage to biomedical samples.

The dedicated X-ray microprobes are situated on a bending magnet port with an energy spectrum extending beyond 30 keV suitable for K-shell excitation to $Z = 50$. There are also wigglers that are insertion devices into straight sections and consist of a series of magnetic poles that induce an oscillation to the electron beam path, producing synchrotron radiation. There is a superconducting wiggler available for nondedicated use with an energy spectrum to well beyond 100 keV and capable of K-shell excitation of all elements. Another type of wiggler is the undulator and has short periods such that interference between the electrons and emitted radiation produces radiation at a fundamental energy and its harmonics with orders of magnitude increased intensity and with a natural collimation in both vertical and horizontal directions. the use of undulators is to be emphasized in the Advanced Light Source at Lawrence Berkeley Laboratory and the Advanced Photon Source at Argonne National Laboratory, yet to be built.

NSLS Microprobe

The NSLS Microprobe Facility uses the two beam lines, A and C, existing on beam port X-26. X-26C was commissioned first and used an unfocused, filtered "white light" spectrum and collimated by using remotely-controlled slits to a minimum size of 20 μm square. A Lawrence Berkeley Laboratory group (2) designed and constructed a Kirkpatrick-Baez optical system in which orthogonal small spherical mirrors focused a collimated 500 μm square source (at 20 m from the electron beam) to a beam spot of 10 μm square. The mirror surface was a synthetic multilayer structure of alternating carbon and tungsten layers to achieve a monochromatic energy of 10 keV with a 1-keV spread. Recently, a platinum-coated toroidal glass mirror was installed to provide 1:1 focusing at 20 m from the source. The mirror provides for an increase of as much as a factor of a few hundred.

Initial experiments on the X-26A beam line utilized a white beam and a collimating system consisting of two pairs of crossed slits of 1-mm thick tungsten each $10 \times 40\ \mu m^2$, at a separation of 2.5 cm. This fixed unit could be rotated around appropriate axes to produce beams down to 5 μm. Future use of the X-26A beam line is based on the use of an ellipsoidal 8:1 platinum-coated mirror machined from an aluminum block. The 30-cm long mirror, placed at a grazing angle of 5 mrad, is designed to focus the source at 900 cm from the source, with a demagnification factor of eight to an area of $\approx 60 \times 200\ \mu m^2$. A pinhole aperture (e.g., 100 μm Au) is then used immediately in front of the target to define the beam to the desired spatial resolution. Care must be taken to placing the pinhole as close to the target as possible since the beam divergence is eight times that at the source. This divergence can be reduced by using slits with greater aspect ratios at the cost of lower intensities. The Pt coating and the 5 mrad grazing angle provide for a high-energy cutoff in the excitation energy spectrum of about 17 keV. X-26A will have provision for a monochromator with either a multilayer plane mirror to produce a wide energy bandwidth monochromatic beam or a four-crystal, channel-cut Si crystal monochromator to provide a narrow bandwidth and a constant height exit beam. The latter will be especially useful in chemical speciation studies at high spatial resolution.

Samples generally are positioned at 45° to the beam on a Klinger X-Y-Z-Θ remotely-controlled stage and viewed by a Nikon SMZ-10 binocular microscope equipped with a high resolution DAGE/MTI TV camera. The fluorescence X-rays have been determined using Si(Li) and intrinsic Ge solid state detectors. A Cameca crystal spectrometer is to be installed on

X-26A and will provide better energy resolution in those regions where peak overlaps are most difficult.

Scanning Techniques

The X-ray microprobe can collect data in a variety of modes. The simplest, and the one for which the calculated and experimentally observed sensitivities are quoted above, analyzes a single small area on a target by collecting an entire fluorescence spectrum for a time period of the order of minutes. The spectra are analyzed for peak areas and the concentrations determined, usually by comparison with appropriate standards.

There are a number of scanning modes that are used on the X-26 microprobe. they can be used to sample one- or two-dimensional arrays of picture elements (pixels). In all scanning modes, the sample stage must be scanned in front of the beam since the beam cannot be rastered, as can charged-particle beams. In the first scanning mode, the entire fluorescence spectrum for each pixel is stored. This procedure requires a large amount of storage for a high resolution scan, as well as an inordinate amount of processing, to extract the desired data. This mode perhaps can be best thought of as an automated procedure for the point analysis described above.

The second mode of scanning limits the amount of information collected by definition of regions of interest (ROI), each of which is a range of channels, including the fluorescence peak for the desired element. At each pixel, the computer program determines the background corrected peak areas for each ROI online and stores only these values. The data are normalized using the integrated ion chamber current at each pixel. As many as 32 ROIs can be defined for each spectrum. Because of the time required to move the stage and process the data, the minimum practical time per pixel is about 1–2 s.

In a third scanning mode, the output of the Si(Li) detector is fed into one or more single channel analyzers (SCA), whose output is then fed into a multichannel scaler (MCS). The window of the SCA is set to the energy range of the fluorescence peak of interest. The scanning is continuous, with the channel advance signal to the MCS being derived from the stepper motor pulses on the scanning stage. The advantage of this scanning mode is speed since the scan rate is limited only by the counting statistics and not the time required for step scanning the stage. However, an extensive amount of hardware is required if a number of elements were to be analyzed simultaneously.

Computerized Tomography

The scanning methods described above have the drawback, particularly in the analysis of thin targets at small spatial resolutions, of having small count rates, thereby making inefficient use of the solid state detector. Gurker (*3*) has applied image reconstruction techniques to X-ray fluorescence analysis and, in effect, has improved the signal to background ratio for a given beam intensity. Initially working with X-ray tube excitation, they scanned a target area with a line source of width corresponding to the spatial resolution desired. This profile of elemental counts, integrated along the line of excitation, across the sample is then repeated for different rotation angles of the sample. Computer filtering and back-projection techniques then provide a two-dimensional distribution of counts for each element determined. The number of linear scans multiplied by the number of rotations is equal to the number of pixels in the map. However, the count rate has been increased by roughly the length of the scan line in pixel length units, thereby improving the signal to noise ratio. They have extended this work to include synchrotron radiation excitation at the HASYLAB facility in Hamburg (*4*).

Their work has also included the use of area irradiations through coded apertures, which at first glance would be an improvement over line scanning. Their analysis shows, however, that area irradiation is superior only at very low concentrations and is decidedly inferior at almost all other concentrations.

Synchrotron radiation, because of its unique properties, can also be used in a transmission mode to perform microtomography with spatial resolutions down to the order of a micrometer. Spanne and Rivers (*5*) used a filtered "white" beam at the NSLS X-26C beam line to produce a microtomogram of a freeze-dried caterpillar with a spatial resolution of 30 μm. The image of a lateral section of the head, approximately at the height of the mouth, consisted of a 177 $\times$ 177 matrix for a diameter of 5 mm. The scan required 50 min and used a gas proportional counter with a maximum count rate of 220,000 pulses/s.

The chemical speciation of trace elements has recently been recognized as being of great importance in biomedical studies. Using X-rays, the technique is based on the shape and energy position of the absorption edge of an element as it is scanned with a narrow energy bandwidth. This technique, as applied to biomedical studies, has been reviewed previously (*6*). Recent advances in detection equipment, including high efficiency intrinsic germanium detectors, lower sensitivity for X-ray absorption studies to $\approx$5 ppm at mm spatial resolution. Beam line X-26A is equipped with a constant exit narrow bandwidth monochromator and, thereby, provides chemical speciation capabilities at microprobe spatial resolutions, albeit at somewhat poorer sensitivity limits.

The properties of a SR beam suggest the possibility of analyzing cells for trace element concentrations in vivo. A wet cell (*7*) was constructed to permit the irradiation of specimen cells, whereas a nutrient solution was circulated through the irradiation cell. The compartment was formed by two 7-μm Kapton films separated by a 15-μm spacer film, so cut to allow for an external circulating fluid. A single myocyte cell was analyzed for trace elements, and the cell shape after irradiation indicated cell survival.

RESULTS

In principle, XRF analysis is an absolute method in which concentrations are determined through knowledge of photoelectric and scattering cross-sections and accurate determination of the experimental parameters. Using SR, Tolkiehn and Petersen (*8*) have determined the mass of a sample from a knowledge of the matrix composition and the appropriate scatter cross-sections. In general usage, however, it is more usual to compare yields with known trace element standards in comparable matrices. Governmental institutions, including the NIST in the US, NIES in Japan, and IAEA in Vienna, produce and distribute trace element standards in a variety of matrices. They are widely used in PIXE although the volume being analyzed is usually much smaller than the amount for which the standard is certified. Standards for trace element microprobe analysis require a criterion of uniformity on a micrometer scale.

NIST (*9*) has recently issued two such standards, with a total of 13 elements sputtered from a glass matrix onto a polycarbonate film. Giauque, et al. (*10*) used these standards to produce a calibration curve for the use of a 100 μm^2 beam spot at the X-26C line using the LBL Kirkpatrick-Baez optics. With a photon fluence of 2×10^9 photons/s, they established detection limits of 20 fg for Ca to 3 fg for Zn in 60-s counts and a 150-mA storage ring current. Themner et al. (*11*) have reported the uniformity of standards produced by MicroMatter Co. using a 200 μm^2 PIXE beam spot. These foils were evaporated layers of elements or binary compounds on polycarbonate films at loadings of 50 $\mu g/cm^2$. Gordon (*12*) has produced a multielemental standard to simulate biomedical thin tissues by depositing a mixed solution of five elements on a Millipore Filter of 1.2 μm pore size. The loadings were ≤ 3 $\mu g/cm^2$ in the filter of surface density 5 mg/cm^2. For a 75 μm square beam spot, the uniformity was better than 2%, whereas a beam spot of 10 μm square yielded a Mn/Cu ratio of better than 1% for seven measured points over a square centimeter. In this latter run, the 10-keV monochromatic beam was below the absorption edge of the other elements present. The center portion of the filter is punched out and analyzed to determine the concentrations of the elements by an appropriate analytical technique. For the experiments simulating tissue analysis, the detection limits for concentration ranged from 0.5 to 10 ppm. These results should be improved considerably with the use of focusing mirrors and more geometrically efficient multilayer monochromators. An X-ray fluorescence spectrum of the Millipore Filter mixed element standard is shown in Fig. 1.

The salt, $Ga(NO_3)_3$, is used therapeutically to aid in the resorption of calcium in several types of bone cancer. In an investigation of the process mechanism and in order to develop better treatment procedures, trace element distributions were made of rat tibia after administration of $Ga(NO_3)_3$ (*13*). This work showed that the gallium concentrated in regions of the bone that were metabolically active. In the course of the

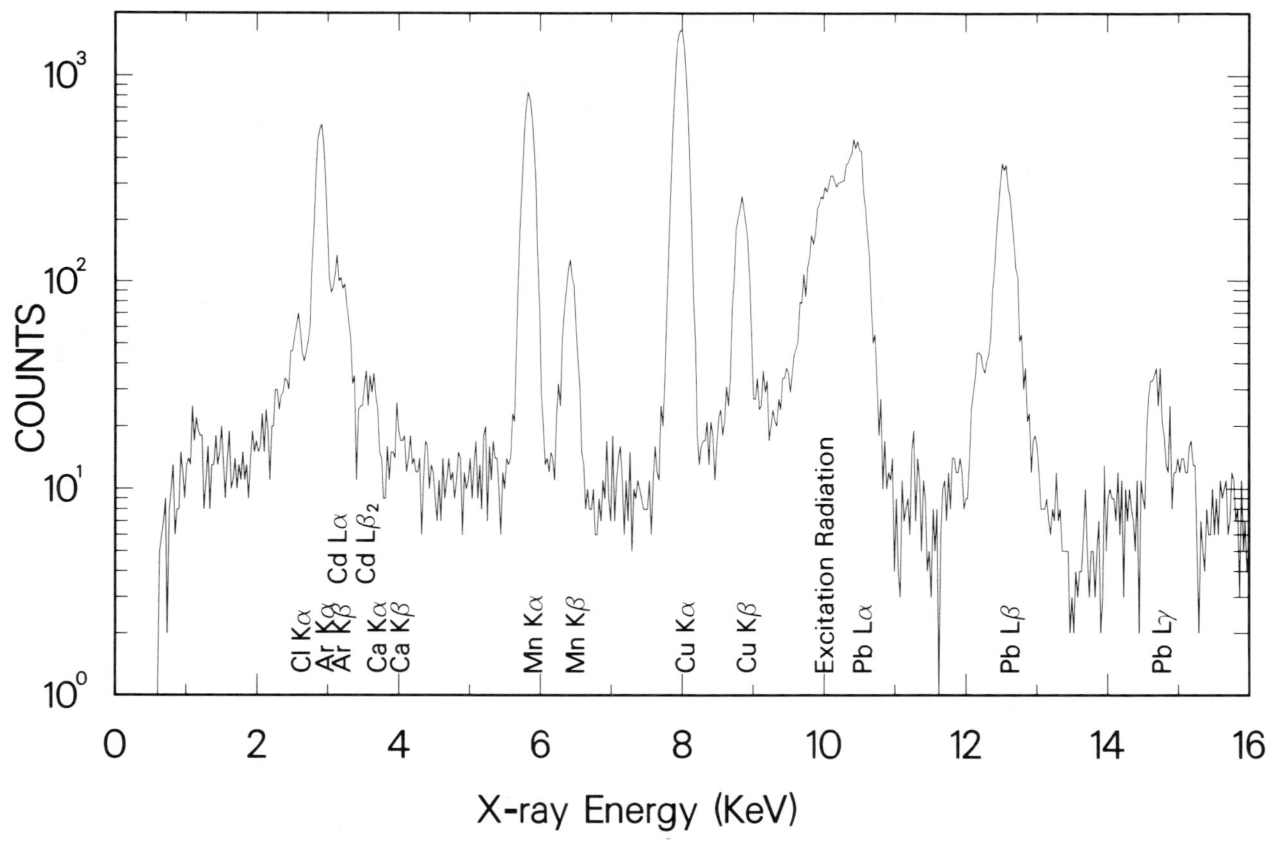

Fig. 1. X-ray fluorescence spectrum for Millipore Filter mixed element standard using a K-B focusing system for a 300-s count at 10×10 μm^2 spatial resolution. The Pb peaks result from excitation of shielding in the hutch.

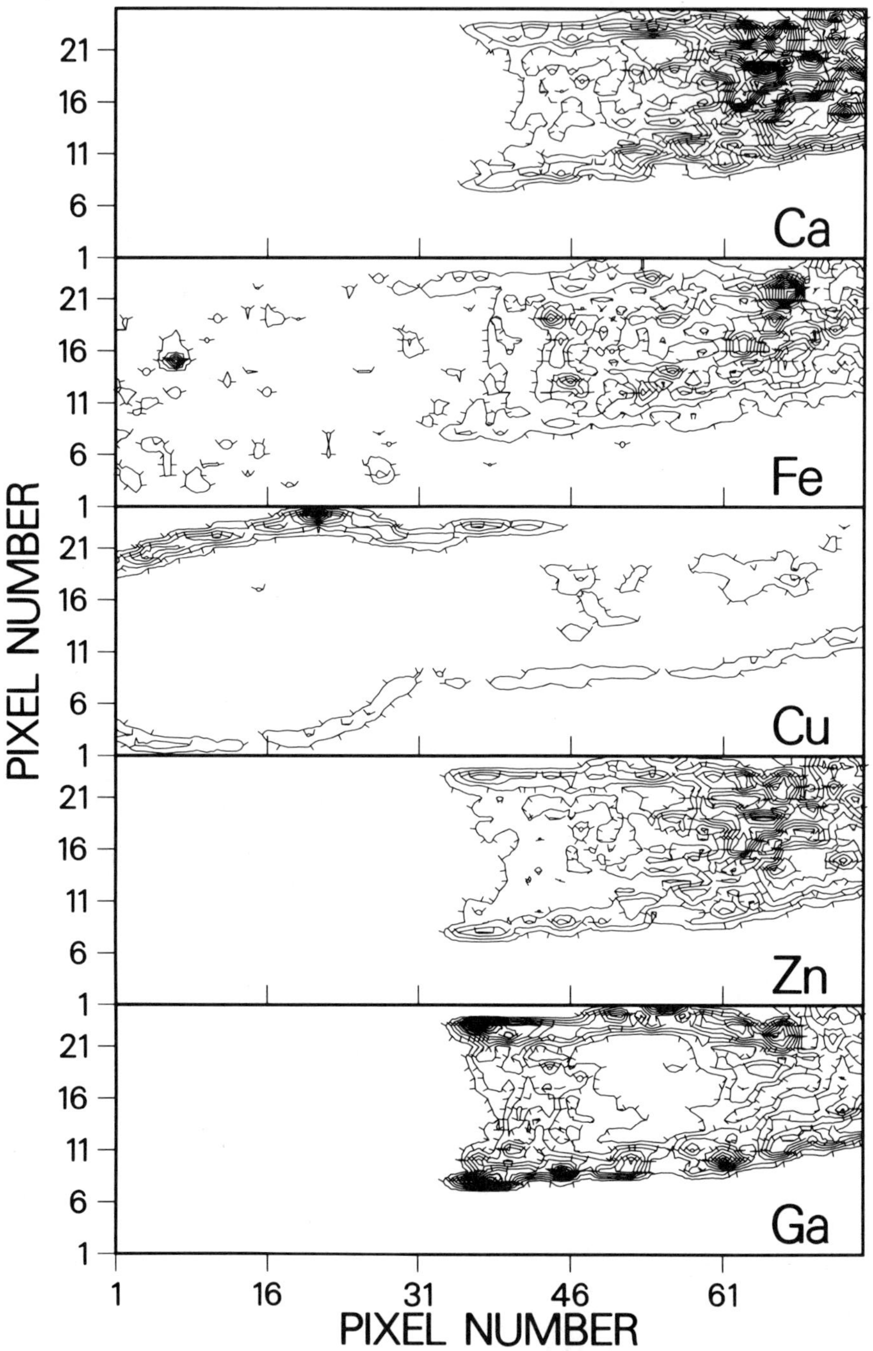

Fig. 2. Elemental distributions obtained by scanning a 1×3 mm^2 area of a fetal rat bone after administration of $Ga(NO_3)_3$. The count interval is 2 s/pixel. The contour intervals are 1500, 20, 50, 50, and 500 for Ca, Fe, Cu, Zn, and Ga, respectively. The tick marks point in the direction of lesser counts.

study, a fetal rat tibia was analyzed after administration of $Ga(NO_3)_3$, and the results of a $1 \times 3\ mm^2$ scan of a thin section is shown in the contour plots of Fig. 2. The area at the right half of the section is calcified.

ACKNOWLEDGMENTS

The development of the X-ray microprobe is supported by the Processes and Techniques Branch, Division of Chemical Sciences, Office of Basic Energy Sciences, US Department of Energy, Contract No. DE-AC02-76CH00016; application to biomedical problems is supported by National Institutes of Health Biotechnology Research Resource Grant No. P41RR01838.

REFERENCES

1. K. W. Jones and B. M. Gordon, *Anal. Chem.* **61,** 341A (1989).
2. A. C. Thompson, J. H. Underwood, Y. Wu, R. D. Giauque, K. W. Jones, and M. L. Rivers, *Nucl. Instr. Methods Phys. Res.* **A266,** 318 (1988).
3. N. Gurker, *X-Ray Spect.* **14,** 74 (1985).
4. A. Knöchel, M. Bavdas, N. Gurker, P. Ketelsen, W. Petersen, M. H. Salehi, and T. Dietrich, *J. Res. Natl. Bur. Stand.* **93,** 379 (1988).
5. P. Spanne and M. L. Rivers, *Nucl. Instr. Methods Phys. Res.* **B24/25,** 1063 (1987).
6. B. M. Gordon, *Biol. Trace Elem. Res.* **12,** 153 (1987).
7. K. W. Jones and J. G. Pounds, *Biol. Trace Elem. Res.* **12,** 3 (1987).
8. G. Tolkiehn and W. Petersen, *Nucl. Instr. Methods Phys. Res.* **215,** 515 (1983).
9. P. A. Pella, D. E. Newbury, E. B. Steel, and D. H. Blackburn, *Anal. Chem.* **58,** 1133 (1986).
10. R. D. Giauque, A. C. Thompson, J. H. Underwood, Y. Wu, K. W. Jones, and M. L. Rivers, *Anal. Chem.* **60,** 855 (1988).
11. K. Themner, N. E. G. Lövestam, U. A. S. Tapper, and K. G. Malmqvist, *Nucl. Instr. Methods Phys. Res.* **B22,** 126 (1987).
12. B. M. Gordon, A. L. Hanson, K. W. Jones, J. G. Pounds, M. L. Rivers, G. Schidlovsky, P. Spanne, and S. R. Sutton, *Nucl. Instr. Methods Phys. Res.* **B45,** 527 (1990).
13. R. S. Bockman, M. Repo, D. Warrell, J. G. Pounds, W. M. Kwiatek, G. J. Long, G. Schidlovsky, and K. W. Jones, *X-Ray Microscopy II*, D. Sayre, M. Howells, J. Kirz, and H. Rarback, eds., Springer-Verlag, New York, 1988, p. 391.

Radioisotope Induced X-Ray Emission (RIXE) Studies in Life Sciences

JOHN J. LABRECQUE

Instituto Venezolano de Investigaciones Científicas (IVIC), Apartado 21827, Caracas 1020A, Venezuela

Received April 17, 1989; Accepted October 10, 1989

ABSTRACT

The inherent advantages of radioisotope induced X-ray emission (RIXE) are briefly presented, with emphasis on their applications to analysis in the life sciences. The following selected applications by RIXE are discussed in general: (1) In vivo analysis: the determination of stable iodine in the thyroid and heavy metals in human bone; (2) the analysis of small tissue samples (about 10 mg) for the simultaneous determination of trace elements between $25 \leqslant Z \leqslant 40$; data are given from a time study of the disease progress of Lewis lung tumor on a rat's liver over a 24-d period; and (3) a homogeneity study of iron, strontium, and zirconium in IAEA's SL-3 Lake sediment standard reference material is presented.

Index Entries: RIXE; life sciences; thyroid; iodine.

INTRODUCTION

X-Ray fluorescence in general has achieved a high degree of success and applications for elemental analysis of geological, industrial, and environmental samples. In the area of biological materials, its application has been more limited. The inherent advantages of radioisotope induced X-ray emission (RIXE) include the following: small compact sources with no external power supply or accessories, flexibility of the acceptance angles and distances between the radioisotope source, the specimen and detector, and low radiation doses. These advantages are favorable for in vivo measurements, as well as for portable instrumentation for on-site analysis. In respect to the use of radionuclides in clinical and investigative applications, RIXE either drastically reduces or eliminates radiation exposure to patients and attending personnel, thus allowing certain tests to be performed in pregnant women and children (*1*). The analysis of heavy metals in tissues is not performed routinely and, in many cases, is bypassed since direct analysis of tissue samples usually requires surgery. Thus, in vivo analysis of tissue samples are welcome especially if the radiation dose can be minimized.

IN VIVO ANALYSIS

Thyroidal Iodine

One of the first reported in vivo applications of RIXE was the determination of the stable iodine content in the thyroid gland (*2*). The first instruments for the stable iodine determination employed high activity (1–10 Ci) sources of Am-241, whereas more recently, lower activities of about 300 mCi of Am-241 have been shown to produce acceptable results (*3*). A typical arrangement of the source, the specimen (thyroid), and the detector is given in Fig. 1. The total cost for a system, including the detector electronics and the multichannel analyzer, can be less than $25,000 when a microprocessor with an appropriate ADC/interfase card is employed.

Theoretical studies (*4*) by Monte Carlo calculations have shown that the maximum excitation of the Iodine K-X-rays at about 5 cm distance is about 45 KeV. Thus, the Am-241 excitation with 59.6 KeV is only about 60% efficiency. To improve upon the excitation efficiency, some workers (*5*) have investigated Am-241 with a Cerium secondary target to produce Cerium K-X-rays (34–39 KeV), and others studied the use of a Gd-153 radioisotope source.

Detection originally was performed with a Si(Li) semiconductor, but more recently have been substituted for a larger, more efficient intrinsic Germanium semiconductor.

With the low activity Am-241 source (300 mCi), a 4 min irradiation is sufficient for determination of stable iodine in the range 0.2–1.0 mg/g. The results from in vivo RIXE correlated well with those from destructive neutron activation analysis; the two methods had a correlation coefficient of 0.93 (*3*).

The major difficulty with quantitation of the iodine content is the attenuation corrections for the tissue overlying the thyroid, but this can be corrected by employing the scattered radiation. It has also been suggested that a transmission measurement also be performed with a second detector to determine the tissue attentuation. Even though there is a large overlap between the iodine content in the thyroid of normal and diseased individuals, in certain cases, useful clinical information can be obtained from the measurement of the thyroidal iodine concentration.

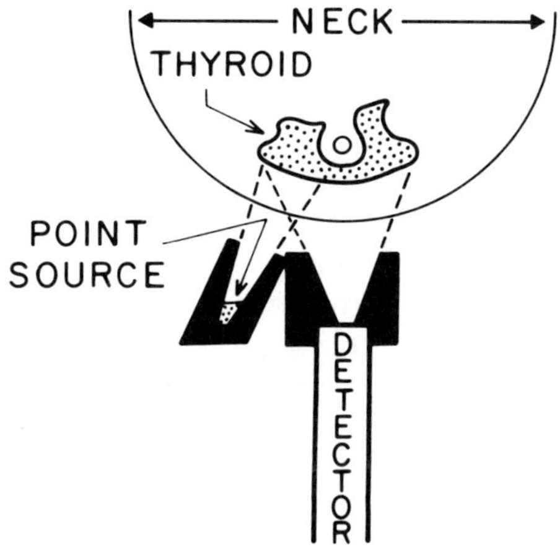

Fig. 1. A typical RIXE setup for in vivo analysis of thyroidal iodine.

Bone Analysis

The correlation between the lead concentration in the blood and bones of occupationally exposed persons have been shown to be poor; thus, there is a real need for analysis of lead in bones, preferably by an noninvasive technique. A simple RIXE system with a 20 mCi Co-57 radioisotope source has been reported (*5*) that has a minimum detectable level of lead in the forefinger bone of about 20 μg/g of bone. The total energy imparted was estimated to be about 0.01 mJ, which is insignificant with respect to an ordinary X-ray picture of the hand.

Both the Cd-109 and I-125 radioisotope have been employed for the determination of strontium in bones (*6*). Again, as in all in vivo studies with RIXE, the attenuation of the X-rays owing to the surrounding tissue needs to be corrected. In this work, besides employing the coherent and incoherent scattered radiation, a determination with a 7 MHz ultrasonic transducer, which has an accuracy of about 0.33 mm, was performed.

The attractive feature of RIXE with external sources is that the withdrawal of the radioisotope source immediately eliminates the radiation dose to the specimen, whereas radiotracer methods leaves activity in the sample for some time; in other words, with RIXE, one can irradiate not only the region of interest, but also only when one wants to. Therefore, the radiation dose can be limited to a certain area for a certain time.

SMALL TISSUE SAMPLES

Most studies of normal trace element concentrations to their disease-related concentrations have been performed by measurement of the healthy tissue and those of the disease-tissue after the specimen had been sacrificed, thus giving no information on the trace element concentration behavior during the disease. In order to study tissue samples as a disease progresses, it is necessary not only to have a low detection limit, but also accept a small sample size in order that the animal can continue to live and give up more portions of the tissue involved.

An optimal source-sample-detector configuration that maximizes the excitation and detection efficiencies, as well as limits the background noise, has been described elsewhere (*7,8*). This system can detect less than 4 nm copper in a synthetic protein. In Figs. 2 and 3, the spectrum of a healthy rat's liver before being subject to Lewis Lung tumor is given, as well as the spectrum from another 10 mg portion of the liver 24 d later. It can be seen that, although the peak intensities of iron and copper are similar in both spectra, the peak intensities have increased relatively for zinc and rubidium and has decreased for bromine. Actually, in this study, other sample portions of about 10 mg were also taken at 7, 14, and 21 d to study the trace element content during the progress of the disease.

HOMOGENEITY STUDIES FOR SRMs

Even though neutron activation analysis and atomic absorption and emission spectrometry have been more widely employed for homogeneity testing and certification of standard reference materials in general, RIXE would probably be the technique of choice if it were readily available for certain materials, such as environmental-geological samples. Soils, sediments, airborne particles, dust, and so on, which contain iron as a minor component (>1%), strontium, and zirconium in concentrations greater than 50 μg/g, are favorable candidates for homogeneity studies by RIXE.

RIXE has been shown (*9*) to be simple, rapid, economical, and truly nondestructive with small sample sizes (less than 50 mg) for IAEA's SL-3 lake sediment standard reference material. Table 1 shows the results when only 20–50 mg of sample from 24 different bottles of SL-3 were placed in separate sample holders without weighing. These samples were then irradiated for only 10 min with a weak Cd-109 (4 mCi) source. Thus, the total time involved can be less than 1 d. In principle, other elements could also be checked too; a good rule would be that their concentration be at least three times their minimiun detection limit.

CONCLUSIONS

Even though conventional RIXE has detection limits in the order of 10–50 times higher than other nuclear and atomic spectrometry techniques in general, it has been shown that, in special cases, it can be considered the technique of choice, such as for in vivo analysis. It should be noted that in the case of solid samples, no sample decomposition is

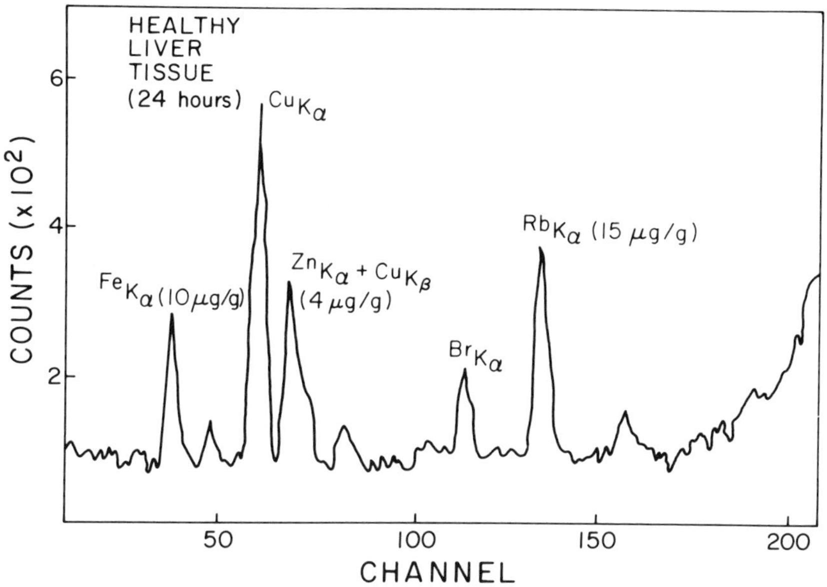

Fig. 2. A spectrum of a healthy rat's liver before being subject to Lewis Lung tumor (*7,8*).

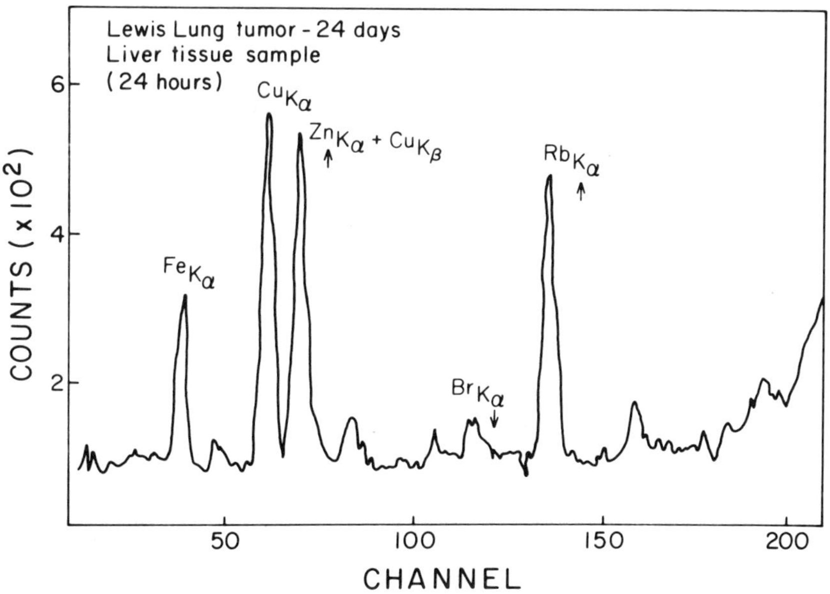

Fig. 3. A spectrum of the same rat's liver 24 d after being subject to Lewis Lung tumor (*7,8*).

Table 1
The Results for the Measurement of Elemental Ratios in IAEA/SL-3 Lake Sediment Standard Reference Material From 24 Different Bottles[a]

Intensity ratio	Means	Relative standard deviation, %	Students's *t*-test, accepted Mos for $\alpha = 0.05$
Fe/Sr	1.34	7.8	>97%
Fe/Zr	1.45	9.2	>96%
Sr/Zr	1.08	6.8	>97%

[a]There was no sample preparation other than thoroughly mixing the contents of the bottles before 20–50 mg of material were placed in the sample holders.

required as in atomic absorption, atomic emission, and some other techniques; thus, with its other inherent advantages, it is comparable to these techniques when the concentrations ranges are a few factors above the detection limits, as is the case for many environmental samples, more specifically soil, sediment, airborne particles, dust, and so on. With respect to conventional neutron activation analysis, the sample results in a radioactive material and, in many cases, the time between the beginning of the irradiation to the final counting is large (many weeks). Finally, RIXE can be truly nondestructive in some cases.

REFERENCES

1. L. Kaufman, D. C. Price, M. A. Holliday, B. Payne, D. C. Camp, J. A. Nelson, and F. Deconnick, *J. Radional. Chem.* **43,** 321 (1978).
2. P. B. Hoffer, W. B. Jones, R. B. Crawford, R. Beck, and A. Gottschalk, *Radiology* **90,** 342 (1968).
3. K. Imamura, Y. Sasaki, N. Sekita, K. Someya, and M. Fujii, *Invest. Radiol.* **14,** 316 (1979).
4. T. Gronberg and S. Mattsson, *Adv. X-Ray Anal.* **24,** 371 (1981).
5. L. Ahlgren, T. Gronberg, and S. Mattsson, *Adv. X-Ray Anal.* **23,** 185 (1980).
6. L. Wielopolski, O. Vartsky, S. Yasumura, and S. H. Cohn, *Adv. X-Ray Anal.* **26,** 415 (1983)
7. J. J. LaBrecque and I. L. Priess, *18th Annual Conference on X-Ray Chemical Analysis,* September 1981, Kyoto, Japan.
8. A. S. Frank, I. L. Priess, A. Adeyemi, and M. K. Schauble, *J. Radional. Nucl. Chem.* **82,** 225 (1984).
9. J. J. LaBrecque and P. A. Rosales, *Adv. X-Ray Anal.* **32,** 191, edited by Charles S. Barrett, John V. Gilfrich, Ron Jenkins, Ting C. Huang, and Paul K. Predecki (Plenum Publishing Corporation, (1989).

Nuclear Analytical Methods for Trace Element Studies in Calcified Tissues

M. A. CHAUDHRI

Department of Medical Physics, Austin Hospital, and the University of Melbourne, Melbourne, 3084, Australia

Received September 12, 1989; Accepted October 10, 1989

ABSTRACT

Various nuclear analytical methods have been developed and applied to determine the elemental composition of calcified tissues (teeth and bones). Fluorine was determined by prompt gamma activation analysis through the $^{19}F(p, \alpha\ \gamma)\ ^{16}O$ reaction. Carbon was measured by activation analysis with He-3 ions, and the technique of Proton-Induced X-ray Emission (PIXE) was applied to simultaneously determine Ca, P, and trace elements in well-documented teeth. Dental hard tissues: enamel, dentine, cementum, and their junctions, as well as different parts of the same tissue, were examined separately. Furthermore, using a Proton Microprobe, we measured the surface distribution of F and other elements on and around carious lesions on the enamel. The depth profiles of F, and other elements, were also measured right up to the amelodentin junction.

Index Entries: Particle-induced gamma emission analysis (PIGE); particle-induced X-ray emission analysis (PIXE); charged-particle activation analysis (CPA); calcified tissues—bones, teeth, enamel, dentine, cementum, and the amelodentin junction; depth profiling; metabolic bone diseases.

INTRODUCTION

The inorganic content of dental hard tissues has been shown to consist not only of Ca, P, O, and H, as indicated by the formula of calcium hydroxyapatite, $Ca_{10}(PO_4)_6(OH)_2$, but also C, Mg, Na, Cl, and

F, as well as a large number of trace elements (*1*). It is also well established that F hardens teeth, and makes them less soluble in acids and more resistant to caries (*2,3*). On the other hand, the role of carbon in dental health is not so well defined, although it is suspected that the introduction of C in the dental matrix increases the solubility of the enamel and makes teeth more prone to decay. There is very little information regarding the role of other trace elements in dental health, either on their own, in relation to F, or to one another. One possible reason for the lack of such information may be the difficulty and inaccuracy associated with chemical analysis of calcified tissues. In order to overcome these difficulties, we developed nuclear analytical methods for investigating the role of different elements in teeth and bones.

The technique of activation analysis with He-3 was developed and applied to determine the carbon content of teeth. For the determination of fluorine and other heavier elements in teeth, the methods of proton-induced gamma emission (PIGE) and X-ray emission (PIXE) analyses were adopted. These nuclear analytical methods are well suited for the analysis of teeth and other calcified tissues since they require little sample preparation and offer the possibility of studying close-lying regions independently.

We have also applied these techniques for nondestructive F-depth profiling, in well-documented teeth, to depths of up to 10 μm in order to study the effect of water fluoridation and topical fluoride applications on F uptake and retention by dental enamel. Moreover, as an extension to these techniques, in the form of a proton microprobe, we have investigated the profiles of F and other elements right across the width of the dental enamel and also measured the spatial distribution of these elements on and around precarious lesions on the enamel surfaces (*4,5*).

These techniques have also been applied to investigate the influence of different trace elements on the fluoride uptake and retention by teeth since there is very little information available on this subject in the literature. In a separate pilot experiment, we also demonstrated the usefulness of PIXE, in the analysis of teeth from mentally handicapped children, for investigating the role of Pb on their mental abilities.

EXPERIMENTAL DETAILS

To avoid any possible error owing to the external effects on teeth (e.g., the environment, topical treatment, varying cleaning habits, and so on), we chose to investigate the inside regions of the teeth, which were specially cross-sectioned for this purpose. All the teeth in this study came from children who have lived in South Australia since birth, where the water supply is generally fluoridated in most areas.

The teeth were carefully cross-sectioned along the vertical median plane and the samples prepared for analysis using the method that has been described in detail, along with the technique for preparing various apatite standards having the same matrix structure as the teeth (*6*). Different dental hard tissues, enamel, and dentine were clearly distinguishable in the sectioned teeth, as shown in Fig. 1, and were investigated independently. The elemental composition of cementum in most of the teeth was also determined.

Fifteen sectioned teeth, and three standards, were mounted in the aluminium target ladder of an insulated stainless steel scattering chamber at an angle of 45 degrees to the incident beam. The target ladder could be manipulated externally, and any one of the teeth or the standards, or any part of them, could be exposed to the incoming proton beam, which was collimated to a 1 mm^2 area. A 40 $\mu g/cm^2$ thin carbon foil was mounted 15 mm in front of the target to eliminate any charging of the teeth during irradiation (*7*). Before entering the scattering chamber, the incident beam had to pass through a 12-in liquid nitrogen trap that eliminated any contamination coming into the chamber along with the incident particle beam.

Teeth were irradiated with 2 MeV protons at intensities varying from 20–100 nA. The characteristic X-rays produced from various elements in teeth passed through a 24-μm Kapton window of the scattering chamber at 90 degrees and were measured with an ORTEC X-ray detector, conventional electronics, and a PDP-11-based analyzer system. Lighter elements were determined using low beam currents (20–50 nA), whereas for elements heavier than Fe, the proton intensity was increased to about 100 nA. A 1.5mm thick perspex filter was placed in front of the X-ray detector in the latter case to reduce the large number of X-rays coming from P and Ca, the main constituents of teeth. The elemental concentrations of different regions of teeth were calculated simply by comparing their X-ray yields with those from the apatite standards having identical matrices (*6*).

The F contents in the teeth were simultaneously determined by observing the 6–7 MeV gammas from the $^{19}F(p, \alpha\ \gamma)^{16}O$ reaction with a stabilized 6 × 5 inch NaI (T1) detector placed in the forward direction. Pulses from an ORTEC Research Pulser were fed through the same electronics as the gamma ray signals and displayed on the multichannel analyzer in order to accurately estimate the dead time of the system and to correct for it. For these measurements, the beam current was kept to 20–25 nA only, which reduced the dead time to no more than a couple of percent. The F standard was also prepared by doping ultrapure hydroxyapatite with analytically pure CaF_2 and shaking the mixture thoroughly for hours with a mechanical device. The uniformity of the standard was found to be very good. It was checked by measuring the 6–7 MeV gamma yields from different portions of the standard.

Since the PIXE and prompt gamma analyses were carried out simultaneously, it was possible to measure the concentrations of F and other trace elements in any selected spot on the tooth and study their interrela-

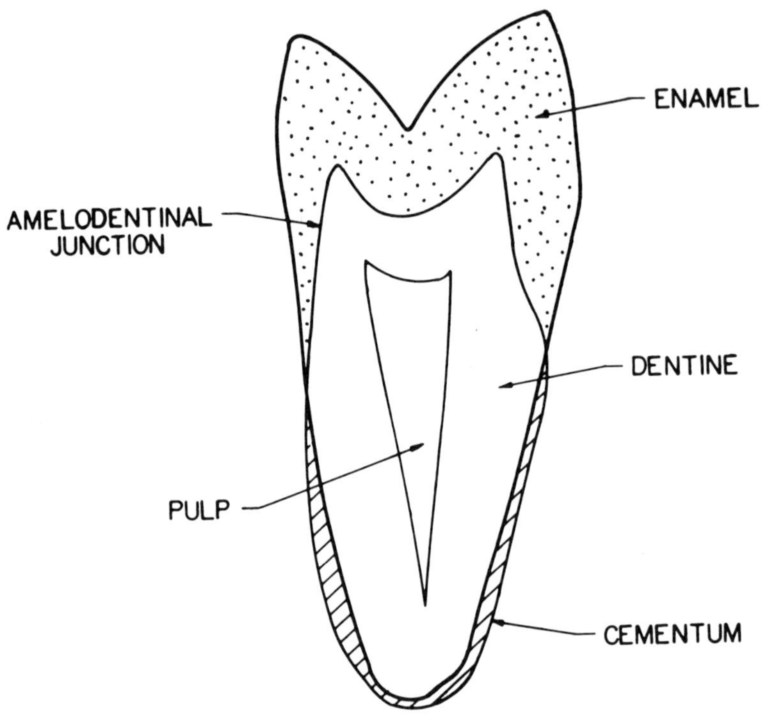

Fig. 1. Schematic diagram of a typical sectioned tooth, clearly showing various dental tissues.

tionship accurately. By irradiating different regions of healthy and decayed teeth, the trace element concentrations of enamel and dentine were measured and the correlations of F with other elements determined.

During irradiation, the beam spot on the tooth was clearly visible owing to the fluorescence. This helped in selecting and identifying various regions of teeth for analysis purposes. Even after the bombardment, the spot on the tooth that was irradiated by the beam could still be identified, owing to slight discoloring of the tooth surface. This helped in selecting the same spot for repeated measurements. It was also established that the slight discoloring of the tooth surface did not change its trace element concentration.

The carbon content of teeth was measured by activation with a 2.7 MeV He-3 beam and counting the 2.3 MeV gamma from the 72 s half-life ^{14}O activity produced through the $^{12}C(^{3}He,n)^{14}O$ reaction with a NaI(Tl) detector (*8*). A piece of silver steel, the carbon content of which was accurately measured by chemical techniques (*9*), was used as the carbon standard. For the C determination, the He beam was focused on or around the same spots on the teeth that had been previously analyzed for F and other trace elements.

For nondestructive F-depth profiling, we measured thick-target yield curves of 6–7 MeV gammas from the $^{19}F(p, \alpha\ \gamma)^{16}O$ reaction, on well-documented teeth, in the energy range of 0.4 to 2.75 MeV and compared them with the calculated yield curve for hydroxyapatite matrix with uniformly distributed fluorine. The difference in the shapes of the yield curves for teeth and the apatite matrices was owing to nonuniform F-distribution in the teeth enamel and, therefore, a comparison of the two curves provided the F-depth distribution in the teeth. In this way, we compared the F-depth profiles in the teeth of experimental rabbits, who were kept on fluoridated water to those of control rabbits, to depths of up to 12 μm, nondestructively.

The Heidelberg Proton Microprobe was used to study the distribution of F and other elements across the entire depth of the dental enamel and investigate the spatial distribution of these elements on and around pre-carious lesions on selected teeth. For depth profiling, the teeth to be investigated were vertically midsectioned and prepared as in ref. (*6*). The section of enamel analyzed was healthy and free from caries. The specimen teeth were mounted, one at a time, in an aluminium holder with steel gripping screws and placed in a scattering chamber. The tooth surface to be investigated was oriented perpendicular to the incident beam direction. The proton beam had a size of 10 microns in width and 50 microns in height on the sample. The 3 MeV proton beam, with an intensity of up to 7 nA, was focused on the edge of the cut surface (surface of the enamel). The 6–7 MeV gammas were detected with a 5-inch × 5-inch NaI (T1) detector, placed directly above the target chamber, while the X-rays were measured simultaneously with a Si(Li) detector mounted at 135 degrees to the beam direction, covered by 1 mm thick plastic and at a distance of 30 mm from the target. To prevent the teeth from charging up during proton bombardment, a small carbon filament was used to spray electrons on to the teeth during irradiation (*5*). After collecting the X- and gamma-ray spectra at the surface/edge, the sample was automatically moved 25 microns and analyzed. In this way, the entire width of the enamel, right up to the amelodentin junction, was studied.

For investigating the spatial distribution of various trace elements on and around pre-carious lesions on teeth, the same experimental setup was used. The teeth used for this experiment were clinically chosen as showing characteristic signs of early caries formation in the form of small, irregularly shaped lesions. With the proton beam, the area on and around the lesion was scanned in steps of 25 microns and the elemental composition determined.

RESULTS AND DISCUSSION

Photon-induced gamma-ray and X-ray spectra from the dentine region of a diseased tooth are shown in Figs. 2 and 3, respectively. It should be mentioned that, although the tooth was diseased, the region investigated was still free from any visible sign of decay. In the gamma-ray spectrum (Fig. 2), the three lines from the reaction $^{19}F(p, \alpha\ \gamma)^{16}O$ are well separated from any other gamma line, with practically no background. These have been labeled accordingly, along with their single and double escape peaks. The total area underneath these peaks can be easily determined, and compared to the corresponding area from the F-standard, to calculate the fluorine concentration in various teeth.

The PIXE spectrum (Fig. 3) was taken using the 1.5 mm perspex filter in front of the detector. The X-ray peaks from Ca, Fe, Ni, Cu, Zn, Br, Sr, Ag, and Pb are easily identified in the spectrum and have been labeled accordingly. From the areas of these peaks, concentrations of different elements have been calculated. In the no filter case, lighter elements (Na, Mg, Al, P, Cl, Ca, and Mn) were also detected and their concentrations determined.

In the gamma spectrum from the activation of C in tooth, through the $^{12}C(^{3}He,n)\ ^{14}O$ reaction, the 2.313 MeV gamma peak and its first escape peak from ^{14}O were very well resolved and, therefore, could be integrated quite easily. The absolute C concentration was obtained by comparing the areas of the 2.313 MeV peaks from the teeth and the C standard. Full details of this project are given in ref. (*8*).

So far, we have investigated healthy and diseased teeth from 25 children and determined the elemental composition of their enamel, dentine, and cementum. Occasionally large variations in the elemental concentrations have been observed among hard tissues from healthy and diseased teeth, in different tissues from the same tooth, and sometimes also in different regions of the same dental tissue. The results are summarized in Table 1, which gives the ranges of concentrations in various dental hard tissues from healthy and diseased teeth. The concentrations of Ca and P for enamel, and Cl for the dentine, of decayed teeth have been shown in Table 1 as the means of two sets of the widely different ranges observed.

From Table 1, certain differences between the elemental compositions of healthy and decayed teeth are obvious. The concentrations of P, Al, and Cl are higher in the enamel of healthy teeth than in the enamel of decayed teeth, whereas for Mg and V, the concentration order is the reverse. The carbon and iron concentrations are higher in dentine from decayed teeth than in the dentine from healthy teeth, whereas there is no obvious difference between the C and Fe concentrations of decayed and healthy enamel. From among the other elements observed, the concentrations of Na, S, Ni, Cu, Zn, Sr, and Pb are found to be higher in both decayed enamel and dentine than in their healthy counterparts, whereas

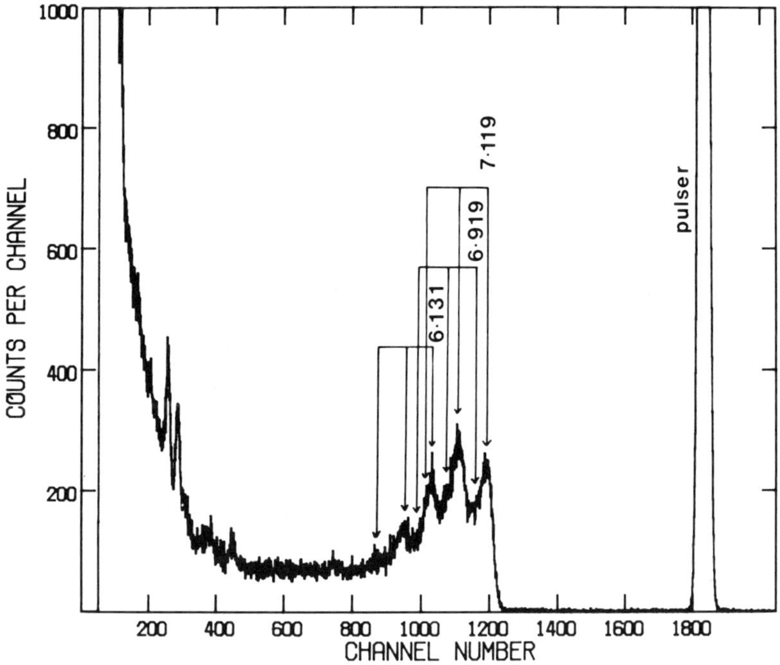

Fig. 2. A prompt gamma ray spectrum from the clean dentine region of a carious tooth. The 6–7 MeV gammas from the reaction ^{19}F (p, α γ) ^{16}O are clearly shown and are marked as such.

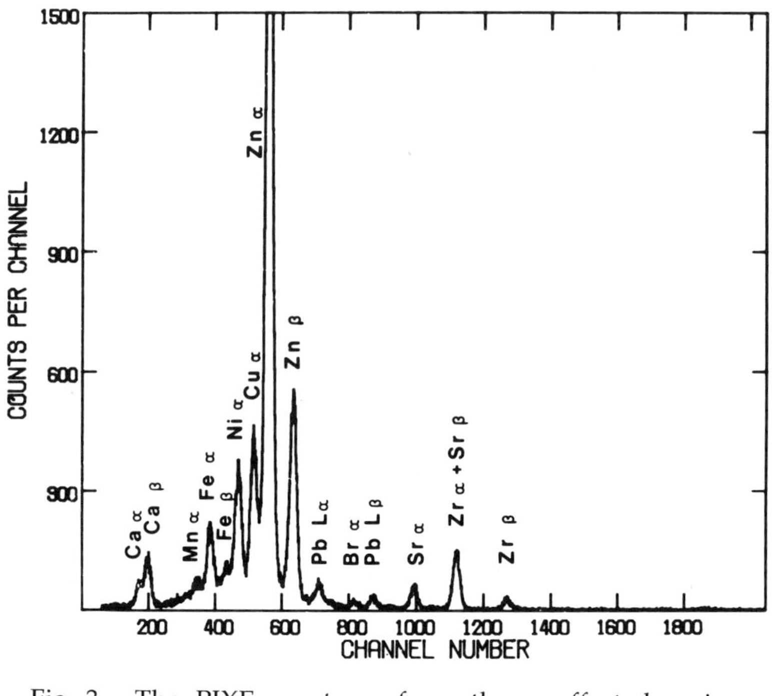

Fig. 3. The PIXE spectrum from the unaffected region of a carious tooth.

Table 1
Trace Elements in Teeth[a]

Element	Enamel		Dentine		Cementum
	Healthy teeth	Decayed teeth	Healthy teeth	Decayed teeth	
C	4.73	4.83	5.88	6.75	ND[b]
F	618	634	1020	1005	ND[b]
P	17.1	8.9, 12.8	14.0	14.4	13.0
Ca	35.7	23.3, 38.4	32.0	31.8	29.5
Na	1700–2600	3400	2000–2300	5700–9400	3000–14,000
Mg	1000–1100	1500	3000–8000	2200–7600	3100–8900
Al	2100–3800	2500	3100–5200	1600–3300	1600–5800
S	1000–1350	4000–5200	706	1400–2300	700–1300
Cl	1600–2800	1000–1500	2500–2600	240, 5000	250
V	10–30	20–50	15–35	15–50	5–30
Cr	20–40	27	15–30	10–25	5–30
Mn	5–25	5–20	5–20	5–10	5–30
Fe	15–100	10–70	10–30	15–130	10–100
Ni	5–15	400–700	5–30	10–200	5–40
Cu	10–30	100–200	5–50	20–500	5–40
Zn	50–150	4500–23,000	200–700	200–3000	700–1400
Sr	100–150	170–250	110–150	120–400	120–190
Pb	2–13	20–50	2–15	10–40	5–50

[a]Concentrations are given in μg/g except for C, P, and Ca, which are given in %.
[b]ND = Not Determined.

F, Ca, Cr, and Mn show no significant difference between healthy and diseased teeth. The Pb concentration of 2–50 ppm is in agreement with other measurements on the teeth of urban children (*10,11*).

In the preliminary study of Pb measurement in the teeth from mentally retarded children, it was observed that the teeth from all children, except those suffering from Down's syndrome (a genetic disease), contained appreciable amounts of Pb (40 ppm and higher). Further work is in progress to investigate the correlation between the Pb content of children's teeth and their mental abilities.

To investigate the effect of different trace elements on the F contents of teeth, linear correlation coefficients (*r*) have been calculated between F and some of the trace elements observed in carious and noncarious teeth. The results are summarized in Table 2.

In the cases of C, Fe, Zn, and Sr, the correlation with F has been determined separately for carious and noncarious teeth. However, both types of teeth have been grouped together for other elements, namely Ni, Cu, Br, Ag, and Pb. As can be seen from Table 2, F correlates moderately well with C in the case of carious teeth, but not at all for noncarious teeth.

A similar degree of correlation is observed in the cases of Cu and Pb for both carious and noncarious teeth. The F correlation is poor to

Table 2
Linear Correlations Between F and Other Trace Elements in Carious and Noncarious Teeth

State of tooth and number of sample regions	Element	*r*
Carious (*10*)	Fe	0.379
Noncarious (*15*)	Fe	0.026
Carious (*10*)	Zn	0.042
Noncarious (*13*)	Zn	−0.394
Carious (*11*)	Sr	−0.252
Noncarious (*14*)	Sr	0.411
Carious (*6*)	C	0.602
Noncarious (*9*)	C	0.007
Carious (*7*); Noncarious (2)	Ni	0.310
Carious (*4*); Noncarious (2)	Pb	0.706
Carious (*5*); Noncarious (2)	Ag	0.385
Carious (*8*); Noncarious (3)	Cu	0.740
Carious (*7*); Noncarious (2)	Br	−0.644

nonexistent for other elements, as clearly shown in Table 2, with the exception of Br, which has a significant negative correlation coefficient. This means that F and Br compete with each other for sites in the dental matrix, and if the concentration of one goes up, that of the other decreases, and vice versa.

The F-depth profiles, measured nondestructively in the teeth of humans and animals, show that the enamel surface has the maximum concentration of F, which goes on decreasing rapidly with depth and becomes a bit more steady at around 3–4 microns below the surface (*12,13*). Furthermore, it was also observed that the teeth of fully-grown monkeys, who were given fluoridated water (2 ppm) since birth, had almost 3 times as much F as the teeth of control animals, right up to 12 microns inside the enamel.

Similar F-depth profiles were also obtained when the depth profiles of F and other elements were investigated right across the width of the dental enamel, using the Heidelberg Microprobe (*4*). The maximum concentration of F was at the enamel surface, which decreased rapidly with depth and then became steady right up to the amelodentin junction. The Zn concentration also peaked at the enamel surface, but dropped slowly for several hundred microns below the surface before leveling out. The concentration of Sr, on the other hand, was low at the enamel surface, but increased steadily with increasing enamel depth. The Ca concentration also dropped rapidly with increasing enamel depth, which is consistent with the surface hardness and density variation cited in ref. (*1*). The depth profiles of some other elements (Fe, Cu, and Pb) displayed uniform distribution throughout the enamel depth.

The concentration of F within a pre-carious dental lesion was found to be approximately 10 times higher than in the surrounding healthy enamel surface. Moreover, this concentration was observed to vary rapidly, within the lesion, in intervals of about 100 microns. However, the F-concentration of the neighboring healthy enamel showed no such spatial distribution (*5*). The Zn concentration, on the other hand, was slightly lower in the lesion, compared to that on the enamel surface outside the lesion, but still displayed more spatial variations inside than outside the lesion. For other elements, such as Fe, Cu, and Pb, patterns similar to F, i.e., that the concentrations peaked at positions within the lesion but remained flat on healthy enamel nearby, were observed.

In this paper, we have discussed the development of various nuclear techniques and their applications, mainly in the field of dental health and, to some extent, in the study of the role of Pb in mental retardation. We have not mentioned other important areas of calcified tissue research where these techniques could play an important role, e.g., in accurate and nondestructive measurements of Ca, P, and other elements in small bone biopsies taken from patients with metabolic bone disorders. There appears to be no suitable chemical method available that can provide accurate assessment of Ca, P, and other trace elements in small bone biopsies. Moreover, the nondestructive nature of the nuclear methods has an extra advantage in that the bone samples, which are normally rather small in quantity, subsequently can be used for histological examination.

Various techniques described here can also be applied to get a better understanding of the role of trace elements in renal patients undergoing dialysis. It has recently been reported that the levels of a number of trace elements, especially Al, in body fluids and tissues of patients undergoing dialysis are altered, causing renal osteodystrophy and other complications (*14*). The elevated level of Al is also thought to be responsible for dialysis encephalopathy (*14*).

The nuclear technique for determining F is one of the easiest and most accurate methods, especially for calcified tissues, that cannot be conveniently and reliably analyzed using chemical procedures. Besides the huge potential of this technique in dental health, as discussed in this paper, it can also provide a great deal of information on the F content of bones and its effect on bone morphology (*15*). Few groups have also used PIXE, XRF, and Synchrotron radiation-excited X-ray emissions analysis of nasal bones for environmental purposes (*16–19*).

REFERENCES

1. J. A. Weatherell and C. Robinson, *Biological Mineralization*, vol. 2, Chapter 3, I. Zipkin, ed., Wiley, New York, 1973, p. 43.
2. *Fluoride, Teeth and Health*, Report to the Royal College of Physicians, Pitman Publishing Ltd, London, 1976.

3. G. N. Jenkins, *Int. Dent J.* **17,** 552 (1967).
4. I. D. Svalbe, M. A. Chaudhri, K. Traxel, C. Ender, and A. Mandel, *Nucl. Instr. Meth. Phys. Res.* **B3,** 648 (1984).
5. I. D. Svalbe, M. A. Chaudhri, K. Traxel, C. Ender, and A. Mandel, *Nucl. Instr. Meth. Phys. Res.* **B3,** 651 (1984).
6. M. A. Chaudhri and T. Ainsworth, *Nucl. Instr. Meth.* **181,** 333 (1981).
7. M. A. Chaudhri and A. Crawford, *Nucl. Instr. Meth.* **181,** 31 (1981).
8. M. A. Chaudhri and A. Crawford, *Biol. Trace Element Res.* **26/7** (1990).
9. B. H. P. Research Laboratories, Melbourne, private communication.
10. P. Bloch, G. Garavaglia, G. Mitchell, and I. M. Shapiro *Phys. Med. Biol.* **20,** 56 (1976).
11. I. M. Shapiro. A. Burke, G. Mitchell, and P. Bloch, *Environ. Res.* **17,** 46 (1978).
12. M. A. Chaudhri and V. West, in press.
13. M. A. Chaudhri and A. Crawford, in press.
14. H. G. Nebeker and J. W. Coburn, *Ann. Rev. Med.* **37,** 79 (1986).
15. M. A. Chaudhri and V. West, in press.
16. U. Lindh, *Nucl. Instr. Meth.* **181,** 171 (1981).
17. M. Jaksic, S. Fazinie, J. Krmpotic-Nemance, M. Bundah, Z. Smit, and V. Valkovic, *Nucl. Instr. Meth. Phys. Res.* **B22,** 93 (1987).
18. M. Jaksic, J. Krmpotic-Nemance, F. Watt, G. W. Grime, and V. Valkovic, *Nucl. Instr. Meth. Phys. Res.* **B33,** 417 (1988).
19. R. M. Wheeler, R. P. Chaturvedi, J. S. Onello, V. Valkovic, and J. Krmpotic-Nemance, *Nucl. Instr. Meth. Phys. Res.* **B24/25,** 658 (1987).

Fluorine Concentrations in Bone Biopsy Samples Determined by Proton-Induced Gamma-Ray Emission and Cyclic Neutron Activation

N. M. SPYROU,*[,1] W. J. ALTAF,[1] B. S. GILL,[1] C. JEYNES,[1]
G. NICOLAOU,[2] R. PIETRA,[2] E. SABBIONI,[2] AND M. SURIAN[3]

[1]University of Surrey, Guildford, Surrey, GU2 5XH, UK;
[2]Radiochemistry Division, Joint Research Centre, CEC, Ispra (Va), Italy;
and [3]Dialysis Service, Maggiore Hospital, Lodi (Mi), Italy

Received April 17, 1989; Accepted September 14, 1989

ABSTRACT

Fluorine concentrations in bone biopsy samples taken from the iliac crest of subjects, divided into four groups depending on the length of dialysis treatment, and aluminium levels in blood and bone pathology, in terms of osteoporosis, were determined by two instrumental methods. Proton-induced gamma-ray emission (PIGE), making use of the resonance reaction of $^{19}F(p,\alpha\gamma)^{16}O$ at 872 keV, and cyclic neutron activation analysis (CNAA), using the $^{19}F(n,\gamma)^{20}F$ reaction in a reactor irradiation facility, were employed. Rutherford backscattering (RBS) was used to calculate the volume, and, hence, mass of the sample excited in PIGE by determining the major element composition of the samples in order to express results in terms of concentration. From this preliminary investigation, a relationship is suggested between fluorine concentrations in bone and aluminium levels in the system.

Index Entries: Bone, fluorine concentrations in renal dialysis patients, fluorine and aluminium in, osteoporosis and renal dialysis; trace analysis, elemental methods of.

*Author to whom all correspondence and reprint requests should be addressed

INTRODUCTION

Fluorine is a trace element essential for animal life. It has a high affinity for bone, with 99% of the human body burden comprising skeletal fluorides (*1*). Traces of fluorine, which is a constituent of osseous and dental tissues, are essential for normal teeth formation (*2*), and therefore, the benefits of fluorine to human health lie in the prevention of dental cavities. However, an excess of 2 μg/g of fluorine in drinking water, which is the main source of fluorine intake in humans, may cause mottled enamel in the teeth of children acquiring permanent teeth (*3,4*). Fluorine compounds also act as anticoagulants, and therefore prevent blood clotting (*2*). In addition, the presence of fluorine was found to have reduced the incidence of osteoporosis among the elderly (*5*). The absorption of fluorides mainly takes place in the stomach and intestine, then circulated by the blood, from which they are either absorbed by the bone (T_{biol} ~30 min) or excreted via the kidneys (T_{biol} ~2–3 h) (*1*).

Concern has been expressed about the uptake of elements, particularly aluminium, in patients receiving kidney dialysis treatment. This is a preliminary investigation to determine the fluorine concentration in bone biopsy samples removed from patients with osteoporosis and on renal dialysis treatment.

SAMPLES

Fresh bone biopsy samples from the iliac crest of the subjects were obtained by the Dialysis Service at the Maggiore Hospital, Lodi, Italy. The subjects were divided by the clinicians into four groups as follows

Group I: eight samples collected from elderly persons who had normal renal function, were affected by osteoporosis, and had undergone no treatment;

Group II: five samples from subjects on dialysis for a period of less than one year, blood plasma concentration of Al less than 80 μg/L. These patients showed no sign of pathology in the bone owing to Al;

Group III: six samples from patients on dialysis for more than 10 years, with high levels of Al in their blood, but showing no sign of pathology in the bone owing to Al; and

Group IV: six samples from dialysis patients affected by osteoporosis owing to Al.

The weights of the bone samples were in the range of 10–140 mg and had an average diameter of 3 mm. No further information concerning the details of treatment and medication were given to the analysts in this preliminary study.

METHODS

Two elemental analysis methods were employed for the determination of fluorine concentrations in the biological specimens.

Proton-Induced Gamma-Ray Emission (PIGE) with Rutherford Backscattering (RBS)

This method is based mainly on the nuclear resonance reactions of the proton that is absorbed by the target nucleus, resulting in the emission of a nucleon and associated gamma rays, or only the emission of gamma rays through the de-excitation process. In other cases, the emission of gamma radiation follows inelastic proton scattering. For fluorine measurement, the $^{19}F(p,\alpha\gamma)^{16}O$ reaction (Table 1) is preferred to the $^{19}F(p,p'\gamma)^{19}F$ reaction ($E\gamma$ = 110keV) (6) since it produces gamma rays in an energy range where the background is low (6.13, 6.92, and 7.12 MeV). A large NaI(T1) detector (130 mm diam × 130 mm) was used to measure the gamma rays in the energy window 4.6 to 7.5 MeV (including the measurement of single and double escape peaks) since there were no interference problems. The largest resonance cross-section of F occurs at a proton energy of 872 keV (661 mb). However, the proton beam from the 2 MeV Van de Graaff accelerator was set at 900 keV (with a resonance width of 4.6 keV) in order to compensate for energy loss with penetration, thus allowing sampling to take place in depth. The beam line is set up to perform analysis of samples using a scanning beam of ~5 to 10 μm diameter. However, for these exploratory investigations, it was decided to use a standard aperture of 1 mm diameter. The gamma ray spectrum was collected for 200 μC of integrated charge (approximately 15 min). The integral number of counts for the set energy window was registered for each sample.

The volume of the material excited depends on the range of the protons and, hence, the elemental composition of the sample. The major elements constituting the sample, therefore, were determined by RBS.

Rutherford backscattering is the elastic scattering of charged particles from the solid matrix, and the yield and energy of the scattered particles are directly related to the elemental composition of the samples (7). The RBS experiments were carried out using a helium ion beam energy of 1.5 MeV. The aperture size was 1 mm in diameter, and by demagnifying the beam by a factor of 5, the beam size was brought down to 0.2 mm in diameter. A silicon surface barrier detector was placed in the vacuum chamber and set at an angle of 165° to the direction of the beam. The backscattered energy spectra were recorded using a 512-channel analyzer.

For PIGE and RBS analysis, eight bone samples and a 7 mm pellet of IAEA (A3/1-1974) calcined animal bone were weighed and mounted on

Table 1
Main Reactions for Fluorine Determination by PIGE and CNAA

Method	Reaction	E(MeV)	Activation x-section, mb	Half-Life	E γ (MeV)
PIGE	$^{19}F(p,\alpha\gamma)^{16}O$	0.872 (resonance)	661	prompt	6.13, 6.92, and 7.12
CNAA	$^{19}F(n,\gamma)^{20}F$	–	9.8	11 s	1.63
	$^{23}Na\ (n,\alpha)^{20}F$ (interference)	4.03 (threshold)	0.8	11 s	1.63

the target holder by sellotape. Target preparation followed standard procedures for charged-particle induced reaction analysis. The sample holder was then placed in the vacuum chamber, where it can be driven horizontally, allowing the samples to be bombarded by protons, in turn, without breaking the vacuum.

Cyclic Neutron Activation Analysis (CNAA)

CNAA is more sensitive than conventional neutron activation methods in the determination of fluorine, owing to the short half-life of ^{20}F, 11s (*8*). Seventeen bone samples were weighed and placed in clean capsules for analysis. No treatment was applied on the samples. The bone samples, along with the standards, were irradiated using the cyclic activation system (CAS) at the Imperial College Reactor Centre in a thermal-neutron flux of 1.2×10^{12} n/cm^2/s^1. The irradiation conditions were 10 s irradiation, 2 s decay, 10 s counting, and 2 s delay for 14 cycles. Gamma ray detection was performed using the Ge(Li) detector associated with the CAS system. The spectra from the samples were recorded on a microcomputer, transferred to a mainframe computer, corrected for dead time, and then analyzed by the SAMPO spectral analysis program.

RESULTS AND DISCUSSION

PIGE and RBS

The counts from the calcined animal bone standard and the bone samples were normalized to counts per μC of charge rather than live-time counting in order to eliminate the proton beam flux variation with time. The detection limit was calculated as 3 (background count of standard)$^{1/2}$ per μC and found for 1 μC of charge to be 23 μg/g $\pm$ 3%. This calculation was deduced from the fluorine concentration of 566 $\pm$ 20 μg/g in calcined animal bone determined by Kerr and Spyrou (*9*). From the RBS spectra of the helium ions and the sizes of the backscatter edge, the average elemental composition of the calcined animal bone and the bone

Table 2
Fluorine Concentrations in Bone Biopsy Samples Using Proton-Induced Gamma-Ray Emission (PIGE), Corrected for Major Element Composition of the Sample Obtained by Rutherford Backscattering (RBS)[a]

Sample	Composition	Sample Mass Excited, μg ± 9.3%	F, μg/g
Calcined Animal Bone	$Ca_3P_2O_7$	37.2	566 ± 20
3	$C_5Ca_3O_{10}P_4$	35.4	272 ± 53
6	C	29.5	96 ± 19
9	C	29.5	246 ± 48
10	$C_{70}P_3Ca$	30.3	240 ± 47
13	$C_{14}P_2Ca$	33.1	427 ± 83
19	C	29.5	288 ± 56
21	C	29.5	82 ± 16
25	C	29.5	110 ± 21

[a]The value for the F concentration in the calcined bone was found by neutron activation analysis (*9*).

samples were determined in order to evaluate the weight of the excited mass of the sample and, hence, the elemental concentrations. Knowing the fluorine counts per μC of charge in the standard, and knowing the counts per μC of charge in the sample, the fluorine concentrations in the samples can be calculated, provided the weight of the excited part of the sample (i.e., the volume of interaction with the beam) is known. The values of the average elemental composition, the sample mass excited, and the fluorine concentration in bone biopsy samples, using PIGE in conjunction with RBS, are presented in Table 2.

CNAA

The bone biopsy samples were found to have high sodium concentrations that, hence, make a significant contribution to the ^{20}F peak via the interference reaction. This contribution was measured as 34 μg apparent F per mg Na, and corrections were applied according to the procedure described in (*9*). The sensitivity of the method, expressed as 3 times the standard deviation of the background in a CaF_2 standard, i.e., with no interference, is 4.6 μg. The fluorine concentrations in the bone samples were calculated and are presented in Table 3. The values obtained by PIGE and RBS are also included in Table 3 and are the ones associated with a standard deviation. The CNAA results have typical standard errors of 20%. The concentrations of the following elements

Table 3
Fluorine Concentrations in Bone Biopsy Samples
Obtained by PIGE and CNAA

Group	Sample number	F, μg/g	$\bar{x}$ ± SD, μg/g
	1F	678	
	2	ND	
	3F	272 ± 53	
I	4F	621	318 ± 250
	5F	386	
	6M	96 ± 19	
	21M	82 ± 16	
	25F	110 ± 21	
	22M	402	
	24M	608	
II	26M	166	515 ± 279
	27M	925	
	11	474	
	7F	1300	
	8F	723	
	9M	246 ± 48	699 ± 505
III	10	240 ± 47	
	12M	1321	
	23M	364	
	13M	427 ± 83	
	14F	916	
IV	15	763	898 ± 522
	16M	1616	
	17M	1380	
	19M	288 ± 56	

were also determined by the technique: Ca, Cl, F, Mg, Na, and Se; these are listed in Table 4.

From the combined CNAA and PIGE results in Table 3, it can be seen that Group IV had the highest level of fluorine concentrations. This is the group of dialysis patients with osteoporosis owing to Al. Group III, which had a high level of Al in blood and a dialytic age greater than 10 yr, had an average value of fluorine concentration that was lower than Group IV. Group II, with values less than 80 μg Al-L plasma, had a lower average fluorine concentration than either Group III or IV; the lowest average fluorine value was obtained for Group I. It should be noted that sample 6M was obtained from a young normal male with no osteoporosis. The highest value of fluorine concentration determined was 2142 μg/g. This value was not included in the set of results because the subject was affected by cirrhosis owing to alcoholism.

Table 4
Values of Elemental Concentrations in Bone Biopsy Samples Obtained by CNAA

Group	Conc. μg/g	Ca[a]	Cl	F	Mg	Na	Se
	4[b]	4	4	3	3	4	2
I	mean	159	1960	562	2928	6680	0.280
	SD	35	1030	126	1397	930	0.040
	5[b]	5	5	5	2	5	2
II	mean	198	2982	515	2523	7751	0.303
	SD	38	1202	279	1328	1596	0.001
	4[b]	4	4	4	4	4	4
III	mean	160	2344	927	1926	6313	0.192
	SD	23	390	464	778	1300	0.06
	4[b]	4	4	4	4	4	3
IV	mean	199	3035	1169	4865	7930	0.342
	SD	60	1570	397	2475	1320	0.310
Range		99–258	1130–4650	166–1616	1175–7831	3300–15,300	0.024–0.352

[a]Values are in mg/g.
[b]Number of samples.

CONCLUSION

Two instrumental methods of elemental analysis were used to determine fluorine concentrations in a preliminary investigation to establish analytical procedures by which bone biopsy specimens from patients on dialysis will be treated. Although CNAA provided additional information about trace element composition in the same experiment, PIGE analysis has the capability of allowing study of variations along the length of the bone specimens, which may prove valuable. However, a direct comparison at this stage cannot be carried out without further work. Results obtained for fluorine concentrations in the four groups of subjects indicate that a positive correlation may exist between fluorine concentrations in the bone and aluminium levels in the system. This was contrary to our expectations and further experiments are planned.

REFERENCES

1. H. C. Hodge and F. A. Smith in *Metallic Containments in Human Health*, D. K. H. Lee (ed), Academic Press, New York, 163, (1972).
2. C. H. Best and N. B. Taylor, *The Living Body*, 4th ed., Chapman and Hall, London, 1959.
3. WHO, *Monograph no. 59*, Geneva (1970).
4. R. C. Weast, *Handbook of Chemistry and Physics*, 67th ed., CRC Press, Boca Raton, Florida, (1987).
5. D. S. Bernstein, N. Sadowsky, D. M. Hegsted, C. D. Guri, *JAMA* **198,** 499 (1966).
6. M. Hyvonen-Debeck, J. Raisanen, J. Dabek, *J. Radioanal. Chem* **63 (1),** 163 (1981).
7. G. Deconninck, *Introduction to Radioanalytical Physics,* Elsevier, Amsterdam, Oxford, New York, (1978).
8. N. M. Spyrou, *J. Radioanal. Chem* **61,** 211 (1981).
9. S. A. Kerr and N. M. Spyrou, *J Radioanal. Chem.* **44,** 159 (1978).

Study of Physiopathological Phenomena in Dental Enamel by Neutron Activation Analysis

M. UNG BAO,[1] V. VERNOIS,[2] N. DESCHAMPS,[2] AND G. REVEL*[,2]

[1]*Laboratoire d'Histologie, Faculté de Chirurgie Dentaire 1, rue Maurice-Arnoux 92120 Montrouge, France; and* [2]*Laboratoire Pierre SUE, CNRS, CEN Saclay, 91191 GIF sur Yvette, Cedex, France*

Received April 17, 1989; Accepted September 14, 1989

ABSTRACT

An epithermal neutron activation method is used to determine the concentration of mineral elements in human dental enamel. A large number (252) of samples from ancient and modern origins are analyzed. The analytical results are mathematically processed using a statistical multivariant method. This allows to differentiate deciduous from permanent teeth and decayed from sound enamel. It is also possible to distinguish the teeth coming from two different necropoles. The origin and the localization of determined elements in the mineralized part, or in the aqueous-organic part, of enamel is suggested. Their role, as witnessed in the physiopathological phenomena of dental enamel, is discussed.

Index Entries: Trace analysis; neutron activation analysis (NAA); dental enamel; deciduous teeth; carious process; archeometry.

INTRODUCTION

Dental enamel is a remnant document of the biological and physiopathological phenomena that follow the life of an individual (*1,2*). Indeed, its privileged peripherical position on the crown part of the tooth

*Author to whom all correspondence and reprint requests should be addressed.

makes it a biological interface in permanent contact with the buccal cavity, so its metabolism depends on both internal and external factors. Dental enamel is a highly mineralized structure. Its composition is made up of a 96% w. mineralized part (principally hydroxyapatite) and of a 4% w. aqueous-organic part. Major elements (Ca, P) constituting the mineralized part have a relatively constant composition, and low variations concerning these elements are difficult to detect.

In opposition, the variations in the composition of trace elements are often manifest and can be associated with the biological and pathological phenomena that occur in dental enamel. So, it is therefore essential to have an analytical method allowing a simultaneous determination of a large number of elements in a large range of concentrations from a same sample. Neutron activation analysis methods make this possible. In order to analyze many samples quickly enough, we have used an epithermal irradiation that reduces the radioactivities induced by $^{32}P^*$, $^{24}Na^*$, and $^{47}Sc^*$ and, consequently, the chemical separations after irradiation. This method is applied to teeth from contemporary and ancient populations. Teeth from ancient necropoles constitute a very interesting group of samples because they come from many sedentary individuals living in definite environments with the same living conditions. Furthermore, different teeth of a same individual can be compared.

MATERIALS AND METHODS

Materials

Different dental samples (252) were prepared from: (a) Contemporary groups: These samples came from individuals living in the district of Paris and consisted of: Permanent sound teeth (25 samples), Deciduous sound teeth (20 samples), and Deciduous decayed teeth (25 samples); and, Ancient groups: The samples came from six French necropoles and from one Swiss necropolis: Vieil-Toulouse (1st century, south of France, 25 samples), Venerque (2nd century, south of France, 35 samples), Montlaur (5th–7th centuries, south of France 40 samples), Mairac (13th–15th centuries, south of France, 6 samples), Ladern (19th century, south of France, 15 samples), Pessat (7th–15th centuries, center of France, 31 samples), and Sezegnin (5th–7th centuries, Switzerland, 30 samples).

Analytical Method

Teeth are cleaned in an ultrasonic bath with distilled water for 10 min. Each one is then divided with an agate pestle and mortar to obtain enamel splinters. In order to have enamel only, each splinter is removed from subjacent tissue using a diamond milling cutter. The enamel samples (50 mg) are then packed in two high purity aluminum foils. The standards are prepared by depositing a known volume of a titrated solution of each considered element on cellulose powder, which is compressed and then packed in aluminum foils, the same as the samples. Samples and standards are put into a cadmium tube crimped into an aluminum cartridge and irradiated for 16 h in the OSIRIS pool reactor at the CEN Saclay. In the irradiation position used for these analyses, the neutron flux is $10^{14}/cm^2/s^1$.

Under these conditions, all the induced radioactivity from $^{24}Na^*$ must be removed before the radioactivity measurements. The separation is obtained using hydrated antimony pentoxide in 12*N* HCl solution. This radiochemical separation method has been previously described and tested (*3*). Radioactivity measurements are made with a high efficiency Ge-Li detector (peak compton ratio = 70; resolution 1.69 KeV at 1333 Kev; relative efficiency = 29.6%).

Data Processing

In each sample, As, Au, Br, Cu, Eu, La, Mn, Mo, Sb, Sr, U, W, and Zn concentrations were always tested. Some of these elements (As, Eu, La, and U) were not always detected in all samples. The results obtained on detected elements were processed using a statistical multivariant method (*4*). This method of factorial analysis allows the simultaneous graphical representation of all samples and elements. Two orthogonal axes are determined. Each sample and element are represented by one vector, which corresponds to the deviation with the average value obtained from the results of the concentrations of each element in all samples. The distance that separates both samples and elements from the origin of the axes is greater according to whether the composition of a given sample differs more from the average composition of all samples and whether the concentration of a given element varies more between the different samples. Thus, it is possible, on the one hand, to bring out the existing correlations between the different samples and, on the other hand, to detect the element that can have the greatest importance in the biological and physiopathological phenomena.

RESULTS AND DISCUSSION

Concentration values of studied elements range from 10^{-4} μg/g to 2 • 10^2 μg/g. These elements can be classified in three groups:

Group A: Elements with concentration values superior to 1 μg/g. These elements are: Zn (30–200 μg/g), Sr (7–110 μg/g), and Cu (1–5 μg/g). Considering the totality of the samples, the variation amplitude in the concentrations of these elements is relatively low for a same group of samples.

Group B: Elements with concentration values from 10^{-2} to 1 μg/g. These elements are Br and Mn. They show considerable amplitudes of variation, in opposition to the elements of Group A.

Group C: Elements with concentration values inferior to 10^{-2} μg/g. The identification of these elements (As, Au, Eu, La, Mo, Sb, U, and W) is largely conditioned by the sensitivity of the analytical method used for their determination. In our experimental conditions, Au, Sb, and W are always detected in all samples. The other elements are only detected in some samples with a concentration significantly higher than the value of their experimental detection limit.

Considering the totality of the samples, without any discrimination between ages, physiological states, or geographical origins, we obtain results, a number of which are illustrated in Fig. 1. These results always allow to distinguish the permanent teeth from the deciduous teeth. The permanent teeth are arranged according to the same direction, forming a kind of axis that we will name "the Zn–Sr axis."

Group A elements are well known to play a fundamental role in growth and mineralization processes, especially Cu and Zn (*5*). The very near position of Cu and Zn in Fig. 1 suggests that they are associated in their biological behavior in the mineralization of dental enamel. Furthermore, this similar and associated role is often suggested for many other biological structures (*6*). Low variations of Group A elements within a same group of samples enhance their constitutive role in dental enamel structure.

On the contrary, Br presents a very eccentric position, drawing away most of the deciduous enamel samples. Its free-ion form is often described in organisms (*7,8*). Furthermore, on the one hand, the aqueous-organic phase is always greater in deciduous enamel than in permanent enamel and, on the other hand, during the maturiation process of dental enamel, the aqueous-organic phase diminishes in parallel to an increase of the mineral phase. In this manner, the more mature the enamel is, the more reduced the aqueous-organic phase is. All these considerations suggest that Br belongs to the aqueous-organic phase of dental enamel and that it could serve as a marker of the physiological state of enamel samples.

In order to study the pathological process of dental enamel, we have applied the statistical analysis method considering only the permanent teeth of all origins. The obtained diagrams show a very clear separation between healthy and decayed teeth. In Fig. 2, the healthy teeth are generally situated above Axis 1 and the decayed teeth below it. In these diagrams, Mn is the most significant element. Indeed, its position at the extremity of the lower part of the diagram attracts most of the decayed samples. The cariogenic role of Mn has often been suggested (*9*). This element, associated with the aqueous-organic phase, could be an important marker of carious disease, and its systematic determination is probably very important for the study of the carious process. Manganese determinations are easy and quick to obtain by neutron activation analysis, using the $^{56}Mn^*$ ($T_{1/2}$ = 2.58 h).

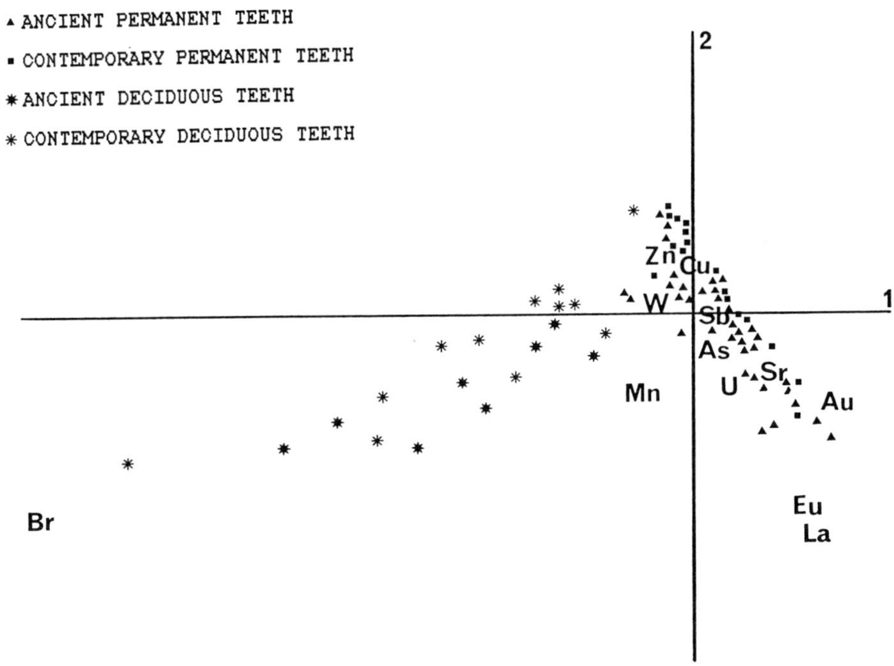

Fig. 1. Differentiation between deciduous and permanent teeth (all teeth are concerned).

It is interesting to note that Sr and Au are situated among sound samples that agree with their cariostatic effect, attributed by Curzon (*10*) and Navia (*11*).

In order to study if individuals coming from a same environment can be characterized, the seven necropoles are analyzed on different correlation diagrams. We have considered only healthy permanent teeth, and Zn and Sr have been removed from the statistical data processing in order to establish these diagrams. Figure 3 represents the result obtained for Pessat and Sezegnin necropoles. Two subgroups of samples are well differentiated by their position in relation to Axis 1. On the left side, Pessat samples are attracted by W, Mo, As, and Sb, whereas on the right side, Sezegnin samples are attracted by La, and Eu.

Analyses of the spring water near these two necropoles were undertaken: the presence of W, Mo, As, and Sb was significantly important in Pessat water, and the presence of La and Eu was significantly important in Sezegnin water. Furthermore, in our study, we had the opportunity of integrating one enamel sample of a tooth coming from a man living today near Sezegnin. In Fig. 3, this enamel sample is well correlated with Sezegnin ancient samples, according to La and Eu influence.

Trace elements pertaining to Group C are specially found in the surrounding where the studied sedentary individuals were living. Con-

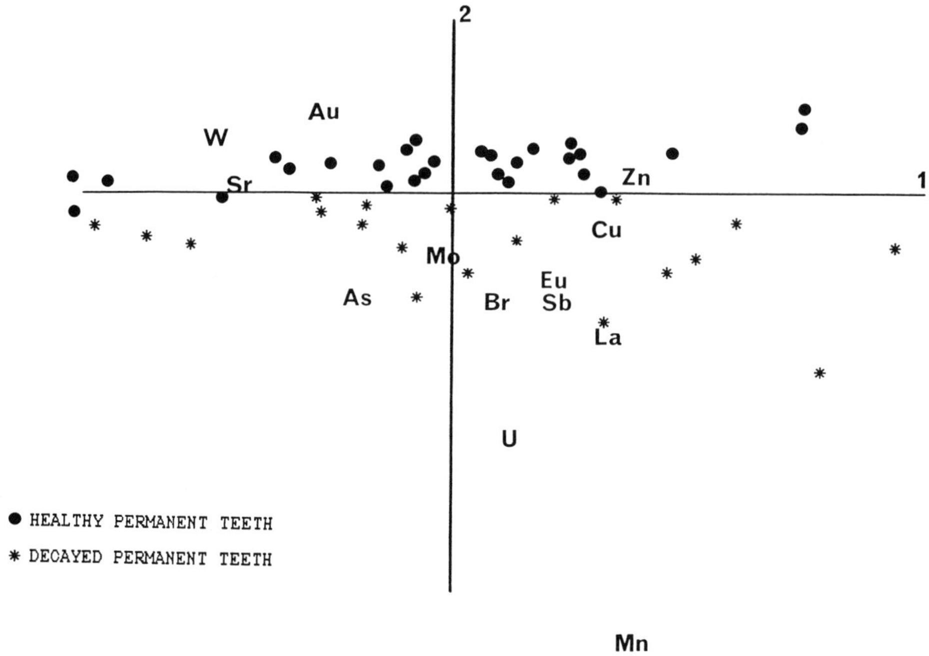

Fig. 2. Differentiation between healthy and decayed teeth (only permanent teeth are concerned).

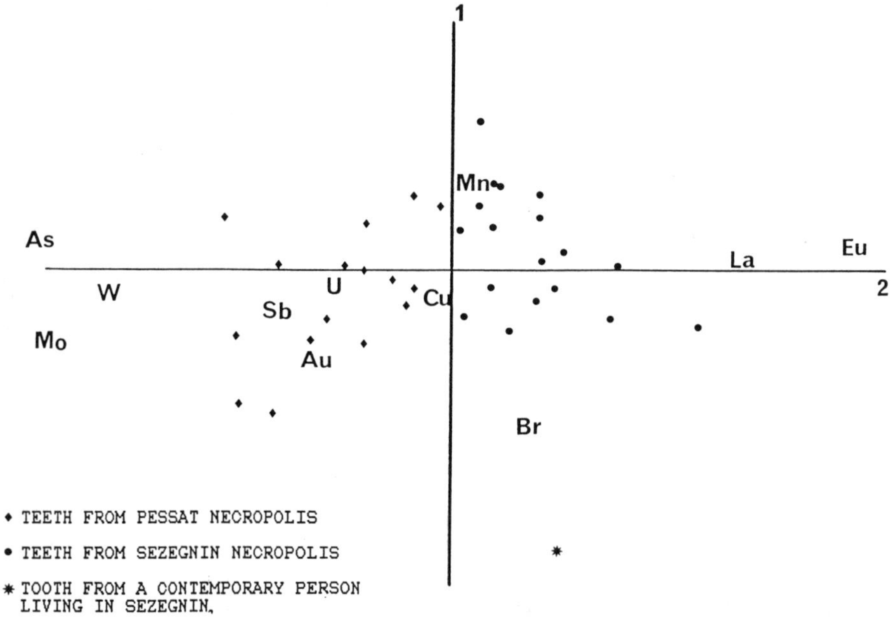

Fig. 3. Differentiation between the teeth from Pessat and Sezegnin necropoles (only healthy permanent teeth are concerned; Zn and Sr are absent from the data processing).

sequently, these elements allow to characterize the geographical origin of samples. We can now better understand the large spreading of their concentration values in our samples: they depend on their environmental concentrations and the space of time during the individual's life, where they are incorporated into the organism. Using percolation tests with radioactive markers, we have checked that the postmortem incorporation of As, Mo, Sb, and W is very small in dental enamel. Lanthanides can be incorporated postmortem in enamel, but experimentations on rats have proven that these elements can also be absorbed during their life (*12*). Furthermore, La and Eu are often present in some contemporary enamel samples, as shown above, for the individual now living nearby Sezegnin. The incorporation of Group C elements during the enamel postmaturation, according to their presence in drinking water or nutritional intakes, suggests that they probably belong to both the mineral and the aqueous-organic phases of dental enamel.

CONCLUSION

Chemical anslysis of dental enamel trace elements shows their important role and offers the means to investigate this role in enamel pathways. The incorporation of trace elements is dependent on the presence of biological systems. Some trace elements could serve as markers of these systems and could help us to have a better understanding of the pathological process and the carious disease of dental enamel. On correlation diagrams, Bromine position appears to be a marker of the maturation process, and manganese position seems to be a marker of the carious process. The systematic determination of trace elements (As, Mo, Sb, W, rare earths, and so on) in dental enamel gives details concerning the life of individuals. These analyses can bring new information in anthropological studies.

REFERENCES

1. V. Vernois, *Les oligoéléments de l'émail dentaire*, Thesis, University of Paris V. 1984.
2. V. Vernois, N. Deschamps, A. Boulay, and M. Goldberg, *Metals in Bone*, N. D. Priest ed., MTP Press Limited, Lancaster, UK, 1984, pp. 407–415.
3. V. Vernois, N. Deschamps, and G. Revel, *J. Trace Elements and Electrolytes in Health and Disease*, **3,** 67 (1989).
4. J. P. Benzecri, *L'analyse des données*, vol. 2, Dunod, Paris, 1980, pp. 18–51.
5. G. Pethes, *I.A.E.A. Technical Report series 197*, Elemental Analysis of Biological Material, Vienna, 1980, pp. 3–19.
6. N. Rao Maturu, *I.A.E.A. Technical Report series 197*, Elemental Analysis of Bilogical Material, Vienna, 1980, pp. 29–39.
7. M. C. Linder, *Seminar in liver's disease*, vol. IV, no. 3, 1984, pp. 264–276.
8. M. C. Linder, *Nutritional Biochemistry and Metabolism*, Elsevier, New York, 1984, pp. 69–81.

9. R. Mejia, F. Espinal, H. Velez, and A. Velez, *Bol. Sanit. Panam.* **66,** 501 (1969).
10. M. E. J. Curzon, P. C. Spector, and H. P. Icker, *Arch. Oral. Biol.* **23,** 317 (1978).
11. J. M. Navia, *Int. Dent. J.* **22,** 427 (1972).
12. V. Vernois, N. Deschamps, G. Revel, D. Septier, and M. Goldberg, *J. Biol. Buccale* **10,** 135 (1982).

Trace Elemental Content of Biological Materials

A Comparison of NAA and ICP-MS Analysis

NEIL I. WARD,* FADI R. ABOU-SHAKRA,
AND STEVEN F. DURRANT

Department of Chemistry, University of Surrey, Guildford, Surrey, GU2 5XH, UK

Received April 17, 1989; Accepted October 10, 1989

ABSTRACT

The advantages and disadvantages of neutron activation analysis (NAA) and inductively coupled plasma-source mass spectrometry (ICP-MS) for the analysis of biological materials is reviewed.

Comparison is made between NAA (instrumental) and ICP-MS (conventional pneumatic solution nebulization and laser ablation) analysis of the biological reference material National Bureau of Standards (NBS) SRM 1577 Bovine Liver. Relatively good agreement is achieved between the results for the 18 elements analyzed by both techniques and those either certified or reported in the literature. Elemental concentrations for Li, Mg, Al, Ca, Cr, Mn, Fe, Cu, Zn, Br, Rb, and Cs are also reported for IAEA Mixed Human Diet (H9), NBS SRM 909 Human Serum, and NBS SRM 1577a Bovine Liver, analyzed by solution nebulization ICP-MS.

Index Entries: NAA; ICP-MS; multielement analysis; biological reference materials.

*Author to whom all correspondence and reprint requests should be addressed

INTRODUCTION

There are many analytical techniques available for measuring trace element levels in biological systems. In recent years, the development of various multielement methods, such as spark-source mass spectrometry, inductively coupled plasma atomic emission spectrometry, NAA, and ICP-MS, have provided the data base of concentrations for a wide range of biological matrices. Since interelemental interactions have recently attracted increasing interest, there is now a tremendous need for multi-element studies at trace levels or lower. Such studies must rely on a high standard of quality control assurance since the elemental results may be the basis upon which future medical, economic, or government regulatory decisions are taken. For most trace analysis measurements, quality control may be based upon interlaboratory and interanalytical method comparisons, or the use of so-called standard reference materials (SRMs). At present, SRMs are often not available for certain matrices (especially biological), or they are limited in the number of certified elements, especially for the majority of ultratrace elements.

Neutron Activation Analysis

NAA is a well-established, multielement analytical technique utilizing instrumental, radiochemical, and prompt gamma ray modes. The basic advantages are the nondestructive, multielement capability, with detection limits for most elements below 1–50 ng/mL, minimal blank ng/mLcorrection, and the simplicity of solids analysis (thereby eliminating the need for pre-analysis sample dissolution steps that may lead to contamination or analyte loss during preparation). NAA also provides good precision at the trace level (*1*), but is severely restricted for small sample masses or liquid volumes, which are not routinely irradiated in thermal reactors owing to fears of radioactive leakage after activation. Such limitations prevent the ultratrace measurement of biological material, such as brain, cerebrospinal fluid, oviduct fluid, or other physiological solutions, and also induce problems relating to sample homogeneity. NAA is also unable to routinely detect such important elements as B, Be, C, H, Li, and Pb, which either have poor cross-sections or do not produce gamma rays. Moreover, long-lived nuclides, for example ^{60}Co ($t_{1/2}$ = 5.27 y), ^{65}Zn ($t_{1/2}$ = 244 d), and ^{59}Fe ($t_{1/2}$ = 45.1 d) require long analysis times. Also, the measurement of many important biological elements, under short-lived nuclide conditions (^{28}Al, ^{49}Ca, ^{66}Cu, ^{20}F, ^{128}I, ^{27}Mg, ^{56}Mn, ^{101}Mo, ^{86m}Rb, ^{77m}Se, and ^{52}V), is normally subject to reduced precision as a result of increased gamma-ray background activity induced by the presence of such major matrix interferences as ^{32}P, ^{24}Na, ^{38}Cl, and ^{80}Br (*2,3*). The necessity for prechemical separation procedures introduces the prospects of analyte loss, contamination, and 'blank correction' problems. Precision problems also arise for cyclic activation analysis where the high elemental content of ^{28}Al ($t_{1/2}$ = 2.246 min) induces pulse pile-up counting errors (especially for soil or sediment samples). In addition, there are 'blank' correction problems induced for ultratrace elements, such as Se and F, owing to impurities in the irradiation vessels or from contamination sources within the rapid sample transfer system. However, acceptable quality control can be achieved by NAA, especially in relation to radiochemical analysis or by the use of Compton suppression for low gamma-ray energies (*4–8*).

Inductively Coupled Plasma-Source Mass Spectrometry

ICP-MS has been successfully applied to the multielement analysis of a wide range of biological materials (*9–12*). Its main features include good sensitivity for undertaking ultratrace element determination with detection limits in aqueous solution generally below 0.1 ng/mL, the ability to determine most elements in the Periodic Table (both in qualitative and quantitative modes), and a dynamic range that extends over six orders of magnitude. Moreover, ICP-MS provides simple spectra, isotopic information, and rapid sample throughput. Generally, signal suppression caused by the major elements present in biological matrices is small and insignificant once the sample is diluted to levels containing less than 1500 μg/mL Na or K (*13,14*). However, a total solids content greater than 1% w/w can lead to injector tube and/or skimmer blockage that reduces instrument sensitivity. The same problem arises when undigested samples containing high levels of protein or fat are nebulized.

Where pre-analysis acid digestion of the biological sample is necessary, the use of HCl, H_2SO_4, or $HClO_4$ as a solvent is undesirable. Oxides and hydroxides of chlorine produce a series of interference peaks, the most important being at m/z 51 (^{35}ClO) and m/z 53 (^{37}ClO), which coincide with monoisotopic vanadium and a small isotope of chromium (9.55% abundance). Similarly, sulfur having a low ionization energy (10.36 eV) yields various oxides and hydroxides of sulfur that occur at m/z of 48, 49, 50, 51, and 52, coincidental with the various isotopes of Ti, V, Ca, and Cr. Since nitrogen has a very high ionization energy, nitric acid causes few problems and is therefore the main acid used (at the 1% v/v level) in the University of Surrey ICP-MS studies. To overcome possible interferences, the spectrum is always 'stripped' using an appropriate matrix matched blank. For the specific analysis of vanadium (^{37}ClO), various correction procedures, based on less naturally-abundant isotopes, are used (*15*).

Another analytical problem that routinely occurs during the analysis of biological samples is the existence of memory effects mainly owing to volatilization of analyte previously deposited upon the walls of the spray chamber or associated nebulizer transfer lines. In general, provided care is taken during the sample throughput run sequence, any memory effects from high elemental levels in standards, sensitivity standards, or samples (especially from B, I, Ba, Hg, and so on) should be minimized.

Basically, a total washout time of only 2–3 min is necessary for routine measurements such that 80–100 samples can be analyzed during a normal 8-h working period.

Numerous sample introduction techniques, including ultrasonic, high-solids nebulizers (*16–19*), and flow-injection methods (*20*) have expanded the analytical scope of ICP-MS, but it is the recent developments in electrothermal volatilization (*21–23*) and laser ablation (*24,25*) that have enabled the direct analysis of solids. Although this is not entirely comparable to the nondestructive nature of NAA, such techniques are already enabling direct multielement and isotope ratio measurements of small sample masses without the concomitant problems of sample dissolution.

MATERIALS AND METHODS

Neutron Activation Analysis

All the elements listed in Table 1 were determined by neutron irradiation and gamma ray spectrometry, utilizing the quoted nuclides and gamma ray energies. Boron was measured using the prompt gamma neutron activation technique and evaluated using a Compton shield (*8*). All other NAA samples were irradiated in the Imperial College Consort MKII reactor (Ascot, UK) at a maximum power of 100 kW. Long irradiations were performed in the 0°(angle) face core tubes in the reactor (thermal flux of 1.6×10^{12} n/cm^2/s) over periods of 1–5 d, with subsequent decay/gamma ray count scheme of 24–72 h/0.5 h for ^{76}As and ^{115}Cd and 7–14 d/1.5 h for ^{60}Co, ^{51}Cr, ^{59}Fe, ^{58}Ni, (^{58}Co), ^{89}Sr, and ^{65}Zn, respectively. Shorter irradiations utilized the In-Core Irradiation System (ICIS) or Cyclic Activation System (CAS), with a thermal flux of 2.3×10^{12} n/cm^2/s. Irradiation schemes involved cyclic mode (t_i = 20 s; t_d = 1 s; t_c = 20 s/10 cycles) for ^{28}Al, ^{77m}Se, and ^{52}V, or rapid transfer mode (t_i = 300 s; t_d = 60–300 s; t_c = 600 s) for ^{49}Ca, ^{66}Cu, ^{27}Mg, ^{56}Mn, and ^{101}Mo. To enhance detection for various short-lived nuclides, a Compton suppression system was also used (*8*). All irradiated samples were counted with Ge(Li) detectors in conjunction with an ND6600 Nuclear Data 4096 channel pulse height analyzer. Sufficient sensitivity for most elements was obtained with 200 mg samples of NBS SRM 1577 Bovine Liver; but for ^{76}As, ^{49}Ca, ^{60}Co, ^{51}Cr, and ^{101}Mo, 500 mg of sample was necessary.

Inductively Coupled Plasma-Source Mass Spectrometry

ICP-MS analysis involving conventional pneumatic solution nebulization was undertaken using the University of Surrey PLASMAQUAD instrument (VG Isotopes Ltd, Cheshire, UK) (*9*). A typical run sequence involved sample solution uptake at a flow rate of ~1.5 mL/min, with MCA scanning over the 5 to 240 m/z mass range using 2048 channels with a dwell time of 250 μs on each channel and 120 sweeps. A multiele-

Table 1
Elemental Concentrations in NBS SRM 1577 Bovine Liver

	NAA			ICP-MS			
Element	Nuclide used	Gamma-ray energy, kev	Value*	Isotope used	ICP-MS**	LA-ICP-MS***	Certified/non Cert.****
Li	—	—	—	7	0.32(0.11)	—	—
B	(prompt, ^{10}B)	472	3.3(0.7)	11	2.8(1.1)	—	3.2(0.9)
Mg	^{27}Mg	843	608(14)	24	608.9(1.3)	—	604(9)
Al	^{28}Al	1779	14.7(1.8)	27	4.5(0.5)	—	/35.50
Ca	^{49}Ca	3083	132(24)	44	116(18)	—	124(6)
V	^{52}V	1434	0.040(0.008)	51	0.061(0.004)	0.021(0.004)	—
Cr	^{51}Cr	320	0.09(0.01)	52	0.125(0.057)	—	—
Fe	^{59}Fe	1292	241(34)	54	262(8)	145(33)	270(20)
Mn	^{56}Mn	846	11.1(1.7)	55	10.9(0.4)	9.2(2.0)	10.3(1.0)
Ni	^{58}Ni	811	0.18(0.02)	60	0.17(0.08)	0.3(0.1)	/0.40
Co	^{60}Co	1332	0.22(0.01)	59	0.04(0.008)	0.01	/0.18
Cu	^{66}Cu	1039	184(16)	63	187.6(4.7)	208(21)	193(10)
Zn	^{65}Zn	1116	132(8)	64	126.6(8.9)	127(9)	130(10)
As	^{76}As	559	0.054(0.002)	75	0.084(0.016)	<0.6	0.055(0.005)
Se	^{77m}Se	161	1.4(0.2)	78	0.72(0.09)	<0.3	1.1(0.1)
Sr	^{89}Sr	910	0.18(0.04)	88	0.21(0.05)	0.47(0.12)	/0.14
Mo	^{101}Mo	307	2.7(0.7)	98	3.1(0.9)	5.9(2.2)	/3.2
Cd	^{115}Cd	528	0.26(0.04)	114	0.42(0.09)	1.1(0.2)	0.27(0.04)
Pb	—	—	—	208	0.25(0.08)	0.29(0.09)	0.34(0.08)

*Neutron activation analysis (instrumental, cyclic, prompt, Compton suppressed); mean (standard deviation).
**ICP-MS, conventional pneumatic solution nebulization.
***Laser ablation.
****Certified or noncertified, literature values.

ment standard solution, containing all the elements of interest, at a final concentration of 100 ng/mL in 1% v/v HNO_3 (Pronalys*ar grade, BDH, UK) was used for calibration. The ICP-MS instrument was optimized using ^{59}Co and ^{209}Bi, giving typical ion count rates of 3–8 × 10^5 counts/s.

Reference materials (NBS SRM 1577, 1577a, 909, and IAEA H9) for conventional pneumatic solution nebulization ICP-MS were prepared by wet digestion with nitric acid. Sample masses ranged between 200–400 mg. Human serum NBS SRM 909 was reconstituted using procedure B, as outlined in the certificate of analysis (*26*). All samples were weighed into 15-mL, metal-free tubes (Elkay Labs, UK), 2 mL concentrated HNO_3 was added, and the tubes set in 500-mL beakers (half full of deionized water). The beaker was transferred into a dessicator and heated in a commercial microwave oven. The microwave was programmed to increase in power step by step (to reach 650 kW in 14 min), such that complete digestion occurred without the samples undergoing rapid frothing or spluttering. The resultant digest was made up to 2.5 mL with double distilled deionized water before introduction to the PLASMA-QUAD.

Laser ablation (LA)ICP-MS was undertaken by the use of the Surrey prototype mass spectrometer (*27*) for the direct analysis of NBS SRM 1577 Bovine Liver, with ablation being undertaken by a ruby laser (*24*). A disk of 1577 Bovine Liver was prepared by mixing 0.96 g of the dried material with 0.24 g of elvasite 2013 (a carbonaceous binder) in a tungsten ball mill. The resulting powder was pressed in an XRF cup (32 mm diameter) to 10 tons. A disk of the Canadian reference material Lobster Hepatopancreas (TORT-1) was similarly prepared to be used as a calibration standard. Five spectra of both the sample and standard were obtained. Five free running shots, each of energy 0.35 J, were employed per integration; the system parameters have been reported elsewhere (*28,29*). Figures 1–4 show the LA-ICP-MS spectra for the analysis of NBS SRM 1577 Bovine Liver over various mass ranges.

RESULTS AND DISCUSSION

Table 1 lists the elemental concentrations in NBS SRM 1577 Bovine Liver analyzed by NAA and ICP-MS (involving both conventional pneumatic solution nebulization and laser ablation). Comparison is made with certified and noncertified/literature values (*30*). Relatively good agreement is achieved between the results for 18 elements (B, Mg, Al, Ca, V, Cr, Fe, Mn, Ni, Co, Cu, Zn, As, Se, Sr, Mo, Cd, and Pb) obtained by both techniques. Li values are also given for ICP-MS. Solid and solution analysis by ICP-MS shows that fair agreement is obtained for most elements. It should be mentioned that precision can be improved by internal standardization for all LA-ICP-MS values (*28,29*).

Table 2 provides the elemental concentrations in IAEA Mixed Human Diet (H9), NBS SRM 909 Human Serum, and NBS SRM 1577a

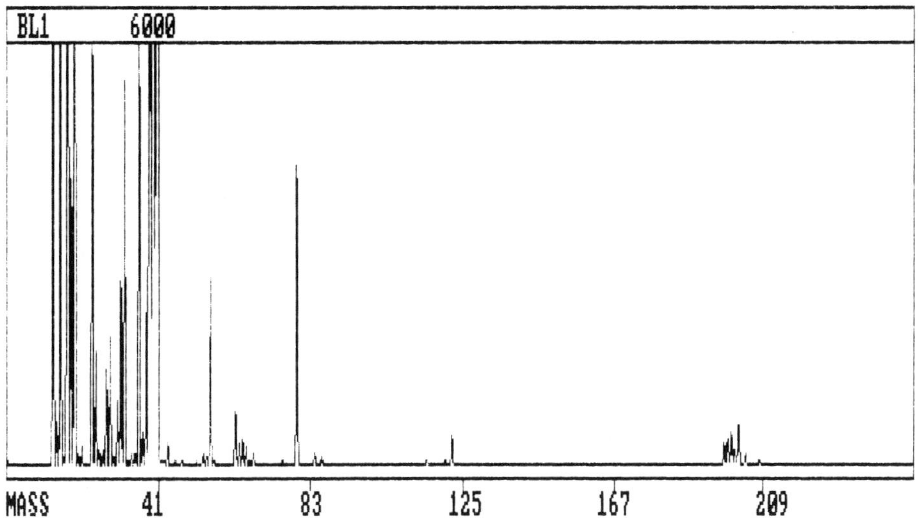

Fig. 1. Analysis of NBS SRM 1577 Bovine Liver by laser ablation ICP-MS. Spectrum showing the full mass range.

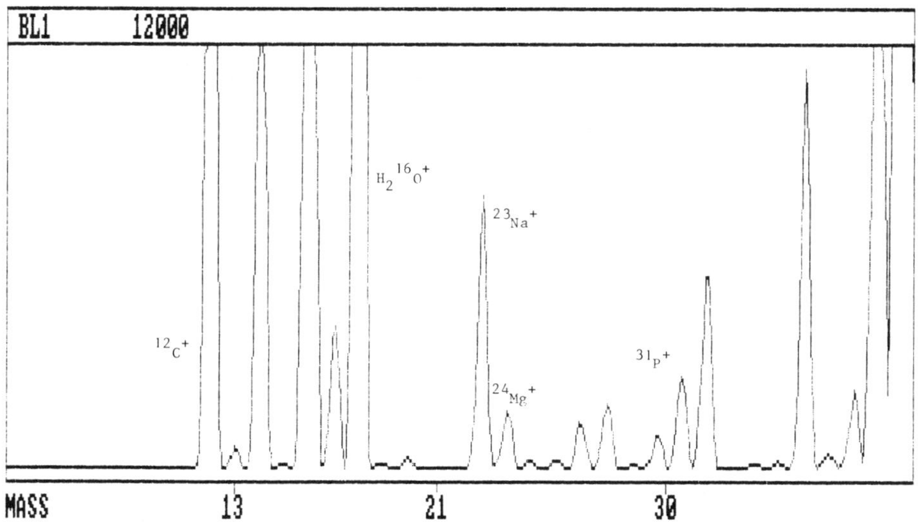

Fig. 2. Analysis of NBS SRM 1577 Bovine Liver by laser ablation. Spectrum showing m/z range 4–40. Integral for m/z 23 was 26,687 counts.

Bovine Liver, evaluated by ICP-MS. Values are compared with certified and literature (*31*) values. Although ICP-MS is still a relatively new analytical method, these results show that it is capable of undertaking both accurate and precise measurements for a range of biological matrices. Now that an increasing number of biological and environmental studies are being undertaken, ICP-MS has the potential of becoming a most important and relevant technique on a level reached by NAA over the past 10 years.

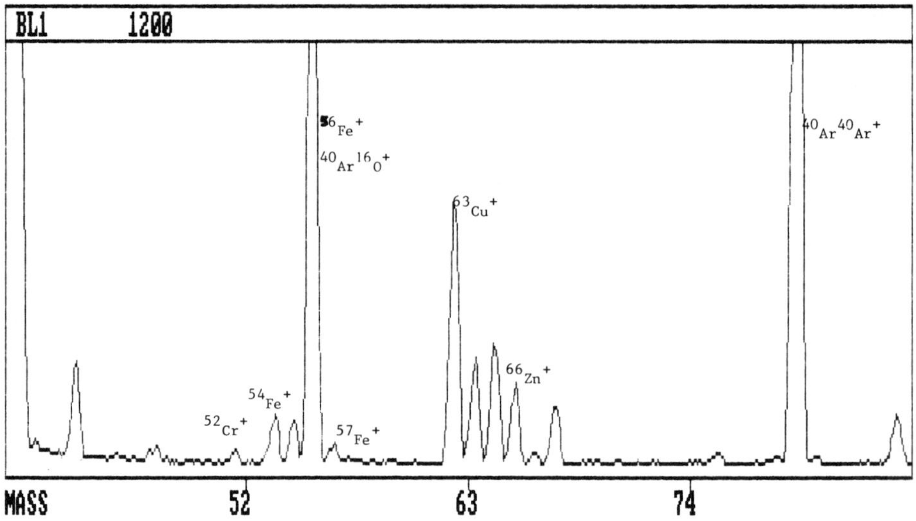

Fig. 3. Analysis of NBS SRM 1577 Bovine Liver by laser ablation ICP-MS. Spectrum showing m/z range 40–85. Integral for m/z 63 was 3,134 counts.

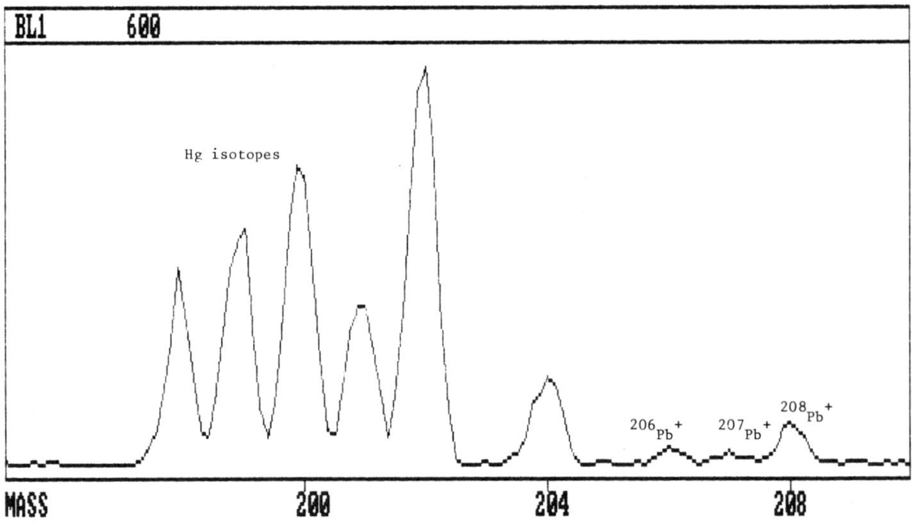

Fig. 4. Analysis of NBS SRM 1577 Bovine Liver by laser ablation ICP-MS. Spectrum showing m/z range 196–210. Integral for m/z 204 was 587 counts.

Table 2
Elemental Concentrations in IAEA Mixed Human Diet (H9), NBS SRM 909 Human Serum, and NBS SRM 1577a Bovine Liver by ICP-MS*

			IAEA:H9		NBS:SRM 909***		NBS:SRM 1577a	
Element	Isotope used	Units**	This work	Cert.(Lit.)****	This work	Cert.(Lit.)†	This work	Cert.(Lit.)
Li	7	$ng g^{-1}$	43.5 ± 4.1	(35.0)	12.1 ± 0.2	11.4 ± 0.3	239 ± 8	(176.5 ± 5)
Mg	26	$\mu g g^{-1}$	645.1 ± 43.3	785	18.7 ± 4.0	29.4 ± 1.2	531 ± 6	600
Al	27	$ng g^{-1}$	9.13 ± 0.86	(16.56)	—	—	1414 ± 69	(2000)
Ca	44	$\mu g g^{-1}$	1705 ± 349	2310	84.3 ± 12.4	120.7 ± 3.5	143 ± 3	120
Cr	52	$ng g^{-1}$	—	150	90.2 ± 12.0	91.3 ± 6.1	826 ± 15	—
Mn	55	$ng g^{-1}$	15.3 ± 0.9	11.8	81.2 ± 9.1	—	13579 ± 161	9900
Fe	56	$\mu g g^{-1}$	49.3 ± 5.2	33.5	1.39 ± 0.45	1.98 ± 0.27	262 ± 8	194
Cu	63	$\mu g g^{-1}$	2.5 ± 0.1	2.9	0.81 ± 0.30	1.1 ± 0.1	155 ± 2	158
Zn	66	$\mu g g^{-1}$	18.09 ± 1.02	27.5	0.54 ± 0.10	(1.09)	92.4 ± 1.2	123
Br	79	$\mu g g^{-1}$	3.76 ± 0.81	7.5	0.74 ± 0.03	—	9.65 ± 0.20	(9)
Rb	85	$\mu g g^{-1}$	3.99 ± 0.29	8	0.0012 ± 0.0001	—	6.4 ± 0.1	12.5
Cs	133	$ng g^{-1}$	35.4 ± 0.93	(25)	<0.4	—	20.8 ± 1.4	—

*Inductively coupled plasma-source mass spectrometry; $n = 5$ replicates.
**μg/g or ng/g for H9 and 1577a; μg/mL or ng/mL for 909.
***Values for reconstituted material (procedure B).
****Certified or Literature values.
†Converted from mmol/L or μmol/L.

ACKNOWLEDGMENTS

The authors wish to thank Alan L. Gray and John Williams for their scientific advice and help in the Department of Chemistry ICP-MS facility. Also, special thanks are given to Efamol, Birthright, and the Hariri Foundation for the assistance in funding and to Alison Gittings for the preparation of this manuscript.

REFERENCES

1. S. Amiel, *Nondestructure Activation Analysis,* Elsevier, New York, 1981.
2. N. I. Ward and D. E. Ryan, *Anal. Chim. Acta* **105,** 185 (1979).
3. N. I. Ward, *Micronutr. Anal.* **2,** 223 (1986).
4. J. Versieck, J. Hoste, F. Barbier, H. Michels, and J. De Rudder, *Clin. Chem.* **23,** 1301 (1977).
5. J. Versieck and R. Cornelis, *Anal. Chim. Acta* **116,** 217 (1980).
6. R. Cornelis, J. Versieck, L. Mees, J. Hoste, and F. Barbier, *Biol. Trace Element Res.* **3,** 257 (1981).
7. J. Versieck, J. Hoste, L. Vanballenberghe, A. De Kesel, and D. Van Renterghem, *J. Radioanal. Nucl. Chem.* **113,** 299 (1987).
8. N. I. Ward, *J. Radioanal. Nucl. Chem.* **110,** 633 (1987).
9. A. R. Date and A. L. Gray, *Applications of Inductively Coupled Plasma Mass Spectrometry,* Blackie, Glasgow, 1988.
10. H. E. Taylor and J. R. Garbarino, *ICP Inf. Newslett.* **11,** 774 (1986).
11. B. T. G. Ting and M. Janghorbani, *Anal. Chem.* **58,** 1334 (1986).
12. J. R. Dean, R. C. Massey, and L. Ebdon, *J. Anal. Atom. Spectrom.* **2,** 369 (1987).
13. A. L. Gray and A. R. Date, *Analyst* **108,** 1033 (1983).
14. A. L. Gray, *Proc. Analytiktreffen,* K. Dittrich, ed., Leipzig, 1987, p. 311.
15. A. L. Gray, *Spectrochim. Acta* **40B,** 1525 (1985).
16. R. S. Houk, V. A. Fessel, G. D. Flesch, H. J. Svec, A. L. Gray, and C. E. Taylor, *Anal. Chem.* **52,** 2282 (1980).
17. A. R. Date and A. L. Gray, *Analyst* **106,** 1255 (1981).
18. M. A. Vaughan, G. Horlick, and S. H. Tan, *J. Anal. Atom. Spectrom.* **2,** 765 (1987).
19. J. G. Williams, A. L. Gray, P. Norman, and L. Ebdon, *J. Anal. Atom. Spectrom.* **2,** 469 (1987).
20. J. R. Dean, L. Ebdon, H. M. Crews, and R. C. Massey, *J. Anal. Atom. Spectrom.* **3,** 349 (1988).
21. D. C. Gregoire, *J. Anal. Atom. Spectrom.* **3,** 309 (1988).
22. C. J. Park and G. E. M. Hall, *J. Anal. Atom. Spectrom.* **3,** 355 (1988).
23. C. J. Park, C. J. VanLoon, P. Arrowsmith, and J. B. French, *Can. J. Spectrosc.* **32,** 29 (1987).
24. A. L. Gray, *Analyst* **110,** 551 (1985).
25. P. Arrowsmith, *Anal. Chem.* **59,** 1437 (1987).
26. S. D. Rasberry, *Certificate of Analysis,* Standard Reference Material 909, Human Serum, National Bureau of Standards, Gaithersburg, MD.
27. A. R. Date and A. L. Gray, *Analyst* **108,** 159 (1983).
28. N. I. Ward, S. F. Durrant, and A. L. Gray, *Mikrochim. Acta,* submitted.
29. S. F. Durrant, A. L. Gray, and N. I. Ward, *Mikrochim. Acta,* submitted.

30. N. I. Ward, *Proc. 2nd Nordic Symp. on Trace Elements in Human Health and Disease*, Odense, Denmark, 1987, p. 118.
31. G. V. Iyengar, W. B. Clarke, R. G. Downing, and J. T. Tanner, *Trace Element Anal. Chem. in Med. and Biol., vol. 5,* P. Brätter and P. Schramel, eds., de Gruyter, Berlin, 1988, p. 267.

Instrumental Neutron Activation Analysis of Geological and Biological Reference Materials Using the k_0-Standardization Method

M. Carmo Freitas

LNETI/ICEN, Departamento de Energia e Engenharia Nucleares, 2686 Sacavém, Codex, Portugal

Received April 17, 1989; Accepted October 10, 1989

ABSTRACT

The k_0-standardization method used in instrumental neutron activation analysis (INAA) was applied to the determination of some elements in geological and biological reference materials. We analyzed NBS SRM 1572 Citrus Leaves and SRM 1645 River Sediment and the CRM materials, IAEA Soil-7, SL-1, and MA-A-2. Comparison is made with reference values whenever available. Good agreement is found. The potential of the k_0-standardization method in reactor INAA is discussed.

Index Entries: Accuracy; activation analysis; calibration standards; neutrons; neutron activation analysis; standardization.

INTRODUCTION

The k_0-standardization method used in reactor neutron activation analysis (INAA) was implemented at LNETI in 1987 (*1*), following the theory established by F. De Corte (*2*). A few papers (*3–5*) have shown the accuracy and precision obtained in our laboratory. It was shown (*3*) that the precision obtained with k_0-standardization and the relative methods are quite similar. No systematic errors were detected in the determina-

Table 1
Experimental Conditions Used in This Work

		Neutron flux (ncm^2/s^1)	
Sample weight, g	Comparator	Thermal	Fast
0.2–1.0	0.112% Au-Al wire	1.4×10^{11}	1.5×10^{10}
Thermal to epithermal ratio	Deviation from the 1/E law	Irradiation time, h	FWHM, keV
66 ± 9	0.05 ± 0.03	14	1.8 (1.34 MeV)

tion of several elements in different matrices (*4*). In this work, some results obtained with the k_0-standardization method are presented for the standard reference materials, (SRM) 1572 Citrus Leaves and 1645 River Sediment, and the certified reference materials, (CRM) Soil-7, SL-1 Lake Sediment, and MA-A-2 Fish Flesh. The results are compared with reference values (whenever available) given by NBS or IAEA, or of Gladney et al.(*6*). The elements presented here and in refs. (*3–5*) have been determined in the same experiment. The potential of the k_0-standardization method to determine a great amount of accurate and precise results in a single irradiation is remarkable.

EXPERIMENTAL

A detailed description of the experimental conditions of preparation, irradiation, and measurement of samples and comparator was made in earlier papers (*1,3,5*). Table 1 shows a summary of those conditions.

RESULTS AND DISCUSSION

Table 2 shows results obtained with the k_0-standardization method for the SRM 1572 Citrus Leaves and 1645 River Sediment and CRM Soil-7, SL-1 Lake Sediment, and MA-A-2 Fish Flesh. The results are compared with reference values (whenever available) given by NBS or IAEA or "consensus" values given by Gladney et al. (*6*). No reference values were found in the literature for Br, Cs, K, Na, and Rb in MA-A-2 Fish Flesh. Good agreement is found, except for Eu and Ta, in SL-1 and La in River Sediment. However, since the reference values for Eu and Ta are noncertified, no errors are assigned to them, and the "consensus" value for La does not agree with ours, but the noncertified value given by NBS agrees quite well.

Table 3 shows the elements determined for a few reference materials (biological and geological), each analyzed in a single 14-h irradiation

Table 2
Results Obtained with the k_O-Standardization Method for Reference Materials

Reference material	Element	Concentration, mg/kg: Reference value[a]	Concentration, mg/kg: This work[b]	Element	Concentration, mg/kg: IAEA/NBS value[a]	Concentration, mg/kg: This work[b]
	Ba	374 ± 26 (n = 4)[c]	321 ± 12	Rb	41 ± 4 (n = 6)	63 ± 8
	Ca (%)	2.65 ± 0.34 (n = 14)	4.0 ± 0.6	Sc	2.6 (n = 2)	2.10 ± 0.05
		2.9			2	
NBS 1645	Cs	2.8 ± 0.5 (n = 3)	2.8 ± 0.1	Sm	1.24(n = 2)	1.1 ± 0.4
	La	24 (n = 2)	7.9 ± 0.3			
		9				
NBS 1572	Ba	23.5 (n = 2)	23 ± 4	Br	8.36 (n = 1)	8.6 ± 0.1
		21 ± 3			8.2	
	Ba	159	143 ± 56	K(%)	1.21	1.02 ± 0.08
IAEA	Br	7	7.6 ± 0.4	Na(%)	0.24	0.21 ± 0.01
Soil-7	Ca (%)	16.3	18.6 ± 0.7	Tb	0.6	0.9 ± 0.1
	Fe (%)	2.57	2.8 ± 0.1	U	2.6	2.1 ± 0.2
	Ga	10	9.6 ± 0.4			
	Eu	1.6	2.8 ± 0.2	Tb	1.4	0.9 ± 0.2
IAEA	K (%)	1.5	1.2 ± 0.5	U	4.02 ± 0.32[b]	6 ± 1
SL-1	Nd	43.8 ± 2.8[b]	47 ± 16	Yb	3.4 ± 0.6[b]	4.2 ± 0.3
	Ta	1.6	1.03 ± 0.08			
IAEA	Br		22.3 ± 0.6	K(%)		1.5 ± 0.1
MA-A-2	Cs		0.18 ± 0.01	Na(%)		0.19 ± 0.03
	Hg	0.47 ± 0.02	0.9 ± 0.6	Rb		8.9 ± 0.5

[a]Uncertainties are standard deviations.
[b]Uncertainties at a significance level of $\alpha = 0.05$.
[c]n indicates the number of data considered (*6*).

Table 3
Elements Determined in a Single 14-h Irradiation at a 1.4×10^{11} ncm^2/s^1 Thermal Neutron Flux with the k_O-Standardization Method for Reference Materials

	SRM NBS			CRM IAEA			
Element	1572 Citrus Leaves	1573 Tomato Leaves	1645 River Sediment	Soil-7	SL-1 Lake Sediment	A-13 Blood	MA-A-2 Fish Flesh
As	X	X	X	X	X		X
Ba	X	X	X	X	X		
Br	X	X		X		X	X
Ca	X	X	X	X			
Cd			X				
Ce	X	X	X	X	X		
Co	X	X	X	X	X		X
Cr		X		X	X		
Cs	X		X	X	X		X
Eu				X	X		
Fe	X	X	X	X	X	X	X
Ga				X			
Hf				X	X		
Hg		X	X				X
K	X	X	X	X	X	X	X
La	X	X	X	X	X		
Mn	X	X	X	X			
Na	X	X	X	X	X	X	X
Nd				X	X		
Rb	X	X	X	X	X		X
Sb	X	X	X	X			
Sc	X	X	X	X	X		
Se							X
Sm	X	X	X	X	X		
Sr	X	X	X				
Ta				X	X		
Tb				X	X		
Th	X	X	X	X	X		
U			X	X	X		
Yb				X	X		
Zn	X	X	X	X		X	X

under the conditions shown in Table 1. Their concentrations, determined with the k_o-standardization method, are given in refs. (*3–5*) and in this work. The values are in good agreement with reference values, except for a few cases discussed in ref. (*5*) and this work (Co and Rb in NBS 1572; Ce and La in NBS 1645, and Eu and Ta in IAEA SL-1). No values were found in the literature for Br, Cs, K, Na, and Rb in MA-A-2.

Table 3 highlights the great advantage of the k_o-standardization method comparatively to the relative method: it would have been impossible to determine so many elements with the relative method, with a single irradiation and so little effort. In the relative method, the preparation of the standards needed for determination of all those elements takes a long time; using available multielement standards would yield results only for those elements for which certified values are available. Besides, both short and long irradiations would have been needed since the amount of sample to be irradiated is limited by the permissible amount of standard because of the condition of equal geometry.

It is remarkable that such a large number of elements was determined accurately in biological reference materials (Citrus Leaves, Tomato Leaves, and MA-A-2) using an instrumental technique with the k_0-standardization method. As shown in ref. (*5*) for Citrus Leaves, Tomato Leaves, and River Sediment, most trace elements whose concentrations have been published were determined accurately with both radiochemical NAA and INAA using the k_0-standardization method and inaccurately with INAA using the relative method. And it should be added that the radiochemical technique is time-consuming and more dangerous than an instrumental one.

CONCLUSION

In this work, we showed results for some elements in SRM NBS 1572 Citrus Leaves and 1645 River Sediment, and CRM IAEA Soil-7, SL-1, and MA-A-2.

These results are a small part of a set of values determined with the k_0-standardization method, using instrumental NAA for a few reference materials (cf. refs. *3–5*). A single 14-h irradiation at a 1.4×10^{11} ncm^2/s^1 thermal neutron flux was used for each reference material. Accurate and precise values were obtained for most of the elements (cf. refs. *3–5* and this work). The number of elements determined was: 19 for 1572 Citrus Leaves, 20 for 1573 Tomato Leaves, 21 for 1645 River Sediment, 27 for Soil-7, 21 for SL-1, 5 for A-13, and 11 for MA-A-2. We conclude that the k_0-standardization method has unique advantages in multielement analysis owing to: the great amount of elements determined per analysis, the precision and accuracy attainable, the short time needed per analysis, and the possibility of avoiding the use of radiochemical separation of some trace elements.

ACKNOWLEDGMENTS

The help from F. De Corte, Research Associate of the National Fund for Scientific Research, Belgium, was invaluable before and during installation of the k_0-standardization method at Sacavém (LNETI); A. De Wispelaere, from the Institute for Nuclear Sciences, University of Ghent, Belgium, gave continued technical assistance. The author is indebted to the IAEA for financial support during a 6-month training period at Ghent.

REFERENCES

1. M. Carmo Freitas, *LNETI/DEEN-A.19*, 1987.
2. F. De Corte, *Agregé Thesis*, Ghent University, Belgium, 1987.
3. M. Carmo Freitas and E. Martinho, *Analytica Chimica Acta*, **219**, 317–322 (1989).
4. M. Carmo Freitas and E. Martinho, *Analytica Chimica Acta*, **223**, 287–292 (1989).
5. M. Carmo Freitas and E. Martinho, *Talanta*, **36(4)**, 527–531 (1989).
6. E. S. Gladney, B. T. O'Malley, I. Roelandts, and T. E. Gills, *Standard Reference Materials: Compilation of Elemental Concentration Data for NBS Clinical, Biological, Geological, and Environmental Standard Reference Materials*, National Bureau of Standards, Washington, DC, 1987.

Nuclear Methods in Environmental Studies

Soluble Metals in the Atmosphere and Their Biological Implications

A Study to Identify Important Aerosol Components by Statistical Analysis of PIXE Data

JOHN W. WINCHESTER

Department of Oceanography, Florida State University, Tallahassee, FL 32306-3048

Received April 17, 1989; Accepted October 19, 1989

ABSTRACT

Multivariate statistical analysis has been applied to time series measurements of aerosol elemental composition from PIXE analysis of filter samples, and principal components have been resolved that represent distinct particle types in an external mixture in the atmosphere. In this study, it is argued that a combination of chemical and statistical analyses of the data may be more powerful in determining chemical species in atmospheric aerosols than studies that employ mainly direct chemical analysis of chemical species in unresolved mixtures of aerosol particle samples. Sulfur is generally associated with mineral dust elements. It is reasoned that the association may represent sulfuric acid coatings on particles that can lead to mineral dissolution and solubilization of significant amounts of aluminum, iron, and other metals.

Upon wet or dry deposition to the surface, the fluxes of these metals in biologically-available form may be sufficient to affect primary productivity in the world ocean and cause ecological damage in lakes. As a consequence, the fluxes of biogenic trace gases to the atmosphere may be changed, possibly leading to changes in the tropospheric concentration of ozone. The inputs to lakes of soluble aluminum, which is toxic to fish, may be partly by deposition directly from the atmosphere, thus not limited to leaching of soils by acid deposition. Human inhalation of soluble aluminum and other solulilized mineral metals may account, in part, for the observed geographic pattern of deaths attributed to chronic obstructive pulmonary disease (COPD) that show high rates in cities of the Western US and the southeast region, but low in most of the midwest and northeast.

Index Entries: Aerosols; analysis, chemical and statistical; atmosphere; ozone.

INTRODUCTION: BIOLOGICAL IMPORTANCE OF SOLUBILIZED TRACE METALS IN THE ATMOSPHERE

Biological Availability of Metals Can Depend on Their Solubility

The biological importance of suspended particulate matter in the atmosphere can depend on its chemical form, as well as on concentrations of its elemental constituents. For example, dry deposition of metals in water soluble forms may impact biota to a greater extent than if they were deposited as insoluble mineral particles. Thus, in assessing atmospheric exposures of either ecosystems or of humans to airborne particulate matter, it is important to have knowledge of both concentrations and chemical forms of its chemical constituents.

Direct determination of chemical species can be labor intensive and force one to compromise the determination of variability in addition to average concentration. But without knowledge of variance, resolution of different principal components that contain the same species may be difficult, and without distinguishing the components, an evaluation of their sources and transport to a receptor may not be possible. An alternative strategy is to infer probable chemical forms from correlations between readily measured elemental concentrations in large data sets with variance sufficient to resolve principal components statistically. By this procedure, the data are fit to a small number of groupings of elements that, when linearly mixed, can account for a high percentage of the variability in concentrations of the elements measured. The internal mixture of elements in a resolved component, together with additional chemical information, may lead to a reasonable estimate of chemical forms of the elements and their biological impacts after aerosol deposition on a receptor.

This approach is illustrated here by two case studies that explore the potential for mineral solubilization in the atmosphere by sulfuric acid and the biological implications for soluble metal deposition. The studies are based on a statistical association of sulfur with elements characteristic of alkaline mineral aerosol and chemical reasoning of its consequences.

Reaction of Sulfuric Acid with Mineral Aerosol and Dissolution of Metals

Sulfuric acid in moist air becomes diluted with water, but under ordinary ambient conditions, the pH remains very low, even if some neutralization by ammonia occurs. Tang et al. (*1*) have constructed a phase diagram for the H_2SO_4-$(NH_4)_2SO_4$-H_2O system, shown in Fig. 1. Measurements of aerosol composition in Ohio (*2*) and northern Florida (*3*) show an average pH<0 in both regions. If dust particles become coated with sulfuric acid while suspended in air, especially in polluted areas where SO_2 is abundant, they may dissolve owing to the very low pH assumed by the coating at ambient relative humidities. From Fig. 1, it can be shown that a droplet of sulfuric acid at equilibrium with water vapor in air at 75% relative humidity will be a 4.5 molal solution with pH $= -1$ and at 90% relative humidity will be 2.2 molal with pH $= -0.65$. Such low pH values in coatings of particles can lead to corrosion of even resistant minerals unless ammonia or other rapidly reacting bases are present. Clay particles often may not resist breakdown during normal atmospheric residence times of a few days, and soluble Al, Fe, and other metals should result and undergo dry or wet deposition in a biologically-available form.

This possibility may be inferred from statistical analysis of aerosol composition measurement data. In a long enough time series of measurements of variable concentrations, such as by aerosol samples collected several times per day for a number of days, correlation between sulfur and the elements typical of soil minerals suggests an internal mixture, e.g., a coating that, under suitable atmospheric conditions, may solubilize even clay minerals in the particles into aqueous ionic species. (An internal mixture, i.e., particles that each contain the different elements in certain proportions, is distinguished from an external mixture, where different kinds of particles are mixed in air. A high degree of correlation among the measured elemental concentrations,which results in a principal component being identified, suggests an internal mixture rather than an external mixture, owing to the latter possibility not being likely to maintain the high correlation over the duration of aerosol sampling.) In this study, we have used time sequence aerosol sampling and PIXE analysis for elemental composition to obtain data for inferring sulfuric acid reaction with mineral matter and the solubilization of metals.

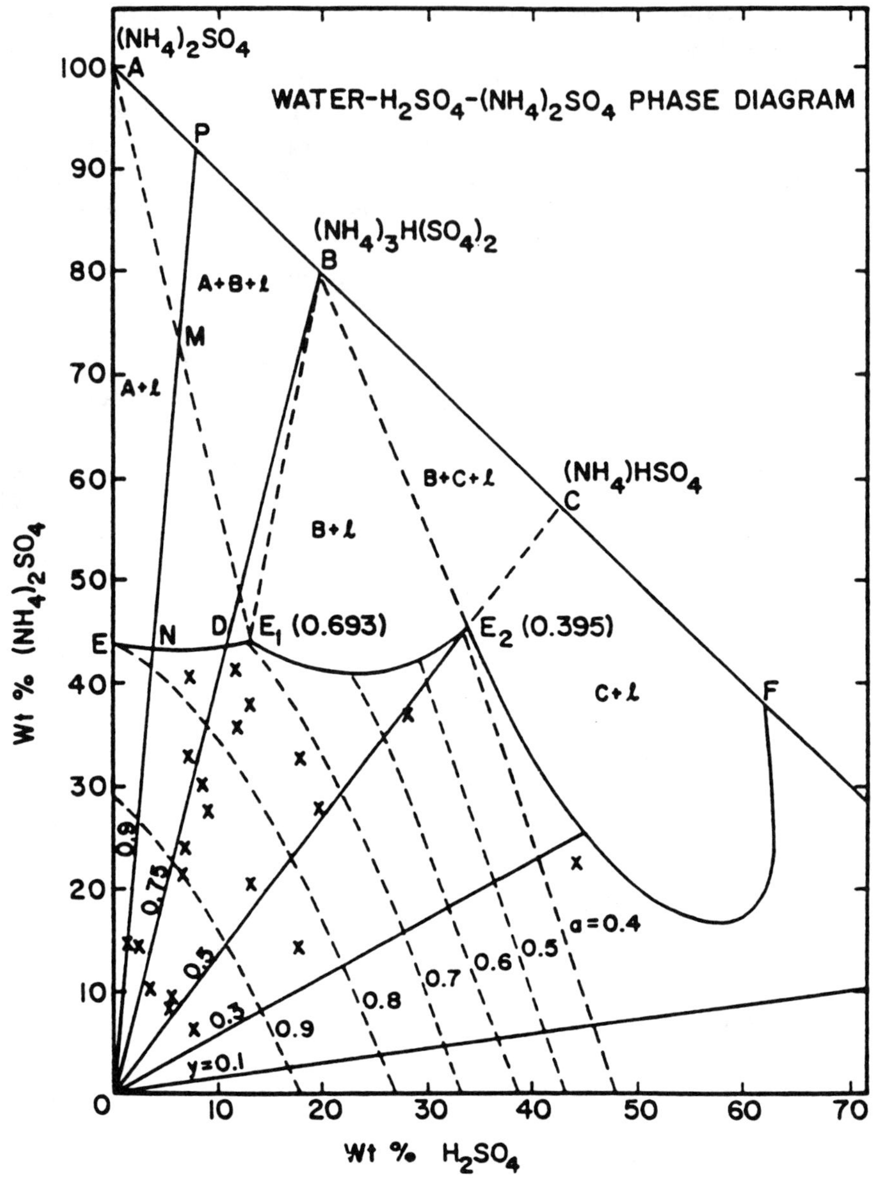

Fig. 1. Phase diagram for aqueous aerosols containing $(NH_4)_2SO_4$, H_2SO_4, and H_2O. Dashed curves for water activity a = 0.9–0.4 (relative humidity 90–40%) indicate compositions in liquid phase region where crystalline phases cannot coexist. Dashed lines above the liquid phase region delineate regions where two of three solid phases can coexist with a eutectic aqueous solution phase. Points "x" represent aerosol composition measured in Tallahassee, FL for aerosol pH calculation. They average pH = −0.17 ± 0.33 Standard Deviation.

STATISTICAL RESOLUTION OF AEROSOL COMPONENTS FROM PIXE ANALYSIS DATA AND INFERRING MINERAL SOLUBILIZATION FROM COMPONENT COMPOSITION INFORMATION

To Characterize Aerosol Mixtures, Balance Analytical Chemistry with Statistical Inference

Building on the distinction made above between internal and external aerosol mixtures, an aerosol is generally an external mixture of components having different chemical properties. Biological effects resulting from aerosol deposition should depend on properties of particles of certain components, not on the mixture of all components. These properties are determined by particle composition, the internal mixture of its chemical constituents. One cannot determine component composition by chemical analysis of the aerosol mixture, for the direct measurement of aerosol chemical species concentrations in an unresolved mixture of different components does not indicate chemical composition of different kinds of particles nor their relative abundances in the external mixture.

But component composition is not easily measured directly. Direct measurement of component properties, e.g., by individual particle analysis, can be time consuming and not allow time for determining the average concentration and time variability of the component, characteristics that may be important for understanding possible biological effects. The dilemma can be resolved by combining statistics, chemical analysis, and chemical reasoning. Principal components can be resolved statistically from large data sets and their elemental compositions calculated. With additional information, the chemical species in the components may then be inferred.

Procedures for Measurement and Data Analysis and an Outline of Some Results

In the present study, sets of aerosol samples were collected on Nuclepore filters by linear streaker (4) in time series of a week to a month, consisting of 60–400 steps each of 2–4 h, and analyzed for 8–20 elements by PIXE (5). Sampling locations were Hawaii, Japan, northwestern China, and nine sites in the northeastern US. Absolute principal component analysis (APCA) (6) of each data set was carried out using IBM PC compatible software (7). Generally, three or four factors were found to be significant and accounted for at least 80% of the total variance. By multiple linear regressions, their relative elemental compositions were calculated and showed patterns that could be recognized, at the different locations, as soil mineral dust, automotive lead with bromine, sulfur with some mineral dust, sea salt, smoke aerosol from burning fuels, and others. Thus, to a good approximation, each data set appears to represent an external mixture of principal components, each component being an internal mixture of elements characteristic of its origin.

In most data sets sulfur is partitioned mainly between two principal components, both containing soil mineral elements together with sulfur, but in different proportions. The data from all sites appear to indicate an active interaction between alkaline clay and carbonate mineral dust and acidic sulfur compounds from SO_2 by several possible mechanisms, leading to sulfuric acid particle coatings. On a chemical equivalent basis, the most sulfur-rich components contain sufficient sulfur which, as H_2SO_4, could react appreciably with their mineral constituents and bring metals into solution. Over the Pacific, the concentrations of metals associated with sulfur are high enough possibly to affect primary production by phytoplankton in oceanic surface waters after aerosol deposition. The absolute concentrations of Al and Fe in the sulfur-rich components at the northeastern US sites are high enough, assuming reasonable deposition velocities, to supply substantial fluxes of these elements in biologically-active ionic forms to lakes and land areas. Human inhalation of solubilized metallic elements may also be biologically significant and help explain geographic patterns of chronic obstructive pulmonary disease (COPD) over the continental US.

CASE 1: MINERAL DUST AEROSOL OVER THE NORTH PACIFIC OCEAN

Resolution and Chemical Significance of Principal Components of Aerosols in Background Air

The objective of the global scale study was to compare aerosol composition measurements made at remote locations and estimate the potential extent of neutralization and solubilization of alkaline aerosol by sulfuric acid through atmospheric uptake of acidic sulfur gases by dust particles. A full account of the study is given elsewhere (*8*). Time sequence samples were collected at five locations in the Asia-Pacific region during 1979 and 1980, as indicated in Table 1. The resulting data sets of 66–229 time steps of 2.4–4.0 h each represented 526 PIXE analyses for 7–12 elements. Principal components, resolved using absolute principal component analysis with varimax rotation and followed by multiple linear regression of measured elemental concentrations vs principal component or factor scores, were evaluated for chemical equivalent concentrations of sulfur and alkaline mineral elements. Two or three factors accounted for 87–97% of the variance of each set, consistent with a model

Table 1
Asia-Pacific Region Streaker Samples Evaluated by Factor Analysis[a]

ID	Location	Dates	Steps Analyzed	Hours per Step	Elements	Percent Variance F1	F2	F3	Sum
XJ	Xinjiang China	1980 Aug 8-24	86	4.0	12	58	37	-	94
XZ	Xizang (Tibet)	1980 Apr 26-May 7	66	4.0	9	68	19	-	87
KN	Niigata Japan	1980 Apr 19-25	59	2.4	9	70	14	12	97
ML	Mauna Loa Observat.	1979 Apr 23-May 6	86	3.68	7	79	18	-	97
HV	Hawaiian Volcano Observat.	1979 Mar 18-28 & Apr 23-May 18	229	3.68	8	55	28	13	96

[a]Elements included in factor analyses from the 5 locations:

XJ: Mg, Al, Si, P, S, Cl, K, Ca, Ti, Mn, Fe, Zn; XZ: Al, Si, S, K, Ca, Ti, Mn, Fe, Zn; KN: Al, Si, S, Cl, K, Ca, Ti, Mn, Fe; ML: Al, Si, S, K, Ca, Ti, Fe; and HV: Al, Si, S, Cl, K, Ca, Ti, Fe. At each location, the concentration data for these elements in the indicated number of sequential time step measurements were resolved into two or three factors, or principal components, that accounted for 87–97% of the variance in the data.

of linear mixing of components for most observed variability. The resulting 12 components from the five locations, when their chemical equivalent compositions are compared, as shown in Table 2, appear to represent a broad range of aerosol aging from recently generated soil dust to particles exposed to trace SO_2 or other acidic sulfur gases for a week or longer. Virtually no aerosol sulfur was found without being associated with soil elements.

All 12 components contained statistically significant amounts of both sulfur and alkaline elements in relative proportions that suggest two rate limitations on acidic gas uptake: neutralization of rapidly reacting bases, mainly $CaCO_3$, and gradual neutralization of slowly reacting bases, mainly clay minerals, by H_2SO_4 after SO_2 uptake. After each stage of titration in the atmosphere, further gas uptake could have been inhibited by an acidic particle surface. The resolved aerosol components show cases approaching both stoichiometric limits, but no cases of excess

Table 2
Summary of Aerosol Aging Categories[a]

	Equivalent Ratios		
Component	S/Σ	S/Ca	$EF_{Al}(S)$
KNF2	0.0014	0.034	4
KNF1	0.0017	0.036	5
HVF1	0.0055	0.072	18
XJF2	0.0069	0.103	21
XZF1	0.0095	0.139	26
Unaged	0.001-0.01	0.03-0.14	4-26
HVF2	0.0236	0.229	105
MLF1	0.0463	0.457	139
XJF1	0.0389	0.719	118
KNF3	0.1166	1.166	482
XZF2	--	*	--
Semiaged	0.02-0.12	0.2-1.2	105-482
MLF2	0.8045	9.709	2357
HVF3	1.079	15.873	3209
Aged	0.8-1.1	10-16	2300-3200

*S/(K + Zn + Mn) = 1.144

[a]The 12 principal components resolved from the five locations are arranged according to the relative magnitudes of the ratios S to alkaline metal concentrations, where Ca represents alkaline mineral, e.g., carbonate, that may react rapidly with sulfuric acid and Σ represents all potentially alkaline mineral metals measured, mainly Ca, Al, Fe, and Si. The enrichment factor of S, relative to Al, compared to that in average earth crust, is also shown. The 12 components are arranged qualitatively into groupings that may represent different degrees of aging of alkaline mineral aerosol in an atmosphere containing acidic sulfur gases.

sulfur beyond the second limit, consistent with these being practical limits in the remote areas investigated. If mineral aerosols are solubilized by strongly acidic surface solutions at ambient relative humidities, as suggested by these results, subsequent deposition to ocean, fresh water, or land surfaces may be more important inputs of nutrient or toxic metal ions than deposition of insoluble mineral particles or of wet or dry deposition of acids followed by dissolution of particles in seawater or leaching of soils.

Biological Implications: Soluble Metals from Aerosol Deposition in Marine Ecosystems Potentially May Affect Marine Primary Production, Release of Reactive Biogenic Gases, and Tropospheric Ozone Levels

A number of lines of evidence point to connections between chemical processes in the atmosphere and biological processes in the ocean. Recently much publicized is the suggestion that cloudiness over the ocean, very important in climate change modeling, is affected by particles coated with sulfates from oxidation of dimethylsulfide (DMS), a marine biogenic trace gas released from certain species of phytoplankton (*9,10*). The result of the present study indicate another possible connection.

Oceanographers have been stirred by a report that iron may be an important limiting nutrient for primary production by phytoplankton in the central North Pacific (*11*). Evidence is growing that an important source for the iron is deposition from the atmosphere, for example, as a constituent of terrestrial dust from Asia (*12*). Measurements of soluble Fe and Al in surface waters indicate higher concentrations at shallower depths than below (*13,14*), as if atmospheric deposition of these metals in an already water soluble form was the source, followed by removal from solution in the water column. It has been reported that Fe and Al in air filter samples over the Pacific are, to a great degree, water soluble, whereas in dust sampled nearer Asia, these metals are soluble only to a limited degree (*15*). This finding supports the concept of particle dissolution during aerosol aging by gradual reaction with sulfuric acid derived from both air pollution SO_2 and natural DMS and other gaseous sulfur precursors. Fe may be a nutrient for phytoplankton, but Al is toxic to some bacteria and fish (*16*) and, perhaps, other biota. Future research will show whether the net effect of atmospheric deposition of soluble mineral elements will be to enhance or depress primary production, either way a result possibly of global importance.

In the Arctic, it has been observed that surface ozone concentrations tend to vary inversely with the concentration of bromide sampled by filters (*17,18,19*). It is hypothesized that bromoform, released from marine organisms to the atmosphere, reacts to destroy ozone and form bromide. Model calculations show that measured concentrations of bromoform in the marine troposphere are sufficient to depress ozone to a measurable degree (*20*). Bromoform has been measured in the atmosphere over a range of low to high latitudes (*21,22,23*), and especially high concentrations have been found in the equatorial Pacific (*24*). Although the organisms that produce the bromoform have not yet been identified, it is known that some species concentrate bromine as organic compounds to a great extent (*25*). Thus, atmospheric chemists believe that bromoform must have a biogenic source not unlike the sources of DMS and other reactive gases from the marine biosphere. The strengths of these marine sources could vary in accordance with primary production and biological activity in general.

To what degree bromoform or any other biogenic gas affects ozone concentrations worldwide has not yet been quantitatively evaluated. In view of the complexity of the chemistry of atmospheric ozone and the presence of a number of pathways for its formation and destruction, it may be bold to suggest here yet another pathway for removal of ozone from the troposphere. However, patterns of tropospheric ozone deduced from satellite data show equatorial concentrations only about one-third those in the mid latitude North Pacific (*26*), a difference that mirrors the pattern of chlorophyll being high in the tropics and low in the mid-ocean gyre regions (*27,28*). Seasonal changes in ozone seem to track those of chlorophyll in the North Pacific (*29*).

Thus, the possible link between marine biological activity and ozone in the troposphere has become an object of scientific attention. If the link, in fact, turns out to be strong, it raises the question of whether the deposition of solubilized mineral Al, Fe, and other metals onto the sea surface, through an impact on the growth of plankton, can affect the release of biogenic trace gases and, thus, ozone concentrations in the troposphere. It will be important to answer this question since the release of SO_2 from fuel combustion in areas with much dust, such as in heavily populated eastern Asia, could increase the solubilization of mineral dust and, thus, the flux of metals to the ocean in biologically-available form. Depending on whether the net effect is to decrease or increase marine life, the impact on the atmosphere could be to raise or lower ozone levels.

CASE 2: AEROSOL COMPOSITION OVER THE EASTERN US

Resolution and Chemical Significance of Principal Components of Aerosols in a Populated Continental Region

The objective of the regional scale study was to examine the association of aerosol sulfur with mineral elements within a region impacted by large amounts of SO_2 from combustion of fossil fuels. Special attention was given to Al, owing to its acknowledged importance in aquatic ecosystems. A full account of the study is presented elsewhere (*30*). Aerosols were sampled at nine classes I sites of the SURE network (Table 3) for the month of October 1977, resulting in 3600 two-hourly time steps analyzed by PIXE for 10 elements. By absolute principal component analysis, 84–91% of the variance at each site could be explained by three or four factors. Using multiple linear regression of element concentrations vs factor scores, the composition of each component could be calculated. One component was mainly sulfur that also contained Al and other soil mineral metals, and the other components were one or two apparently from soil dust and one from automotive Pb and Br emissions.

The first of these components is interpreted to represent sulfuric acid taken up by airborne alkaline mineral dust in cloud or clear air processes. Al, Fe, Si, and K are present in the component, but Ca is generally absent or negatively correlated, suggesting separate atmospheric interactions of SO_2 or other acidic sulfur species with clay and carbonate mineral particles. At all nine sites, as indicated in Table 3, the chemical equivalents of S exceed those of Al + Fe + Si, indicating excess sulfuric acid over the particle alkaline capacity, so that at ambient humidities, the particles could have been rendered completely soluble by strongly acidic surface coatings. Subsequent wet or dry deposition would bring these metals to the surface in biologically-available form.

At two sites in Ohio and Indiana, the Al concentrations in the S-rich component are higher than elsewhere, areas where especially high dust fluxes by wind erosion of soils have been estimated (*31*). This suggests that dust transported upward to heights where interaction with SO_2 may be favored, by either cloud or clear air processes, becomes internally mixed with H_2SO_4 coating of particles. At two sites in Pennsylvania and Massachusetts, the Al concentrations in the S-rich component are high enough to account for much of the soluble Al reported to be present in Adirondack lakes. In fact, their equivalent ratios Al/S = 0.043 ± 0.006 and 0.038 ± 0.010 are not far from the geometric mean of 0.072(SD × 3.1) of 21 lakes monitored for 3 years in the early 1980s (*32*). Deposition of both sulfate and soluble Al from the atmosphere could account for this similarity. At all the sites, the atmospheric deposition fluxes of metals, after solubilization by reaction of airborne soil mineral particles with sulfuric acid, may be an important input to aquatic and, perhaps, terrestrial ecosystems, a possibility examined further in the next section.

Biological Implications: Soluble Al Aerosol Deposition May Affect Aquatic and Terrestrial Ecosystems

Al is prominent in the literature on the effects of acid deposition on ecosystems (*33,34*) since soluble Al in lakes and soils can be toxic to their organisms, including bacteria and fish (*16*), although the biochemical literature rates Al low among trace metals in biological importance (*35*). Water-soluble Al is reported to occur in fresh water of eastern North America at concentrations higher than considered to be natural (*32*) and is attributed to acid deposition processes, but by mechanisms that are still unclear. The cause of high aquatic concentrations of Al generally

Table 3
Statistical Association of Sulfur and Aluminum in the Eastern US Composition of Aerosol Sulfur Component by Factor Analysis of Element Concentration Data at 9 SURE Sites, October 1977

		Data & Factor Analysis Characteristics				Sulfur Factor Composition			
						Conc., ng m^{-3}		Equivalent Ratio	
SURE Site		n+1	#F	%Var	$\Sigma r^2(S)$	S	Al	Al/S±SD	(Al+Si+Fe)/S±SD
Duncan Falls	OH (4)	414	4	87.9	1.242	2136.	389.	0.324±0.012	0.911±0.071
Fort Wayne	IN (7)	412	3	84.0	1.517	2682.	364.	0.242±0.010	0.771±0.024
Rockport	IN (5)	413	3	91.0	1.101	3489.	191.	0.097±0.010	0.186±0.030
Lewisburg	WV (9)	437	3	84.1	1.126	2252.	112.	0.089±0.009	0.245±0.017
Giles County	TN (6)	413	3	88.7	1.060	3121.	33.	0.019±0.009	0.054±0.027
Montague	MA (1)	412	4	91.6	1.177	1332.	25.	0.033±0.003	0.296±0.011
Scranton	PA (2)	373	4	90.8	1.103	2158.	52.	0.043±0.006	0.137±0.017
Indian River	DE (3)	408	4	88.9	1.035	1682.	36.	0.038±0.010	0.190±0.023
RTI, RTP	NC (8)	425	4	90.8	1.098	2439.	72.	0.052±0.008	0.402±0.022

[a]Aerosol sampling time, 2 h per analyzed step for 31–36 d. Number of step samples, $372 < n < 436$. Absolute factor analysis used an additional zero sample.

[b]Factors were defined using eight elements—S, Al, Si, K, Ca, Fe, Pb, and Br—and three or four factors (#F) explained 84.0–91.6% of the variance (% Var), with a satisfactory significance of the mainly S factor ($\Sigma r^2(S)$).

[c]Multiple linear regressions used 10 elements, including Cl and P, and led to determining the average concentration of each element in each factor.

[d]The contents of S and Al in the mainly S factor and the ratios, in chemical equivalent units, of Al/S and (Al + Si + Fe)/S with their standard deviations are shown here.

seems to be considered the leaching of soils and sediments by acidified water. However, at commonly measured pH > 4 in rainwater and lakes, the observed micromolar Al concentrations are in excess of those expected by soil mineral dissolution at chemical equilibrium.

In the results of the present study, based on multivariate statistical analysis of data from aerosol samples analyzed by PIXE for elemental concentrations, we find evidence for another possible source of soluble Al in surface waters of air polluted regions: direct deposition from the atmosphere of soil dust aerosol that has already undergone dissolution by strong sulfuric acid attack while airborne. Although oceanographers have suggested dustfall as contributing to Al concentrations measured in surface seawater (*14*), such a process apparently has not been previously reported in the acid deposition literature.

It is now important to obtain independent evidence that soluble aerosol Al, Fe, and other metals occur in polluted areas at concentrations high enough to be biologically significant. In one test (*36*), where filter samples of total suspended particulate matter were extracted and analyzed for soluble Al, concentrations were greater in urban Denver than at a nonurban site 50 km distant. The urban concentrations approximated those of Al in S-rich principal components at the eastern US SURE sites. This test should be followed by others to verify the inferences made by the statistical analysis of PIXE data.

POTENTIAL HEALTH EFFECTS MAY RESULT FROM INHALATION OF SOLUBILIZED MINERAL AEROSOL

Most attention given to the possible consequences of acid air pollution has been to effects on ecosystems and materials. But direct health effects also deserve consideration. Now that we have evidence that Al, Fe, and other mineral constituents of aerosois may be rendered soluble by acid reactions in the atmosphere, the deposition of soluble metals in the respiratory tract potentially may endanger human health. As reported elsewhere (*37*), a possible link of soluble Al and other metals to chronic obstructive pulmonary disease (COPD) can be argued.

Age-adjusted 1979–1981 mortality rates for COPD as the principal cause of death of males and females in 305 cities (standard metropolitan statistical areas) are well correlated with each other and, for white males, range for about 15–50 deaths per year per 100,000 population. White female rates are about 1/3 as high. Figure 2 shows the areas of high, intermediate, and low white male rates as contours of 305 data points by a kriging procedure using IBM PC software (*38*). The rates are highest in two broad regions of the US. One is in the southeast, with age-adjusted rates high in Georgia and northern Florida, but decreasing toward southern Florida. The other is the western plains, with rates high in Colorado

A

B

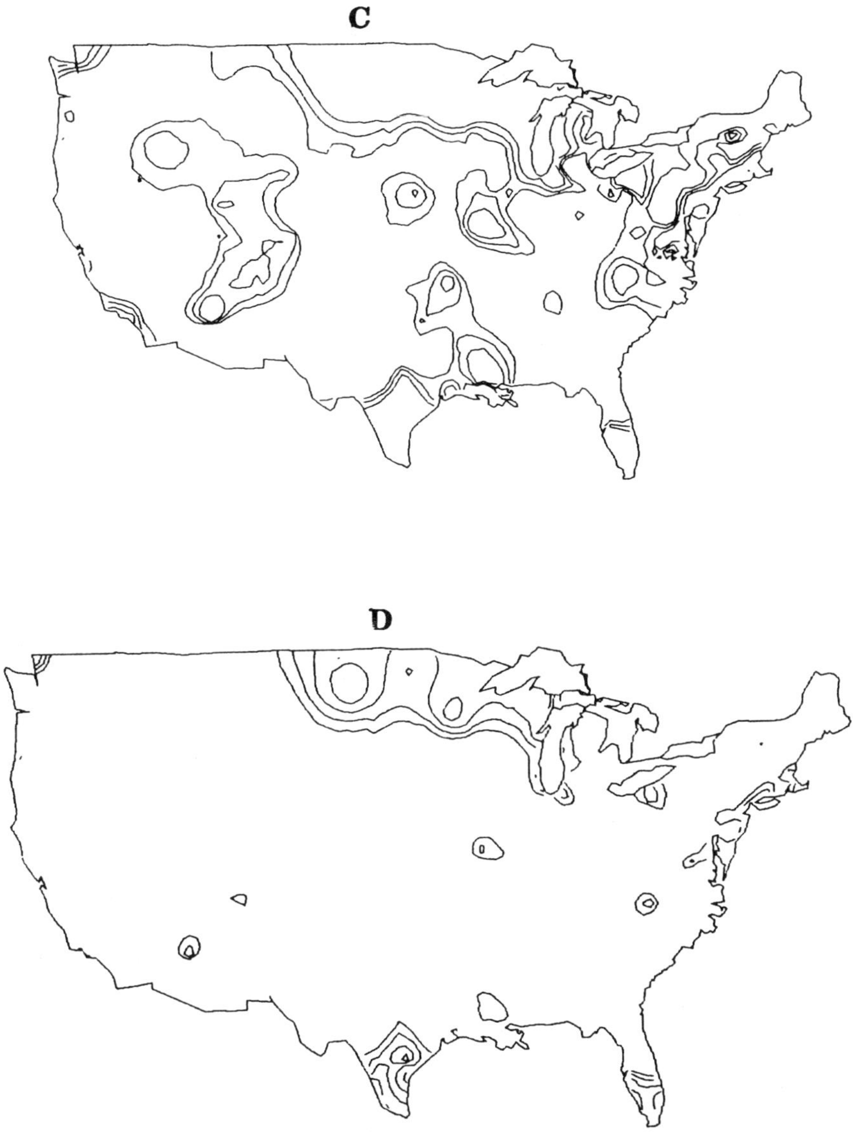

Fig. 2. Pattern of 1979–1981 white male age-adjusted mortality attributed to chronic obstructive pulmonary disease (COPD) as principal cause of death in urban areas, classified into 305 standard metropolitan statistical areas (SMSA). Data were obtained from the US Public Health Service and summarized by Winchester. The patterns are presented as contours by a kriging procedure calculated by Bryan Paul at NCAR. (**A**) Locations of SMSA in 48 states; (**B**) Regions with cities having the highest third; (**C**) middle third; and (**D**) lowest third of mortality rates, where for ease in visualization, the contours of the urban data over the 48 states were subdivided into thirds. Thus, in (**B**) all areas covered by contours, i.e., not blank, lie in the highest third of urban COPD mortality rates, in (**C**) the second third, and in (**D**) the lowest third of urban US rates.

and northern Texas, but decreasing toward southern Texas. Rates are generally low in the northeast, upper midwest, and far west, as well as in the largest cities of these regions. These geographic patterns suggest that atmospheric environmental conditions may contribute, along with other factors, to the risk of COPD.

Based on measured aerosol characteristics and atmospheric chemical reasoning, it should be pointed out that ambient air in cities of the high COPD regions exposes people to fine particles that may contain high concentrations of the reaction products of acid air pollutants. These may be especially irritating to the respiratory tract. In the southeast, sulfuric acid aerosol is abundant, apparently because of a sunny warm humid climate that favors rapid oxidation of sulfur dioxide, as well as the region's proximity to large primary air pollution sources further north. Particulate sulfur is also associated with soil mineral constituents. In the western plains, concentrations of alkaline dust are high because of soil erosion during windy dry conditions. Acid air pollutants in western cities can be efficiently scavenged by mineral particle surfaces and form chemically reactive coatings that lead to solubilized mineral Al and other metals. These may be inhaled and deposited in the respiratory tract and contribute to COPD mortality risk. To test this possibility, rigorously appropriate studies should be carried out jointly between environmental chemists and medical experts. Such collaboration can achieve a more thorough understanding than studies conducted by either scientific discipline alone.

CONCLUDING REMARKS

In this study, we have combined pattern recognition and chemical reasoning to infer the chemical nature of aerosol particles, with specific reference to their potential biological effects. Statistical relationships by themselves do not prove cause and effect, nor does any observation of attributes prove the essence of things, a philosophical issue pondered since ancient times. But they provide clues and aid understanding. In the present study, the combination of PIXE analysis, where large data sets can be acquired, the principal component analysis of the data has provided insights into likely chemical forms of aerosol constituents. These insights may now be subjected to direct test in suitable experiments. If verified, the approach taken here may be given greater confidence as a means of assessing biological exposures to important components of airborne particulate matter.

ACKNOWLEDGMENT

Stimulation for exploring chemical measurement data by multivariate statistical analysis methods was provided by Paulo Artaxo during an extended visit in 1985 to the University of Sao Paulo, Brazil, and Lloyd

Currie during the author's sabbatical at the National Bureau of Standards in 1987. The present study was continued during a sabbatical at the NOAA GMCC and completed during a special assignment at the National Center for Atmospheric Research. Permission by Florida State University for these leaves of absence and the generous support by the USP, NBS, NOAA and NCAR are gratefully acknowledged.

REFERENCES

1. I. N. Tang, H. R. Mukelwitz, and J. G. Davis, *J. Aerosol Sci.* **9,** 505 (1978).
2. R. J. Ferek, A. L. Lazrus, P. L. Haagenson, and J. W. Winchester, *Environ. Sci. Technol.* **17,** 315 (1983).
3. Fan Song-Miao, *Acidity of Atmospheric Aerosols in North Florida,* M. S. Thesis, Florida State University, Tallahassee, FL (1985).
4. J. W. Nelson in *X-Ray Fluorescence Analysis of Environmental Samples,* T. G. Dzubay, editor, Ann Arbor Science Publishers, Ann Arbor, MI (1977).
5. T. T. Johansson, R. E. Van Grieken, J. W. Nelson, and J. W. Winchester, *Anal. Chem.* **47,** 855 (1975).
6. G. D. Thurston, and J. D. Spengler, *Atmos. Environ.* **19,** 9 (1985).
7. Statgraphics, Statistical Graphics System by Statistical Graphics Corporation, User's Guide, Version 2.6, STSC, Inc., Rockville, MD (1987).
8. J. W. Winchester, and M. X. Wang, *Tellus* **41B,** 323 (1989).
9. R. J. Charlson, J. E. Lovelock, M. O. Andreae, and S. G. Warren, *Nature* **326,** 655 (1987).
10. B. C. Nguyen, B. Bonsang, and A. Gaudry, *J. Geophys. Res.* **88,** 10,903 (1983).
11. J. H. Martin, and R. M. Gordon, *Deep-Sea Research* **35,** 155 (1987).
12. H. B. Maring, and R. A. Duce, *Earth Planet. Sci. Lett.* **84,** 381 (1987).
13. W. M. Landing, and K. W. Bruland, *Geochim. Cosmochim. Acta* **51,** 29–43 (1987).
14. K. J. Orians, and K. W. Bruland, *Earth Planet, Sci. Lett.* **78,** 397 (1986).
15. S. Kanamori, N. Kanamori, M. Nishikawa, and T. Mizoguchi, Paper 11-1, The 2nd International Conference on Atmospheric Sciences and Applications to Air Quality, Tokyo, October (1988).
16. E. I. Ochiai, *General Principles of Biochemistry of the Elements,* Plenum, New York, NY (1987).
17. L. A. Barrie, J. W. Bottenheim, R. C. Schnell, P. J. Crutzen, and R. A. Rasmussen, *Nature* **334,** 138 (1988).
18. S. J. Oltmans, R. C. Schnell, P. J. Sheridan, R. E. Peterson, S. M. Li, J. W. Winchester, P. P. Tans, W. T. Sturges, J. D. Kahl, and L. A. Barrie, *Atmos. Environ.* **23,** 2431 (1989).
19. S. M. Li, J. W. Winchester, J. D. Kahl, S. J. Oltmans, R. C. Schnell, and P. J. Sheridan, *J. Geophys. Res.* **95,** (1989) (in press).
20. N. D. Sze, Personal communication from Atmospheric and Environmental Research, Inc., Cambridge, MA (1989).
21. W. W. Berg, L. E. Heidt, W. Pollock, P. D. Sperry, R. J. Cicerone, and E. S. Gladney, *Geophys. Res. Lett.,* **11,** 429 (1984).
22. R. J. Cicerone, L. E. Heidt, and W. H. Pollock, *J. Geophys. Res.* **93,** 3745 (1988).
23. R. J. Cicerone, *Rev. Geophys.* **19,** (1981).
24. L. E. Heidt, personal communication from the National Center for Atmospheric Research (1989).

25. D. W. Carey, and J. W. Farrington, *EOS* **69(44),** 1119 (1988).
26. J. Fishman, personal communication from NASA Langley Research Center (1989).
27. J. H. Ryther, *Science* **166,** 72 (1969).
28. J. B. Garver, The World and Pastures of the Sea map supplement, *National Geographic* **174,** 910, (1988).
29. W. Esaias, personal communication from NASA Goddard Research Center (1989).
30. J. W. Winchester, in *Acidic Precipitation, Vol. 3, Sources, Deposition, and Canopy Interactions,* (Advances in Environmental Science), S. E. Lindberg, A. L. Page, and S. A. Norton, editors, Springer-Verlag, New York, NY (1990).
31. D. A. Gillette, and K. J. Hanson, *J. Geophys. Res.* **94,** 2197 (1989).
32. C. T. Driscoll, and R. M. Newton, *Environ. Sci. Technol.* **19,** 1018 (1985).
33. D. W. Schindler, *Science* **239,** 149 (1988).
34. *The Environmental Chemistry of Aluminum,* G. Sposito, editor, CRC Press, Boca Raton, FL (1989).
35. E. Frieden, in *Biochemistry of the Essential Ultratrace Elements,* E. Frieden, editor, Plenum, New York, NY (1984).
36. J. W. Winchester, and G. L. Kok, paper ENVR-6, American Chemical Society 198th National Meeting, Miami, September (1989).
37. J. W. Winchester, *Arch. Environ. Contam. Toxicol* **18,** 291 (1989).
38. Golden Software Inc., Surfer Version 3.00 Manual, Golden, CO 80402 (1987).
39. National Center for Health Statistics, Mortality Statistics Branch, Public Health Service, U.S. Dept. of Health & Human Services, Hyattsville, MD 20782, *Average age-adjusted death rates for chronic obstructive pulmonary diseases and allied conditions 490-496 1979–81,* Table 803, pages 1313–1340, dated 85/08/09, for 305 Standard Metropolitan Statistical Areas (unpublished document, 1985).

Major and Trace Elements in Spruce Needles by NAA

ARMIN WYTTENBACH,* SIXTO BAJO, AND LEONHARD TOBLER

Paul Scherrer Institute, CH-5232, Villigen PSI, Switzerland

Received April 17, 1989; Accepted October 19, 1989

ABSTRACT

Concentrations of 23 elements in needles of Norway spruce (*P. abies*) have been determined at 47 sites. It is shown that a thorough removal of the aerosols sitting on the needles surface is necessary in order to get the inherent needle concentrations. Neutron activation was used to determine concentrations from 10^{-9} to 10^{-2} g/g. Irradiation and counting conditions are given. The essential elements, Ca, Cl, Cu, Fe, K, Mg, Mn, P, Zn, and the nonessential elements, Al, As, Ba, Br, Co, Cs, Hg, La, Na, Rb, Sb, Sc, Sr, and V, could be determined. The concentrations of most elements are about a factor of 6 smaller than the mean concentrations in land plants. Analytical reproducibility was much better than the variation among individual trees, and the variation within sites is smaller than among sites. In general, essential elements have smaller variations than nonessential elements. For some elements, variations between sites are owing to differences in the soil pH or the emission situation.

Index Entries: *P. abies*, elemental concentrations in needles; neutron activation analysis; surface contamination on needles, removal; variation of concentrations, intra- and intersite; distribution functions of concentrations.

Abbreviations: NAA, neutron activation analysis; CV, coefficient of variation; DF, degrees of freedom; LSD, least significant difference.

*Author to whom all correspondence and reprint requests should be addressed.

INTRODUCTION

Elemental contents in plants are often determined with one of the following scopes: assessment of the nutritional status, detection of deficiency of essential elements or excess of toxic elements, detection of effects induced by immission in order to achieve early recognition of latent or already established pollution damage (biomonitoring), and investigation of the influence of soil parameters, such as pH, metal content, or geology on the plant (biogeochemistry).

In order to get valid results, at least the following criteria must be met: the sample has to be free of surface contaminants; the analytical technique must be reliable and sensitive enough to determine traces, and it should preferably be a multielement method; and the sample analyzed must be representative for the site investigated, and its error must be known.

The present work was undertaken in order to establish the necessary parameters for the analysis of needles of Norway spruce and the discussion of the obtained results. Interest in Norway spruce (*P. abies*) has grown over the past few years since this tree has, in many places, shown damage that might be connected to environmental influences and to its nutritional status (*1*). Unfortunately, needles of Norway spruce are difficult objects to work with. This arises from the following facts: spruce needles are covered by a natural wax that tenaciously retains surface contaminants. The inherent elemental content of spruce needles is much smaller than the values generally assumed for land plants (*2*), and this places some restraint on suitable analytical methods. Lastly, Norway spruce retains up to seven successive years of needles that differ among each other and that furthermore show marked seasonal variations (*3*).

MATERIALS AND METHODS

Sites, Trees, and Needles

The investigated sites are distributed over an area of 150 km^2 in the vicinity of the city of Winterthur, Switzerland. The whole area has a rather uniform geological setting (moraines containing debris of crystalline and sedimentary origin). The mean distance between sites is about 2 km. A total of 47 sites were sampled, mostly in the center or at the edge of forests in moderately polluted rural regions, but some sites were chosen on the border of a motorway.

A site was considered to be an area of about 100 m^2. At each site, twigs from four trees standing a few meters apart were sampled. At five sites, these samples were treated individually, whereas at the remaining 42 sites, the twigs from the four trees were combined into one mixed sample.

The trees sampled had an age of 20–40 yr. Sampling was done at the beginning of April before sprouting, taking only terminal twigs. Therefore, the needles had an age just short of 1 yr.

Sample Preparation

The individual twigs of one sample were immersed into a mixture of toluene and tetrahydrofuran (1:1) and mechanically shaken. This treatment dissolves the natural wax on the surface of the needle and efficiently removes any surface contamination, but it does not leach elements from the interior of the needles. The material removed from the surface of the needles was recovered by taking the solvent to dryness and analyzed separately (*4*).

The washed twigs were then kept at 65°C. After 4 d, the needles were separated from the twig axis and ground with a commercial coffee grinder. Blanks were run by substituting sugar for the needles, and they revealed no detectable traces originating from grinding. The resulting needle powder was directly used for analysis. Heating aliquots to 85°C for 24 h reduced their weight by 5.7%. All quoted concentrations refer to the weight at 85°C.

Neutron Activation Analysis

A total of 23 elements were determined by NAA using irradiation positions in the SAPHIR research reactor. From these elements, 20 were measured directly by γ-spectroscopy, Cu and As were radiochemically separated before γ-spectroscopy, and P was determined by measuring its bremsstrahlung.

Elements Determined Directly by γ-Spectroscopy

Irradiation and counting conditions for these elements are given in Table 1. The complete analysis uses three irradiations and four measurements. The total sample mass is about 2 g, and the total counting time about 14 h. Adequate standards were prepared by pipeting calibrated solutions on charcoal. The analysis is straightforward, but the following points should be noted

1. Counting conditions can usually be kept constant for the individual samples since most elements do not show large variations (Table 2). Exceptions are Cl, Co, Cs, and Rb (max/min = 10–100) and Br, Mn, and Na (max/min = 150–600). Problems arise only from high concentrations of Mn and Na since these necessitate a longer decay time before counts B and C.
2. Interfering nuclear reactions are important only with the determination of Al. For the mean concentrations as found in the present work (Al 15, P 1450, Si 1420 μg/g) and the irradiation conditions used, $^{31}P(n,\alpha)^{28}Al$ contributes 4% and $^{28}Si(n,p)^{28}Al$ 13% to the registered ^{28}Al activity.

Table 1
Irradiation and Counting Conditions for INAA

#	irradiation (a) time	sample mass mg	#	counting (b) decay time	counting time	measured nuclides
1	20 s	270	A	3 min	8 min	Al-28, Ca-49, Cl-38, Mg-27, V-52
2	30 min	270	B	10 h	17 min	Mn-56, K-42, Na-24
			C	3 d	3 h	Br-82, K-42, Na-24
3	2 h	1200	D	12 d	10 h	Ba-131, Co-60, Cs-134, Fe-59, La-140, Rb-86, Sb-122,124, Sc-46, Sr-85, Zn-65

(a)Irradiations at 2.2*10**13 n/cm**2, s (SAPHIR reactor).
(b)Counting on Ge(Li)-detector (0.015 counts/γ at 662 keV).

P was determined by measuring the bremsstrahlung of ^{32}P in the samples from irradiation #3 (Table 1). Counting was done for 1 min in a well-type NaI-detector between 12 and 30 d after irradiation (*7*).

As and Cu can only be determined after their radiochemical separation. Five hundred milligrams of the needle powder were irradiated for 15 min, cooled for 1 d, and then destroyed by heating with $Mg(NO_3)_2$ and HNO_3 (raising the temperature from 20 to 550°C within 3 h). The residue was taken up in a mixture of HCl, KI, and ascorbic acid, and Cu and As were extracted with Zn-diethyldithiocarbamate in $CHCl_3$ (*5,6*). The extract was measured for 4 h in a well-type Ge(Li) detector.

RESULTS AND DISCUSSION

All results are summarized in Tables 2 and 3 and will be discussed in the following sections.

Analysis of Washed Needles vs Unwashed Needles

In working with conifer needles, it is often overlooked that surface contamination by adhering aerosols is difficult to remove and can invalidate the results. This is shown by the results in Table 2, col. 7, which

Table 2
Results of the Analysis of 1-Yr-old Needles of Norway Spruce from 47 Sites in Switzerland

(1)	(2) units	(3) mean	(4) range min	max	(5) max/min	(6) LSD (0.05)	(7) fraction in the aerosol %
Al	μg/g	15	7	35	4.8	6	62
As	ng/g	15	9	23	2.5	6	62
Ba	μg/g	32	19	64	3.3	17	0.8
Br(a)	ng/g	407	119	711	6.0	154	35
Br(b)	ng/g	685*	119	18600	156		30
Ca	μg/g	5623	3019	9191	3.0	1700	0.3
Cl(a)	μg/g	420	204	600	3.0	160	1.1
Cl(b)	μg/g	544*	204	3392	17		1.3
Co	ng/g	20*	4	149	35		32
Cs	ng/g	12*	4	120	32		22
Cu	μg/g	3	2	3	1.7	0.6	n.a.
Fe	μg/g	26	17	40	2.3	6	45
Hg	ng/g	31	18	64	3.5	14	38
K	μg/g	5298	2754	8155	3.0	1360	0.2
La	ng/g	13	6	21	3.5	8	57
Mg	μg/g	1042	698	1460	2.1	205	0.8
Mn	μg/g	74*	4	1445	321		0.7
Na(a)	μg/g	4	2	5	2.3	1.6	46
Na(b)	μg/g	18*	2	1356	528		30
P	μg/g	1449	819	2120	2.6	219	n.a.
Rb	ng/g	1920*	240	17000	71		1.9
Sb	ng/g	19	5	34	6.8	9	70
Sc	ng/g	2	1	4	4.1	0.9	73
Sr	μg/g	41	10	90	9.0	22	n.a.
V	ng/g	78	40	113	2.8	n.a.	62
Zn	μg/g	34	16	58	3.7	9	1.8

(1) Br, Cl, and Na show large influences of environmental contamination. (a) denotes the relatively clean sites only according to the following criteria: Br(a): <800 ng/g (n = 33), Cl(a): <650 μg/g (n = 32), and Na(a): interior of forests only (n = 14); (b) denotes the whole population (n = 47).

(2) These units apply to columns 3, 4, and 6.

(3) Arithmetic means for elements with max/min <10, geometric means for other elements. Geometric means are marked by *.

(6) Least significant difference at the 0.05 level between individual sites. LSD = $t \cdot (2 \cdot s^2/n)^{1/2}$, with n = 4 and s^2 = pooled error variance. s^2 is calculated from the data in Table 3, col. 3 and has DF = 15.

(7) The value given is 100% · b/(a + b), with a = mass inherent in the needle (true concentration), b = mass owing to the aerosol on the surface of the needle. (a + b) is the apparent concentration of unwashed needles. b was not measured for Cu, P, and Sr, and the fraction in the aerosol is not available (n.a.).

Table 3
Statistics for the Concentrations of Elements in Needles of Norway Spruce

(1)	(2) analyt. reprodu-cibility	(3) variation among trees	(4) expected error of a composite sample (4 trees)	(5) variation among sites
	%	%	%	%
Al	3.1	25	13	45
As	5.5	26	13	18
Ba	1.0	36	18	29
Br(a)	1.9	25	13	39
Ca	1.4	20	10	25
Cl(a)	2.7	25	13	24
Cu	3.2	13	7	12
Fe	1.6	16	8	22
Hg	5.0	29	15	30
K	2.5	17	8	22
La	3.5	40	20	34
Mg	2.2	13	7	18
Na(a)	5.0	27	14	25
P	1.0	10	5	23
Sb	3.1	34	17	38
Sc	3.8	30	15	40
Sr	5.0	35	18	38
V	2.1	n.a	n.a	30
Zn	1.0	17	9	27
	%	factor	factor	factor
Co	10	1.59	1.26	2.59
Cs	10	1.52	1.23	2.10
Mn	2.7	1.50	1.22	4.16
Rb	4.1	1.69	1.14	2.40
Br(b)	1.9	1.25	1.12	3.13
Cl(b)	2.7	1.25	1.12	1.70
Na(b)	5.0	1.25	1.12	4.80

(1) The upper part of the table contains elements that have a max/min <10; in these cases, normal distributions were assumed. In the other cases (lower part of the Table), lognormal distributions were assumed. (a) and (b) are as in Table 2.

(2) Experimental reproducibility of the analytical determination obtained from 4 to 8 independent analyses.

(3) Variation among individual trees at a given site. Values calculated from five different sites, each with four individual trees ($DF = 20 - 5 = 15$). The values given are relative standard deviations (CV) for normal and geometric standard deviations (GSD) for lognormal distributions.

(4) Estimated error of a composite sample from 4 trees per site, with respect to the mean of all trees from this site. This value was calculated as $CV\sqrt{4}$ or as $(GSD)^{1/2}$.

(5) Experimental variation between the 47 sites investigated. A composite sample from 4 trees was analyzed per site. The values given are CV (upper part of table) or GSD (lower part).

gives the relative distribution of the individual elements between their inherent needle concentration and their surface contamination. Although the effect is negligible for some elements (Ba, Ca, Cl, K, Mg, Mn, Rb, and Zn), it is important for all other elements investigated. Thus, neglecting proper washing will render a meaningful discussion of the concentration of Fe (an essential element) or Al (which is suspected to be toxic to the plant) impossible. It must be stressed that the figures given represent relatively clean sites from the interior of forests. Loading of the needle surface with aerosols was found to be enhanced at the forest edge, for aerodynamic reasons, by a factor of 2.6, with respect to the interior of forests (*8–10*), and therefore makes the problem more complicated. The same is true for sites with increased immission. The composition of the aerosols residing on the surface of needles has been discussed in detail elsewhere (*8*).

Determinable Elements, Sensitivities, Analytical Reproducibilities, and Accuracy

The list of elements that can be determined (Table 2) includes nine elements that are essential to plants, and 14 nonessential elements that occur mostly as traces. Essential elements that can't be determined by the present procedure are B, Mo, N, and S. Most elements are present in spruce needles in concentrations that are about six times smaller than the generally assumed values for land plants (*2*). Exceptions are Ba and Sr, which occur in about the expected concentrations.

The sensitivity of the analysis was found to be adequate for all occurring concentrations, with the exception of Cs (8 samples out of 62 below the detection limit of 4 ng/g) and Sr (7 samples below 10 μg/g). Mn concentrations >100 μg/g make the determination of V, which is present in the 50–100 ng/g range, impossible.

The analytical reproducibility was determined by analyzing 4–8 aliquots of one sample. The results are given in Table 3, col. 2. Fourteen elements have CV ≤ 3.5%, 21 elements ≤5%, and two elements (Co, Cs) 10%. These reproducibilities are negligibly small compared with the other variations. Accuracy was checked by analyzing NBS SRM 1571 (orchard leaves); no significant deviations from certified or literature values exceeding 10% were found.

Distribution Functions

The values of all 47 sites were checked for their distribution function visually (plots of cumulative probability distribution) and numerically (test of skewness and kurtosis). The results show that most elements can be approximated reasonably well by normal distribution functions. Exceptions are Co, Cs, Mn, and Rb, which are lognormally distributed. Al can be described by both functions about equally well. The elements Na,

Cl, and Br show increased values at sites with increased immission (*11*); although they are normally distributed at clean sites, lognormal distributions are more appropriate for the whole population. The corresponding distribution functions were used for the calculation of means and variations with the individual elements.

Variations Within and Among Sites

The variation among trees within a given site was evaluated from the results of the five sites, at which four individual trees each were analyzed. The pooled CV have 15 DF and are given in Table 3, col. 3. The smallest values (CV from 10 to 20%) are shown in order by the essential elements P, Mg, Cu, Fe, K, Zn, and Ca. The other elements have CV from 20 to 40%. These variations are much larger than the analytical reproducibilities and, therefore, are real.

The variation among trees can be used to estimate how well the mean value of all trees of a given site is represented by the analysis of just four trees per site, either separately or as a mixed sample. If the site is composed of only the four trees analyzed, the result will be exact. On the other hand, if the site is composed of more than the four trees analyzed, the result will deviate from the true mean. This deviation is approximated by dividing the variation among trees by $\sqrt{n}$ (with n = number of sampled trees per site = 4) and is called the expected error of a composite sample. Its value is given in Table 3, col. 4.

The variation among all 47 sites is given in Table 3, col. 5. The smallest variation among sites is shown by Cu (CV = 12%) and the largest by Mn and Na (geometric standard deviation ~4). Again, the essential elements—with the exception of Mn—have the smallest variations (CV from 12 to 27%, corresponding to max/min ~2.5). The largest variations are shown by Cl, Br, and Na (when sites showing influences of immission are included) and by Cs, Rb, Co, and Mn with a max/min from 30 to 300. It has been shown that the concentration of Cs, Rb, Co, and Mn are dependent on the soil pH (*12*), and the large variations found in this work partially reflect the large range of soil pH (from 3.2 to 7.8) between the different sites.

Comparison of the variation among sites, with the expected error of a composite sample, shows that the latter is always smaller. This means that real differences exist between the individual sites within the region investigated. This is also shown by the values of LSD for sites (Table 2, col. 6), which for most elements is at least 3 times smaller than the range of all sites. As mentioned before, these differences can be attributed for some elements to differences in soil pH or in the immission situation, but for most elements the reasons are not clear at the present state of the work. It also remains to be seen if there are significant differences between regions with different geological settings.

ACKNOWLEDGMENTS

This study was done at PSI. It was supported by the Swiss National Foundation (grant 4.831) and the Swiss Federal Institute for Forestry Research. We want to thank V. Furrer and R. Brütsch for their help with the collection and the analysis of the samples and R. Lorenzen for the manuscript.

REFERENCES

1. V. A. Mohnen, *Sci. Am.* **259,** 14 (1988).
2. B. Mason, C. B. Moore, *Principles of Geochemistry,* Wiley, New York, 1982, p. 258.
3. A. Wyttenbach, and L. Tobler, *Trees* **2,** 52 (1988).
4. A. Wyttenbach, S. Bajo, and L. Tobler, *J. Radioanal. Nucl. Chem. Articles* **114,** 137 (1987).
5. S. Bajo, and A. Wyttenbach, *Anal. Chem.* **48,** 902 (1976).
6. S. Bajo and A. Wyttenbach, *Extraction liquide-liquide de As, Sb, Se et Te par le diéthyldithiocarbamate de Zn,* #EIR 336, Würenlingen, Switzerland, 1978.
7. S. Bajo, and A. Wyttenbach, *Talanta* **35,** 747 (1988).
8. A. Wyttenbach, L. Tobler, and S. Bajo, *J. Aerosol Sci.* **18,** 609 (1987).
9. B. Hasselrot and P. Grennfelt, *Water, Air, Soil Poll.* **34,** 135 (1987).
10. G. P. J. Draaijers, W. P. M. Ivens, and W. Bleuten, *Water, Air, Soil Poll.* **42,** 129 (1988).
11. A. Wyttenbach, L. Tobler, and S. Bajo, *Toxicol. Environ. Chem.* **19,** 25 (1989).
12. A. Wyttenbach, L. Tobler, S. Bajo, and H. Conradin, in preparation.

Transfer of Toxic Elements from the Atmosphere to Cattle and Game Animals

C. RONNEAU,* J. PATIGNY-EVERS, AND J. CARA

Université Catholique de Louvain, Louvain-la-Neuve, Belgium

Received April 17, 1989; Accepted October 19, 1989

ABSTRACT

Neutron activation analysis was used for determining trace element content in hair of herbivorous animals. Pollutant deposition from the air onto foliage must be considered as the main cause of contamination of the animals.

Index Entries: Neutron Activation Analysis (NAA); air pollutant deposition; herbivorous animals, domestic and wild; roe deer; red deer; boar.

INTRODUCTION

As a general rule, the quality of the atmosphere at forest and rural sites may be considered as satisfactory, even in heavily industrialized countries such as Belgium. However, problems might appear at the level of grass foliage where deposition (wet and dry) reconcentrates pollutants originally dispersed into hectometers of air layers. The problem is even more acute under trees since they very efficiently accumulate air pollutants during dry periods: these pollutants are then deposited on soil by the washing action of rain (*1,2*). As a consequence, forest soil is a significant sink for air pollutants and especially for potentially toxic heavy elements.

*Author to whom all correspondence and reprint requests should be addressed.

The point is to know whether this reconcentration is a tolerable phenomenon as far as cattle and game herbivorous animals are concerned. In the cases studied by us in a few farms in Belgium, trace element concentrations observed in cows' hair and milk are traced back to elemental deposition from air onto the pastures.

The situation is less clear-cut in the case of game animals, where observations are still in progress. However, significant differences in contamination are observable between different sites, depending on their remoteness from industrial plants; differences are also observable between different animal species.

EXPERIMENTAL

The experimental procedure used in this study has been thoroughly described in previous papers. (*3,4*). Hair was selected as the most interesting indicator of internal contamination because of the ease of sampling and it is a way of excretion for toxic elements (*5*).

Cow's hair (5-yr-old animals) was sampled at the end of three grazing seasons at 12 farms in Belgium (3 farms during the third season). These farms were selected on the basis of their distance (close or remote) from industrial and urban zones of the country. As a consequence, their pastures are exposed to more or less intense elemental deposition fluxes, as ascertained by measurements made on deposition samples collected during the grazing periods.

Hair from game animals was sampled during hunting periods. Deposition fluxes at the hunting territories were also determined by means of bulk deposition gages. Determinations were made by atomic absorption spectrometry (AAS) for Pb and Cd and by instrumental neutron activation analysis (INAA) for As, Ba, Br, Ce, Co, Cs, Fe, K, La, Na, Rb, Sb, Sc, Se, Sm, Th, U, and Zn. All these elements were not systematically detected in all hair samples.

RESULTS AND DISCUSSION

Elements in Cows' Hair

Figure 1 shows the example of the correlation between the concentrations of As in the hair of cattle and the fluxes of As on the pastures during the grazing season (each point is a mean of at least 10 determinations made in a single farm). Part of these results have been published already (*6*). More recent unpublished measurements (x on the graph) have confirmed the previous observations.

Significant correlations between elemental concentrations in cows' hair and deposition fluxes on the pastures are observed for As, Ba, Ce, Fe, La, Se, and Sm. A clear distinction should be made between nones-

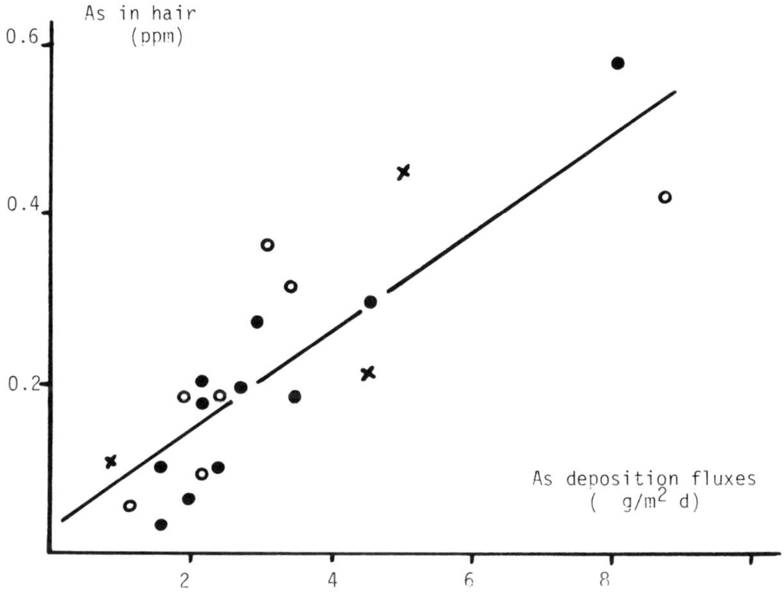

Fig. 1. As in cows' hair as a function of mean deposition fluxes onto the pastures during three grazing seasons (○, 1983; ●, 1984; X, 1987).

sential elements (As for example) and trace essential elements, such as Se and Zn, which are generally more poorly correlated with deposition fluxes because of a significant positive intercept of the regression line, as shown in Fig. 2, in the case of Zn. Even in this case, the influence of deposition is clearly perceptible.

Note that no correlation was found between the elements in cows' hair and in pasture soil (after extraction with ammonium acetate 1*N*): this indicates that deposition from the air directly affects grass foliage, not soil.

Elements in Game Hair

Hair from game animals has been sampled at seven sites, of which three are relatively close to industrial plants. However, because large hunting territories (and big game) are restricted to relatively remote and less polluted zones of the country, it was not possible to systematize the correlations between elemental deposition fluxes and hair concentrations, as was done for cows. Furthermore, it was not possible to take into account the age of the animals because of the relatively limited number of samples.

Hair was collected from: rabbit (*Orictolagus cuniculus*); hare (*Lepus europaeus*); roe deer (*Capreolus capreolus*); red deer (*Cervus elaphus*); and wild boar (*Sus scrofa*).

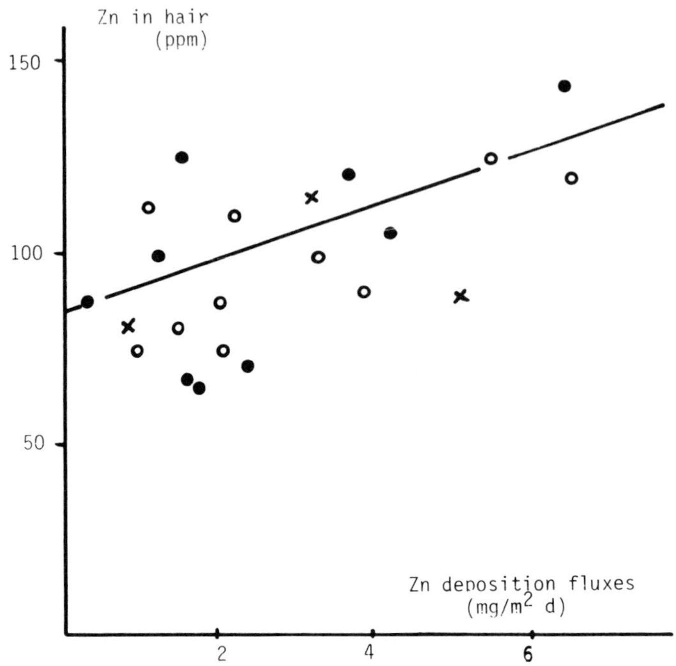

Fig. 2. Zn in cows' hair as a function of mean deposition fluxes onto the pastures.

After analysis, it appeared necessary to distinguish three groups of elements

1. Essential elements: K, Na, Co, Cr, Fe, Se, and Zn that are observed at rather constant concentrations in hair.
2. Terrigenous elements: Au, Ce, La, Sc, Sm, Th, and U that, when detected, do not seem to respond to any systematic behaviour.
3. Potentially toxic elements: As, Cd, Pb, and Sb that are more closely related to deposition fluxes.

Differences Between Territories

Hare was the only animal to be systematically sampled at each hunting site. Trends in elemental hair concentrations are observable as a function of the remoteness from industrial plants: this is most easily perceptible for toxic elements. Figure 3 shows as an example the relation between As in hair and rain water collected at five sites. The correlation is of poor quality, supposedly because, in two cases, rain was collected *after* the hunting period. The trend is nonetheless evident: deposition may be considered as the main cause of hare contamination by toxic elements.

Essential elements are less dependent upon deposition and here again we come to the conclusion made about cows. Terrigenous ele-

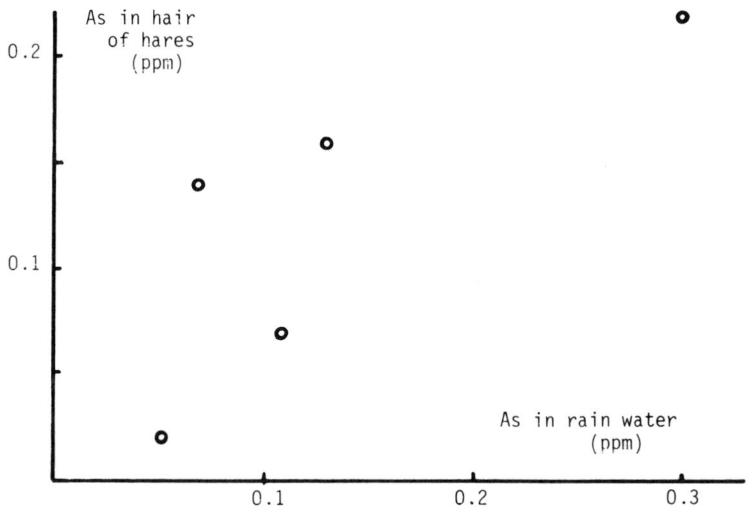

Fig. 3. As in the hair of hares as a function of deposition fluxes onto the hunting sites.

Table 1
Mean Concentrations Measured at Anlier (ppm), with Standard Deviation

Element	Roe deer (12 samples)	Red deer (12 samples)	Boar (12 samples)
Na	62 (31)	87 (55)	180 (132)
K	377 (160)	824 (1210)	298 (200)
Fe	16 (9)	71 (50)	334 (120)
Zn	38 (5)	42 (4)	63 (7)
Se	0,06 (0,05)	0,13 (0,06)	0,30 (0,07)
Co	0,04 (0,01)	0,05 (0,04)	0,13 (0,04)
As	0,02 (0,02)	0,11 (0,10)	0,35 (0,17)
Pb	6,2 (2,7)	6,4 (2,3)	12,9 (1,5)
Cd	0,63 (0,11)	0,68 (0,17)	1,6 (0,53)

ments, on the other hand, show no definite trend from one site to another.

Differences Between Animals

We shall restrict the discussion to hair from big game at a single remote site (Anlier in the Ardenne, 6500 hectares). Table 1 shows the mean concentrations measured for a few significant elements in the hair of roe deer, red deer and boar at Anlier.

As can be seen from the table, the hair of boar is systematically enriched in all the elements (except K). The high potassium levels in deers, as well as its high variability, is attributable to the production of

Table 2
Comparison of Our Own Measurements of Toxic Elements At Anlier (Remote Forest) with the Results of Holm (7) in Low-Saxony (D)[a]

Element		Our determination means (ppm)	Range of concentrations found by Holm
Pb	RoD	6,2	1,13 - 5,86
	ReD	6,4	0,48 - 3,92
	B	12,9	2,86 - 8,42
Cd	RoD	0,63	0,041 - 0,109
	ReD	0,68	0,034 - 0,346
	B	1,6	0,102 - 0,251
As	RoD	0,02	0,031 - 0,051
	Red	0,11	0,09 - 0,208
	B	0,35	0,12 - 0,27

antlers. The omnivorous diet of the boar and its foraging habits explain the contaminations observed for a majority of elements.

CONCLUSIONS

Activation analysis appears to be a very fruitful method to investigate the contamination suffered by domestic and wild herbivorous animals. Furthermore, hair is confirmed to be excellent indicator of this contamination.

We have shown that air pollutant deposition may be considered as the main route of transfer of elemental contamination toward living organisms: this has been evidenced for cows; it remains to be unequivocally demonstrated for game animals.

Although we did not determine elements in internal organs, we are inclined to recommend cautiousness in wild meat consumption (mainly liver and kidney). This is based on comparisons with the observations made by Holm (7) in Low-Saxony (Germany). These comparisons are reported in Table 2.

With the exception of As, the values observed in Belgium are systematically higher. Holm advises against the consumption of liver and kidney from hares whose hair contains more than 0.043 ppm of As: such a level is exceeded in all but one site studied by us (see Fig. 3). We suggest that this kind of measurement be made on a systematic mode at least at exposed sites.

ACKNOWLEDGMENTS

Thanks to the staff of the BR-I, CEN-Mol (Belgium), who performed the irradiations. The study was sponsored by the IISN and the GEBEG (Brussels).

REFERENCES

1. A. Chamberlain, *Atmospheric Environ.* **4,** 57 (1970).
2. B. Coenen, C. Ronneau, and J. Cara, *Water, Air, and Soil Pollution* **36,** 231 (1987).
3. C. Ronneau, M. Detry, J.-Ph. Hallet, and P. Lardinois, *Agric. Ecosyst. and Environ.* **10,** 285 (1983).
4. C. Ronneau, M. Detry, and J. Cara, ESNA XVI Annual Meeting, Warsaw, September 1985.
5. M. Anke and M. Risch, *Haaranalyse und Spurenelement Status,* Fischer Verlag, Jena, East Germany, 1979, pp. 164–171.
6. C. Ronneau and J. Cara, *Sci. Total Environ.* **39,** 135 (1984).
7. J. Holm, *Fleischwirtsch* **63(11),** 1764 (1983); **64(5),** 613 (1984); **64(8),** 970 (1984).

Determination of Trace Elements in Aerosol Samples by Instrumental Neutron Activation Analysis

R. P. Paiva,[1,*] C. J. S. Munita,[1] I. I. L. Cunha,[1] J. Romano,[2] and C. D. Alonso[2]

[1]*Instituto de Pesquisas Energéticas e Nucleares, IPEN-CNEN/SP, Radiochemistry Division, CEP 05508, São Paulo-SP, Brazil; and*
[2]*Companhia de Tecnologia de Saneamento Ambiental, Av. Professor Frederico Hermann Jr., 345 CEP 05459, São Paulo-SP, Brazil*

Received April 17, 1989; Accepted October 19, 1989

ABSTRACT

Two nuclear techniques, Energy-Dispersive X-Ray Fluorescence Analysis (EDXRF) and Instrumental Neutron Activation Analysis (INAA), were used to analyze aerosol samples collected in the city of São Paulo, Brazil. Na, Cl, Mn, V, Al, Sm, Mo, W, La, As, Br, Sb, K, Ba, Se, Th, Cr, Rb, Ca, Fe, Ce, and Sc were determined by INAA, and Al, Si, P, S, Cl, K, Ca, Ti, V, Cr, Mn, Fe, Ni, Cu, Zn, Ga, As, Se, Br, Rb, Sr, Hg, and Pb were determined by EDXRF. A preliminary identification of the main source of the atmospheric aerosol was performed based on enrichment factor and correlation coefficient calculations.

Index Entries: Aerosol; EDXRF; INAA; enrichment factor; correlation coefficient; source identification.

INTRODUCTION

In recent years, air pollution has become a critical problem, with effects on atmospheric properties, human health, soil, water, vegetation,

*Author to whom all correspondence and reprint requests should be addressed.

animals, and ecosystems. The aerosols in urban and industrialized areas are clearly mutagenic and, most probably, traffic and certain industrial sources contribute to this occurrence (*1*).

São Paulo, the largest industrialized city in Latin America, has approximately 12 million people within the city and 14 million if we consider the surrounding zones. There are approximately 150,000 industries of all types and sizes and 4 million vehicles, of which 80% use only gasoline blended with 24% ethanol and the other 20% use pure alcohol fuel.

To protect the environment against pollution, it is necessary to have techniques that not only monitor the air pollution, but also clarify its origin. Nuclear techniques, in particular Instrumental Neutron Activation Analysis (INAA) (*2–4*) and Energy-Dispersive X-Ray Fluorescence Analysis (EDXRF) (*5,6*) are useful and have been widely used to nondestructively analyze the elemental composition of aerosols.

This work is part of the "Co-ordinated Research Programme on Use of Nuclear and Nuclear-Related Techniques in the Study of Environmental Pollution Associated with Solid Wastes" initiated by the IAEA in December 1987, and it is being carried out in cooperation with the CETESB—Companhia de Tecnologia de Saneamento Ambiental-Governmental Institution responsible for environmental control in the state city of São Paulo.

This investigation has the purpose of obtaining more information about the aerosols of the city of São Paulo and to perform, by means of the trace element results obtained by EDXRF and INAA, a preliminary identification of the main sources of the atmospheric aerosol based on enrichment factor (EF) and correlation coefficient (r) results.

EXPERIMENTAL

Air Sampling

The collection of the aerosols was made by CETESB using a Sierra Instruments model 244 Dichotomous Sampler (Virtual Impactor) that fractionates suspended particles into two size fractions, 2.5–15 μm (coarse, noninhalable particles) and less than 2.5 μm (fine, inhalable particles). The two particle fractions were collected uniformly on two 37 mm Teflon membrane filters. The virtual impactor head has a single stage design with a cut point of 2.5 μm and a flow rate of 1 m^3/h. The sampling site was on the western periphery of the city, 2 m above ground level. The time of collection was 24 h.

Analytical Procedures

The filters were analyzed first at CETESB by EDXRF and after that at IPEN-CNEN/SP by INAA.

EDXRF

The sample was analyzed on a TEFA system that has a dual anode Mo/W-X-ray tube. Mo and Cu primary X-ray filters can be used to obtain monochromatic excitation.

The X-ray spectrometer consists of a Si(Li) X-ray detector, amplifier, and pulse-height analyzer that measures the energy and intensity of characteristic X-rays generated in the excitation procedure.

INAA

A nondestructive neutron activation analytical procedure has been established. Samples and standards were irradiated for 5 min using a pneumatic rabbit station of the swimming pool type research reactor, whose flux is about 10^{12}/n/cm^{2}s, for the analysis of elements that give rise to short-lived radionuclides. For the analysis of long-lived radionuclides, samples were irradiated for 24 h on a neutron flux of 10^{13}/n/cm^{2}s.

Standards were prepared by pipeting suitable aliquots of standard solutions, obtained by dissolution of high degree of purity metal or oxide, of elements on pieces of Whatman no. 41 filter paper and drying under an infrared lamp. The measurements were carried out after suitable cooling times with Ge(Li) detector Ortec model 8001-1022V, resolution of 2.6 keV at the 1332 keV gamma peak of ^{60}Co, coupled to a 4096 channel gamma spectrometer Ortec model 6240B.

RESULTS AND DISCUSSION

In order to test the accuracy of the INAA method, about 50 mg of two USGS rock standards (W-1 and AGV-1) and 2 filters of the IAEA reference material (AIR 3/1) were analyzed. Table 1 gives a comparison between the results obtained and the certified values (*8,9*). The relative standard deviations obtained for most elements were smaller than 10%, which is generally considered as a good result in trace analysis. For the elements V in W-1 and As in AIR 3/1, the reproducibility was not so good owing to their low activities for Ge(Li) counting.

The described analytical procedures were applied to the analysis of aerosols collected on 24 filters (fine and coarse particulate). Twenty-two elements could be determined by INAA: Na, Cl, Mn, V, Al, Sm, Mo, W, La, As, Br, Sb, K, Ba, Se, Th, Cr, Rb, Co, Fe, Ce, and Sc, whereas by means of EDXRF, Al, Si, P, S, Cl, K, Ca, Ti, V, Cr, Mn, Fe, Ni, Cu, Zn, Ga, As, Se, Br, Rb, Sr, Hg, and Pb could be determined. The distribution of the results obtained was considered log-normal (*4*).

Geometric means, geometric standard deviation, and median of elemental concentration obtained by EDXRF and INAA and the number of data used for calculation of the means (*n*) are given in Tables 2 and 3, respectively. In the calculation of the means only the number of samples

Table 1
Results Obtained by INAA for Some Certified Reference Materials in ppm, Unless Indicated

ELEMENT	W - 1		AGV - 1		AIR 3/1 (μg/filter)		
	This Work mean ± SD[(a)]	Certificate[(8)] mean ± SD	This Work mean ± SD[(a)]	Certificate[(8)] mean ± SD	This Work mean ± SD[(a)]	Certificate Content	Certificate Confidence limit(0.05)[(9)]
Na (%)	1.55 ± 0.16[(4)]	1.58 ± 0.08	3.10 ± 0.04[(3)]	3.15 ± 0.09	104 ± 8[(2)]	(b)	(b)
Al	7.80 ± 0.53[(4)]	7.93 ± 0.14	8.50 ± 0.09[(2)]	9.07 ± 0.18	92.0 ± 3.7[(2)]	(b)	(b)
Sc	36.2 ± 0.5[(5)]	35 ± 2	11.8 ± 0.09[(3)]	12.1 ± 0.9	(c)	(b)	(b)
V	282 ± 43[(4)]	260 ± 25	140 ± 6[(2)]	123 ± 12	8.42 ± 0.24[(2)]	7.8	5.8 ÷ 8.3
Cr	119 ± 2[(4)]	120 ± 14	(c)	12 ± 3	5.04 ± 0.01[(2)]	5.0	4.5 ÷ 5.1
Mn	1314 ± 38[(4)]	1300 ± 125	733 ± 28[(4)]	740 ± 60	31.76 ± 0.20[(2)]	30	28 ÷ 32
Fe (%)	8.27 ± 0.52[(5)]	7.79 ± 0.16	4.93 ± 0.08[(4)]	4.73 ± 0.13	226 ± 3[(2)]	204	196 ÷ 210
Co	46 ± 2[(4)]	46 ± 4	14.8 ± 0.7[(3)]	15.1 ± 1.2	1.18 ± 0.03[(2)]	1.1	0.8 ÷ 1.4
As	(c)	2.2 ± 0.3	(c)	0.84 ± 0.27	4.98 ± 0.58[(2)]	5.0	4.6 ÷ 5.8
Se (ppb)	(c)	113 ± 7	(c)	(b)	0.93 ± 0.02[(2)]	1.0	0.8 ÷ 1.2
Ba	162 ± 10[(2)]	162 ± 5	(c)	(b)	(c)	(b)	(b)
La	11.9 ± 0.7[(3)]	10.9 ± 1.3	40.7 ± 0.8[(3)]	38 ± 3	(c)	(b)	(b)
Sm	3.3 ± 0.2[(3)]	3.5 ± 0.3	5.5 ± 0.3[(3)]	5.9 ± 0.5	(c)	(b)	(b)
Th	2.8 ± 0.1[(3)]	2.4 ± 0.4	7.4 ± 0.2[(3)]	6.50 ± 0.37	(c)	(b)	(b)

[(a)]Number of determination used in the mean calculation.
[(b)]Certified value not presented.
[(c)]Element not determined in this work.

Table 2
Geometric Means, Geometric Standard Deviation, and Median from EDXRF Analysis In ng/m^3

Element	Fine Particles			Coarse Particles		
	n	GEOMETRIC MEANS ± SD	MEDIAN	n	GEOMETRIC MEANS ± SD	MEDIAN
Al	12	256,4 ± 1,7	260,5	12	2307,4 ± 1,6	2391,0
Si	12	401,6 ± 1,8	399,5	12	3629,6 ± 1,5	3747,5
P	12	134,2 ± 1,6	142,0	12	162,1 ± 1,4	172,5
S	12	2693,3 ± 1,6	3195,5	12	1220,5 ± 2,0	1383,0
Cl	11	72,1 ± 3,1	78,0	12	533,3 ± 1,7	521,0
K	12	311,9 ± 2,1	355,5	12	342,6 ± 1,6	338,0
Ca	12	115,2 ± 2,5	161,0	12	1617,9 ± 1,8	1537,0
Ti	6	14,7 ± 1,6	12,9	9	161,3 ± 2.1	203,4
V	7	13,3 ± 1,6	14,8	8	27,0 ± 1,8	35,1
Cr	11	7,4 ± 2,3	6,3	12	25,3 ± 1,4	23,5
Mn	12	28,9 ± 3,8	21,6	12	59,8 ± 2,1	53,8
Fe	12	196,6 ± 1,7	191,0	12	1583,8 ± 1,5	1560,5
Ni	10	6,4 ± 2,1	8,1	10	7,6 ± 2,4	9,6
Cu	12	9,7 ± 2,7	13,0	12	29,7 ± 1,7	30,5
Zn	12	94,4 ± 2,0	79,5	12	69,6 ± 1,6	71,5
Ga	11	3,9 ± 2,4	4,9	12	3,8 ± 1,9	4,5
As	4	13,4 ± 1,2	13,4	4	7,6 ± 1,1	7,9
Se	11	2,8 ± 2,9	3,5	10	2,2 ± 2,3	2,6
Br	12	14,2 ± 2,0	15,8	12	3,2 ± 3,4	4,2
Rb	10	5,1 ± 2,9	6,0	12	3,2 ± 4,8	5,3
Sr	10	6,0 ± 4,0	8,6	12	11,8 ± 2,0	12,1
Hg	3	4,5 ± 1,6	5,6	4	4,3 ± 1,9	3,6
Pb	12	88,6 ± 1,8	88,5	12	58,2 ± 2,0	73,0

(*n*) where the determination of the considered element was possible were used.

In order to obtain information about aerosol sources, the enrichment factor (EF) and correlation coefficients (*r*) were calculated:

$$EF = \frac{(E/R)\ \text{aerosol}}{(E/R)\ \text{reference}}$$

where *E* and *R* are geometric mean concentrations of any element *E* normalized to a selected element *R*, and *r* is the usual linear correlation coefficient.

Table 3
Geometric Means, Geometric Standard Deviation, and Median, from INAA Analysis in pg/m³, Unless Indicated

Element	Fine Particles			Coarse Particles		
	n	GEOMETRIC MEANS ± SD	MEDIAN	n	GEOMETRIC MEANS ± SD	MEDIAN
Na (ng/m^3)	12	305,7 ± 1,9	296,5	12	245,0 ± 2,4	237,0
Al (ng/m^3)	11	154,6 ± 1,8	179,0	12	1480,0 ± 1,8	1642,0
Cl (ng/m^3)	10	74,5 ± 9,6	90,5	11	351,4 ± 1,8	314,0
K (ng/m^3)	9	476,6 ± 2,7	593,0	9	340,4 ± 2,7	475,0
Sc	8	12 ± 2	12	9	183 ± 2	252
V (ng/m^3)	12	15,0 ± 1,6	17,3	12	6,4 ± 1,6	6,4
Cr (ng/m^3)	9	5,3 ± 3,8	12,3	9	9,1 ± 2,1	11,8
Mn (ng/m^3)	12	36,5 ± 3,6	23,3	12	38,1 ± 2,4	29,6
Fe (ng/m^3)	9	171,5 ± 2,6	251,0	9	947,2 ± 2,3	1211,0
Co	3	265 ± 3	221	9	207 ± 4	220
As	8	2480 ± 2	1975	4	388 ± 2	367
Se	8	1718 ± 3	1482	8	357 ± 2	391
Br	8	8852 ± 2	9147	7	1010 ± 3	1289
Rb	6	3779 ± 2	4209	6	2236 ± 2	2300
Mo	7	1033 ± 2	952	6	1332 ± 2	1200
Sb	9	4018 ± 2	4872	9	1848 ± 2	1600
Ba (ng/m^3)	7	13,2 ± 1,3	12,6	7	42,0 ± 1,8	42,3
La	9	259 ± 2	279	9	1021 ± 2	1300
Ce	8	255 ± 4	342	9	2429 ± 2	3175
Sm	9	51 ± 2	48	9	157 ± 2	172
W	9	493 ± 2	409	9	755 ± 3	840
Th	8	60 ± 2	54	9	200 ± 2	246

The EF were calculated using soil dust and sea salt as reference composition (7), owing to the fact that they are well known. Si and Cl were used for normalization of the results obtained by EDXRF (Tables 4 and 5), whereas for INAA results, Fe and Na were used (Tables 6 and 7).

Although a small difference has been verified for EF results obtained by both techniques (EDXRF and INAA), the final conclusions are practically the same. The differences in EF values were probably owing to different elements being chosen as reference. For one to suggest that the

Table 4
Enrichment Factor (EF), Relative to Soil Dust, and the Correlation with Reference Element, r, Calculated from EDXRF Results

		Reference Element	Al	K	Ca	Ti	Mn	Fe
Fine	EF	Si	1.6	17	3.3	1.5	16	2.2
	r (a)		0.94 (1)	0.92 (2)	0.60 (3)	0.20 (4)	0.84 (2)	0.73 (3)
Coarse	EF	Si	1.5	2.1	4.9	1.8	3.7	2.0
	r (a)		0.99 (1)	0.92 (2)	0.55 (3)	0.91 (2)	0.84 (2)	0.96 (1)
Soil dust weight ratio element/Si .			0.411	0.0444	0.0911	0.0249	0.0045	0.222

[a]Significance level: (1) $p < 0.001$; (2) $0.001 < p < 0.01$; (3) $0.01 < p < 0.1$; and (4) $p > 0.1$.

Table 5
Enrichment Factor (EF), Relative to Sea Salt, and the Correlation with Reference Element, *r*, Calculated from EDXRF Results

		Reference Element	Al	S	K	Br
Fine	EF	Cl	85	466	211	58
	r (a)		0.23 (4)	- 0.15 (5)	0.23 (4)	- 0.10 (4)
Coarse	EF	Cl	99	28	30	1716
	r (a)		0.59 (3)	- 0.32 (4)	0.57 (3)	- 0.03 (4)
Sea salt weight ratio element/Cl			0.0435	0.0827	0.0209	0.0352

[a]Significance level: (1) $p < 0.001$; (2) $0.001 < p < 0.01$; (3) $0.01 < p < 0.1$; and (4) $p > 0.1$.

Table 6
Enrichment Factor (EF), Relative to Soil Dust, and the Correlation with Reference Element, *r*, Calculated from INAA Results

		Reference Element	Na	Al	K	V	Cr	Mn
Fine	EF	Fe	3.4	0.5	14	33	11	10
	r (a)		0.61 (4)	0.48 (4)	0.64 (4)	0.94 (3)	0.41 (4)	0.61 (4)
Coarse	EF	Fe	0.5	0.8	1.8	2.5	3.6	2.0
	r (a)		0.80 (3)	0.97 (1)	0.51 (4)	0.70 (3)	0.34 (4)	0.39 (4)
Soil dust weight ratio element/Fe .			0.528	1.850	0.200	0.0027	0.0027	0.0203

[a]Significance level: (1) $p < 0.001$; (2) $0.001 < p < 0.01$; (3) $0.01 < p < 0.1$; and (4) $p > 0.1$.

Table 7
Enrichment Factor (EF), Relative to Sea Salt, and the Correlation with Reference Element, *r*, Calculated from INAA Results

		Reference Element	Al	K	Fe	Cl	Br
Fine	EF	Na	6.1	40	51	0.1	4.3
	r (a)		0.46 (4)	0.58 (4)	0.61 (4)	0.55 (4)	0.04 (4)
Coarse	EF	Na	73	35	355	0.75	16000
	r (a)		0.75 (3)	0.34 (4)	0.80 (3)	0.92 (2)	0.82 (3)
Sea salt weight ratio element/Na .			0.0825	0.0397	0.0109	1.90	0.0067

[a] Significance level: (1) $p < 0.001$; (2) $0.001 < p < 0.01$; (3) $0.01 < p < 0.1$; and (4) $p > 0.1$.

trace elements involved have a common origin, the EF must be close to unity, and it is also expected that *r* values be close to unity too.

This is observed for fine and coarse particulates (Tables 4–7) relative to soil dust. Besides, high *r* values are obtained for Al, K, Ti, Mn, and Fe, showing that soil dust is the dominant source of fine and coarse particulates. On the other hand, EF values show an enrichment relative to sea salt for all trace elements considered, except for Cl that is depleted. This can denote a complete disappearance of any marine contribution.

Also, a high correlation between Fe and the elements Al, Mn, Se, Rb, La, and Ce (>0.80) is observed in coarse particulate, which suggests a common source, perhaps soil dust or a new aerosol source, different from sea salt and soil dust.

CONCLUSION

In this work, the results obtained by means of the two techniques applied, EDXRF and INAA were considered separately. But, since some important elements, such as Si, Ti, Cu, P, and S, could not be determined by INAA, it has been more interesting to combine the results obtained by both techniques.

Considering that the aerosols of São Paulo is one of the most complex of the world, because it includes a large urban-industrial area, and for this reason, its characterization is very difficult, also, considering the small number of samples, obtained from a single point of sampling that was analyzed and submitted to a simple mathematical treatment, it was possible to perform a preliminary investigation of the main sources of aerosol in the city of São Paulo.

ACKNOWLEDGMENTS

The authors wish to acknowledge financial support from FINEP and the International Atomic Energy Agency (IAEA), Vienna (Research Contract no. 4957/RB).

REFERENCES

1. B. G. Bennett, J. G. Kretzschmar, G. G. Akland, and H. W. de Koning, *Environ. Sci. Technol.* **19,** 298 (1980).
2. S. Habib and M. J. Minski, *J. Radioanal. Chem.* **63,** 379 (1981).
3. A. Albini, A. Cesana, and M. Terrani, *J. Radioanal. Chem.* **34,** 185 (1976).
4. A. Alian and B. Sansoni, *J. Radioanal. Nuclear Chem. Articles* **89,** 191 (1985).
5. B. W. Loo, W. R. French, R. C. Gatti, F. S. Goulding, J. M. Jaklevic, J. Llacer, and A. C. Thomson, *Atmos. Environ.* **12,** 759 (1978).

6. G. S. Kowalczyk, G. E. Gordon, and S. W. Rheingrover, *Environ. Sci. Technol.* **16,** 79 (1982).
7. P. N. Artaxo, C. Q. Orsini, *Doctoral dissertation* (in Portuguese), Institut of Physics, University of São Paulo, Brazil, 1985.
8. F. J. Flanagan, *Geochim. Cosmochim. Acta* **37,** 1189 (1973).
9. International Atomic Energy Agency, *Certified Reference Material, Air 3/1* (1982).

Production of ^{134}Cs Thermo-Generated Aerosols

Study of Their Behavior After Deposition on Spruce Trees

C. RONNEAU,* K. FONSNY, AND C. MYTTENAERE

Université de Louvain, Louvain-la-Neuve, Belgium

Received April 17, 1989; Accepted October 19, 1989

ABSTRACT

This paper presents observations made on the behavior of radioactive cesium in laboratory conditions. ^{134}Cs-loaded aerosols are generated from UO_2 pellets in an electric oven and allowed to deposit on small spruce trees. The transfer of radioactivity to the plants has been studied. In another series of experiments, trees are submitted to artificial rain contaminated by the radioactive aerosols. The two modes of contamination are compared.

Index Entries: Radioactive aerosol generation; ^{134}Cs aerosols; aerosols, physicochemical structure of; deposition on plants; thermo-generated ^{134}Cs, solubility of.

INTRODUCTION

The assessment of radiological consequences of nuclear accidents usually emphasizes the contamination of edible plants, neglecting the fact that trees are powerful accumulators of radioactivity (*1*). Observations made after the Chernobyl accident (*2,3*) suggested that airborne

*Author to whom all correspondence and reprint requests should be addressed.

radioactive cesium could be partly incorporated into insoluble submicron aerosols, leading to a behavior different from what could be observed when using totally soluble cesium salts, for example, enhanced mobility in soils.

Toward the end of 1985, we decided to build a small laboratory simulator capable of producing aerosols from heated UO_2 pellets contaminated by ^{134}Cs. Initially, we studied the physicochemical characteristics of thermo-generated radioactive aerosols, as well as their deposition behavior on plant surfaces. We found that a small, but significant fraction of the cesium released by the pellets at about 2000°C was insoluble in water at pH 6. Soon after this study, on May 3–4 1986, the same observation was made for rain water contaminated by the Chernobyl accident: this was, however, nonconclusive because of the many other chemical pollutants (organic and inorganic) present in rain that were likely to alter the intrinsic solubility of the radiocontaminants. The present paper is a summary of the results obtained in our laboratories during simulation experiments in which small spruce trees were exposed to deposition (wet and dry) of thermo-generated ^{134}Cs-loaded aerosols emitted from UO_2 pellets.

EXPERIMENTAL

Figure 1 schematizes the apparatus used for producing ^{134}Cs-tagged aerosols. UO_2 pellets containing Cs (inactive and active) are heated up to 2000°C in a graphite oven, as used in an Atomic Absorption Spectrometer (AAS). The aerosols released during the thermal phase (30 s) are swept away by a current of argon (99.90% purity) and pass into a quartz oven heated to about 700°C, where they are, if necessary, mixed with CO_2, H_2O, and air to simulate some kind of "maturation" in the plume released from a damaged reactor.

At the exit of the quartz oven, the aerosols enter a small cylindrical room (diameter, 1.5 m; height, 2 m), where various surfaces (plants, soils, and so on) are exposed to dry deposition. Another possibility is to trap the aerosols in water at the exit of the generator and to perform wet deposition experiments in a rain simulator. In the lower part of the deposition chamber, the aerosols are eventually sampled by filtration or impaction for physicochemical analyses.

The radioactivity of the various samples is determined by γ-spectrometry on a Ge(Li) detector coupled with a 4096 channel analyzer. Correction is made for counting geometry of the samples. Radioactivity in plant material was determined by a NaI well-type counter: NaI and Ge(Li) counters were cross-tested at regular intervals. X-ray photoelectron spectroscopy (XPS) was performed on inactive samples to determine the chemical characteristics of the aerosol surface. XPS analyses allow determination of the atomic composition of the aerosol surface down to a

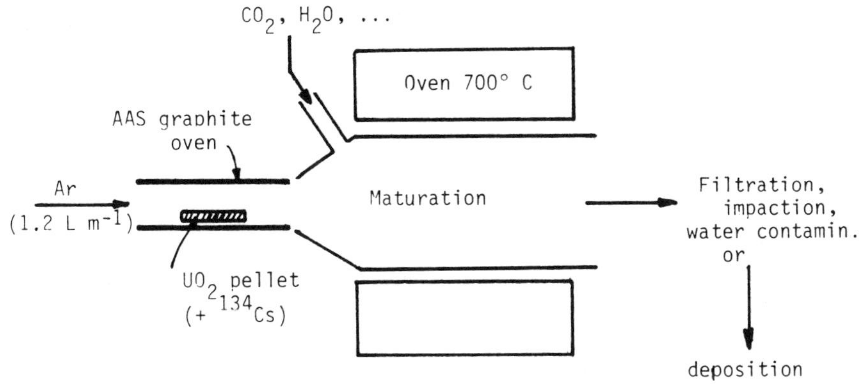

Fig. 1. Scheme of the laboratory aerosol generator.

depth of 2 nm. Radioactivity measurements give the global composition of the samples before and after the aerosol generation. The solubility of cesium was assayed on PVC or Teflon membrane filters (pore size 0.2 μm).

EXPERIMENTAL RESULTS

Structure of the Aerosols

Results concerning the physicochemical properties of the aerosols have already been partially published (*4,5*) and are summarized below.

Because of its relatively high volatility (bp 678°C), cesium is enriched in the aerosols, compared with the composition of the starting pellets. Moreover, the surface of the aerosols is enriched in cesium, compared with their bulk composition. This may be schematized as follows (results obtained during a typical experiment; the composition balance is UO_2).

Starting pellet	0.54% Cs	(atomic composition)
Aerosols	4.5% Cs	(global atomic composition)
Aerosols surface	71% Cs	(atomic composition in 2 nm surface layer).

The surface enrichment is very important and could explain the solubility of the major part of the cesium deposited after the Chernobyl event. On the other hand, a fraction of the cesium remains trapped in the UO_2 core of the aerosols, and this fraction might constitute the insoluble (or less soluble) part of deposited ^{134}Cs.

Given that the mass-median equivalent (aerodynamic) diameter of the aerosols is of the order of 0.5 μm, their mean physical diameter should be about 60 nm, provided the volumic mass of airborne UO_2 remains equal to 11 g/cm^3. A 2 nm surface layer corresponds to ~10% of the volume of the aerosol sphere, and this layer accumulates 7% of the Cs content of the particle.

The deposits of aerosols on the impactor stages (Teflon filters: pore size 0.2 μm) were progressively washed with 3 × 50 mL of deionized water (which is a large volume, compared with the sample quantity). The solubilization of Cs is gradual: 2–5% of the radioactivity remains on the filter. Since ^{134}Cs is not carrier-free, it seems logical to attribute this insolubility to retention into UO_2 aerosols.

Dry Deposition on Spruce Trees

Dry deposition on small trees was performed without any turbulence, giving way to Stokes' sedimentation ($\sim 1 \cdot 10^{-4}$ cm/s). These, of course, are unnatural conditions; however, we consider that the transfer behavior of radioactive cesium into plant material is not affected by these experimental conditions. It appears that deposited radioactivity is homogeneously distributed in tree foliage; small filter papers located inside and outside the trees showed the same homogeneity, indicating that diffusion is the main driving mechanism for deposition (in our experimental conditions). Comparisons were made between the activity intercepted by the trees and the activity deposited on an inert surface (filter paper). The interception has been estimated at about 80%, which is a high figure compared with the 40 to 60% given in the literature (6,7); once more, this is attributable to the mode of deposition.

Radioactivity Distribution in Tree Canopy

As a general rule, needles of the same age display no difference in contamination with regard to the position in the tree. For a given weight, young needles are more numerous than older ones; consequently, they develop a greater surface for aerosol capture. Twigs accumulate five times less activity (Bq/g dm) than needles. After contamination, the trees are placed into a greenhouse; no significant variation of contamination with time occurs after 37 d.

Rain Washing

Some contaminated trees were exposed to natural rain. On the average, radioactivity exported by rain amounts to 60% of the deposited contamination (160 L rain water/m^2 in a 3-mo period).

Mechanism of Retention of Cs by the Needles

Contaminated twigs were washed with deionized water. Up to 90% of the radioactivity is eliminated by water. The washing water is then filtered on Teflon membranes (to avoid parasitic adsorption of ^{134}Cs). We observed that 70–80% of the extracted Cs remains trapped on the particles retained by the filters. In other instances, the insoluble fraction was isolated by centrifugation. Figure 2 shows the amount of radioactivity extracted in a typical experiment as a function of the number of washing

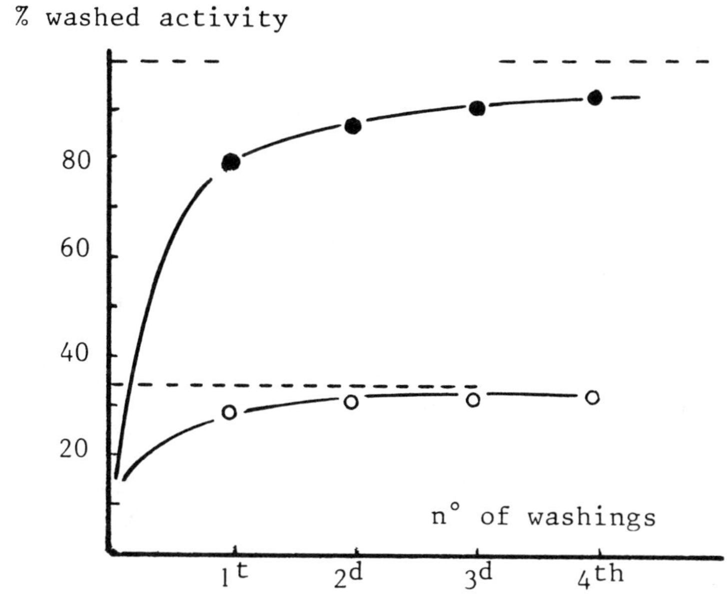

Fig. 2. As a function of washing operations (2 g of needles washed each time by 20 mL water): ●, fraction of total extracted radioactivity; and ○, fraction of centrifugated radioactivity.

operations (from 2 g of needles); the insoluble fraction separated by centrifugation is also shown.

These results agree with conclusions drawn by other authors (*8,9*). Note that, after examination by electron microscopy and electron probe, particles extracted from needles or filtered from rain water present the same physicochemical characteristics.

Wet Deposition on Spruce Trees

In some experiments, ^{134}Cs-loaded aerosols from the laboratory generator were allowed to contaminate deionized water (~2000 Bq/L). Trees were submitted to artificial rain produced from this water. In this case, we observed contamination gradients into trees: the radioactivity decreased by a factor of about 1.5 from the top to the base and from the exterior of the canopy to the trunk.

The ratio between radioactivity deposited on the needles (Bq/g dm) and radioactivity deposited on passive wet gages outside the canopy (Bq/cm^2) is about 14. In the case of dry deposition, the same ratio, with respect to deposition onto filter paper, is five times higher (this was also observed in the deposition of "classical" pollutants (*10*)). This indicates that rain is, at the same time, a deposition and a washing agent. After the Chernobyl accident, however, the major part of the radioactivity was deposited by rain, at least onto the Belgian territory.

CONCLUSIONS

Under controlled conditions and in the experimental procedure developed in our laboratories, the following conclusions may be drawn

1. The interception factor of spruce trees is very high (0.80) when dry deposition occurs via Stokes' sedimentation, with a probable significant contribution of diffusion effects.
2. Canopies are contaminated in a uniform way, and contamination rates are directly linked to the weight of intercepting matter.
3. Deposited radioactivity is easily washed by rain water; a significant fraction of radioactive cesium is bound to natural particles on leave surfaces and in dripping water.

Further experiments are in progress to determine the transfer of radioactive cesium into the different parts of the plants, as well as the relative role of root uptake in the internal contamination of trees.

ACKNOWLEDGMENTS

This work was sponsored by the CEA (Cadarache). Thanks to J. Cara for valuable technical assistance. ESCA measurements were made by W. Stone at the University of Louvain.

REFERENCES

1. G. Desmet and C. Myttenaere, *J. Environ. Radioactivity* **6,** 197–202 (1988).
2. E. A. Bondietti, J. N. Brantley, and C. Rangarajan, *J. Environ. Radioactivity* **6,** 99–120 (1988).
3. H. Maubert and M. Grauby, personal communication.
4. A. H. Al Rayyes, C. Ronneau, D. Apers, and C. Myttenaere, *Dokl. Akad. Nauk Ukr. SSR, ser. B* **8,** 76–79 (in Russian) (1988).
5. C. Ronneau et al., *IVe Symposium International de Radioécologie de Cadarache,* tome 1, B18–B25 (1988).
6. S. E. Lindberg, G. M. Lovett, D. D. Richter, and D. W. Johnson, *Science* **231,** 141–145 (1985).
7. P. J. Coughtrey and M. C. Thorne, *Radionuclide distribution and transport in terrestrial and aquatic ecosystems. A critical review of data,* vol. I, ANS, A. A. Balkema Publishers, Rotterdam, Netherlands.
8. L. Tobler, S. Bajo, and A. Wyttenbach, *J. Environ. Radioactivity* **6,** 225–245 (1988).
9. Y. Takashima, N. Momoshima, and M. R. Hamidian, *Radiochem. Radioanal. Letters* **36(1),** 69–72 (1978).
10. B. Coenen, C. Ronneau, and J. Cara, *Water, Air, Soil Pollution* **36,** 231–237 (1987).

Whole Body Content and Turnover of Cs and K

D. BERG,*[,1] W. E. KOLLMER,[1] K. HENRICHS,[2] AND G. VOIGT[2]

[1]*Inst. für Strahlenbiologie; and* [2]*Inst. für Strahlenschutz, Gesellschaft für Strahlen, u. Umweltforschung, 8042 Neuherberg, West Germany*

Received April 17, 1989; Accepted October 23, 1989

ABSTRACT

In persons who had accumulated radiocesium from the fallout of the Chernobyl nuclear reactor accident, the amount of radiocesium and radiopotassium was measured in the total body and in urine. The stable Cs content in the urine was determined at the same time by instrumental nuclear activation analysis. From the data obtained, the total body Cs pool was calculated to be 1500 μg, that of K 110 g. The Cs turnover rate was found to be 0.64%/d for men and 0.81 for women; the K turnover 2.4 and 2.7, respectively. The Cs:K ratio in the total body was 4 times higher than that in urine, demonstrating that, in relation to the intake, the body is able to accumulate more Cs than K.

Index Entries: Whole body Cs and K; Cs and K turnover; instrumental neutron activation; humans.

INTRODUCTION

The nuclear reactor accident at Chernobyl resulted in a radionuclide fallout consisting mainly of ^{131}I, ^{134}Cs, and ^{137}Cs. Although the intake of radioiodine by humans was mainly through inhalation, radiocesium entered the human food chain and was ingested. We have examined the

*Author to whom all correspondence and reprint requests should be addressed.

radioactivity in the local community for more than 2.5 years by carrying out whole body measurements (*9–11*).

The excretion rate of stable Cs, published by the Task Group on Reference Man (*1*), was based on tracer experiments, because data on this were not available. The present paper deals with some findings in connection with the determination of stable Cs and K in the total body and in urine. These findings constitute part of a larger investigation in which the impact of a heavily contaminated test meal was also examined.

MATERIALS AND METHODS

The study was begun 1 1/4 years after the Chernobyl accident. At that time, there were practically no fluctuations in the whole body radiocesium. Five female and five male volunteers (Table 1) ate a test meal containing approximately 300–700 Bq ^{137}Cs and 150–350 Bq ^{134}Cs. The bulk of the radioactivity was contained in venison from roe heavily contaminated by Chernobyl fallout. A whole body meaurement was made before the meal. Within the week following the meal, the measurements were repeated every day, and later on at wider intervals, until 200 d after the start of the study. Starting 1 d before the test meal, samples of urine were collected every 24 h for a week, and then at longer intervals, until 35 d had elapsed since the meal.

The whole body counter was equipped with four 5-inch by 4-inch (12.7 cm $\times$ 10.2 cm) sodium iodide crystals and shielded by quartz sand, steel, and lead. The measurements were performed using a bed-like geometry. With a measuring time of 25 min, the detection limit for an adult person was approximately 70 Bq ^{137}Cs. Because of the low count rate, the low resolution of the sodium iodide crystals, and the poor reproducibility of the positioning of the persons, the measurement error is estimated to be 10–20%.

The radioactivity in the urine was measured with a low-level intrinsic Ge detector. Optical reproducibility was ensured by using a Marinelli beaker with a volume of 1000 cm^3. The counter was calibrated. For a 12 h measurement period, it had a sensitivity limit of 1 Bq/L. After the radioactivity had been measured, 10 g of each of the urinary samples were freeze-dried and the ratio of fresh to dry weight calculated. Approximately 5–80 mg of the dry matter were transferred to quartz vials. These samples, together with a standard preparation containing 5.159 μg of stable Cs in the form of chloride, were irradiated for 24 h in the nuclear reactor (FRM-Garching) using a fishing line irradiation procedure. The neutron flux was $1.3 \times 10^{13}/s/cm^2$. After an 8-mo period of radioactive decay, the vials were rinsed with 7M HNO_3, distilled water, ethanol, and acetone. Then they were transferred to a counter equipped with a planar Ge(Li) detector for γ-spectroscopy. Stable Cs was determined using the 796.3 and 801.3 keV emission of ^{134}Cs. The 604 keV emission was not used because of high background emission in this region owing to the radioactive products of uranium–radium decay.

RESULTS

Figure 1 demonstrates the course of the whole body retention of the sum of ^{134}Cs and ^{137}Cs in one of the volunteers. Before the test meal, it had been nearly constant at 1500 Bq. After the meal, an increase to approximately 2500 Bq was recorded. Within the following 4 to 5 d, it dropped to approximately 2100 Bq. Only a minor decrease was observed after the eighth day.

The amounts of radiocesium (^{134}Cs, ^{137}Cs) excreted daily in the urine were also measured and recorded. In addition, the specific activities of urinary Cs were calculated from the measurements of fallout radiocesium in urine, combined with the analyses of stable Cs via NAA. The specific activities are not influenced by incorrect timing of the urine sampling since this is equally reflected in the content of fallout radiocesium and stable Cs. In Fig. 2, these data are shown for the same volunteer as the data in Fig. 1. A very similar course of excretion was also observed in all the other volunteers.

In the whole body counter, ^{40}K was also determined at the same time as ^{134}Cs and ^{137}Cs. Because of the constant specific activity (30.9 Bq/g stable K), this measurement also reflects the whole body pool of stable K. In Table 1, the K pools of the volunteers, based on the whole body measurements, are compared with those calculated for each of them, according to the formula of Ellis et al. (*12*), on the basis of their body weights, body heights, age, and sex. This comparison shows that two of the volunteers exhibit a K retention below normal. The normally lower retention in women, compared to men, can also be seen in Table 1.

The urinary excretion of K by the volunteers, determined by ^{40}K activity measurements, is shown in Table 2. For the female volunteers, it was approximately 2 to 3 g/d, for the males 3 to 4 g.

The amount of stable Cs excreted, calculated on the basis of the neutron activation analyses, was 9.3 $\pm$ 1.2 μg/d (arithm. mean $\pm$ standard error) in the case of the female volunteers and 11.1 $\pm$ 2.7 in the case of the males. The difference between females and males was not statistically significant.

For calculating the whole body pool of Cs, it was assumed that the transient increase in urinary Cs owing to fallout in the test meal had become negligible after the eighth day following the meal, and thus, the specific activity in the urine reflects that in the whole body. The whole body pool of stable Cs, therefore, was calculated by dividing the ^{137}Cs activity measured in the total body by the specific activity observed in the urine later than 8 d after the test meal. A numerical simulation using published data (2) shows that the error, owing to the deviation from

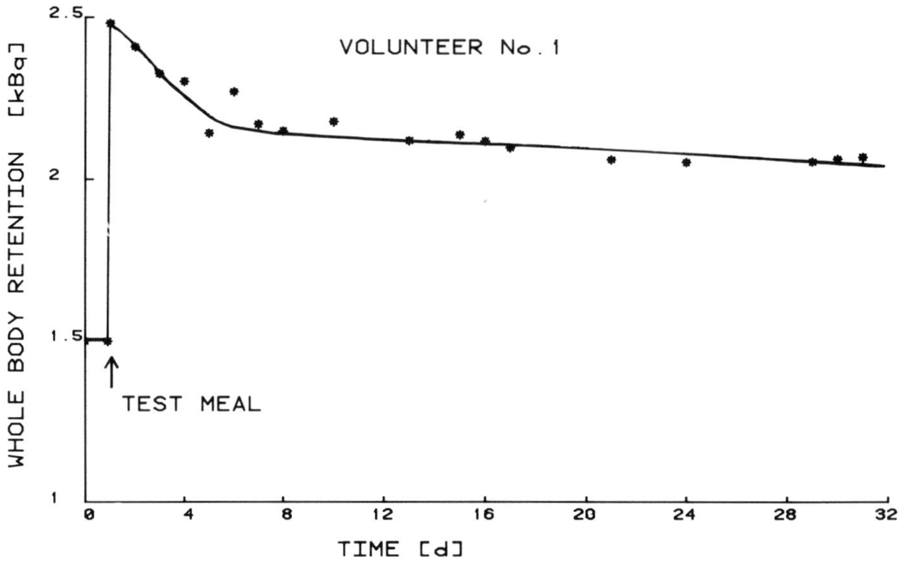

Fig. 1. Whole body retention of radiocesium (^{137}Cs + ^{134}Cs) before and after the test meal containing fallout cesium from the Chernobyl reactor accident.

equilibrium, may lead to an underestimate of the total body Cs well below 10%. The whole body pool of stable Cs was found to vary between 800 and 1900 μg for the female volunteers and between 1600 and 2100 for the males.

DISCUSSION

Since the chemical properties of Cs are very similar to those of K, it is not surprising that it exhibits a similar fate in biological systems, including the human organism. One of the aims of this study was to establish the physiological relations between the two elements.

By analogy with whole body measurements made during the course of another study (*9,13*), the dynamic equilibrium of radiocesium in the body was assumed to have exhibited only minor fluctuations prior to the start of the present study. The radioactivity in the test meal was exceedingly high by comparison and, thus, caused an increase in the whole body retention, followed by a rapid decrease during the first week. The amounts of the stable elements Cs and K excreted remained unchanged. In the case of the element K, it is known that the amounts resorbed and excreted are homeostatically controlled, ensuring a constant level of K in the various tissues, especially the muscle. No homeostatic regulating mechanism has been described in the case of Cs. Its content in normal food is approximately one millionth that of K. The dietary supply of Cs

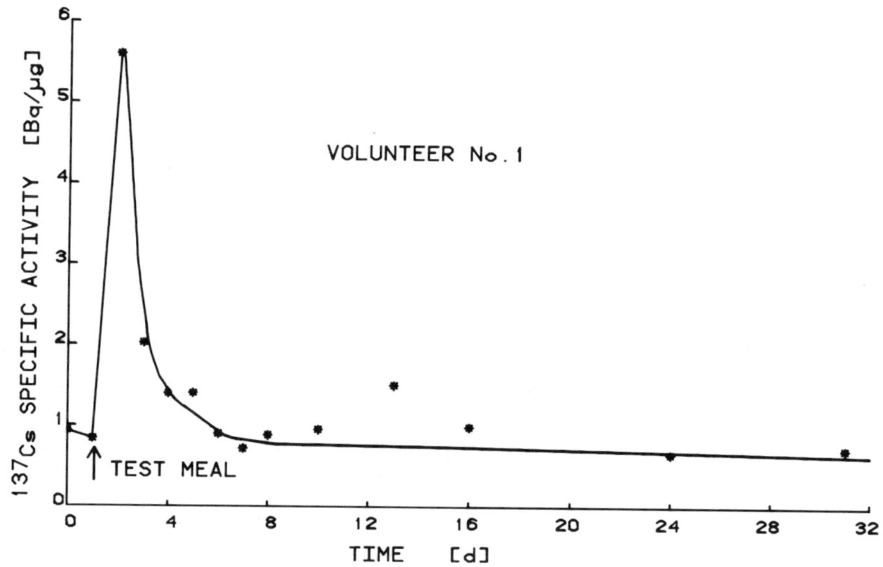

Fig. 2. Specific activity of ^{137}Cs in urine before and after the test meal containing a high load of fallout radionuclides from the Chernobyl accident.

Table 1
Total Body Pools of Cesium and Potassium (Mean ± SEM)

Subject No	Weight	Height	Age	Body Cesium	Body Potassium measured	Body Potassium calculated[a]
	cm	kg	y	µg	g	g
1[c]	170	79	45	1690 ± 140	132 ± 2	126
2[c]	187	83	36	1680 ± 170	129 ± 2[b]	160
3[c]	187	78	36	2060 ± 130	117 ± 2[b]	155
4[c]	181	95	43	1590 ± 100	161 ± 2	157
5[c]	183	85	31	1830 ± 130	147 ± 2	157
6[d]	159	54	46	930 ± 60	77 ± 2	77
7[d]	170	62	49	1860 ± 40	100 ± 2	93
8[d]	158	56	28	1010 ± 100	100 ± 2	80
9[d]	165	58	38	780 ± 50	85 ± 2	87
10[d]	165	61	35	1700 ± 150	94 ± 3	90

[a]After Ellis et al. (*12*): $(\alpha_1 - \alpha_2 \cdot \text{Age}) \cdot \text{Height}^2 \cdot \sqrt{\text{Weight}}$; men: $\alpha_1 = 5.52$ and $\alpha_2 = 0.014$; women: $\alpha_1 = 4.58$ and $\alpha_2 = 0.010$.

[b]Suboptimal K status.

[c]Man.

[d]Woman.

and K varies with the composition of the diet. In volunteer No. 1, this led to a variation in the urinary excretion rate of between 3 and 4.5 g of K and between 5.8 and 12.4 μg of Cs per day. Investigations with radiocesium in another study (*13*) demonstrate that the intestinal resorption of Cs is nearly complete over the whole range of "normal" dietary content. The same is true for K (*1*).

The mean rate of excretion for each person, obtained by calculating the average 24 h excretion over at least 11 d, demonstrates a variation between individuals of 2.0 to 3.7 g for K and 5.1 to 14.9 μg for Cs. The data for K are well in agreement with the figures of 2.8 and 3.5 g per day reported by others (*1,2,8*). The significantly lower rate of excretion in the women (Table 2) of our study may be explained by their lower food consumption owing to their lower body weights. In the case of Cs excretion, no difference between man and woman was apparent. Cs excretion is not well documented in the literature. In the compilation of Iyengar et al. (*3*), four references are cited (*4–7*): three between 9.9 and 13, and one of 50 μg/d. An estimate of 9 μg/d has been published in the ICRP publication on Reference Man (*1*). Only a few data concerning the ingestion of Cs have been published. The intake of dietary Cs in Japan has been found to range between 8.6 and 11 μg/d (*14,15*). Moreover, it was assumed in the ICRP publication on Reference Man (*1*) that the ingestion of Cs in the US and Europe should be somewhat higher than in Japan. Assuming that about 80% of dietary Cs is absorbed in the intestine (*13*) and that 90% of this is excreted in the urine (*1*), the estimated daily intake of stable Cs by our volunteers, based on the urinary excretion, is 12.9 to 15.4 μg/d, supporting the assumption made in the ICRP publication (*1,2*). Since, with a very few exceptions, the organism and, thus, the kidney is unable to differentiate between the isotopes of an element, the specific activity of Cs in urine ought to be the same as in the blood. In blood, it is a mixture of the activity of Cs fed back to blood from all other compartments of the body and that newly absorbed in the intestine.

Assuming a steady state between intake and excretion of stable Cs in all healthy adults, it may be concluded that the specific activity in all compartments of the body is the same if there is no simultaneous change in the whole body radiocesium and in the specific activity of the urinary Cs. When such conditions prevail, the total body stable Cs may be calculated from the known parameters. During an equilibrium, the differences in the urinary activity between various individuals must therefore be attributed to differences in their dietary habits. In particular, the variation of specific activity of dietary constituents from regions with different contaminations from fallout causes the differences in urinary activity. From calculations performed after the amount of radioactivity from the test meal had stabilized in both the female, as well as the male volunteers, a mean of 21.3 μg Cs was obtained per kg body weight. Because of the somewhat larger variation in women, 1490 ± 510 μg Cs was calculated for a 70 kg woman and 1490 ± 240 for a 70 kg man. This

Table 2
Rate of Excretion, Turnover, Ratio in Urine and Total Body of Stable Cesium and Stable Potassium (Mean ± SEM)

Subject No	Rate of Excretion in Urine		Rate of Turnover in Whole Body		Ratio Cs/K x10⁻⁶	
	K g/d	Cs μg/d	K %/d	Cs %/d	Urine	Total Body
1[a]	3.0 ±0.4	8.6 ±0.8	2.3	0.51	2.86	12.8
2[a]	3.0 ±0.6	8.5 ±0.7	2.3	0.51	3.13	13.0
3[a]	3.0 ±0.4	11.7 ±1.0	2.6	0.57	3.85	17.5
4[a]	3.5 ±0.6	14.9 ±2.9	2.2	0.94	3.85	9.90
5[a]	3.7 ±0.3	11.9 ±0.6	2.5	0.65	3.13	12.5
Mean[a] ± SEM	**3.2 ±0.2**	**11.2 ±2.7**	**2.38 ±.07**	**0.64 ±.08**	**3.36 ±0.21**	**13.1 ± 1.2**
6[b]	2.2 ±0.3	8.9 ±0.8	2.9	0.97	3.85	12.1
7[b]	2.6 ±0.3	10.5 ±1.0	2.6	0.57	4.00	18.5
8[b]	2.6 ±0.4	8.8 ±1.0	2.6	0.87	3.33	10.1
9[b]	2.0 ±0.3	7.8 ±0.8	2.4	1.00	3.33	9.2
10[b]	2.6 ±0.3	10.6 ±0.9	2.8	0.62	3.85	18.2
Mean[b] ± SEM	**2.4 ±0.2**	**9.3 ±1.2**	**2.66 ±.09**	**0.81 ±.09**	**3.67 ±0.14**	**13.6 ± 2.0**

[a]Man.
[b]Woman.

finding is in accordance with the 1500 μg estimate for the 70 kg reference man (*1,2*).

The daily turnover of stable Cs was found to be 0.81 ± 0.02%/d in women and 0.64 ± 0.18 in men. Accordingly, the biological halftime for the women was 86 d and for men 108 d. No statistically significant difference between the sexes could be established. Differences between the sexes, however, was substantiated by our observation that, with a change in the dietary intake of radiocesium, a new dynamic equilibrium is reached more quickly in women than in men. This observation is based on more than 3000 whole body measurements within 3 yr following the Chernobyl accident.

The ratio stable Cs:stable K in the whole body was approximately 4 times higher than in the urine (Table 2). This difference was found to be statistically significant at the level of $2p < 0.005$. Although the concentration of K in the body is much higher than Cs, the body is able, in relation to the intake, to accumulate more Cs than K.

ACKNOWLEDGMENTS

The authors would like to thank H. Ruhstorfer, Ch. Üblacker, and M. Reithmaier for their technical assistance.

REFERENCES

1. Report of the Task Group on Reference Man, *ICRP 23* (1975).
2. Limits for Intakes of Radionuclides by Workers, *ICRP 30*, Part 1, 91 (1979).
3. G. V. Iyengar, W. E. Kollmer, and H. J. M. Bowen, *The Elemental Composition of Human Tissues and Body Fluids*, Verlag Chemie, Weinheim, 1978, p. 127.
4. H. Bostroem and P. Wester, *Acta Med. Scand.* **183,** 209 (1968).
5. H. Bostroem and O. Wester, *Acta Endocrinol.* (Copenhagen) **60,** 380 (1969).
6. R. Cornelis, A. Speche, and J. Hoste, *Anal. Chim. Acta* **78,** 317 (1975).
7. P. O. Wester, *Acta Med. Scand.* **194,** 505 (1973).
8. K. Lux-Sparschuh, *Kalium, Biochemie-Klinik-Therapie*, Hippokrates Verlag, Stuttgart, West Germany, 1978, p. 33.
9. D. Berg, W. E. Kollmer, H. Kriegel, and K. Henrichs, *Proc. 21 Conf. on Trace Substances in Environmental Health XXI*, D. D. Hemphill, ed., Univ. of Missouri, Columbia, MO, 1987, pp. 219–225.
10. D. Berg, W. E. Kollmer, and H. Kriegel, *Der Nuklearmediziner* **10,** 87–92 (1987).
11. D. Berg, L. Bogner, H. Czempiel, W. E. Kollmer, H. Kriegel, and P. Mühle, *Der Nuklearmediziner* **10,** 353–358 (1987).
12. K. J. Ellis, K. K. Shukla, S. H. Cohn, and R. N. Pierson, *J. Lab. Clin. Med.* **83,** 716 (1974).
13. K. Henrichs, G. Paretzke, G. Voigt, and D. Berg, *Health Phys.*, **57,** 571 (1989).
14. N. Yamagata, *J. Rad. Res.* **2,** 158–169 (1962).
15. N. Yamagata, *Nature* **196,** 83, 84 (1962).

Studies of Cadmium Uptake in Bone and Its Environmental Distribution

S. S. KRISHNAN,*[1,2] S. M. W. LUI,[1] R. E. JERVIS,[1] AND J. E. HARRISON[2]

[1]*Department of Chemical Engineering, University of Toronto, Toronto, Ontario, Canada; and* [2]*Toronto General Hospital, Toronto, Ontario, Canada*

Received April 17, 1989; Accepted November 3, 1989

ABSTRACT

Rat experiments indicate that oral ingestion of cadmium through drinking water leads to an accumulation of cadmium in bone, in addition to liver and kidney. After five weeks of cadmium intake in drinking water (50 to 100 mg/L), the bone cadmium levels increased in proportion to the intake concentration. Bone and kidney histology showed no signs of bone or kidney damage up to 5 wk of cadmium ingestion. Cadmium accumulation in bone was a primary phenomenon and not secondary to renal failure.

In addition, cadmium levels have been estimated in a variety of sources, e.g., foodstuff, fertilizer, and sewage sludge, using neutron and proton activation analyses and atomic absorption spectrophotometry. Cadmium levels of Canadian foods are in the range of 0.002–0.07mg/kg, and soils are in the range of 0.55 to 1.72 mg/kg. Fertilizers contain cadmium from 0.3 to 1.25 mg/kg, whereas sewage sludge contains up to 122 mg/kg.

Index Entries: Cadmium; bone; cadmium pollution; itai-itai disease; trace cadmium levels.

*Author to whom all correspondence and reprint requests should be addressed.

Biological Trace Element Research Editor: G. N. Schrauzer

INTRODUCTION

Cadmium is toxic to living systems. Itai-itai, a painful bone disease, has been associated with cadmium poisoning (*1,2*). Previous investigations of cadmium toxicity to bone has indicated that the bone lesions may be secondary to kidney damage. We felt, as a result of our preliminary work, that cadmium toxicity to the bone may be a primary effect at certain levels of cadmium exposure and accumulation. In order to investigate this hypothesis, experiments were carried out with rats, the cadmium being administered through drinking water.

We also investigated the cadmium levels of food, fertilizer, and soil, which are potential sources of this element in the human life cycle. In addition to this, preliminary experiments indicated that sewage sludges may emit cadmium to the environment during their treatment process, which includes incineration. Since this may be a potential source of inhaled cadmium, we also analyzed sewage sludge samples for their cadmium levels.

METHODS

Animal Experiments

Three groups of female Wistar albino rats, each containing six animals, were given drinking water containing 0, 50, and 100 mg/L of cadmium. The period of ingestion was five weeks. These cadmium levels and the experimental period were chosen on the basis of preliminary experiments. At these cadmium levels, the animals did not get sick or suffer acute toxicity. Beyond the 5-wk period, the uptake of cadmium by bone did not significantly increase.

At the end of the experimental period, these animals were sacrificed, and the bones were removed. The bones were cleaned free of tissue, dissolved in 6M hydrochloric acid, and analyzed for cadmium.

Samples

Various food samples were obtained from across Canada. They were freeze-dried for 24 h prior to analysis.

Soil and fertilizer (used by the farmers) samples were collected from the Holland Marsh area, which is a market gardening tract in Southern Ontario. Sewage sludge samples were collected from the sewage treatment plants of the Municipality of Metropolitan Toronto. These samples were air-dried prior to analysis.

Analytical Methods

Food, soil, fertilizer, and sewage sludge samples were analyzed by radiochemical neutron activation analysis (RNAA) and/or instrumental proton activation analysis (IPAA). These are described in detail elsewhere (2). The acid-dissolved bone samples were analyzed by graphite furnace atomic absorption spectrophotometry. In this procedure, no matrix effects were found, as seen from repeat analysis of diluted samples and analysis using the method of additions.

The accuracy of cadmium assay methods has been checked and found correct by intercomparison of analysis data using more than one method on the same sample. Furthermore, NBS Orchard Leaves Standard Reference Material (1571) and multielement gelatin standard (TEG-50-A) of the Eastman Kodak Co., and other coal and fly ash intercomparison standards of the US EPA and NBS have all been analyzed by the method used here to check its accuracy.

RESULTS AND DISCUSSION

Cadmium Content of Food

The cadmium contents of Canadian food samples (3) have been expressed on an "as received basis" in Table 1. That means dry items, such as wheat, tea, and so on, are given in mg/kg dry weight, whereas with meat, the cadmium level is expressed in mg/kg wet weight. The cadmium levels in nonmeat dry items varied from 0.007 to 0.29 mg/kg. Meat such as beef, chicken, and lamb contained 0.002–0.021 mg/kg cadmium in their muscle tissues. In these samples, specific organs, such as kidney and liver, which are known to accumulate cadmium, contained much higher levels, and these are given in brackets.

Cadmium Content of Environmental Sources

The cadmium levels in samples collected from the environmental sources, e.g., soil, fertilizer, and sewage sludge, are high compared to those in food. These are given in Table 2. The cadmium level in Holland Marsh soil varied from 0.5 to 1.72 mg/kg. Similar concentrations were found in fertilizer samples. Sewage sludges, however, contained cadmium at much higher levels, indicating input from industrial sources.

Intake of Cadmium by Living Systems

One source of intake of cadmium by living systems is the food chain. The average cadmium intake by Canadians through this source is about 50–90 μg/d (350/700 mg/week), compared with the FAO/WHO recommended "Provisional Tolerable Weekly Intake" of 400–500 μg. Less than 10% of dietary cadmium is absorbed by humans (4).

Sewage sludges indicate very high levels of cadmium. Our preliminary experiments indicate that it is possible that, in these treatment plants, large amounts of cadmium are emitted from incinerator stacks.

Table 1
Cadmium Content of Certain Canadian Foods

Sample	Cadmium level (mg/Kg)*
Wheat	0.29
Tea	0.004, 0.015
Coffee	0.007, 0.017
Beef	0.002 to 0.021 (0.061, 0.115)**
Lamb	0.003 to 0.016 (0.253)
Chicken	0.002 to 0.003 (0.063)

*Dry samples are expressed as mg/kg dry wt.; meat samples are expressed as mg/kg wet wt. The concentrations of individual samples are shown in the first three items (where enough samples to give a range had not been analyzed).

**The amounts in parentheses are levels beyond the range quoted and are mostly for organs such as liver and kidneys.

Table 2
Cadmium Content of Canadian Environmental Samples

Sample	Cadmium level (mg/Kg)
Fertilizer	0.302, 1.25 (n = 2)
Soil	0.55 to 1.72 (n = 8)
Sewage sludge	7 to 122 (n = 8)

This is a significant source of inhaled cadmium. Up to 50% of cadmium can be absorbed by humans when it is inhaled (*4*).

Cadmium in the Bone

The bone levels of rats given cadmium through drinking water are given in Table 3. The bone cadmium levels increased in these animals as a function of ingested dose. For example, the right femur of control rats contained 0.16 ± 0.14 mg/kg dry wt. On the other hand, the rats given 50 mg/L of cadmium for 5 wk showed a bone cadmium level of 1.09 ± 0.34 mg/kg. The rats fed 100 mg/L cadmium showed a bone level of 2.6 ± 0.54 mg/kg. Similar results were obtained when the right tibias of these animals were assayed. Histological examination of the bone samples

Table 3
Cadmium Content of Femurs of Rats Fed Cadmium[a]

Cd intake Conc.(mg/L)	Femur Cd (mg/Kg dry) (mean ± S.D.; n = 6)
0	0.16 ± 0.08
50	1.09 ± 0.34
100	2.60 ± 0.54

[a]Mode of ingestion: drinking water; period of ingestion: 5 wk.

from these animals showed no unmineralized osteoid surfaces. The kidney samples showed no lesions either. This indicates that cadmium directly goes to the bone even when the kidney is normal. Therefore, it appears possible that the toxic effect may also be primary to the bone rather than solely secondary to renal failure.

SUMMARY

Rat experiments show that intake of cadmium through drinking water results in the accumulation of cadmium in the bone. This is a primary phenomenon and not secondary to renal failure. The bone cadmium level was directly proportional to the ingested dose.

Cadmium is widely distributed in food, fertilizer, and soil. The highest amount of cadmium was found in sewage sludges, which are suspected to emit cadmium to the atmosphere during incineration.

ACKNOWLEDGMENT

The authors wish to acknowledge the financial support from the Natural Sciences and Engineering Research Council of Canada.

REFERENCES

1. N. Yamagata and I. Shigematsu, *Bull. Inst. Pub. Health* **19,** 1 (1970).
2. L. Friberg, T. Kjellstrom, G. F. Nordberg, and M. Piscator, *US EPS Document,* Environmental Protection Technol. Ser., Washington, DC, Rep. EPA-650/2-75-049 (1975).
3. R. E. Jervis, B. Tiefenbach, and A. Chattopadyay, *Can. J. Chem.* **52,** 3008–3020 (1974).
4. Effect of Cadmium in the Canadian Environment, *Document No. NRCC No. 16743,* NRC Canada (1979).

Elemental Concentrations in Hair of Inhabitants of a Cadmium-Polluted Area

T. TAKEUCHI,*[,1] Y. NAKANO,[1] A. AOKI,[2]
S. OHMORI,[3] AND M. KASUYA[4]

[1]Research Reactor Institute, Kyoto University, Kumatori-cho, Sennan-gun, Osaka 590-04, Japan; [2]Kyoto Prefectural University, Kyoto 606, Japan; [3]Osaka Prefectural Institute of Public Health, Osaka 537, Japan; and [4]Toyama Medical and Pharmaceutical University, Toyama 930-04, Japan

Received April 17, 1989; Accepted October 24, 1989

ABSTRACT

To investigate the present situation of inhabitants living in the cadmium-polluted area of Toyama Prefecture, 95 hair samples were analyzed instrumentally by the neutron activation method. The concentrations were compared with nonpolluted samples. The concentrations of Cd, Cu, Zn, and Cu/Zn in hair were compared with those in some reported organs. The values of Cu/Zn were significantly lower ($p < 0.001$) in hair of both female groups, but only somewhat lower ($p < 0.05$) in hair of the male group. Causes of these facts are discussed. Results obtained from principal component analysis are also mentioned.

Index Entries: Instrumental neutron activation analysis; human hair; cadmium-polluted area; Toyama Prefecture; copper to zinc, ratio of; Itai-itai disease; principal component analysis.

*Author to whom all correspondence and reprint requests should be addressed.

INTRODUCTION

An outbreak of a disease with severe bone pain (Itai-itai disease, abbreviated throughout as I-disease) occurred in the Jintsu River basin of the Toyama Prefecture in Japan early in this century. The disease was caused mainly by Cd, stemming from a mine located in the upper area of the river. Cadmium incorporated from drinking water and eating food containing large amounts of this element causes renal tubular dysfunction (characterized by the excretion of low molecular weight protein), followed by osteomalacia, together with osteoporosis.

Since the 1970s, several measures to protect human health from Cd pollution have been undertaken to cope with this difficult problem. At this stage, it is important to evaluate precisely the present status of those who live in the polluted area and to what extent they are still contaminated with Cd, as well as to study the effects of Cd and other elements on the metabolism of the human body.

METHODS

Hair samples were collected in 1982 from the inhabitants of the Toyama district, where it has been designated officially as a Cd-polluted area. The area has been polluted with Cd, Cu, and Zn by factors of about 4, 1.5, and 5 times respectively, compared with nonpolluted areas (*1*). The results of a health survey of the inhabitants of Toyama Prefecture showed that the incidence of people who show low molecular weight protein in their urine is about 3 times higher and that the incidence of people whose renal tubular reabsorption of P is less than 80% is over 10 times higher than those of people in nonpolluted areas (*2*).

Sample analysis was carried out with the same procedure that was used previously for the samples of nonpolluted Japanese (abbreviated throughout as "Normal") (*3,4*). Since it is well known that the elemental concentrations in hair vary with age, sex, and hair treatment, the obtained data were analyzed statistically in the same way mentioned in ref. (*5*).

RESULTS AND DISCUSSION

Concentrations of 27 elements in hair of the Cd-polluted area were compared in refs. (*6,7*) with those of Normal. In the present paper, attention will be paid to the concentrations of Cd, Cu, and Zn. It has been pointed out that not only the absolute concentration of Cu is important to human health, but also the concentration ratios of Cu to other elements, especially to Zn, have very important meanings from the viewpoint of antagonism between these two elements (*8*).

Medians and their geometric standard deviations of Cu/Zn ratios in hair obtained from three adult groups of the Cd-polluted area are compared with those of Normal in Table 1. The concentration ratios of Cd, Cu, Zn, and Cu/Zn in hair of inhabitants of the Cd-polluted area to those of Normal are compared with those ratios in some organs calculated from the reported values (*9*) in Fig. 1. From the Fig. 1, the following may be concluded about the concentrations in hair of the inhabitants of the Cd-polluted area, compared with those of nonpolluted inhabitants: (1) The concentrations of Cd and Zn in hair are significantly higher in both the female and male groups, but for Cu, significantly lower concentrations were obtained only in the female group. (2) The values of Cu/Zn in hair are most significantly lower ($p < 0.001$) in both of the female groups, but in the male group a significantly lower value ($p < 0.05$) is obtained, (3) The concentration ratio pattern seen in hair is quite similar to those seen in liver and pancreas although the age range and the extent of Cd exposure are largely different from those of people whose organs were analyzed and (4) The concentration of Cu in organs is significantly lower only in liver, but the values of Cu/Zn are significantly lower not only in liver, but also in pancreas.

These facts are consistent with the results reported by Niimura et al. (*10*) that significantly higher and lower excretions of Cu and Zn in urine have been seen as the result of renal tubular and glomerulus dysfunctions, respectively, which may be caused by excess incorporation of Cd. Furthermore, one of the reasons why the values of Cu/Zn are significantly lower in hair of the female groups, but why in hair of the male group the value is only somewhat significantly lower, can be considered from the fact that the amount of Cu lost from a mother is very large (about 0.4 mg/d) during the lactation period (*8*). This phenomenon is also consistent with the fact that the conditions of I-disease patients have been heaviest in women who had experienced more childbirths.

From principal component analysis of the correlation coefficient matrix, the factor loadings of elements were estimated. In Fig. 2, the first and second factor loadings obtained for 13 elements (the main 12 elements + Cd) in the Normal group are compared with those of the Normal + Toyama group. It can be seen in Fig. 2 that the second factor loading of Cd is large, but the first one is small in the Normal group, whereas the first factor of Cd becomes very large in the Normal + Toyama group. It may be concluded from this fact that Cd is related closely to the second factor in the Normal group, but that a close correlation to the first factor may occur by the excess incorporation of Cd. It is also shown in Fig. 2 that in the Normal + Toyama group, elements such as Br, Mn, and Zn, which have significant positive correlation coefficients to Cd (*5*), move in the same direction as Cd, whereas Cu moves perpendicular to Cd.

In order to evaluate the extent of contamination by Cd on individuals, the first and second factor scores are shown in Fig. 3 for each sex,

Table 1
Comparisons of Cu/Zn Ratios in Hair
of Inhabitants of the Cd-Polluted Area with Those of Nonpolluted Areas

Group	Nonpolluted (n)	Polluted (n)
Male	0.058 $\overset{\times}{\div}$ 1.4 (108)	0.049 $\overset{\times}{\div}$ 1.4 (22)[a]
Female without hair waving	0.068 $\overset{\times}{\div}$ 1.4 (39)	0.030 $\overset{\times}{\div}$ 1.4 (11)[b]
Female with hair waving	0.064 $\overset{\times}{\div}$ 1.6 (80)	0.037 $\overset{\times}{\div}$ 1.4 (51)[c]

[a]$p < 0.05$.
[b]$p < 0.001$.
[c]$p < 0.001$.

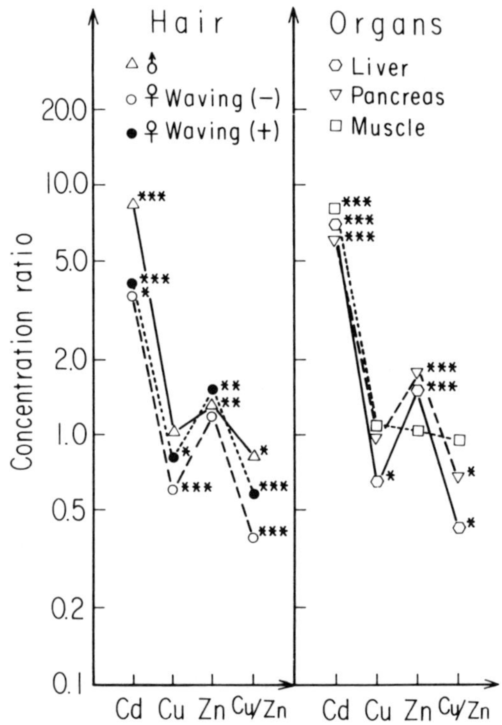

Fig. 1. Concentration ratios in hair of inhabitants of the Cd-polluted area to inhabitants of nonpolluted areas, compared with those in organs calculated from the arithmetic means reported in ref. (9), which were obtained from 49 cases of the Cd-polluted area (7 males + 42 females; 17 designated + 32 suspected as I-disease patients; age = 80 ± 6) and 27 cases of nonpolluted areas (11 males + 16 females; age = 79 ± 12).

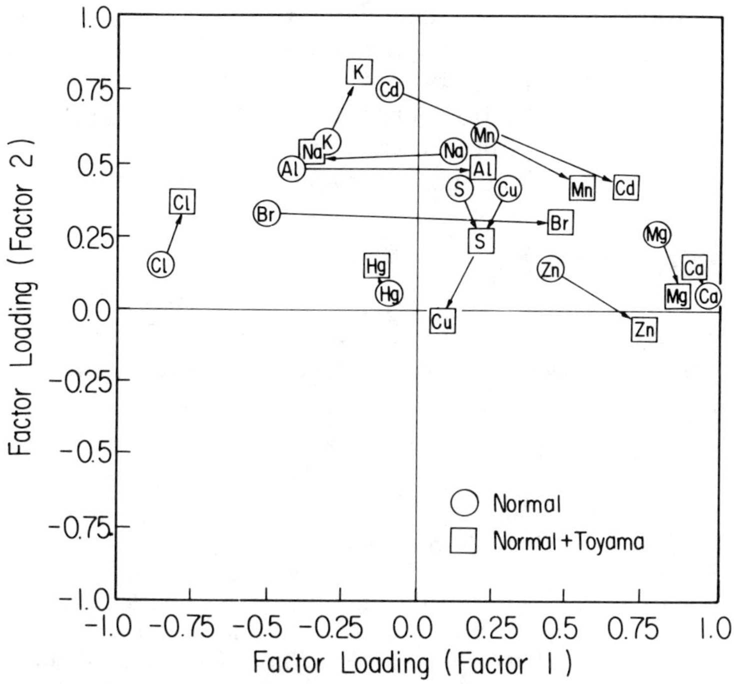

Fig. 2. Factor loadings calculated for 13 elements (the main 12 elements + Cd) in the Normal and Normal + Toyama groups, excluding females with waved hair.

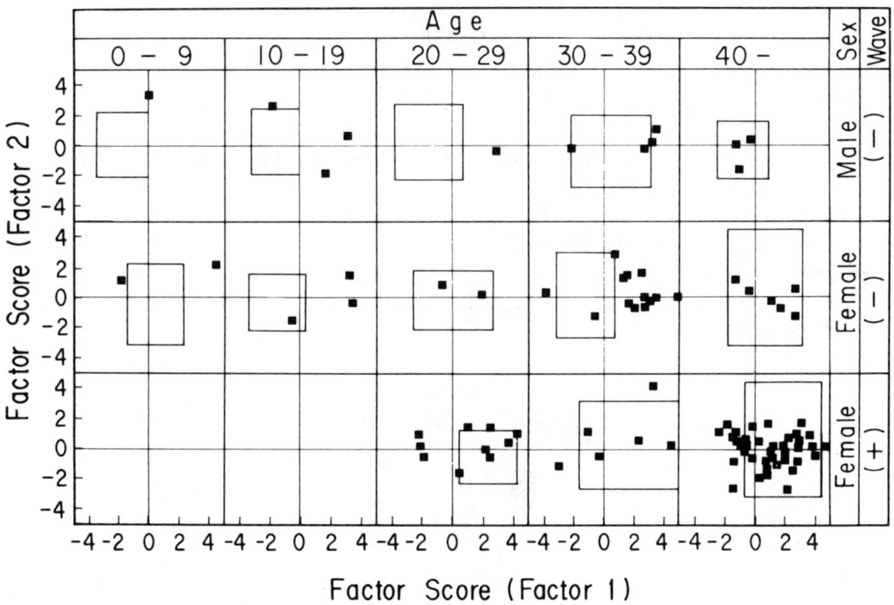

Fig. 3. Factor scores for each sex, generation, and hair treatment. The solid lines indicate the arithmetic mean ± twice the standard deviation calculated from those of the Normal.

generation, and hair treatment calculated from the main twelve elements in the Normal + Toyama group. It is shown in Fig. 3 that in the Toyama group, many samples show abnormal values not only in the older, but also in the younger generation. From this fact, it may be concluded that children living in the polluted area are also affected by Cd. This is consistent with the results obtained from Cd concentrations in urine (*11*).

ACKNOWLEDGMENT

The authors are grateful to H. Kozuka of the Toyama Medical and Pharmaceutical University for his support and interest in this study.

REFERENCES

1. M. Yanagisawa, Y. Niimura, N. Yamada, A. Segawa, and K. Kida, *Bull. Toyama Agric. Exp. Stan.* **15,** 8 (1984).
2. Dept. Public Welfare, Toyama Pref., private communication.
3. T. Takeuchi, T. Hayashi, J. Takada, M. Koyama, H. Kozuka, H. Tsuji, Y. Kusaka, S. Ohmori, M. Shimogi, A. Aoki, K. Katayama, T. Tomiyama, *Nuclear Activation Techniques in the Life Sciences,* IAEA ed., Vienna, 1979, pp. 545–561.
4. T. Takeuchi, T. Hayashi, J. Takada, Y. Hayashi, M. Koyama, H. Kozuka, H. Tsuji, Y. Kusaka, S. Ohmori, M. Shimogi, A. Aoki, K. Katayama, T. Tomiyama, *J. Radioanal. Chem.* **70,** 29 (1982).
5. T. Takeuchi, Y. Nakano, A. Aoki, S. Ohmori, and T. Tsukatani, *J. Radioanal. Nucl. Chem. Articles* **112,** 259 (1987).
6. T. Takeuchi, Y. Nakano, A. Aoki, S. Ohmori, and T. Tsukatani, *Annu. Rep. Res. Reactor Inst.* **19,** 89 (1986).
7. T. Takeuchi, Y. Nakano, A. Aoki, S. Ohmori, and T. Tsukatani, *Annu. Rep. Res. Reactor Inst. Kyoto Univ.* **20,** 159 (1987).
8. E. J. Underwood, *Trace Elements in Human and Animal Nutrition,* Academic, New York, 1975.
9. Y. Kuzuhara, M. Fijita, C. Hayashi, K. Sumino, and S. Kitamura, *Jpn. J. Public Health* **33 (10),** 817 (1986).
10. T. Niimura, T. Tanaka, and K. Shiroishi, *Jpn. J. Hyg.* **43,** 129 (1988).
11. K. Nogawa, E. Kobayashi, T. Honda, and I. Tsuritani, *Jpn. J. Public Health,* **28 (10),** 364 (1981).

Determination of Manganese in Airborne Particulates in a Dry-Cell Battery Factory Using X-Ray Fluorescence Technique

T. SUMITRA* AND S. KONGSOMBATSUK

Department of Nuclear Technology, Chulalongkorn University, Bangkok, Thailand

Received April 17, 1989; Accepted October 19, 1989

ABSTRACT

Feasibility of using X-ray fluorescence techniques for the determination of airborne manganese in dry-cell battery factories was investigated. Both Energy Dispersive (EDX) and Wavelength Dispersive (WDX) X-ray fluorescence techniques of analysis were used. The minimum detectable quantity of Mn by the EDX method was found to be 50 μg for membrane filter (AA type) and 100 μg for cellulose filter (pore size 2.5 μm), respectively, with a counting time of 2000 s. The most suitable exciting source was 1.11 GBq (30 mCi) Pu-238. The minimum detectable quantity of Mn by the WDX method was 10 μg, with a counting time of 100 s. Results of the X-ray fluorescence and the AA methods were found to be in good agreement. Field measurements were carried out in a dry-cell battery factory. Concentrations of airborne manganese in the factory were found to be 0.02–41.1 mg/m^3.

Index Entries: X-ray fluorescence technique; airborne manganese, determination of; dry-cell battery factories, airborne manganese in; Energy Dispersive (EDX), Wavelength Dispersive (WDX) X-ray fluorescence; Mn, minimum detectable quantity of; the X-ray fluorescence and AA methods, results of; airborne manganese, concentration of.

*Author to whom all correspondence and reprint requests should be addressed.

INTRODUCTION

In 1964, the first case of "manganese poisoning" was reported at a dry-cell battery factory in Thailand. About 41 workers in the milling section suffered from chronic manganese poisoning, resulting in permanent disability (*1*). Since then, the dry-cell battery factories have been regularly inspected by the Division of Occupational Health, Department of Public Health. Therefore, determination of Mn in airborne particulates in the working environment is very important.

The most commonly used method of determining Mn is Atomic Absorption Spectrophotometry (AAS). This method was used from 1982–1984 to determine the concentration of Mn in air (*2,3*) when the Environmental Research Institute, Chulalongkorn University, had undertaken a coordinated research program on the hazard of Mn in some dry-cell battery factories in Bangkok and Samutprakarn, 25 km east of Bangkok. The method is relatively simple and fast, and the sensitivity is high. However, other analytical methods may be useful and should be encouraged. In this study, feasibility of using the X-ray fluorescence method (XRF) was investigated. X-ray fluorescence has been applied to directly determine airborne metallic dusts and fumes collected on filter paper (*4,5*). The method may provide a means for nondestructive and faster analysis of Mn. It may be cheaper when a simple EDX system could be used.

METHODOLOGY

Air samples were taken in a dry-cell battery factory in the Samutprakarn Province. The milling section and the mixing section have been previously identified as areas with high concentrations of MnO_2 particulates in air (*3*). However, in this study, air samples were taken in three work sections, i.e., milling, mixing, and tamping. The samplings took place in August and October 1987 and February 1988. Personal air samples were taken on workers in high risk groups. The duration of each sampling was about 15–30 min, depending on the work cycle of that particular worker. Two types of filters were used, i.e., the standard 37 mm membrane filters (pore size 0.8 μm, Type AA) and the 37 mm cellulose filters (pore size 2.5 μm). Standard three-piece cassette filter holders were employed. The flow rate used was about 1.5 L/min. The pesonal air samplers employed the low-cost, in-house developed pumps.

The analytical methods used in this study were the XRF and the standard AAS methods. Both the energy dispersive (EDX) and wavelength dispersive (WDX) X-ray fluorescence techniques were used. The EDX system consisted of an isotopic exciting ring source, a HpGe detector, a preamplifier, an amplifier, and an MCA. The system was arranged coaxially, with the X-ray source in the middle between the detector and the sample to be measured. The source was placed directly on the detector, but for Am-241 and Cd-109, a Cu ring of 5 mm thick was placed between the source and detector to avoid "detector shock." The detection system was cooled with liquid nitrogen. Three isotopes were used, i.e., a 1.11 GBq (30 mCi) Pu-238, a 1.11 GBq (30 mCi) Am-241, and a 0.74 GBq (20 mCi) Cd-109. The distances between these sources and the samples were 7, 7, and 5 mm, respectively, which were found to give a maximum number of counts. Acrylic rings of corresponding thickness were used as spacers. The measuring chamber was made of acrylic and was at atmospheric pressure. The filter was excited on the reverse side. This arrangement resulted in less fluorescent X-rays to the detector, but was necessary to avoid possible Mn dust contamination of the detection system. The characteristic 5.9 keV K X-ray of Mn was counted.

For the WDX method, the JEOL mod. JSX-60PA system was used. Since the measuring chamber must be operated in vacuum, the filters were wrapped with 0.5 mL polyethylene foils to avoid possible contamination of the detection system. In the case of AAS, the PERKIN ELMER 4000 was used.

RESULTS

It was found that, using a counting time of 2000 s, the simple EDX method, with no vacuum chamber, could analyze Mn particulates collected on a membrane filter to a concentration of about 50 μg, with 0.81% standard deviation. The minimum detectable amount of Mn on a cellulose filter was found to be about 100 μg with the same counting time. When standard addition technique was used, the calibration curves for the EDX method using Pu-238, Am-241, and Cd-109 were as shown in Fig. 1. With the same technique, the calibration curves for the WDX method were as shown in Fig. 2. The minimum detectable amount of Mn by the WDX method was found to be 10 μg with a counting time of 100 s.

A total of 83 samples were collected and analyzed by the EDX method. The results of this preliminary analysis in the three work sections are summarized as in Table 1. The airborne Mn concentrations in the tamping section were consistently lower than the detection limit; therefore, samples from this section were not analyzed further. Later on, when WDX and AAS equipment became available, comparison between the three analytical methods were performed. Analysis of all samples were not possible because of budget limitations and limited equipment time available for this project. Only 32 samples were randomly selected from the milling and mixing sections. The samples were analyzed first by EDX, then WDX, and finally by the AAS method. The concentrations of Mn in airborne particulates were determined, with the results for cellu-

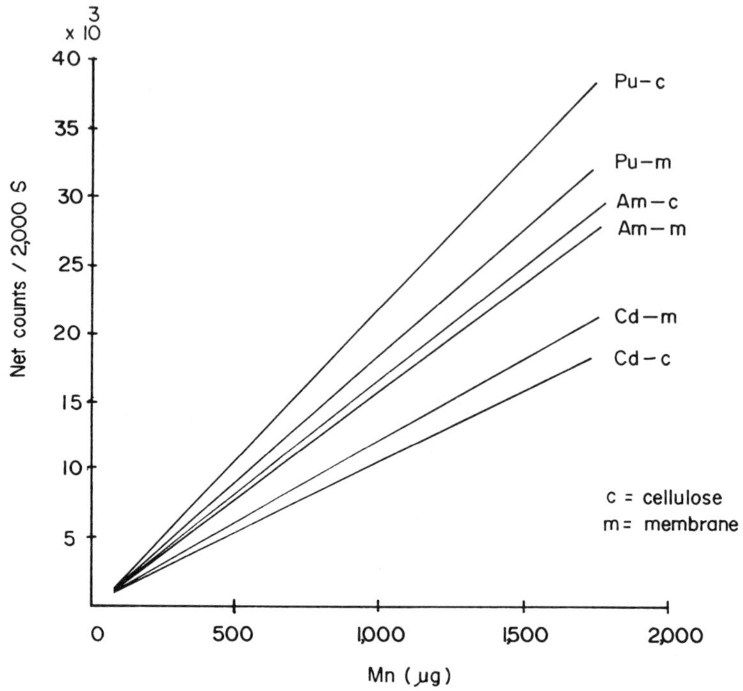

Fig. 1. Calibration curves for the EDX method.

lose filters shown in Table 2 and for membrane filters in Table 3. Concentrations of Mn in air were found to vary from 0.02 to 19.66 mg/m^3 when using WDX and 0.2 to 19.60 mg/m^3 when using AAS. For the EDX method, the maximum concentration of Mn in air was found to be 20.29 mg/m^3, using Pu-238. The method was not sensitive for Mn concentrations in air of less than 1.6 mg/m^3. Good agreement between the results, obtained by the WDX and AAS methods, can be seen in Fig. 3. In the case of cellulose filters, the correlation coefficient was found to be 0.989, and for membrane filters, the correlation coefficient was 0.987. Therefore, the two methods could be said to yield comparable results. Agreement between the results of the EDX method with the other two methods could be noticed only for Mn concentrations in air exceeding 3 mg/m^3.

From the 83 samples analyzed by the EDX method, with Pu-238 as exciting source, the maximum concentration of Mn in air was found to be about 41.1 mg/m^3 in the milling section and 20.3 mg/m^3 in the mixing section. It should be noted that, when air samples were taken in this same factory in 1984, the maximum Mn concentration in air in these two sections were found to be 17.7 and 15.4 mg/m^3, respectively (3). It can be concluded that the risk of having concentrations of Mn in air exceeding the ceiling value of 5 mg/m^3, in these two sections, is still very high; thus, care, as well as personal protection gears, should always be applied.

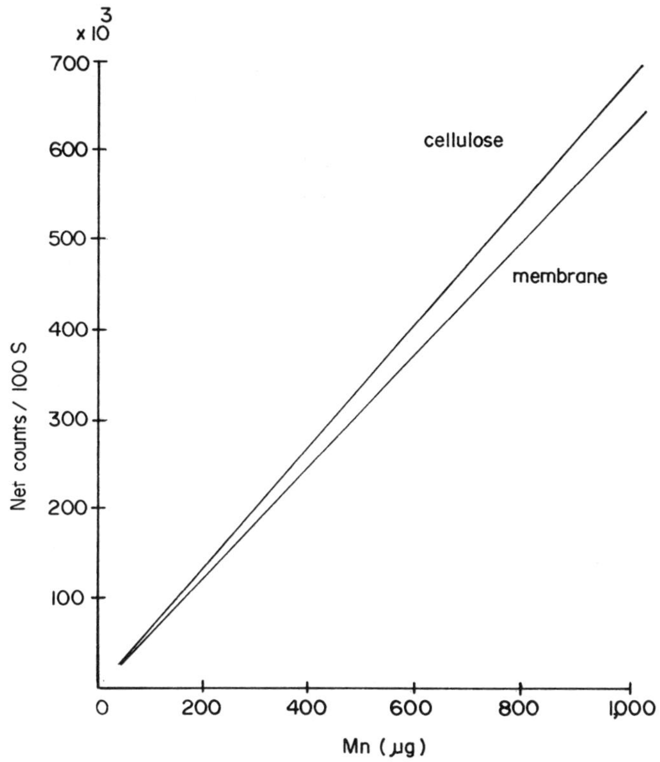

Fig. 2. Calibration curves for the WDX method.

DISCUSSION

The determination of airborne Mn concentrations to identify the ceiling level by the XRF method is feasible with the existing equipment. The EDX method is comparable to the AAS and WDX methods for concentrations greater than 3 mg/m^3. In this case, the method would be fast, economical, and sufficiently reliable. It is convenient because samples can be analyzed directly without preparation. The advantage of the nondestructive method is evident in this case. If extensive checking of Mn concentrations in air is required, the EDX method could be used as a screening method to check against the ceiling level. Samples with high concentrations of Mn may be re-analyzed more carefully with other analytical methods, e.g., AA, WDX, and so on. For the WDX method, it could be considered comparable to the AAS method, even for concentrations as low as 0.4 mg/m^3. The method may be used when the AAS equipment is not available. In this case, polyethelene wrap should be used to avoid contamination of the detection system. Care must also be taken when analyzing membrane filters with the WDX method, because

Table 1
Maximum Concentration of Mn in Airborne Particulates in 3 Working Sections, as Determined by the EDX Method Using Pu-238[a]

Section	No. of Samples	Maximum Concentration (mg/m^3)
Milling	17	41.1 ± 0.68
Mixing	50	20.3 ± 1.21
Tamping	16	ND

[a]ND = Not Detectable (<1.6 mg/m^3).

Table 2
Concentrations of Mn in Airborne Particulates, as Analyzed by the EDX, WDX, and AA Methods (Cellulose Filter)[a]

Analytical Method	XRF Method (mg/m^3)				AA Method (mg/m^3)
	EDX			WDX	
	Pu-238	Am-241	Cd-109		
Sample No.					
1	ND	ND	ND	0.42 ± 0.002	1.41 ± 0.012
2	2.36 ± 0.70	2.05 ± 0.96	2.16 ± 0.54	3.01 ± 0.015	3.06 ± 0.007
3	ND	2.51 ± 1.17	2.03 ± 0.65	2.91 ± 0.015	3.51 ± 0.008
4	7.24 ± 0.70	6.40 ± 1.63	7.17 ± 0.91	6.40 ± 0.025	7.36 ± 0.012
5	ND	ND	ND	0.91 ± 0.010	1.04 ± 0.010
6	11.67 ± 0.78	15.79 ± 1.82	11.50 ± 1.01	15.94 ± 0.042	13.55 ± 0.013
7	ND	ND	ND	2.71 ± 0.022	2.49 ± 0.017
8	ND	ND	ND	0.57 ± 0.008	0.50 ± 0.008
9	6.20 ± 0.70	8.81 ± 1.64	ND	8.56 ± 0.029	7.57 ± 0.012

[a]ND = Not Detectable.

a long exposure time may result in damage to the filters because of heat imparted. This is no problem for cellulose filters.

SUMMARY

A method for determining Mn concentration in air has been investigated and field tested. The technique employs both the energy-dispersive (EDX) and the wavelength-dispersive (WDX) X-ray fluorescence of Mn collected on membrane and cellulose filters. The EDX method

Table 3
Concentrations of Mn in Airborne Particulates, as Analyzed by the EDX, WDX, and AA Methods (Membrane Filter)[a]

Analytical Method	XRF Method (mg/m^3)				AA Method (mg/m^3)
	EDX			WDX	
	Pu-238	Am-241	Cd-109		
Sample No.					
1	5.69 ± 0.44	6.11 ± 0.59	4.96 ± 0.78	4.11 ± 0.013	3.74 ± 0.005
2	ND	ND	ND	0.27 ± 0.002	0.26 ± 0.002
3	4.39 ± 0.61	4.51 ± 0.80	3.58 ± 1.07	4.57 ± 0.016	3.78 ± 0.006
4	ND	ND	ND	0.30 ± 0.004	0.46 ± 0.006
5	ND	ND	ND	0.27 ± 0.004	0.40 ± 0.005
6	3.40 ± 0.40	3.29 ± 0.53	2.92 ± 1.42	3.14 ± 0.011	2.64 ± 0.004
7	7.91 ± 0.43	8.05 ± 0.57	7.38 ± 0.75	6.91 ± 0.017	5.91 ± 0.004
8	ND	ND	ND	0.20 ± 0.003	0.33 ± 0.004
9	ND	2.33 ± 0.51	ND	1.42 ± 0.007	1.47 ± 0.004
10	ND	ND	ND	1.96 ± 0.015	3.00 ± 0.012
11	7.37 ± 0.72	7.57 ± 0.95	4.72 ± 1.27	7.40 ± 0.022	5.06 ± 0.007
12	4.53 ± 1.05	5.76 ± 1.38	2.28 ± 1.84	2.46 ± 0.016	2.26 ± 0.011
13	ND	ND	ND	0.91 ± 0.009	1.04 ± 0.010
14	16.51 ± 1.15	17.90 ± 1.53	13.06 ± 2.03	17.40 ± 0.043	NM
15	ND	ND	ND	1.22 ± 0.012	1.47 ± 0.012
16	1.57 ± 0.68	2.49 ± 0.89	ND	1.08 ± 0.021	1.26 ± 0.007
17	4.51 ± 0.83	4.52 ± 1.09	2.30 ± 1.44	3.42 ± 0.016	3.63 ± 0.009
18	11.96 ± 1.16	10.54 ± 1.52	7.60 ± 2.03	11.84 ± 0.034	9.73 ± 0.012
19	ND	ND	ND	1.16 ± 0.010	1.21 ± 0.010
20	20.29 ± 1.21	22.30 ± 1.70	11.82 ± 2.26	19.66 ± 0.043	19.60 ± 0.013
21	ND	ND	ND	1.50 ± 0.015	1.65 ± 0.017
22	ND	ND	ND	0.02 ± 0.002	0.51 ± 0.009
23	2.75 ± 1.12	4.11 ± 1.52	1.70 ± 2.01	2.44 ± 0.019	3.21 ± 0.012

[a]ND = Not Detectable; NM = Not Measured.

employs a coaxially-arranged detection system. Pu-238 was found to be better than Am-241 and Cd-109. The minimum detectable quantity of Mn was found to be 50 μg for membrane filters and 100 μg for cellulose filters. The counting time in both cases was 2000 s. For the WDX method, the counting time was 100 s, and the minimum detectable quantity of Mn was found to be about 10 μg. The method was applied to the situation of a dry-cell battery factory, where the ceiling level of Mn in air was to be determined. The concentrations of Mn in air were found to have a maximum of about 41.1 mg/m^3 in the milling section and about 20.3 mg/m^3 in the mixing section of the factory.

Results of field tests were compared to those obtained by the standard AAS method. The results of the WDX and AAS methods were in

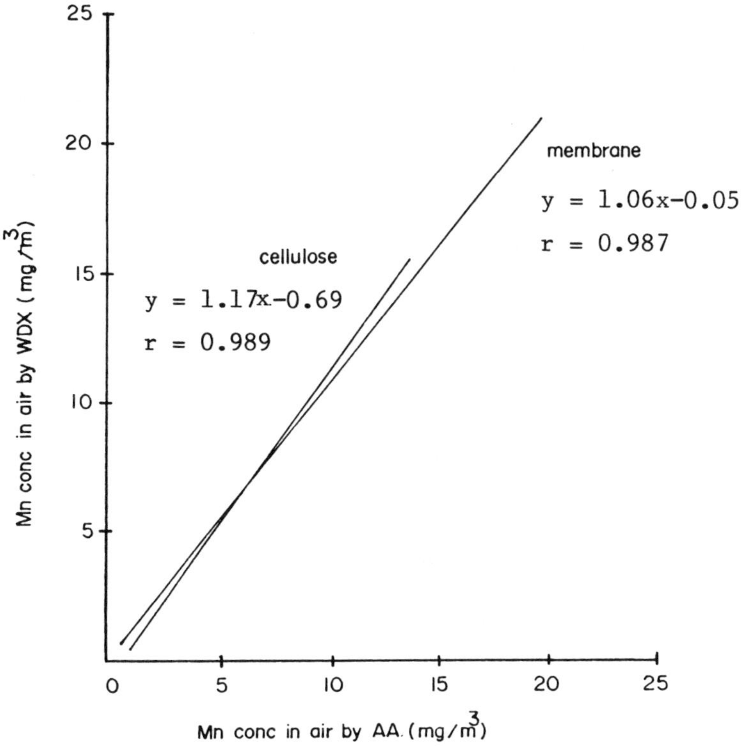

Fig. 3. Correlation between Mn concentration in air, determined by the AAS and WDX methods.

good agreement in the whole range of concentrations. In contrast, the results of the EDX and AAS methods were in agreement only for concentrations greater than 3 mg/m^3.

ACKNOWLEDGMENTS

The authors are deeply indebted to Nares Chankow and Suvit Punnachaiya, assistant professors of the Department of Nuclear Technology, Faculty of Engineering, Chulalongkorn University, who provided them with valuable advice and suggestions about the XRF detection systems. Acknowledged are Thirapol Kangkaket, scientist of the Environmental Research Institute, Chulalongkorn University, who was responsible for the analysis by the AAS method and Wanida Chinnasart, who was responsible for the analysis by the WDX method at the Scientific and Technological Research Equipment Center, Chulalongkorn University. Acknowledgement is also due to Somyot Srisatit, Bancha Unpanich, and Achara Sakulman for the preparation of this manuscript.

REFERENCES

1. Department of Public Health, *J. Public Health* **1,** 178–183 (in Thai) (1975).
2. P. Krieosakul, *J. Environ. Res.* **5,** (in Thai) (1983).
3. T. Sumitra, *J. Environ. Res.* **8,** 23–38 (in Thai) (1986).
4. H. E. Bumsted, *Am. Ind. Hyg. Assoc. J.* **25** (1964).
5. J. W. Cares, *Am. Ind. Hyg. Assoc. J.* **29** (1968).

Environmental Impact of the Gold Mining Industry in Ghana

S. AKOTO BAMFORD,*[1] E. OSAE,[1] I. ABOH,[1]
C. A. BINEY,[2] AND L. A. ANTWI[2]

[1]*Physics Department, National Nuclear Research Institute, Kwabenya, Ghana; and* [2]*Institute of Aquatic Biology, Accra, Ghana*

Received April 17, 1989; Accepted October 23, 1989

ABSTRACT

X-ray Fluorescence (XRF) Analysis and Atomic Absorption Spectrophotometry (AAS) have been used in assessing heavy metal pollution from some gold mines in Ghana. The presence and levels of heavy metals in gold ore, gold tailings, inland waters, and river sediments have been determined. Using these techniques, the heavy metals: Cr, Mn, Fe, Cu, Zn, As, Pb, Rb, Sr, Y, Zr, and Nb were identified in some of the solid samples within a concentration range of 0.08 ppm—4.9%. However, the inland waters showed the presence of only Fe and Zn at levels of 0.08–2.4 μg/mL.

Index Entries: X-ray fluorescence analysis; atomic absorption spectrophotometry; heavy metal pollution.

INTRODUCTION

For a long time, gold mining has been the most important mining industry in Ghana, and Ghana is one of the major gold producing countries in the world (*1*). There are now five producing gold mines in Ghana, and more licenses are currently being granted to private firms for gold prospecting and processing. However, the nature of the gold depos-

*Author to whom all correspondence and reprint requests should be addressed.

its in Ghana (1) and the process of gold mining have been contributing toward the pollution of the environment.

Figure 1 is a schematic diagram showing the essential principles involved in the processing of gold, from the ore to the gold bullion. From a study of the said diagram, the following potential pollution sources can be identified in the gold mining industry: gold ore, dust particulates, chemical reagents used in gold processing, gaseous emissions, and gold tailings. In this work, attention has been focused on the environmental impact of the gold ore and gold tailings in two of the gold mines (Tarkwa and Prestea mines), with special reference to their effect on surrounding inland waters—namely streams, ponds, rivers receiving effluent from the mines, underground water sources, and tap water.

The techniques of X-ray Fluorescence and Atomic Absorption Spectrophotometry have been used to analyze the presence and concentration levels of heavy metals in gold ore, gold tailings, and inland waters, including river sediments.

METHODS

Analytical Methods

In the XRF analysis, Cd-109 radioisotope and a molybdenum X-ray tube with a Zr secondary target were the excitation sources used for irradiating the samples. The fluorescent X-rays produced were detected with a Canberra Si(Li) detector having a resolution of 186 eV at 6.4 KeV, and analysis was done with a Canberra 35+ multichannel analyzer linked to a mini-computer.

The AAS facility used was a Varian 1275. The metals Cr, Fe, Cu, Zn, As, Pb, and Cd were determined by flame atomization and Hg by cold vapor atomic absorption spectrophotometry.

Sampling and Sample Preparation

Gold ore samples from the Prestea mines and gold tailing samples from the Prestea and Tarkwa mines were collected, sun-dried, and ground to a very fine powder. "Thin" and "infinitely thick" samples (2) of each were prepared for XRF analysis. The fundamental parameters method (3) was used in the quantitative analyses. For AAS, the solid samples were wet-digested in a teflon vessel.

The inland water samples were collected in previously cleaned polyethylene containers, and the water samples were acidified to a pH of about 2 at site (4). For XRF analysis, 100 mL portions of the water samples were adjusted to a pH of about 4 and then preconcentrated onto 0.45 μm Millipore filters, using ammonium pyrolidinedithiocarbamate (APDC) as the precipitant (5,6). For AAS, the water samples were shaken, allowed to settle, and about 2 mL of the clarified portion was sucked into the atomizer of the AAS for analysis.

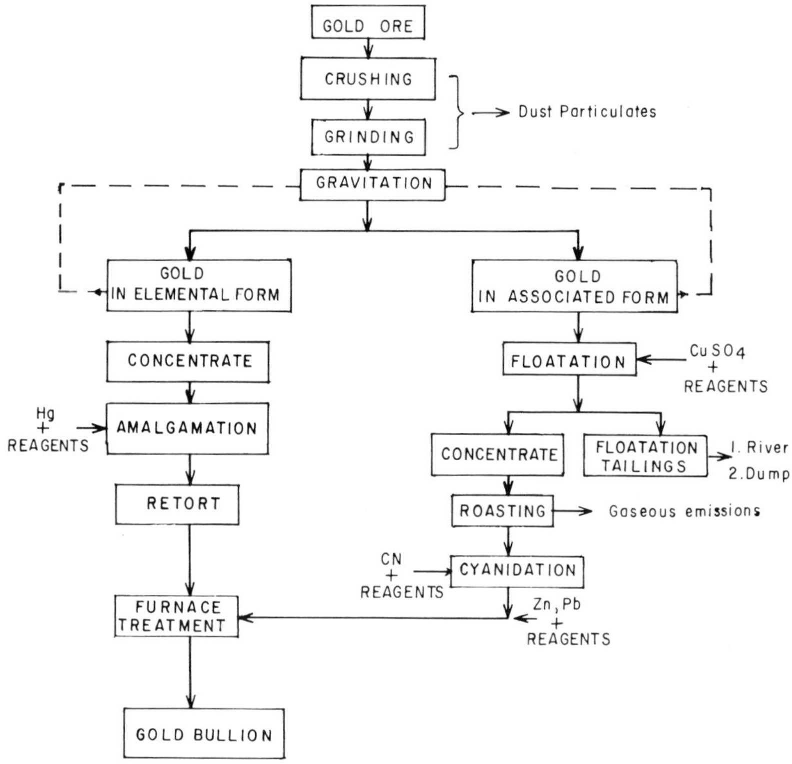

Fig. 1. Schematic diagram showing essential principles in gold processing.

For the inland water analyses, standard samples of the metal ions were prepared from stock solutions of Fluka metal ion standard solutions.

RESULTS AND DISCUSSION

Analytical quality control tests were applied to the AAS technique using certified reference materials of mussel and fish muscle (7). Figures 2 and 3 are spectra of gold tailing samples from the Prestea and Tarkwa mines, respectively. The metals Mn, Fe, Zn, Rb, Sr, Y, and Zr were found in both tailings, whereas Cu, As, and Pb were identified in the Prestea tailing only. This confirms the difference in the nature of the ores from the two mines, with the Prestea ore having a relatively greater proportion of arsenopyrites. This difference also dictates the type of processing used in extracting the gold ore. Thus, the Prestea mines tend to introduce more pollution sources than the Tarkwa mines.

Tables 1 and 2 give the concentration of the metals in gold ore and gold tailings, as measured using XRF and also AAS. Both techniques identified most of the elements, but XRF could not identify Hg and Cd,

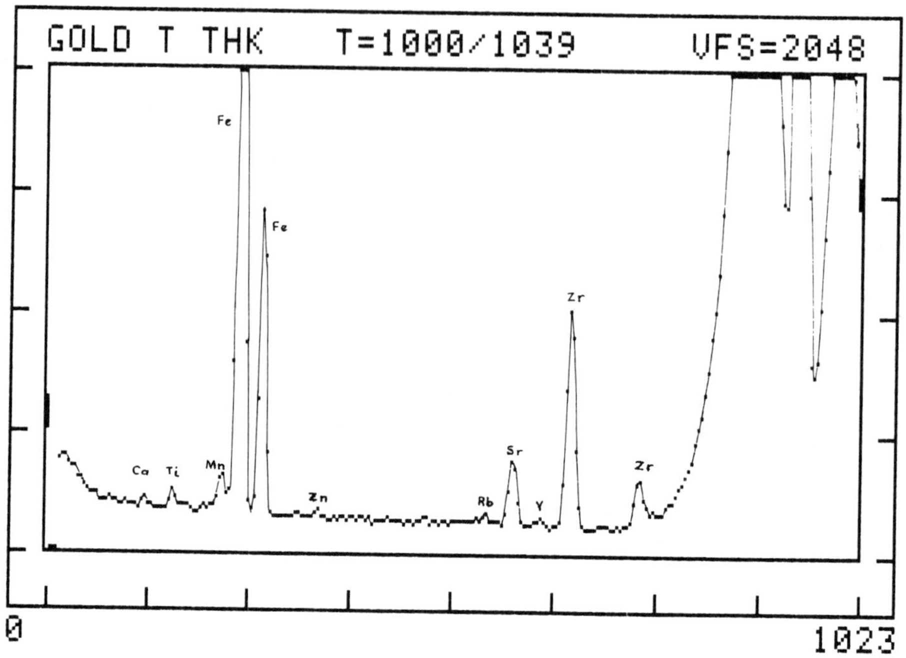

Fig. 2. Typical X-ray spectrum of Tarkwa gold tailings.

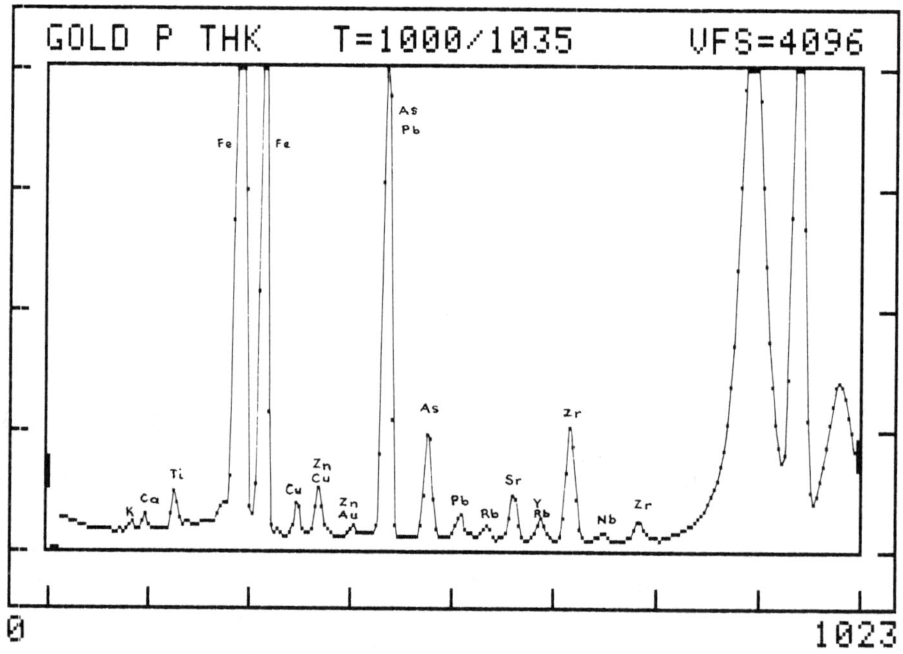

Fig. 3. Typical X-ray spectrum of Prestea gold tailings.

Table 1
Metal Concentrations (% by wt.) in Gold Ore and Tailings Using XRF

Location	Hg	Cu	Pb	Zn	Cd	Fe	Mn	Cr	As	Rb	Sr
Gold Ore - Prestea	+	0.011 ±.002	++	0.011 ±.004	+	4.89 ±.09	0.109 ±.030	1.12 ±.05	2.54 ±.09	0.01 ±.004	0.009 ±.
Tailings - Prestea	+	0.08 ±.01	++	0.03 ±.01	+	3.6 ±.06	++	++	0.06 ±.005	0.01 ±.002	0.02 ±.003
Tailings - Tarkwa	+	+	++	++	+	2.9 ±.05	0.07 ±.02	++	+	++	0.05 ±.008

+ = Element not identified; ++ = element identified, but not quantified [<LLD].

Table 2
Metal Concentrations (% by wt.) in Gold Ore and Tailings Using AAS

Location	Hg	Cu	Pb	Zn	Cd	Fe	Mn	Cr	As
Gold Ore - Prestea	0.16(-4) ±.24(-5)	0.010 ±.004	0.005 ±.001	0.009 ±.4(-3)	0.15(-3) ±.2(-4)	4.00 ±.32	0.08 ±.008	0.80 ±.24	1.53 ±.20
Tailings - Prestea	0.17(-4) ±.32(-5)	0.05 ±.003	0.002 ±.5(-3)	0.01 ±.006	.90(-4) ±.12(-4)	2.4 ±.2	0.06 ±.007	0.70 ±.18	0.08 ±.01
Tailings - Tarkwa	0.08(-4) ±.12(-5)	0.001 ±.7(-3)	0.001 ±.2(-3)	0.003 ±.2(-3)	.51(-4) ±.07(-4)	2.0 ±.24	1.10 ±.13	0.60 ±.16	+

Table 3
Trace Metal Concentrations in Water Samples, Using the AAS Technique[a]

Location	Fe	Zn
Prestea		
River Ankobra - Upstream	2.4 ± .2	
- Confluence	1.0 ± .08	
- Downstream	4.4 ± 0.3	
Stream	1.9 ± 0.1	
Tarkwa		
Pond 1	0.5 ± .04	
Pond 2	0.1 ± .07	
Stream 1 - Upstream	3.6 ± 0.3	
- Confluence	0.08 ± .006	
- Downstream	0.3 ± .02	
Underground water	0.006 ± .005	
Stream 4	1.0 ± .08	0.5 ± .03

[a]Concentrations in μg/mL.

and AAS could also not analyze for Rb, Sr, and Zn. Analysis of the inland water samples identified only Fe and Zn, whereas the sediments of these waters showed the presence of Hg, Cu, Pb, Zn, Cd, Fe, Mn, Cr, and As.

SUMMARY

Gold ore and gold tailings are some of the identified pollution sources in the gold mining industry. Using the combined techniques of XRF and AAS, the following heavy metal elements were identified: Cr, Mn, Fe, Cu, Zn, As, Pb, Hg, Rb, Sr, Y, Zr, and Cd in a concentration range of 0.08 ppm to 4.9%. Even though some of the tailings are dumped into ponds, streams, or rivers, the analysis of these inland waters showed the presence of only Fe and Zn at concentration levels of 0.08–2.4 μg/mL. However, the sediments of these waters showed the presence of most of the heavy metals. This indicates that most of the heavy metal pollutants exist in particulate form and do not go into solution, but rather settle into the sediments of the inland waters.

ACKNOWLEDGMENTS

The above work is part of an ongoing coordinated research project of the International Atomic Energy Agency. The authors acknowledge the sponsorship of this project by the Agency. The authors also wish to thank the collaborating institutions taking part in this project and also express their appreciation of the cordiality and help being given by the State Gold Mining Corporation in the collection of samples from the mines.

REFERENCES

1. G. O. Kesse, *Proc. 11th GSA Biennial Conf.*, Kumasi, Ghana, 1979.
2. J. R. Rhodes et al., *Applications of Low Energy X-ray and Gamma Rays*, C. Zeigler, ed., Gordon and Breach, New York, 1971, p. 317.
3. R. D. Giauque and J. M. Jaklevic, *Advances in X-ray Analysis*, vol. 15, K. Heinrich et al., eds. Plenum, 1971, pp. 164–174.
4. J. Toussant and R. Borniforti, *Int. J. Environ. Anal. Chem.* **6,** 217 (1979).
5. D. E. Leyden et al., *Int. J. Environ. Anal. Chem.* **7,** 85 (1979).
6. M. Nagj et al., *Nucl. Instr. Meth.* **A236,** 563 (1985).
7. C. Biney, "Monitoring of Pollution in the Marine Environment of West and Central African Region", *Report, Workshop WACAF/2*, Accra, 1988.

Trace Element Determination in Scalp Hair of People Working at a Copper Smelter

M. D. Tshiashala,* K. Kabengele, and B. M. Lumu

Laboratoire Central d'Analyse, Centre Régional d'Etudes Nucléaires de Kinshasa, BP 868, Kinshasa XI, Zaïre

Received April 17, 1989; Accepted October 27, 1989

ABSTRACT

Instrumental neutron activation analysis was performed to assess the exposure degree of a worker group from a copper smelter. The results were compared with data of inhabitants of Kinshasa and Bandaka in Zaire. Excessive concentrations of Mg, Ca, V, Co, Cu, As, and Cd were observed in the investigated group. The distribution patterns of trace elements are shown. The plots of Cu concentration relations to selected elements, i.e., As, Co, and Zn, showed a plateau at about 300 ppm of Cu.

Index Entries: Cu smelter pollution; hair analysis; instrumental neutron activation analysis; occupational exposure; trace element determination.

INTRODUCTION

The processing of ores for the recovery of metals leads to emission of quantities of smog and dust in the atmosphere that can cause both occupational health hazards to smelter workers and environmental health problems to people living in the vicinity of the plants. It can also alter the equilibrium of environmental systems if their trace element contents exceed certain limits.

*Author to whom all correspondence and reprint requests should be addressed.

Biological Trace Element Research Editor: G. N. Schrauzer

The Lubumbashi copper smelter in the southern part of Zaire produces more than 125,000 tons of Cu a year (*1*). The bulk of treated materials consists of sulfide concentrates grading about 30–40% Cu (*2*). An independent monitoring of dust from the plants exhibited a high degree of Zn (18.20%), Cu (15.20%), Pb (1.90%), Cd (0.19%), and As (0.17%) (*3*). The trace element burden of people working in this environment could be largely influenced by these metals.

The assessment of exposure risk was studied by hair analysis. This tissue is the most inviting in epidemiological studies because it is easily sampled, shipped, and analyzed, and most of the trace elements are present at concentrations higher than those in body fluids or other accessible tissues (*4*). The aim of the present study is to make more information available on trace element levels in hair of people groups with different types of environmental and occupational exposure (*5–11*).

EXPERIMENTAL

Scalp hair samples were collected by the medical staff of the company, using stainless steel scissors. A total of 34 samples were collected, stored in plastic bags, and shipped to our laboratory for analysis. A questionnaire submitted to donors for background information revealed the homogeneity of the group with untreated hair and a similar diet. Their ages varied between 24 and 56 yr and their exposure history between 3 and 28 yr.

To remove external adherent contaminations, the hair was washed following the IAEA procedure (*12*), consisting of a washing with 20 mL acetone, followed by 3 successive washings with 20 mL water each and a final washing with 20 mL acetone. The contact time was 10 min, with mechanical shaking. Afterwards, the washed hair samples were dried in a glove box at room temperature.

Hair analysis was performed by instrumental neutron activation analysis. About 20–100 mg of washed hair samples were weighed in polyethylene vials and submitted to irradiation in the CREN-K Triga Mark II reactor at a flux of 10^{12} neutrons/cm^2/s. The irradiation times were 10 min, 5h, and 30 h, depending on the half-lives of the products formed (listed in Table 1). For 10-min irradiations, an Al flux monitor was used.

RESULTS AND DISCUSSION

Results of scalp-hair analysis are given in Table 2. They include the concentration range, the median values, and the arithmetic and geometric means, along with their correlated standard deviations calculated on the basis of the number of individual determinations (*n*).

Table 1
Analytical Characteristics of Detected Elements

Element	Radionuclides	Half-life	Gamma energies used (in KeV)
Na	^{24}Na	15.0 h	1368
Mg	^{27}Mg	9.46 m	1014
Al	^{27}Al	2.24 m	1779
K	^{42}K	12.4 h	1525
Ca	^{48}Ca	8.7 m	3084
V	^{52}V	3.8 m	1434
Mn	^{56}Mn	2.6 h	887;1181
Co	^{60}Co	5.26 y	1173;1332
Cu	^{64}Cu	12.8 h	511
	^{66}Cu	5.1 m	1039
Zn	^{65}Zn	245.0 d	1115
As	^{76}As	26.3 h	559
Br	^{82}Br	35.9 h	554;777
Cd	^{115}Cd	53.5 h	528
Sb	^{122}Sb	2.8 d	564
	^{124}Sb	60.3 d	603
Cs	^{134}Cs	2.05 y	605;795
Hg	^{203}Hg	46.8 d	279

Because of the small quantities of collected hair samples, a number of determined elements were found to be present below their detection limits, calculated on the basis of the Currie criteria (*13*).

The median values were found to be closer to the geometric means than the arithmetic means for all determined elements. The geometric standard deviation must be lower than 1.6 for essential trace elements of the human body (*14*). But all the determined elements in this study, except Mg, have higher values; thus, they do not distribute normally in the investigated population. Accordingly, the distribution patterns to be expected are log-normal and are plotted in Fig. 1. Those of the patterns with more than one maximum probably indicate the presence of several kinds of distributions in the investigated population (*14*). Neither the age nor exposure history classification showed significant differences between groups.

Table 2
Trace Element Content in Scalp Hair of Workers from Cu-Smelter Plants†

Element	Exposed group n	range	median	arith. mean	geom. mean	Non-exposed group (14) geom. mean
Na	30	43.3 - 1425	219.5	288.9 (279.2)	202.2 (2.0)	61.6 (5.5)
Mg	19	217.0- 1475	555.0	665.0 (332.5)	597.0 (1.6)	93.3 (3.0)
Al	32	4.5 - 162.1	39.4	45.2 (36.1)	29.4 (2.7)	72.4 (3.2)
K	16	3.9 - 54.7	10.2	16.3 (14.4)	17.9 (4.9)	87.1 (3.7)
Ca(%)	29	0.07 - 11.1	2.3	2.9 (2.2)	2.4 (3.9)	0.84 (2.6)
V	28	0.02 - 1.9	0.56	0.64 (0.5)	4.45(2.7)	0.39 (2.0)
Mn	17	1.2 - 18.0	3.4	5.7 (5.5)	3.8 (2.4)	4.3 (2.4)
Co	34	0.2 - 27.3	6.7	9.4 (10.8)	5.0 (3.3)	0.25 (3.3)
Cu	34	9.3 - 530.0	161.4	200.4 (148.3)	142.2 (2.6)	11.3 (1.9)
Zn	29	78.8 - 570.1	188.0	239.1 (133.2)	207.0 (1.8)	132.0 (1.5)
As	28	1.6 - 42.8	11.2	15.2 (17.0)	10.5 (2.3)	0.12 (0.11)*
Br	31	1.04 - 26.2	6.6	7.8 (5.8)	5.9 (2.2)	3.9 (3.3)
Cd	10	9.2 - 55.6	25.6	28.9 (15.8)	24.8 (1.8)	0.77 (0.25)*
Sb	28	0.01 - 0.84	0.1	0.18 (0.2)	0.1 (3.5)	1.4 (3.7)
Cs	19	0.08 - 1.5	0.22	0.37 (0.3)	0.27 (2.3)	0.09 (2.3)
Hg	33	0.3 - 6.28	1.10	1.9 (2.0)	1.4 (2.2)	2.35 (2.04)*

*Values computed from IAEA report 50/RL (*12*).
†(in ppm, unless otherwise noted)

An independant monitoring of dust from the smelter revealed the following concentration sequence for the major elements: Zn (18.20%) > Cu (15.20%) > Cd (0.19%) > As (0.17%) (*3*). A similar concentration sequence is observed in scalp hair from workers at the copper smelter. This finding is the proof of scalp hair ability to reflect the environmental status.

In Fig. 2 the concentration relations of Cu to As, Cd, and Zn are plotted. The copper concentrations show a tendency to attain a maximum concentration at about 300 ppm.

Trace element contents in hair of inhabitants of Kinshasa and Bandaka in Zaire have been previously reported (*15*). Results are given in the last column of Table 2 as comparative data. The As, Cd, and Hg values reported herein are the mean values computed from IAEA Report 50/RL (*12*). The elements As and Cd were not detected in these nonexposed groups, whereas Hg had high values owing to the use of antiseptic soaps containing Hg compounds.

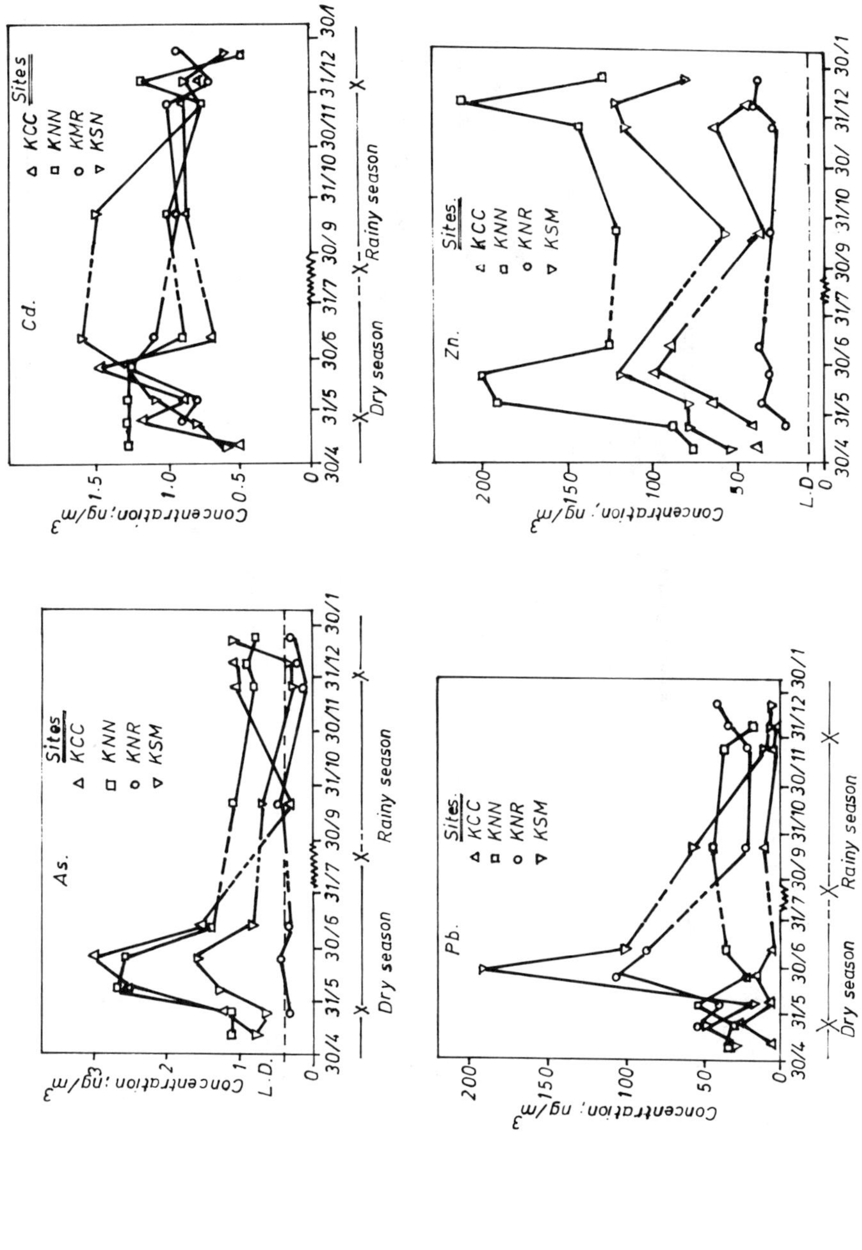

Fig. 1. Variation of As, Cd, Pb, and Zn concentration in airborne particulates in Kinshasa vs time.

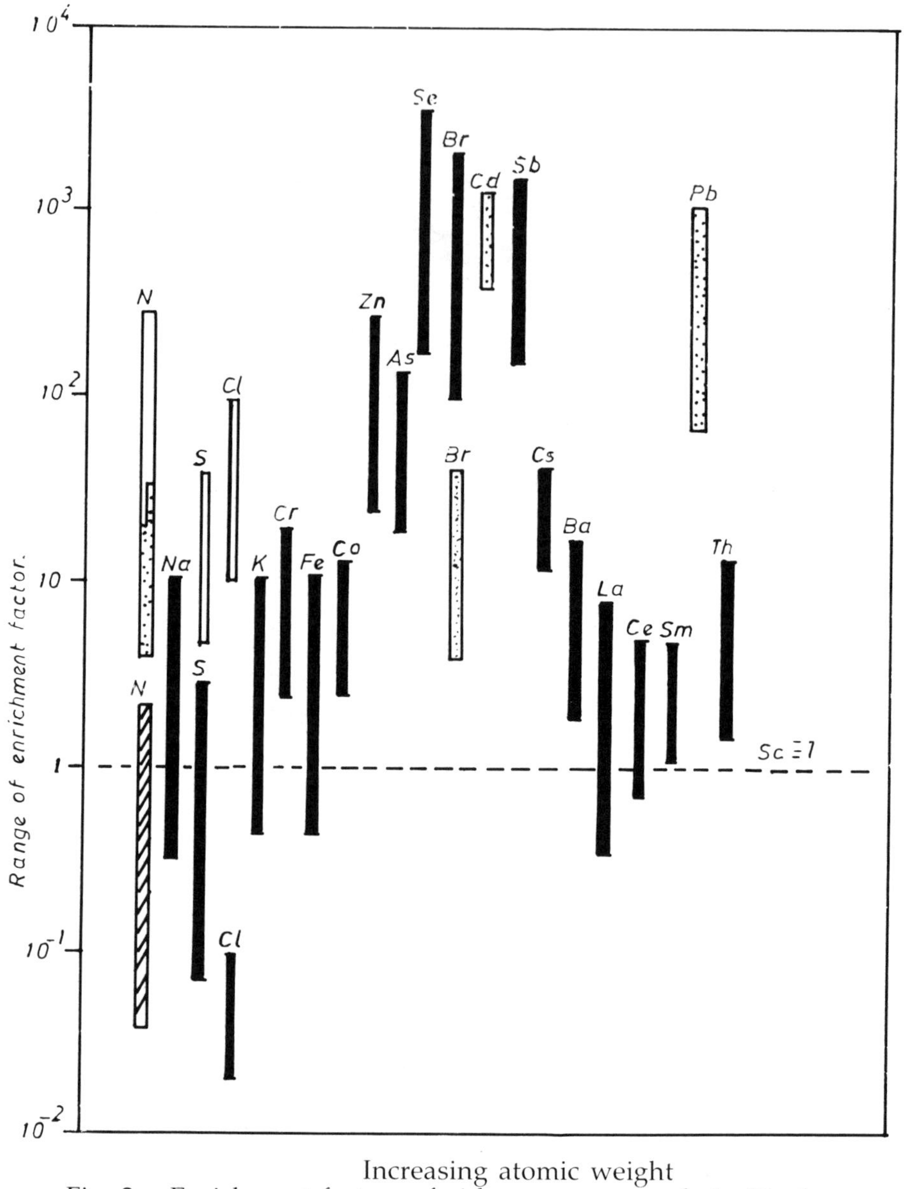

Fig. 2. Enrichment factors of airborne compounds in Kinshasa.

■ Particulate elements, □ Gaseous elements,

⁙ Water soluble partic., under NO_2^- form, ▨ Particulate N determined

The elements Al and K exhibited lower values than those in nonexposed groups. This could be attributable to variations in results of this study and those quoted for nonexposed groups. Sb, Mn, and Br values are reasonably similar in both groups. Mercury contents were found to be similar to those in normal populations (*12*).

The zinc content increased only by a factor 1.6 despite its high concentration in the plant's dust. This shows a tendency of this element to keep constant its concentration both in hair of the normal and occupational-exposure population groups.

Large increases are those exhibited by Co (20-fold), Cu (12.6-fold), V (11.4-fold), and Mg (6.4-fold), compared with nonexposed groups. Arsenic exhibited values higher than those in normal populations by a factor of 80. Relatively similar values were observed by Tomza et al. (*5*) in hair of a worker group from a non-ferrous smelter.

The cadmium content exceeded normal population levels by a factor of over 30. Similar data were reported by Tomza et al. (*5*). Recently, Kucèra reported a severe exposure of workers in a factory producing alkaline storage batteries (Ni-Cd alkaline cells) (*11*).

CONCLUSIONS

This study has demonstrated the reliability of instrumental neutron activation analysis in the determination of trace elements in biological materials. It has also shown the usefulness of hair as a good monitor of environmental exposure to trace elements.

The high exposure of the investigated population group to toxic trace elements such as As, Cd, and others can, over certain limits, become hazardous for human health. Nevertheless, we think that in an environment highly contaminated with elements, the external contamination is the main way of trace element incorporation into scalp hair. To confirm this assumption, further studies are to be performed to find out the relationship between trace element concentrations in body fluids, such as blood and urine, and in hair of occupational-exposed communities. These studies will be extended to people living near Cu smelters.

ACKNOWLEDGMENTS

The authors wish to thank the Gecamines medical staff for their contributions to the sample collection. They gratefully acknowledge Prof. Malu, and Messrs. Matamba and Tshibengabo both for discussing, reviewing, or final reading of the manuscript.

REFERENCES

1. La Gécamines, **rapport annuel**, Lubumbashi, Zaire, 1975, 17 pp.
2. K. Akilimali, Y. Mambo, B. M. Lumu, and K. S. Mandungu, *Rev. Zaïr. Sci. Nucl.* **1,** 31 (1980).
3. La Gécamines. *Coll. Natl. Environ. et Pollut. Industr.*, Dept. Environ. Conser. Nature, Lubumbashi Zaïre, 1986, unpublished paper.
4. M. Limić and V. Valkovic, *Biol. Trace Elem. Res.* **12,** 363 (1987).
5. U. Tomza, T. Janicki, and S. Kossman, *Radiochem. Radioanal. Letters* **58,** 209 (1983).
6. R. E. Jervis and B. Tienfenbach, *Proc. Symp. Nucl. Act. Techn. in the Life Sciences*, IAEA, Vienna, 1978, 627 pp.
7. J. P. W. Houtman, M. De Bruin, J. J. M. Goeij, and P. S. Tjie, *Proc. Symp. Nucl. Techn. in the Life Sciences*, IAEA, Vienna 1978, 599 pp.
8. I. Obrusnik and V. Bencko, *Radiochem. Radioanal. Letters* **38,** 751 (1979).
9. J. Akashi, I. Fukushima, A. Imahiro, S. Shiohara, Y. Takahashi, and T. Tumura, *J. Radioanal. Chem.* **68,** 59 (1982).
10. A. Chattopadhayay, T. M. Roberts, and R. E. Jervis, *Arch. Environ. Health* **34,** 226 (1977).
11. J. Kucèra, V. Senft, P. Huzl, and L. Soukal, *J. Radioanal Chem.* **122,** 361 (1988).
12. IAEA, *report 50/RL*, Vienna (1978).
13. L. A. Currie, *Anal Chem.* **40,** 586 (1968).
14. J. M. Lenihan and H. Smith, *Proc. Symp. Nucl. Act. Techn. in the Life Sciences*, IAEA, Vienna, 1967, 601 pp.
15. M. D. Tshiashala and B. M. Lumu, *Rev. Zaïr. Sci. Nucl.* **6,** 205 (1985).

Air Quality Status in Kinshasa as Determined by Instrumental Neutron Activation Analysis, Atomic Absorption Spectrometry, and Ion-Exchange Chromatography

K. K. Lobo,*[,1] M. D. Tshiashala,[2] B. M. Lumu,[2] K. Matamba,[2] and C. Ronneau[3]

[1]*Faculte des Sciences, Universite de Kinshasa, Kinshasa, Zaire;* [2]*Laboratoire Central D'Analyse, Centre Regional D'Etudes Nucleaires de Kinshasa, Kinshasa, Zaire; and* [3]*Faculte des Sciences, Universite Catholique de Louvain, Louvain, Belgium*

Received May 22, 1989; Accepted October 27, 1989

ABSTRACT

Three independent analytical techniques: instrumental neutron activation analysis, atomic absorption spectrometry, and ion-exchange chromatography were applied to airborne particulates collected on filters, and to atmospheric acid gases collected in carbonate buffer solutions. Twenty trace elements and seven acid gases and acid aerosols were determined.

Results were compared with those observed elsewhere and showed that air pollution is low in Kinshasa and should not cause anxiety. The main known sources of pollutants are vehicle exhaust and aeolian processes on stripped soils.

Index Entries: Aerosol trace elements; aerosol acid compounds; air quality; atmospheric acid gases determination; atomic absorption spectrometry; detection limits; instrumental neutron activation analysis; ion-exchange chromatography; trace element determination.

*Author to whom all correspondence and reprint requests should be addressed.

INTRODUCTION

During the last few decades, studies have been initiated on atmospheric aerosols in the tropical and equatorial zones. An important one is the study of the atmospheric low layer in the Amazon basin (Brazil) initiated by the NAS (*1*). Some subprojects of this program concerned the chemical characterization of aerosol particulates.

Studies in progress in Africa are timely, but limited to the northern hemisphere (*2–5*). To our knowledge, the only investigation on aerosols in Zaire is that performed by Maenhaut et al. (*6*). Today, there is a necessity for Africa to establish the impact of nascent industrial pollution sources on the environment and to know the aerosol background levels in the rural environment as reference areas. From a fundamental point of view, these studies could contribute significantly to the understanding of geological cycles of more elements, to the study of long-distance transport of toxic compounds, and to the identification of unknown natural sources of aerosol constituents.

The study of aerosols in Zaire is interesting because of its climatic particularities and contrasts between rural, urban, and mining regions. It aims at the determination of the aerosol composition of important cities to

(1) Obtain information on typical urban aerosol composition and its potentially toxic element composition;
(2) Evaluate the impact of man-made activities on its composition;
(3) Inventory the different sources of atmospheric particulates, their elementary profiles, and their contributions to the natural contents of various elements; and
(4) Establish its evolution with time.

This paper is a part of a research project on the air quality status in Kinshasa, initiated in connection with the Laboratory of Inorganic and Nuclear Chemistry of Louvain-la-Neuve University in Belgium. It deals with the preliminary results obtained by instrumental neutron activation analysis (INAA), atomic absorption spectrometry (AAS), and ion-exchange chromatography (IEC).

EXPERIMENTAL

Samples herein were collected during 1985–1986 at four pilot stations installed throughout Kinshasa and whose characteristics were:

(1) KCC (Kinshasa Campus Centre Nucléaire): site located on a ventilated hill, the Mont Amba, with scattered population, low vehicle traffic, and absence of industries;

(2) KNR (Kinshasa Ndjili Regideso): site located beside industrial area with busy vehicle traffic and middle population density;
(3) KNN (Kinshasa Ngiri-Ngiri): site with high population density, low vehicle traffic, and absence of industries;
(4) KSM (Kishasa 20 May Stadium): populous quarter located beside industrial area with busy vehicle traffic.

Sample methodology is described elsewhere (*7*). Samples analyzed in the framework of this study are airborne particulates collected on Whatman 41 and TFA 41 filters and atmospheric acid gases trapped in carbonate buffer solutions. The sampling duration was, on an average, 14 d.

Sample Pretreatment

Sample filters were cut into four equal parts. One quarter was submitted to direct INAA. Another one was dipped into 5 mL deionized water and submitted to a 15-min ultrasonification shaking. The water solution was then analyzed by AAS for the determination of Cd and Pb and by IEC for the determination of acid aerosols. The last method has also been applied to the bubbling carbonate buffer soltutions for the determination of atmospheric acid gases.

ANALYTICAL TECHNIQUES

Instrumental Neutron Activation Analysis

Quarters of sample filters were irradiated in the thermal column of the CEN-Mol BR-1 reactor at an average integrated flux of 2.9×10^{16} neutron cm^2. The counting system comprises a Ge(Li) detector with associated electronics coupled to a 4096 Canberra Series 35 PLUS analyzer. Two sets of counting conditions (600 s after 1–2 d decay time for short-lived radioisotopes and 6000 s after 10–15 d decay time for long-lived radioisotopes) were used. Quantitation was based on a comparison with standards prepared by dropping 20μL of solutions, containing elements to be analyzed, at known concentration on Whatman 41 that were irradiated along with the sample filters.

Atomic Absorption Spectrometry

The water soluble Cd and Pb were determined, respectively, at λ = 283.3 and 228.8 nm, using a graphite furnace Perkin-Elmer Atomic Absorption Spectrometer 303. The measurement device was calibrated with Cd and Pb solutions prepared from standard solutions.

Ion-Exchange Chromatography

The concentration of water soluble airborne particulates and those atmospheric acid gases trapped in carbonate buffer solutions were determined using a Dionex IC 10 ion-exchange chromatograph, using their anionic forms. Concentrations of prepared standards were in the same range as the observed sample concentrations. All the analyses with INAA, AAS, and IEC were performed in the analytical laboratory of the faculty of sciences at Louvain-la-Neuve University in Belgium.

RESULTS AND DISCUSSION

More than 60 samples of airborne particulates collected on Whatman and TFA 41 filters, and atmospheric acid gases trapped in carbonate buffer solutions, were analyzed in the course of this study. The analytical characteristics of detected radioisotopes are summarized in Table 1, along with the detection limits of corresponding elements, calculated following the Currie method (*8*). The Cd and Pb analytical characteristics and their detection limits are given in Table 2. The detection limits of anions from airborne particulates or atmospheric acid gases were calculated on the basis of 2 times the value of the background observed in the spectrum. The values found are given in Table 3. Experimental results are reported in Tables 4–6, respectively, for 18 trace elements determined in airborne particulates by INAA, nine constituents from water soluble airborne particulates, and four anions from atmospheric acid gases.

Owing to insufficient data to perform a statistical treatment, only the concentration ranges of determined atmospheric constituents are given. Some of them were found to be below their detection limits in some samples, as shown in Fig. 1. These data show an abnormal spatial distribution of constituents through the different sampling sites.

The anthropogenic compounds show the following characteristics

(1) Arsenic concentration is below its detection limit level at the KNR site. Concentration similarities are found at sites KCC and KNN, with relatively higher values than those found at KSM.
(2) Bromine has two distinct distributions: its concentrations are similar at KNN and KSM sites, with higher values than those at KCC and KNR, which also are found to be similar.
(3) Cadmium concentration shows a constancy in time. It varies only about 1 μg through the four sampling sites.
(4) Atmospheric Pb concentration is higher at busy traffic sites KRN and KSM than at other relatively low traffic sites.
(5) Antimony is one of the elements whose determination was not easy. Its concentration lies at its detection limit level at the

Table 1
Analytical Characteristics and Detection Limit (DL) of Radioisotopes

Element	Radioisotope	Half-life	γ Energy, (KeV)	DL, ng/m^3
As	^{76}As	26.4 h	559	0.4
Ba	^{131}Ba	11.5 d	496	8.2
Br	^{82}Br	35.3 h	779	1.1
Ce	^{141}Ce	32.5 d	145	0.6
Co	^{60}Co	1934 d	1332	0.3
Cr	^{51}Cr	28 d	320	2.0
Cs	^{134}Cs	730 d	796	0.4
Fe	^{59}Fe	45 d	1099	166
K	^{42}K	12.5 d	1524	150
La	^{140}La	40.3 d	1595	0.2
Na	^{24}Na	15 h	1368	2.8
Rb	^{86}Rb	19 h	1077	3.3
Sb	^{124}Sb	61 d	603	0.4
Sc	^{46}Sc	84 d	889	0.1
Se	^{75}Se	121 d	265	0.2
Sm	^{153}Sm	47 h	103	0.1
Th	^{233}Pa	27 d	312	0.2
Zn	^{65}Zn	245 d	1115	9.4

KCC and KRN sites, whereas high values are observed at the KNN and KSM sites.
(6) The spatial Zn distribution in Kinshasa shows the following concentration sequence: KRN < KCC < KSM < KNN.

Similar characteristics are shown by the crust-derived elements. For instance, Ce has high values at the KNN site, whereas Co and Cr show low values at the KRN site. Stripped soils is one of the causes that could explain the trace element behavior throughout Kinshasa, but it is uncertain.

Owing to the lack of data on interelementary concentration relations in the soil of Kinshasa, the enrichment factors were calculated on the basis of Rahn ore values (*9*) of Sc as the reference element. Figure 2 illustrates the determined factors.

Table 2
Characteristics and Detection Limits of Cd and Pb, Determined by Atomic Absorption Spectrometry

Element	(nm)	Furnace Temperature (°C) Dryness 30 s	Thermal decomposition 30 s	Atomisation 10–30 s	D.L. (μg/L)
Cd	228.8	100	350	1800	1
Pb	283.3	100	550	2040	10

Table 3
Detection Limits of Ions Determined by Ion-Exchange Chromatography

Anions	Detection Limit μg/L	μg/m³ air
F^-	0.07	0.032
Cl^-	0.13	0.046
Br^-	0.7	0.334
NO_2^-	0.04	0.02
NO_3^-	1.03	0.367
PO_4^{3-}	1.13	0.51
SO_4^{2-}	0.98	0.35

For some selected elements, concentrations are plotted vs time. The graphs show clearly the weather influence on trace element behavior. During the rainy season, which lasts from October through May and is interrupted by a break in December and January, airborne compounds are largely washed out. The concentrations increase during the dry season, which is from June through September. The highest concentra-

Table 4
Trace Elements Determined in Airborne Particulates in Kinshasa by Instrumental Neutron Activation Analysis

	Concentration range (ng/m^3)			
Element	KCC	KNN	KNR	KSM
As	0.6 - 3.0	0.8 - 2.7	0.4 - 0.5	0.5 - 1.6
Ba	15.7 - 44.4	32.6 - 123.0	18.4 - 32.7	13.4 - 57.7
Br	3.8 - 36.4	31.5 - 72.8	16.7 - 36.0	16.4 - 79.5
Br	0.2 - 1.5	0.4 - 1.6	0.4 - 1.0	0.2 - 1.1
Ce	0.6 - 4.8	1.1 - 4.8	0.7 - 2.1	0.7 - 2.8
Co	0.6 - 1.4	0.5 - 1.2	0.5 - 0.5	0.6 - 2.1
Cr	4.6 - 10.5	7.0 - 18.5	2.3 - 5.6	3.4 - 19.1
Cs	0.4 - 0.5	0.5 - 1.4	0.4 - 0.5	0.5 - 0.6
Fe	446.0 - 5112.0	1037.0 - 4069.0	223.0 - 694.0	828.0 - 2779.0
K	451.0 - 3969.0	1136.6 - 3764.0	147.0 - 740.0	407.0 - 2116.0
La	0.2 - 4.3	0.4 - 3.8	0.2 - 0.5	0.2 - 2.1
Na	159.0 - 1512.0	742.0 - 3653.0	118.0 - 345.0	164.0 - 1193.0
Rb	4.3 - 12.0	5.4 - 21.1	3.7 - 4.7	4.0 - 4.6
Sb	0.4 - 2.1	1.3 - 3.9	0.5 - 0.9	0.6 - 2.8
Sc	0.1 - 1.0	0.2 - 0.9	0.1 - 0.2	0.1 - 0.6
Se	0.2 - 0.3	3.4 - 3.8	0.3 - 0.4	0.3 - 2.1
Sm	0.2 - 0.4	0.1 - 0.4	0.1 - 0.2	0.2 - 0.3
Th	0.2 - 1.3	0.3 - 1.8	0.3 - 0.4	0.4 - 0.7
Zn	37.0 - 98.9	76.2 - 214.0	19.9 - 44.6	53.6 - 123.0

tions of atmospheric constituents determined in Kinshasa are compared with the values found elsewhere at Louvain-la-Neuve, a subrural area of Belgium (*10*), at some rural areas of Great Britain (*11*), in the Osaka District of Japan (*12*), and in the U.S. (*13*). These results emphasize the following characteristics

(1) The acid gaseous oxides NO, NO_2, and SO_2, characteristic of industrial areas, are present at much lower concentrations than in rural areas in Europe. This fact could be attributable to the absence of industries using fossil fuels.
(2) This study reveals the presence of NO_2^-, PO_4^{-3}, and Br^- in the atmosphere of Kinshasa, compounds not detected at Louvain-la-Neuve.

Table 5
Water Soluble Airborne Particulates Determined by Atomic Absorption Spectrometry and Ion-Exchange Chromatography

Constituent	Concentration range (ng/m^3) KCC	KNN	KNR	KSM
Br^-	0.2 - 1.5	0.4 - 1.6	0.4 - 1.0	0.2 - 1.1
Cd	0.5 - 1.5	0.5 - 1.3	0.7 - 1.3	0.5 - 1.6
Cl^-	0.1 - 1.2	1.6 - 4.1	1.1 - 1.8	0.7 - 2.1
F^-	0.1 - 0.3	0.1 - 0.3	0.1 - 0.2	0.1 - 0.3
NO_3^-	1.0 - 8.0	2.5 - 8.9	2.5 - 5.7	1.4 - 6.3
NO_2^-	0.07 - 0.6	0.01 - 0.2	0.01 - 0.03	0.02 - 0.6
Pb	11.9 - 27.6	17.7 - 54.4	20.2 -124.0	16.5 -190.0
PO_4^{3-}	0.2 - 0.9	0.1 - 0.3	0.2 - 0.3	0.1 - 1.7
SO_4^{2-}	0.3 - 1.6	1.4 - 12.3	3.0 - 10.2	2.3 - 10.0

Table 6
Anions from Atmospheric Acid Compounds Determined by Ion-Exchange Chromatography

	Concentration range, ng/m^3			
Cl^-	950–420	67–72	48–119	45–54
F^-	8–24	6–19	4–48	6–13
NO_3^-	5–12	55–76	5–20	15–28
SO_4^{2-}	32–164	39–108	15–82	27–90

(3) The high Pb concentrations found at the busy traffic sites, KRN and KSM, are similar to the average found in rural areas in Europe, but much lower than those found in Japan.
(4) Ba, Ce, Cs, K, La, and Na are the same order of magnitude as those in Japan, but much higher than the observed values both in rural and urban areas in Europe and in rural areas in the US.
(5) The elements Zn, Cr, Sb, and Cd exhibit lower values in Kinshasa than in Japan.

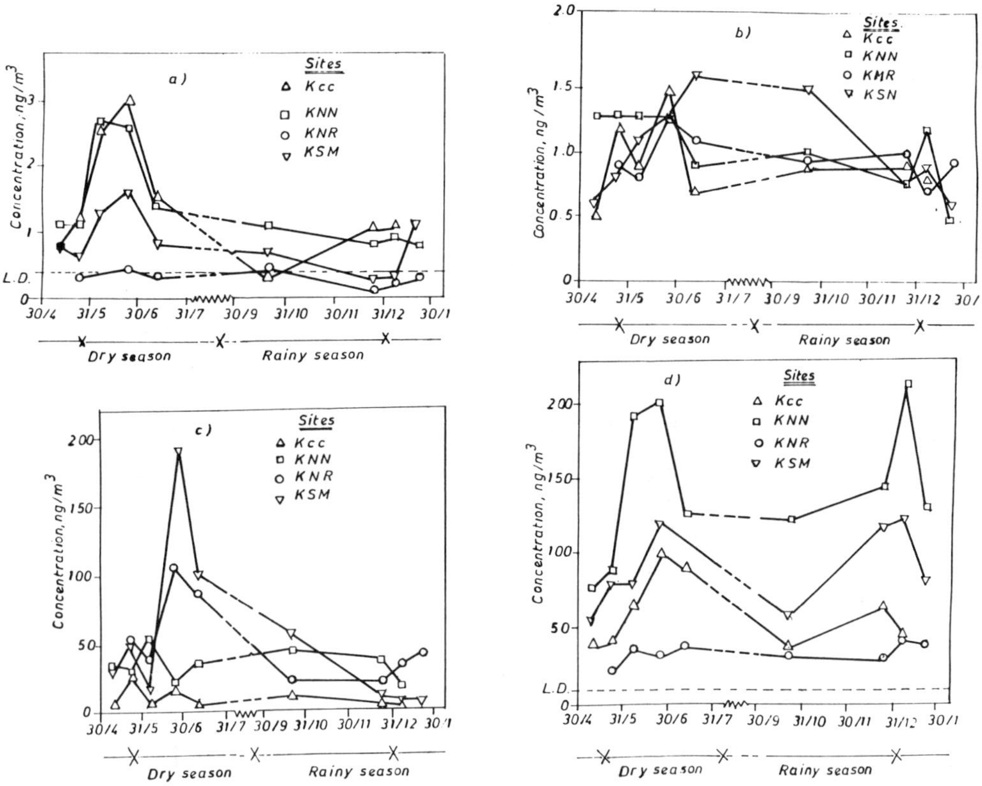

Fig. 1. Variation of As, Cd, Pb, and Zn concentrations in airborne particulates in Kinshasa vs time.

(6) Other elements are the same order of magnitude as those found elsewhere, both in rural and urban areas.

CONCLUSIONS

These preliminary studies allow the establishment of the analytical conditions for the determination of more than 26 components in the atmosphere of Kinshasa. Their levels emphasize that air pollution in Kinshasa is low and should not cause anxiety.

The main activities known to contribute to the perturbation of the quality of air in this city are vehicle exhaust and aeolian processes on stripped soils. Further measurements and a good knowledge of parameters, such as particulate sizes, wind speed and direction, and so on, could allow a good understanding of the spatial distribution of atmospheric constituents throughout Kinshasa and of their natural and pollution sources.

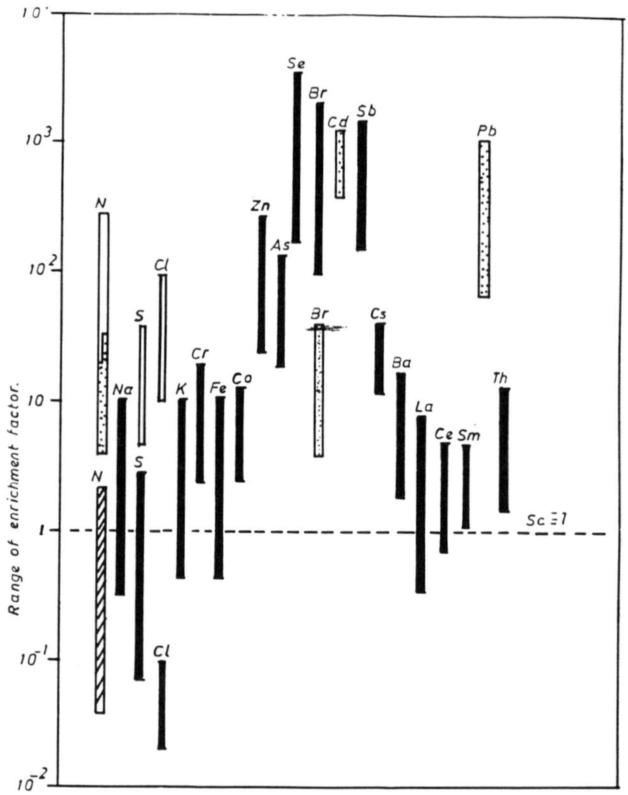

Fig. 2. Enrichment factors of airborne compounds in Kinshasa: ☐ Particulate elements; ☐ Gaseous elements, ▦ Water soluble particulate, ▨ Particulate N determined under NO_2^- form.

ACKNOWLEDGMENTS

The financial support of the International Atomic Energy Agency under Research Contract No. 5130/RB is gratefully acknowledged.

REFERENCES

1. National Academy of Sciences (NAS), *The Tropospheric Transport of Pollutions and Other Substances to the Oceans*, Printing and publishing Office, NAS, Washington, DC, 1978.
2. G. Grozat, *Thèse de Doctorat ès Sciences Physiques nr 844*, Université Paul Sabatier de Toulouse, 1978.

3. J. L. Domergue, *Thèse de Doctorat ès Sciences Physiques*, Université Paul Sabatier de Toulouse, 1980.
4. S. A. Penkett, D. H. F. Atkins, and M. H. Unsworth, *Tellus* **31**, 295 (1979).
5. P. Pélassy, *Atmospheric Environment* **18**, 2245 (1984).
6. W. Maenhaut and K. Akilimali, *Nucl. Instr. Meth. B* **32**, 254 (1987).
7. M. D. Tshiashala, *I.A.E.A. report NAHRES-1*, Vienna, annex 21, 1988.
8. L. A. Currie, *Anal. Chem.* **40**, 586 (1968).
9. K. A. Rahn, *Technical Report*, Graduate School of Oceanography, University of Rhode Island, 1976.
10. J. P. Hallet, *Thèse de Doctorat*, Université Catholique de Louvain, 1983.
11. P. A. Cawse, *AERE report R 7669*, London, 1974.
12. T. Mamura, Y. Matsuda, A. Mizohata, and T. Matsunami, *Ann Rep. Cent. Osaka Pref.* **13**, 1 (1972).
13. L. A. Rancitelli, J. A. Cooper, and R. W. Perkins, *Proc. Symp. Nucl. Tech in the Life Sciences* (Otaniemi, Japan, Aug. 27–31, 1973), IAEA, Vienna, 1974, pp. 431–454.

Mapping Technique Based on Elemental Hair Composition Data

L. I. ZHUK AND A. A. KIST*

Institute of Nuclear Physics, Academy of Science of Uzbek SSR, Tashkent, USSR

Received April 17, 1989; Accepted October 31, 1989

ABSTRACT

The content of 24 elements has been determined in 2000 hair samples of the inhabitants of Uzbekistan. INAA was used for analysis. Reference material IAEA HH-1 and laboratory standards were used. Measurements were done using a Ge(Li) detector and a multichannel analyzer.

No correlation was found between the element content and hair color and ethnic group, whereas the content of some elements depended on sex, hemoglobin content, blood group, and occupation. Arithmetical and geometrical means and medians were calculated. Cumulative histograms show mainly lognormal distributions. Maps of selected regions, based on the elemental hair composition data, were made. The element content was shown to depend on the location of large cities and biogeochemical anomalies. The relationship between the death rate and the element content in hair in some countries has been shown.

Index Entries: Elemental human hair composition mapping; hair samples, analysis of; lognormal distribution; elemental hair composition and death rate; Instrumental Neutron Activation Analysis (INAA).

*Author to whom all correspondence and reprint requests should be addressed.

Fig. 1. Geographical situation of Uzbek SSR.

INTRODUCTION

Attempts to study the changing nature of the geochemical situation in vast areas have a long history in geohygiene, medical geography, teaching on geochemical provinces, soil science, and other disciplines (*1*). Today, this problem acquires special importance in terms of environmental contamination. Its solution is connected with certain difficulties, in particular, with the difficulty to select the objects to be analyzed. The usage of soil, plant and animal tissues, aerosols and atmospheric fallouts, surface waters, and so on for this purpose is difficult, owing to the exclusive spatial mosaic pattern, as well as to the changes in their elemental composition in time. For large areas, this requires a great number of analyses, which is hardly possible with the analytical and technical means available. Thus, human hair may be considered a more suitable object for this type of study. This is because of the simple and harmless way the samples can be collected, and the fact that they reflect the status of the final object of the environment protection undertakings—a human being. Also, the existing methods allow one to determine many elements in a sample. One more advantage lies in the fact that the information on the body status appears to be recorded by the hair length covering a long enough period of time. Keeping all of this in mind, one may consider human hair and animal wool to be highly generalized systems that permit elemental mapping of large areas.

This mapping procedure gives the possibility to find the natural and anthropogenic biogeochemical anomalies and trace the impact of large industrial clusters on the environment. The comparison of such maps with those of the distribution of diseases may raise a question on the prognosis of the human body status in a certain area. By considering frequency histograms, one can evaluate the degree of sampling homogeneity and find anomalous sample sets. Besides, since this work requires a large number of analyses, one has the possibility to obtain statistically reliable data on the average element content in hair within a population. In its turn, this allows one to consider the regularities in human hair elemental composition. One can evaluate the comparability of the averaged data obtained by different techniques of statistical processing. All the above mentioned provoked us to consider the informativity of human hair elemental composition maps of the Uzbek SSR and then compare the data obtained with the information available on other regions of the Earth.

The Uzbek SSR is located in the south of the Soviet Union (Fig. 1). Its area is 447,000 km^2, and the population is about 18 million. The region is quite attractive in terms of climate, geography, economy, and ecology.

METHODS

Sampling Procedure and Preparation of Samples

The territory of the republic was divided into 71 zones, from which the samples of whole human blood and hair were collected. The total made 2000 hair samples and 2500 blood samples. This is about 0.01% of the population and corresponds to previous recommendations (2) considering 0.01–0.001% to be enough for the determination of elemental hair composition in the population. Particulars of a donor, such as age, sex, resident time in the given area, ethnic group, occupation, hemoglobin content, blood group, and hair color, were recorded.

Hair samples were collected from five places on the head, and the surface washing, according to (2), was performed. Blood samples were placed in sterile vials that were flared and immediately sent for lyophilization (3,4).

Elemental Analysis

Instrumental neutron activation analysis was applied to study the hair samples collected. Samples were irradiated at a flux of $5 \cdot 10^{13}$ n/cm^2/s for 15 s in polyethylene capsules. The measurements were carried out with a Ge(Li)-detector-based computerized gamma-ray spectrometry system. Immediately after irradiation, the induced activities of ^{37}Cl and ^{128}I were measured. One hour after irradiation, the induced activities of ^{24}Na, ^{64}Cu, and ^{56}Mn were determined. One week later, the samples were repacked in pure aluminum foil and irradiated at the same flux of 15 h. A week later, gamma-ray spectrometry was used to measure the induced activities of ^{47}Sc (from calcium), ^{82}Br, ^{76}As, ^{99}Mo, ^{115}Cd, ^{140}La, ^{187}W, ^{198}Au, and ^{239}Np (from uranium). The induced activities of ^{46}Sc, ^{51}Cr, ^{56}Mn, ^{60}Co, ^{65}Zn, ^{75}Se, ^{110m}Ag, ^{115}Cd, ^{124}Sb, ^{134}Cs, and ^{203}Hg were measured 1 mo after irradiation. The number of elements to be determined may be increased with a longer than 15-h irradiation (Sr, Ba, Hf, Tl, and Te) by activation using fast neutrons from a neutron generator to determine F, and for analysis via short-lived nuclides after irradiation by thermal neutrons (V, Al, Ti, and Mg).

Standards

Standard samples were made by placing a certain amount of element solutions on strips of ashless filter paper. At the same time, a laboratory reference material was prepared. The quality of standards and that of the laboratory reference material were evaluated by comparing the IAEA HH-1 reference material. The agreement obtained was quite satisfactory.

RESULTS AND DISCUSSION

Mean Values

Mean values may be characterized by arithmetical means by excluding the data by the criteria 2.5 or 5 d, or using all the results (those beyond the detection limit were given the value of 0.5 of the detection limit), geometrical means with the same additions, mode, and median. Characteristic cumulative histograms in Fig. 2 are given in order to evaluate the applicability of different ways of data presentation. For most of the elements, the histograms may be considered to be close to lognormal. For Cu, an approximation by a normal distribution is possible, whereas Fe, Co, Ag, and Cd may have bimodal distributions.

Table 1 lists arithmetical and geometrical means and medians, as well as the percentage of samples in which the element content was within the detection limit or above it. It follows from Table 1 that, taking into account the considerable spread in the data, all given ways of analytical information presentation are comparable if the sensitivity of the method used allows one to determine the element content in more than 70% of the cases. This conclusion is very important although it contradicts the recommendation (*6*) to use the analytical techniques with the detection limit at least an order of magnitude less than the minimum content. It seems acceptable to compare the results of different investigators even if the mean element content values have been obtained by different statistical techniques. This conclusion can be made on the basis of other sufficiently large data sets (e.g., *7,8*).

Correlation

As a whole, a reliable correlation between the content of elements in hair and blood has not been found. For some elements a relationship has been observed between age and elemental hair composition (Fig. 3). Reliable data on the relationship between the elemental hair composition and color, as well as ethnic group, has not been found. The difference in the elemental hair composition of males and females is also statistically unreliable, except for the cases of Cl, Ca, I, Ag, and Au. A very weak relationship has been found between elemental hair composition and hemoglobin content in the blood. It appeared to be positive for Ca, As, Mo, and La and negative for Sc, Se, Br, Cs, And U. Quite unexpectedly, some elements showed a relationship between the hair composition and blood group (Fig. 4).

A positive correlation was observed within the whole set for the following pairs: Na–Cl, Mo–As; Mo–Cr, Mo–Co, Co–Cr, Co–Fe, Co–Cs, Se–La, Sb–Cd, and U–Cs. In the case of Sb–Cd, the correlation coefficient reached the value of 0.98. Most of the remaining correlations are connected, we believe, with the state of the environment, since one can find predominantly the priority contaminators there. A negative correlation was found only for Ag–W.

Maps

From reviewing the maps, different conclusions can be drawn. For example, the regions with elevated contents of mercury or tungsten (Fig. 5) may be clearly correlated with the known biogeochemical anomalies. Iodine anomalies may be referred to the zones with different levels of thyroid gland affliction, although such a tendency may not exist since iodine-treated salt is widely consumed in the territory of the republic as a prophylaxis.

Elements such as chlorine, sodium, and so on are distributed more uniformly in the territory studied. Some of the anomalies may be explained by salinization and by the effect of the Aral Sea, which is drying out. Scandium is also characterized by a rather uniform distribution pattern. Anomalies may be connected with the production of phosphorus fertilizers. The more or less uniform distribution of scandium allows one to use it as a reference element in order to calculate the enrichment factor, e.g., for iodine. In this case, anomalies become more pronounced.

One element maps are characterized by the overlapping of many processes that simultaneously influence the elemental hair composition. Here, it is of interest to apply the methods of multifactor analysis in order to compare the elemental hair composition with other characteristics of the area. To obtain more information from the maps, one may use extremely simplified techniques. Thus, Fig. 6 shows the distribution of the coefficient, K_a, equal to

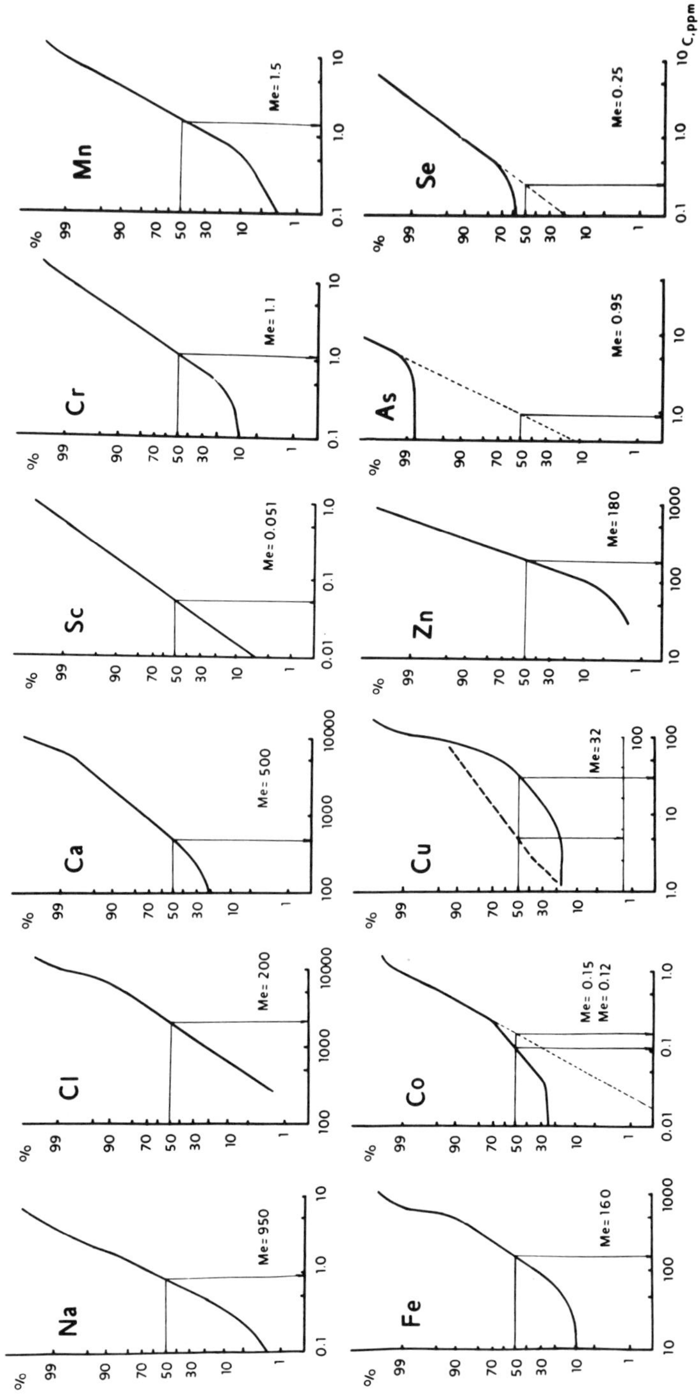
Na
Me= 950
Cl
Me= 200
Ca
Me= 500
Sc
Me= 0.051
Cr
Me= 1.1
Mn
Me= 1.5
Fe
Me= 160
Co
Me= 0.15
Me= 0.12
Cu
Me= 32
Zn
Me= 180
As
Me= 0.95
Se
Me= 0.25
%
99
90
70
50
30
10
1
C, ppm

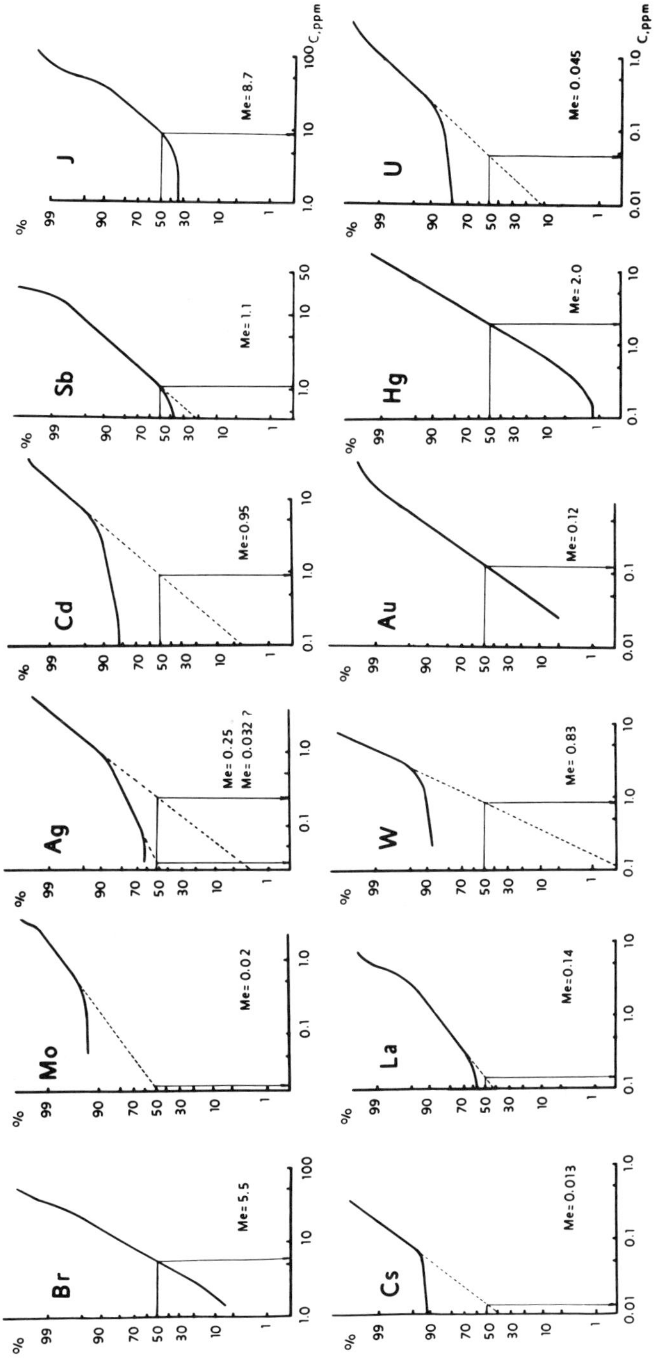

Fig. 2. Cumulative frequency distributions.

Table 1
Concentrations of Elements in Hair of Uzbek SSR Inhabitants (in ppm)

Element	% of Sample in which the element was detected	Arithmetic mean[a]		Geometric mean	Median
		1[b]	2[c]		
Na	98	1080 ± 156	1077 ± 156	944 ×÷ 2.4	950
Cl	99	2870 ± 52	2870 ± 52	2163 ×÷ 2.6	2000
Ca	85	857 ± 18	694 ± 16	497 ×÷ 3.2	500
Sc	95	0.06 ± 0.001	0.057 ± 0.001	0.055 ×÷ 2.8	0.051
Cr	90	2.0 ± 0.05	1.8 ± 0.05	1.39 ×÷ 4.0	1.1
Mn	98	2.2 ± 0.04	2.1 ± 0.04	1.66 ×÷ 2.6	1.5
Fe	90	200 ± 4	179 ± 3.6	203 ×÷ 3.4	160
Co	75	0.47 ± 0.02	0.36 ± 0.02	0.11 ×÷ 5.1	0.13
Cu	82	72 ± 1.4	65 ± 1.4	36.5 ×÷ 5.3	32
Zn	98	200 ± 2.4	200 ± 2.3	203 ×÷ 1.9	180
As	2	13.1 ± 1.7	0.55 ± 0.08	0.12 ×÷ 2.1	0.95
Se	40	1.0 ± 0.05	0.49 ± 0.03	0.25 ×÷ 3.6	0.25
Br	95	7.7 ± 0.15	7.5 ± 0.14	5.9 ×÷ 2.7	5.5
Mo	20	0.81 ± 0.07	0.13 ± 0.005	0.13 ×÷ 2.6	0.02
Ag	35	1.18 ± 0.07	0.46 ± 0.02	0.2 ×÷ 3.1	0.25
Cd	15	8.1 ± 0.4	1.46 ± 0.1	0.38 ×÷ 4.5	0.95
Sb	50	2.1 ± 0.07	2.08 ± 0.07	1.23 ×÷ 3.2	1.1
J	65	25 ± 0.6	16.9 ± 0.47	6.7 ×÷ 7.3	8.7
Cs	8	0.12 ± 0.01	0.018 ± 0.0008	0.013 ×÷ 2.3	0.013
La	40	1.1 ± 0.1	0.53 ± 0.03	0.23 ×÷ 3.4	0.14
W	15	3.6 ± 0.2	0.69 ± 0.033	0.44 ×÷ 2.8	0.83
Au	95	0.24 ± 0.01	0.24 ± 0.007	0.15 ×÷ 2.9	0.15
Hg	99	2.6 ± 0.1	2.56 ± 0.05	2.22 ×÷ 2.8	2.0
U	18	0.59 ± 0.04	0.13 ± 0.01	0.022 ×÷ 4.8	0.045

[a]The arithmetic mean is given as X ± m.
[b]The results excluding the content below the determination limit and high values.
[c]All the results.

$$K_a = C_{Cr} \times C_{Cd} \times C_{Hg} \times 10C_{Co} \times 0.01C_{Cu} \times 0.01C_{Zn} \times C_{Mn}$$

In any case, such a plot leaves the possibility for comparatively reliable evaluation of the environmental state from the elemental hair composition data.

Many studies confirm the relationship between the changes in elemental hair composition and the occurrence of various diseases. More often, such regularities are found if the changes in the concentration of several elements are measured by many factors although a more simple correlation is possible. We do not discuss the reasons for such cases here, however, it should be noted that it may be important to consider the relationship between the occurrence of the diseases and the elemental composition from the point of view of geography. It has been shown above how different information may be used (comparison of different characteristics of mean values is acceptable) to make worldwide distribution maps for various elements. As an example, Fig. 7 shows a mercury distribution map based on the data obtained by different authors. Although there are no data for most of the countries, the overlapping of zones speaks for the possible regularities of element distribution in different regions.

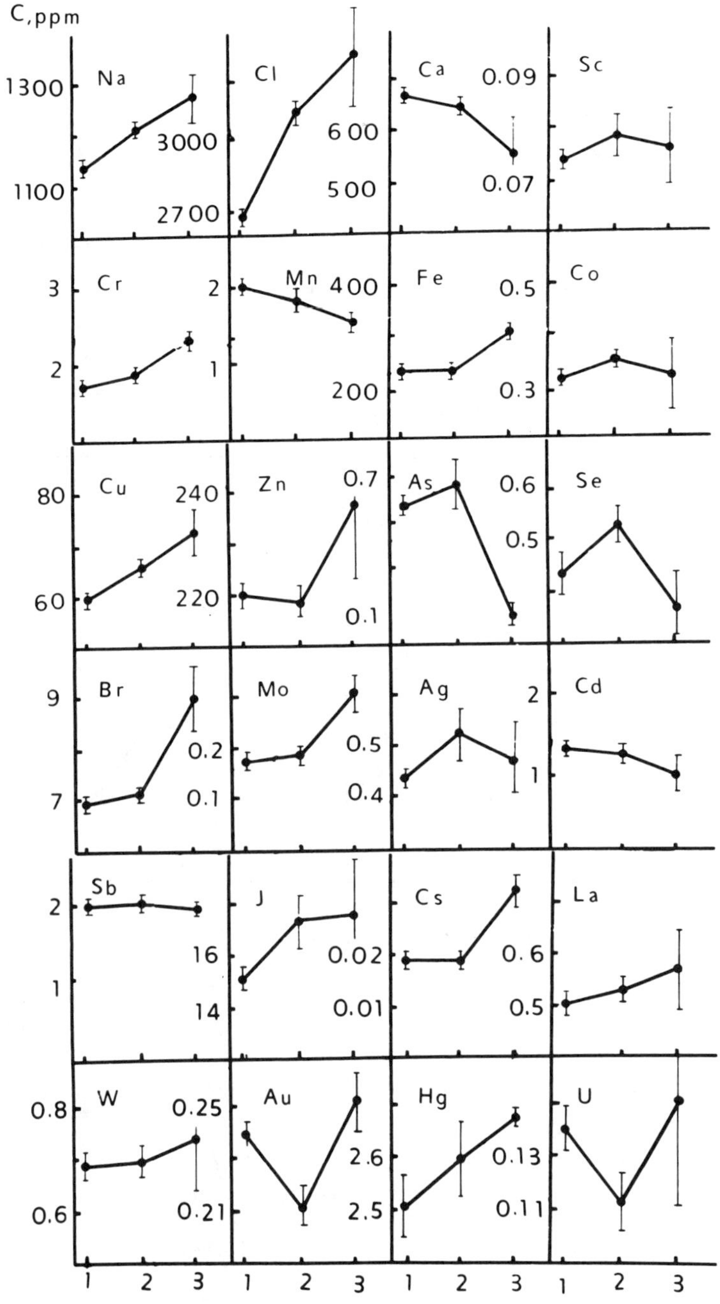

Fig. 3. Concentrations vs age (1, less than 30 yr; 2, 30–50 yr; and 3, more than 50 yr).

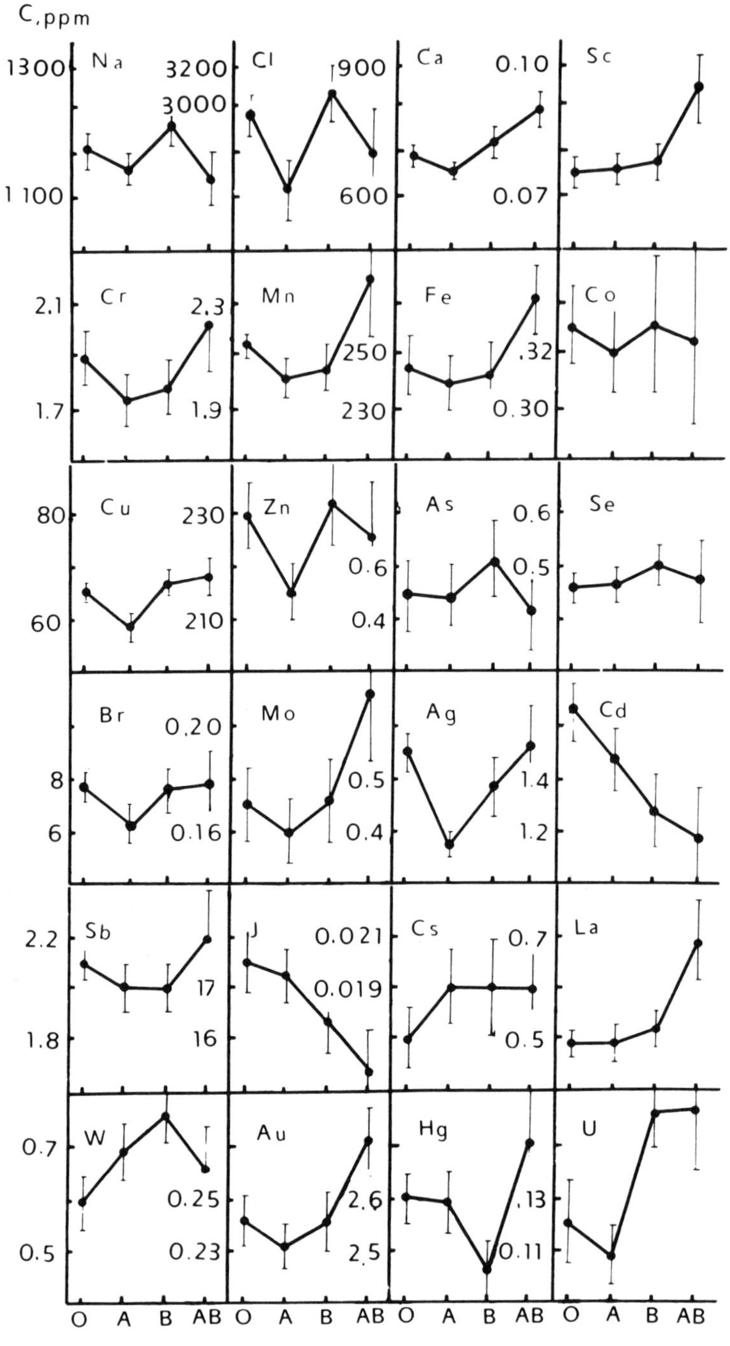

Fig. 4. Concentrations vs blood group.

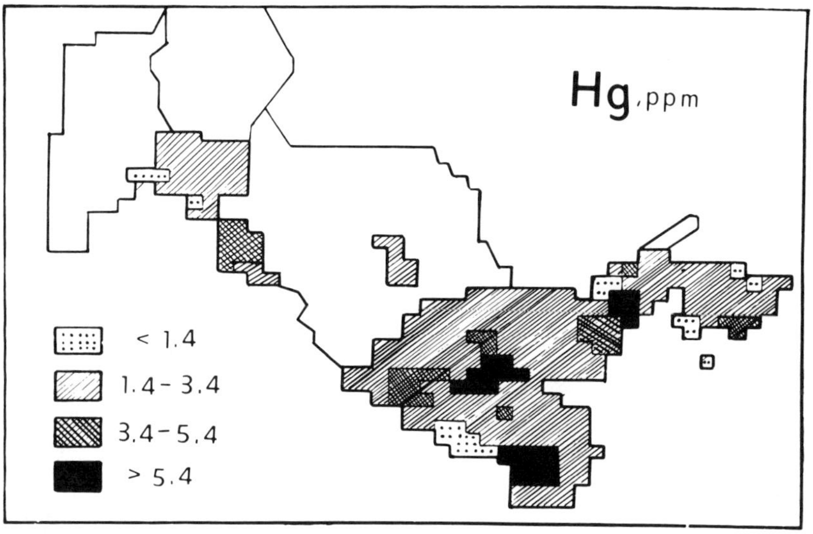

Fig. 5. Mercury distribution in Uzbek SSR.

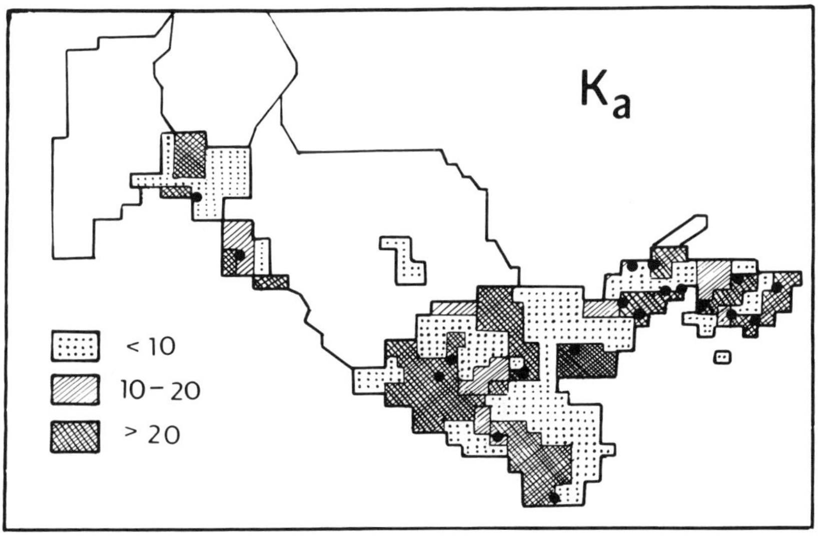

Fig. 6. Factor K_a distribution (●, cities with more than 50,000 inhabitants).

WHO-provided data allowed us to calculate the pair correlation coefficients for hair composition and death rate on the basis of 24 cases (9). Despite the fact that the data on both hair composition and death rate are not complete, being considered together, they allowed us to find the correlation between the average death rate values and elemental composition, with the pair correlation coefficients in some cases not exceeding 0.9 by the absolute value. As an example, Fig. 8 shows the relationship

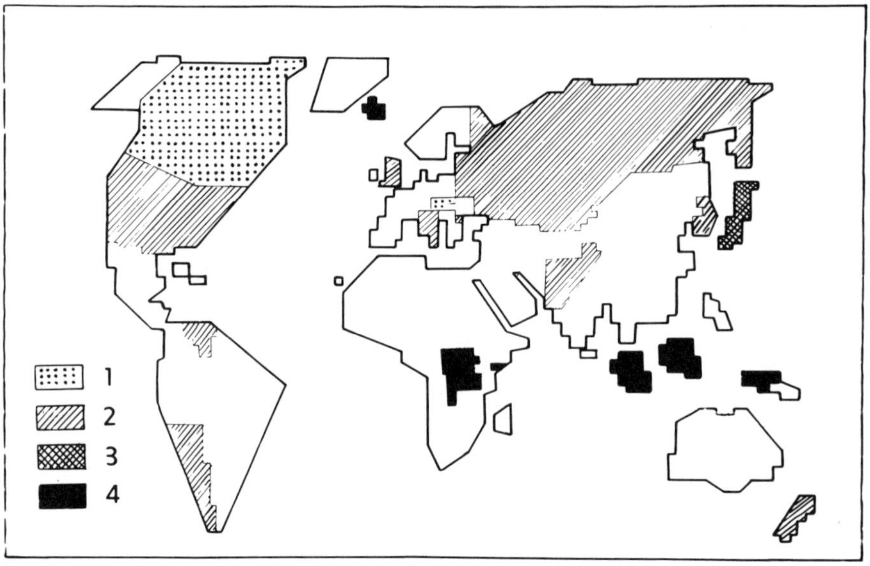

Fig. 7. Mercury distribution (1, less than 1 ppm; 2, 1–4 4 ppm; 3, 4–6 ppm; and 4, more than 6 ppm).

between the rate of death caused by breast cancer and the total cadmium and mercury content in hair. Despite a certain incorrectness of such a comparison (the first group naturally, includes only women, whereas the second group includes men as well), the correlation found seems to be quite illustrative. A similar relationship is also shown in Fig. 8 for the death rate caused by lung cancer vs the total amount of cadmium and mercury.

It is obvious that the usage of more complete and correct information, as well as the application of multifactor analysis, should make these dependencies more accurate. The regularities found speak for the fact that, despite the possible differences in diagnostics level, efficiency of prophylaxis and treatment, the peculiar living conditions, and so on, the search for the dependence of the elemental hair composition in different countries and regions on the distribution pattern of various diseases may be of great interest for making a prognosis of the health status. It may be looked upon as an evidence of the environmental influence on the state of health or vice versa; this may point to the changes in the trace element status of an organism at the early stages of a disease. In the latter case, it seems reasonable to use these data for the identification of risk groups.

SUMMARY

These data provide exciting evidence of how informative elemental hair composition may be. Here, the geographic approach to the solution of the problem seems to be very attractive.

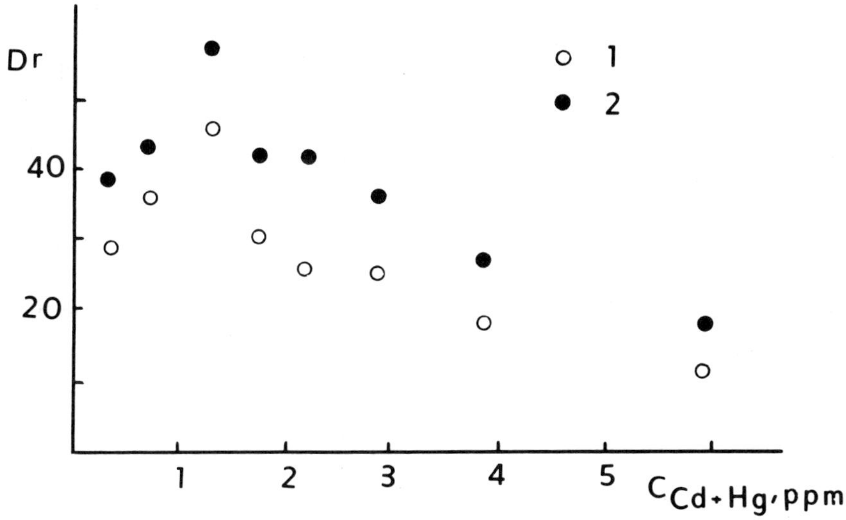

Fig. 8. Mortality vs sum of Cd and Hg concentrations in hair for different countries (1, mammary carcinoma; and 2, lung carcinoma).

As the biogeochemical situation undergoes rapid global changes, it appears timely to make worldwide maps of elemental hair composition, based on common methodological approaches, possibly in one laboratory.

The information obtained would allow one to estimate the state of the biogeochemical environment on the Earth, and repeated series of such studies, after certain periods of time, may appear as a powerful means of monitoring and prognosis of the environmental state. Besides, the results of these studies allow the evaluation and prognosis of the health status in various regions and countries. Taking into account that different investigators carry out simultaneous and independent studies of elemental hair composition relative to various diseases, the cumulative data will allow the total examination of the world population in order to identify the risk groups and ensure early diagnosis. It should be noted that our laboratory registered cases in which the characteristic changes in elemental hair composition were observed one or two years before the patient consulted a doctor. Now we can only speculate on the additional information that may be obtained. This makes the whole subject very attractive and challenging.

In any case, an attempt to draw worldwide maps of the elemental hair composition seems to be quite justified and important. This can be done within a reasonable period of time, involving reasonable resources.

REFERENCES

1. A. I. Perelman, *Geokhimia,* M. Vishaya shkola, 1979.
2. Yu. S. Ryabukhin, *J. Radioanal. Chem.* **60,** 7–30 (1980).
3. *Activation analysis of hair as an indicator of contamination of man by environmental trace element pollutants,* IAEA/RL 50, 1978.

4. L. I. Zhuk, A. A. Kist, T. Tillayev, S. M. Bakhramov, et al., *Med. J. Uzbekistan* **6,** 28, 29 (1986).
5. L. I. Zhuk, A. A. Kist, I. N. Nikolskaya, N. S. Osinskaya, T. Tillayev, S. Tursunbaev, and S. V. Agzamova, *J. Radioanal. Nucl. Chem.* **120,** 369–377 (1988).
6. S. B. M' Baku and R. M. Parr, *J. Radioanal. Chem.* **69,** 171–180 (1982).
7. I. Takeuchi, T. Hayashi, J. Takada, M. Koyama, H. Kozuka, H. Tzuji, Y. Kusaka, S. Ohmori, M. Shinogi, A. Aoki, K. Katayama, and T. Tomiyoma, *Nuclear activation techniques in the life sciences*, IAEA, Vienna, 1985, pp. 545–561.
8. A. A. Kist, Fenomenologiya biogeokhimii i bioorganicheskoi khimii, T, FAN, 1987.
9. Ezhegodnik mirovoi sanitzrnoi statistiki, 1984, WHO report, Geneva, 1988.

Environmental Specimen Banking

A Complement to Environmental Monitoring

ROBERT E. LEE

US Environmental Protection Agency, Atmospheric Research and Exposure Assessment Laboratory, Research Triangle Park, NC 27711

Received June 5, 1989; Accepted October 23, 1989

ABSTRACT

The risk of environmental damage from the countless chemicals and chemical combinations is estimated by monitoring the air, water, soil, and biota and comparing the findings to known risks. Environmental monitoring involves the collection and chemical analysis of selected samples to determine pollutant trends, forecast damage to the environment, and develop control strategies. However, past experience has taught us that there is always a level of uncertainty associated with the chemical analysis of complex environmental samples. This uncertainty arises from insufficient analytical sensitivity to detect trace levels of damaging pollutants, unsuspected analytical inaccuracies associated with unknown interferences, and limited resolution to identify and quantify all pollutants of interest.

Environmental specimen banking can serve as a complement to environmental monitoring by allowing future chemists to analyze preserved samples retrospectively with emerging analytical methodologies. In other words, specimen banking will give later environmentalists the opportunity to characterize chemicals whose hazards were not recognized or for which sufficiently accurate analytical methods were not available. This will enable us to determine whether

Mention of trade names or commercial products does not constitute endorsement or recommendation for use.

a newly recognized pollutant is truly a new environmental threat or whether it has always been with us.

Although environmental specimen banking is still in its infancy, we now have many experiences that clearly demonstrate that valuable information can be produced. In all environmental monitoring programs, some consideration should be given to preserving selected samples. Ideally, specimen banking will involve; (1) a stable funding commitment, (2) a cryogenic storage facility, (3) development of selection criteria to collect the most valuable environmental specimens, and (4) a continuing research program to develop optimal procedures for collecting, handling, and preserving specimens.

Index Entries: Tissue archive; pollutant trend monitoring; long-term storage; pollutants, inorganic and organic.

INTRODUCTION

Some of us remember the early environmental scare of the 1960s when mercury was found in swordfish and other aquatic organisms. The question was asked: Is mercury a newly introduced contaminant in the environment, or has it always been with us? To answer this question, preserved samples of aquatic organisms were located in museum archives, but were found unsuitable because of contamination during the preservation process. Had carefully collected and cryogenically-preserved fish specimens been available, the above asked question would have been quickly and reliably answered.

There are countless chemicals and chemical combinations that may produce harmful effects in the environment, sometimes at concentrations below our ability to analyze or detect. The way we assess the environmental risk from chemical contamination is through monitoring to identify pollutant hazards and to determine trends. These data are used to forecast damage to the environment and develop control strategies. However, the level of uncertainty in assessing the impact of long-lived biologically-active chemicals is high. We repeatedly confront unexpected consequences of our technologies. As R. Lewis has written, "Uncertainty is the constant companion of risk assessment" (*1*).

A way to cope with the uncertainty of hazard prediction is to preserve selected environmental specimens for chemical (and perhaps microbiological) analysis, using improved analytical procedures developed in the future. Consequently, specimen banking can serve as a "safety net" in the assessment of chemical risk. New technologies almost always produce unintended and unpredicted impacts, but by environmental monitoring, complimented by a specimen banking program, the accuracy of risk assessment will almost certainly be improved.

WHAT IS ENVIRONMENTAL SPECIMEN BANKING?

An Environmental Specimen Bank will allow environmentalists to do something that cannot be done any other way: to characterize chemicals whose hazards were not recognized earlier or for which sufficiently accurate analytical methods were not available.

Monitoring alone will not allow us to go back in time to do retrospective analysis, nor allow us to take advantage of newly emerging analytical methodologies that are not available today. The capabilities of analytical chemistry have increased rapidly over the past 40 years with the development of neutron activation and radiochemical techniques in the 1940s, gas chromatography and other separation techniques in the 1950s, X-ray fluorescence and atomic absorption techniques in the 1960s, and the rapid proliferation of combined ("hyphenated") techniques, such as high resolution gas chromatography/mass spectrometry and gas chromatography/matrix-isolation infrared spectroscopy in the 1970s and 1980s. New capabilities can be expected at an even faster rate in the 1990s and beyond with the development of microelectrosensor technology, improved chemical separation techniques such as supercritical fluid separations, computer enhancement of data output, and other electronic improvements, and the continuing trend of coupling totally different separation techniques and detectors that began in the 1970s. All of these developments will enable the future analyst to determine chemical constituents with a degree of sensitivity and specificity that is unavailable today. By preserving environmental samples that are collected as part of the expanding ecological monitoring networks, we shall be able to reanalyze these specimens with future technologies to answer questions such as the one posed above: Is this pollutant a new environmental threat or has it always been with us?

VALUE OF ENVIRONMENTAL SPECIMEN BANKING

There already exists a number of environmental specimen banks and other types of systems for sample preservation in the US and several other countries, including Germany, Canada, and Sweden. The US Environmental Protection Agency, in conjunction with the National Institute of Standards and Technology (formerly the National Bureau of Standards), has supported the collection and preservation of human liver samples collected from autopsies since 1980. The National Cancer Institute has supported the preservation of human serum samples, The National Oceanic and Atmospheric Administration has supported the preservation of marine samples under its National Status and Trends Network, and the US Department of Agriculture and the Food and Drug

Administration has supported the preservation of food samples. Experiences from these and other specimen banking programs have already given us valuable information on the true worth of specimen banking, illustrated in the following examples.

Kepone Episode in the James River/Chesapeake Bay

In 1975, extensive contamination by the pesticide Kepone was discovered in the James River and the Chesapeake Bay, which resulted from unscrupulous dumping of waste from chemical production (*2,3*). This episode is a prime example where ecological samples of aquatic life, shellfish, or other specimens from the Chesapeake Bay or the James River could have been used, if available, for determining Kepone levels before the start-up of Kepone production in that area. These specimens would have been extremely useful to assess the change in levels of that pollutant in the environment. Fortunately, limited samples dating back 3–4 years before the discovery of Kepone contamination were available from the Virginia Institute of Marine Sciences. These samples proved to be invaluable in establishing the extent to which Kepone affected marine life. Had earlier samples been available from a banking system, a much better assessment of Kepone baseline levels prior to the dumping episodes could have been made, thereby allowing environmental officials to assess environmental damage more quickly and implement enforcement actions.

Introduction of Chlorinated Dioxins in the Environment

A classic example of the value of stored samples was presented at the 6th International Symposium on Chlorinated Dioxins held in Fukuoka, Japan in 1986. Although it is well known that chlorinated dioxins are highly toxic environmental contaminants introduced through the production of herbicides and other synthetic compounds, it has been proposed that these materials may also be formed in trace amounts by the combustion of untreated wood (the trace chemical theory of the origin of dioxins). This theory suggests that chlorinated dioxins would have existed prior to the synthesis of dioxin-containing chemicals in the early part of the century, thereby indicating that humans and animals might have evolved in the presence of these compounds. Schecter and coworkers (*4*) found lower levels of dioxin in human tissue in nonindustrialized countries than in persons living in industrialized countries, suggesting an anthropogenic origin. To explore this hypothesis further, they analyzed human liver and lung tissue from two female Eskimos frozen over 100 years ago and meat found in the Eskimos' home. In addition, charred wood from an ancient California Redwood Tree was analyzed. Nondetectable levels of dioxins were found, further supporting the theory that dioxins are a recent introduction into the environment and are of an anthropogenic origin.

Contamination of Bird Tissue

In the 1960s, increased levels of pollutants in bird tissue were measured by researchers in Sweden through comparison of recent bird tissue samples with those of bird skins preserved in academic and museum collections dating as far back as 1940 (*5*). This study underscored the value of retrospective analysis for comparing chemical constituents in preserved environmental materials with chemical constituents in freshly collected samples.

Archived Wildlife Samples in the Great Lakes Area

Several Canadian agencies, including the Canadian Wildlife Service, have determined pollutant trends in the Great Lakes area through retrospective analysis of preserved wildlife samples (*6*). Contaminant trends for three major chlorinated organics (PCBs, DDE, and oxychlordane) were studied in herring gull eggs preserved since the 1970s from Lake Ontario and Lake Michigan. Using improved methods that rendered some of the earlier data questionable, comprehensive retrospective analysis for these chlorinated organic compounds clearly demonstrated declining trends for PCBs and DDEs, with a relatively constant level of oxychlordane throughout the study period from 1970–1982. This study demonstrated the effectiveness of controls imposed in Canada and the US on the use and disposal of PCBs and many other organochlorine pesticides. Chlordane, of which oxychlordane is a major metabolite, was used extensively in the United States after restrictions on DDT were imposed in the late 1970s.

In an effort to study the decline in the population of northern gannet in the Gulf of St. Lawrence, adult and juvenile bird samples and eggs were collected and preserved in a specimen bank from 1968 to 1976. These samples were analyzed for organochlorine contaminants and compared with recently collected egg samples in 1984. High levels of DDE found in the earlier samples and the thinning of egg shells were associated with the period of reduced reproductive success. PCB levels also decreased in gannet eggs throughout the study period, but at a slower rate than DDE levels. Since the analytical methods used in 1969 seriously underestimated the levels of PCBs, reanalyses of the preserved specimens with modern analytical technology again underscored the value of retrospective analysis.

CONCLUSIONS

The preservation of ecological samples, collected in conjunction with a monitoring program, can be an invaluable asset. The examples noted above clearly show the value of retrospective analysis of preserved samples as more accurate analytical techniques are developed. In the past, and perhaps even today, available analytical techniques can often lead to inaccurate and misleading data. However, through retrospective analysis of carefully preserved samples, using more sensitive and precise techniques, more accurate assessments of pollutant trends can be made.

Four elements of a viable ecological specimen banking program can be identified

1. A stable funding commitment;
2. A central cryogenic storage facility, ideally operating at liquid nitrogen temperature;
3. Establishment of selection criteria to collect the most valuable environmental specimens; and
4. A continuing research and development program to develop methods and procedures for the collection, handling, and preservation of environmental specimens.

It is absolutely critical to make a commitment for stable funding to establish and maintain a successful specimen banking program. Without such a commitment, all earlier efforts to collect and store environmental samples will be lost.

Much is known about the collection, handling, and preservation of biological specimens through work done at the National Institute of Standards and Technology and similar institutes in other countries. Procedures have been developed for the collection and preservation of human liver specimens, marine sediments and tissues, terrestrial wildlife, and vegetation. However, very little is known about optimal methods for collecting and preserving air pollution samples for specimen banking. Airborne pollutants may consist of nonvolatile components that can be collected on filters (and perhaps cryogenically preserved), semivolatile components that require both filter media and absorption devices for collection, and volatile components (especially organic compounds) that may require whole air sampling. As a first step, it should be possible to cryogenically preserve filter samples containing nonvolatile organic and inorganic constituents for retrospective chemical analysis. A much more difficult task involves devising appropriate methods for collecting and preserving semivolatile and volatile compounds. Means might also be developed for collecting and preserving rainwater samples as part of a future environmental specimen bank.

In conclusion, environmental specimen banking is still in its infancy. Much more needs to be done to develop the collection, handling, and preservation techniques for air samples, rainwater samples, and environmental biota. Although long-term funding must be assured to maintain a successful specimen banking program, the rewards to future generations can be great.

REFERENCES

1. R. A. Lewis, *Proceedings of the 11th US-German Seminar of State and Planning on Environmental Specimen Banking*, Bayneuth, Bavaria, May 1–3, 1988, in press.
2. R. J. Huggett, M. M. Nichols, and M. E. Bender, *Contaminants and Sediments*, vol. I, R. A. Baker, ed., Ann Arbor Science, Ann Arbor, MI, 1980, pp. 33–52.
3. G. M. Goldstein, *EPA Report No. 600/1-78-022*, Health Effects Research Laboratory, Research Triangle Park, NC, 1978.
4. A. Schecter, M. Gross, and A. Dekin, *Abstracts of Dioxin 1986, 6th Int. Symposium on Chlorinated Dioxin and Related Compounds*, Fukuoka, Japan, Sept. 16–19, 1986.
5. M. Olsson, *International Review of Environmental Specimen Banking*, S. A. Wise and R. Zeisler, eds., NBS Spec. Publ. 706, US Government Printing Office, Washington, DC, 1985, pp. 26–32.
6. J. E. Elliott, R. J. Norstrom, S. W. Kennedy, and G. A. Fox, *Progress in Environmental Specimen Banking*, S. A. Wise, R. Zeisler, and G. Goldstein, eds., NBS Special Publication 740, US Government Printing Office, Washington, DC, 1988 pp 131–142.

Alaskan Marine Mammal Tissue Archival Project

PAUL R. BECKER,*[,1] BARBARA J. KOSTER,[2] STEPHEN A. WISE,[2] AND ROLF ZEISLER[2]

[1]*Ocean Assessments Division, National Oceanic and Atmospheric Administration, Anchorage, Alaska 99513; and* [2]*Center for Analytical Chemistry, National Institute of Standards and Technology, Gaithersburg, MD 20899*

Received June 5, 1989; Accepted October 23, 1989

ABSTRACT

A project to establish an archive of Alaskan marine mammal tissues was conceived in 1987 to be a part of the National Biomonitoring Specimen Bank (NBSB). Protocols and field collection of marine mammals, long-term storage, and analysis are summarized in this paper. Instrumental neutron activation analysis has been used for an initial evaluation of trace element content in samples of northern fur seal (*Callorhinus ursinus*) from the Pribilof Islands. The findings agree with previously observed trace element levels in northern fur seals. The archived specimens can be used in future studies when comparisons of past and present pollution levels are needed.

Index Entries: Specimen bank; long-term storage; marine mammals; Alaska and marine mammals; trace element analysis and marine mammals; neutron activation analysis; northern fur seal (*Callorhinus ursinus*); ringed seal (*Phoca hispida*).

*Author to whom all correspondence and reprint requests should be addressed.

Certain commercial equipment, instruments, or materials are identified in this paper in order to specify adequately the experimental procedure. Such identification does not imply recommendation or endorsement by the National Institute of Standards and Technology, nor does it imply that the materials or equipment identified are necessarily the best available for this purpose.

INTRODUCTION

Within the last decade, Alaska has experienced a tremendous increase in industrial activities, such as mineral extraction and offshore oil exploration. Concerns have been raised as to what impact these industries have, and will have, on the Alaskan environment. To address some of these concerns, the Minerals Management Service (MMS) and the National Oceanic and Atmospheric Administration (NOAA) initiated the Alaskan Marine Mammal Tissue Archival Project (AMMTAP) in 1987 (*1*).

Most marine mammals are considered top predators in the marine environment. Chemical analysis of their tissues can be particularly useful in determining whether bioaccumulation of contaminants is occurring in marine food chains. Not only is the impact on the marine food chain important, but of equal importance are the human health implications. Marine mammals are a prime food source for many of the native Alaskan people. Additionally, fish taken from areas such as the Bering Sea are exported all over the world, yet there is no nearby human population to easily test for contaminants derived from these fish. Hence, the Alaska marine mammal tissues can provide information that may be of worldwide interest. The collection of marine mammal tissues, over a period of several years, will provide an archive of samples for the determination of baseline contaminant levels and against which future contaminant measures can be compared.

Only in the past decade has the concept of an environmental specimen bank for the archiving of biological and environmental samples for retrospective analysis been recognized as an important component to systematic environmental monitoring. For 10 years, the National Institute of Standards and Technology (NIST) has been involved in environmental specimen banking activities. These activities resulted in the development of collection, storage, processing, and analysis procedures for long-term archiving of a variety of environmental specimens, including human liver, fish muscle and liver, marine bivalves, sediment, and marine mammal tissues. Although the total scientific value of the banked samples is not fully known at this time, the current uses of the banked samples, and the implementation of the concept, have already contributed to major monitoring programs in the US and abroad. The National Biomonitoring Specimen Bank (NBSB) can serve as a valuable resource for the assessment of long-term trends of pollutants affecting human and environmental health, in particular for those pollutants that have been unnoticed thus far, or could not be measured in the past (*2*).

Several species of marine mammals have been selected for study in the Alaskan Marine Mammal Tissue Archival Project. The species are selected on the basis of criteria, such as availability, migration habits, feeding habits, reference information, and animal and organ size. The animals are collected with the cooperation of NIST, the National Marine Fisheries Service (NMFS), the US Fish and Wildlife Service (USFWS), the

Alaska Department of Fish and Game (ADF&G), and the North Slope Borough Department of Wildlife Management, as well as many individual Alaskan native hunters and hunting organizations. Sample collections have been taken during the Pribilof Island subsistence hunt of the northern fur seals (*Callorhinus ursinus*) and at Barrow, Alaska during the summer ringed seal (*Phoca hispida*) hunt. Animals considered for future sampling include: the harbor seal (*Phoca vitulina*), the bearded seal (*Erignathus barbatus*), the Pacific walrus (*Odobenus rosmarus divergens*), belukha whale (*Delphinapterus leucus*), and polar bear (*Ursus maritimus*).

MATERIALS AND METHODS

One of the primary concerns of the AMMTAP is to avoid contaminating the tissues during the collection and handling process, thereby assuring the integrity of the sample. Therefore, detailed sampling protocols have been established for handling the animal from the time it is sacrificed until the time the tissue sample is permanently archived in the NBSB. The principles of such procedures have been evaluated previously with the human liver project (3). These protocols cover: the use and cleaning of specific equipment, such as the titanium instruments and Teflon bags (Clean Room Products, Ronkonkoma, NY) or Teflon jars (Savillex, New Prague, MN), the cleanliness of the physical area where the sample is packaged, sample handling procedures, and the procedure for quick freezing and shipment of the samples to the NBSB. Additionally, sample information forms are filled out for each animal from which samples are taken. The forms include information such as, location and time of sampling, size and sex of animal, weight of the tissues processed, weather conditions, handling times, and personnel responsible. Tissue histology and age determination are added to the form later. This information is stored at NBSB in hardcopy and on computer.

Because laboratory facilities are not always available at the sampling site, a completely self-contained sampling kit was designed for use in field situations. The kit included all supplies and implements necessary for sampling one animal. The kit has been evaluated twice during field collection and was found to be a viable solution for sampling in the field.

Four types of tissue were sampled from each animal during the Pribilof hunt: liver, kidney, muscle, and blubber. For future collections, the muscle tissue has been eliminated from the process owing to homogenization difficulties and low levels of analytes. Three hundred grams of each tissue are considered essential for archiving. Each 300 g tissue sample is divided into two replicate samples: one section (A) is placed in the specimen bank at liquid nitrogen temperatures for long-term storage, and the second section (B) is homogenized and part of the tissue is analyzed to provide baseline data.

To evaluate the significance of the various tissues for environmental studies, three tissue types were selected for analysis. Initial determinations of both inorganic and organic constituents were performed on cryogenically-prepared homogenates (*4*) from two northern fur seals from the 1987 Pribilof Islands hunt. The contaminants of interest included both organic compounds (polycyclic aromatic hydrocarbons, chlorinated pesticides, and polychlorinated biphenyls) and trace elements.

The determination of 23 trace elements was performed on three tissue types from two northern fur seal samples using instrumental neutron activation analysis (INAA), and the results are reported in this paper. The procedure followed the approach developed for analyzing human tissue from the NBSB (*5*). The tissue homogenates were freeze-dried, pressed into 200 mg pellets, and irradiated. The blubber tissue could not be transformed into a sample suitable for the INAA procedure and, therefore, was not analyzed. The INAA procedure consisted of a 120-s irradiation at a fluence rate of $2 \cdot 10^{17}$ neutron/m^2/s^1, followed by two counts on a high resolution gamma spectrometer, and a second 16-h irradiation followed by two counts after appropriate decay times. Element standards for quantitation by the comparator method and certified reference materials (NIST SRM 1577 Bovine Liver, NIST SRM 1572 Citrus Leaves, and IAEA MA-A-2/TM Fish Flesh) for quality assurance were included in the irradiation sets. A special high-count-rate gamma spectrometer, combined with loss-free counting techniques (*6*), was used to lower the instrumental detection limit for vanadium.

RESULTS AND DISCUSSION

The results of the initial INAA studies are listed in Table 1. Replicate samples from the various tissues did not deviate more than 5% relative if the analytical uncertainty, owing to counting statistics, was sufficiently small. This result is in agreement with earlier findings on human liver tissue in the NBSB (*7*). The analysis of the certified reference materials did not reveal any bias. The following elements were also included in the quantitative evaluation, but were below their detection limit in all investigated tissues: Sr, La, Ce, Sm, Eu, Tb, Hf, Ta, Au, and U.

Both the human and seal liver tissues have comparable element concentrations that allow for a similar suite of elements to be detected with the established procedure. The seal kidney and muscle show significantly different trace element concentration profiles for some elements. For example, arsenic is highly enriched in muscle tissue, whereas other elements, such as silver and cadmium, were below our detection limit. Therefore, as in human monitoring, the liver tissue may be the most suitable tissue for the determination of both inorganic and organic constituents. These limited results are comparable to results reported by

Table 1
Trace Element Concentrations in Northern Fur Seal Tissues[a]

	LIVER		KIDNEY		MUSCLE	
	Seal 1	Seal 2	Seal 1	Seal 2	Seal 1	Seal 2
Na	760.4±6.0	699.8±5.5	1305.4±9.8	1382.3±10.2	505.4±3.7	479.4±3.6
Mg	194±4.5	213±4.4	189±5.4	168.4±4.2	231±3.9	272±4.5
Al	0.533±0.081	0.522±0.042	1.599±0.085	0.780±0.053	0.313±0.041	0.712±0.049
Cl	1016±13	891±12	1578±18	1697±16	538.9±8.0	470.0±6.0
K	3316±50	3155±59	2610±49	2464±33	3350±33	3615±38
Ca	38.3±4.4	30.6±3.8	38.1±4.4	53.1±4.0	31.5±3.7	22.1±2.5
Sc	≤0.00012	≤0.00013	≤0.00011	≤0.00009	0.000143±0.000007	0.000087±0.000006
V	0.0843±0.0062	0.1088±0.0044	≤0.013	≤0.008	≤0.0073	≤0.0073
Cr	0.0251±0.0044	0.0551±0.0050	0.0390±0.0036	0.0106±0.0032	0.0612±0.0030	0.0386±0.0030
Mn	4.179±0.045	3.185±0.036	1.157±0.022	1.260±0.015	0.1630±0.0058	0.1857±0.0063
Fe	171.8±1.0	118.8±0.8	74.00±0.49	51.70±0.37	75.12±0.48	70.58±0.45
Co	0.01884±0.00016	0.01624±0.00016	0.02501±0.00017	0.01733±0.00014	0.005094±0.000068	0.003641±0.000057
Cu	56.34±0.82	17.03±0.49	7.06±0.52	8.71±0.35	2.20±0.45	2.07±0.30
Zn	56.53±0.42	57.89±0.43	63.28±0.47	37.18±0.28	35.00±0.26	33.43±0.25
As	≤0.092	0.187±0.020	0.164±0.021	0.503±0.023	0.174±0.012	0.311±0.014
Se	6.289±0.052	5.450±0.045	3.856±0.032	3.066±0.026	3.109±0.026	2.147±0.018
Rb	1.874±0.047	1.759±0.049	1.368±0.033	1.038±0.030	1.518±0.028	1.408±0.028
Mo	0.38±0.10	0.65±0.11	≤0.40	≤0.29	≤0.25	≤0.28
Ag	0.1721±0.0023	0.3815±0.0037	≤0.0044	0.0129±0.0011	≤0.0030	≤0.0029
Cd	18.51±0.29	8.01±0.19	66.66±0.79	29.02±0.39	≤0.26	≤0.39
Sb	0.00105±0.00027	0.00375±0.00033	≤0.00069	≤0.00054	≤0.00048	0.00060±0.00016
I	0.447±0.071	1.74±0.12	≤0.32	≤0.16	≤0.20	≤0.21
Cs	0.01691±0.00048	0.02880±0.00053	0.02474±0.00050	0.02312±0.00044	0.04045±0.00046	0.04321±0.00045

[a]INAA results (mg/kg), errors represent uncertainty of a single determination and are comprised of counting statistics and estimated uncertainty owing to standards and irradiation and counting geometries.

others for trace element levels in this species (*8,9*). More detailed baseline studies, as well as values for other trace elements, can be obtained from those samples in the same collection, which have not yet been analyzed. The long-term stability of samples being stored in the NBSB has been recently confirmed (*10*).

Although the emphasis is on the collection of tissues for the determination of contaminants that may be associated with industrial activities, it is also recognized that the development of an archive of marine mammal tissues, which have been collected and stored using carefully controlled procedures, provides a resource that goes beyond the immediate needs of the program. Such an archive, developed over several years, provides a resource of materials for future investigators addressing questions concerning the transport of elements and compounds throughout the polar ecosystem, regardless of source. It is hoped that this resource will gain wide support from the many agencies involved in marine mammal research and management and environmental assessment and management, as well as organizations and individuals with interests in the polar ecosystem.

ACKNOWLEDGMENT

This work was supported by the Minerals Management Service, Department of the Interior and the National Oceanic and Atmospheric Administration. Special acknowledgment is extended to the seal harvest crews of St. Paul, Pribilof Islands, Alaska, TDX Corporation, NMFS, and the North Slope Borough Department of Wildlife Management, Barrow, Alaska, for working with us to obtain samples.

REFERENCES

1. P. R. Becker, S. A. Wise, B. J. Koster, R. Zeisler, *Alaskan Marine Mammal Tissue Archival Project: A Project Description Including Collection Protocols,* NBSIR 88-3750, 1988, 46 pp.
2. S. A. Wise, B. J. Koster, R. M. Parris, M. M. Schantz, S. F. Stone, and R. Zeisler, *Intern. J. Environ. Anal. Chem.*, **37,** 9–106 (1989).
3. S. A. Wise and R. Zeisler, *Environ. Sci. Technol.* **18,** 302A–307A (1984).
4. R. Zeisler, J. K. Langland, and S. H. Harrison, *Anal. Chem.* **55,** 2431–2434 (1983).
5. R. Zeisler, R. R. Greenberg, and S. F. Stone, *J. Radioanal. Nucl. Chem.* **124,** 47–63 (1988).
6. G. P. Westphal, *J. Radioanal. Chem.* **70,** 387–410 (1982).
7. R. Zeisler, S. H. Harrison, S. A. Wise, *Biol. Trace Elem. Res.* **6,** 31–49 (1984).
8. C. J. Goldblatt and R. G. Anthony, *J. Environ. Quality,* **12(4),** 478–482 (1983).
9. D. R. Smith, *Report to NMFS,* Juneau, Alaska by Creative Educational Consultants, Monterey, California, 1986, 5 pp. plus 4 pp. appendix.
10. R. Zeisler, R. R. Greenberg, S. F. Stone, and T. M. Sullivan, *Fresenius Z. Anal. Chem.* **332,** 612–615 (1988).

A Comprehensive Study on the Contents and Leaching of Trace Elements from Fly-Ash Originating from Polish Hard Coal by NAA and AAS Methods

RAJMUND DYBCZYNSKI,*[1] KRZYSZTOF KULISA,[1] MARIA MAŁUSECKA,[2] MARIA MANDECKA,[1] HALINA POLKOWSKA-MOTRENKO,[1] SŁAWOMIR STERLINSKI,[1] AND ZYGMUNT SZOPA[1]

[1]Department of Analytical Chemistry, Institute of Nuclear Chemistry and Technology, Warsaw, Poland; and [2]Department of Technology for Power Station Waste Utilization, Katowice, Poland

Received April 17, 1989; Accepted October 27, 1989

ABSTRACT

In order to assess the environmental risks associated with the emission of fly-ash into the atmosphere and its storage on waste heaps, the trace element contents of fly-ashes from burning Polish hard coal were determined by a newly developed INAA method. Leaching of trace elements from the fly-ash by water and H_2SO_4 solution (pH $\approx$ 2.5) simulating acid rain, respectively, was studied using AAS and spectrophotometric methods. Analogous experiments were done with neutron-irradiated fly-ash, following the composition of the eluate gamma-spectrometrically. The new fine fly-ash (CTA-FFA-1) candidate reference material was prepared, and the certification was undertaken on the basis of an international intercomparison run. Preliminary evaluation of results shows that at least 38 elements will be certified and, in addition, the "information values" for at least 12 elements will be given.

Index Entries: Polish hard coal; fly-ash; INAA; trace elements, leaching of; H_2O; H_2SO_4; environmental risks.

*Author to whom all correspondence and reprint requests should be addressed.

INTRODUCTION

The power engineering industry being an important element of the national economy, also constitutes a serious problem for the natural environment. Burning of coal releases large amounts of sulfur and nitrogen oxides, giving rise to acid rain and, in addition, produces huge amounts of solid wastes in the form of slag, bottom ash, and fly-ash, part of which is emitted into the atmosphere in the form of aerosols. In Poland, ca. 80% of power production comes from burning solid fuels, i.e., hard and brown coals (*1*).

The chemical composition of fly-ash, which is known to contain a number of essential and toxic elements (*2–4*), as well as the susceptibility of these elements to leaching by water and acid rain, respectively, are of primary importance for the evaluation of the environmental impact of storage of large masses of fly-ash on waste heaps, as well as the deposition of fly-ash from the atmosphere. In particular, leaching studies can throw some light on the passage of the trace elements from the fly-ash into soil, vegetation, and, consequently, to the food chain.

A comprehensive approach used in this work included:

1. Elaboration of a suitable procedure, based on instrumental neutron activation analysis (INAA), for the determination of several trace elements in fly-ash and geological materials of similar matrix, as well as checking its reliability with the use of suitable certified reference materials.
2. A study on leaching of trace elements from fly-ash of known composition by water and simulated acid rain, respectively, with the use of atomic absorption spectrometry and spectrophotometry.
3. Column studies on leaching of trace elements from neutron-irradiated fly-ash samples by water and simulated acid rain.
4. Preparation of a fine fly-ash candidate reference material and its certification by international intercomparison and our own method of data evaluation.

METHODS

INAA Procedure

One hundred milligram samples of fly-ash of known moisture contents (as determined in separate subsamples by drying to constant weight at 105°C) were weighed into circular polyethylene (PE) foil bags (Ø ≈ 2.5 cm, sample thickness 2–3 mm) and heat sealed. Mixed standards of the elements being determined, as well as Au neutron flux monitors, were prepared by dropping the aliquots of standard solutions onto filter paper disks that were subsequently dried and individually heat sealed in PE bags. CRMs (IAEAs Soil-5 and Lake sediment SL-1) were prepared in the same manner as the samples.

Samples, mixed standards, CRMs and flux monitors, separated by Al foil disks, were tightly packed in a few cylindrical PE containers that were assembled together, wrapped, and heat sealed into PCV foil. A typical irradiation package consisted of 8–12 samples, 8–12 mixed standards, 2–3 CRMs, and 8–9 Au flux monitors distributed uniformly along the vertical axis of the package. The package was irradiated in a vertical channel of the EWA reactor at a flux of $\varphi \approx 1.8 \cdot 10^{13}$ n/cm^2/s^1 for 1 h and cooled for 3–4 d. After that time, the first gamma-spectrometric measurement was done using a 55 cm^3 Ge(Li) detector (ORTEC), with ORTEC analog line (resolution 3.2 keV for 1332.5 keV ^{60}Co line), and Didac-4000 PHA (Intertechnique), equipped with a 9-track magnetic tape station (MICRO-DATA), for the determination of As, La, Na, U, and W. The second measurement for other elements was performed after a cooling time of 20–30 d. Measurement time for samples, standards, and flux monitors amounted to 1 h, 10 min, and 5 min, respectively. The spectra were processed with the aid of NAGRAJ, SAWAPS, and MAZYG computer programs (*5*), which enable qualitative and quantitative analysis to be performed by a computer CYBER 73 (operation system SCOPE 3.4.4) at CYFRONET (Swierk). The results were further corrected for microgradient of the neutron flux, as determined with the aid of Au monitors.

Leaching of Trace Elements from the Fly-Ash

Ten grams of fly-ash was placed as a thin (ca .5 mm) layer in a funnel with a sintered glass disk (ϕ = 3.5 cm) between two filter paper disks. Either 1800 mL of distilled water or solution simulating acid rain (H_2SO_4, pH $\approx$ 2.5) was passed through the funnel under suction during the time of ca. 6h. Trace elements in the filtrate were determined as follows: Cd, Cr, Co, Cu, Mn, Ni, Pb, and Zn by FAAS after 25-fold preconcentration of the sample by evaporation; As by ETA-AAS after separation of AsI_3 by extraction with chloroform; Mo, and V by ETA-AAS after chloroform extraction of their complexes with BFHA; Se by ETA-AAS after its separation as a complex with 3, 3-diaminebenzidine by extraction with toluene; and F, spectrophotometrically as a triple complex: La-alizarin-complexone-diphenylguanidine after separation of fluorides by distillation with water vapor.

Column Studies After Neutron Irradiation

Eight hundred milligrams of fly-ash (Kozienice) were placed in a quartz ampule and irradiated together with standards of several elements in the EWA reactor at a flux of $1 \cdot 10^{13}$ n/cm^2/s for 15 h. After cooling for 6 d, the sample was quantitatively transferred to a glass column (ϕ = 10 mm) with a sintered glass disk at the bottom. The column was eluted with 150 mL of water or H_2SO_4 (pH ≈ 2.5) solution, respectively, at a linear flow rate of ca. 0.3 cm/min., collecting 7 mL fractions into standard Erlenmeyer flasks. Gamma-ray spectra of the fractions were measured as described above and compared with those of standards (in identical geometrical conditions). The pH of the solution was measured with a pH-meter (Radiometer, Copenhagen, Denmark).

RESULTS AND DISCUSSION

Composition of Fly-Ashes from Burning Polish Hard Coal, as Determined by INAA

The INAA method devised in this work enables the determination of 22 elements, with the precision (relative standard deviation) for most elements better than 15%. Results of analysis of fly-ashes from the three different power stations (Rybnik, Skawina, and Kozienice) are shown in Figs. 1 and 2. Comparison of our results obtained for Soil-5 and SL-1 CRMs with the recommended values (*6,7*), as shown at the bottom of Figs. 1 and 2, confirms generally good or at least satisfactory accuracy of the method. It should be mentioned that the conditions used here are not optimal for all elements, but constitute rather a reasonable compromise.

The interesting conclusion that follows from Figs. 1 and 2 is that concentrations of trace elements, e.g., rare earth elements, thorium, chromium, hafnium, rubidium, cobalt, barium, and tantalum, as well as that of the macro-constituent iron, in the fly-ashes coming from the three different power stations are very similar.

Larger differences were observed for As, for which the ratio of extreme concentrations amounts to 2.6, Sb (ratio = 3.0), U (5.0), Cs (2.1), and Na (4.2).

Generally speaking, concentrations of trace elements in fly-ashes from Polish hard coals studied in this work are closer to the lowest, rather than highest, extreme values of the concentration ranges reported for Chinese and Canadian ashes (*8*).

Leaching of Trace Elements

The results of leaching of trace elements from "Kozienice" fly-ash with water and simulated acid rain (H_2SO_4, pH ≈ 2.5), performed according to methodology similar to that described in Polish Norm (*cf.* Experimental section), are summarized in Fig. 3. This methodology simulates leaching of fly-ash deposited on soil, vegetation, and so on by water or acid rain over the period of ca. 3.5 yr (assuming the mean precipitation in Poland of 550 mm H_2O per year (*9*).

The pH of the filtrate amounted to 8.6 and 2.9 for water and simulated acid rain, respectively. As can be seen from Fig. 3, leaching of trace elements that normally occur as divalent cations (Cd, Co, Cu, Mn, Ni,

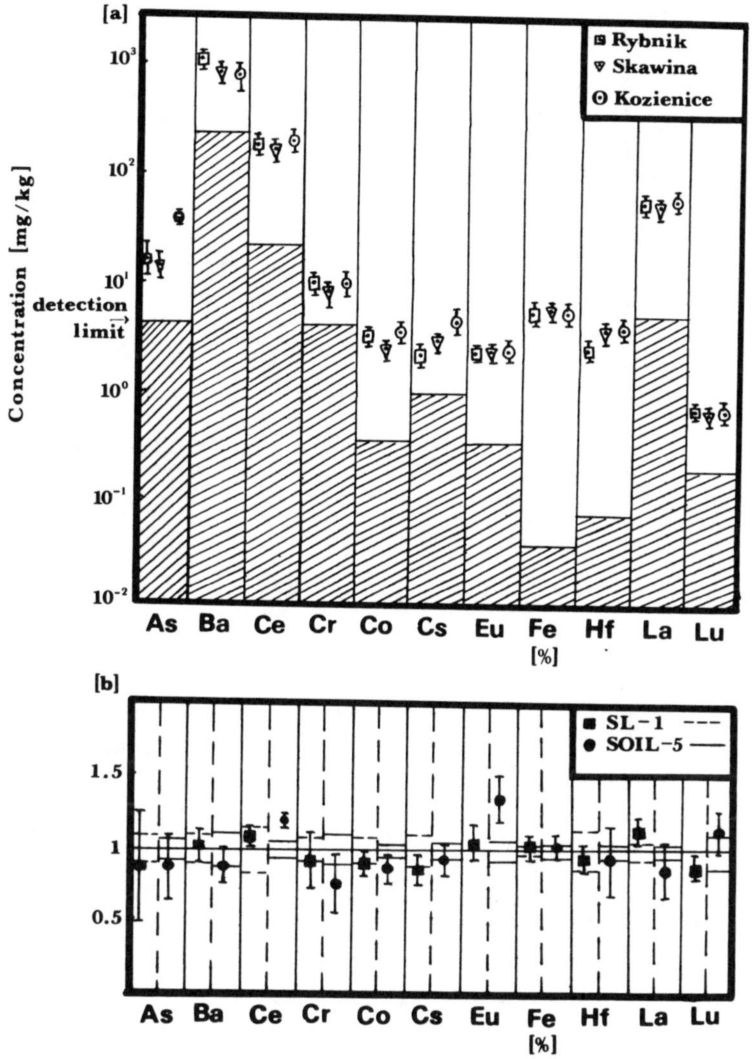

Fig. 1. Trace element content of fly-ashes from three different power stations, as determined by INAA, and results of control analyses of two certified reference materials. Detection limits cross-hatched in Fig. 1(a).

Pb, and Zn) by pure water is small or negligible. On the other hand, nonmetals and elements that occur mostly as anions (F, As, Se, and Mo), and to a smaller extent V and Cr, are eluted at much faster rates. Apparently, these elements are present in the fly-ash as easily soluble salts with alkali metals, or as oxides. Especially striking is the case of selenium, which, owing to volatility of its oxide, is probably condensed on the outer surface of the fly-ash particles. Thanks to that and to good solubility of the oxide, over 80% of Se is eluted with pure water.

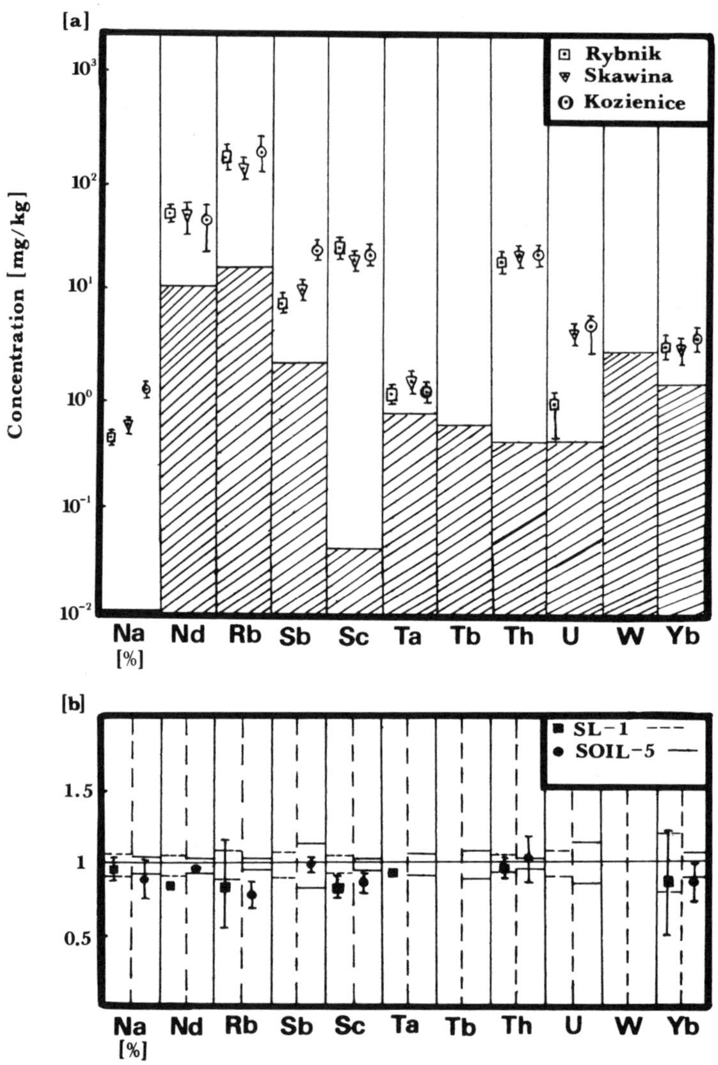

Fig. 2. Trace element content of fly-ashes from three different power stations, as determined by INAA, and results of control analyses of two certified reference materials. Detection limits cross-hatched in Fig. 2(a).

The leaching behavior of elements in simulated acid rain is different from that in water. Transition metals (Cd, Co, Cu, Mn, Ni, and Zn) are already eluted to an appreciable extent. On the other hand, elements that probably are present as oxyanions (As, Mo, Se, and V) are solubilized by dilute H_2SO_4, but often to a smaller extent than by pure water.

As was found by scanning electron microscopy and X-ray diffraction methods, the investigated fly-ash contained a considerable amount of amorphous glassy spheres ($\phi \leq 30$ μm). From a chemical point of view,

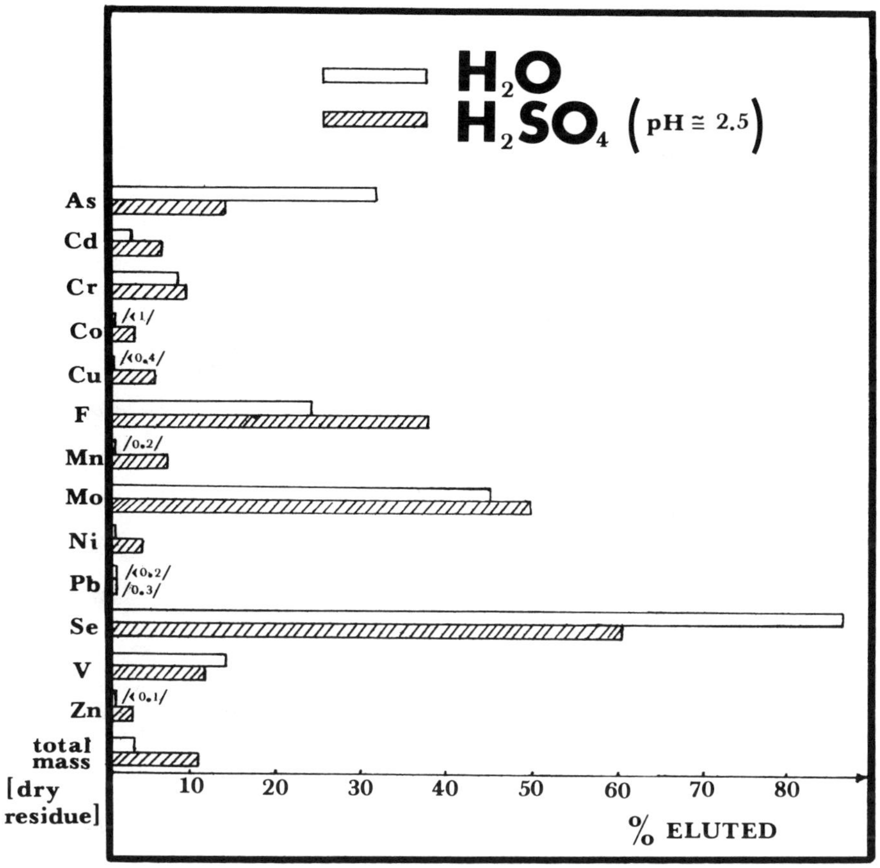

Fig. 3. Results of leaching studies of the "Kozienice" fine fly-ash, as determined by AAS and spectrophotometry.

this is a material of a typical alumino-silicate matrix, and as such, it should exhibit cation exchange properties. This assumption well explains the elution behavior of the elements. Relatively strongly-held divalent transition metal cations cannot be removed by pure water, but are gradually displaced from the matrix by hydrogen ions from acid rain. The lower percentage of elution of elements occurring as oxyanions by acid rain, in comparison to water, may be interpreted as a result of shifting the chemical equilibrium from anionic to cationic form, which can be subsequently retained by a cation exchange mechanism. It cannot be excluded, however, that fly-ash may behave as an amphoteric ion exchanger and retain anions in acidic medium via an anion exchange mechanism. An important finding from the ecological point of view is that the highly toxic element lead is eluted only to a negligible extent both by water and acid rain (in the latter case probably owing to insolubility of its sulfate).

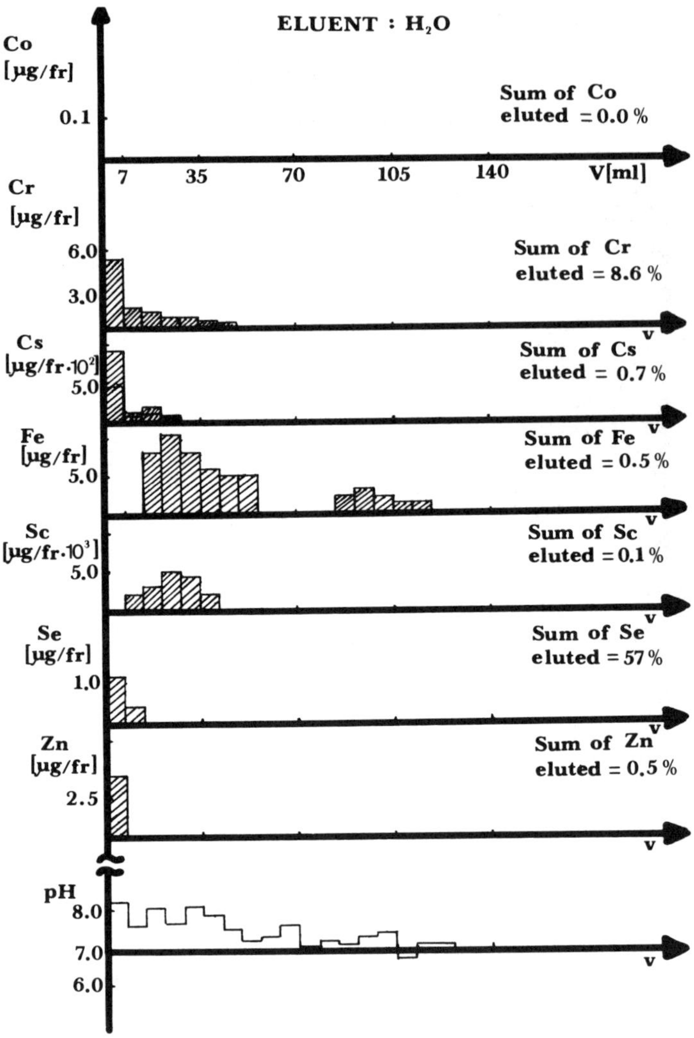

Fig. 4. Elution of trace elements from the neutron-irradiated "Kozienice" fine fly-ash (column: 0.8 cm × 0.78 cm²) by water.

Column Studies With Neutron-Irradiated Fly-Ash

Observations on elution of trace elements from the columns filled with neutron-irradiated fly-ash (Figs. 4 and 5) are in qualitative agreement with the conclusions drawn in the preceding section. In particular, the exceptionally high percentage of elution of selenium, both by of water and dilute H_2SO_4, has been confirmed. Only small amounts of Fe and Sc pass into the solution, and it is characteristic that they are better eluted by water than by acid rain, probably owing to peptization of their hydroxides. The elution of Co and Zn with H_2SO_4 (pH ≈ 2.5) starts only after the pH of the effluent drops below a certain value. The elution

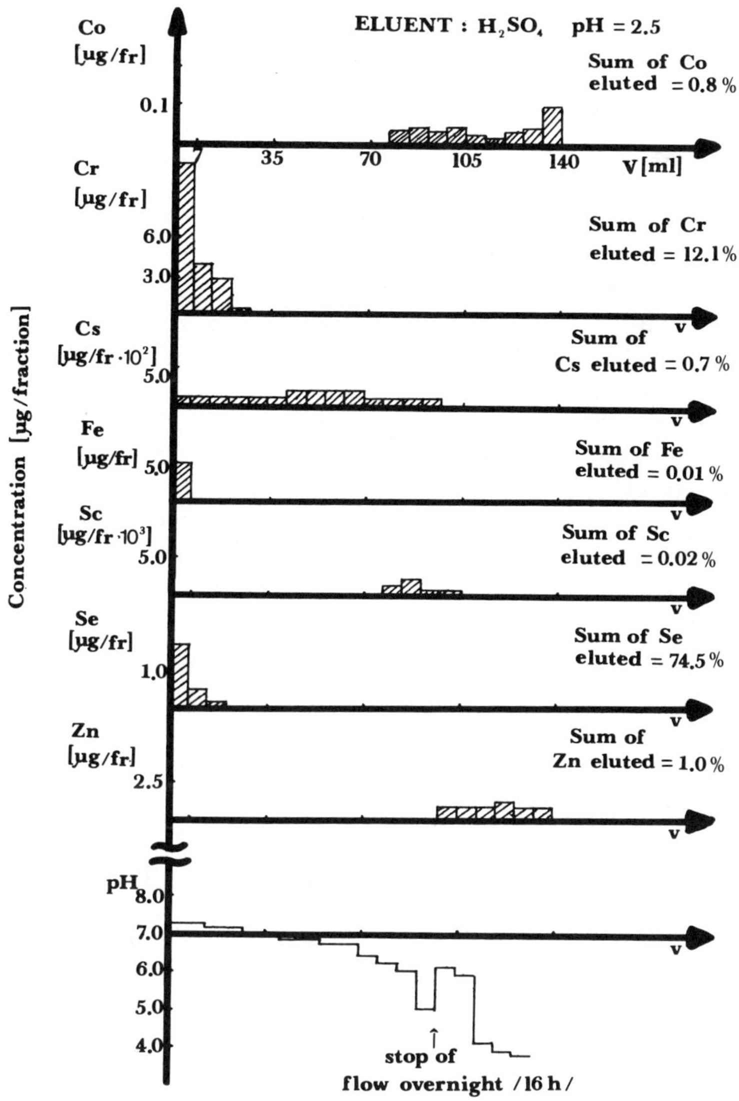

Fig. 5. Elution of trace elements from the neutron-irradiated "Kozienice" fine fly-ash (column: 0.8 cm × 0.78 cm²) by solution imitating acid rain (H_2SO_4, pH ≈ 2.5).

pattern of Cs is different in water and dilute H_2SO_4, in agreement with the expected differences in the elution mechanism, i.e., solubility of salts (or oxides), and ion exchange reaction, respectively.

Fine Fly-Ash (CTA-FFA-1): New Certified Reference Material

Studies on the effect of fly-ash on pollution of the environment require appropriate CRMs to check the accuracy of the analytical methods used. Although a number of geological and environmental CRMs is available from various suppliers (*10,11*), it was felt that, in specific Polish conditions, a new fly-ash CRM certified for a possibly greater number of trace elements is urgently needed.

About 50 kg of fine fly-ash, collected from the third zone of electrofilters of the Kozienice Power Station working on hard coal, was chosen as a candidate CRM. The material had a density of 2.53 g/cm^3 and a rather uniform particle size (>93% passed through the 0.06 mm sieve).

An international intercomparison run was organized to certify the material. So far, 57 laboratories, employing a variety of analytical methods, participated, sending 3828 bits of chemical data (915 laboratory means) on 65 elements. Although several laboratories are still expected to send their results and the final report will not be issued earlier than in a few months, the preliminary evaluation of results has shown that it will be possible to certify at least 38 elements (Al, As, Ba, Ce, Co, Cr, Cs, Cu, Dy, Eu, F, Fe, Hf, La, Li, Lu, Mg, Mn, Na, Nd, Ni, P, Pb, Rb, Sb, Sc, Se, Si, Sm, Sr, Ta, Tb, Th, U, V, W, Yb, and Zn) and give "information values" for an additional twelve (Be, Ca, Cd, Er, Ga, Gd, In, K, Mo, Ti, Tm, and Y) (*12*).

The certification criteria were the same as formulated previously (*13*), with some recent modifications (*14*). It may be stated already now that CTA-FFA-1 can be ranked among the best environmental CRMs with a mineral matrix ever produced.

ACKNOWLEDGMENTS

This work has been supported, in part, by the International Atomic Energy Agency Contract No 4906/R1/RB, as well as the Polish Central Research Programmes CPBR 5.8, cel 1.06, and CPBP 01.17.

REFERENCES

1. R. Ney, *Energetyka,* **1,** 17–22 (1989).
2. J. M. Pacyna, *Scientific Papers of the Institute of Environmental Protection Engineering of Wrocław Technical University,* No. 47, Monograph Series No. 17, Wydawnictwo Politechniki Wroclawskiej, Wroclaw, 1980.
3. D. J. Swaine, *Bull. New Zealand DSIR,* 127–131 (1977).
4. D. J. Swaine, *Colloquium on Combustion of Pulverized Coal: The Effect of Mineral Matter,* I. M. Stewart and T. F. Wall, eds., Univ. of Newcastle, Newcastle NSW, Australia, 1979.
5. Z. Szopa, M. Plejewska, and J. Staszelis, *Institute of Nuclear Research Report,* INR/1945/VIII/E/B, Warsaw (1982).
6. R. Dybczyński, A. Tugsavul, and O. Suschny, *Geostandards Newsletter* **3,** 61 (1979).
7. R. Dybczyński, O. Suschny, *Final Report on the Intercomparison Run SL-1 for the Determination of Trace Elements in a Lake Sediment Sample,* IAEA/RL/64, IAEA, Vienna, 1979.
8. Sun Jingxin and R. E. Jervis, *J. Radioanal. Nucl. Chem. Articles* **114,** 89 (1987).

9. P. Bieńkowski, Z. Fisher, and J. Godzikiewicz, *Ekol. Pol.* **34(1),** 75 (1986).
10. F. J. Flanagan, *Geostandards Newsletter* **10,** 191 (1986).
11. K. Govindaraju, *Geostandards Newsletter* **8,** (special issue) (1984).
12. R. Dybczyński, H. Polkowska-Motrenko, Z. Samczyński, and Z. Szopa, in preparation.
13. R. Dybczyński, *Anal. Chim. Acta* **117,** 53 (1980).
14. R. Dybczyński, H. Polkowska-Motrenko, Z. Szopa, and Z. Samczyński, Certified Reference Materials, *Proc. ISCRM '89,* May 15–18, 1989, Beijing, China, International Academic Publishers-Pergamon Press, Beijing, 1989, p. 312–321.

An Environmental Research on Trace Elements in Freshwater Fish by Neutron Activation Analysis

K. S. Park,* N. B. Kim, H. J. Woo,
Y. Y. Yoon, and K.Y. Lee

Korea Institute of Energy and Resources, 219-5 Garibong-dong, Guro-gu, Seoul 152, Korea

Received April 17, 1989; Accepted October 27, 1989

ABSTRACT

For a case study of environmental contamination, radiochemical activation analysis has been applied to the crucians collected in the Han River. Sixteen trace elements (Hg, Cd, As, Br, Cu, Na, K, Se, Cr, Hf, Rb, Fe, Zn, Co, La, and Cs) were separated into three groups using distillation and diethyldithiocarbamate extraction methods, and their contents were determined by a single comparator method. Compared with the values 15 years ago, the values for mercury and cadmium have been drastically decreased at the middle and lower part of the river, but no typical change is found in other elements.

Index Entries: Han River; environmental contamination; trace elements; neutron activation analysis (NAA).

INTRODUCTION

Widespread contamination caused by industrialization and urbanization makes it desirable to have a reliable picture of the present and future background of the environment, especially the extent of specific pollutants, such as mercury, cadmium, and arsenic. Related to this, the

*Author to whom all correspondence and reprint requests should be addressed.

necessity for a suitable monitor for controlling and assessing the influence of toxic technogenic pollutants is inevitably required.

Reactor neutron activation analysis is a suitable method for determining trace elements in biological materials and, with the advent of sophisticated instruments, is often used for instrumental determination of elements such as As, Br, Cu, Na, K, Zn, Se, Cr, Hf, Rb, Fe, Co, La, Cs, Sc, and so on. However, some toxic elements of interest, such as Hg, Cd, and As, cannot be determined purely by instrumental neutron activation analysis. This is mainly owing to the presence of predominant radionuclides in the gamma-ray spectrum. Therefore, a radiochemical separation is required for determining these elements. The most commonly used separation methods are ion exchange (*1–3*), solvent extraction methods (*4,5*), or combinations of these two methods.

For this case study of the environment, the Han River is one of the most suitable areas in Korea, because it flows from east to west through Seoul, the most urbanized and industrialized city. The river was investigated by measuring mercury, cadmium, arsenic, and four other elements in crucian as a monitor 15 years ago (1973) (*2,3*). In that experiment, a distillation and an ion exchange method were applied for the group separation of trace elements. That method was a somewhat time-consuming one.

In the present work, the Han River has been investigated again for the past three years (1986–1988). Crucians were collected from 10 sampling sites and analyzed to evaluate the change of the contamination level during the past 15 years. A simple and rapid solvent extraction method has been used to separate trace elements into three groups. It is based on the extraction of cadmium and copper and diethyldithiocarbamic acid (*6*).

EXPERIMENTAL

Sample Collection and Pretreatment

Freshwater fishes (crucians) were collected from 10 points along the Han River (as shown in Fig. 1) in the years 1986–1988. Among them, only crucians of 18–20 cm length were used for this investigation. Samples were cut into pieces with a plastic knife. Axial muscles were taken. Skin and bones were removed as completely as possible. The pieces of muscles were washed several times with redistilled water. They were transferred into polyethylene bags and dehydrated by drying for 3 d in a thermovac freeze unit (Labconco Model 5, Kansas City, MO, US).

Neutron Irradiation

One gram of the dehydrated crucian was weighed and put into a pre-decontaminated silica ampule. The ampule was irradiated in the rotary specimen rack of the TRIGA Mark III reactor with two comparators

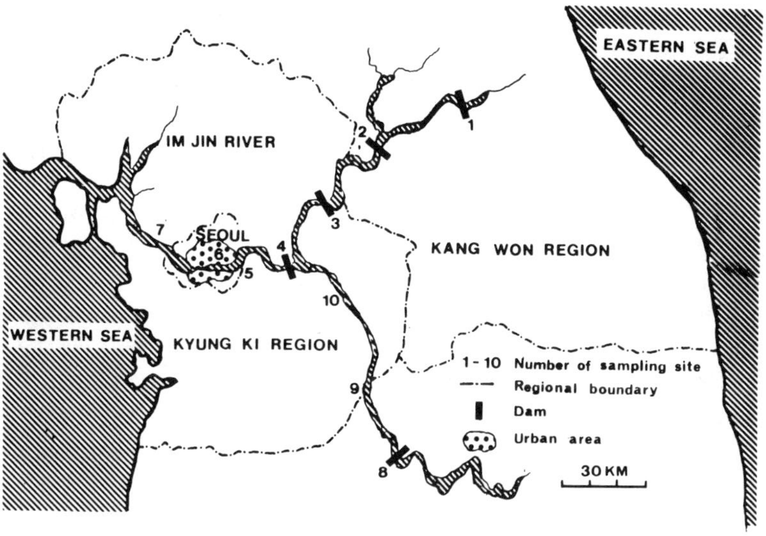

Fig. 1. The sampling sites along the Han River.

of gold and cobalt for 3 d. The neutron flux was 5 × 10^{12}n/cm^2/s at the irradiation site.

Radiochemical Separation

After the two comparators were taken off, the irradiated silica ampules were thoroughly cleaned to remove surface contamination and dipped into liquid nitrogen to prevent possible loss of mercury and bromine by volatilization (*1*). Each ampule was inserted into a piece of tygon tubing that was previously sealed at one end. The tubing was then filled with 3 mL of an oxidation mixture of 90% concn. nitric acid and 10% concn. sulfuric acid in order to cover the ampule completely and also with 1 mL of mercuric chloride solution containing 10 mg of mercury as a carrier. The tygon tubing was sealed at the other end with a Hoffman clamp. Pressure was applied to crush the ampule. The sample was then immediately mixed with the oxidation mixture. The gases were absorbed by shaking the tubing. The tygon tubing was then opened.

The irradiated sample and broken pieces of the ampule were quantitatively transferred into the distillation flask of a modified Bethge distillation apparatus (*7*), using a few milliliters of concentrated nitric acid for washing. Ten milliliters of oxidation mixture was added to the distillation flask with 1 mL cadmium carrier solution containing 10 mg of cadmium. The flask was heated until the nitric acid was distilled into the reservoir. The temperature was allowed to rise to 320°C. After cooling, the distillate was run back into the flask. The heating cycle was repeated until the solution was completely clear and colorless. Then, a mixture of 5 mL 70%

perchloric acid and 0.5 g of glycine, dissolved in 5 mL of water, was added to the digested sample solution. The flask was heated again. The mercury fraction was collected with distillate in the reservoir. The distillation was interrupted when the temperature rose up to 250°C. The distillate in the reservoir was collected in a 50 mL volumetric flask for counting the activities of mercury and bromine.

The digested solution in the flask was transferred into a 250 mL separatory funnel. Forty milliliters of tartaric acid solution (50 mg/mL) and 60 mL of 10*M* sodium hydroxide solution were added in the funnel. Thirty mililiters of chloroform, containing 3.4×10^{-3} mol HDDC, was added in the funnel as an organic phase. The separatory funnel was shaken for 1 min. Cadmium and copper were extracted into the organic phase. These elements were extracted again with 10 mL of chloroform. The HDDC and the chloroform phases were collected together into a 50 mL volumetric flask for counting. The aqueous phase was evaporated by heating to reduce the volume and transferred into a 50 mL volumetric flask to count the activities of other elements. The HDDC solution was prepared as follows (*6,8*): the precipitate ($Zn(DDC)_2$), which was produced by adding NaDDC to zinc chloride solution, was dissolved in chloroform to give a 4×10^{-2} *M* solution. This solution was stable for at least 5 mo if kept in a dark bottle. HDDC was then prepared for every analysis by shaking equal volumes of $Zn(DDC)_2$/chloroform and 1 *M* hydrochloric acid. The shaking time must not be longer than 20 s. The resulting organic phase contains HDDC free of zinc.

Activity Counting and Determination of Each Element

Three volumetric flasks were counted by a HPGe detector (CANBERRA, active volume = 70 cm^3 FWHM at 1.332 MeV of ^{60}Co = 1.9 keV) connected to a 4096-channel analyzer (CANBERRA Model 90). The photopeak counting rate of each nuclide in the gamma-ray spectrum was calculated with SPECTRAN-F (*9*) and PIAD (Peak Identification and Activity Determination) (*10*) computer codes. The radionuclides selected in this work are shown in Table 1. The content of each element in a sample was calculated by the single comparator method, using two comparators (*11,12*). The analytical results are given in Table 2.

RESULTS AND DISCUSSION

The method for pretreatment and digestion used in this work was the same as in the previous papers (*2,3*). This paper reveals that about 70% of the mercury in a biological sample was vaporized during the reactor neutron irradiation, but by using a silica ampule and the condensing and distilling apparatus, almost all of the vaporized mercury can be condensed. Tracer experiments were carried out to determine the accu-

Table 1
Radionuclides Selected for This Experiment

Radio-nuclide	Half-life, hours	Energy, keV	Radio-nuclide	Half-life, hours	Energy, keV
Hg-203	1122	279.2	Cr-51	664.8	320.0
Cd-115	53.38	336.2	Hf-181	1019	482.0
As-76	26.32	559.1	Rb-86	451.7	1077.2
Br-82	35.34	554.3	Fe-59	1069	1099.2
Cu-64	12.70	1345.6	Zn-65	5856	1115.5
Na-24	15.03	1368.6	Co-60	46180	1332.5
K-42	12.36	1524.6	La-140	40.27	1596.5
Se-75	2843	264.7	Cs-134	17897	795.8

racy of this experiment. A mixed solution of known amounts of mercury, cadmium, arsenic, selenium, and zinc was irradiated in a silica ampule. The same experimental procedure was applied to the solution with a nonirradiated crucian sample. The recoveries of these elements, obtained by the tracer work, were over 98% for all these elements. To verify the reliability of this experiment, two NBS biological reference materials, SRM 1570 (Spinach) and SRM 1571 (Orchard Leaves) were analyzed by this experiment, as shown in Table 3. Our values for most of the elements agreed with the NBS certified values (*13*) within a 10% deviation.

Figure 1 shows the middle part of the Korean peninsula, where the Han River flows from east to west through Seoul, the most industrialized and urbanized city in Korea, and the 10 sampling sites: Soyang (site no. 1), Euiam (site no. 2), Chungpyung (site no. 3), Paldang (site no. 4), Kwangnaru (site no. 5), Yeoido (site no. 6), Haengju (site no. 7), Chungju (site no. 8), Yeoju (site no. 9), and Yangpyung (site no. 10). Sites 8–10 were not included in the 1973 investigation, but are included for the two most recent investigations (1987 and 1988). The analytical results for crucians collected in the past three years are shown in Table 2 for 16 elements (Hg, Cd, As, Br, Cu, Na, K, Se, Cr, Hf, Rb, Fe, Zn, Co, La, and Cs). The contents of 7 elements (Hg, As, Br, Cu, Na, and K) determined in the previous work are also included in Table 2 to be compared with the recent one.

The mercury and cadmium contents are graphically shown in Fig. 2. It is easily seen in Fig. 2 that mercury has such a regional trend that the

Table 2
Analytical Results of Trace Elements in Crucians (ppm)

Element	Year	Soyang	Euiam	Chung-pyung	Pal-dang	Kwang-naru	Yeoido	Haeng-ju	Chung-ju	Yeoju	Yang-pyung
Hg	73	0.17	0.23	0.18	0.77	0.83	2.7	2.9			
	86	0.13	0.16	0.17	0.26	0.29	0.31	0.57			
	87	0.081	0.13	0.14	0.19	0.21	0.21	0.27	0.16	0.53	0.71
	88	0.14	0.20	0.21	0.31	0.32	0.32	0.34	0.39	0.42	0.59
Cd	73	0.21	0.23	0.25	0.62	0.71	1.1	1.2			
	86	0.009	0.004	0.007	0.007	0.032	0.052	0.037			
	87	0.010	0.003	0.007	0.028	0.019	0.017	0.031	0.006	0.001	0.042
	88	0.005	0.010	0.004	0.016	0.004	0.005	0.039	0.003	0.003	0.047
As	73	0.30	0.59	0.62	0.32	0.20	16	0.13			
	86	0.35	0.29	0.20	0.24	0.35	0.41	0.56			
	87	0.15	0.15	0.17	0.56	0.27	0.50	0.65	0.66	0.11	1.2
	88	0.24	0.17	0.13	0.26	0.28	0.40	0.55	0.23	0.34	0.65
Br	73	5.5	1.4	4.1	5.9	5.9	3.5	6.8			
	86	1.5	1.4	0.90	3.2	1.6	2.5	4.4			
	87	1.7	2.7	3.6	1.4	1.5	1.7	1.8	1.3	2.3	1.5
	88	1.4	1.1	1.8	2.5	6.0	1.5	3.9	5.2	2.7	1.6
Cu	73	0.071	2.2	4.2	3.5	5.5	7.2	3.5			
	86	0.49	0.66	0.46	0.62	0.78	1.5	1.1			
	87	0.62	0.72	0.30	0.55	0.24	0.64	1.2	0.23	1.1	0.50
	88	0.95	0.70	1.3	0.61	0.59	0.73	1.6	1.2	1.4	0.52
Na(%)	73	0.15	0.14	0.20	0.16	0.21	0.13	0.24			
	86	0.056	0.089	0.055	0.083	0.084	0.086	0.12			
	87	0.094	0.11	0.14	0.10	0.065	0.10	0.088	0.11	0.12	0.073
	88	0.13	0.11	0.10	0.17	0.16	0.10	0.16	0.13	0.16	0.066
K(%)	73	0.67	0.87	1.0	0.75	0.47	1.9	1.1			
	86	1.2	1.0	1.0	0.96	1.2	1.3	1.4			
	87	0.95	1.2	1.1	1.2	1.3	1.2	1.2	1.2	1.3	1.2
	88	1.2	1.2	1.5	1.1	1.2	1.2	1.2	1.4	0.86	1.2
Se	86	0.69	0.61	0.87	0.37	0.46	0.51	0.27			
	87	0.72	0.98	0.59	0.57	0.62	0.64	0.57	0.47	0.64	0.47
	88	0.68	0.46	0.92	0.50	0.51	0.74	0.35	0.86	0.85	0.57
Cr	86	0.38	0.37	0.36	0.31	0.39	0.43	1.0			
	87	0.22	0.33	0.51	0.44	0.38	0.68	0.60	0.79	0.64	0.70
	88	0.21	0.24	0.73	0.53	0.41	0.82	0.68	0.36	0.59	0.10
Hf	86	0.008	0.007	0.005	0.007	0.010	0.016	0.014			
	87	0.010	0.004	0.005	0.006	0.009	0.014	0.023	0.004	0.005	0.006
	88	0.011	0.013	0.012	0.015	0.014	0.015	0.018	0.004	0.005	0.007
Rb	86	5.3	9.9	5.2	9.6	9.1	11	13			
	87	4.8	5.8	7.3	9.6	7.5	13	8.6	11	8.3	13
	88	6.6	6.1	6.5	6.9	7.5	8.5	9.2	11	4.7	8.1
Fe	86	17	27	29	25	15	24	35			
	87	14	12	6.3	14	16	18	20	20	19	7.7
	88	34	13	15	10	11	12	25	25	25	13
Zn	86	65	87	35	55	100	36	55			
	87	52	49	92	100	59	54	76	70	36	39
	88	55	48	55	71	50	28	88	31	66	42
Co	86	0.015	0.029	0.021	0.009	0.025	0.027	0.024			
	87	0.012	0.011	0.011	0.019	0.006	0.012	0.008	0.006	0.012	0.005
	88	0.014	0.006	0.013	0.015	0.017	0.019	0.014	0.015	0.015	0.007
La	86	0.022	0.017	0.014	0.036	0.051	0.050	0.058			
	87	0.008	0.017	0.032	0.040	0.023	0.031	0.026	0.033	0.022	0.036
	88	0.011	0.015	0.022	0.016	0.021	0.029	0.031	0.026	0.019	0.028
Cs	86	0.011	0.015	0.027	0.026	0.032	0.034	0.027			
	87	0.012	0.016	0.022	0.019	0.028	0.032	0.035	0.019	0.006	0.063
	88	0.009	0.020	0.016	0.025	0.037	0.039	0.030	0.056	0.010	0.036

Table 3
Trace Element Concentrations in Spinach and Orchard Leaves (ppm)

Element	SRM 1570 (Spinach)		SRM 1571 (Orchard leaves)	
	Certified value	This work	Certified value	This work
Hg	0.030 ± 0.005	0.028 ± 0.005	0.155 ± 0.015	0.156 ± 0.018
Cd	1.5	1.55 ± 0.15	0.110 ± 0.010	0.114 ± 0.012
As	0.15 ± 0.05	0.16 ± 0.02	10 ± 2	9.80 ± 0.41
Br	54	49.8 ± 1.2	10	9.3 ± 0.3
Cu	12 ± 2	12.3 ± 2.0	12 ± 1	10.9 ± 1.7
Na(%)	1.41	1.35 ± 0.05	0.0082 ± 0.0006	0.0080 ± 0.0003
K(%)	3.56 ± 0.03	3.22 ± 0.14	1.47 ± 0.03	1.39 ± 0.05
Se	0.160	0.154 ± 0.009	0.064	0.062 ± 0.004
Cr	4.6 ± 0.3	4.25 ± 0.29	2.6 ± 0.3	2.52 ± 0.16
Hf	0.04	0.050 ± 0.014	0.03	0.028 ± 0.007
Rb	12.1 ± 0.2	12.9 ± 0.7	12 ± 1	11.6 ± 0.8
Fe	550 ± 2	550 ± 29	300 ± 20	309 ± 17
Zn	50 ± 2	53.1 ± 1.9	25 ± 3	27.2 ± 1.6
Co	1.5	1.40 ± 0.03	0.200	0.186 ± 0.016
La	0.370	0.35 ± 0.11	1.16	1.10 ± 0.03
Cs	0.061	0.065 ± 0.025	0.040	0.040 ± 0.016

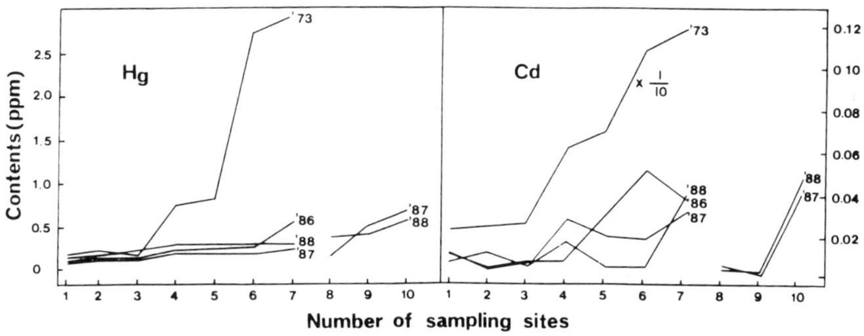

Fig. 2. Trends of mercury and cadmium contents along the Han River.

content increases as the river flows down. At the first three sites (upper stream), the mercury contents for the most recent three years have similar values, in the range of 0.08–0.20 ppm, which are very close to a natural background level and are slightly lower than the previous value. The ratio of the contents at the middle part, located around Seoul (site nos. 4 and 5), to those at the upper part is drastically decreased compared with those of the previous values. This discrepancy between the previous and the recent values is increased by a factor of 4. After the river passes through Seoul (site nos. 6 and 7), the discrepancy reaches a factor of 8. Cadmium also has a similar regional trend with mercury and a similar discrepancy between the previous and recent three values. But, even at the first three sites, the previous values are 20 times larger than the recent values, and that phenomenon still remains a question. It can be concluded from the above facts that the upper part of the Han River has been a little contaminated with those two toxic elements and that the middle and lower parts, which were seriously contaminated, have been decontaminated drastically. The reason seems to be that several sewage treatment facilities were installed, and contaminated sediment has been cleaned by means of the project on the redevelopment of the Han River (1983–1985).

All other elements show a slightly increasing trend as the river goes down; however, any significant meaning cannot yet be found from these trends. Otherwise, only selenium shows a different trend of decreasing, but the reason has not yet been found.

REFERENCES

1. N. B. Kim, C. Lee, and H. I. Bak, *J. Korean Nucl. Soc.* **8**, 41 (1976).
2. C. Lee, N. B. Kim, and E. B. Park, *Korean Nuclear Soc.* **5**, 279 (1973).
3. C. Lee, N. B. Kim, and I. C. Lee, *Report of IAEA Research Contract No. 1136/RB (1974).*
4. G. Zhuang, Y. Wang, M. Zhi, W. Zhou, J. Yin, M. Tan, and Y. Cheng, *J. Radioanal. Nucl. Chem.* **129**, 459 (1989).
5. E. Taskaev, I. Penev, and L. Kinova, *J. Radioanal. Nucl. Chem.* **120**, 83 (1988).
6. S. Bajo and A. Wyttenbach, *Analytical Chem.* **49** 158 (1977).
7. B. Sjoestrand, *Anal. Chem.* **36**, 814 (1964).
8. A. Wyttenback and S. Bajo, *Anal. Chem.* **47**, 1813 (1975).
9. *Technical Reference Manual for SPECTRAN-F*, Version 2, Canberra Industries Inc., Meriden, CT, 1981.
10. N. B. Kim, Y. S. Kim, J. K. Kim, and K. S. Park, *J. Radioanal. Nucl. Chem.* **125**, 403 (1988).
11. J. I. Kim, *J. Radioanal. Chem.* **63**, 121 (1981).
12. N. B. Kim, K. S. Park, and H. I. Bak, *J. Korean Nucl. Soc.* **18**, 85 (1986).
13. E. S. Gladney, B. T. O'Malley, J. Roelandts, and T. E. Gills, *Standard Reference Materials Compilation of Elemental Concentration Data for NBS Clinical, Biological, Geological, and Environmental Standard Reference Materials*, NBS Special Publication 260-111, NBS, Washington, DC, 1987.

Identifying Sources of Groundwater Pollution Using Trace Element Signatures

ILHAN OLMEZ* AND MARILYN J. HAYES

Nuclear Reactor Laboratory, Massachusetts Institute of Technology, 138 Albany Street, Cambridge, MA 02139

Received April 17, 1989; Accepted October 23, 1989

ABSTRACT

A simple receptor modeling approach has been applied to groundwater pollution studies and has shown that marker trace elements can be used effectively in source identification and apportionment. Groundwater and source materials from one coal-fired and five oil-fired power plants, and one coal-tar deposit site have been analyzed by instrumental neutron activation analysis for more than 20 minor and trace elements. In one of the oil-fired power plants, trace element patterns indicated a leak from the hazardous waste surface impoundments owing to the failure of a hypolon liner. Also, the extent and spatial distribution of groundwater contamination have been determined in a coal-tar deposit site.

Index Entries: Groundwater, trace elements; contamination sources; source identification.

INTRODUCTION

Recently, groundwater pollution has been recognized as one of our most critical environmental problems. Although groundwater contamination has been going on for decades, we are just now beginning to observe the effects. The flow rate of groundwater is very slow, a few inches to a few feet per day; therefore, it may take years or even decades

*Author to whom all correspondence and reprint requests should be addressed.

for pollutants to be detected in a nearby water supply. Also, unlike the turbulent flow of surface water systems, groundwater flow is laminar, i.e., packets of water move along separate paths, with little mixing occurring as it moves (*1*). For this reason, a point source may not contaminate the entire body of groundwater in the aquifer. Therefore, it is necessary to sample many wells, at various locations and depths in the aquifer, to determine if the groundwater has been contaminated.

To determine whether groundwater samples have been contaminated, it is necessary to detect and identify the contamination sources. Chemicals that enter the soil can be attenuated through adsorption and chemical reactions with other organic or inorganic species existing in the soil matrix, or they may be enhanced through scavenging. Thus, the original composition of the pollutant may be greatly altered at the sampling point (wells), which makes it almost impossible to predict the pollution and/or the pollution source. Therefore, it is desirable to find and utilize certain signatures that are not modified during the transport between the source and the receptor. Inorganic elements are much better candidates for this signature approach, because even if their absolute concentrations change as they move away from the source, individual elements with similar chemical characteristics are expected to be affected equally. Organic compounds usually change their identity totally because of their chemical characteristics.

In recent years, "marker trace elements" have been successfully used in atmospheric studies to identify individual pollution sources. For example, V and rare earth elements (REE) are indicators of oil combustion; most of the airborne Pb in urban areas originates from motor vehicles; S, Se, and As are almost always associated with coal-related pollution. These and some other elements have been used extensively to assess the impact of different sources on atmospheric particle loadings at given receptor sites and have also been used to understand regional and even global movements of air masses (*2–4*). This methodology has been modified and used in the past few years by our group to study the sources of groundwater pollution. The methodology has been found to be as useful as in atmospheric studies (*5,6*).

EXPERIMENTAL

Site Descriptions

During the last two years, we have studied one coal- and five oil-fired power plants and one coal-tar deposit site. All of the hazardous waste surface impoundments around the oil-fired power plants were located on old coal ash deposits produced from previous operations. At one of the oil-fired power plants, waste water was discharged into unlined earthen basins. The remaining plants utilized hypolon liners and waste treatment plants.

The coal-fired power plant was located in an area where the bedrock composition was almost identical to the coal. Previous studies conducted at this particular site had failed to separate coal and bedrock contributions to the groundwater contamination. Coal-tar, which is a waste product of old coal gasification plants, is usually deposited within the plant site boundaries and is covered with a layer of soil. There are numerous known and perhaps as many unknown sites of this kind around the country.

Sample Collection and Analysis

Samples of groundwater, surface water, and source materials, such as coal, fuel oil, ash, bedrock, coal-tar, and soil, were collected from each site. Liquid samples were stored in clean polyethylene bottles and transferred to the laboratory. These samples were then filtered through 0.2 μm Teflon filters to separate soluble and insoluble fractions.

To increase the analytical sensitivity and eliminate geometry factors, 20–30 mL of the water samples were freeze-dried in small, acid-cleaned polyethylene bags ($\approx$60 mg), using successive additions of 5 mL aliquots following each drying. This procedure, in contrast to freeze-drying large volumes of water in a bigger bag at one time, was found to be effective in preventing the introduction of high activities and blank concentrations from the sample container.

Elemental concentrations were determined by instrumental neutron activation analysis (INAA). The INAA procedure applied is very similar to that reported previously (*7*). Samples were subjected to two different irradiations: first, for a few minutes to observe radioisotopes with short half-lives; and second (after they were allowed to decay for several days), for several hours to observe radioisotopes with long half-lives. Both irradiations were performed at a neutron flux of 8×10^{12} n/sc/cm^2 in the MIT Research Reactor. For some of the samples (e.g., basins), the activity following the short irradiation was very intense owing to the high concentrations of Na in the matrix. Because of these high activities, we were unable to determine the concentrations of elements, such as V, which is an important marker element for oil-fired plants. Therefore, we performed a preirradiation chemical separation using an ion exchanger to eliminate alkaline metals from these samples. About 1.0 g of Chelex-100 resin (200–400 mesh size) was cleansed with 2.5*M* HNO_3 and converted to NH_4^+ with 2.0*M* NH_4OH. A 10 mL water sample was then passed through the resin at a very slow rate. The alkaline metals were eluted by washing the resin with 1.0*M* NH_4COOCH_3. We obtained more than 90% recovery for V in this study using this technique.

RESULTS AND DISCUSSION

In this study, more than 30 elements (Na, Mg, Al, Cl, K, Ca, Sc, V, Cr, Mn, Fe, Co, Zn, As, Se, Br, Rb, Sr, Sb, Cs, Ba, La, Ce, Nd, Sm, Eu, Gd, Yb, Lu, Hf, Th, and U) were observed by INAA. On the average,

Table 1
List of Elements that are Found to be Marker Elements in Groundwater Studies for Specific Pollution Sources

Source	Marker Elements
Coal-fired Power Plants	As, Se, Sr, Ba
Oil-fired Power Plants	V, Sb, Fe, REE
Coal-tar Deposits	Mg, Cl, Ba
Photographic Industry	Br, Ag
Soil	Al, Sc

however, concentrations of typically 20 elements were determined in any one of the 65 individual monitoring wells.

To identify marker trace elements for groundwater contamination sources, we examined elemental ratio patterns for source and groundwater compositions and calculated enrichment factors (EF). In environmental studies, the absolute concentrations of elements provide limited information about the sources and source-receptor relations. It is common practice to normalize observed concentrations of the different elements to an element with a known source or calculate enrichment factors with respect to a normalizing matrix, such as the composition of average crustal material. In our groundwater studies, we used Sc, an element representing the crustal material, as a normalizing element. We also determined EFs with respect to background well compositions, which represent elemental concentrations in unpolluted aquifer water. Using these normalization techniques, we were able to identify marker trace elements for various sources encountered in our field studies. These marker trace elements are listed in Table 1.

Examination of the EFs, elemental ratio patterns and absolute concentration variations of signature elements in monitoring wells and hazardous waste surface impoundment samples can provide much information about groundwater contamination. Indeed, using this methodology, we discovered that at one of the oil-fired power plants, groundwater was being polluted by leakage of hazardous waste surface impoundments. At this particular site, the compositions of individual monitoring wells indicate contamination from oil- [high EF(V) and La/Sm] and coal-related [high EF (As)] sources. The coal-related contamination is probably owing to leaching of elements by impoundment water passing through old coal ash deposits (under the basin) during leakage.

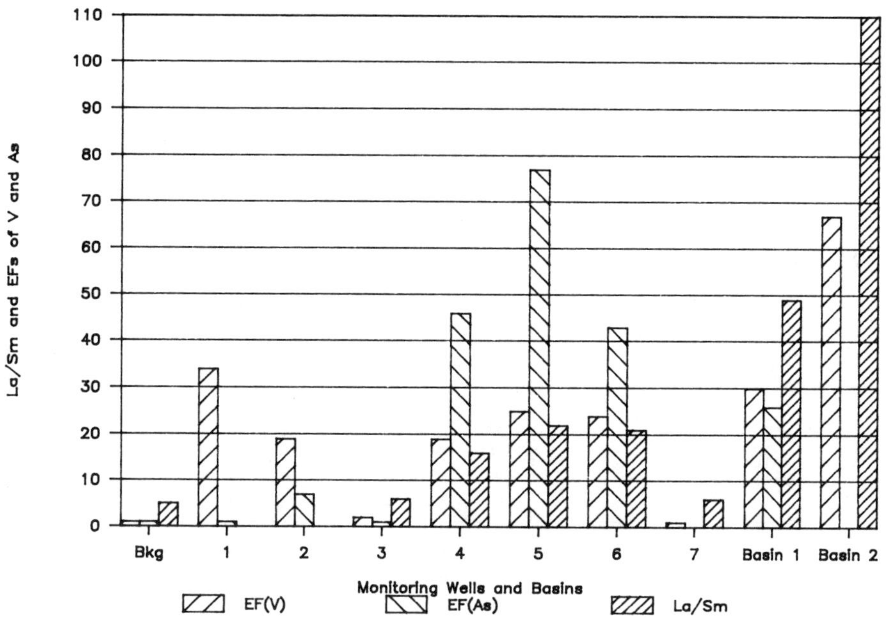

Fig. 1. The La/Sm ratio and enrichment factors observed in an oil-fired power plant, indicating leakage from surface impoundments.

The EFs for V and As and La/Sm ratios for the monitoring wells and basins at the oil-fired power plant are shown in Fig. 1. All of the wells, except Background, 3, and 7, show high degrees of enrichment in V and As, oil-, and coal-related elements, respectively. Also, the La/Sm ratio reaches the 20s for the basins and wells 4, 5, and 6. For the remaining wells, La/Sm ratios range from 5 to 6. Since the observation of unusual rare earth patterns in oil-related emissions, the La/Sm ratio became a unique signature for oil in source apportionment studies (*8*). The La/Sm ratio is found to be more than 20 in fuel oil-related contaminations, while being 5-6 in other source materials, such as coal and soil. The high La/Sm ratios, and the elevated EFs for V and As for a majority of the wells at the oil-fired power plant, show that groundwater has been contaminated by oil- and coal-related sources.

Trace marker elements were also used to detect and determine the extent of groundwater pollution around a coal-tar deposit site. Elemental distribution among the monitoring wells has revealed that several elements, including Mg, Cl, and Ba, had high relative concentrations at medium depth (M) and/or deep (D) monitoring wells, relative to shallow (S) and background (Bkg) wells (Fig. 2). This observation has allowed us to construct spatial distribution patterns and determine the degree of impact of pollution in the area.

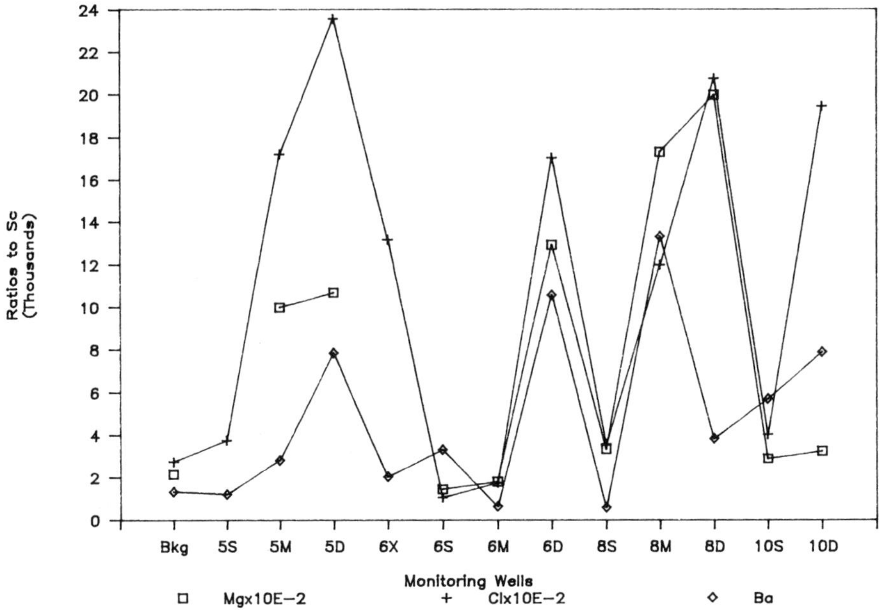

Fig. 2. Relative concentrations (normalized to Sc) of several elements observed in monitoring well waters in a coal-tar deposit site.

CONCLUSION

It has been shown that "marker trace elements" can be very effective tools in source attribution in groundwater, as well as in atmospheric studies. The success of this new approach in groundwater studies can be outlined as follows: unlike organic compounds, inorganic elements carry exactly the same signatures as the source material regardless of the distance between the source and the monitoring wells. Most of the organic compounds are so chemically active that they change their chemical compounds completely after they travel through just a few centimeters of soil. Also, routine analytical techniques used in groundwater contamination are not sensitive enough to detect marker trace elements at 10^{-9} or, in certain cases, even 10^{-12} g levels. Although INAA is one of the most sensitive, selective, and accurate multielement analytical techniques available in environmental studies, it is not as widely available to the research community as other techniques. After the marker element methodology has been established by INAA, however, other common analytical techniques can be utilized, e.g., by preconcentration of samples, with respect to the marker elements that are specific to that or similar sites.

REFERENCES

1. C. H. Ward, W. Giger, and P. O. McCarty (eds.) *Ground Water Quality*, Wiley, New York, NY (1985).
2. S. G. Tuncel, I. Olmez, J. R. Parrington, G. E. Gordon, and R. K. Stevens, *Environ. Sci. Technol.* **19**, 529 (1985).
3. K. Rahn and D. H. Lowenthal, *Science* **223**, 132 (1984).
4. T. G. Dzubay, R. K. Stevens, G. E. Gordon, I. Olmez, A. E. Sheffield, and W. Courtney, *Environ. Sci. Technol.* **22**, 46 (1988).
5. I. Olmez, *Trace Element Signatures in Wastewater Treatment Facilities and Underground Water Relations*, MIT Report No. MITNRL-017 (1987).
6. I. Olmez, *Receptor Models in Air Resources Management*, J. G. Watson, ed., APCA, Pittsburgh, PA, 3 (1989).
7. M. S. Germani, I. Gokmen, A. C. Sigleo, G. S. Kowalczyk, I. Olmez, A. M. Small, D. L. Anderson, M. P. Failey, M. C. Gulovali, C. E. Choquette, E. A. Lepel, G. E. Gordon, and W. H. Zoller, *Anal. Chem.* **52**, 240 (1980).
8. I. Olmez and G. E. Gordon, *Science* **229**, 966 (1985).

An Application of INAA and PIXE on the Analysis of Nutritional and Toxicological Elements in Samples of Drinkable Water

P. Blasi,*[1,2] G. Capannesi,[3] A. Cecchi,[2] F. Lucarelli,[1,2] and F. A. Sedda[3]

[1]Dipartimento di Fisica dell'Università, Largo E. Fermi 2, 50125 Florence, Italy; [2]Istituto Nazionale di Fisica Nucleare, Largo E. Fermi 2, 50125 Florence, Italy; and [3]Laboratori ENEA, 0060 Casaccia, Rome, Italy

ABSTRACT

Multielemental detection using Instrumental Neutron Activation Analysis (INAA) and Proton-Induced X-Ray Emission (PIXE) of drinkable water of Florence (Italy) is described. The concentrations of 52 elements were measured. No chemical treatment was performed on the samples, before or after irradiation. On the basis of an appropriate combination of the two analytical techniques, a procedure for aluminum determination was developed. Results are briefly discussed from a methodological point of view.

Index Entries: Drinkable water; trace elements; target preparation; INAA; PIXE.

INTRODUCTION

Trace elements play a key role in a variety of processes necessary for life. Various studies have shown that a definite correlation exists between trace elements and many common diseases (*1–4*). Furthermore, it is likely that some of the trace elements are more important in human

*Author to whom all correspondence and reprint requests should be addressed.

nutrition than is now generally acknowledged. Since trace elements are incorporated into the human through food, the first step in the study of their behavior is the knowledge of the concentration of these elements in that which is commonly ingested (*5,6*).

Since drinkable water is a fundamental factor in human nutrition, the presence of as many elements as possible in drinkable water was investigated. Two highly sensitive nuclear techniques, INAA and PIXE, were used to determine the multielemental content of drinkable water. The aim of our work was to provide an analytic procedure that could yield accurate results, relative to the highest number of elements, using the most efficacious resources of INAA an PIXE.

EXPERIMENTAL

Sampling

The drinkable water distributed to Florence (Italy) is drawn from the Arno River and is treated at the municipal aqueduct to make it potable. Figure 1 shows the steps carried out in the potabilization process.

The sampling sites were selected from seven private homes located in various urban areas to get an overall picture of the quality of the water supply. All samples from the private homes were taken after extended flow. In addition, two sampling sites were chosen at the aqueduct: one at the point of entry (raw water) and the other at the outlet (distribution water).

A part of the sample collecting was carried out using polyethylene vials, which, because of their high degree of purity and resistance to radiation damage, are suitable for irradiation. The quantity of water in each of these vials for INAA was 20 g per sample (8 g for short irradiations). At the same time, 1-L bottles of selected polyethylene were employed for collecting samples for both the PIXE target preparation (35 mL) and other INAA targets (250 mL). The latter were prepared using the freeze-drying method. All plastic containers were rinsed with the drinkable water itself before the samples were collected.

The samples from private homes refer to single samplings taken in December 1988; the samples from the aqueduct refer to samplings in September, October, and December 1988.

Target Preparation, Irradiations, and Counting

INAA

All samples were irradiated at 1 mW TRIGA MKII reactor of the ENEA (Casaccia, Roma). Eight grams of each sample of drinkable water were irradiated for 10 min in an incore system with pneumatic transfer. The flux was 1.1×10^{13} neutron/cm^2/s^1. Short-living radionuclides were produced. Gamma spectrometry was done by means of a hyperpure

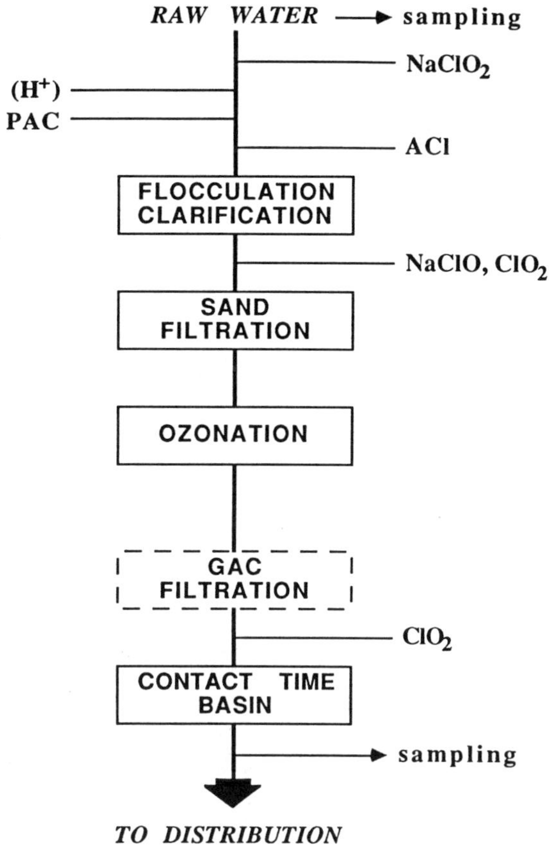

Fig. 1. The potabilization process at the Municipal Aqueduct.
PAC: *powder activated carbon.*
PAC1: *poly-aluminum cloryde.*
GAC: *granular activated carbon.*

Ortec Ge detector (25% efficiency, 1.7 KeV full width at half maximun at 1.3 MeV), storing 8K spectra on a multichannel analyzer (MCA).

Since it is well known that ^{41}Ar interferes with the measurement of these radionuclides, an N_2 flow was introduced into the container immediately after irradiation. A first spectrum was recorded, 4 min after irradiation, with a counting time of 15 min: Mg, Cl, Ca, V, Cu, and I were determined. A second spectrum, 60 min after irradiation, with a counting time of 30 min, was recorded: K, Mn, Sr, In, Ba, and Dy were detected.

Twenty grams of the samples in appropriate vials were irradiated for 30 h in a neutron flux of $2.6 \cdot 10^{12}$ n/cm^2/s^1. Long-living radionuclides were obtained. The sample holder had a capacity of 40 vials and was kept in constant rotation so that the homogeneity of the flux was better than 99.85%, and the maximum temperature was 40°C. The measurements were taken using a hyperpure Ge detector (29% efficiency, 1.7 keV FWHM at 1.3 MeV) with the abovementioned electronic equipment. The samples were allowed to decay 3–5 d before a spectrum (lasting 1–3 h) was recorded. Further measurements (lasting 5–10 h) were taken 7–10 and 30–40 d after irradiation.

Long irradiations were also performed on the solid product, obtained by freeze-drying (at a pressure of 5 Pa) 250 g of drinkable water. The preparation of each sample was carried out in the same vials used for the liquid sample irradiations, through successive steps of freeze-drying. The time required for the complete operation was 1 wk. The counting of these samples, by reducing the self-shielding, allowed for the detection of radionuclides that have low-energy gamma emission. The detection of some radionuclides that were present in drinkable water at very low levels was also possible using these targets, because the detectable quantity had been increased.

Finally, INAA was carried out using a thermalized column where the neutron flux was $4 \cdot 10^9$ n/cm^2/s^1. The thermalized column consisted of a 70 cm long pipe, filled with D_2O. The value of R_{Cd} was >1000. The irradiation time was 5 min, the cooling time 1 min, and the counting time 10 min. The weight of each sample was 1000 g. Each sample was distributed along a large area (400 cm^2) of a 1-mm thick polythene container to minimize the stopping power effect. For both short and long irradiations, solutions of properly mixed spectroscopically pure elements (Johnson-Matthey, UK) were employed for calibrations in the same geometry used for the samples.

PIXE

Thirty-five milliliters of each sample of the drinkable water were evaporated on a thin Mylar backing at 30°C by means of a lamp. The water was dropped onto the Mylar foil pressed by a plastic ring on a flat polyethylene container. The evaporation was performed in a glove box with an Argon atmosphere to avoid external contamination. The target preparation method was tested by processing standard solutions of some trace elements, in particular volatile elements. The analysis carried out on these targets has revealed no environmental contamination, nor any effect such as selective evaporation.

The solid targets from drinkable water thus obtained showed a thickness that was sufficiently thin for the analysis; however, it was not uniformly distributed on the backing. On the other hand, the preparation of the target, according to the procedure described above, is exclusively instrumental. It has the advantage of almost totally reducing the number of operations, such as chemical procedures, that other kinds of target preparation perform (*7*). We did not have to detect concentrations below tens µg/L by PIXE because it was used in combination with INAA, which is characterized by elevated sensitivity.

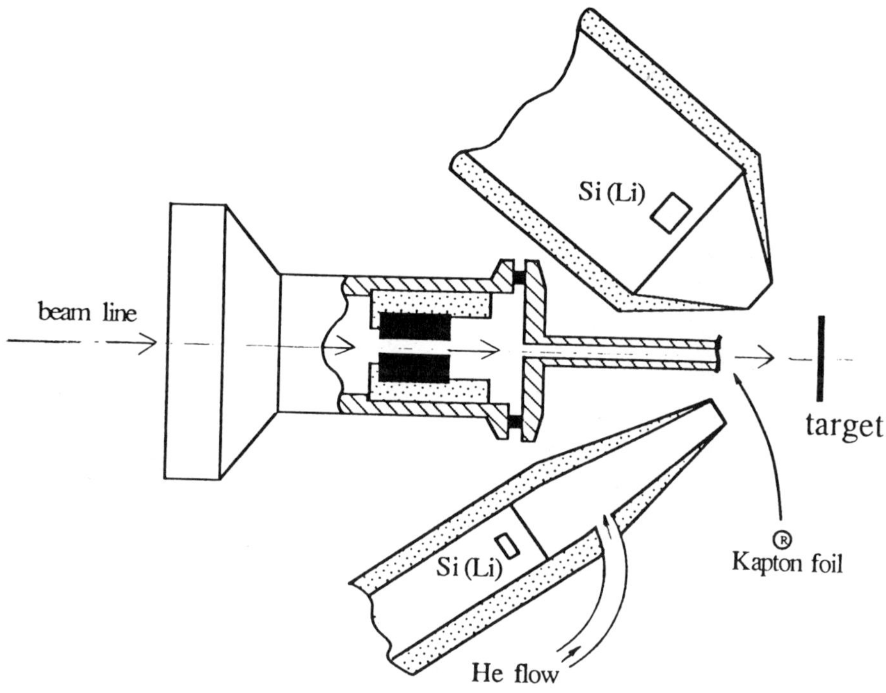

Fig. 2. Experimental set-up of the PIXE system.

In Fig. 2, a scheme of our PIXE system is reported. We used the 3 mV Van de Graaff accelerator at the Physics Department of the University of Florence. The proton beam is extracted from the vacuum system of the accelerator through a thin (20 μm) Kapton ® foil into air at atmospheric pressure. Under these conditions, it is possible to overcome or reduce the damages associated with physical instabilities in vacuum (e.g., selective sublimation of some components) or connected with the ion bombardment in vacuum (e.g., local heating of the sample on the beam spot).

Two problems arise with the presence of air. One is the strong Argon peak that appears in the spectrum when the beam passes through the air; the other is that air itself absorbs low energy X-rays going from the sample to the detector. Both problems can be avoided by making measurements in a Helium atmosphere. Under these conditions, we were able to detect the X-rays for elements with $Z > 10$.

X-rays were detected with two Si(Li) detectors placed at backward angles (135°) with respect to the beam direction. One detector has a surface area of 13 mm^2, a thickness of 3mm, and an energy resolution of 155 eV. This one was placed 8 cm from the sample. The other detector has a surface area of 30 mm^2, a thickness of 5 mm, and an energy resolution of 180 eV. It was placed as close as possible to the sample (about 3.5 cm) in order to maximize the acceptance solid angle. It was

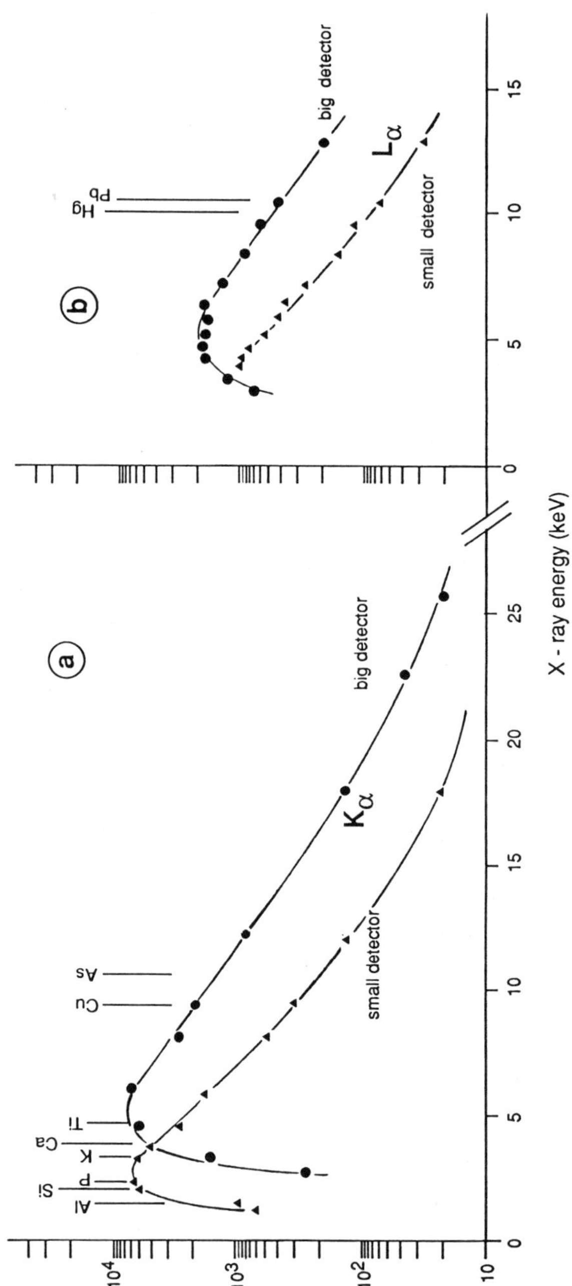

Fig. 3. The yields obtained by bombarding the thin standards in the same geometry of the samples.

covered with a 300 μm layer of Mylar to absorb the intense low energy X-rays and prevent too high a counting rate.

The two detectors were complementary in the sense that the latter was optimized for the detection of X-rays from higher Z elements, whereas the former was more sensitive to elements of lower atomic numbers. To illustrate this point, Fig. 3 shows the thin target yields of the two detectors, as deduced in our experimental set-up from thin calibrated layers of many standards.

In our experiment, a proton beam of 2.5 MeV on the target and 1 mm of diameter was used. A current of about 3 nA (measured by a Faraday cup of spectroscopically pure carbon behind the sample) was found to provide counting statistics for the X-rays sufficient to examine a spot in 10 min, even for elements of relatively low concentrations. At this beam currents, no evaporation or decomposition was observed on the bombarded samples. Blanks were prepared and run in the same way as the irradiated samples, both by INAA and PIXE.

RESULTS AND DISCUSSION

The instrumental sampling and analysis of drinkable water were basic aspects of our research. In addition, the combination of INAA and PIXE allowed for notable advantages in the analysis of the samples studied. In fact, INAA (*8*) is one of the most sensitive, selective, and accurate multielemental analytical techniques used in environmental studies; PIXE (*9*) has good detection limits for most elements and can be rapidly carried out.

Using PIXE, one can have information about the concentrations of most elements by means of an initial screening. Furthermore, three elements, Si, P, and Pb, which are not easily identified using INAA, can be detected. Fig. 4 shows a PIXE spectrum of a sample of drinkable water. As mentioned above, the targets prepared using our method do not have uniform thickness. The concentration values of our samples were obtained by comparing the peak areas of drinkable water with those of thin standards (Micro Matter Co., USA) of composition and quantity known (Fig. 3). Table 1 shows that the values of concentration, obtained using INAA and PIXE, are comparable. Table 2 reports the concentrations of all elements, determined using the INAA and PIXE methods, in the samples taken from private homes. Table 3 shows the concentration values of the more significant elements of the samples, taken from the aqueduct before and after the potabilization process.

The reproducibility of the INAA results from different measurements of the short- and long-living radionuclides showed differences $\leq$ 3%. Errors in the concentration values of the elements determined by INAA were principally owing to the presence of some elements in the vials used. Some blanks, randomly chosen from each batch, were analyzed by means of irradiations and gamma spectrometry in the same

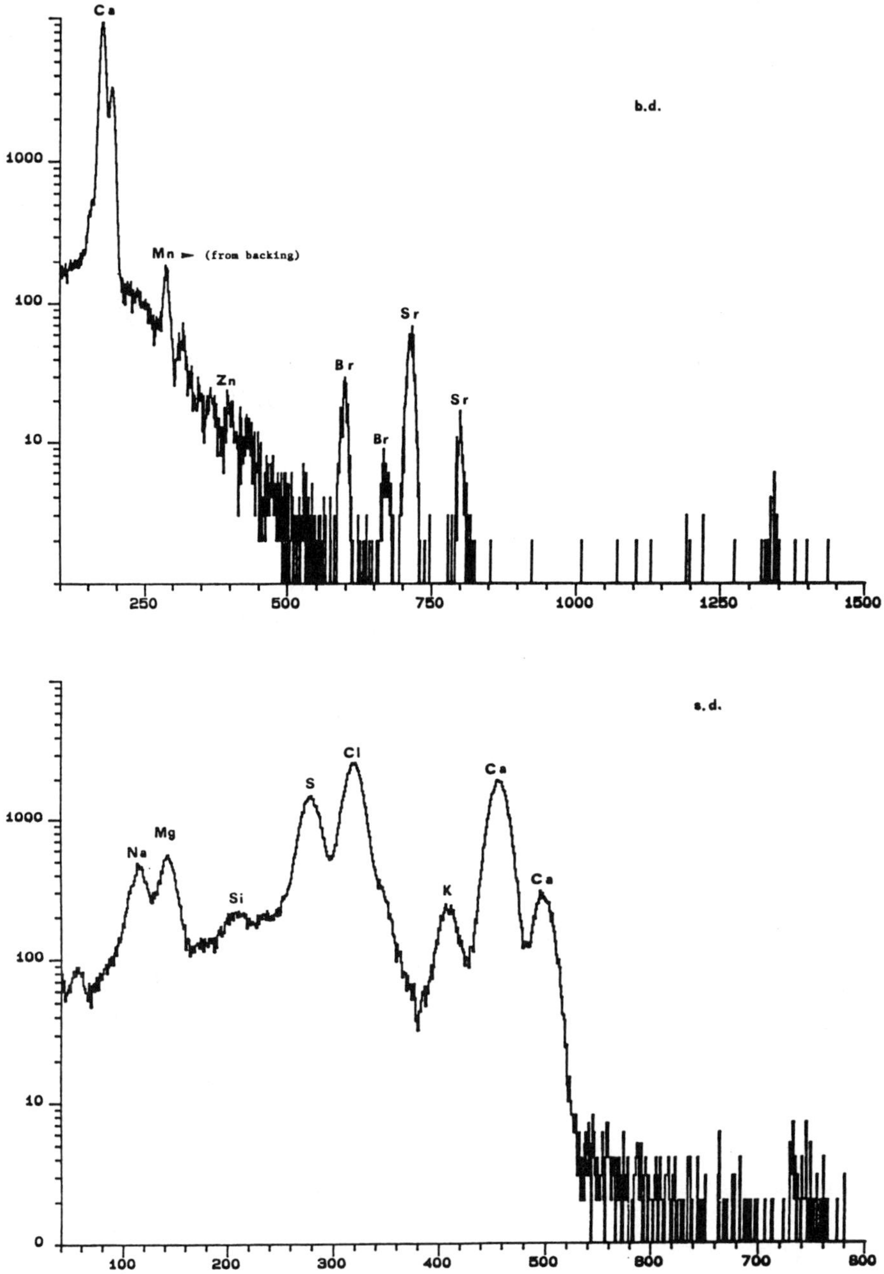

Fig. 4. PIXE spectrum of a drinkable water sample.

Table 1
Comparison Between INAA and PIXE
for the Concentration (μg/L) of Br, Zn, and K[a]

	Br		Zn		K	
Sampling sites	INAA	PIXE	INAA	PIXE	INAA	PIXE
1	36	41	18	29	2800	2600
2	42	54	64	81	3400	3200
3	39	59	9	14	3200	3400
4	44	28	5	9	2100	2000
5	50	54	1.2	0.8	3200	2600
6	26	38	1.5	1.7	4800	4300
7	15	8	9.4	8.6	1500	1100
8	125	104	0.7	≤0.6	5100	4600
9	32	30	2.3	2.6	2500	2300

[a]Sites 1–7: private homes, sites 8 and 9: aqueduct.

geometry used for the samples. The results showed that the content of the elements in these blanks was rather constant, and the standard deviation was calculated to be 8%. The presence of Cr, Co, Zn, Au, and Hg in the blanks may contribute up to 70% of the values found. Therefore, in this case, we used the results obtained from the measurements of the freeze-dried samples instead of the liquid ones. The same results obtained from the freeze-dried samples were also used to evaluate the concentrations of Mo, some earth rare elements (Gd, Ce, and Sm), and Ni, Sn, Nd, Lu, and Ta. The total error on the concentrations of the elements determined by INAA was estimated to be around 10%.

The concentrations of the trace elements Si and P were determined using PIXE. Silicon and Phosphorus have different origins in drinkable water. Silicon comes primarily from the sand filters used during the potabilization process. Phosphorus, however, is present in compound forms in water from the Arno River. Silicon concentrations varied between a minimum value of 30 and a maximum of 280 μg/L, whereas P concentrations were ≤100 μg/L.

Using PIXE, we defined a detection limit of 80 μg/L in the samples analyzed, regarding Al. On the other hand, Al was determined by INAA in samples taken from the aqueduct using the thermalized column. The values are reported in Table 4. The possibility of using a similar facility is especially interesting when there are high enough concentration values of Si and P. In that case, the reactions generated by the fast neutrons, ^{28}Si (n,p) ^{28}Al and ^{31}P (n,α) ^{28}Al, are in competition with ^{27}Al (n,γ) ^{28}Al. The

Table 2
Concentrations of Elements Detected in Private Homes[a]

Element	Min.	Max.	Element	Min.	Max.
Na	20	40 mg/l	In	0.05	0.12
Mg	18	35 mg/l	Sn	0.2	0.6
Al	60	120	Sb	0.08	0.38
°Si	$\leqslant$ 30	284	Te	0.09	0.36
°P	-	$\leqslant$100	I	0.04	1.0
°S	1.9	11 mg/l	Cs	0.005	0.018
Cl	50	80 mg/l	Ba	60	130
K	2.5	3.5 mg/l	La	0.01	0.03
Ca	65	85 mg/l	Ce	0.1	0.4
Sc	$2.5 \cdot 10^{-3}$	$4.5 \cdot 10^{-3}$	Nd	0.06	0.15
V	0.14	0.65	Sm	0.04	0.14
Cr	0.1	0.4	Eu	$0.8 \cdot 10^{-3}$	$2.2 \cdot 10^{-3}$
Mn	3.6	10	Gd	0.008	0.024
Fe	14	110	Tb	0.001	0.006
Co	0.03	0.45	Dy	0.001	$2.4 \cdot 10^{-3}$
Ni	0.6	2.4	Yb	$0.8 \cdot 10^{-3}$	0.003
Cu	12	36	Lu	0.004	0.012
Zn	1.2	64	Hf	$0.6 \cdot 10^{-3}$	$1.6 \cdot 10^{-3}$
As	0.05	0.65	Ta	0.008	0.04
Se	0.08	0.19	W	0.2	1.2
Br	15	50	Ir	$5 \cdot 10^{-5}$	$1.5 \cdot 10^{-3}$
Rb	2.0	12	Au	0.006	0.011
Sr	200	500	Hg	0.05	0.11
Mo	0.15	1.8	°Pb	$\leqslant$ 4	18
Ag	0.06	0.25	Th	0.001	0.002
Cd	0.04	0.47	U	0.5	0.9

[a]Quoted values are in µg/L, unless differently noted. The concentrations are the minimum and the maximum values found at the seven sampling sites.
○Elements determined by PIXE.

sensitivity of a thermalized column at the Triga Mark II reactor for the determination of Al in drinkable water is shown in Table 5.

In the specimens of drinkable water examined, the Al content could be determined using short irradiations. In fact, the levels of P and Si in our samples were rather low. In any case, in those samples where P and Si were identified at higher levels through PIXE screening, the use of the thermalized column was absolutely necessary for Al determination.

Table 3
Concentrations of More Significant Elements Detected at the Aqueduct[a]

Element	Raw water	Treated water
Al	10	80
°Si	≤40	122
Cr	1.4	0.45
Co	0.08	0.14
Ni	1.8	0.7
Cu	9.0	2.0
Zn	0.7	2.3
As	0.45	0.65
Se	0.21	0.09
Br	125	32
Mo	0.8	0.9
Cd	0.07	0.09
Sb	0.28	0.08
La	0.02	0.02
Hg	0.10	0.05
°Pb	≤4.0	≤4.0

[a]Quoted values are in μg/L. The concentrations are given as averages of the values found for the three samplings taken in September, October, and December 1988.

°Elements determined by PIXE.

Table 4
Concentration of Al (μg/L) Detected by INAA[a]

Sampling	Raw water	Treated water
September 88	10	66
October 88	8.0	74
December 88	10	96

[a]The irradiations were performed on samples of 1000 g of drinkable water in the thermalized column.

Table 5
Detection Limits of Thermalized Column at the Casaccia Reactor for the Al Determination in Drinkable Water

Single irradiation/sample		Ten cycles of irradiation/sample	
Sample weight (g)	Minimum detectable concentration (ppm)	Sample weight (g)	Minimum detectable concentration (ppm)
1	1	1	0.2
10	0.16	10	0.03
50	0.06	50	0.001
1000	0.008	1000	0.0016

The Se and Mg contents were determined with good resolution at ultratrace levels using INAA. The evaluation of the Se contribution to the 279 keV peak was obtained by determining the ratio of the 279 and 264 keV gamma emissions from a primary standard of Se that was measured using the same geometry as for the samples analyzed. Under our experimental conditions, we determined the Hg content in the presence of Se when the concentration ratio Hg:Se was above 1:20.

During the flocculation step of the water potabilization process at the aqueduct, the concentrations of the heavy elements were reduced to very low levels. The results of the analysis of the water samples from private homes are affected by the corrosive effect of the water itself on the pipelines. This effect is increased by the addition of chlorinated compounds during the treatment process at the aqueduct. Consequently, the concentration values of Cu, Zn, Cd, and Pb were higher (Table 2) in comparison with those obtained in samples at the aqueduct distribution point (Table 3). An increase by factor 10 was noted in the concentration of these four elements after a period of no flow.

CONCLUSIONS

The contemporaneous application of INAA and PIXE in the determination of trace and ultratrace elements in drinkable water is reliable and efficient. The instrumental procedure applied during this study allowed for a complete picture of the concentration levels of 52 elements. These elements are commonly found before and after the potabilization process at the aqueduct and in private homes in Florence.

REFERENCES

1. M. Simonoff, Y. Llabador, C. Hamon, B. Berden, G. Simonoff, C. Conri, B. Fleury, P. Couzigon, and A. Lucena, *MTAA* **1**, 517 (1986).
2. T. Kouris, C. Papandopoulou, G. D. Kanias, E. Bairaktori-Kouri, and G. Theodossiadis, *MTAA* **2**, 893 (1986).
3. S. S. Krishnan, K. G. McNeill, A. J. W. Hitchmau, J. R. Mernagh, S. C. Linn, and J. E. Harrison, *Proc. of Fifth Int. Conf. Puerto Rico*, CONF-840408, 861 (1984).
4. S. Ahmad, *J. Radioanal. Chem.* **120**, 89 (1988).
5. L. S. Mc Dowell, P. R. Giffen, and A. Chatt, *J. Radioanal. Chem.* **110** 519 (1987).
6. A. Chatt, H. S. Dang, B. B. Fang, C.K. Jayawickreme, L. S. Mc Dowell, and D. L. Pegg, *J. Radioanal. Chem.* **124,** 257 (1988).
7. E. M. Johansson and S. A. E. Johansson, *Nucl. Instr. Meth.* **B3**, 154 (1984).
8. J. E. Milley and A. Chatt, *J. Radioanal. Chem.* **110**, 345 (1987).
9. R. E. Jervis, *Int. J. Environ. Anal. Chem.* **15**, 89 (1983).

Plastics from Household Waste as a Source of Heavy Metal Pollution

An Inventory Study Using INAA as the Analytical Technique

P. Bode,*[,1] M. De Bruin,[1] Th. G. Aalbers,[2] and P. J. Meyer[2]

[1]Delft University of Technology, Interfaculty Reactor Institute, Mekelweg 15, 2629 JB Delft, The Netherlands, and [2]National Institute of Public Health and Environmental Protection, PO Box 1, 3720 BA Bilthoven, The Netherlands

Received April 17, 1989; Accepted October 23, 1989

ABSTRACT

An inventory study to the levels of cadmium in the plastic component of household waste was carried out utilizing INAA as the analytical technique. In a 2-h irradiation, 2-d decay, and 1-h measurement, protocol adequate sensitivities could be obtained for Cd, but also for a group of other metals: Cr, Co, Ni, Cu, Sr, Zn, As, Se, Mo, Sn, Sb, Ba, and Hg. Red-, orange-, and yellow-colored plastics either contain Cd at high levels (over 1000 mg/kg) or have relatively low Cd concentrations (<50 mg/kg). High concentrations were also occasionally found for Sr, Se, Ba, Sb, and Hg. INAA appeared very well to be routinely usable for such analysis because of the absence of a destruction step, adequate sensitivity, high accuracy, and multielement results.

Index Entries: INAA; plastic household waste; environmental pollution; Cd levels; Cr, Zn, Se, Sb, Ba, and Hg, high levels of; plastic color, relation with; original use, relation with.

*Author to whom all correspondence and reprint requests should be addressed.

Biological Trace Element Research Editor: G. N. Schrauzer

INTRODUCTION

The relationship between (trace) element concentrations in household waste and in incinerator residues of these wastes has been the subject of study by the Dutch National Institute of Public Health and Environmental Protection. One of the results of this study was that cadmium predominantly occurs in the component "plastics" in such a waste stream. As can be seen from Table 1, as much as 2000 mg Cd/ton of untreated waste was found, which corresponds to about 110 g Cd/ton of electrofilter ashes.

The presence of Cd in plastic materials has to do with the use of Cd compounds as a pigment to obtain red, orange, and yellow coloring and sometimes as a stabilizer agent in polyvinylchloride. New plastic products for use in daily life (e.g., packing materials of food and detergents) are regularly being introduced. Many of these products have short life cycles since they can be used only once. Therefore, a regular flow to the waste stream occurs, and an environmental problem may be foreseen, owing to the high Cd levels.

Such a situation might be tackled by focusing on the raw materials and reducing the use of Cd compounds therein. Still, information is needed on the levels of Cd in many plastics first and answers have to be given to questions, such as "Are the levels of the same order of magnitude for different products, or are there any products with substantially higher concentrations?" It was recognized that, for such an inventory study, INAA would be the most appropriate technique. Its nondestructive character easily overcomes the cumbersome problems that occur when the samples have to be brought into a solution prior to analysis. Moreover, the high accuracy of INAA is extremely important for a National Institute that plays a role in formulating and controlling legal pollution standards. Since quick answers are often asked for, attention had to be paid to develop an analysis protocol that shared adequate sensitivity, optimal multielement information with a short turn-around time.

METHODS

Sampling

Sampling was carried out by the National Institute of Public Health and Environmental Protection at their special waste stream facility. Both colored (red, orange, yellow, green, blue, brown, white, and matte-white) and transparent materials were sampled. The materials were selected from the following original fields of application: cleaning agents, personal care articles, soft-drink bottles, food packages, foils (plastic bags), florist articles, toys, hobby articles, and utensils.

Table 1
Mean Metal Concentrations (in mg per ton; 1000 kg) of Fresh Household Waste and its Relation to Different Components of the Waste

component	As	Cd	Cr	Cu	Hg	Pb	Zn
vegetables, fruits, garden waste	447	110	37930	25670	53	53100	37080
paper, cardboard	77	43	3242	8725	18	5822	37730
plastics	22	2014	3952	2372	2.8	14880	10490
glas	3744	41	36860	1860	<1	59030	6435
ferrous metals	1185	<14	22970	17590	n.d.	2117	5923
non-ferrous metals	22	11	412	1044000	n.d.	30950	157200
textile	11	123	5752	3459	1.0	2881	3547
bread	<0.5	1.4	151	73	0.7	76	350
ceramics	241	19	991	266	<0.2	26570	6322
carpets, mats	1.3	8.6	222	57	0.3	555	4673
leather, rubber	5.6	229	33630	360	0.5	5149	32350
wood	1.5	7.1	137	145	0.7	574	2784
total	5757	2606	146251	1104322	76	201708	304917

Sample Preparation

Samples were thoroughly cleaned from adjacent dirt by washing with distilled water. Drying occurred at room temperature. A sample of 5 × 5 cm was taken; soft materials were cut using titanium scissors. Hard materials were embrittled after immersion in liquid nitrogen. The reduced sample (400 mg) was packed in a polyethylene vial.

Plastic Characterization

The type of plastic (polyvinylchloride, polyethylene, polypropylene, polystyrene, or others) was determined by the Plastic and Rubber Research Institute of the Netherlands Organization for Applied Scientific Research. A distinction could be made between the pe and pp fractions, and the pvc, the ps, and other fractions.

INAA

Irradiations were carried out in the pneumatic irradiation facilities at the Hoger Onderwijs Reactor of the Interfaculty Reactor Institute. Two protocols were tried. In the first, samples were irradiated for 30 s at 1.3×10^{13} n/s^1/cm^2 and measured during 300 s after a 30-s decay period at a 10 cm distance from a 10% Ge detector. In the second protocol, irradiations were performed during 2 h at 5.0×10^{12} n/s^1/cm^2, and after a 2–3 d decay period, the samples were counted during 1 h at a 10 cm distance from a 17% Ge(Li) detector. Aspects of the IRI system for routine INAA have been extensively described previously (*1*).

Table 2
Sensitivities for a Group of Elements as Found by Two Analysis Protocols and the Relation with the Legal Standard for Chemical Waste

	analysis protocol(*)		legal standard for chemical waste
	I	II	
	sensitivity in mg/kg		mg/kg
Cr	20000	10	50 (#)
Co	2000	10	5000
Ni	10000	200	5000
Cu	25	10	5000
Zn	3500	25	20000
As	50	0.5	50
Se	0.5	10	50
Sr	200	125	(**)
Mo	200	1	5000
Cd	150	1	50
Sn	75	200	5000
Sb	200	1	50
Ba	200	100	20000
Hg	600	5	50

*I: T(irr), T(decay), and T(meas): 30 s, 30 s, and 300 s, respectively; II: T(irr), T(decay), and T(meas): 2 h, 2–3 d, and 1 h, respectively.

**Not yet defined, depends on compound, e.g., 20,000 for sulfides and 50,000 for oxides and #50 for Cr(VI) and 5000 for Cr(III).

RESULTS

Table 2 lists sensitivities for some elements, as found in the two analytical protocols. In Table 3, element concentration levels are given for some plastic materials of different color, as found by the 2-h irradiation protocol. Frequency distributions of Cd, with respect to color, field of application, and type of plastic, are given in Table 4.

DISCUSSION

INAA Procedure

Very quick answers can be obtained by the analysis protocol in which the fast rabbit facility was used (30-s irradiation). However, the ultimate sensitivity for Cd, based on the detection of ^{111m}Cd or ^{117m}Cd, is

Table 3
Some Typical Results from a Routine Analysis[a]

	colour of plastic					
	red	red	red	yellow	yellow	green
Cr	2.8	<6	<6	<5	<1.5	860
Fe	<350	<350	<500	<300	<400	<500
Ni	<70	<70	<90	<75	<60	<100
Zn	<0.4	30	<30	904	17.5	<35
As	<0.01	<0.02	<0.06	<0.03	<0.03	<0.2
Se	<1.9	409	<5.4	16.5	<2.4	<6.5
Br	0.03	0.06	<0.03	0.3	0.08	32
Sr	<160	<150	<200	<180	<150	<600
Cd	<0.6	2490	1598	8050	<0.9	<7
Sb	0.15	<0.03	<0.05	1.3	<0.02	317
Ba	<42	265	218	628	4720	<200
La	<0.02	<0.03	<0.04	<0.04	<0.04	2.1
Au	0.004	0.002	0.002	0.002	0.025	0.025
Hg	<0.7	<4.4	396	<2.7	<0.9	<4

[a]t(irr), 2 h; t (decay), 3 d; and t (meas) 1 hr. All concentrations are in mg/kg.

not good enough; the demands set were 50 mg/kg, and the experiment learned that the method is useful to determine Cd levels higher than 150 mg/kg. Still, this protocol might have some value for a quick reference for cases in which a high Cd level is suspected. For Co, Cu, Zn, As, Se, Sr, Mo, Sn, and Ba, the observed sensitivities were adequate compared to the limiting values set for chemical waste.

The protocol in which the samples are being irradiated for 2 h appears to be very suitable not only for the determination of Cd, but also for many other (trace) elements of environmental interest—all better than the demands set with respect to sensitivity. The involved turn-around time is still acceptable. In general, the entire analysis protocol in this application of INAA is such that large series of samples can be analyzed easily. The absence of laborious and sample type to sample type varying destruction steps also add up to a favorable and attractive cost of the analysis.

Cd and Other Trace Element Concentration Levels

At present, no laws exist in the Netherlands with respect to the trace element content of household waste. The outcome of this study has led to a comparison with the limitation values for some elements, as set by

Table 4
Frequency Distributions for Observed Amounts of Cadmium in 90 Samples with Relation to Plastic Color, Field of Application, and Type of Plastic

	colour										application										plastic type			
mg/kg Cd	red	orange	yellow	green	blue	brown	white	matt white	transparent	other	cleaning agents	pers. care art.	soft drink bottles	food	foils	florist art.	toys	hobby art.	utensils	others	pe.+pp.	pvc.	ps.	others
0 - 1	3	1	5	2	-	-	2	1	3	1	8	1	1	2	1	-	-	-	5	-	14	3	1	-
1 - 5	6	3	11	2	11	7	5	1	-	5	17	4	1	5	8	1	1	3	6	5	36	6	9	-
5 - 10	-	1	1	-	2	-	1	-	-	-	1	1	-	-	2	-	-	-	-	1	3	1	1	-
10 - 25	-	1	3	-	-	-	1	-	-	-	1	1	-	-	2	-	-	-	1	-	3	1	1	-
25 - 50	-	-	-	-	-	1	-	-	-	-	-	-	-	-	-	-	-	-	1	-	1	-	-	-
50 - 100	-	-	2	-	-	-	-	-	-	-	1	-	-	-	1	-	-	-	-	-	2	-	-	-
100 - 200	-	-	-	-	-	-	-	-	-	-	-	-	-	-	-	-	-	-	-	-	-	-	-	-
200 - 300	-	-	-	-	-	-	-	-	-	-	-	-	-	-	-	-	-	-	-	-	-	-	-	-
300 - 400	-	-	-	-	-	-	-	-	-	-	-	-	-	-	-	-	-	-	-	-	-	-	-	-
400 - 500	-	-	-	-	-	-	-	-	-	-	-	-	-	-	-	-	-	-	-	-	-	-	-	-
500 - 600	-	-	-	1	-	-	-	-	-	-	-	-	-	-	-	1	-	-	-	-	1	-	-	-
600 - 700	-	-	-	-	-	-	-	-	-	-	-	-	-	-	-	-	-	-	-	-	-	-	-	-
700 - 800	-	-	-	-	-	-	-	-	-	-	-	-	-	-	-	-	-	-	-	-	-	-	-	-
800 - 900	1	-	-	-	-	-	-	-	-	-	-	-	-	-	-	-	-	-	-	1	1	-	-	-
900 -1000	-	-	-	-	-	-	-	-	-	-	-	-	-	-	-	-	-	-	-	-	-	-	-	-
> 1000	1	1	2	1	-	1	-	-	-	-	2	-	-	-	-	-	-	-	4	-	4	1	1	-
total	11	7	24	8	11	10	8	2	3	6	30	7	2	7	14	2	1	3	17	7	65	12	13	-

laws for chemical waste (<50 mg/kg). It can easily be seen (Tables 2 and 3) that when the same legal regulations were valid for household waste, in many cases this waste should be treated as chemical waste.

Cd occurs in high to very high concentrations only in a limited number of thoroughly colored products. In the majority of the products analyzed, the levels are well below 50 mg/kg/ There seems to be no correlation between color of the plastic and Cd concentration. Red-, orange-, and yellow-colored material apparently contain the higher amount of Cd, but also in many plastics of this color, low Cd levels were found.

Nearly all products with a Cd concentration of more than 50 mg/kg originate from the short lifecycle stock. Half of these products have been used for food packing. No correlation seems to exist between the type of plastic and high Cd concentrations. Therefore, it can be concluded that Cd is primarily added as a pigment.

Multielement INAA clearly revealed that a low or acceptable Cd level by no means implies that the material wouldn't contain potential environmental contaminants. Many other heavy metals appear to occur in sometimes alarmingly high concentrations.

SUMMARY

INAA has proven to be a very valuable aid for the routine determination of trace elements in plastics from household waste. The absence of a destruction step, the high accuracy, multi-element results, and adequate sensitivity have contributed to this. The method easily allows for analysis of both large series of samples for inventory work as for rapid analysis for specific cases.

Red-, orange-, and yellow-colored plastics variably contain high Cd amounts since material of this color often appears to be low in Cd. Efforts have to be put in the development of usable, harmless organic pigments. Meanwhile, Cd, Hg, and Cr pigments may be replaced by pigments of other metals, such as Sb, Sr, Ni, Ba, and Fe. As a result, the release of these trace elements may increase. A decision will have to be made as to which alternative priority has to be given.

Recently, the Dutch cabinet has decided that governmental agencies should pursue a policy focused on the purchase of goods preferentially "Cd-free," i.e., with a Cd content lower than 5 mg/kg. Maybe this initiative will work as a snowball to other institutions and industries. For the enforcement of this act, as for random checks on products being already in the market, all (multi)element analyses will be carried out using INAA.

REFERENCES

1. M. de Bruin, P. J. M. Korthoven, and P. Bode, *J. Radioanal. Chem.* **70,** 497–512 (1982).

In Vivo Analysis

Reference Man and Woman More Fully Characterized

Variations on the Basis of Body Size, Age, Sex, and Race

KENNETH J. ELLIS

USDA/ARS Children's Nutrition Research Center, Department of Pediatrics, Baylor College of Medicine and the Texas Children's Hospital, Houston, TX 77030

Received June 14, 1989; Accepted November 3, 1989

ABSTRACT

Total body neutron activation analysis, prompt-gamma neutron activation analysis, and whole body counting have been used to determine the elemental composition of the human body. The total body elements measured were potassium, nitrogen, calcium, sodium, chlorine, and phosphorus. Total body water was also determined by the dilution principle using tritiated water. Observations were made in an adult US population that totaled 1374 and ranged in age from 20 to 90 yr. The dataset for the white population consisted of 175 males and 1134 females observations; for the black population, it consisted of 30 male and 35 female observations. The variation in the elemental composition of both males and females in any 5-yr age group was large and ranged up to 20% (SD). Age-, race-, sex-, and size-specific differences were evident. When equations were developed that predicted the elemental composition of the adult on the basis of age, weight, and height, the variation in the age groups was reduced approximately in half. Age-specific mean values for the 20- to 29-yr-old white population were also compared with values for the International Commission on Radiological Protection (ICRP)-23 Reference Man. The "average" young adult male was larger than Reference Man; the in vivo data also indicated a larger skeletal mass, more lean tissues and body water, but lower body sodium. When in vivo prediction equations were used to adjust for size differences, good agreement was found between the expected values and those for Reference Man. The ICRP-23 does not contain elemental data for Reference Woman; therefore, the in vivo data in the present study provide the first estimates of body composition for Reference Woman.

Index Entries: Whole-body activation analysis; total body (inorganic composition); body composition; age, sex, and race variations; reference man and woman.

INTRODUCTION

A task group of the International Commission on Radiological Protection (ICRP) has developed the concept of a Reference Man (a.k.a. Standard Man) for use in calculating permissible levels of exposure for a typical radiation worker (*1*). The Reference Man has been defined as one who is 20 to 30 years of age, 70 kg in weight, 170 cm in height, and lives in a climate with an average temperature of 10 to 20°C. He is Caucasian and a Western European or North American in habitat and custom. Although the general characteristics of Reference Man were not intended to predict any particular individual or class of individuals, their use has been extended for this purpose in recent years, especially with the resurgent clinical interest in body composition assessment (*2–4*).

Our knowledge of the elemental composition of the human body has increased significantly since the development of in vivo neutron activation analysis. This nuclear analytical technique has enabled a direct quantitative chemical profile of the human body (*5*). Extensive in vivo measurements of the bulk elements in the body (i.e., calcium, sodium, chlorine, phosphorus, nitrogen, and potassium) have been performed in adults. The values of ICRP-23 Reference Man can now be compared with those from the in vivo observations to provide an indication of the "normal" range of these elements in this population.

Total body neutron activation analysis (TBNAA), prompt-gamma neutron activation analysis (PGNAA), and whole body counting (WBC) procedures have been performed on patient populations, with a broad spectrum of metabolic and endocrine disorders in which body composition could be altered. To evaluate the degree of abnormality among the patient populations, it was necessary to establish a normal reference range from observations of a control group of healthy subjects from the general US population. The data collected from the control group have been used to establish the extent of individual variation in body elements among normal subjects, including the effects of age, sex, body size, and race. On the basis of these interrelationships, prediction equations were established for the individual. The in vivo elemental data were compared with values for the ICRP-23 Reference Man. These findings will be used to investigate the concept of a Reference Woman.

METHODS

Subject Populations

Body elemental composition measurements were performed over a 15-yr period, from 1972 to 1987. Black and white adult volunteers were recruited from the general US population who lived in the New York City-Long Island, New York metropolitan area. All the subjects were considered physically active for their age and had no prior history or current symptoms of metabolic, renal, or cardiovascular disease at the time of the procedures. Only subjects whose weights were within 25% of that expected for their height and sex were included. Many of the older women were also participants in a prospective study of the incidence of postmenopausal osteoporosis in aging. No subjects received medications known to influence calcium or potassium metabolism. The project was approved by the Human Studies Research Committee at the Brookhaven National Laboratory and an informed consent was obtained from each subject.

Whole Body Counting (WBC)

The absolute measurement of total body potassium (TBK) was obtained by counting the natural ^{40}K content of the whole body. The counter is housed in a shielded room that contains 54 NaI detectors arranged in two 3 $\times$ 9 configurations. The counters are positioned above and below the subject. For a 15-min count, the accuracy and precision for measurements of TBK are reported to be $\pm$3.3% using an anthropomorphic adult phantom (*6*). Reproducibility of the TBK value in a normal healthy adult was found to be $\pm$2.8%.

Total Body Neutron Activation Analysis (TBNAA)

The design of the irradiation facility used exclusively for human TBNAA studies has been described previously (*7,8*). Fourteen 50-Ci ^{238}Pu, Be neutron sources are arranged in two rows above and below the midline of the subject. Fast neutrons are moderated by surrounding the subject with a 3/4 in layer of polyethylene. The exposure time is fixed at 5 min and delivers a total neutron (QF = 10) and gamma dose measured at less than 2.8 mSv (278 mrem).

The induced activity is measured in the whole body counter for 5 and 15 min, starting 3 min postirradiation. The first 5-min count is optimized for the rapid decay ($t_{1/2}$ = 2.8 min) of the induced signal from body phosphorus generated by the fast reaction $^{31}P(n,\alpha)^{28}Al$. The 15-min spectra contains the Ca, Na, and Cl peaks induced by thermal neutron activation. The accuracy and precision for TBNAA, using an anthropomorphic phantom (same model as used for TBK), are reported as $\pm$1% for total body calcium (TBCa), $\pm$2% for total body chlorine and sodium (TBCl, TBNa), and $\pm$4% for total body phosphorus (TBP) (*7,8*).

Prompt-Gamma Neutron Activation Analysis (PGNAA)

A detailed description of the facility designed for prompt-gamma neutron activation analysis (PGNAA) has been presented previously (*9,10*). In vivo measurements are performed using two large volume NaI detectors (6 in dia × 6 in thick) that are positioned above the subject, just out of the direct view of the neutron beam. An 85-Ci ^{238}Pu, Be source is located approximately 50 cm below the subject and is shielded to provide a rectangular beam (30 × 60 cm) at the level of the bed. The subject is scanned in prone and supine positions for a total measurement time of 30 min, with a total body dose less than 0.40 mSv (40 mrem). Although most elements in the body emit prompt gammas after thermal neutron capture, body nitrogen and hydrogen are particularly well suited for this type of in vivo analysis. Gamma signals from body chloride, carbon, and calcium are also detectable in the prompt gamma spectra; their precision, however, is less than the TBNAA procedure, and the dosage must be increased significantly.

The calculation of total body nitrogen (TBN) is based on an "internal standardization" relative to the body hydrogen gamma signal (*9*). An independent estimate of body hydrogen is obtained using a tracer dose of tritiated water (30 μCi); the total body dose is less than 0.03 mSv (3 mrem). The HTO is administered orally, and a blood sample is collected 3 to 4 h later. The HTO content of 1 mL plasma is counted using standard liquid scintillation spectroscopy, and total body water (TBW) is determined by the dilution principle. The precision for TBW has been reported at ±1% in normal healthy adults (*11*). Reports indicate that a more accurate measurement of body nitrogen is obtained when it is combined with the measurement of body water (*10*). The precision and accuracy for the in vivo measurement of TBN, using the anthropomorphic phantom, is ±3% (*9,10*). Repeated measurements in a normal healthy adult male over a 6-mo interval showed a ±3.5% variation in the value of TBN.

RESULTS

Body Composition of the Adult White Population

A total of 1309 body composition observations were made in an adult white population. The sampling between the sexes was not uniform, because many of the women also served as controls for clinical protocols involved in the study of osteoporosis. Thus, the number of older females in our sample was larger than younger women or younger and older men. The distribution of males and females is given in Table 1 for each 5-y interval, beginning with age 20. The mean age-specific anthropometric values and standard deviations for age, weight, and height are given in Table 1. The body mass index (BMI), defined as weight/height^2,

Table 1
Distribution of Age, Weight, Height, and Body Mass Index (BMI) in an Adult White Population (1134 Females, 175 Males) in the US

Age Group	Subjects (n)	Age (yr)	Weight (kg)	Height (m)	BMI* (kg/m^2)
			Females		
20–24	18	22.1 ± 1.4†	59.2 ± 6.7	1.62 ± .06	22.3 ± 1.9
25–29	14	27.3 ± 1.4	58.9 ± 5.9	1.63 ± .06	21.9 ± 1.7
30–34	41	32.1 ± 1.4	58.2 ± 10.9	1.62 ± .08	22.1 ± 3.8
35–39	53	36.6 ± 1.4	62.6 ± 10.8	1.63 ± .06	23.2 ± 3.3
40–44	64	42.4 ± 1.3	66.3 ± 13.4	1.62 ± .06	25.0 ± 4.5
45–49	118	47.3 ± 1.4	66.2 ± 10.2	1.63 ± .04	24.7 ± 3.4
50–54	243	52.2 ± 1.3	68.4 ± 12.9	1.63 ± .06	25.7 ± 4.5
55–59	276	56.8 ± 1.4	67.1 ± 10.4	1.62 ± .05	25.5 ± 3.6
60–64	169	61.6 ± 1.4	65.8 ± 10.4	1.62 ± .06	24.8 ± 3.4
65–69	80	66.8 ± 1.3	65.5 ± 9.2	1.58 ± .05	26.0 ± 4.1
70–74	58	72.0 ± 1.9	61.6 ± 6.4	1.58 ± .05	24.5 ± 2.9
			Males		
20–24	18	22.4 ± 1.4	79.9 ± 14.2	1.78 ± .04	24.9 ± 4.1
25–29	8	27.1 ± 1.2	71.7 ± 9.3	1.74 ± .07	23.3 ± 1.7
30–34	20	31.6 ± 1.5	72.1 ± 11.5	1.74 ± .08	23.5 ± 2.9
35–39	17	37.4 ± 1.4	77.7 ± 8.1	1.76 ± .07	25.0 ± 2.6
40–44	20	41.9 ± 1.4	79.9 ± 13.8	1.76 ± .07	25.6 ± 3.8
45–49	20	46.9 ± 1.4	76.1 ± 9.3	1.75 ± .04	24.7 ± 2.7
50–54	14	51.8 ± 1.4	82.9 ± 14.2	1.76 ± .07	26.3 ± 3.0
55–59	15	56.6 ± 1.4	79.1 ± 10.9	1.72 ± .07	26.6 ± 3.2
60–64	15	61.8 ± 1.7	80.1 ± 9.8	1.77 ± .04	25.4 ± 2.3
65–69	6	66.3 ± 0.8	80.1 ± 15.6	1.74 ± .03	26.2 ± 4.7
70–74	14	72.7 ± 2.4	76.6 ± 11.2	1.71 ± .06	26.2 ± 4.0
75–90	8	85.0 ± 4.4	72.5 ± 10.7	1.63 ± .07	26.9 ± 2.4

*BMI ($weight/height^2$).
†Mean ±SD.

has also been calculated and included in Table 1. Full stature appeared to have been achieved by age 20–25 for both sexes; the mean values were 1.62 m for females and 1.76 m for males. The mean values for height remained relatively stable through ages 60–64, after which a decrease occurred in the older age groups. As expected, males also weighed more (an average of 14.2 kg) than females at all ages. Mean weights were less stable than mean heights, varying between 59.2 to 68.4 kg in females and 71.7 to 82.9 kg in males. Body weight tended to increase for both sexes through ages 60–64, followed by a moderate loss in the older ages. The trend for body mass index increased throughout life for both sexes. The mean values for BMI ranged from 23 to 27 kg/m^2 for males and 22 to 26 kg/m^2 for females. Although males had higher BMI values at all ages, the rate of increase in BMI with age was more rapid in females.

The mean values for total body calcium (TBCa), phosphorus (TBP), sodium (TBNa), and chlorine (TBCl) in adult white males and females as a function of age are tabulated in Table 2. The variability in values for each element in each age group was large; the standard deviations ranged from 8 to 14% for both sexes. The skeleton is the major body compartment represented by TBCa and TBP; > 98% of body calcium, > 85% of body phosphorus are contained in bone. The mean value of TBCa for females is 20–35% lower than the mean for each age group for males. The mean TBCa value for the young male began at 1230 g at 20–24 yr and decreased uniformly to 1096 g by 70–74 yr, a difference of 134 g in 50 yr. When a similar comparison was made among the female groups, a slightly different pattern emerged. The mean TBCa value for ages less than 40 was relatively constant and averaged approximately 900 g, after which a reduction of 175 g occurred over the next 30–35 y. Between the ages of 20 and 70, the mean TBCa value for females decreased significantly in absolute amount and percentage change per year. The difference between the sexes was even more dramatic for ages greater than 45 yr. The body phosphorus pattern was similar to body calcium, although the reduction with age was less because of the higher fraction of body phosphorus in the soft tissues. Total body sodium and chlorine values, however, appeared relatively constant during adult life in healthy individuals. The 175 males had mean values of 83 g sodium and 77.5 g chlorine. The 1134 females had mean values of 63.5 g sodium and 61.5 g chlorine.

The results for TBK, TBW, and TBN, obtained by whole body counting, prompt-gamma neutron activation, and tritiated water dilution measurements, respectively, are tabulated in Table 3. The mean values for TBK, TBW, and TBN decreased with increasing age for both males and females. Other investigators have reported similar observations (*13,16*). The females at ages 70–74 had approximately 11 g less potassium, 3 L less water, and 334 g less nitrogen than the 20- to 24-yr-old subjects. The relative effects were a 13% reduction for TBK, a 9% decrease for TBW, and 22% lower TBN levels. The older males also had reduced values for

Table 2
Total Body Elemental Composition Measured by In Vivo Neutron Activation Analysis in an Adult White Population in the US*

Age Group	Calcium (g) TBCa	Phosphorus (g) TBP	Sodium (g) TBNa	Chlorine (g) TBCl
		Females		
20–24	905.2 ± 124.0†	439.3 ± 70.4	62.9 ± 7.6	60.1 ± 7.0
25–29	905.8 ± 78.9	428.3 ± 53.4	62.8 ± 7.6	62.9 ± 8.2
30–34	870.3 ± 125.4	424.0 ± 63.2	62.3 ± 7.8	60.0 ± 7.7
35–39	915.7 ± 116.6	433.2 ± 59.9	63.8 ± 7.2	60.6 ± 8.4
40–44	859.8 ± 108.2	394.1 ± 55.2	63.9 ± 7.8	61.8 ± 7.7
45–49	892.2 ± 87.9	412.6 ± 51.3	65.4 ± 6.3	61.8 ± 7.8
50–54	858.8 ± 109.8	400.3 ± 63.3	65.1 ± 6.8	62.8 ± 7.6
55–59	829.0 ± 111.8	391.0 ± 67.2	63.3 ± 6.3	61.3 ± 7.3
60–64	799.1 ± 99.9	382.1 ± 61.9	62.2 ± 6.3	61.0 ± 8.4
65–69	767.5 ± 93.1	374.3 ± 59.5	60.3 ± 5.5	58.9 ± 6.8
70–74	731.1 ± 102.1	380.8 ± 76.6	59.4 ± 5.0	59.6 ± 8.0
		Males		
20–24	1229.8 ± 101.1	561.4 ± 68.6	84.0 ± 7.2	82.1 ± 7.9
25–29	1157.0 ± 101.5	515.7 ± 15.5	79.7 ± 5.0	73.1 ± 4.6
30–34	1151.1 ± 145.0	508.7 ± 76.9	81.4 ± 9.7	76.6 ± 11.6
35–39	1174.5 ± 143.9	531.3 ± 56.0	84.4 ± 9.5	76.7 ± 7.6
40–44	1189.5 ± 117.1	504.1 ± 72.9	85.2 ± 9.9	74.9 ± 9.8
45–49	1107.5 ± 142.3	481.7 ± 69.8	81.2 ± 5.6	75.6 ± 7.9
50–54	1144.4 ± 108.4	468.2 ± 48.9	86.5 ± 8.9	78.5 ± 11.1
55–59	1103.3 ± 150.9	463.7 ± 84.8	84.1 ± 9.2	76.4 ± 12.2
60–64	1147.7 ± 160.0	492.6 ± 95.8	89.2 ± 9.7	79.5 ± 10.1
65–69	1029.6 ± 122.8	460.8 ± 81.8	79.0 ± 5.4	77.9 ± 8.0
70–74	1096.2 ± 131.8	538.0 ± 64.8	82.5 ± 8.1	79.1 ± 5.6
75–90	1010.8 ± 179.1	466.3 ± 62.7	82.6 ± 9.7	77.8 ± 9.8

*Delayed Neutron Activation Analysis (total body dose <2.8 mSv).
†Mean ± SD.

Table 3
Elemental Composition Obtained by Whole Body Counting, Prompt-Gamma Neutron Activation, and Tritiated Water Measurements in an Adult White Population in the US

Age Group	Potassium (g)* TBK	Water (L)† TBW	Nitrogen(g)‡ TBN
		Females	
20-24	85.8 ± 12.9§	30.5 ± 2.9	1525.8 ± 174.7 (7)‖
25-29	91.2 ± 12.8	32.4 ± 2.3	1424.5 ± 155.1 (5)
30-34	85.6 ± 14.9	32.0 ± 2.6	1429.4 ± 181.7 (7)
35-39	87.3 ± 12.8	33.3 ± 1.4	1556.3 ± 23.3 (3)
40-44	85.0 ± 12.2	26.6 ± .8	1308.4 ± 127.9 (2)
45-49	87.7 ± 9.8	32.2 ± 2.8	1430.7 ± 149.6 (9)
50-54	86.5 ± 11.4	31.2 ± 2.3	1326.6 ± 62.3 (5)
55-59	83.2 ± 8.8	31.6 ± 1.1	1340.9 ± 126.2 (4)
60-64	82.1 ± 10.9	28.6 ± 2.6	1372.2 ± 125.9 (4)
65-69	78.1 ± 9.8	27.7 ± 3.9	1225.3 ± 150.5 (9)
70-74	74.5 ± 7.8	27.7 ± 2.3	1192.3 ± 104.0 (9)
		Males	
20-24	157.2 ± 18.4	46.7 ± 5.6	2153.2 ± 207.0 (17)
25-29	141.2 ± 17.2	41.9 ± 6.3	1899.2 ± 258.3 (6)
30-34	142.9 ± 17.9	43.0 ± 5.0	1925.6 ± 230.0 (9)
35-39	142.1 ± 19.8	40.7 ± 4.6	1846.6 ± 98.2 (7)
40-44	145.8 ± 20.2	42.9 ± 4.1	1899.0 ± 101.7 (4)
45-49	135.6 ± 12.0	43.4 ± 3.3	1890.8 ± 240.7 (8)
50-54	142.4 ± 20.4	44.9 ± 7.8	1898.0 ± 320.9 (8)
55-59	131.7 ± 17.0	42.5 ± 4.1	1861.1 ± 197.2 (6)
60-64	135.5 ± 10.7	43.6 ± 2.9	1972.0 ± 180.0 (6)
65-69	121.4 ± 11.6	40.4 ± 5.0	1721.1 ± 120.0 (5)
70-74	115.4 ± 15.7	39.7 ± 3.1	1832.0 ± 144.2 (9)
75-90	95.6 ± 13.5	37.1 ± 1.2	1504.7 ± 30.8 (2)

*Whole body ^{40}K counting (natural radioactivity, 0.0 mSv).
†Dilution of tritiated water (total body dose <0.03 mSv).
‡In vivo prompt-gamma neutron activation (total body dose <0.30 mSv).
§Mean ±SD.
‖Subset from the total population in which both total body nitrogen and water measurements were performed (87 males, 64 females).

the same 50-y age span: 42 g (26%) for TBK, 7 L (15%) for TBW, 321 g (15%) for TBN.

Body Composition of the Adult Black Population

The mean values for the 35 black females and 30 black males are summarized in Table 4. Only the means and SD for the total groups are presented, because there were too few subjects for adequate representation of each 5-yr age group, as used to describe the white population. The age-specific trends observed in the black female and male groups, however, were similar to those observed in the white population, i.e., decreased absolute levels of TBK, TBCa, TBP, and TBN in the older subjects. The TBNa in black women also appeared to decrease slightly with age, whereas TBCl remained relatively constant. Both TBNa and TBCl tended to increase slightly with age in the black males.

Comparison of Body Composition of Black and White Groups

Although both the white and black subjects in the present study were of similar mean stature, the black males and females had higher mean body weights. The mean values for TBK, TBN, TBCa, TBP, TBNa, and TBCl in the black groups were also higher than those for the white groups. Body potassium levels were approximately 10% higher for both sexes, whereas body calcium was increased 7–11%. TBN was increased 17% in black women, but not in black males. Body sodium was increased by 6% in both sexes, whereas the increase in body chlorine was different between the sexes (13% in females, 3% in males).

Four-Compartment Model of Body Composition

Various models of body composition have been proposed. The simplest model consists of a body fat compartment and a nonfat or lean body mass (LBM) compartment. LBM contains the body cell mass (BCM) that is the more metabolically active component of the lean tissues. The model used in ICRP-23 to describe the composition of Reference Man is based on four compartments: protein, water, ash, fat, and carbohydrates (*1*). Total body protein can be obtained from TBN (protein $= 6.25 \times$ nitrogen). Total body water is directly measured by the tritiated water technique. Extracellular water (ECW) has been shown to be directly proportional to TBCl (*11*). Intracellular water (ICW) is the difference between total body water and the extracellular component. The ash compartment is dominated by bone mineral; total body calcium is approximately 34% of total ash (*1*). The relatively invariant size of the carbohydrate compartment averages approximately 0.5 kg in the adult.

The results for the four-compartment model are given as a function of age and sex for the adult white population (Table 5). Body cell mass

Table 4
Subject Characteristics and Absolute Levels of Elemental Composition in Black and White Adults in the US

	Females		Males	
	White	Black	White	Black
No. of Subjects	1134	35	175	30
Mean age (yr)*	53.5 ± 10.5	49.3 ± 18.2	47.2 ± 16.9	48.2 ± 13.1
Weight (kg)	65.9 ± 11.2	72.1 ± 10.7[a]	77.6 ± 11.8	85.5 ± 15.7[a]
Height (m)	1.62 ± .06	1.60 ± .05	1.75 ± .07	1.75 ± .06
Potassium (g)	84.0 ± 11.3	92.0 ± 12.7[c]	137.0 ± 21.5	150.5 ± 23.9[b]
Nitrogen† (g)	1362 ± 174	1603 ± 135[c]	1921 ± 237	1940 ± 248
Calcium (g)	838 ± 115	899 ± 140[b]	1137 ± 142	1265 ± 158[c]
Phosphorus (g)	397 ± 64	402 ± 57	501 ± 75	527 ± 71
Sodium (g)	63.4 ± 6.7	66.8 ± 5.7[c]	83.9 ± 8.8	9.1 ± 7.4[c]
Chlorine (g)	61.6 ± 7.9	69.5 ± 7.4[c]	77.3 ± 9.4	79.6 ± 8.8

*Age range = 20–75 yr.

†Prompt-gamma nitrogen measurements were obtained for a subset of 72 females (64 white, 8 black) and 94 males (87 white, 7 black). Results of *t*-test for black vs white ([a]$p < 0.05$, [b]$p < 0.01$, [c]$p < 0.001$).

(BCM) is defined on the basis of the equation of Moore et al. (*12*), BCM = k × TBK. ECW is defined as 0.9 × (TBCl(mEq)/plasma Cl). Fat is defined as the difference between body weight and lean body mass, which is the sum of protein, water, and bone ash. The absolute amount of fat is similar between the sexes. When expressed as a percent of body weight, however, a higher proportion of fat is found in females. The variability of body fat within age groups is large; although, not surprisingly, body fat increases with age in both sexes. Women appear to reach a maximum body fat content during the 6th decade of life, after which a decline occurs. Body fat content in males, however, appears to increase gradually throughout adult life.

The composition of the lean tissues also changes during the process of aging. There is a monotonic decrease of total body protein in both sexes that averages 32–36 g/yr. Body cell mass, however, is relatively constant for females younger than 59 yrs of age; a decrease is evident only in older age groups. The male BCM pattern shows the highest value in the youngest age group, followed by a continual decline with advancing age. By age 70, for example, males have a mean body cell mass (major component is muscle mass) that is 14% lower than the 20- to 29-yr-old age group; a decline approximately half this size (*8*) occurs in females.

Table 5
Mean Estimates of Body Composition in an Adult US White Population[a]

Age Group	Protein (kg)	BCM (kg)	ECW (l)	ICW (l)	Ash (kg)	Fat (kg)
			Females			
20–29	9.27 ± 1.04*	21.0 ± 3.0	15.5 ± 1.9	15.4 ± 3.0	2.66 ± .30	15.9 ± 3.4
30–39	9.19 ± .98	20.6 ± 3.2	15.3 ± 2.0	15.7 ± 2.5	2.63 ± .35	21.7 ± 6.4
40–49	8.80 ± .92	20.6 ± 2.5	15.6 ± 1.9	15.6 ± 2.4	2.59 ± .28	21.7 ± 7.5
50–59	8.33 ± .57	20.2 ± 2.4	15.7 ± 1.9	15.5 ± 1.6	2.47 ± .32	30.6 ± 4.8
60–69	7.94 ± .96	19.2 ± 2.5	15.3 ± 2.0	13.2 ± 2.7	2.32 ± .29	24.0 ± 6.6
70–79	7.45 ± .65	17.7 ± 1.8	15.1 ± 2.0	12.6 ± 1.9	2.15 ± .30	21.6 ± 3.8
			Males			
20–29	13.07 ± 1.50	36.2 ± 4.5	20.1 ± 2.0	24.9 ± 4.8	3.55 ± .30	19.8 ± 13.0
30–39	11.80 ± 1.12	33.9 ± 4.4	19.4 ± 2.4	23.5 ± 3.9	3.42 ± .42	16.1 ± 8.8
40–49	11.83 ± 1.21	33.5 ± 4.1	19.1 ± 2.2	24.7 ± 3.2	3.38 ± .39	19.8 ± 8.3
50–59	11.76 ± 1.66	32.6 ± 4.5	19.6 ± 2.9	24.5 ± 4.5	3.30 ± .38	21.0 ± 8.4
60–69	11.61 ± 1.23	31.2 ± 2.9	20.0 ± 2.4	22.0 ± 3.4	3.28 ± .46	22.5 ± 6.6
70–79	11.45 ± .90	27.5 ± 3.7	20.0 ± 1.5	20.3 ± 3.0	3.22 ± .38	25.6 ± 10.6
80–89	9.40 ± .20	22.7 ± 3.2	19.8 ± 2.4	16.8 ± 2.0	2.97 ± .52	27.2 ± 2.5

*Mean ± SD.

[a]Protein = 6.25 × TBN (kg); BCM = Body cell mass = 0.00932 × TBK (mEq); ECW = Extracellular water = 0.9 × (TBCl (mEq)/plasma Cl); ICW = Intracellular water = TBW − ICW; Ash = TBCa (kg)/0.34; and Fat = Weight − LBM, where LBM = Protein + TBW + Ash.

Extracellular water is relatively constant in both sexes, whereas intracellular water declines with advancing age. The decline in ash content for both sexes reflects the changes already noted for total body calcium.

DISCUSSION

A major focus of this study has been to provide elemental and body compositional data that describe the general population of the US. Age-, race-, and sex-specific differences in elemental composition are evident in the data presented in Tables 2–5. The variation in the elemental composition of both males and females in any age group, however, is large and ranges up to 20% (SD). In a cross-sectional study such as this, the variability can reasonably be attributed to differences in body size and habitus. Subjects whose weights and heights deviate from the mean also could be expected to have elemental data that are different from the age-specific mean, as well. For example, one would not expect a smaller healthy male (165 cm, 65 kg) to have the same skeletal mass (TBCa) as a taller larger male (187 cm, 90 kg) of the same age. The mean values based on sex and age alone are inadequate as references for the evaluation of an individual's status, particularly if body composition measurements are to be of clinical value. Body weight and body surface area have been used as reference parameters in an effort to normalize differences in body size. Height, however, is an important index of size and should be used in any formulation that attempts to describe body compositional differences between groups or within the individual.

Using the database for the white population, a set of equations was developed that related variations in the elemental composition of the body to differences in age, weight, and height for each sex (Table 6). The model allowed for noninteger exponents of weight, height, and sex differences. Each element was first fitted to a function of height and weight, and then adjusted, using a linear relationship, for changes resulting from age. In all cases, height was the dominant anthropometric parameter. Although similar equations have been reported previously (*13–16*), they were derived for different populations and may not be adequate to describe the interrelationships between body elements in an individual. The present set of equations for TBK, TBCa, TBNa, TBCl, TBN, TBW, and body fat were derived from a single group and provide estimates of the elemental composition in the individual relative to age, weight, and height. The absolute value in the total body can then be expressed as a ratio to the calculated or predicted value using the age, weight, and height equations. If an individual's elemental data are exactly the same as predicted, then the ratio is 1.00. When this approach is used to examine the variation seen in each age group, the standard deviation is generally reduced approximately by half.

A second objective of this study was to compare the data now available from in vivo neutron activation studies in the general US adult population with the elemental and body composition values compiled for Reference Man. Although the ICRP-23 states that Reference Man was not intended to be predictive for any particular individual or class of individuals, it has clearly been used for this purpose in recent years as a result of the increased clinical interest in body composition assessment (*2–4*). In Table 7, two sets of values are given for comparison with Reference Man, one for the age-specific "average" person and the other derived on the basis of the prediction equations using the same age, weight, and height as Reference Man. It is evident that the average 20- to 30-yr-old male in the US is larger than Reference Man, as judged by weight and height alone. He is approximately 7.5 kg (10%) heavier and 8 cm (5%) taller. All of the additional body weight is not the result of an increase in body fat. The mean or "average" values for 20- to 30-yr-old males indicated a

Table 6
Prediction Equations using Age, Weight, and Height for the Adult White Population in the US[a]

Body Element	Sex	n	a	b	c	d	r*	SEE*
TBK (g)	F	1134	69.7	0.46	-0.24	-46.8	.74	7.5
	M	175	102.2	0.68	-0.62	-65.5	.84	11.7
TBN (g)	F	61	1481.4	-†	-3.60	-853.5	.75	116
	M	88	834.7	7.60	-5.63	105.3	.72	168
TBCa (g)	F	1133	861.1	1.20	-3.18	-469.0	.62	90
	M	175	1165.2	1.63	-†	-1025.4	.67	106
TBNa (g)	F	1133	55.5	0.17	-.048	-35.5	.67	4.9
	M	175	56.5	0.29	0.10	-42.5	.73	6.1
TBCl (g)	F	1130	44.0	0.19	-†	-22.9	.52	6.7
	M	175	24.8	0.36	-†	6.0	.58	7.6
TBW (l)	F	64	27.3	0.17	-0.045	-21.9	.84	1.9
	M	87	28.4	0.25	-0.092	-21.9	.86	2.8

Nonlinear equations for estimation of total body elemental composition‡

Potassium (M) (F):
$$K_p = \begin{Bmatrix} 12.33 - 0.043 \times age \\ 11.83 - 0.028 \times age \end{Bmatrix} Wt^{.325} \; Ht^{1.52 + .57 * sex}$$

Calcium (M) (F):
$$Ca_p = \begin{Bmatrix} 251.4 - 0.18 \times age \\ 291.2 - 0.90 \times age \end{Bmatrix} Wt^{.077 + .036 * sex} \; Ht^{1.89}$$

Sodium (M) (F):
$$Na_p = \begin{Bmatrix} 13.57 - 0.014 \times age \\ 14.86 - .011 \times age \end{Bmatrix} Wt^{.185 + .10 * sex} \; Ht^{1.48 - .52 * sex}$$

Chlorine (M) (F):
$$Cl_p = \begin{Bmatrix} 13.80 \\ 13.80 \end{Bmatrix} Wt^{.21 + .11 \times sex} \; Ht^{1.20 - .66 * sex}$$

Nitrogen (M) (F):
$$N_p = \begin{Bmatrix} 480.9 - 1.09 \times age \\ 468.2 - 0.85 \times age \end{Bmatrix} Wt^{.167 * sex} \; Ht^{2.41 - 1.05 * sex}$$

TBW (M) (F):
$$TBW_p = \begin{Bmatrix} 3.41 - .005 \times age \end{Bmatrix} Wt^{.35 + .03 * sex} \; Ht^{1.63}$$

Fat (M) (F):
$$Fat = \begin{Bmatrix} 0.90 + .007 \times age \\ 0.99 + .005 \times age \end{Bmatrix} Wt^{2.85 - .08 * sex} \; Ht^{-4.31}$$

[a] $Y = a \times Ht + b \times Wt + c \times Age + d$.
*r = multiple r value; SEE = standard error of the estimate.
†Stepwise multiple regression analysis indicated no significant relationship to age at the 0.05 level.
‡An age-dependent coefficient is given for each sex. The exponent value for weight and height are obtained using sex = 0 for females and sex = 1 for males.

Table 7
Comparison of ICRP-23 Reference Values with IVNAA Data for the Adult US White Population

	ICRP-23 Reference Male	ICRP-23 Reference Female	IVNAA Average Male*	IVNAA Average Female*
Age	20–30	20–30	23.9	24.4
Wt (kg)	70	58	77.4	59.1
Ht (cm)	170	160	177.7	163.3
Body Composition Values				
TBW (L)	42	29	45.5	31.3
ICW (L)	23.8	17.4	24.9	15.5
ECW (L)	18.2	11.6	20.2	15.6
LBM (kg)	57.4	39.6	64.4	44.0
Fat (kg)	13.5	16	19.9	15.9
BCM (kg)	41.2	26.9	36.3	21.0
Total Body Elements				
Ca (g)	1000 (1079)†	-(896)†	1209	906
P (g)	580	-	549	435
Na (g)	100 (75)	-(62)	83	63
Cl (g)	95 (73)	-(58)	80	61
K (g)	140 (138)	-(85)	152	88
N (g)	1800 (1907)	-(1407)	2093	1484

*Values are for the age range 20–29 yr.

†The values in parentheses are those predicted for an individual with the same age, weight, and height as reference man and woman. The prediction equations are given in Table 6. Elemental data for reference woman are not available in ICRP-23.

larger skeletal mass (TBCa), more lean tissues (TBK, TBN), and more body water (TBW). Body levels of sodium and chlorine (TBNa, TBC), however, are approximately 25% lower than the reference values; these differences, however, have been noted previously (*15*). The predicted absolute elemental values for a male or female in the US population with the same weights and heights as the Reference Man and Woman are in good agreement (3–8% differences in absolute amounts) with those given in the ICRP-23.

As an increasing number of women enter the radiation workforce, the ICRP has begun the development of data for a Reference Woman. The development of these data will also enable the expansion and application of the reference concept to various studies of body composition. At present, no elemental data are tabulated in the ICRP-23 for Reference Woman. Thus, it is difficult, if not impossible, to compare the in vivo data presented in this paper with reference values. In fact, the data presented in this paper may serve as a comprehensive reference for the US adult white female population. Some estimates of body composition, however, are given in the ICRP-23 and have been included in Table 7. The "average" female is only slightly taller and heavier than reference woman; the differences are not significant. Using the ICRP values of 58 kg for body weight, 160 cm for height, and a mean age of 25 yrs for Reference Woman, the associated elemental values have been calculated for TBCa, TBK, TBNa, TBCl, and TBN.

SUMMARY

Body composition analysis, although not a new concept, can be investigated more fully because of the considerable advantages offered by in vivo techniques. Unique human data are obtained by the TBNAA, PGNAA, and WBC techniques, which continue to prove that they are valuable in the evaluation of clinical diagnostic procedures, normal physiological aging, and the efficacy of therapeutic regimens. Investigation of the effects of dietary intake and nutritional support during growth, aging, or malnutrition can be examined. The in vivo data presented for the US population and the prediction equations provide a comprehensive reference of the elemental composition of the adult human. This reference information also provides a basis for comparisons with data developed for other countries or different ethnic groups. The present data confirm the ICRP task group's recognition that Reference Man is not the "average" or "median" adult male. What is equally important is that the values for Reference Man agreed with the in vivo data when adjusted for age, weight, and height. In addition, the in vivo data for the US female population can be used to define the body composition of the Reference Woman more accurately.

ACKNOWLEDGMENTS

In vivo measurements were performed between 1972 and 1987 at the Brookhaven National Laboratory. Some of the volunteers in this study also served as controls for the following investigators: J. Aloia, S. H. Cohn, A. Vaswani, S. Yasumura, and I. Zanzi. Editorial assistance was provided by E. R. Klein and J. D. Eastman. This work is a publication of the USDA/ARS Children's Nutrition Research Center, Department of Pediatrics, Baylor College of Medicine and Texas Children's Hospital, Houston, TX. This project has been funded, in part, by the USDA/ARS under Cooperative Agreement No. 58-7MN1-6-100 and by the USDOE under Contract No. DE-AC02-76CH00016. The contents of this publication do not necessarily reflect the views or policies of the USDA or USDOE, nor does mention of trade names, commercial products, or organizations imply endorsement by the US Government.

REFERENCES

1. W. S. Snyder, M. J. Cook, E. S. Nasset, L. R. Karhausen, G. P. Howells, and T. H. Tipton (eds.), *Report of the Task Group on Reference Man (ICRP 23)*, Pergamon, New York, 1984.
2. K. J. Ellis, S. Yasumura, and W. D. Morgan (eds.), *In Vivo Body Composition Studies*, Institute of Physical Sciences in Medicine, London, 1987.
3. A. F. Roche (ed.), *Body Composition Assessments in Youth and Adults*, Sixth Ross Conference on Medical Research, Ross Laboratories, Columbus, OH, 1985.
4. W. J. Klish, and N. Kretchmer (eds.), *Body Composition Measurements in Infants and Children*, Ninety-eighth Ross Conference on Pediatric Research, Ross Laboratories, Columbus, OH, 1989.
5. K. J. Ellis and S. H. Cohn, *Proceedings of the Fifth International Conference on Nuclear Methods in Environmental and Energy Research*, DOE Conf-840408, J. R. Vogt, ed., 1984, pp. 729–741.
6. S. H. Cohn and C. S. Dombrowski, *J. Nucl. Med.* **11,** 239 (1970).
7. S. H. Cohn, K. K. Skukla, and R. G. Fairchild, *J. Nucl. Med.* **13,** 487 (1972).
8. S. H. Cohn, C. S. Dombrowski, and R. G. Fairchild, *Int. J. Appl. Radiat. Isot.* **21,** 127 (1970).
9. D. Vartsky, K. J. Ellis, and S. H. Cohn, *J. Nucl. Med.* **11,** 239 (1970).
10. D. Vartsky, K. J. Ellis, A. N. Vaswani, S. Yasumura, and S. H. Cohn, *Phys. Med. Biol.* **29,** 209 (1984).
11. S. Yasumura, S. H. Cohn, and K. J. Ellis, *Am. J. Physiol.* **244,** R36 (1983).
12. F. D. Moore, K. H. Olesen, J. D. McMurrey, H. V. Parker, M. R. Ball, and C. M. Boyden, *The Body Cell Mass and Its Supporting Environment: Body Composition in Health and Disease*, Saunders, Philadelphia, PA, 1963, p. 21.
13. S. H. Cohn, K. K. Shukla, and K. J. Ellis, *J. Nucl. Med. Biol.* **1,** 131 (1974).
14. K. J. Ellis, K. K. Shukla, S. H. Cohn, and R. N. Pierson, *J. Lab. Clin. Med.* **83,** 716 (1974).
15. K. J. Ellis, A. Vaswani, I. Zanzi, and S. H. Cohn. *Metabolism* **25,** 645 (1976).
16. K. J. Ellis, S. Yasumura, D. Vartsky, A. Vaswani, and S. H. Cohn. *J. Lab. Clin. Med.* **99,** 917 (1982).

In Vivo Neutron Activation Analysis of Organ Cadmium Burdens

Referent Levels in Liver and Kidney and the Impact of Smoking

D. M. FRANKLIN,[1] C. J. G. GUTHRIE,[1,2] D. R. CHETTLE,*[,1] M. C. SCOTT,[1] H. J. MASON,[3] A. G. DAVISON,[4] AND A. J. NEWMAN TAYLOR[4]

[1]School of Physics and Space Research, University of Birmingham, Birmingham B15 2TT, England; [2]Present address: Communicable Diseases (Scotland) Unit, Ruchill Hospital, Glasgow, G20 9NB, Scotland; [3]Health and Safety Executive, Occupational Hygiene and Medicine Laboratories, 403 Edgware Rd., London, NW12 6LN, England; and [4]Cardiothoracic Institute Brompton Hospital, London, SW3 6HP, England

Received April 17, 1989; Accepted November 3, 1989

ABSTRACT

In vivo neutron activation measurements of liver and kidney cadmium have been made in 77 exposed workers and 101 referents. Cadmium levels were higher in exposed workers than in referents; both in liver, 25.7 cf. 0.6 μg/g, and kidney, 17.9 cf. 2.7 mg. The 19 referents who never smoked had lower mean organ cadmium burdens than the other referents, the difference achieving statistical significance in the kidney, $p < .01$. Cigaret smoking was estimated to increase cadmium body burden by 370 ± 140 μg/pack year. These referent cadmium levels are similar to, although slightly below, previous in vivo and autopsy data.

*Author to whom all correspondence and reprint requests should be addressed.

Biological Trace Element Research Editor: G. N. Schrauzer

Index Entries: in vivo neutron activation; liver cadmium; kidney cadmium; industrial cadmium exposure; cadmium and smoking; referent cadmium levels.

INTRODUCTION

Excessive exposure to cadmium can be harmful, so when it is used industrially, members of the exposed work force are routinely monitored to ascertain levels of cadmium absorbed into the body and detect early signs of any harm it may cause. In 1975, McLellan et al. (*1*) reported the use of in vivo neutron activation analysis to measure organ cadmium levels noninvasively. Such measurements complement those made in blood and urine samples, because they look at stored rather than circulating cadmium.

In a collaborative survey, a group of cadmium workers, together with age and socioeconomically-matched referents, were studied. The impact of cadmium on the workers' lung (*2*) and kidney functions (*3*) has already been reported. As a part of the survey, in vivo neutron activation measurements were made of cadmium in liver and left kidney. The contrast between organ cadmium levels in exposed and referent groups is presented, as is the impact of smoking on stored cadmium levels within the referent group.

SUBJECTS AND METHODS

The study population comprised 77 men who had been exposed to cadmium fumes and dust at a factory producing copper/cadmium alloys. Of these, 11 were exposed at the time of study, another 32 were still employed but no longer exposed to cadmium, and the remaining 34 had ceased employment. The referent group comprised 101 men because, for each of those cadmium workers who had ceased employment before the age of 65, two referents were sought, one who had also ceased working before the age of 65 and another who had not retired early.

Among the other measurements made, each subject had the cadmium content of their liver and left kidney measured by neutron activation analysis. Transportable instruments based on ^{238}Pu-Be neutron sources were used since the survey was conducted on-site at the factory. Neutrons are thermalized in the subject's body and then can be captured by ^{113}Cd. The most prominent of the resulting prompt γ-rays, at 559 keV, is then detected in germanium spectrometers, and the cadmium content is assayed by comparison with appropriate standard phantoms (*4,5*).

The duration of a person's exposure to cadmium while in employment was assessed by reference to the factory records. Also made available were the results of cadmium-in-air sample measurements over a 30-yr period. Smoking history was determined by a questionnaire. Retinol-binding protein and creatinine were among the parameters measured in urine samples; these were used to detect signs of renal dysfunction.

RESULTS

The mean liver cadmium concentration in the 77 exposed workers was 25.7 μg/g, whereas in the 101 referents, it ws 0.6 μg/g. The mean cadmium content in left kidney was 17.9 mg in exposed and 2.7 mg in referent workers. The mean duration of exposure in the cadmium workers was 8.2 yr.

Of the cadmium-exposed workers, 33 were classified as exhibiting renal dysfunction in that their urinary excretion of retinol-binding protein, normalized to creatine, exceeded the upper 95th percentile of this parameter in the referent group, which, in this case, was 35 μg/mmol creatinine. Among the 44 people without raised urinary levels of retinol-binding protein, there was a strong correlation between liver cadmium and kidney cadmium ($r = 0.783$ $t = 8.171$, $p < .00001$). An estimate of the true underlying relationship between these parameters was derived using Quenouille's method (*6*), resulting in the equation

$$\text{kidney} = 1.19 \text{ liver} - 0.13$$

This relationship weakened when people with raised retinol-binding protein excretion were added to the data set ($r = 0.482$, $t = 4.769$, $p < .00005$).

Liver cadmium correlated with the duration of exposure in people with normal kidney function ($r = 0.487$, $t = 3.612$, $p < .001$), those exhibiting dysfunction ($r = 0.444$, $t = 2.758$, $p \approx .01$), and in the whole group ($r = 0.582$, $t = 6.203$, $p < .00001$). The mean differences in liver cadmium between exposed and referent groups corresponded to annual increases of 2.5 μg/g and 3.3 μ/g in the subgroups with normal kidney function and dysfunction, respectively, and to 3.0 μg/g/yr exposure for the whole group. Kidney cadmium also correlated with the duration of exposure in people with normal kidney function ($r = 0.594$, $t = 4.791$, $p < .0001$), but only weakly in those with renal dysfunction ($r = 0.292$, $t = 1.700$, $p \approx .1$). The relationship was again more significant in the combined data set ($r = 0.458$, $t = 3.737$, $p < .0005$). Increases in kidney cadmium were 2.5 mg/yr and 1.5 mg/yr for the subgroups with normal renal function and dysfunction, respectively, and 1.8 mg/yr for the whole group. In the referent group of 101 subjects, kidney cadmium correlated weakly with liver cadmium ($r = 0.307$, $t = 3.214$ $p \approx .002$), and the underlying relationship was:

$$\text{kidney} = 0.73 \text{ liver} + 2.31$$

Nineteen of the referent group had never smoked; the remaining 82 had a mean smoking index of 26.3 pack years, where the unit pack years is the product of packs of 20 cigarets/d smoked and the duration of smoking in years. The mean cadmium levels in nonsmokers were 0.0 μg/g in liver and 0.6 mg in kidney; those in smokers were 0.7 μg/g in liver and 3.2 mg in kidney. The difference in mean liver cadmium concentrations was not significant (t = 0.765 $p > .4$); however, in kidney cadmium levels, it was significant (t = 3.104 $p < .005$).

The whole-body burden of cadmium can be estimated from the organ burdens. Nordberg et al. (*7*) report liver cadmium to comprise 14–18% of the total and in the two kidneys to comprise 50–56% of the total. If the right kidney is assumed to contain as much cadmium as the left, and the average liver mass for adult males is taken to be 1.8 kg (*8*), then body burden is given by

$$(1.8 \times \text{liver concentration} + 2 \times \text{kidney burden})/(0.16 + 0.53)$$

On this basis, the mean cadmium body burden of the nonsmoking referents was 1.65 mg, whereas that of the smokers was 11.34 mg. The difference corresponds to an increase in body burden of 368 ± 140 μg/pack year. For comparison, the mean increase in body burden in the cadmium-exposed workers corresponded to 13.4 mg/yr. However, cadmium-in-air levels had decreased markedly during the working lives of these subjects, so liver and kidney levels were correlated with a cadmium exposure index derived from cadmium-in-air measurements and factory records (*2,3*). From these relationships, the increase in body burden, corresponding to mean cadmium-in-air levels of 25 μg/m^3, was found to be 3.0 mg/yr.

DISCUSSION

These data from referent subjects can be compared with previously published results. The effect of smoking on organ cadmium accumulation was studied by Ellis et al. (*9*), who also used neutron activation analysis. They found mean increases of 1.8 μg/g in liver, and 2.7 mg in kidney of their smokers compared to their nonsmokers, and the mean smoking index was 38.7 pack years. Using Ellis' data to compute body burden according to Norberg's data, as above, leads to an estimated increase of body burden of 12.5 mg or 324 μg/pack year. These two estimates of the changes in cadmium body burden are clearly consistent.

More generally, however, some data sets covering organ cadmium levels in population groups not occupationally exposed are summarized in Table 1. The ICRP data (*8*) and those of Iyengar et al. (*10*) are based on in vitro measurements, as are those from the recent extensive survey by Scott et al. (*11*). The kidney concentrations quoted by Iyengar (*10*) and Scott (*11*) have been converted to masses by assuming a single kidney mass of 155 g, of which 70% is cortex (*8*) and the rest medulla. The ICRP liver data have been converted from cadmium mass to concentration by

Table 1
Organ Cadmium Levels

Source of data	population group	mean kidney burden (mg)	n	mean liver concentration ($\mu g\ g^{-1}$)	n
Snyder et al, 1975	referent	5.0	145	2.2	150
Iyengar et al, 1978	referent	3.3	98	3.1	96
Ellis et al, 1979	referent, non-smokers	3.1	8	2.3	8
	referent, smokers	5.8	12	4.1	12
Cummins et al 1980	referent, non-smokers	3.9	13		
	referent, smokers	4.2	15		
	hypertensive, non-smokers	2.4	15		
	hypertensive, smokers	4.0	15		
Scott et al 1987†	referent, non-smokers	1.8	467		
	referent, light-smokers	2.2	159		
	referent, heavy-smokers	2.6	175		
present study	referent, non-smokers	0.6± 0.8	19	0.0± 0.9	19
	referent, smokers	3.2± 0.4	83	0.7± 0.5	83

†Smoking levels at time of death, rather than pack years.

assuming a liver mass of 1.8 kg (*8*). The data of Cummins et al. (*12*) were obtained in in vivo neutron activation analysis. The ICRP data (*8*) and those of Iyengar et al. (*10*) were not subdivided according to smoking history; those of Scott et al. (*12*) were broadly categorized regarding smoking habits shortly before their death, rather than pack years.

All three sets of in vivo data are in moderately close agreement with the in vitro studies of autopsy material. The data from the present study appear somewhat low, but the kidney levels are very comparable with those observed by Scott, especially when it is considered that they were unable to take full smoking histories that will have tended to reduce the differences among their subgroups.

In conclusion, the agreement between the present data set and that of Ellis regarding the quantitative effect of smoking is reassuring, as is the general agreement between data derived from in vivo neutron activation analysis and those obtained in vitro at autopsy. These observations tend to confirm the reliability of the in vivo neutron activation analysis technique. It is also clear that the effect of smoking on cadmium body burden, at 0.3–0.4 mg/pack year, is minor compared to the industrial exposure considered here, which historically was equivalent to 13.4 mg/yr, which would amount to 3.0 mg/yr at a cadmium-in-air level of 25 $\mu g/m^3$.

ACKNOWLEDGMENTS

We thank H. Holden and D. O'Malley for their help during this work. Most of the funding for this work was provided by the Department of Health and Social Security and the Health Safety Executive; a travel grant from The Wellcome Trust made it possible for one of us (D. R. Chettle) to present this paper; this financial support is gratefully acknowledged.

REFERENCES

1. J. S. McLellan, B. J. Thomas, J. H. Fremlin, and T. C. Harvey, *Phys. Med. Biol.* **20,** 88 (1975).
2. A. G. Davison, P. M. Fayers, A. J. Newman Taylor, K. M. Venables, J. Darbyshire, C. A. C. Pickering, D. R. Chettle, D. M. Franklin, C. J. G. Guthrie, M. C. Scott, D. O'Malley, H. Holden, H. J. Mason, A. L. Wright, and D. Gompertz, *Lancet* **i,** 663 (1988).
3. H. J. Mason, A. G. Davison, A. L. Wright, C. J. G. Guthrie, P. M. Fayers, K. M. Venables, N. J. Smith, D. R. Chettle, D. M. Franklin, M. C. Scott, H. Holden, D. Gompertz, and A. J. Newman Taylor, *Br. J. Ind. Med.* **45,** 793, (1988).
4. M. C. Scott, D. R. Chettle, S. A. Coward, M. J. Faddy, and J. G. Fletcher, *Trans. Am. Nucl. Soc.* **44,** 33 (1983).
5. D. M. Franklin, *Ph.D. thesis,* Department of Physics, University of Birmingham, England, 1986.
6. M. H. Quenouille, *Associated Measurements,* Butterworths, London, 1952, p. 71.
7. G. F. Nordberg, T. Kjellström, and M. Nordberg, *Cadmium and Health,* vol. 1, L. Friberg, C. G. Elinder, T. Kjellström, and G. F. Nordberg, eds., CRC Press, Boca Raton, FL, 1985, pp. 103–178.
8. W. S. Snyder, M. J. Cook, L. R. Karhausen, E. S. Nasset, G. Parry Howells, and I. H. Tipton, *Report of the Task Group on Reference Man,* ICRP23, 1975.
9. K. J. Ellis, D. Vartsky, I. Zanzi, S. H. Cohn, and S. Yasamura, *Science* **205,** 323 (1979).
10. G. V. Iyengar, W. E. Kollmer, and H. J. M. Bowen, *The Elemental Composition of Human Tissues and Body Fluids,* Verlag Chemie, Weinheim, FRG, 1978.
11. R. Scott, E. Aughey, G. S. Fell, and M. J. Quinn, *Human Toxicol.* **6,** 111 (1987).
12. P. E. Cummins, J. Dutton, C. J. Evans, W. D. Morgan, A. Sivyer, and P. C. Elwood, *Euro. J. Clin. Invest.* **10,** 459 (1980).

In Vivo Measurements of Cadmium and Lead in Occupationally-Exposed Workers and an Urban Population

W. D. MORGAN,*[,1] S. J. S. RYDE,[1] SARAH J. JONES,[1] RUTH M. WYATT,[1] I. R. HAINSWORTH,[2] STEPHANIE S. COBBOLD,[3] C. J. EVANS,[4] AND R. A. BRAITHWAITE[5]

Departments of [1]Medical Physics, [2]Chemical Pathology, and [3]Medicine, Singleton Hospital, Swansea SA2 8QA, Wales, UK; [4]Department of Physics, University College of Swansea, Swansea SA2 8PP, Wales, UK; and [5]Regional Laboratory for Toxicology, Dudley Road Hospital, Birmingham B18 7QH, UK

Received April 17, 1989; Accepted November 3, 1989

ABSTRACT

This paper reports the preliminary findings of a survey of lead and cadmium body burdens in a nonoccupationally exposed population in Swansea, Wales, using the techniques of in vivo neutron activation and X-ray fluorescence analysis. Some measurements on an occupationally cadmium-exposed group are also included. The results confirm the association between cadmium and smoking and bone lead and age. The in vivo measurements demonstrate a degree of comparability with other data, which supports the further detailed analysis of the relationships between body burden and exposure, on the one hand, and possible health effects on the other.

Index Entries: In vivo neutron activation analysis; liver Cd; kidney Cd; Pb and Cd, body burden of; in vivo X-ray fluorescence analysis; tibia Pb; blood Pb and Cd; environmental and occupational exposure; smoking.

*Author to whom all correspondence and reprint requests should be addressed.

INTRODUCTION

Human exposure to heavy metals may induce a variety of health effects, such as kidney dysfunction, hypertension, impairment of the central nervous system, and anemia. Both lead and cadmium are dispersed in the general environment, from where the metals may be inhaled or ingested by humans. Some population groups may be more exposed than others, e.g., workers in metal industries or those whose environment may have been particularly contaminated as a result of the mining or smelting of metalliferous ores.

Biological monitoring of persons who may be considered to be at risk involves blood and urine sampling for metals and urinary proteins and enzymes. These procedures adequately monitor current exposure to the metals, but since both lead and cadmium are retained in the body organs and tissues and released only very slowly over many years, other methods are required for assessing the body burden of either metal so that the risk to a particular organ (e.g., Cd in the kidney), or systemic harm if a large deposit is subsequently mobilized (e.g., Pb release from bone), can be more accurately determined.

Two techniques are suitable for noninvasive measurement of body composition. In vivo neutron activation analysis (NAA) has been used for many years to measure the cadmium content of the liver and kidneys, which together account for more than half of the body burden of Cd; whereas in vivo X-ray fluorescence analysis (XRF) is ideally suited to the measurement of lead in superficial bony sites, such as the tibia and finger. The role of these measurement techniques in the field of occupational medicine has recently been reviewed (*1*), and their wider application to studies of environmental exposure to heavy metals has also been considered (*2*).

Previous work by the Swansea In-Vivo Analysis Research Group (SIVARG) has focused on the role of cadmium in the etiology of essential hypertension (*3*) and on the nephrotoxicity of cadmium in occupational exposure (*4*). This paper reports the preliminary findings of a survey of lead and cadmium body burdens in a cross-section of Swansea residents and compares these data with other measurements made in Swansea and with data reported by other laboratories using a variety of techniques.

METHODS

In Vivo Neutron Activation Analysis of Liver and Kidney for Cadmium

This method has been widely used by several laboratories during the last 10 years (*1,2*). The method involves irradiating the organ of interest with a collimated neutron beam and on-line counting of the prompt 559 keV gamma rays emitted during neutron capture in the stable isotope of cadmium, ^{113}Cd (12.3%, σ_c = 20,000 barns).

The current Swansea instrument (*5*), which uses a 200 μg (4 GBq) Cf-252 neutron source and two high-efficiency (20%) HpGe detectors, achieves a detection limit (2 SD of the background) of 2.4 mg cadmium in the kidney and 2.5 ppm in the liver, for a dose equivalent to the skin of 3 mSv, assuming a quality factor of 10 for neutrons.

The use of ultrasound to locate the organ enables corrections to be made for differences in anatomy and ensures that an accuracy of ±10% can be achieved (*6*). This was further tested in Swansea by measuring a set of anthropomorphic phantoms (*7*) whose composition was not known at the time of measurement. Measured kidney cadmium values (in mg) of 3.3 ± 1.3 and 4.1 ± 1.5 were, respectively, 6 and 32% higher than the known content of 3.1 mg, but were within the errors of measurement.

In Vivo X-Ray Fluorescence Analysis of Tibia Lead

This technique, which uses a 200 mCi (7.4 GBq) annular Cd-109 source, and a planar HpGe detector in a 180° scattering geometry, was first described by the Birmingham University Group (*8*) and subsequently validated by cross-comparison of the same bone samples measured by atomic absorption spectrometry (AAS) at the University of Minnesota, Duluth (*9*). Two of these tibia specimens have since been measured in Swansea, and the values of 26.5 ± 1.9 and 85.5 ± 2.7 μg Pb/g ash bone compared with the figures of 23.9 ± 1.8 and 80.4 ± 1.7 and 24.7 ± 2.6 and 79.2 ± 6.6 previously reported using XRF and AAS, respectively (*9*). The minimum detectable concentration (2 SD of net peak counts) with this method is approximately 11 μg Pb/g wet bone (corresponding to ~ 19 μg Pb/g ash bone) for a maximum skin dose of 0.6 mGy and a mean absorbed dose to a 20 cm section of leg of only 0.13 mGy.

The Swansea Survey: Recruitment and Study Protocol

A 1983 UK survey of heavy metal content of household dust and garden soil included 100 dwellings in Swansea, a South Wales city with a history of metallurgical industry, but whose boundaries include large areas of nonindustrial development. Therefore, it represents a suitable population in which to examine the relationships between environmental exposure, body burden, and possible health effects. A total of 74 adult volunteers were recruited from 57 dwellings. All subjects provided written, informed consent in accordance with the protocol approved by the District Ethical Committee.

In addition to the in vivo measurements of kidney and liver Cd and tibia Pb, blood and urine samples were also taken for heavy metal analysis and an assessment of kidney function. Blood pressure and radial bone mineral concentration were also measured and a structured questionnaire completed to obtain relevant information on family and medical history, occupation, residency, smoking habit, and diet.

Table 1
Results of Cadmium Measurements
in Nonoccupationally Exposed Subjects[a]

Parameter	All Subjects	non-smokers	ex-smokers	smokers	Reference
		Group			
n	69	21	23	25*	
Blood Cd ($\mu g\ l^{-1}$)	1.04(0.65)	0.70(0.40)	0.86(0.42)	1.48(0.75)	this work
n	69	20	23	26*	
Liver Cd (ppm)	1.3(2.0)	1.0(2.1)	1.4(2.1)	1.6(1.9)	"
n	70	20	23	27*	
Kidney Cd (mg)	2.9(2.0)	1.9(1.4)	2.7(2.3)	3.9(1.7)	"
n	60	26	15	19+	
Kidney Cd (mg)	3.7(2.5) (p>0.05)	3.0(2.0) (p<0.03)	3.7(3.1) (p>0.05)	4.7(2.3) (p>0.05)	3 (1980)
n	130	46	38	46*+	combined data 1980 & 1989
Kidney Cd (mg)	3.3(2.3)	2.5(1.9)	3.1(2.7)	4.2(2.0)	
n	130	46	38	46	
Cortex Cd ($\mu g g^{-1}$)	30.5(20.9)	24.6(18.4)	27.7(24.0)	38.6(18.1)	"
G.M.	24.8	16.1	23.3	33.9	

[a]Data for smokers include 2 (*), 4 (+), and 6 (*+) pipe smokers, respectively. Values are quoted as arithmetic means (±SD).

RESULTS

Cadmium

The results of cadmium measurements are shown in Table 1. Also included in Table 1 are the values of kidney Cd obtained in an earlier study (*3*). At that time, liver Cd was not measured, so the kidney values now reported in Table 1 have been adjusted for interference between liver and kidney using the data from the current survey. This results in an average reduction in kidney Cd of 5%.

Although the 1980 data are generally higher than the current values, the difference is only significant ($p < 0.03$) for the nonsmokers. Therefore, the two sets of data have been combined, as shown in the penultimate line of Table 1. The kidney Cd data have also been converted to renal cortex concentrations, assuming a ratio of 1.25 between cortex and whole kidney concentrations (*10*) and a kidney weight of 145 g for males and 124 g for females (Table 1). This enables the data to be directly compared with results from other laboratories throughout the world, as shown in Fig. 1.

In addition to the measurements on nonoccupationally exposed persons, 13 ex-employees of a nickel-cadmium battery factory were stud-

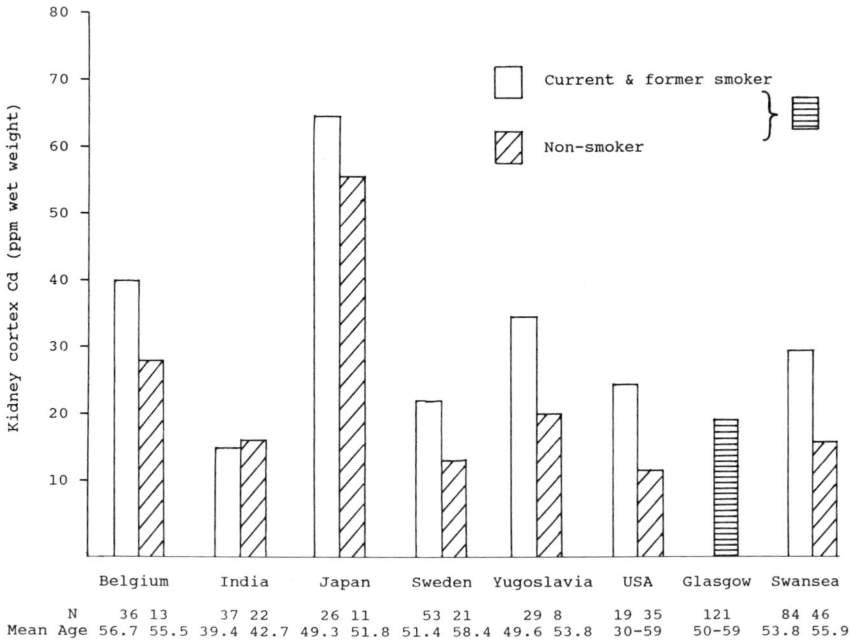

Fig. 1. Concentration of cadmium in kidney cortex (geometric mean values) from different countries. Data from Elinder (*10*), Scott (*12*), and this work.

ied after an interval of approximately five years since leaving the factory. These data are typically an order of magnitude higher than the environmentally-exposed group (Fig. 2).

Lead

Blood lead values in the Swansea survey were all within the normal range, with a mean of 1.04 ± 0.65 μg/L and a geometric mean of 0.88 μg/L. Tibia lead values in 59 subjects ranged from −10 to 42 μg/g wet bone, with a mean of 7.2 ± 10.8 μg/g. There was only a poor correlation between tibia lead and blood lead, but the in vivo results showed a significant correlation with the age of the subject, as shown in Fig. 3.

DISCUSSION

The association between cadmium and smoking is well known, and for every parameter in Table 1, except the liver, the difference between smokers and nonsmokers is highly significant ($p < 0.02$). The ratio of kidney cadmium in 46 smokers (mean age 52.2 y) to 46 nonsmokers (mean age 55.9 yr) is 1.66, which compares very well with the ratios of 1.5 and 1.8 in males and females, respectively, reported by Hahn et al. (*11*). An approximate cumulative measure of smoking exposure is the pack

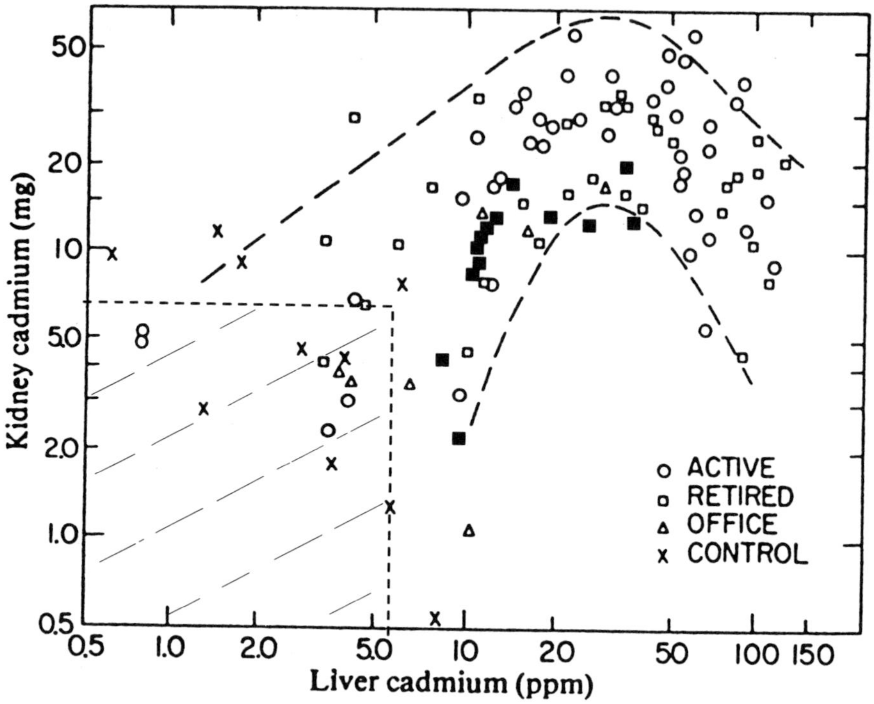

Fig. 2. Measurements of kidney and liver cadmium in ex-employees of a South Wales nickel-cadmium battery factory (■) superimposed on similar data from a cadmium smelter work force in the US (*14*). The shaded area represents the boundaries of the nonoccupationally exposed Swansea survey.

year, the product of packs of 20 cigarets per day smoked and the duration in years. The values for smokers, exsmokers, and all smokers, respectively, were 21.9, 25.1, and 23.3.

Age is also an important parameter when comparing different population groups, because kidney Cd is known to increase gradually from birth and reach a maximum in the fifth and sixth decades before declining (*10*). Thus, age and smoking can probably account for the generally higher values of kidney Cd observed in the 1980 study (*3*), compared with the more recent survey, since the mean ages of the two groups were, respectively, 58.5 ± 9.7 and 51.4 ± 14.2 overall and 61.1 ± 9.7 and 49.8 ± 14.8 among the nonsmokers. The smoking pack years were 29.9 ± 24.5 and 18.9 ± 21.6, respectively.

Figure 1 shows that the kidney cortical Cd content of Swansea residents is comparable with the values found in other European countries, although somewhat higher than the results obtained by Scott et al. in Glasgow (*12*). This may indicate a real difference in exposure to Cd between the two cities. It is interesting to note that a similar survey to Scott's (i.e., atomic absorption analysis of autopsy tissues) by Drasch et al. (*13*) in Munich reported geometric mean cortex concentrations of 17.9 μg/g in nonsmokers and 52.1 μg/g in heavy smokers. In terms of occupa-

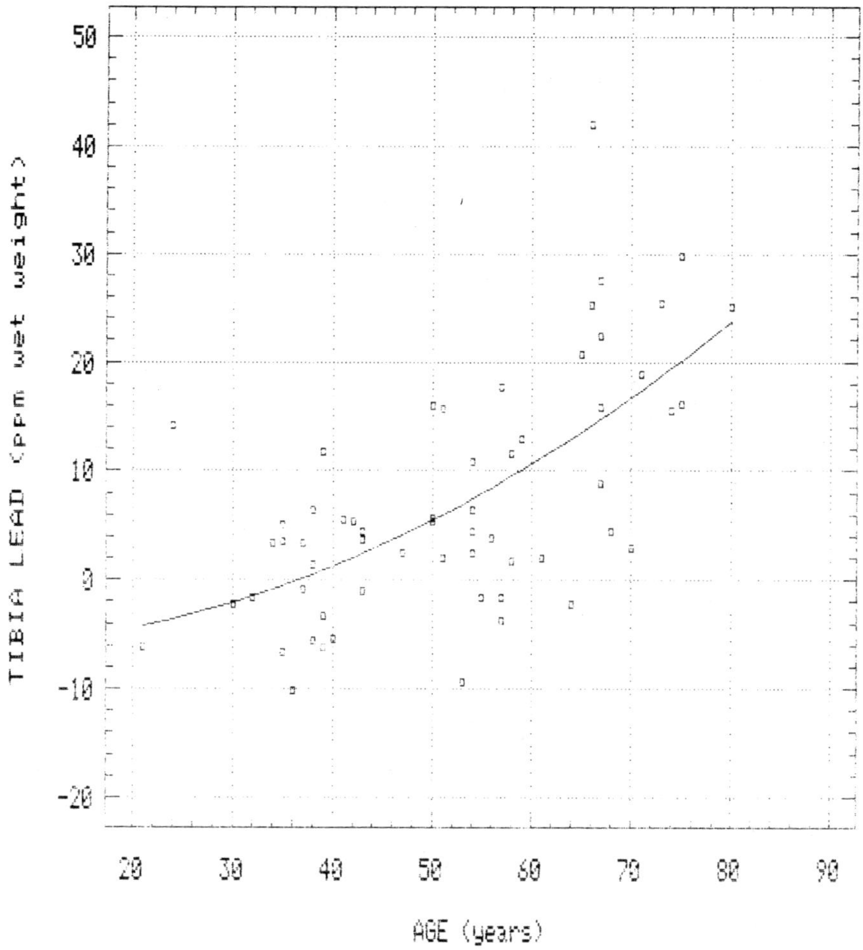

Fig. 3. Tibia lead (μg/g wet bone) vs age of subject. The fitted curve is of the form: tibia Pb (ppm) = $-6.24 + (0.0047) \times (\text{age})^2$ (r = 0.65).

tional exposure to cadmium, the Swansea in vivo measurements of kidney and liver coincide very well with the pattern of distribution between the two organs, reported by Ellis et al. (*14*), as depicted in Fig. 2.

Lead measurements are available only in persons who are not known to have been occupationally exposed to the metal. The mean value of tibia lead concentration is similar to that reported by Somervaille et al. (*8*), although the age dependence is slightly more marked in the Swansea group.

SUMMARY

Preliminary results of a survey to investigate the body burden of lead and cadmium in an environmentally-exposed population have demonstrated a degree of comparability with other data that supports the further detailed analysis of the relationships between body burden and exposure, on the one hand, and possible health effects on the other.

ACKNOWLEDGMENTS

Much of this work was supported by research grants from the Welsh Office and the Medical Research Council. One of the authors (W. D. Morgan) received a Wellcome Foundation Travel Grant to attend this conference, and another author (S. J. Jones) is fully supported by the Tenovus charity.

REFERENCES

1. M. C. Scott and D. R. Chettle, *Scand. J. Work Environ. Health* **12,** 81–96 (1986).
2. D. R. Chettle, M. C. Scott, K. J. Ellis, and W. D. Morgan, *In Vivo Body Composition Studies,* K. J. Ellis, S. Yasumura, and W. D. Morgan, eds., IPSM, London, 1987, pp. 300–310.
3. P. E. Cummins, J. Dutton, C. J. Evans, W. D. Morgan, A. Sivyer, and P. C. Elwood, *Eur. J. Clin. Invest.* **10,** 459–461 (1980).
4. R. R. Ghose, W. D. Morgan, and P. E. Cummins, *Br. J. Ind. Med.* **38,** 185–186 (1981).
5. S. J. S. Ryde, W. D. Morgan, A. Sivyer, C. J. Evans, and J. Dutton, *Phys. Med. Biol.* **32,** 1257–1271 (1987).
6. W. D. Morgan, K. J. Ellis, D. Vartsky, S. Yasumura, and S. H. Cohn, *Phys. Med. Biol.* **26,** 577–590 (1981).
7. D. K. Bewley, *Phys. Med. Biol.* **33,** 805–813 (1988).
8. L. J. Somervaille, D. R. Chettle, and M. C. Scott, *Phys. Med. Biol.* **30,** 929–943 (1985).
9. L. J. Somervaille, D. R. Chettle, M. C. Scott, A. C. Aufderheide, J. E. Wallgren, L. E. Wittmers, Jr., and G. R. Rapp, *Phys. Med. Biol.* **31,** 1267–1274 (1986).
10. C. G-Elinder, *Cadmium and Health: A Toxicological and Epidemiological Appraisal,* vol. 1, L. Friberg, C. G-Elinder, T. Kjellstrom, and G. F. Nordberg, eds., CRC Press, Boca Raton, FL, 1985, pp 81–102.
11. R. Hahn, U. Ewers, E. Jermann, I. Freier, A. Brockhaus, and H. W. Schlipkoter, *Int. Arch. Occup. Environ. Health* **59,** 165–176 (1987).
12. R. Scott, E. Aughey, G. S. Fell, and M. J. Quinn, *Human Toxicol.* **6,** 111–119 (1987).
13. G. Drasch, G. Kauert, and L. von Meyer, *Int. Arch. Occup. Environ. Health* **55,** 141–148 (1985).
14. K. J. Ellis, W. D. Morgan, D. Vartsky, I. Zanzi, and S. H. Cohn, *Cadmium '81,* D. Wilson and R. A. Volpe, eds., Cadmium Association, London, 1982, pp 161–165.

Recent Developments in In Vivo Neutron Activation Analysis Facilities

S. S. KRISHNAN,*[,1] K. G. MCNEILL,[2] J. R. MERNAGH,[2] AND J. E. HARRISON[1,2]

[1]*Toronto General Hospital; and* [2]*University of Toronto, Toronto, Ontario, Canada*

Received April 17, 1989; Accepted November 3, 1989

ABSTRACT

Two new facilities for in vivo activation analysis of patients have been designed, developed, and constructed at Toronto General Hospital. One of these is for the determination of body calcium for the diagnosis of osteoporosis and other diseases associated with bone loss. The other is for the measurement of total body nitrogen for the determination of protein status.

These facilities replace old university facilities and take into account the comfort and management of patients. In addition, in the case of the calcium facility, the precision of the measurements has been improved because of larger detector volume and increased neutron source strength. Both the facilities are now in routine hospital clinical use.

Index Entries: Neutron activation analysis; in vivo NAA; total body calcium; total body nitrogen; osteoporosis; malnourished state.

INTRODUCTION

In vivo activation analysis for the measurement of body calcium and body nitrogen in humans has been in use in this hospital for many years.

*Author to whom all correspondence and reprint requests should be addressed.

For the measurement of body calcium, which is a measure of bone mineral mass, approximately a third of the body, viz. from the neck to mid thigh, is irradiated by a neutron flux. Plutonium-beryllium sources are now used as the neutron source. The induced ^{49}Ca ($T_{1/2}$ = 8.8 min, $E\gamma$ = 3.1 MeV) activity provides a measure of the body calcium. The ratio of this activity, corrected for body size, to a normal person of similar size gives what we define as the calcium body index (CaBI) of the patient. The data for normal persons have been measured using healthy volunteers. This procedure has been described elsewhere (*1*).

The calcium bone index is a measure of the bone mineral mass of the patient. This index is used as a diagnostic indicator of osteoporosis (*2*). The CaBI values of 38 healthy volunteers (20–40 years of age) gave a mean value of 0.97 ± 0.12 (SD). The CaBI values of 61 older volunteers (age 68.3 ± 6.3 (SD)) without any bone problem gave a mean CaBI value of 0.83 ± 0.11. In contrast, the mean CaBI value of 71 patients (age 68.7 ± 7) with vertebral fractures was 0.67 ± 0.12 (SD).

In the case of body nitrogen measurement, approximately a third of the body, as in the calcium measurement, is irradiated with Pu-Be neutron sources placed above and below the patient. The prompt gamma rays from the induced ^{15}N ($E\gamma$ = 10.8 MeV) are measured during the irradiation by four sodium iodide detectors placed at the sides of the patient. Thus, this measurement is a prompt-gamma assay of nitrogen. As in the case of calcium, appropriate corrections are made for body thickness, and a nitrogen index is calculated for the patient. The nitrogen index is the ratio of measured nitrogen activity of the patient to a volunteer of the same size. Here, we also have nitrogen data for healthy volunteers. This procedure has been described elsewhere (*3*).

The nitrogen index is a measure of the protein status of the patient and is useful in examining the nutritional status of patients, e.g., those with malabsorption syndrome. Clinical uses of the nitrogen index for a variety of diseases have been described elsewhere (*4*).

The above in vivo calcium and nitrogen facilities have been in use for clinical work. Although these pilot facilities were technically effective, they were not appropriate for routine hospital use. These facilities have now been redesigned with the cooperation and under contract with Monserco Ltd., Mississauga, Ontario, Canada. They are described in this paper.

IMPROVED IN-HOSPITAL IN VIVO FACILITIES

Calcium System for Bone Mineral Mass Assay

The new calcium facility for CaBI measurement is located in the outpatient area of the Toronto General Hospital (Fig. 1). The irradiation

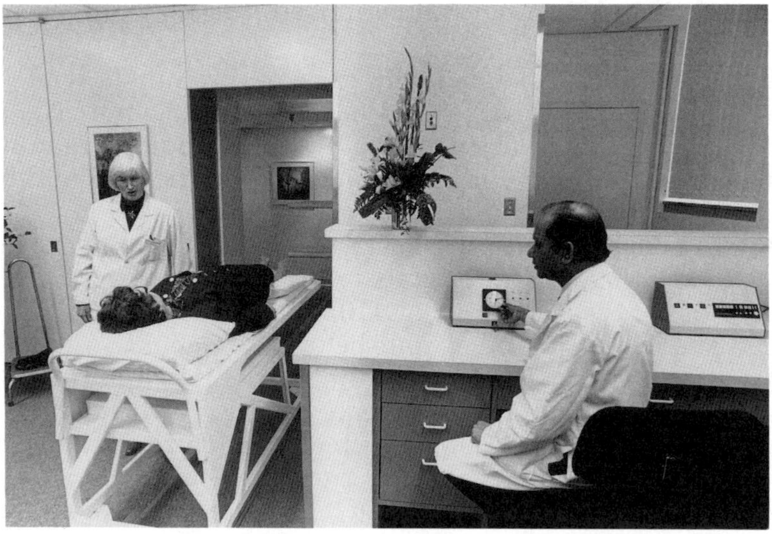

Fig. 1. In vivo body calcium irradiation facility.

sources and control equipment occupy an area of 19 ft × 37 ft (5.8m × 11 m). The neutron exposure is carried out in a cubicle room of 8 ft (2.4 m) sides and height. The walls of the irradiation room are made up of a 4 feet (1.2 m) thick wall of concrete blocks. The 16 in (41 cm) steel door contains borated wax and moves on railings on the top.

The irradiation sources consist of a total of 110 curies of Pu-Be neutron sources. These are in 5- or 10-Curie sealed containers. During neutron irradiation, these sources occupy positions such that 55 curies are placed above and below the patient. The sources are placed so that the area from the sternal notch to mid-thigh area is irradiated, similar in configuration to the previous arrangement (*1*) except for the higher source strength.

The neutron sources are brought from their shielded location within the wall to the irradiation position hydraulically. They travel inside aluminum tubes positioned above and below the patient (Fig. 1). The patients, therefore, do not see any movement of sources themselves.

There are many safety features incorporated in this facility. The door must be closed before the sources can be moved out of the shielding. In emergency situations, the sources can be withdrawn and the door opened by emergency buttons accessible to the patient inside and another for the operator outside the irradiation room. In case of a power failure, the sources will be withdrawn and the door opened automatically by the reserve hydraulic pressure. If everything including the hydraulics fail, the sources and the door can be moved manually after removing a few screws and pins.

The entire facility is painted light in color, brightly lit with reflected light, and decorated. Music is piped in during irradiation. The patients are on closed-circuit television and a two-way radio communication, which allow us to observe and communicate with the patient during the irradiation procedure.

The ^{49}Ca activity is measured in a whole-body counter using eight sodium iodide crystals. The crystals are the extruded type, each with dimensions of 4 in. × 4 in. × 16 in. (10 × 10 × 4 cm). Four of these are positioned above and the other four below the patient so that the ^{49}Ca activity is effectively measured from the sternal notch to the mid-thigh area of the patient. The output from each detector is directed by a mixer router to a 256-channel area of a multichannel analyzer (Canberra Model 35). Thus, the gamma-ray spectrum from each detector is separately displayed, covering an energy range of 0 to 5 MeV. Therefore, the Cl-38, Na-24, and Ca-49 activities are displayed individually from each detector on the analyzer. The operator then locates the appropriate peak areas for these three peaks in each detector. Once that is done, the data of all the 24 peak areas are transferred automatically to an IBM-PC clone computer. The computer is programmed to apply the appropriate corrections and calculate the Ca-49 activity from each of the eight detectors. It then adds these to obtain the total Ca-49 activity of the patient and calculates the CaBI, as in the previous procedure (*1*). The computer is programmed to produce a suitable report. Thus, after 20 min counting of the patient, the report is obtained immediately from the computer (Fig. 3).

The precision of CaBI measurement with this facility is of the order of ±3.5% (CV) compared to about ±5.5% in the previous facility. This is achieved by increased neutron flux and detector volume. The radiation dose is about 550 mR, compared to about 300 mR in the previous facility.

Nitrogen Assay

The in vivo nitrogen facility is also used for routine clinical measurement (Fig. 2). The neutron sources are located above and below the patient, and the detectors are placed on the sides. The configuration is similar to the previous set-up (*3*). The neutron sources being used, so far, are two 10-curie Pu-Be sources. Recently, we have changed to ^{252}Cf sources, which yield a lower background signal, so that the nitrogen signal-to-background ratio is higher (*5*).

The prompt-gamma ^{15}N signals are collected by four 5 in. × 4 in. (12.7 × 10.2 cm) sodium iodide detectors. A mixer router then routes these to four sections (of 512 channels each) of a 2048-channel Nuclear

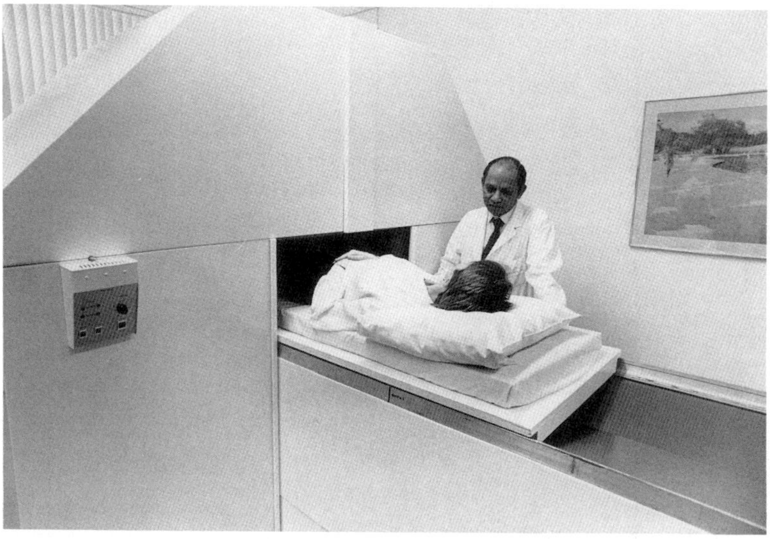

Fig. 2. Prompt-gamma in vivo nitrogen measurement facility.

Data analyzer, as in the case of the calcium system. The N and H signals are measured for each detector, and the nitrogen amount is calculated as the percent of body weight, as described elsewhere (*3*). The H signal is used as an internal standard. The %N valves from all the four detectors are added, and an average value is calculated from the sum. The nitrogen index is then calculated for the patient as in the previous case (*3*).

The facility is motorized so that the sources can be moved as close to the patient as possible. This allows us to obtain the maximum nitrogen signal with patients of different sizes. The detector signals are handled by a multichannel analyzer (ND66) connected to an IBM PC computer that calculates the nitrogen index and produces a report immediately after the operator has manually identified the N and H peak areas.

SUMMARY

Two facilities, one suitable for routine measurement of bone mineral and another for protein status of patients, have been designed and constructed for routine hospital use. The equipment is moved automatically for measurement configurations. The data and report preparation are done with suitable computer programs. The outer appearance of the equipment is attractive (see Figs. 1 and 2), in keeping with commercial facilities.

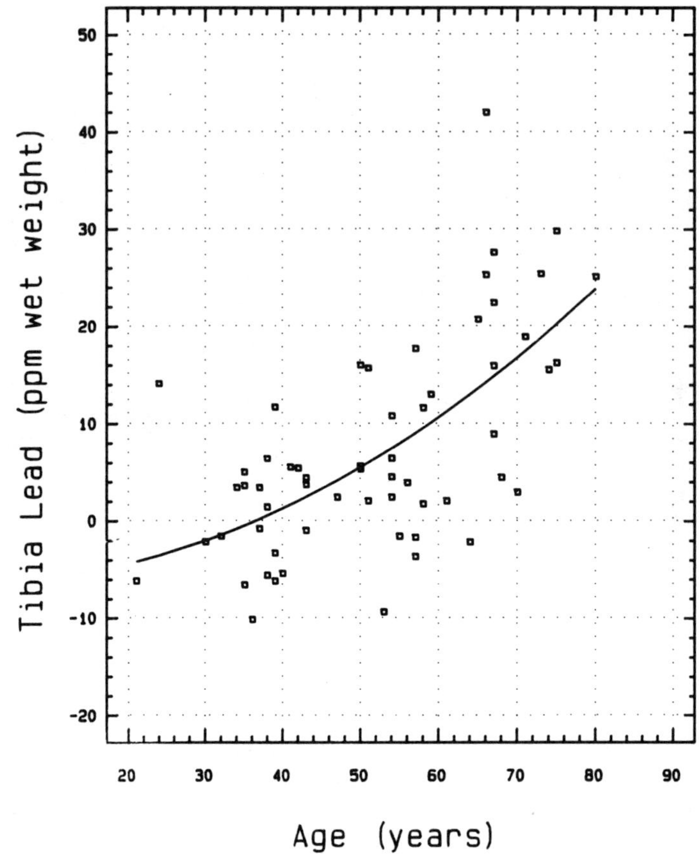

Fig. 3. Plot of tibia lead against age.

ACKNOWLEDGMENTS

The financial assistance from the Womens Auxiliary of the Toronto General Hospital for the Calcium Facility is gratefully acknowledged.

REFERENCES

1. J. E. Harrison, C. Williams, J. Watts, and K. G. McNeill, *J. Nucl. Med.* **16,** 116–122 (1975).
2. J. E. Harrison, K. G. McNeill, N. Patt, C. Mueller, J. B. Houpt, J. C. Kirsh, R. Chow, and the Bone and Mineral Group, *In-Vivo Body Composition Studies,* Chapter 40, K. J. Ellis, S. Yasumura, and W. D. Morgan, eds., The

Institute of Physical Sciences in Medicine, London, England, 1986, pp. 243–249.
3. J. R. Mernagh, J. E. Harrison, and K. G. McNeill, *Phys. Med. Biol.* **22,** 831–835 (1977).
4. K. G. McNeill, J. E. Harrison, and S. S. Krishnan, *J. Radioanal. and Nuclear Chem.* **110,** 655–661 (1987).
5. K. G. McNeill, D. J. Borovnicar, S. S. Krishnan, H. Wang, C. Waana, and J. E. Harrison, *Phys. Med. Biol.* **34,** 53–59 (1989).

A Proposed Three-Phase Counting System for the In Vivo Measurement of the Major Elements Using Pulsed 14 MeV Neutrons

S. MITRA,* J. F. SUTCLIFFE, AND G. L. HILL

University Department of Surgery, Auckland Hospital, Auckland, New Zealand

Received April 17, 1989; Accepted November 3, 1989

ABSTRACT

It is proposed to employ a pulsed source of 14 MeV neutrons for in vivo activation analysis. This would permit the differentiation, with time, of the resulting gamma ray emission into three separate spectra, according to the type of nuclear reaction and mode of decay. The three-phase counting has been divided into ~ 10 μs during "beam-on," and 200 and 800 μs during "beam-off." Measurement of the major elements, C, N, O, Cl, and P, to give nutritionally-important body compartments of total body fat, protein, water, minerals, and extracellular water, thus is expected with a single scan.

Index Entries: Body composition; pulsed 14 MeV neutron activation; gamma-ray spectra, time differentiation of.

INTRODUCTION

In vivo neutron activation analysis for the study of human body composition is an established technique. The variations in the design of facilities, neutron sources used, the associated detection systems, and so

*Author to whom all correspondence and reprint requests should be addressed.

on, for the study of various elements are dependent on the varying requirements of each center. These have been reviewed earlier (*1–3*).

Prompt-gamma in vivo neutron activation analysis with $^{238}Pu/Be$ sources has been used to measure total body nitrogen, for protein, in seriously ill patients in the Department of Surgery at Auckland Hospital since 1982. The $^{14}N\ (n, \gamma\ )^{15}N$ reaction is utilized, and the 10.83MeV gamma rays emitted by the deexcitation of ^{15}N are measured. The precision of this measurement, i.e., the SD of 20 measurements of an anthropomorphic phantom, is $\pm$ 48 g. Total body water (TBW) is measured by dilution of tritiated water (SD $\pm 1.5\%$, amounting to 0.6 kg of Reference Man), and total body mineral and glycogen are estimated from a measure of skeletal size (*4*). Total body fat (TBF) as derived, i.e., "by difference" between body mass and the sum of the masses of the other compartments, has a precision of 0.7 kg. "Snapshots" of the patient's body composition, in terms of water, protein, and fat at various intervals, can show changes that can reveal the underlying trends in the metabolic response to illness and surgery.

A disadvantage of our present procedure in clinical applications is that TBW is measured by dilution of tritiated water. This technique incurs an additional radiation dose to the patient, and the associated equilibration times and chemical analysis procedures preclude the benefit to the clinician of immediate results for TBW and TBF. Therefore, it is desirable to measure fat directly from total body carbon by utilizing the fast-neutron inelastic-scattering reaction $^{12}C\ (n,n'\gamma)\ ^{12}C$ (*5*) and TBW "by difference." Alternatively, a direct measurement of TBW can be made from total body oxygen (*6,7*); however, this measurement, either by the $^{16}O\ (n,p)\ ^{16}N$ or the $^{16}O\ (n,n'\gamma)\ ^{16}O$ reactions, is subject to interference from the 6.1 MeV gamma ray emission from thermal neutron capture in chlorine. Also, the mean neutron energy from the Pu/Be sources ($\sim$5 MeV (*8*)) is appreciably lower than the reaction thresholds required for the above reactions. These are in excess of $\sim$6.6 MeV in the case of oxygen (*9*).

The measurement of total body chlorine itself is desirable as a means of dividing the water compartment into intracellular and extracellular components. The measurement of total body calcium, the main component of the mineral compartment, is prevented in the existing facility by the high emission from this element in the structure of the building, owing to the continuous irradiation from the Pu/Be neutron sources.

To measure the four major body compartments (water, fat, protein, and minerals) more precisely, to provide clinically relevant diagnostic information, we plan to develop multielement analysis by employing a pulse of 14 MeV neutrons and an associated three-phase counting system. This extends the pulsed-beam method developed at the Brookhaven National Laboratory, Upton, NY, US (*10*) to include measurement of delayed emission. This new technique is expected to improve on existing methods that utilize delayed gamma-ray emission (*11*) in two ways: first, it will deliver a significantly smaller dose of fast neutrons to the subject; second, the subject will not need to be transferred quickly after irradiation to a separate low background counting facility; this rapid transfer is not practicable for the majority of seriously ill patients in this department.

METHODS

A block diagram of the proposed experimental facility is shown in Fig. 1. Central to the facility is a pulsed 14 MeV neutron generator system. The four NaI (Tℓ) detectors have energy resolutions better than ±3% at 6.1 MeV.

Each neutron pulse will be approximately 10 μs long. The pulsing circuit will provide an external trigger pulse at the beginning of each neutron burst to generate gates for three periods of counting. The first phase (10 μs) will coincide with the 10 μs neutron pulse. The second phase (~200 μs) will immediately follow the end of the neutron pulse. The third phase will cover the remaining ~800 μs until the start of the next neutron pulse. This counting cycle can be maintained by a 1 kHz neutron pulse repetition frequency. The signals from these three phases will be analyzed and stored in separate parts of the memory of a multichannel pulse-height analyzer.

Table 1 shows the body elements of interest and their abundances, based on Reference Man (*12*), along with the counting regimen for their detection and analysis. During the neutron pulse of approximately 10 μs, fast-neutron inelastic-scattering reactions, predominantly with carbon and oxygen, will be detected. For approximately 200 μs following each neutron pulse, the prompt gamma ray emission owing to thermal neutron capture will be observed. This emission will consist mainly of 2.223 MeV quanta from hydrogen, the element with the greatest nuclear abundance in the human body (*3*). The emissions from nitrogen, chlorine, calcium, and phosphorus would also be observed. Zamenhof et al. (*13*) report that, with the exception of the hydrogen peak, the entire capture gamma-ray spectrum is dominated by chlorine, thus making it possible to precisely measure total body chlorine. Corrections for chlorine contamination in the calcium peaks can then be made.

If the neutron generator is operated at a pulse repetition frequency of 1 kHz, delayed emission alone would be observed in the approximately 800 μs before the next pulse. In this phase, the gamma ray emissions of interest are at 6.1 MeV from ^{16}N (half-life, 7.2 s), following an (n,p) reaction with ^{16}O, and those owing to the production of ^{38}Cl, ^{49}Ca, and ^{28}Al (from ^{31}P). Thus, the third phase of counting would allow independent measurements of oxygen, chlorine, calcium, and phosphorus and help in some of the spectral interference corrections required in the second phase of counting.

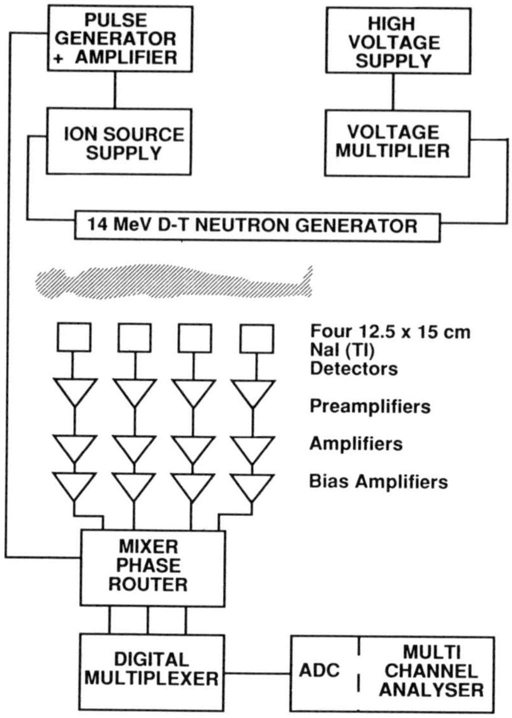

Fig. 1. Schematic diagram of a pulsed neutron system.

Table 1
Body Elements of Interest and Analysis

Reference man (*12*),		Energy of Gamma-Ray Emission (MeV), "Beam-on"	"Beam-off"	
Element	Amount in Body, kg	Phase I, ~10 μs	Phase II, ~200 μs	Phase III, ~800 μs
O	42.7	6.1 (^{16}O)		6.1 (^{16}N, $t_{1/2}$ = 7.2 s)
C	16.1	4.43 (^{12}C)		
H	7.0		2.223 (^{2}H)	
N	1.82		10.83 (^{15}N)	
Ca	0.98		6.42 (^{41}Ca)	3.08 (^{49}Ca, $t_{1/2}$ = 8.7 m)
P	0.77		3.9 (^{32}P)	1.78 (^{28}Al, $t_{1/2}$ = 2.24 m)
Cl	0.084		6.1 (^{36}Cl)	2.2 (^{38}Cl, $t_{1/2}$ = 37.3 m)

DISCUSSION

In addition to the greater signals resulting from the inelastic scattering of 14 MeV neutrons with carbon and oxygen, compared with neutrons from Pu/Be sources, a switchable source of neutrons would reduce the activation of the environment and present a lower background against which the subject is measured. This would result in better precision in the measurement of all elements, and, hence, the magnitude of the major body compartments, at acceptable dose levels to the subject. The dose to the subject in the present facility is approximately 0.3 mSv. On the basis that 14 MeV neutrons each carry three times the energy of neutrons from a Pu/Be source, on average, for the same neutron fluence per scan, this would give a dose to the subject of about 0.9 mSv. However, the reduced background activation should allow a lower neutron fluence per scan to achieve a similar signal-to-background ratio. Therefore, 0.9 mSv would represent an upper limit of the dose likely to be incurred.

The pulse regimen proposed in this paper is largely based on the Monte Carlo simulation and experimentation of Zamenhof et al. (*13*) on the time dependence of the total neutron-induced gamma ray flux incident on the detectors. A test rig is currently being built, and the most suitable pulse protocol will be decided by experiment.

ACKNOWLEDGMENTS

The authors are grateful to L. Burkinshaw for his constructive criticism of this manuscript. This work is supported by the Medical Research Council of New Zealand.

REFERENCES

1. A. H. Beddoe and G. L. Hill, *J. Parenteral Enteral Nutrition* **9,** 504 (1985).
2. S. H. Cohn and R. M. Parr, *Clin. Phys. Physiol. Meas.* **6,** 275 (1985).
3. D. R. Chettle and J. H. Fremlin, *Phys. Med. Biol.* **29,** 1011 (1984).
4. A. H. Beddoe, S. J. Streat, and G. L. Hill, *Metabolism* **33,** 270 (1984).
5. K. Kyere, B. Oldroyd, C. B. Oxby, L. Burkinshaw, R. E. Ellis, and G. L. Hill, *Phys. Med. Biol* **27,** 805 (1982).
6. E. D. Williams and K. Boddy, *Int. J. Appl. Radiat. Isot.* **29,** 281 (1978).
7. B. W. East, T. Preston, I. Robertson, D. L. Davies, and G. Herd, *In Vivo Body Composition Studies*, K. J. Ellis, S. Yasumura, and W. D. Morgan, eds., IPSM, London, 1987, pp. 154–162.
8. A. H. Beddoe, H. Zuidmeer, and G. L. Hill, *Phys. Med. Biol.* **29,** 371 (1984).
9. J. A. Auxier, W. S. Snyder, and T. D. Jones, *Radiation Dosimetry*, vol. 1, 2nd edition, F. H. Attix and W. C. Roesch, eds., Academic, New York, 1968, p. 282.

10. J. J. Kehayias, K. J. Ellis, S. H. Cohn, S. Yasumura, and J. H. Weinlein, *In Vivo Body Composition Studies*, K. J. Ellis, S. Yasumura, and W. D. Morgan, eds., IPSM, London, 1987, pp. 427–435.
11. A. Sharafi, D. Pearson, C. B. Oxby, B. Oldroyd, D. W. Krupowicz, K. Brooks, and R. E. Ellis, *Phys. Med. Biol.* **28,** 203 (1983).
12. ICRP 1975, *Report of the task group on reference man*, Publication 23, Pergamon, Oxford.
13. R. G. Zamenhof, O. L. Deutsch, and B. W. Murray, *Med. Phys.* **6,** 179 (1979).

Measurements of Total Body Calcium by Prompt-Gamma Neutron Activation Analysis Using a ^{252}Cf Source

S. J. S. RYDE,*[,1] W. D. MORGAN,[1] J. COMPSTON,[2]
C. J. EVANS,[3] A. SIVYER,[1] AND J. DUTTON[3]

[1]Department of Medical Physics and Clinical Engineering, Singleton Hospital, Swansea SA2 8QA, Wales, UK; [2]Department of Pathology, University of Wales College of Medicine, Heath Park, Cardiff CF4 4XN, Wales, UK; and [3]Department of Physics, University College of Swansea, Swansea SA2 8PP, Wales, UK

Received April 17, 1989; Accepted November 3, 1989

ABSTRACT

A clinical neutron activation instrument has been developed for in vivo elemental analysis. Utilizing the prompt-capture gamma ray technique, simultaneous total body (TB) measurements of primarily Ca, but also Cl, N, C, and H are routinely performed. This paper describes a technique for the measurement of TBCa (g) that relies on the use of TBCl as an internal standard. The method has been tested with four anthropomorphic phantoms covering a range of body habitus. The mean discrepancy between the measured and known Ca contents was 3.6%. The technique has been applied to two patient groups, and encouraging results were obtained.

Index Entries: In vivo neutron activation analysis; Ca measurement; neutron activation analysis; body composition analysis; Ca, in vivo measurement of; ^{252}Cf; prompt gamma ray neutron activation; total body calcium.

*Author to whom all correspondence and reprint requests should be addressed.

INTRODUCTION

In vivo neutron activation analysis has been used for nearly 20 years for measurements of total and partial body calcium (*1*). A variety of neutron sources and geometries have been used, but in all cases, a delayed activation technique, using the reaction ^{48}Ca (n,γ) ^{49}Ca, has been employed. The 3.1 MeV gamma ray from ^{49}Ca (half-life, 8.8 min) is measured in a low-background counter after a period of neutron irradiation (*2*). The alternative prompt technique requires the detection of gamma rays resulting from neutron capture in ^{40}Ca during irradiation.

Previous work indicated the possibility of measuring total body calcium (TBCa) by the prompt-gamma technique using a pulsed neutron beam (*3*), and the feasibility of partial body measurements was experimentally demonstrated by the Swansea group in 1983, using a ^{252}Cf source (*4*). It had also been suggested that total body measurements might be achieved using ^{252}Cf because the thermal neutron depth distribution generated in the body by this source is not too dissimilar from that of ^{238}Pu,Be, used by other workers for the measurement of nitrogen (*5*). The suitability of ^{252}Cf for these measurements has been recently investigated and confirmed (*6*).

This paper describes a ^{252}Cf-based instrument that is in regular clinical use for sequential measurements of total body calcium (TBCa); it also reports recent progress in achieving a calibration for TBCa so that single measurements can be of diagnostic value on their own.

THE INSTRUMENT

A schematic diagram of the instrument is shown in Fig. 1. The main parts comprise shielding material, a safe storage position, a pneumatic transport system to deliver the neutron source to one of two collimator positions, a scanning couch, and a gamma ray detection system. In order to achieve adequate shielding for the nominally 4 to 5 GBq neutron source and a proper height for the patient couch, the greater part of the main instrument is housed below floor level. A detailed description of the instrument is given elsewhere (*7,8*).

The gamma ray spectroscopy system comprises two *n*-type hyperpure Ge detectors, each of 20% relative efficiency and 1.9 keV energy resolution at 1.33 MeV. These are connected to independent spectroscopy channels comprising amplifier (Silena 7611/L), 3 μs successive approximation ADC (Silena 7423), and an 8k PC-based multichannel analyzer (System 100, Canberra Instruments, Meriden, CT). The performance of the instrument was assessed by investigating the measurement uniformity with depth and width of subject, and the precision of repeated measurements (*6*).

The anterior-posterior uniformity (the root-mean-square deviation

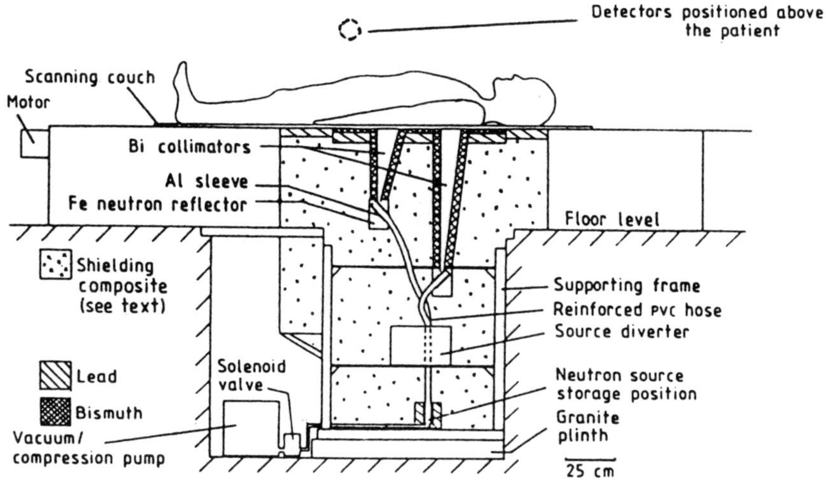

Fig. 1. Schematic diagram of the IVNAA instrument.

about the mean) varied from ±17.1% for a phantom section of 120 mm thickness to ±6.7% for a section of 200 mm thickness. Measurements were made to within 10 mm of the upper and lower surfaces. The lateral uniformity was ±23.7% to within 20 mm of the surface of a 300 mm-wide phantom. The uniformity of activation and detection was further assessed by scanning a spine and rib phantom in which the mass and position of calcium could be varied. Linear regression analysis of the calcium content against calcium gamma ray counts gave a correlation coefficient of 0.986 ($p < 0.0001$) for bilateral irradiation. The measurement precision was determined from several scans of a Bush-type phantom filled with a tissue-equivalent solution. The precision (expressed as the coefficient of variation) for calcium and other simultaneously determined elements is given in Table 1.

CALCIUM MEASUREMENT TECHNIQUE

At the start of the clinical program it was intended to measure changes in calcium counts from the same individual over a period of time. Provided that the subject's body habitus does not change significantly from one measurement to the next, the difficulties associated with whole-body measurements thereby can be avoided. However, it was recognized from the outset that an independent Ca measurement would be very desirable, principally to obtain a value of measured-to-expected TBCa in an individual measurement, but also to overcome the difficulties associated with those patients who undergo a significant body habitus change between measurements. The method that has been adopted relies on the use of total body chlorine (TBCl) as an "internal standard" for calcium. A similar approach was previously used by Vartsky et al. (*9*)

Table 1
Counting Precision for Homogeneously-Distributed Elements Measured Simultaneously in an Anthropomorphic Phantom Using a Bilateral Scanning Geometry[a]

Element	Gamma rays measured (keV)	Counts (g^{-1})	Phantom/ Reference Man content	Precision (% CV)
Ca	6 420	8.55	1.00	2.6
	5 909			
	5 398			
	4 419			
Cl	6 111	430.8	1.13	1.5
	5 600			
	5 089			
P	3 900	0.69	3.60	5.1
C	4 945	0.44	0.77	4.2
	4 434			
	3 923			
N	10 828	0.98	0.99	3.6
	10 317			
	9 806			
H	2 223	321.1†	0.95	0.4

†Includes contribution from polythene shell of phantom.
[a]The total body mean skin dose equivalent was 6.4 mSv.

for the evaluation of total-body nitrogen, using hydrogen as the internal standard.

The internal standardization technique requires that chlorine should be measured or assessed independently. Chlorine is found predominantly in the extracellular fluid volume (ECFV), and therefore, the TBCl should be related to the size of this compartment. In practice, the ECFV can be determined by dilution of a suitable tracer (*10*). Assuming that the TBCl is known, then TBCa is calculated from

$$\text{TBCa(g)} = \text{TBCl(g)} \times \frac{\text{Ca(cts)}}{\text{Cl(cts)}} \times \text{K} \quad (1)$$

where factor K is the ratio of Cl counts/g to Ca counts/g and is obtained by phantom measurements.

In order for chlorine to be a suitable internal standard for calcium, it should ideally fulfill the following criteria: (1) the gamma rays of chlorine and calcium should be detected with the same efficiency; (2) the gamma rays should suffer the same attenuation in the body; (3) the calcium and chlorine neutron reaction cross-sections should vary in a similar manner; and (4) the distribution of the two elements in the body should be the same.

It is evident that criteria (1) and (2) are satisfactorily fulfilled because the separation of the respective gamma ray energies is only 309 keV

Table 2
Anthropomorphic Phantoms Used to Test the Internal Standardization Method of Calibration

Phantom	Height (m)	Weight (kg)	Ca (g)	Cl (g)	Reference
1	1.50	41.4	581±6	47.1±1.6	11
2	1.82	66.8	967±10	113±2.9	11
3+	1.75	94.4	1213±13	105.4±4.8	11
4	1.70	70.2	1010++	97++	12

+The abdomen section was omitted and the values adjusted accordingly.

++The errors in these quantities are unknown.

(Table 1). Criterion (3) is satisfied since both reactions proceed via thermal neutron activation. However, criterion (4) is not easily met since calcium resides mostly in bone, whereas chlorine is found mostly in soft tissue. This dissimilarity of distribution places great importance on there being an adequate uniformity of measurement response throughout the body. Although such variations have been assessed and found to be satisfactory, further studies using nonhomogeneous phantoms were needed to validate the approach.

These studies were based upon the measurement of four tissue-equivalent anthropomorphic phantoms containing simulated skeletons and soft tissue in which the contents of chlorine and calcium were known (*11,12*). The phantoms covered a wide range of height, weight, and body habitus (Table 2).

Each phantom was measured bilaterally from head-to-toe, receiving a mean radiation dose equivalent (QF = 10) to the "skin" of either 3.1 mSv (phantoms 1, 2, and 3) or 4.0 mSv (phantom 4). The resulting 6.420 and 6.111 MeV prompt-gamma peaks from calcium and chlorine, respectively (Table 3), and their single (5.909 and 5.600 MeV) and double (5.398 and 5.089 MeV) escape peaks were analyzed by a spectrum-fitting program. The escape peaks from another gamma ray of chlorine (6.620 MeV) were also included.

For each phantom, the ratio of calcium to chlorine counts was multiplied by the known chlorine content and plotted against the known calcium content (Fig. 2). These quantities should be linearly related if the internal standardization calibration technique is valid. The data in Fig. 2 were analyzed by linear regression analysis. The correlation coefficient of 0.984 ($p < 0.02$) implies that 96.8% of the variation is explained by the content of calcium in the phantoms.

The slope of the line in Fig. 2, i.e., the inverse of the factor K (Eq. (1)) provided a value for K of 58.9 ± 7.5 (SEM), compared with a calculated figure of 67.9 ± 7.5. Using the measured value, the calcium content of

Table 3
Calculation of Calcium Content in the Phantoms

Phantom	Ca/Cl counts ratio	Measured Ca (g)	$\frac{\text{Ca(measured-known)}}{\text{Ca(known)}} \times 100\%$
1	0.191±0.017	566±53[+]	-2.6%
2	0.150±0.009	1034±67	+6.9%
3	0.187±0.012	1197±89	-1.3%
4	0.164±0.008	973±47	-3.7%
		mean =	$\overline{3.6\%}$

[+]Errors are based on counting statistics (phantom 4) and the known error in the chlorine content (phantoms 1–3).

each phantom was calculated and the discrepancy of the measured-to-known calcium determined (Table 3). The discrepancy is comparable with measurements by the delayed activation method of phantoms 1–3 at other centers (*11*), where calibrations were achieved by other means.

CLINICAL APPLICATION

Several clinical studies are in progress to determine changes in TBCa with time. Data from patients with Crohn's disease, and women receiving estrogen implant therapy, have been retrospectively analyzed to obtain a measure of TBCa. For each patient, a predictor equation based on sex, height, and weight was used to determine the TBCl (*13*). Equation 1 was then used to calculate TBCa. The values obtained were compared with those expected on the basis of sex, height, weight, and age (*13*) (Fig. 3).

Patients were admitted to the implant group on the basis of menopausal symptoms and previous hysterectomy. The TBCa measurements were undertaken before implant therapy commenced. The Crohn's patient group had a number of risk factors, including corticosteroid therapy, malabsorption, malnutrition, and amenorrhea. This latter group also had been measured using quantitative computerized tomography of the lumbar spine and single-photon absorptiometry of the wrist, both of which measures have, in general, been below the values observed in age- and sex-matched normal subjects (*14*).

Therefore, it is not unreasonable that the TBCa of both groups might be below that expected in a normal population, as indicated in Fig. 3. However, some caution is necessary since the use of predictor equations for TBCl and TBCa may only be strictly applicable to the North American

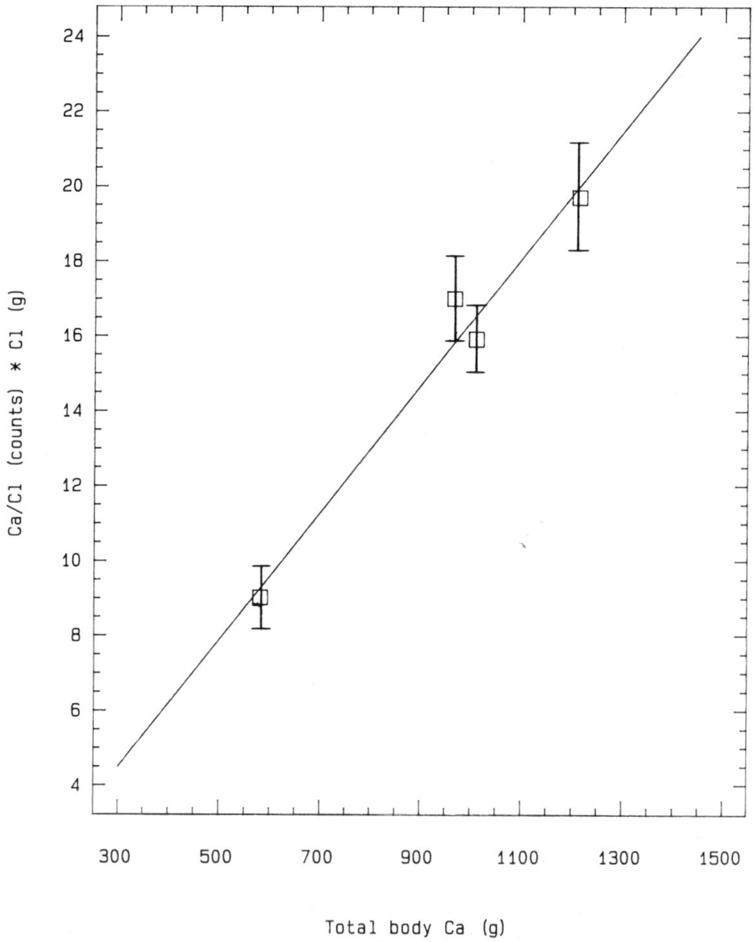

Fig. 2. The regression of the measured Ca-to-Cl counts ratio, times the known Cl content, on the known Ca content for four anthropomorphic phantoms.

population from which they were derived. Therefore, work is in progress to measure the TBCl directly, using a stable bromide dilution technique, and to consider further the effects of body habitus upon the Ca/Cl ratio.

SUMMARY

A technique has been proposed for the absolute measurement of TBCa based on the use of TBCl as an internal standard. The technique has been shown to be applicable over a range of body habitus, as determined from phantom measurements. Further work is needed to establish the validity of the method, although initial results are very encouraging.

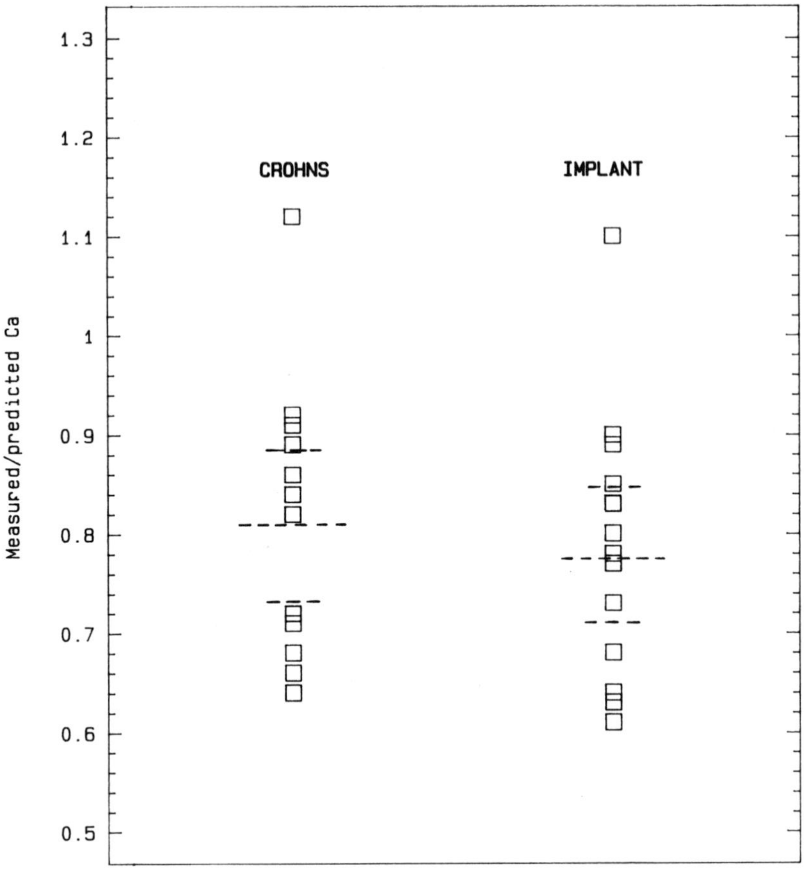

Fig. 3. Comparison of measured-to-predicted TBCa in two patient groups. Dashed lines indicate the mean ±95% confidence interval.

ACKNOWLEDGMENTS

We are grateful to the Medical Research Council for supporting the construction and calibration of the instrument through Project Grants G82 0801353 and G86 00030CA. The support of the National Kidney Research Fund, the Wellcome Trust, and Organon is also gratefully acknowledged.

REFERENCES

1. S. H. Cohn, *Atom. Energy Rev.* **18,** 599 (1980).
2. D. R. Chettle and J. H. Fremlin, *Phys. Med. Biol.* **29,** 1011 (1984).
3. R. G. Zamenhof, O. L. Deutsch, and B. W. Murray, *Med. Phys.* **6,** 170 (1979).
4. W. D. Morgan, S. J. S. Ryde, J. Dutton, C. J. Evans, and A. Sivyer, *Proc. Vth Int. Conf. on Nucl. Meths. In Environ. and Energy Res.*, J. R. Vogt, ed., US DOE Conf-840408, 1984, pp. 751–758.

5. W. D. Morgan, D. Vartsky, K. J. Ellis, and S. H. Cohn, *Phys. Med. Biol.* **26,** 413 (1981).
6. S. J. S. Ryde, W. D. Morgan, S. Cobbold, A. Sivyer, J. Dutton, C. J. Evans, J. E. Compston, and R. Motley, *Osteoporosis and Bone Mineral Measurement* (IPSM 4), E. F. J. Ring, W. D. Evans, and A. S. Dixon, eds., IPSM, York, UK, 1989, pp. 105–115.
7. S. J. S. Ryde, *Ph.D. Thesis*, University of Wales, 1988.
8. S. J. S. Ryde, W. D. Morgan, A. Sivyer, C. J. Evans, and J. Dutton, *Phys. Med. Biol.* **32,** 1257 (1987).
9. D. Vartsky, N. W. Prestwich, B. J. Thomas, J. T. Dabek, D. R. Chettle, J. H. Fremlin, and K. Stammers, *J. Radioanal. Chem.* **48,** 243 (1979).
10. S. Yasumura, S. H. Cohn, and K. J. Ellis, *Am. J. Physiol.* **244,** R36 (1983).
11. D. K. Bewley, *Phys. Med. Biol.* **33,** 805 (1988).
12. C. B. Oxby and K. Brooks, *Phys. Med. Biol.* **24,** 440 (1979).
13. K. J. Ellis, private communication (1989).
14. J. E. Compston, W. D. Evans, E. O. Crawley, D. Judd, R. Motley, C. Evans, and J. Rhodes, *In Vivo Body Composition Studies*, K. J. Ellis, S. Yasumura, and W. D. Morgan, eds., IPSM, London, 1987, pp. 226–233.

Applications in Medical Research

Combination of Neutron Activation Analysis, Tracer Techniques, and Biochemical Methods in the Investigation of Selenium Metabolism

D. BEHNE,* S. SCHEID, H. HILMERT, H. GESSNER, D. GAWLIK, AND A. KYRIAKOPOULOS

Hahn-Meitner-Institut Berlin, D-1000 Berlin 39, FRG

Received June 5, 1989; Accepted November 16, 1989

ABSTRACT

In several studies on rats, the metabolism of selenium was investigated. The quantitative determination of the element was carried out by instrumental neutron activation analysis. For in vivo tracer experiments, ^{75}Se-labeled selenium compounds were used. In addition to these methods, procedures for the measurement of the selenoenzyme glutathione peroxidase, and for the investigation of other selenoproteins, were applied. In this way, information on the specific pools and sites of action of the element, on biologically important selenoproteins and the regulation of the selenium metabolism, was obtained.

Index Entries: Selenium metabolism; glutathione peroxidase; selenoproteins; rat.

INTRODUCTION

The discovery that selenium is essential to animals (*1*) and that it is a biologically active component of the enzyme glutathione peroxidase (*2*) has led to widespread interest in the metabolism and biological functions

*Author to whom all correspondence and reprint requests should be addressed.

of this element. This interest in selenium was further enhanced by the finding that a low selenium status may play a role in several pathological states (3). Diseases that are discussed in connection with selenium deficiency are, for instance, sterility, muscular dystrophies, cardiomyopathy, cardiovascular disease, and cancer.

In selenium research, analytical methods have been used to determine the element level in various tissues. The kinetics of the metabolism, such as uptake and excretion, have been studied by means of tracer experiments. In this paper, it will be shown, with some examples, how considerably more detailed information can be obtained if these methods are applied simultaneously and combined with biochemical procedures.

METHODS

In these studies, rats were fed either a selenium-deficient diet, or diets with sufficient amounts of the element, for several generations. For the tracer experiments, selenium compounds labeled with ^{75}Se were used. After sampling, several biochemical parameters were determined, such as the enzymatic activity of glutathione peroxidase or the protein concentrations. For the identification and investigation of the biologically active selenium compounds, protein separation methods were applied. The analytical procedures included the measurement of the selenium content and the ^{75}Se tracer in the tissues or tissue fractions.

For the determination of the selenium levels, instrumental neutron activation analysis was used, either via the long-lived ^{75}Se or the short-lived ^{77m}Se. The data for these procedures are listed in Table 1. Both methods have advantages and disadvantages. With the short-lived radionuclide, the information is obtained almost immediately, but the detection limit is relatively high. In several cases, especially in the investigation of extremely selenium-depleted animals, methods with lower detection limits were needed. In these cases, selenium determination via the long-lived ^{75}Se, after long-time irradiation at high neutron fluxes, was used. Here, long decay periods of about 2 to 3 mo were necessary in order to reduce the interfering phosphorus bremsstrahlung sufficiently. However, since very large numbers of samples had to be analyzed in these experiments and the time factor was not so important, we preferred the instrumental method to the quicker, but more laborious, radiochemical separation.

STUDIES ON SELENIUM METABOLISM

The first step in our investigation of selenium metabolism was the determination of the selenium contents in the different tissues of the rat, carried out in order to obtain information on specific pools and sites of action (4). In this study, it was found that the element is unevenly

Table 1
Determination of Selenium in Biological Materials by Means of Instrumental Neutron Activation Analysis[a]

Radionuclide	^{75}Se	^{77m}Se
Half-life	120 d	17.5 s
Activation	10 d at $7x10^{13} cm^{-2} s^{-1}$	20 s at $2x10^{13} cm^{-2} s^{-1}$
Decay time	60 - 90 d	0.4 s
Counting time	2 h	48 s
Detection limit	0.15 ng	5 ng
Precision	2 %	8 %
Accuracy	1.14 ± 0.11	1.1 ± 0.1 µg/g

[a]The accuracy was tested by using the standard reference material NBS-SRM 1577 with a selenium content of 1.1 ± 0.1 µg/g.

distributed among the tissues. The highest levels found in the testes (6.7 mg Se/kg dry mass) and the kidneys (6.0 mg Se/kg) differed from the lowest in the muscle and brain (0.6 mg Se/kg) by one order of magnitude.

Further important information was obtained by the simultaneous determination of the glutathione peroxidase activity in the tissues. This enzyme, at present, is the only selenium compound with known biological functions in mammals. From the selenium content and the glutathione peroxidase activity, the percentage of the tissue selenium bound to the enzyme was calculated (*4*). In this way, it was possible to estimate whether the element in a tissue was mainly present as glutathione peroxidase or in the form of other selenium compounds. It was found that the enzyme-bound fraction of the element was particularly high in the erythrocytes and liver and low in the muscle, brain, and testes.

By calculating the total amount of the enzyme-bound element, it was estimated that the majority of the selenium in the body of the rat is not contained in the glutathione peroxidase, but in other compounds instead. This is of great interest since after the discovery that selenium is an essential part of glutathione peroxidase, most of the studies in selenium research have concentrated on the enzyme. This finding was the basis of our research on new selenoproteins, discussed later.

This example shows that, by combining the values for the selenium content and the glutathione peroxidase activity, much more detailed information can be obtained than by independent measurement of the two parameters. The same is true if one combines the analytical values for the selenium content with the data of tracer experiments with ^{75}Se.

This approach was used, for instance, in the investigation of the

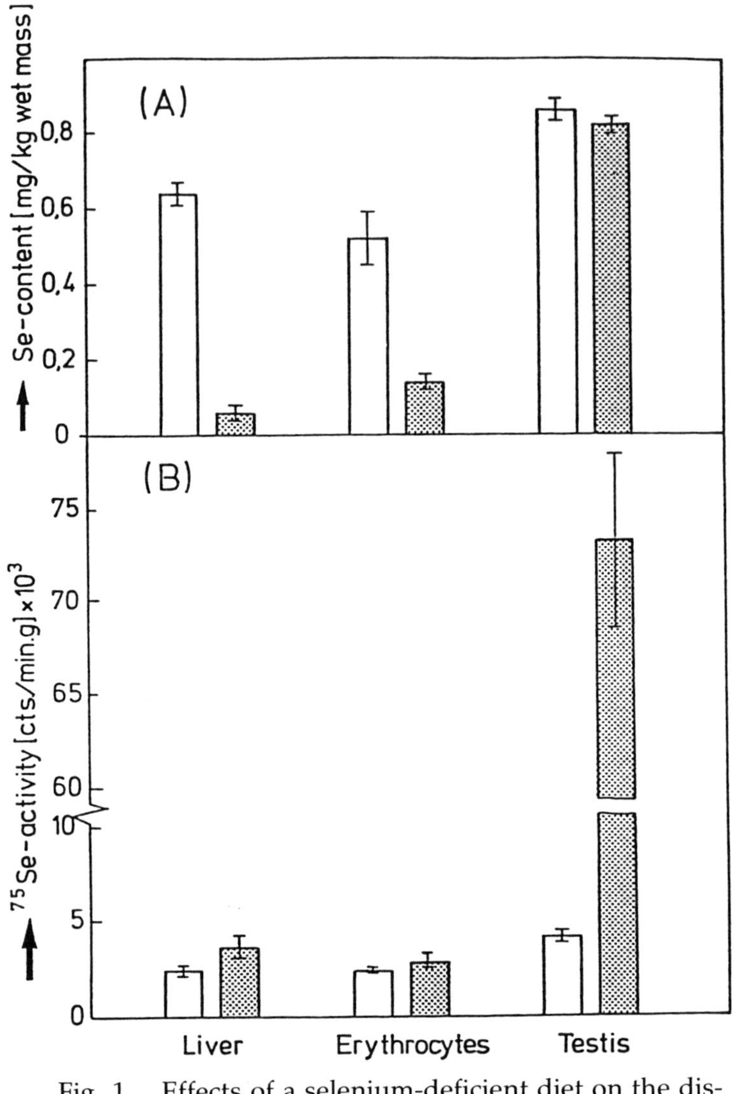

Fig. 1. Effects of a selenium-deficient diet on the distribution of selenium in male rats. (A) Selenium content (mean ± SD, $n = 4$) in animals fed on a low selenium diet with <0.015 mg Se/kg (shaded bar), or the same diet with 0.25 mg Se/kg added as selenite (open bar) for 70 d. (B) ^{75}Se activity (mean ± SD) in animals fed on the selenium-adequate ($n = 5$), or the selenium-deficient diet ($n = 4$) for 95 d, 35 d after the injection of ^{75}Se-selenite (*5*).

selenium metabolism in rats during selenium depletion (*5*). In Fig. 1A, the changes in the selenium content in some tissues are shown after the selenium-deficient diet had been fed for 70 d. Compared with sufficiently-fed controls, the greatest decreases were found in the liver and the erythrocytes. In the testes, however, the selenium content did not differ significantly from the value of the control animals. This phenomenon cannot be explained only by consideration of a very slow turnover of the element in the male gonads and, thus, a slower loss than in the other tissues, since, as a result of the growth of the testes during the depletion period, the selenium levels would already have been reduced to a much greater extent.

When the animals were then given a small amount of labeled selenium (Fig. 1B), we found a very interesting effect. It was not the tissues with a low selenium status that retained a large proportion of the dose, but retention was highest in the testes even though the selenium level had remained more or less unchanged.

The results of some of our later studies on the functions of the element suggested, however, that the testes are not the only organ in which the selenium level is regulated. Based on these findings, we then performed a tracer experiment similar to the one described before, in which the retention of the element in Se-deficient rats was investigated in a larger number of tissues (*6*). Since it was expected that the differences in the retention would be more clearly noticeable, the lower the selenium status of the animals, extremely selenium-depleted rats were used in this study. These animals and selenium-supplemented controls were given a very small dose of labeled selenite. From the ^{75}Se-distribution 6 wk after administration of the tracer, retention factors were calculated by relating the specific activity in each tissue in the deficient rats to the corresponding value in the controls. Table 2 shows the highest and lowest retention factors thus obtained.

From these results, it can be seen that the severely-depleted animals retained about 20–25 times more of the dose in the brain, the reproductive organs, and several endocrine organs than the selenium-adequate controls. As in the experiment mentioned above, retention was lowest in the tissues that, during depletion, had the greatest decreases in their selenium contents. It is interesting to note that the heart, muscle, and liver, which are in this group, are the tissues in which pathological changes first appear in selenium deficiency. The differences in the maintenance of the selenium levels, therefore, might explain the occurrence of lesions in these tissues. The results also show that, in animals with a low selenium status, there is a priority transport of the element to certain target tissues. Since it can be assumed that selenium is preferentially supplied to the sites in which it is most needed, these findings indicate important functions of the element in these organs.

STUDIES ON SELENOPROTEINS

In the experiments discussed so far, the selenium metabolism was investigated without consideration of the chemical forms of the element in the tissues. As mentioned above, we calculated that less than half of the selenium in the rat was bound to glutathione peroxidase. Therefore, it was of interest to find out to what extent other selenoproteins, besides

Table 2
Effects of the Selenium Status on the Selenium Retention in Rats[a]

Tissue	RF	Tissue	RF
Brain	51.9 ± 2.6	Heart	1.9 ± 0.2
Thyroid	46.9 ± 3.9	Muscle	1.8 ± 0.2
Pituitary	42.0 ± 3.6	Plasma	1.7 ± 0.2
Testes	22.5 ± 2.6	Liver	0.9 ± 0.05
Adrenals	21.9 ± 2.4	Erythrocytes	0.6 ± 0.04
Ovaries	24.2 ± 3.0		
Corpora lutea	42.7 ± 9.1		

[a]A small dose of selenite, labeled with ^{75}Se, was injected into rats that had been kept on either a selenium-deficient or selenium-adequate diet. Six weeks after the administration of the tracer, the tissue ^{75}Se activities were determined. For each tissue, the retention factor, RF, was calculated as a ratio of the specific ^{75}Se activity (% of dose/g tissue) in the depleted rats to that in the control animals. The data (*6*) represent the mean ± SEM for five male animals in each group. Corpora lutea and ovaries were taken from similarly-treated females.

this enzyme, are present in the organism and, therefore, should be included in studies on the metabolism and functions of the element.

For these investigations, the tissue selenoproteins were labeled in vivo by injecting rats with ^{75}Se-selenite several times during a period of 5 wk (*6*). The animals were then killed and the proteins separated, using SDS-polyacrylamide gel electrophoresis. The ^{75}Se-containing compounds in the gel were identified by means of autoradiography. By investigating a large number of rat tissues, a total of 13 selenium-containing proteins or protein subunits have been detected in this way. Their molecular weights are shown in Table 3. Most of the selenoproteins were found in all the tissues. With protein nos. 9 and 13, which were found in nearly all of the tissues, it can be assumed that they too occur in all tissues, but were not sufficiently labeled in some to be detected. Protein no. 6 is the glutathione peroxidase subunit. Protein no. 7 seems to be a special case. It is one of the main selenoproteins in the thyroid, but was not present in most of the other tissues investigated.

After these selenoproteins had been detected, we wanted to obtain some information on their tissue levels and the selenium distribution among these compounds in the normal state. Direct measurement of the selenium concentration in these proteins was not possible since, in the separation methods with sufficiently high resolution, only very small amounts of the proteins could be separated, and their selenium concen-

Table 3
Selenium-Containing Proteins or Protein Subunits in Tissues of the Rat[a]

Protein		M_r	Protein		M_r
1	+++	12100 ± 500	8	+	33300 ± 1300
2	+++	15600 ± 300	9	++	55500 ± 1300
3	+++	18000 ± 300	10	+++	59900 ± 2200
4	+++	19700 ± 400	11	+++	64900 ± 1200
5	+	22200 ± 600	12	+	70100 ± 1300
6	+++	23700 ± 700	13	++	75400 ± 2300
7	+	27800 ± 400			

[a]The ^{75}Se-labeled proteins in homogenates of adrenals, brain, corpora lutea, heart, kidneys, liver, lungs, muscle, ovaries, pituitary, prostate, spleen, testes, and thyroid were separated by SDS-polyacrylamide gel electrophoresis and identified by autoradiography (6). Their relative molecular weights, M_r, are the mean ± SD of five determinations. The selenoproteins were numbered in the order of their increasing molecular masses. The plus signs indicate that the protein was detected either in a few (+), nearly all (+ +), or all (+ + +) of the tissues investigated.

trations were below the limit of detection. We could, however, quantitatively determine the ^{75}Se tracer. Therefore, an experiment was carried out in which we tried to match the distribution of the tracer with the native selenium (6). For this purpose, we repleted the selenium pools in selenium-deficient rats by feeding a selenium-adequate diet and injected the tracer at regular intervals during the repletion period. We then determined the distribution of the tracer in the tissues and compared it with the distribution of the native selenium in adequately-supplied control animals. The highly significant correlation ($r = 0.98$) found between the selenium content in the controls and the tracer activity in the repleted animals showed that, under these conditions, the tracer was distributed in the same way as the native selenium, and the ^{75}Se activity, therefore, reflected the selenium content.

Table 4 shows the relative distribution of the main selenoproteins found in this way in the repleted animals and, thus, in the normal state in the brain, liver, and thyroid. In most of the tissues, the element was present in seven main proteins, including the glutathione peroxidase. Only in the thyroid was there an additional main selenium compound, and it can be assumed that selenoprotein no. 7 has specific biological functions in this organ.

Table 4
Selenium Distribution Among Selenoproteins in Rat Tissues[a]

Protein	Relative protein ^{75}Se-activity (%)		
	Adrenal cortex	Brain	Thyroid
1	5	15	7
2	14	13	16
3	3	9	7
4	11	15	7
6 (GSH-Px)	44	15	23
7	--	--	13
10+11	15	20	11

[a]Selenium-deficient male rats were fed a selenium-adequate diet for 4 wk. ^{75}Se-selenite was administered repeatedly during repletion. The ^{75}Se activity in the main labeled proteins was measured and expressed as a percentage of the total protein-bound tissue ^{75}Se (*6*).

CONCLUSIONS

By combining neutron activation analysis, tracer methods, and biochemical procedures, detailed information on the metabolism of selenium was obtained. In this way, several new selenoproteins were detected in the tissues of the rat. The studies also provide evidence for the existence of regulatory mechanisms that, during an insufficient dietary selenium intake, maintain the supply of the element, above all to the brain and the organs with reproductive and endocrine functions. Liver, heart, and muscle are last in this hierarchy, which might be the reason for the occurrence of the diseases found in these tissues during selenium deficiency.

REFERENCES

1. K. Schwarz and C. M. Foltz, *J. Am. Chem. Soc.* **79,** 3292 (1957).
2. J. T. Rotruck, A. L. Pope, H. E. Ganther, A. B. Swanson, D. G. Hafeman, and W. G. Hoekstra, *Science* **179,** 585 (1973).
3. O. A. Levander, *Trace Elements in Human and Animal Nutrition*, vol. 2, W. Mertz, ed., Academic, Orlando, FL, 1986, pp. 209–279.

4. D. Behne and W. Wolters, *J. Nutr.* **113,** 456 (1983).
5. D. Behne, T. Hoefer, R. von Berswordt-Wallrabe, and W. Elger, *J. Nutr.* **112,** 1682 (1982).
6. D. Behne, H. Hilmert, S. Scheid, H. Gessner, and W. Elger, *Biochim. Biophys. Acta* **966,** 12 (1988).

Speciation of Chromium in Plasma and Liver Tissue of Endstage Renal Failure Patients on Continuous Ambulatory Peritoneal Dialysis

FRÉDÉRIQUE BORGUET,[1] RITA CORNELIS,*,[1] AND NORBERT LAMEIRE[2]

[1]*Laboratory of Analytical Chemistry, Rijksuniversiteit Gent, Proeftuinstraat 86, B-9000 Gent, Belgium; and* [2]*Renal Division, University Hospital, Gent, Belgium*

Received April 17, 1989; Accepted November 16, 1989

ABSTRACT

This work is a first attempt to determine the speciation of Cr in human plasma. With the aid of in vitro and in vivo ^{51}Cr-labeled experiments, it was possible to develop the necessary biochemical techniques for the separation of the plasma proteins. Further work will use real samples, taking care to avoid contamination of the various fractions and to preserve the original binding of the Cr to the specific plasma compounds.

In a first attempt on the distribution of Cr over the different organelles of liver tissue, work will be restricted to in vivo labeled experiments with rats. The procedure to do the speciation work seems so elaborate that it may be impossible ever to achieve the contemplated speciation of Cr in human liver tissue by subcellular fractionation.

Index Entries: Chromium in human plasma, speciation of; chromium in human liver, speciation of; patients on continuous ambulatory peritoneal dialysis (CAPD); endstage renal failure, chromium in; gel permeation chromatography and ion exchange chromatography, chromium speciation with.

*Author to whom all correspondence and reprint requests should be addressed.

INTRODUCTION

This research was initiated by our finding of extremely high Cr levels in serum of patients with terminal renal failure and on chronic hemodialysis or continuous ambulatory peritoneal dialysis (CAPD) (*1–3*). A mean serum value of 2.22 ± 1.33 ng/mL (mean ± SD), range 0.70–4.34 ng/mL (n = 6), was found for hemodialysis patients and 4.30 ± 1.92 ng/mL, range 1.80–7.05 ng/mL (n = 5) for CAPD patients, whereas the normal Cr content is 0.160 ± 0.083 ng/mL (*4*). These observations were confirmed by a research group in Glasgow, who obtained 0.2 ± 0.2 ng/mL in controls, 6.2 ± 2.1 ng/mL in patients on hemodialysis, and 7.5 ± 1.7 ng/mL on CAPD (*5*). The underlying reason for this increased concentration is that the small amount of Cr, 1 ng/mL, present as an impurity in the dialysate liquid, is irreversibly taken up by certain substances in the serum. This occurs against a concentration gradient. It is the purpose of this study to detect to which substances in serum or plasma this Cr is bound.

On the basis of kinetic modeling studies of Cr in the human body, an accumulation of Cr in liver tissue of CAPD patients may be anticipated. Therefore, the second aim of this research is directed to the speciation of Cr in liver cell organelles (*6*).

MATERIALS AND METHODS

Plasma

The experiments to speciate the Cr in plasma started with in vitro and in vivo labeled ^{51}Cr plasma samples. This approach reduced the detection of the element Cr to a mere radioactivity measurement since only ^{51}Cr is to be detected, thus circumventing the immense problem of possible contaminations of the samples with exogenous Cr. This methodology has two major shortcomings. First, there is no certainty that the nonradioactive or "cold" Cr present in the plasma has exchanged with the "warm" ^{51}Cr spike. Second, the Cr added may be easily bound to one or the other of the many ligands present in plasma. This may lead to the finding of "artefacts." The experiments were carried out under as much noncontaminating conditions as possible for Cr. Blood samples were collected from healthy individuals and CAPD patients. Heparin was added to the samples, and the plasma was separated from the packed cells after 20 min centrifuging at 3000 rpm. The plasma was kept in the refrigerator for no longer than 5 d. The in vitro labeled human plasma was obtained after a 2-h incubation with ^{51}Cr(III) (Amersham, UK 0.1*M* HCl solution; carrier 3 pg/Bq) under continuous shaking.

Sprague Dawley rats were injected intraperitoneally (ip) on two consecutive days, each with 10 mL ± 0.4 Bq ^{51}Cr-labeled hypertonic glucose dialysate (3.86% glucose) (Baxter Co., Deerfield, IL). The animals were anesthetized with ether, and the blood was taken by venous puncture with a heparinized needle. This procedure induces less artefacts, compared with in vivo labeling.

Liver

Liver tissue was obtained from a female patient (age 39 yr), who died after a trauma, and a male patient of 61 yr with terminal renal failure owing to chronic pyelonephritis and treated for 7 yr with CAPD.

Big lumps of those livers, obtained at autopsy, were immediately stored in a freezer at −26°C. All subsequent manipulations were done under clean room circumstances. After partial thawing, the outer layer of the liver tissue was removed with a polyethylene knife in order to eliminate contaminated surfaces. Liver samples of 2.5 g wet tissue were stored in clean polyethylene vials.

Methods

The techniques used to separate the proteins were gel permeation chromatography (GPC) and cation and anion exchange chromatography (GPC gels: Sephacryl S200 and Fractogel TSK HW55. Eluents: 0.1*M* NH_4Ac at pH 5 and 7; 1.39% $(NH_4)_2CO_3$ pH 7, and 2.3% lactate solution, pH 5.5). The NH_4Ac buffer is made from ultrapure NH_4OH, HAc, and H_2O. This was considered the buffer of choice because of its absence of measurable amounts of Cr impurities. A second advantage is that this buffer can be sublimated through lyophilization. Column characteristics: height, 1 m; diameter, 2.4 cm; sample volume: 5 mL plasma; the column was eluted at 0.7 mL/min; duration: 10 h. A high capacity column was required, in view of the 5 mL plasma to be separated. This quantity is required for the detection of Cr in the separated fractions.

The cation exchange chromatography column was: carboxymethylgel CM-Sephadex C-50 and the eluting agents were: 0.025*M* Tris-HCl pH 8; 0.01*M* NH_4Ac pH 8. The column characteristics were: height, 15 cm; diameter, 1.5 cm; sample volume: 1 mL plasma; the column was eluted at 1 mL/min; duration: 30 min. The eluted proteins were detected with a UV monitor (wavelength 280 nm).

The anion exchanger was diethylaminoethyl (DEAE) gel (DEAE tris-acryl® M). The eluting agent was a NaCl gradient (0–0.25*M*) in 0.025*M* Tris-HCl, pH 8, or in 0.01*M* NH_4Ac, pH 8 (column characteristics: sample volume: 1–10 mL; the column was eluted at 1 mL/min; duration: 2h).

The identification of the proteins (essentially transferrin and albumin) was done with gel electrophoresis (PhastSystem™ Pharmacia, Uppsala, Sweden). The separation consisted of an isoelectric focusing in the pH interval 3–9 (Gel: Phastgel IEF 3–9 Pharmacia). The concentration of the transferrin and albumin was determined by nephelometry, using the Behring Nephelometer.

The activity of the ^{51}Cr present in the different fractions was measured by NaI(Tl) scintillation spectrometry.

Neutron activation analysis was used to determine the total concentration of Cr in the liver samples. The lyophilized liver samples were irradiated for 28 h at a flux of $2 \cdot 10^{12}$ n/cm^2.s in the Thetis reactor of the University of Gent. Radiochemical separation of the Cr followed irradiation. This necessitated the destruction of the sample in 6 mL 1:1 14*M* HNO_3: 70% $HClO_4$ and distillation of the Cr as CrO_2Cl_2 in the presence of 5 mg CrO_3 carrier and with the aid of a continuous flow of dry HCl gas. The ^{51}Cr activity was measured for 24 h by Ge(Li) spectrometry. The Cr standard (Cr metal 99.999% Cr UCB, Brussels, Belgium) was dissolved in HCl and measured under identical conditions as the samples.

RESULTS AND DISCUSSIONS

Chromium in Plasma

Previous studies (*7–9*) have already shown that Cr is mainly bound to transferrin and, to a lesser degree, albumin. Therefore, a method was developed to separate transferrin and albumin from human plasma. In the past, the investigations simply aimed at the qualitative speciation of the Cr in their relation to particular molecules. The present research pursues a quantitative speciation in human plasma. Many problems may be anticipated, as described by Gardiner (*10*).

The GPC-DEAE Experiments

The first strategy consisted of separating the human plasma proteins with GPC. Owing to their relatively small difference in molecular mass, transferrin (± 80,000 dalton) and albumin (± 70,000 dalton) are eluted together from the GPC column. These proteins are then separated on the basis of their different number of negative charges by passing them through a DEAE ion exchange chromatography column. These experiments were associated with insurmountable problems.

Elution of 5 mL ^{51}Cr-labeled human plasma of healthy individuals on the Sephacryl S200 column with 0.1*M* NH_4Ac at pH 7 resulted in a 77% absorption of the ^{51}Cr on the gel. The recovered 23% Cr appeared to be split into two regions: about 78% of this Cr eluted in the albumin + transferrin region and 20% showed up at a mass of ± 5,000 dalton (*See,* Fig.1). These experiments point to a fundamental disturbance in the original Cr-protein binding. A major part of the Cr-protein has probably been subject to hydrolysis phenomena. Notwithstanding this major shortcoming, work continued on the very low yield (Cr − transferrin + albumin) fraction collected from the GPC column. For further separation of this fraction, its volume had to be reduced. Therefore, the fraction was first lyophilized and then suspended in 2 mL buffer. The attempt to separate transferrin and albumin on the DEAE ion exchange column with

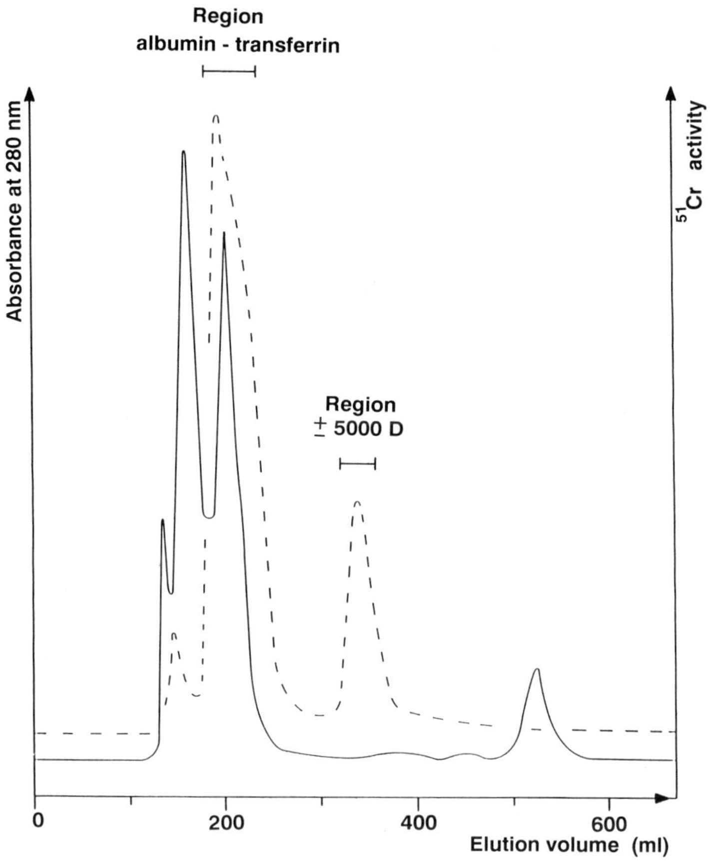

Fig. 1. UV chromatogram (-) and ^{51}Cr elution pattern (---) of ^{51}Cr-labeled human plasma after separation over the GPC column Sephacryl S 200 with 0.1*M* NH_4Ac, pH 7 eluent.

a 0–0.25*M* NaCl gradient in 0.025*M* Tris-HCl at pH 8 failed completely, since the proteins and the ^{51}Cr eluted together with the void volume.

These experiments were repeated on a different type of GPC column, Fractogel, but the results were similar. The elution with 0.1*M* NH_4Ac at pH 7 resulted in an improved recovery of 35% of the Cr from the column. The distribution of the eluted ^{51}Cr over the different fractions was similar, including the appearance of a ghostpeak at ± 5000 dalton. The explanation for the major retention of the Cr on the column was thought to be the cleavage of the original Cr-protein binding. The absence of complexing agents (such as lactate, carbonate and so on) gives rise to the hydrolysis of the Cr and formation of polymers that cannot be eluted anymore. Apparently, acetate ions cannot fulfill the role of a complexing agent. Actually, the tendency to hydrolyze already starts at pH 4. These hydrolysis phenomena can be counteracted by elution at a lower pH or through the addition of complexing agents, thus allowing any free Cr present to elute as a mobile complex. It is not yet possible to predict if the complexing agents may compete with the original ligands of the proteins and thus disrupt the original bindings.

Experiments with ^{51}Cr-labeled human plasma on Fractogel GPC and 0.1*M* NH_4Ac at pH 5 instead of pH 7 showed far less retention of the ^{51}Cr on the column. It decreased to 15% at pH 5 v 65% at elution with NH_4Ac at pH 7. Recovery became 100% when eluting with 1.39% $(NH_4)_2CO_3$ and 2.3% lactate solution. The ^{51}Cr activity was not only traced back in the albumin-transferrin region, but also on a molecule with a mass of ±5000 dalton. The presumption that this 5000 dalton peak is an artefact was proven by eluting the lyophilized albumin-transferrin fraction once more on the GPC column. This resulted in a 5000 dalton peak, with Cr associated to it, splitting off from the originally solely albumin + transferrin. The Cr was roughly divided as 80 and 20% over the two fractions.

Through these consecutive failures, it became clear that it is impossible to keep the original Cr-protein binding intact under these experimental conditions (GPC gel: Fractogel or Sephacryl; column, 1 m high; diameter, 2.4 cm; elution time, 10 h). A gradual cleavage of the Cr-protein binding occurs. The idea occurred that this phenomenon might be directly related to the long residence time of the compounds on the resin. Therefore, a much shorter separation scheme was designed to diminish possible changes in the protein structure, including interactions of the protein and the Cr with the separating medium.

Cation and Anion Exchange Separations Combined

A preliminary experiment consisted in subjecting the plasma, without any previous separations, to anion exchange chromatography. One milliliter of ^{51}Cr-labeled human plasma was eluted on a DEAE column with a 0–0.25*M* NaCl gradient in 0.025*M* Tris-HCl, pH 8. No adequate separation between transferrin and albumin could be achieved. The fact that a major part of the proteins and the ^{51}Cr activity eluted in the void volume made it clear that such a procedure is pointless. Isoelectric focusing of this fraction revealed that this void volume fraction not only contained albumin and transferrin, but also many more positive ions.

The procedure was adapted to the new observations. The plasma was first eluted over the cation exchanger column with 0.025*M* Tris-HCl at pH 8. (The use of 0.01*M* NH_4Ac at pH 8 as an eluting agent proved unsatisfactory.) The least positive fraction was then eluted over the DEAE column, with a 0–0.25*M* NaCl gradient in 0.025*M* Tris-HCl at pH 8. The UV absorption diagrams and the ^{51}Cr elution patterns are given in Figs. 2 and 3. Qualitative analysis of the fractions, with the aid of isoelectric focusing (pH 3–9), showed that fraction C_2, obtained from the cation exchange, contained transferrin and albumin, together with some positive proteins. Peaks A_2 and A_3 from the DEAE separation, respectively, carried the transferrin and albumin. Peak A_1 showed positive proteins, but neither transferrin nor albumin appeared to be present.

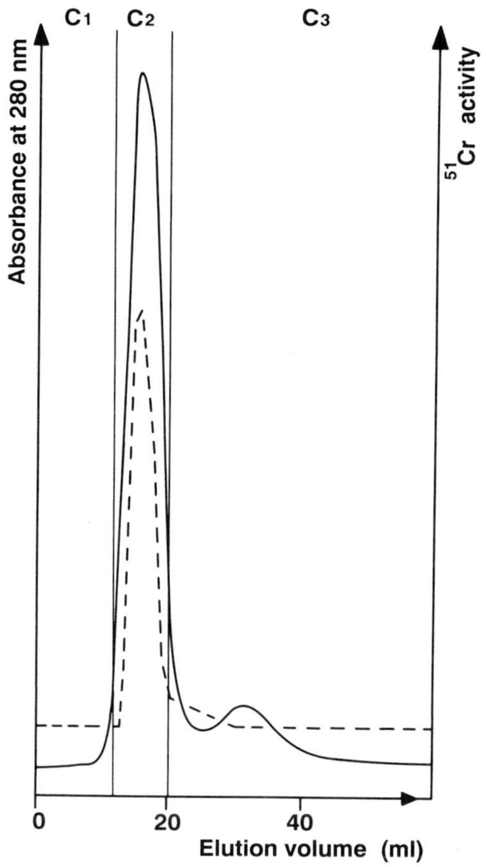

Fig. 2. UV chromatogram (-) and ^{51}Cr elution pattern (---) of ^{51}Cr-labeled human plasma after separation over the cation exchanger with 0.025*M* Tris-HCl, pH 8 eluent.

Anion exchange chromatography of the least positive fraction from the cation exchange proved adequate to materialize the separation of transferrin and albumin. The recovery of transferrin and albumin in the isolated fractions vs the "original content in the plasma" was done with the aid of nephelometry.

Table 1 gives the transferrin and albumin concentrations in the fractions C_1, C_2 and C_3, and A_1, A_2, A_3, and A_4. It appears that, after cation exchange, the transferrin and albumin are quantitatively collected. Fraction C_2, which was used for anion exchange, contained 92% of the albumin and 78% of the total transferrin. After anion exchange, all of the transferrin could be located in peak A_2. The albumin was distributed over different fractions: 56% in A_3, 5% spread over the other fractions, and 31% retained on the column.

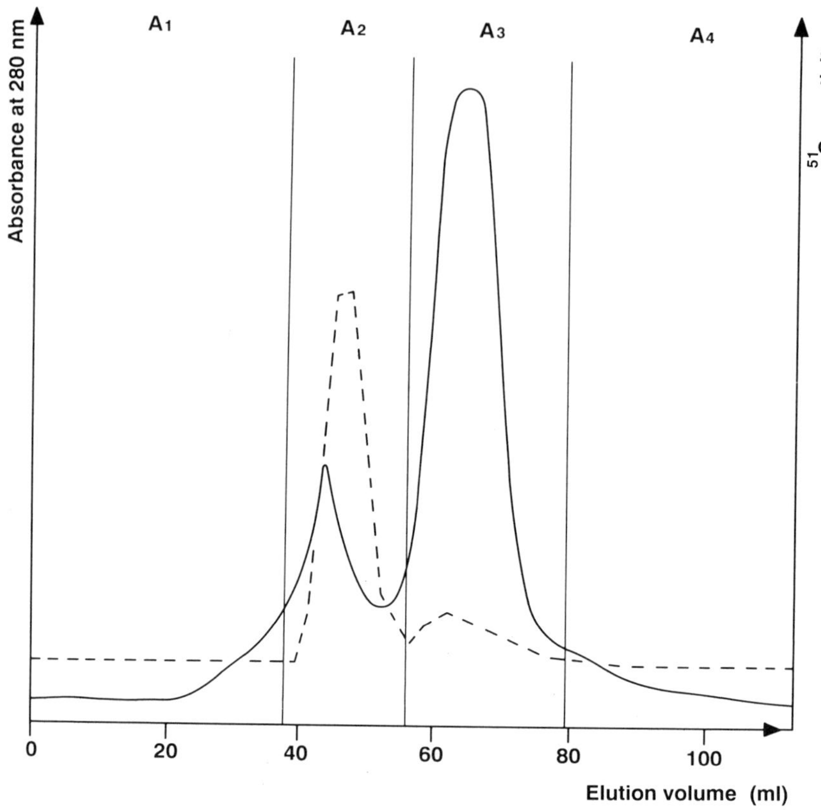

Fig. 3. UV chromatogram (-) and ^{51}Cr elution pattern (---) of ^{51}Cr-labeled human plasma after separation over the DEAE anion exchanger of fraction C_2 of Fig. 2 (^{51}Cr-labeled human plasma after separation over the cation exchanger) with 0–0.25*M* NaCl gradient in 0.025*M* Tris-HCl, pH 8.

Measurement of the ^{51}Cr radioactivity of the isolated fractions allowed calculation of the recovery of the procedure. It proved to be quantitative after cation exchange. Fraction C_2 submitted to the anion exchange carried 90% of the initial Cr activity. The transferrin fraction A_2 retained 62% of the Cr. Furthermore, 13% was present in A_3 (albumin), 6% was divided over the other fractions, and 10% was sticking to the column.

The ^{51}Cr distribution over the different protein fractions has been studied in in vitro ^{51}Cr-labeled plasma of healthy individuals and of CAPD patients. The pattern was similar in both cases.

The behavior of Cr in the in vitro incubated plasma may, however, be different from that of the Cr that is normally carried by the bloodstream. This prompted experiments with in vivo labeled rat plasma. It was submitted to a preliminary cation exchanging, followed by anion exchange of the least positive fraction. The UV chromatograms of the proteins were similar to human plasma. The recovery of the Cr in the

Table 1
Distribution of Albumin and Transferrin over the Different Fractions After Separation of Human Plasma Proteins over Cation and Anion Exchangers[a]

Concentration	Transferrin mg/100 mL	Albumin mg/100 mL
start	242	4200
cation exchange		
C_1	9	–
C_2	190	3 847
C_3	29	366
anion exchange		
A_1	–	70
A_2	186	44
A_3	–	2343
A_4	–	82

[a]0–0.25*M* NaCl gradient in 0.025*M* Tris-HCl, pH 8.

different fractions is given in Table 2. From these data, it can be seen that, upon elution of in vivo incubated rat plasma over a cation exchanger, 84% of the Cr is eluted. The fraction (C_2) submitted to anion exchange contained 77% of the Cr. After separation of this C_2 fraction, all activity was found in the transferrin fraction (A_2). Contrary to the in vitro incubated human and rat plasmas, no ^{51}Cr activity was found in the albumin fraction (A_3).

Chromium in Liver

The Cr content in the liver of the healthy individual amounted to 3 ng/g wet weight. This is in agreement with the previous findings at the laboratory by Lievens and coworkers (*11*) and Vuori and coworkers (*12*) in Finland. The Cr concentration in the liver of the CAPD patient was 100 ng/g wet tissue, i.e., 30 times higher than the normal value. Chromium is known to exert mutagenic effects when it succeeds in penetrating the cell nucleus. Therefore, it was considered very important to learn in which subcellular fractions Cr resides.

The procedure to be followed is the subcellular fractionation method. The liver cells are first homogenized by hand with a Potter-Elvehjem homogenizer (glass) in a 0.25*M* sucrose solution, pH 7.4, at 4°C and passed through a nylon sieve. The homogenate is next submitted to

Table 2
Recovery of ^{51}Cr in the Different Fractions Obtained After Cation and Anion Exchange of In Vivo ^{51}Cr-Labeled Rat Plasma

cation exchange	% recovery of the ^{51}Cr
C_1	–
C_2	76.73
C_3	7.03
anion exchange	
A_1	0.73
A_2	75.97
A_3	1.36
A_4	–

differential centrifugation. Five fractions (four precipitates and the supernatant) are separated at 600, 10,000, 37,000, and 100,000*g*, respectively. Although the nuclei, mitochondriae, lysosomes, and microsomes will be relatively enriched, a fraction never consists of a single organelle. The distribution of an organelle over the different fractions can be learned by the analysis of its specific enzymes, so-called marker enzymes. For instance, the mitochondriae contain cytochrome-C oxidase and succinate dehydrogenase, whereas the lysosomes contain hexosaminidase. The determination of those enzymes allows one to conclude about the presence of the different types of organelles. The determination of the enzymes occurs through the interaction with specific substrates. The intensity of the colored complex formed is measured spectrophotometrically.

The total protein content of each fraction is determined by the Lowry method. The RSA (relative specific activity of a particular enzyme) equals the ratio of the percent of the activity of the enzyme in the fraction to the percent of the protein content of the fraction. It is a measure of the enrichment of a certain organelle. The original idea was to determine the Cr distribution over the different fractions and then, knowing the relative distribution of the organelles by the enzyme analyses of a small, independent aliquot, to deduce the distribution of the Cr over the organelles. Immense problems may be anticipated as far as contamination of the cell fractions is concerned. To this, the lack of knowledge on the preservation of the original Cr binding must be added. It is also a fact that not all of the liver cells present in the homogenate will have been homogenized. Yet, the only feasible experiments seem to be an attempt to learn the subcellular Cr distribution on in vivo labeled ^{51}Cr rat livers. The results will be indicative only since the authenticity of the original Cr binding cannot be ascertained. Experimental work on the liver has now started along the lines described above.

CONCLUSIONS

It is the purpose of this research to apply the above-outlined procedure of ion exchange chromatography to isolate the main Cr-binding proteins from the plasma of healthy individuals and CAPD patients and assess their Cr contents. It is evident that all this should be done in contamination-free circumstances, or with a blank value that is negligible in comparison with the Cr concentration present. This will be a most difficult endeavor. In addition, the challenge not to disturb the original Cr binding with the plasma components should be considered.

Since plasma of CAPD patients carries abnormal lower molecular weight molecules, it is feasible that some of the Cr may be bound to those components. The research on the Cr speciation in liver cells by use of subcellular fractionation seems practicable only in in vitro labeled rat tissue.

ACKNOWLEDGMENTS

The authors thank the National Fund for Scientific Research in Belgium and the Research Board of the University of Gent for their financial support. They are indebted to H. Hilderson and A. Lagrou of the RUCA, University of Antwerpen, for their scientific advice concerning the subcellular fractionation work, and they express their thanks to J. Delanghe for enabling the nephelometric measurements. The authors acknowledge, with gratitude, the technical assistance of L. Mees and D. Kesteloot.

REFERENCES

1. R. Cornelis, S. Ringoir, L. Mees, N. Lameire, B. Wallaeys, and J. Hoste, *Trace Element Metabolism in Man and Animals (TEMA-4)* J. McC. Howell, J. M. Gawthorne, and C. L. White, eds., Australian Academy of Science, Canberra, Australia, 1981, pp. 530–533.
2. R. Cornelis, D. Van Renterghem, B. Wallaeys, L. Mees, N. Lameire, and S. Ringoir, *Trace Elements in Man and Animals (TEMA-5)* C. F. Mills, I. Bremmer, and J. K. Chesters, eds., Commonwealth Agricultural Bureaux, Farnham Royal, Slough, UK 1985, pp. 819–822.
3. B. Wallaeys, R. Cornelis, L. Mees, and N. Lameire, *Kidney Int.* **30,** 599 (1986).
4. J. Versieck, J. Hoste, F. Barbier, and H. Steyaert, *Clin. Chem.* **24,** 303 (1978).

5. D. J. Halls, A. C. T. Leung, I. S. Henderson, G. S. Fell, J. W. Dobbie, and A. C. Kennedy, *Trace Elements in Man and Animals (TEMA-5)*, C. F. Mills, I. Bremmer, and J. K. Chesters, eds., Commonwealth Agricultural Bureaux, Farnham Royal, Slough, UK 1985, pp. 816–819.
6. B. Wallaeys, R. Cornelis, N. Lameire, and M. Van Landschoot, *Proc. 5th Benelux Symposium*. N. H. Lameire and S. M. Ringoir, eds., University of Gent, Belgium 1986, pp. 61–73.
7. L. L. Hopkins and K. Schwarz, *Biochim. Biophys. Acta* **90,** 484 (1964).
8. D. C. Harris, *Biochemistry* **16,** 560 (1977).
9. B. Wallaeys, R. Cornelis, F. Belpaire, and N. Lameire, *Trace Element Analytical Chemistry in Medicine and Biology*, vol. 4, P. Brätter and P. Schramel, eds., de Gruyter, Berlin, FRD 1986, pp. 451–461.
10. P. H. E. Gardiner, *Topics in Current Chemistry* **141,** 145 (1987).
11. P. Lievens, J. Versieck, R. Cornelis, and J. Hoste, *J. Radioanal. Chem.* **37,** 483 (1977).
12. E. Vuori and J. Kumpulainen, *Trace Elements in Medicine* **4,** 88 (1980).

A Search for Longitudinal Variations in Trace Element Levels in Nails of Alzheimer's Disease Patients

D. E. VANCE,[1,2] W. D. EHMANN,*,[1] AND W. R. MARKESBERY[1]

[1]*Department of Chemistry and Sanders Brown Center on Aging, University of Kentucky, Lexington, KY 40506*
[2]*Department of Chemistry, Eastern Kentucky University, Richmond, KY 40475*

Received April 17, 1989; Accepted November 16, 1989

ABSTRACT

Levels of Br, Hg, K, and Zn were determined by INAA in the nails of Alzheimer's disease (AD) patients at 6-mo intervals for up to 3 yr. These elements have been shown to be imbalanced in AD nail. Bromine showed no trends. Mercury tended to decrease in nail with increasing age of patient, and with the duration and severity of the dementia. Potassium and Zn tended to increase with these same factors. Hence, progressive changes in trace-element levels do occur in AD nail, although imbalances are detected even in the earliest sampled stages of the disease.

Index Entries: Alzheimer's disease, INAA, nails, trace element levels, radiochemical analysis, longitudinal study, myasthenia gravis, bromine, mercury, potassium, zinc.

INTRODUCTION

Alzheimer's disease (AD) is a neurological disorder characterized by progressive dementia. A number of recent reviews of the current status of AD research may be found in the literature (*1–3*). The cause of the

*Author to whom all correspondence and reprint requests should be addressed.

disease is unknown. Hypotheses regarding its etiology include genetic factors, infectious agents, and environmental toxins, especially certain trace elements. Markesbery and Ehmann have recently reviewed the possible role of trace elements in dementing disorders (*4*).

A large part of the current AD research effort has involved the postmortem characterization of pathological and biochemical alterations in the brains of AD patients. Much less work has been directed at describing elemental alterations in nonneural tissues of living AD patients (*5–8*). Longitudinal studies have been conducted on such factors as clinical and neuropsychological progression (*9,10*), but these authors are not aware of any previous studies that have monitored trace element alterations in AD patients over time. The present study was initiated to examine possible longitudinal trace element variations in nail. Our previous work has shown that Br, Hg, K, and Zn are imbalanced in AD nail compared to controls (*8*). Therefore, these were the 4 elements of greatest interest, although other elements were also determined.

METHODS

Twenty AD patients participated in the longitudinal study. One patient was followed for 3 yr, 9 patients for 2½ yr, 2 for 2 yr, 5 for 1½ yr, and 3 for 1 yr. The patient population consisted of 8 males and 12 females, with ages ranging from 56–84 yr. Extensive medical histories were available for each patient, including relative severity of dementia, whether or not the AD was familial, year of onset, other medical conditions, and drug therapies. Subjects who had AD complicated by another disease were excluded from the study. Seven control subjects (3 males and 4 females, ages 60–72 yr) were also followed. The controls were generally close relatives of the AD patients, living in the same household.

Detailed descriptions of the INAA procedures have been published previously (*11*). Briefly, fingernail samples were obtained from the AD patients at 6-mo intervals, using stainless-steel scissors. The nails were scraped with quartz knives to remove external contamination. They were washed, using the IAEA acetone-water method (*12*), and air-dried. The nails were sealed in *Suprasil* quartz vials and irradiated for 40 h at the Missouri Universities Research Reactor. Upon return to the laboratory, samples were counted for 15 min for As, Au, Br, K, and Na. Two weeks later, a 3-h count was done for Ag, Cs, Co, Cr, Fe, Hg, Rb, Sb, Sc, Se, and Zn. Only the results for Br, Hg, K, and Zn are presented here. The detector–analyzer system consisted of ORTEC Ge(Li) and HPGe detectors coupled to a Nuclear Data ND680 analyzer. Appropriate corrections for spectral interferences were applied where required; e.g., Se with Hg.

Comparator standards used in this work were NIST Standard Reference Materials SRM 1566, SRM 1577, and SRM 1571. Analyses of Bowens' Kale as a secondary standard were in good agreement with accepted values (*13*). The Bowen's Kale used in this study has also been analyzed using primary standard solutions (*14*).

RESULTS

There are difficulties associated with attempts to perform statistical inference tests on data sets of this type, as discussed by Berg et al. (*9*). These include the choice of a zero reference point, patient attrition, missing data resulting from patient scheduling, and selection of appropriate ways to group patients based on clinical criteria.

The time of entry into this study was used as the zero point. The patients were not all at the same stage of the disease at that point or at succeeding 6-mo intervals. However, all did show progressively worsening dementia throughout the study. Attrition and missing data led to a data set that was extremely unbalanced, and had a very small number of points for some time intervals. Because of this and the clinical heterogeneity of the groups at each time interval, application of exacting statistical tests was of questionable validity. Therefore, it was decided to use only some simple descriptive measures to look for longitudinal changes. These included:

1. The calculation of elemental means for each time period
2. Plotting trace element levels vs time for each patient and determining a Pearson correlation coefficient and
3. The preparation of CUSUM statistical plots (*15*) to show cumulative changes in trace element levels for the AD patient groups.

Plots of the mean values for Br, Hg, K, and Zn at each 6-mo interval for AD and control subjects are shown in Fig. 1. Mean Br values for AD patients are higher at all points than those for controls, but no apparent trend with time toward significantly higher Br levels is observed. For Hg, AD levels are below those of controls at all but the last two points. Unfortunately, only two controls are included for these last two time periods, so these control means may not be representative for Hg. Potassium levels in AD patients are higher than in controls, but there is again an ambiguous trend toward longitudinal accumulation of K in AD nails. The Zn levels in AD patients do show a tendency to increase with time and are generally higher than the control means.

In a second attempt to characterize the data, individual plots of the changes in each of the 4 elements of interest over time were made for each AD patient. A Pearson correlation coefficient was calculated for each, using the SAS CORR procedure. It is recognized that the small number of points for most plots makes the validity of the Pearson correlations questionable. Each plot was also visually examined. For Br, only 2 of the 20 AD patients showed a significant ($p < 0.05$) trend toward increased values over time. For Zn, only 1 subject showed a significant

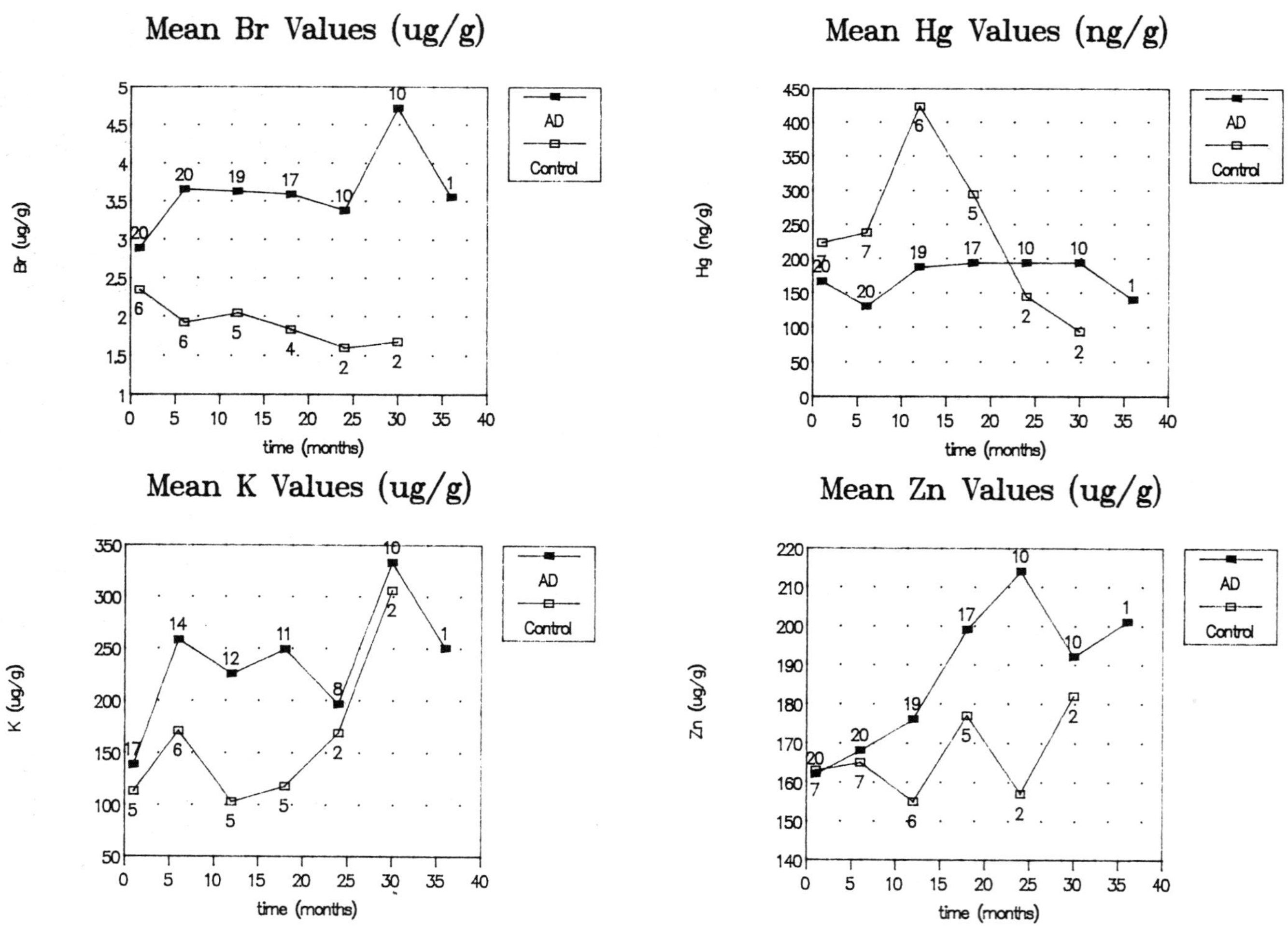

Fig. 1. Mean elemental values for AD and control groups at each time interval (number of samples indicated near each point).

increase, and 1 patient a decrease. For Hg, 3 AD patients showed a significant trend toward lower Hg values, but 2 showed a significant upward trend. For K, no significant trends were apparent in any subject. The elemental plots were also made for the controls. No trends for any of the 4 elements were observed in controls. Generally, examination of data on nail trace element levels for a single individual is not very useful, because of the natural variability of the elemental concentrations.

The third approach used to extract information from the data was to analyze it using CUSUM plots, similar to those used by Krolner and Nielsen (*15*). In this method, the patients were ranked using a criterion appropriate for the study (e.g., age, time since onset, and so on). The overall change in the dependent variable for each subject over the time of the study was found by taking the difference between the first and last values. A plot was then made of the cumulative change in the dependent variable vs the total time. Patients were ranked according to three criteria:

1. Age at start of study
2. Number of years since disease onset and
3. A test evaluation of the severity of the dementia.

Figure 2 shows the set of CUSUM plots obtained when patients were ranked by chronological age, from youngest to oldest. Bromine and Hg show apparently random scatter. For K and Zn, however, there may be a tendency for increasing elemental levels with increasing age of the patient.

CUSUM plots obtained when patients were ranked by years since reported onset of the disease are shown in Fig. 3. The range was from 1–15 yr duration. The Br plot again shows only random scatter. The K and Zn plots, however, indicate some tendency for increasing levels for patients who have had the disease the longest. The Hg plot shows an opposite tendency.

Figure 4 shows CUSUM plots obtained when testable patients were ranked according to scores on the Mini-Mental State Examination (MMSE). Patients who were not testable were ranked on the same plots according to a second criterion—years since onset of disease. Considering the testable group only, no trends were obvious for Br, Hg, and K. However, in the case of Zn, there did appear to be a tendency towards higher Zn levels with decreasing test scores. In the nontestable group, Br, K, and Zn tended to increase, and Hg tended to decrease with years since onset.

Nail trace element levels can reflect metabolic intake, as shown by our longitudinal analyses of nail from a myasthenia gravis patient. The patient showed nail Br values of 32–68 μg/g, much higher than the normal control values of under 4 μg/g observed for her sister living in the same household. In this case, the patient was taking pyridostigmine bromide as a treatment for the disease.

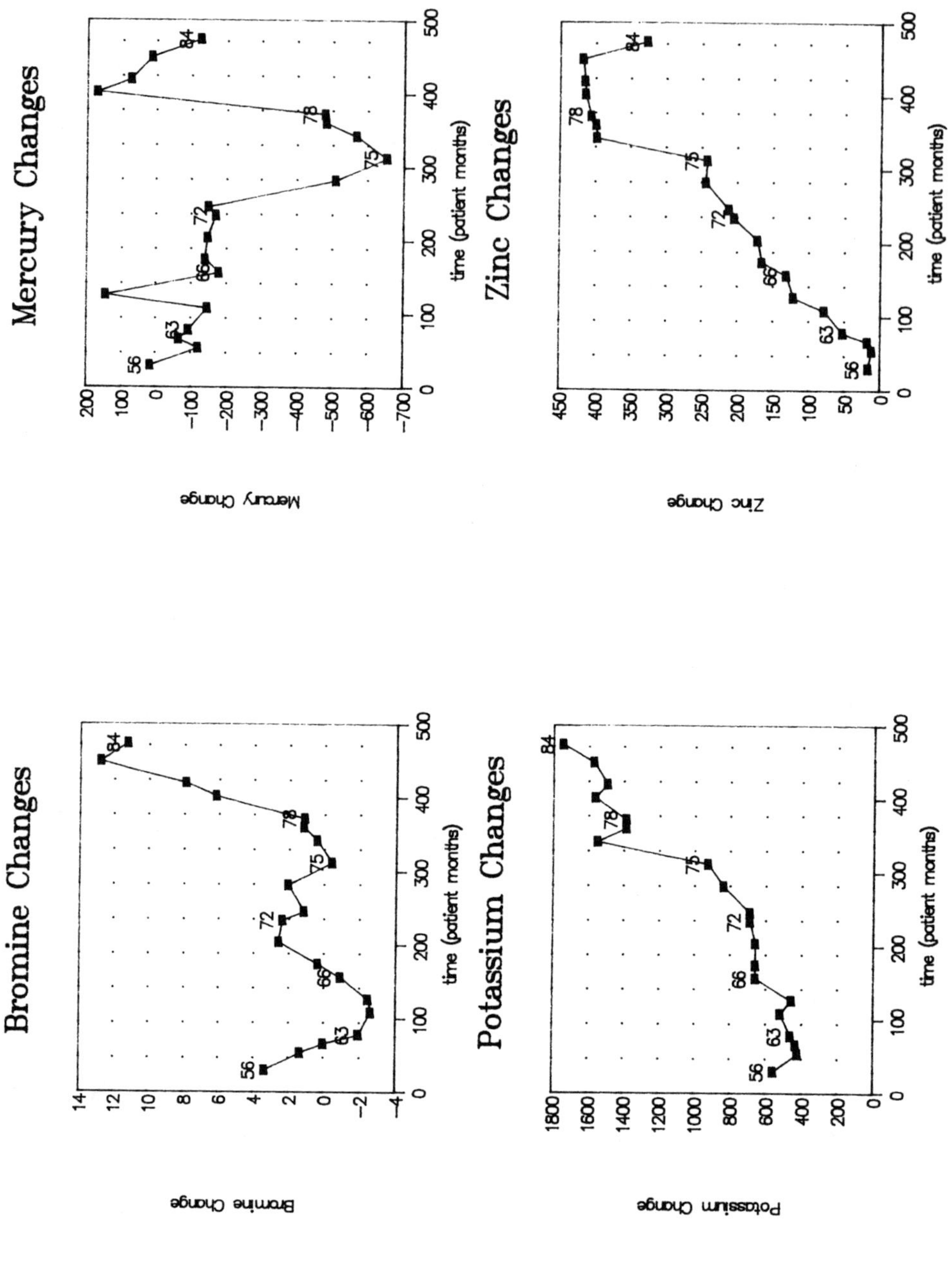

Fig. 2. Cumulative change in elemental levels for patients ranked by age in yr (age range 56–84 yr).

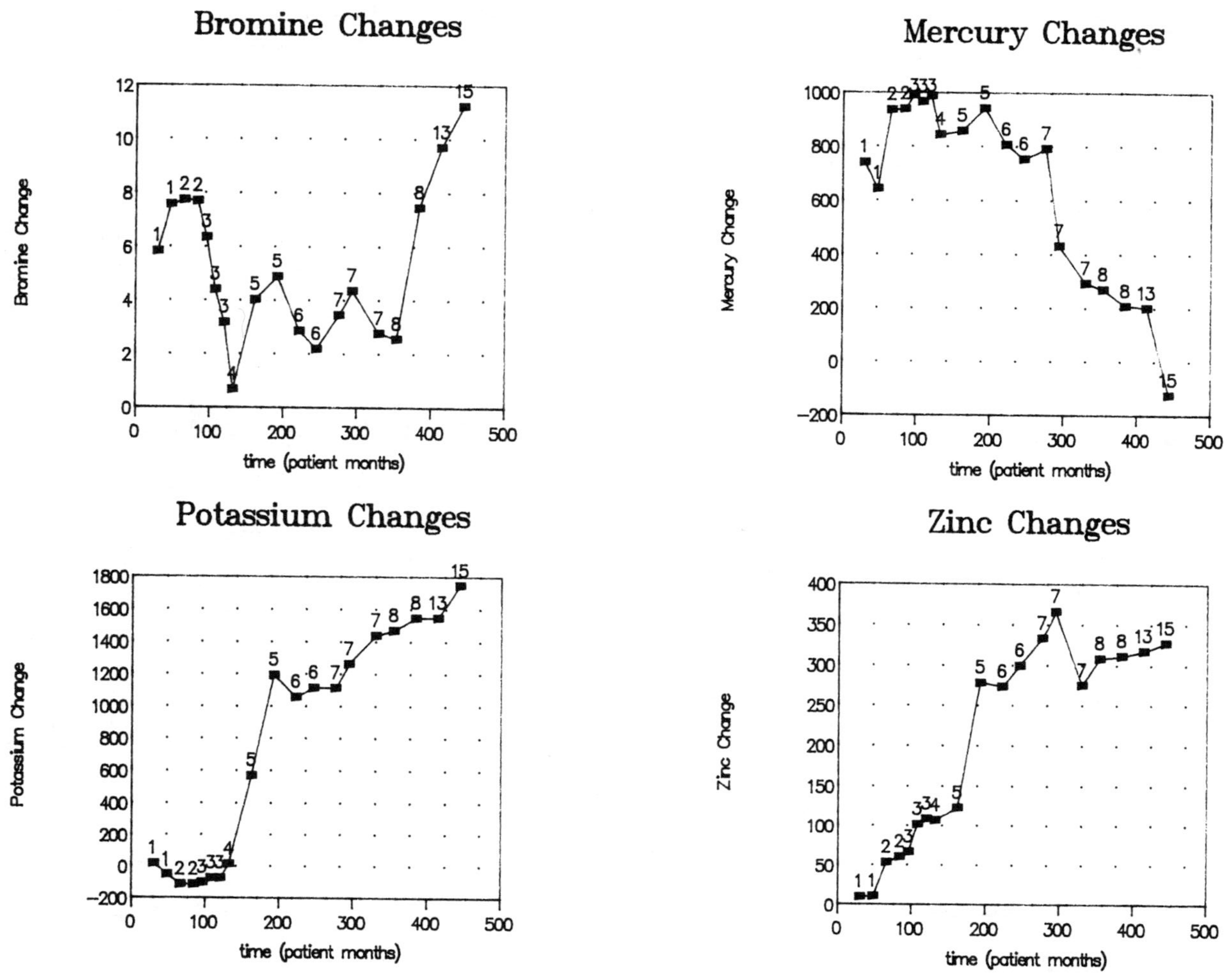

Fig. 3. Cumulative change in elemental levels for patients ranked by yr since onset of disease (range 1–15 yr).

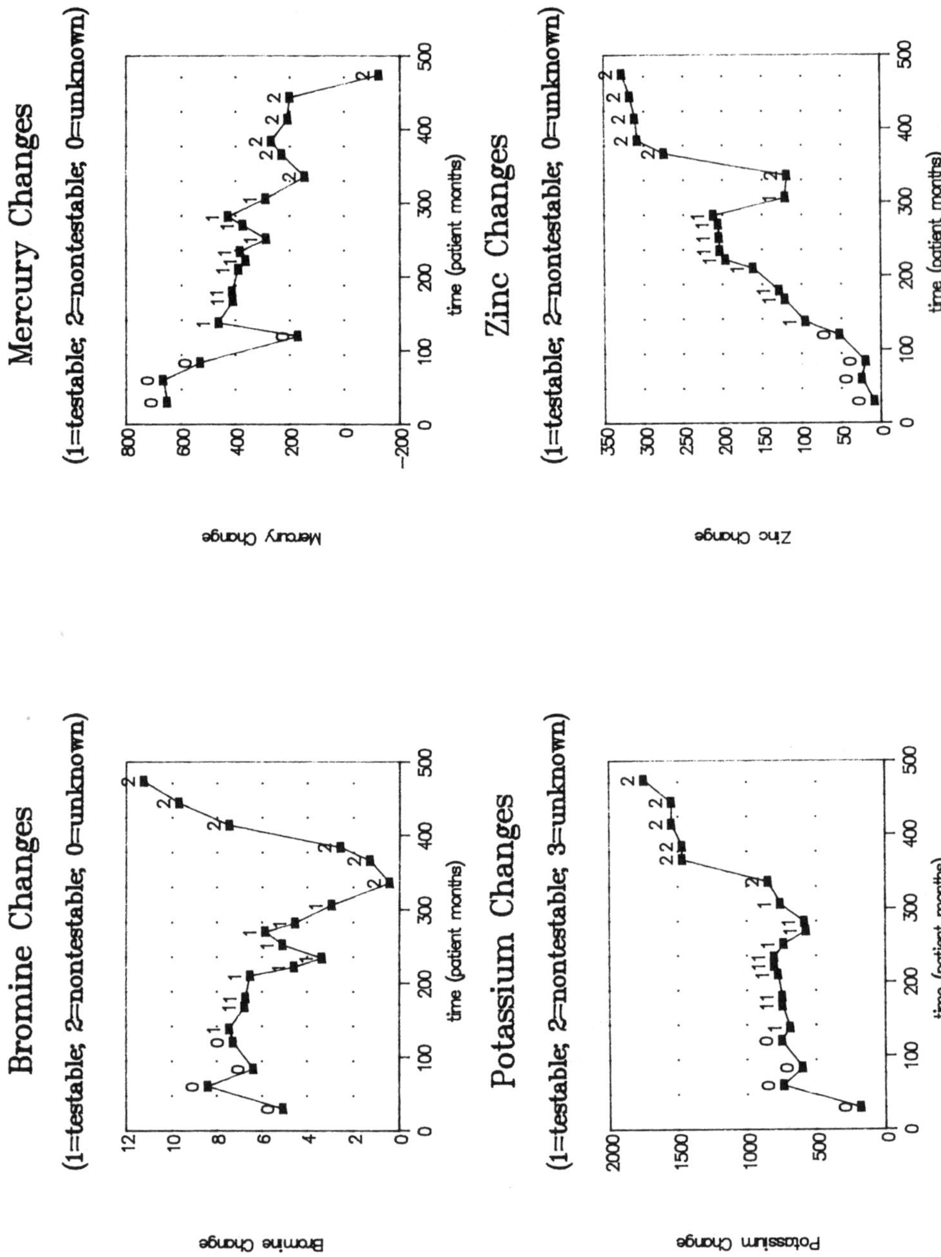

Fig. 4. Cumulative change in elemental levels for patients ranked by severity of disease.

DISCUSSION

This study suggests that longitudinal changes in trace element levels in nail of AD patients do occur. However, it is also clear that the observed trace element imbalances are detectable even at the earliest sampled stages of the disease. These observations are based on cumulative changes in element levels for a group of 20 AD patients observed over a 3-yr period. Statistical tests for significant differences among succeeding time intervals could not be meaningfully performed, because of the unbalanced data set and the small number of data for some intervals. Trends toward elevated or decreased elemental levels are generally not apparent for individual subjects.

Bromine levels do not appear to vary consistently over time, except in the case of nontestable, late-stage patients when ranked according to years since onset of disease. Bromine has been shown to be elevated in bulk AD brain samples as compared to age-matched controls (*16*), but in control brain, Br levels exhibit no trend with age (*17*). The levels of Hg in the nails of AD patients appear to decrease with increasing age, increased duration of disease, and increased severity of dementia. This decrease is counter to the elevated levels of Hg observed in AD brain, as compared to age-matched controls (*16*). The Hg level reduction in nail could merely reflect recent reduced environmental exposure, because these patients would be the ones most likely to be institutionalized, whereas the elevated brain levels reflect prior exposure and Hg retention. Control bulk brain Hg levels exhibit no significant trend with age (*17*).

Potassium and Zn levels in AD nail tend to increase with increasing age, increased duration of disease, and increased severity of dementia. For Zn, the increased levels could be a reflection of a lowered metabolic rate, or an age effect. It has previously been noted that Zn levels are higher in nails of control subjects over 45 than in those younger (*11*). No variations in control brain Zn levels with age were observed in a previous work (*17*). No differences in bulk brain Zn levels were noted in the AD neocortex (*18*), but Zn was found to be elevated in the hippocampus and amygdala of AD patients, compared to age-matched controls (*19*). The K increase is clearly not an age effect, because it has been observed that control subjects over age 45 have lower K nail levels than younger subjects (*11*). Potassium has been observed to decrease with age in adult control brain (*17*), but bulk brain K levels in AD and age-matched control brains are not significantly different (*16*).

ACKNOWLEDGMENTS

This work was supported in part by NIH Grants 1-PO1-AG05119, 5-P50-AG05144, and the Reactor Sharing Program at the Research Reactor, University of Missouri. We wish to thank Dr. Richard Kryscio

(Dept. of Statistics, University of Kentucky) for advice on statistical data processing.

REFERENCES

1. G. K. Wilcock, *Age Ageing* **17,** 73 (1988).
2. J. A. Hardy and D. C. Davies, *Br. J. Hosp. Med.* **39,** 372 (1988).
3. A. Cronin-Golomb, *Neuropsychologia,* **26,** 187 (1988).
4. W. R. Markesbery and W. D. Ehmann, in *Nutritional Modulation of Neural Function,* J. E. Morley, M. B. Sterman, and J. H. Walsh, eds., Academic Press, New York, 1988, pp. 179–190.
5. J. R. Atack, E. K. Perry, R. H. Perry, I. D. Wilson, M. J. Bober, G. Blessed, and B. E. Tomlinson, *J. Neurol. Sci.* **70,** 1 (1985).
6. N. R. Sims, J. M. Finegan, and J. P. Blass, *N. Engl. J. Med.* **313,** 638 (1985).
7. G. S. Zubenko, B. M. Cohen, C. F. Reynolds, F. Boller, I. Malinakova, and N. Keefe, *Am. J. Psychiatry* **144,** 860 (1987).
8. D. E. Vance, W. D. Ehmann, and W. R. Markesbery, *NeuroToxicology* **9,** 197 (1988).
9. L. Berg, J. P. Miller, M. Storandt, J. Duchek, J. C. Morris, E. H. Rubin, W. J. Burke, and L. A. Coben, *Ann. Neurol.* **23,** 477 (1988).
10. W. J. Jagust, R. P. Friedland, T. F. Budinger, E. Koss, and B. Ober, *Neurology* **38,** 909 (1988).
11. D. E. Vance, W. D. Ehmann, and W. R. Markesbery, *Biol. Trace Element Res.* **17,** 109 (1988).
12. International Atomic Energy Agency, *IAEA/RL/50,* IAEA, Vienna, Austria, 1978.
13. D. E. Vance, *Trace Element Relationships in Alzheimer's Disease: Hair and Nail Analysis by INAA,* Ph.D. Dissertation, University of Kentucky, 1986.
14. R. A. Nadkarni and W. D. Ehmann, *J. Radioanal. Chem.* **3,** 175 (1969).
15. B. Krolner and S. P. Nielsen, *Clin. Sci.* **62,** 329 (1982).
16. W. D. Ehmann, W. R. Markesbery, M. Alauddin, T. I. M. Hossain, and E. H. Brubaker, *NeuroToxicology* **7,** 197 (1986).
17. W. R. Markesbery, W. D. Ehmann, M. Alauddin, and T. I. M. Hossain, *Neurobiol. Aging* **5,** 19 (1984).
18. W. D. Ehmann, W. R. Markesbery, and M. Alauddin, in *Neurobiology of Zinc,* C. J. Frederickson, G. A. Howell, and E. J. Kasarskis, eds., Alan R. Liss Inc., New York, 1984, pp. 329–342.
19. C. M. Thompson, W. R. Markesbery, W. D. Ehmann, Y. X. Mao, and D. E. Vance, *NeuroToxicology* **9,** 1 (1988).

Whole Body Elimination Routes of Gold in Humans After a Single-Dose Application of the Antirheumatic Auranofin

C. SCHNIER*,[1] AND H. P. BENN[2]

[1]Institute of Physics, GKSS Research Centre Geesthacht Max-Planck-Str., D-2054 Geesthacht, FRG

[2]Second Department of Internal Medicine, University of Kiel Metzstr. 53-57, D-2300 Kiel, FRG

Received April 17, 1989; Accepted November 16, 1989

ABSTRACT

After removal of the gallbladder (cholecystectomy) 7 patients were administered the antirheumatic agent Auranofin in its commercially available tablet form (Ridaura) 3 d postoperative. The single dose consisted of 5 tablets (4.35 mg of gold). Gold was determined in samples of blood, plasma, urine, bile, and feces (1–2 mL and 2.5 g, respectively). The specimens were drawn 10 min to 8 d p.a. and on d 14 p.a. The determinations were performed by instrumental neutron activation analysis (INAA). The limit of detection was less than 1 ng of gold. The courses of the mean gold concentrations show maxima after 1.9 h in blood and plasma, and after 16 h in urine and bile. The mean half-lives (terminal phases) are 7.6 d (blood), 23 d (plasma), and 6.5 d (bile).

Index Entries: Auranofin; gold in blood, plasma, bile, urine, feces; instrumental neutron activation analysis (INAA).

*Author to whom all correspondence and reprint requests should be addressed.

INTRODUCTION

Oral and parenteral gold therapies have proven to be beneficial in the management of rheumatoid arthritis, but substantial adverse effects must be taken into account (*1*). Long-term oral gold therapy and its possible adverse reactions require detailed knowledge about pathways and mechanisms of gold elimination. So far, no data exist on the biliary gold elimination in humans that may be harmful to the liver metabolism.

A single dose of the antirheumatic agent Auranofin, containing triethylphosphin (TEP)-gold, was administered to 7 patients 3 d after removal of the gallbladder (cholecystectomy). The renal, fecal, and biliary gold excretions were investigated by means of instrumental neutron activation analysis (INAA) of feces, urine, and bile, as well as the whole blood and plasma up to 13 d after application (p.a.). A multielement analysis by INAA of the bile samples will be published elsewhere (*2,3*). INAA was chosen because it is a simple and sensitive method for the determination of low concentrations of gold in biological materials. No preconcentrations or radiochemical procedures are required. This method is free from reagent contamination (*4*).

SAMPLE COLLECTION

Seven patients who only suffered from cholecystolithiasis were involved in this study. The medical parameters are described by Benn et al. (*1*). All 7 fasting (n.p.o.) patients (Table 1) were administered a single dose of Auranofin (5 tablets Ridaura®), containing 4.35 mg of gold in the commercially available tablet form, 3 d postoperative. None of the patients experienced gastrointestinal side effects. The dose was chosen as a result of previous investigations, which yielded measurable amounts of gold determined by means of INAA in blood, urine, and feces.

Samples of blood, plasma, bile, urine, and feces were collected successively 8 d post oral administration of Auranofin. Except for bile, additional samples were obtained on d 14 p.a. Drawn venous blood was divided into whole blood and plasma, won immediately by the centrifugation of heparinized blood. The samples were collected according to the following schedule:

1. Blood and plasma were drawn immediately before dosing, and at 10 and 30 min; 1, 2, 4, 6, 12, 24, 36, and 48 h; and 3, 4, 5, 6, 7, 8, and 13 d following drug administration.
2. Bile samples were drawn from 6 complete 4-h fractions during the first day p.a. (blocked choledochus) (*1*). On each of the following 7 d, 2 samples were obtained by pooling 4 complete 2-h fractions (blocked choledochus) and 4 4-h fractions (spontaneous bile flow) (*1*), respectively.
3. Urine samples were obtained from 6 4-h portions entirely

Table 1
Personal Data of the 7 Investigated Patients

Patient	A	B	C	D	E	F	G
Age (yr)	64	63	64	48	50	52	67
Sex	f	f	f	m	f	m	f
Height (cm)	154	160	163	166	169	176	160
Body weight (kg)	80	69	50	74	63	59	65
Blood vol (L)	4.2	4.0	3.3	4.8	4.0	4.4	3.8

collected during the first day p.a., from 2 12-h portions on the second day, and 24-h portions at 3–8 d and 13 d after administration.

4. Samples of feces were obtained according to spontaneous defecation in 24-h portions.

METHODS

After homogenization of the fecal samples, weighed aliquots of about 2.5 g were transferred into polyethylene (PET) vials and oven-dried (24 h at 70°C). Two milliliters of the liquid samples blood, plasma, and urine were pipeted into vials and dried as mentioned above. All samples were hermetically sealed in the PET vials and irradiated for 2 h in the nuclear reactor of the GKSS research center in Geesthacht. The thermal-neutron flux F_{th} measured for each irradiation with an Al/Au-monitor foil was about 8×10^{12} cm^{-2} s^{-1}. The epithermal and the fast fraction in the neutron spectrum amounted to $F_{epi}/F_{th} \approx 1.6\%$ and $F_f/F_{th} \approx 15\%$. The waiting times after irradiation until measurement were 5–10 d, and the counting time was 2.5 h. For the gold determination, the 411.8 keV γ radiation of ^{198}Au (half-life 2.697 d) was used.

Under these conditions, the detection limit for gold was ca. 1 ng, and the precision varied from less than 1% for gold contents of more than 20 ng up to 10% for less than 10 ng of gold. The accuracy, which was checked using self-produced reference materials because no appropriate standard reference material was available, was about 25%. Besides gold, the elements Na, K, Br, and Fe (only in whole blood) were determined (2).

Aliquots of 1.5 mL from the bile samples were pipeted into quartz ampules and freeze-dried. After sealing the ampules,the samples were irradiated for 3 d in a thermal-neutron flux F_{th} of about 5×10^{13} cm^{-2} s^{-1} ($F_{epi}/F_{th} \approx 4.5\%$, $F_f/F_{th} \approx 30\%$). The neutron flux was measured for each irradiation with Fe- and Au-monitors. The irradiation was followed by 3 counting periods, after waiting times varying from

1. 5–10 d
2. 14–23 d
3. 30–40 d.

This complicated procedure was chosen in order to determine as many elements as possible besides gold. The detection limit for gold was about 30 pg, the precision about 2%, and the accuracy about 30%. In the bile samples, 12–23 elements could be determined (*2,3*).

The irradiated samples and the neutron-flux monitors were counted on a high-purity germanium detector (19.7% efficiency. 1.69 keV resolution FWHM, and a peak/compton ratio of 56.6:1, all relative to ^{60}Co 1.332 MeV peak), coupled to a 4096-channel pulse-height analyzer. For the evaluation of the spectra, the computer program system AKAN (*5,6*), developed in the GKSS research center, was applied. It calculates the absolute amounts of the elements in the samples using the nuclear data, the neutron fluxes in the reactor during the irradiations, and the previously determined parameters of the detector device.

RESULTS

More than 500 samples were analyzed by means of INAA (*2*). The course of the mean gold concentrations with time is shown in Figs. 1 and 2. After the single administration of the Auranofin tablets, a fast rise of the gold content is observed in all matrices. It is followed by a decline that is nearly linear in the semilogarithmic plot. A biological half-life can therefore be calculated. For bile, 2 linear parts are found.

The maximum gold concentration in whole blood (140 ± 42 ng/mL) and plasma (173 ± 54 ng/mL) is reached 1.9 ± 0.7 h p.a. (*1*), whereas the maximum value in urine (43 ± 28 ng/mL) coincides with the maximum in bile (65 ± 50 ng/mL) after 16 h p.a. The mean half-lives evaluated from the decline part of the curves are given in Table 2.

The mean urine values are almost constant from 1 d until 8 d p.a. A decrease seems to start later, so that at 13 d p.a. the mean concentration of 13 ± 8 ng/mL is measured.

Because feces could not be collected each day, and because the individual gold contents in the feces of the 7 patients differed sometimes by more than 2 orders of magnitude, the results were not plotted. For example, the gold concentration in feces of patient A attained its maximum (3600 ng/g wet wt) after about 2 d p.a., and decreased with a half-life of 2 d until the end of the study (13 d p.a.).

DISCUSSION

A balance study for the elimination of TEP-gold in the 7 patients was performed in reference (*1*). Here, only a short summary is given. Within 8 d only 1.6 ± 0.7%, corresponding to (70 ± 32 μg) of the administered 4.35 mg of gold, was recovered in bile, whereas 4 ± 1% or 174 ± 42 μg was recovered in urine, and 40 ± 8% in feces. So absolute biliary gold

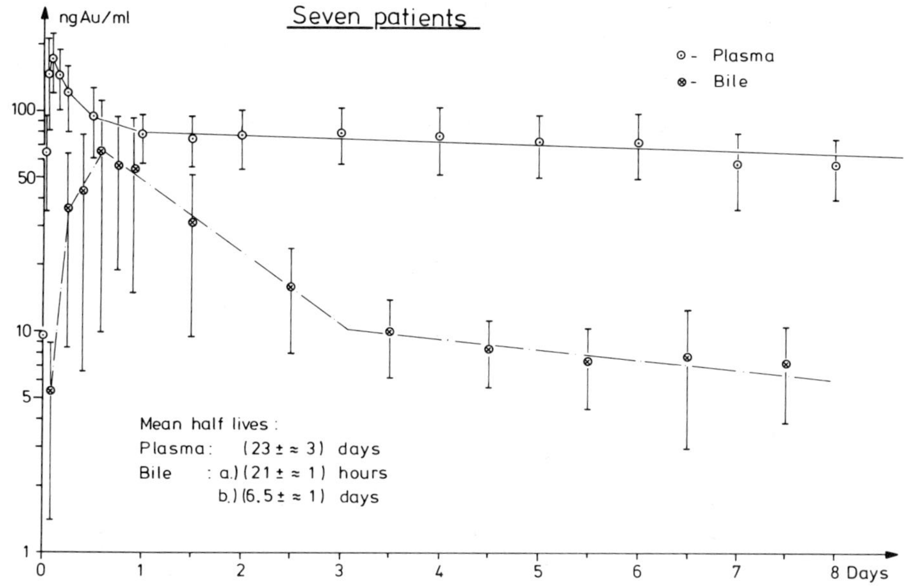

Fig. 1. Mean concentration courses ($\bar{x} \pm$ SD) of gold in human plasma and bile (ng/mL) of 7 individuals over the period of 8 d postingestions of 5 Auranofin tablets (4.35 mg of gold).

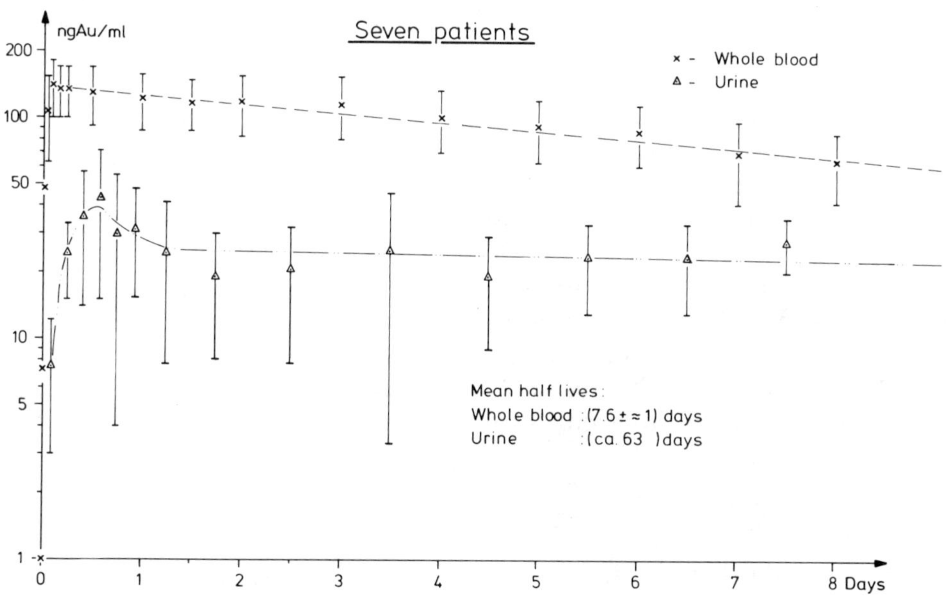

Fig. 2. *See* text of Fig. 1, but for whole blood and urine.

Table 2
Mean Half-Lives for the Terminal Part of the Mean Gold Concentration Courses

Matrix	Mean half-life	Range of validity
Whole blood	7.6 ± = 1 d	0.5–13 d
Plasma	(a) 23 ± = 1 d	2–13 d
	(b) 15 ± = 2 d	2–8 d
Urine	ca. 63 d	1–8 d
Bile	(a) 21 ± = 2 h	0.5–3 d
	(b) 6.5 ± = 1 d	3–8 d

excretion within 8 d amounted to 40% of the renal gold elimination and about 4% of the fecal excretion. This difference still increased with time up to the 14th d after drug ingestion.

The individual biliary gold concentration curves show considerable differences during the first 72 h after administration of Auranofin. The measured peak bile values range from 14–157 ng/mL. Nevertheless, a significant correlation (correlation coefficient $r = 0.864$) can be established between AUC-plasma and AUC-bile (AUC = area under curve) (*1*).

The results show that biliary gold elimination ceased 23 d p.a. and that in total 2.2% of the administered single dose of 4.35 mg TEP-gold was eliminated via bile. About 60% of the gold was retained 8 d p.a. (corresponding to the complement of fecal gold recovery).

Based on their bile/plasma concentration ratio, Brauer (*7*) defined 3 groups of biliary excreted agents: Class A substances, including electrolytes and glucose, appear in the bile in the same concentration as in the blood. Class B substances appear in the bile enriched by a factor of 100, indicating an active transport mechanism. This category comprises most metals, nonionized substances, and cations and anions, which are most pronounced with compounds of mol wt of 300–500. Macromolecules, such as phospholipids and mucoproteins, are excreted into bile only in very low concentrations, if at all (ratio <1). These substances have been designated class C. Surprisingly, our investigations indicate that orally administered TEP-gold belongs to the class C substances, because the bile/plasma concentration ratio or AUC (gold) bile/AUC (gold) plasma (compare Fig. 1) is <1 throughout the 2 wk of this study for all 7 patients, irrespective of the individual differences in plasma gold levels.

CONCLUSIONS

Our single-dose balance study indicates considerable mean intestinal gold retention (ca. 60%), when the antirheumatic agent Auranofin is administered in the original tablet form. In addition, differing quantitative biliary and renal gold eliminations are found. Although renal gold loss maintains a considerable nondeclining urinary concentration level for more than 1 wk after drug ingestion, steadily decreasing biliary gold excretion ceases after 3 wk. Measured cumulative amounts of gold excreted by different pathways (gut:kidney:liver corresponding to 40:4:1.6%) correlate remarkably with the incidence of reported and clinically assessed adverse events related to gut, kidney, and liver (gastrointestinal: 47% total; kidney 3%; liver 0.6%) (*8*).

ACKNOWLEDGMENTS

The authors wish to thank Professor Michaelis (GKSS Research Center Geesthacht) for supporting this study. We express our gratitude to I. Bendler and T. Wedderien for their help in performing the analyses, and to R. Niedergesaess for his help in all questions concerning INAA. We are grateful to Dr. K. Doden for his advice on mathematical operations, and to H. Elhoeft for sample preparation.

REFERENCES

1. H. P. Benn, C. Schnier, K. U. Seiler, E. Bauer, and H. Löffler, *J. Rheumatol.* (in preparation) (1989), and A. Doenhardt, ed., *Nordwestdeutsche Gesellschaft für Innere Medizin, Kongressbericht 109,* Tagung, Hamburg 21.-23 (Jan. 1988). pp. 47, 48. Hansisches Verlagskontor Lübeck, 1988.
2. C. Schnier, H. P. Benn, T. Wedderien, and R. Niedergesaess, The concentration of gold in blood, plasma, bile, urine and feces after a single oral administration of the antirheumatic Auranofin, determined by means of INAA. GKSS research report, 90/E/.
3. C. Schnier and H. P. Benn, Multielement analysis of human bile by means of instrumental neutron activation analysis (INAA) *J. Radioanal. Nucl. Chem.* 1990.
4. E. S. Waller et al., *Res. Commun. Chem. Pathol. Pharmacol.* **37,** 33–47 (1982).
5. L. Greim, R. Niedergesaess, and K. Motamedi, *Seminar "Aktivierungsanalyse,"* **Jülich, 3./4.** (Nov. 1975).
6. L. Greim and R. Niedergesaess, *GKSS-Report* **80/I/28** (1980).
7. R. W. Brauer, *JAMA* **169,** 1462–1466 (1959).
8. M. A. Heuer et al., *Auranofin. J. Rheumatol.* **12,** 695–699 (1985).

Trace Element Distribution in the Hair of Some Sickle Cell Anemia Patients and Controls

A. F. OLUWOLE,*[,1] O. I. ASUBIOJO,[2] A. D. ADEKILE,[3]
R. H. FILBY, A. BRAGG,[4] AND C. I. GRIMM[4]

Departments of Physics,[1] Chemistry,[2] Pediatrics and Child Health,[3] Obafemi Awolowo University, Ile-Ife, Nigeria, and Nuclear Radiation Center Washington State University, Pullman, WA 99614-1300

Received April 17, 1989; Accepted November 16, 1989.

ABSTRACT

Hair samples of some young sickle cell anemia (SCA) and control patients in Nigeria were analyzed for 12 elements, viz, Se, Hg, Cr, Fe, Zn, Co, Cu, Br, As, Sb, Na, and Sc, using Instrumental Neutron Activation Analysis (INAA).

With the exception of Cu, which was found to be significantly higher in the hair of SCA patients (at the 0.05 level of the t-test), there were generally no significant differences in elemental concentrations within the two groups. A preliminary study of the elemental contents of the fingernails of the same subjects showed a higher abundance of most of the elements in nail than in hair. These preliminary results were compared with similar studies from some other parts of the world.

Index Entries: Hair samples; sickle cell anemia patients; t-test; elemental concentrations.

INTRODUCTION

Sickle Cell Anemia (SCA) is the commonest inherited hemoglobinopathy of clinical importance among the Black race. It is particularly common in Nigeria, where about 2% of the population are affected. It is

*Author to whom all correspondence and reprint requests should be addressed.

now well known that trace elements play a significant role in health and disease (*1*). Recently, aluminum was implicated in Alzheimer's disease (*2*), and calcium deficiency in hair of osteoporotic patients (*3*). There have been some inconclusive reports on the role of some trace elements in North American and West African SCA patients (*4–6*). Detailed trace-element data on Nigerian SCA patients and control subjects should provide some useful information in efforts to combat the disease.

Hair and nail have been found to be very useful in the study of trace element accumulation in the body, because both are keratinized tissues with low metabolic turnover. Hair, in particular, has been shown to be a reliable and atraumatic method of assessing body trace element stores (*7,8*). Olatunbosun et al. (*6*) have determined the concentrations of 12 trace elements in the hair of some Nigerian subjects by INAA and AAS. However, only elements producing short-lived radioisotopes on neutron irradiation were determined.

This article reports the concentrations of trace elements (measuring both intermediate and long-lived radioisotopes by INAA) in the hair of some young Nigerian SCA patients and control subjects.

EXPERIMENTAL

Scalp hair was obtained from 10 SCA and 11 normal young outpatients in the Pediatrics Clinic of the Obafemi Awolowo University Teaching Hospital. The age of the patients ranged between 3–12 yr.

The hair samples were washed, in stages, with acetone and deionized water in an ultrasonic water bath, and dried overnight at 65°C in specially cleaned polyethylene vials. They were then irradiated, along with NBS standards, for 8 h at a thermal-neutron flux of 6×10^{12} neutrons $cm^{-2} s^{-1}$ at the Washington State University TRIGA III—fueled research reactor. The irradiated samples were counted on Ge(Li) detectors interfaced to an ND 6620 spectrometer for 4000 and 80,000s, after decay times of 30 h and 21 d, respectively, as shown in Table 1. Details of the γ-ray spectrometry and data reduction procedures have been presented elsewhere (*9*).

RESULTS AND DISCUSSION

The results of the elemental concentrations of the hair samples are presented in Table 2. There is an unusually high concentration of Hg(22.2 ppm) in the hair of 1 of the SCA patients. This patient, besides being a sickler, has no other unusual medical record, and there is no corresponding Hg enrichment in his nail. The high level of Hg in his hair is suspected to be the result of environmental contamination. Previous studies (*10*) have shown that most of these elements follow a lognormal

Table 1
Irradiation and Counting Procedure for the Determination of Trace Elements by Instrumental Neutron Activation Analysis

Element determined	Irradiation time	Decay time	Counting time
Cu, Br, As, Sb, Na,	8 h	36 h	4000 s
Se, Hg, Cr, Fe, Zn, Co, Sc	8 h	21 d	80,000 s

Table 2
Elemental Concentrations (in ppm) of the Hair of SCA Patients and Controls

	SCA Patients		Controls	
Element	Range	Geom mean	Range	Geom mean
Se	0.19–0.97	0.40 × 1.61	0.15–1.03	0.38 × 2.00
Hg	0.10–22.2	0.14 × 13.0	0.09–0.33	0.16 × 1.54
Cr	0.48–3.22	0.96 × 1.76	0.34–1.97	0.88 × 2.05
Fe	33–222	91.4 × 1.72	27–302	85.8 × 1.91
Zn	27–220	88.7 × 2.02	33–143	70.3 × 1.64
Co	0.09–0.29	0.14 × 1.44	0.06–0.38	0.17 × 1.68
Cu	4.13–23.3	10.3 × 1.74	2.12–12.8	6.12 × 1.79
Br	0.44–9.35	1.74 × 2.97	0.73–5.90	2.39 × 2.19
As	0.04–0.32	0.09 × 2.33	0.03–0.43	0.09 × 2.07
Sb	0.08–3.69	0.23 × 4.42	0.04–0.50	0.13 × 2.17
Na	4.94–56.2	13.2 × 2.08	4.38–29.8	11.4 × 1.75
Sc	0.005–0.026	0.015 × 0.017	0.003–0.050	0.013 × 0.019

distribution in hair and nails, in which case the median is closer to the geometric mean than to the arithmetic mean. Hence, the geometric means and standard deviations of the 2 groups of subjects are compared for any significant differences. Using the *t*-test, there were no significant differences (at the 0.05 level) in the elemental concentrations in hair of SCA patients, compared to normal patients, the only exception being Cu (Table 3). The higher levels of Cu in the hair of SCA patients is in agreement with the previous finding of Olatunbosum et al. (6), but the medical implications of this are not clear at the present time. A preliminary study of the elemental contents of the fingernails of the same subjects shows that, with the exception of Hg, for which the values are comparable, all the trace elements determined are more abundant in nail than in hair, in some cases by as much as 2 orders of magnitude. This is expected because most of these patients are children of low-income people whose fingernails are prone to considerable contamination

Table 3
t—Test for the Significance of Difference between Mean of Elemental Concentrations in Hair of Sicklers and Control Subjects

Element	n	t_c	$t_{0,5}$	t_{0-10}
Zn	21	0.018	2.08	1.72
Cu	21	2.08	2.08	1.72
Sb	21	1.26	2.08	1.72
Br	21	1.25	2.08	1.72
Hg	11	0.998	2.20	1.78

through occupational exposures, as has been observed in a previous IAEA report (*11*).

Since there are very few differences in the elemental concentrations in the hair of SCA compared to control subjects in Nigeria, the two groups are merged together and compared with similar data from other parts of the world, as shown in Table 4.

In Table 4, it is observed that the ranges of elemental concentrations in the hair of the Nigerian subjects are within the same order of magnitude as those reported in a recent study of the Malaysian population (*12*). A similar comparison with average US data reveals higher values of Se, Hg, Cr, Zn, and Cu, but lower values of Fe and Co in the US subjects (*13*). Higher values of Se and Hg in the US subjects are not surprising, because they would be more prone to environmental exposure resulting from industrial effluents than the rural Nigerian subjects investigated here.

In summary, the following are the finding of this study:

1. In general, there are no significant differences in the elemental contents of the hair of young Nigerian SCA patients and control subjects.
2. The trace element contents of the fingernails are generally higher than the hair samples for the same group of subjects.
3. The range of concentrations of trace elements in the hair of the Nigerian subjects is in agreement with similar studies elsewhere, although lower values of some toxic elements were observed, compared to US results.

A comprehensive program is underway to determine the trace elements in the whole blood, serum, and hair of SCA patients and controls. This will take into account the different age groups and social (nutritional) gradients.

Table 4
Intercomparison of Trace Element Concentrations in Human Hair (in ppm)

Element	Nigeria		Malaysia		U.S.A.	
	Range	G.M.	Range	G.M.	Range	G.M.
Se	0.15–1.03	0.37 × 1.79	0.08–1.10	0.45 × 1.88	NA	1.15 × 4.7
Hg	0.09–22.2	0.30 × 0.005	0.20–16.40	2.97 × 6.57	NA	1.8 × 1.5
Cr	0.34–322	0.92 × 1.89	0.37–4.03	1.08 × 1.59	NA	1.5 × 2.7
Fe	27–302	88.4 × 1.80	36–253	65.2 × 1.6	NA	30 × 1.7
Zn	27–220	78.5 × 1.83	33–407	138 × 1.9	NA	164 × 1.4
Co	0.06–0.38	0.15 × 1.58	0.01–0.19	0.05 × 2.32	NA	0.03 × 1.6
Cu	2.12–23.3	7.84 × 1.85	NA	NA	NA	151 × 1.4
Br	0.44–9.35	2.05 × 2.52	0.32–25.4	3.77 × 3.42	NA	NA
As	0.03–0.43	0.09 × 2.15	0.05–1.7	0.28 × 2.04	NA	0.13 × 4.6
Sb	0.04–3.69	0.17 × 3.22	0.09–0.80	0.26 × 1.86	NA	0.167 × 2.1
Na	4.38–56.2	11.7 × 1.89	NA	NA	NA	NA
Sc	0.003–0.050	0.014 × 0.0017	NA	NA	NA	NA

REFERENCES

1. A. S. Prasad, ed., *Trace Elements in Human Health and Diseases,* Vol. 1,2, Academic Press, New York, 1976.
2. D. Crapper, S. Krishnam, and A. Dalton, *Science* **180,** 511 (1973).
3. L. G. Stephens-Newsham, M. J. M. Duke, T. R. Overton, and D. Ng, Proc. 7th MTAA Conf., 909 (1976).
4. G. J. Brewer, F. J. Oelshlegel, A. S. Prasad, C. A. Knutsen, and N. Meyers, *1st Nat. Symp. on Sickle Cell Disease (Abs.),* N. I. H. Bethesda, MD, 1974.
5. A. S. Prasad, E. B. Schoomaker, J. Ortega, G. J. Brewer, D. Oberleas, and F. J. Oeschlegel, *Clin. Chem.* **21,** 582 (1975).
6. D. Olatunbosun, S. Morris, F. Lichte, S. R. Koirtyohann, and J. R. Vogt, *Trace Substances in Environmental Health* **10,** 383 (1977).
7. W. H. Strain, L. T. Steadman, C. A. Lankan, et al. *J. Lab. Clin. Med.* **68,** 244 (1966).
8. I. Othman and N. M. Spyrou, in *Proc Int. Symp on Nuclear Activation Techniques in the Life Sciences,* IAEA, Vienna, p. 583, 1979.
9. R. H. Filby, W. A. Haller, and K. R. Shah, *J. Radioanal. Chem.* **5,** 277 (1970).
10. G. J. Evans and R. E. Jervis, *J. Radioanal. Nucl. Chem.* **110,** 613 (1987).
11. IAEA Report CVD\73.4, Trace Elements in Relation to Cardiovascular Diseases, Vienna, 19-23 February, 1973 and Geneva (WHO), 2-6 April, 1973.
12. S. Sarmani, *J. Radioanal. Nucl. Chem.* **110,** 627 (1987).
13. Y. S. Ryabukhin, Report IAEA/RL/50, 1978.

Minor and Trace Elemental Contents of Cancerous Breast Tissue Measured by Instrumental and Radiochemical Neutron Activation Analysis

AMAR N. GARG,*[,1] RAJIV G. WEGINWAR,[1] AND VARSHA SAGDEO[2]

[1]*Department of Chemistry, Nagpur University, Nagpur-440 010, India*
[2]*Department of Surgery, Government Medical College, Nagpur-440 014, India*

Received April 17, 1989, Accepted November 16, 1989

ABSTRACT

Many minor and trace elements influence the permeability of cell membranes by competing for binding sites, and exert direct or indirect action on the carcinogenic process. Instrumental and radiochemical neutron activation analysis has been employed for the determination of more than 20 elements in normal and cancerous breast tissues of 6 patients. Most trace elements, viz., Zn, Cu, Mn, Co, Se, Br, As, Sb and Cd, are elevated in cancerous tissue, whereas lower levels are observed for Fe, Cs, I, and Sr. Similarly, concentrations of minor constituents, such as Na, K, P, Cl, and Mg, are enhanced compared to normal tissue. Several elements incorporate into the normal cell, change its enzymatic activity, and accelerate the growth of tumor.

Index Entries: Trace elements; neutron activation analysis; breast cancer; neoplastic tissue; malignancy; carcinogenic; solvent extraction; gamma spectrometry.

*Author to whom all correspondence and reprint requests should be addressed.

Biological Trace Element Research Editor: G. N. Schrauzer © 1990 by The Humana Press Inc.

INTRODUCTION

The essentiality of trace elements in a number of biological processes, through their action as activator or inhibitor of enzymatic reactions, is now well established. Furst (*1*) suggested that perhaps metals, aided by viruses, may penetrate living cells and either enhance or retard the kinetics of anabolic or catabolic enzymes. Versieck (*2*) has critically reviewed the trace elemental contents in human body fluids and various tissues, especially with regard to reference materials. During the last few years, several attempts have been made to correlate trace metal analysis data with the tumor tissues and serum from cancerous patients. Mega et al. (*3*) found a positive correlation between mortality from stomach cancer and zinc content of ground water and rice. Berg and Burbank (*4*) have compared trace metal concentration with the death rate from cancer, for major water basins in the US. Wright and Dormondy (*5*) have found elevated zinc levels in cancerous liver tissue. Some workers (*6–10*) have found elevated copper levels in serum of patients suffering from acute leukemia, lung cancer, and oral cavity.

There has been growing recognition that metal compounds are an important class of environmental and occupational carcinogens. Many compounds of Cd, Co, Cr, Ni, Zn, and Pb have been used to induce cancer in experimental animals, indicating that metal ions can interact with nucleic acids to influence base pairing and conformation (*11,12*). Such effects have been known to cause somatic mutations leading to cellular transformations. Cr, Ni, As, and Be have been shown as human carcinogens, and exhibit genetic toxicity in a number of test systems, suggesting that metagenesis is involved in the initiation of cancer by these metals.

The literature contains several studies of trace metal concentrations in paired samples of cancerous and normal tissues from the same human breast of an individual. Choice of paired samples avoids variations resulting from differences in age, diet, sex, hormonal status, medication, genetics, and other environmental factors. De Jorge et al. (*13*) were the first to employ colorimetric methods for determining Cu, Ca, Mg, and P in 13 sample pairs, from which Cu was found to be significantly higher in the tumors. Mulay et al. (*14*) reported the measurement of Al, Ba, Ca, Cr, Cu, Fe, Mg, Mn, Sn, Sr, Te, and Zn in 8 paired samples, by emission spectroscopy. In recent years, neutron activation analysis (NAA), an ideal technique for trace elemental analysis of biological samples, has been widely used by several workers (*15–18*). This technique has an advantage of having simultaneous multielemental capability, freedom from reagent and laboratory contamination, and an adequate sensitivity for most elements of interest (*19,20*). Schwartz et al. (*15*) first used NAA for the determination of 17 elements in 9 paired samples, and observed substantial differences for certain trace elements in malignant and normal breast tissues. Increase in the concentration of essential trace elements has been attributed to enhancement in the metabolism and enzymatic activity of cancerous tissue. Santoliquido et al. (*16*) have observed significantly higher concentrations of Mg and Zn in cancerous tissue. Most recently, Rizk and Sky-Peck (*21*) have reported the determination of 16 elements in 25 patients, using the EDXRF method, and found a characteristic pattern of elevation for Ca, V, Cu, Zn, Se, and Rb in neoplastic tissues. Also, Mangal and Kumar (*18*) have employed INAA for the determination of 10 elements in 17 paired samples, and observed a positive correlation of Fe, Zn, Cr, and so on, with the corresponding change in neucleocytoplasmic ratio.

In the present work, we have analyzed histologically normal and neoplastic human breast tissues obtained from 6 patients, by instrumental and radiochemical neutron activation analysis. The elements determined were Na, K, P, Cl, Zn, Cu, Mn, Co, Se, Br, As, Cd, Fe, Cs, Sr, Mg, I, and Sb. Ga, Rb, and Ag were also detected in some samples. An attempt has been made to understand the degree of disease relationship with the concentration of elements and their role in regulating tumor growth.

EXPERIMENTAL

Sample Collection

Paired samples of normal and malignant tissues were obtained from 6 patients who had undergone radical mastectomy for breast cancer. In each case, two specimens were removed from the same breast, using tungsten carbide scissors so as to avoid any metal contamination. A piece of cancerous breast tissue, weighing about 10–15 g, and an equal weight of normal breast tissue from the diagonally opposite site to the cancerous mass at a maximum distance were taken from the same breast. Microscopic examination of the malignant tissue showed infiltrating duct carcinoma, and care was taken to separate the two types of tissues. The excised surgical specimens were washed thoroughly to remove any blood, and kept in isotonic saline water (0.9% sodium chloride solution) in polythene containers.

Sample Preparation

After bringing the samples to the chemical laboratory on the same day, they were cut into pieces and thoroughly washed in boiling double-distilled water, to dissolve most of the low-melting fats. At least 10–15 washings were performed in each case, and all handling was done in Corning™ petridishes, employing a glass rod and a Teflon™-coated spatula. Afterwards, all water was removed by decantation, and then samples were dried under an infrared lamp at ~70°C, avoiding any charring. After complete drying, tissues were powdered in an agate mortar and irradiated in a cobalt-60 Gamma Chamber-900 to a dose of 2.5 Mrads.

Irradiation and INAA Counting Procedure

About 50 mg each of samples, synthetic standard (prepared by impregnating an appropriate mount of pure salt solution in nitrate form on Whatman filter paper), and comparators (NBS SRM 1577a Bovine Liver and IAEA H-4 Animal Muscle) were packed in pure aluminium foil (INDAL). For INAA, irradiations were carried out for 10 min, 1 h, and 10 h at a thermal-neutron flux of 10^{12}n cm^{-2} s^{-1} in the APSARA reactor at BARC, Bombay. For RNAA work, however, 80–100 mg each of samples and standards were encapsulated in quartz ampules and irradiated in the CIRUS reactor at a thermal-neutron flux of 10^{13}n cm^{-2} s^{-1} for 2 wk.

After a delay of 40 min samples were opened, decontaminated, and counted on a 45 cm^3 HPGe dector (EG & G ORTEC) and Canberra Series 35 Multiparameter MCA system in a fixed geometry at the radiochemistry laboratories of BARC, Bombay. Various nuclides were identified by their γ-ray energies. At least 4 countings were performed at successive intervals to follow the half-lives of short-lived nuclides (*22*). Phosphorus was determined by β^- counting ^{32}P after a delay of 3–4 wk, employing a gas-flow proportional counter (ECIL, Hyderabad) and Al filter (28 mg cm^{-2}), as already described (*23*).

RNAA Procedure

After a delay of 2 wk, samples were transported to our laboratories and decomposed in 5 mL of aquaregia and 2 mL of perchloric acid. Carrier solutions containing about 5–10 mg each of Fe, Zn, and Co from high-purity chemicals (AR, more than 99%) were added and heated to dryness. The contents were dissolved in 10 mL of 6 *M* HCl, and the solution was made to 50 mL. From a 10-mL aliquot of the stock solution, Fe was extracted with diethyl ether, followed by a 6% aqueous solution of cupferron and $CHCl_3$ at pH 4, from which rediochemically pure fractions were obtained. To the aqueous phase, 5% KCN was added to mask Co, and then Zn was extracted using 0.1 *M* 2-thenoyltrifluoroacetone (TTA) in isobutyl methylketone (IBMK) at pH 6. Finally, Co was extracted using α-nitroso β-naphthol in chloroform at pH 5 (*24*).

In another radiochemical procedure for Sb and Se, respective carriers were added before dissolution, and a stock solution was prepared in 0.5 *M* HCl. To a 10-mL aliquot, 2 mL of 2% citric acid solution was added for masking of Sb. Selenium was complexed with o-phenylenediamine, using a 0.1% solution in absolute alcohol. Two extractions were carried out, each with 10 mL of benzene. The aqueous phase was reduced to 2 mL and added 6 mL of 6 *M* H_2SO_4, 1 mL of 1 *M* KI, and 1 mL of 0.1% aqueous solution of 4-(2-pyridylazo)-resorcinol (PAR). Finally, Sb was extracted with 10 mL of benzene (*25*). In each of these cases an aliquot of the organic phase was counted on a well type 2 × 1.75 in (5.1 cm × 4.4 cm) (Tl) detector, coupled to a Nuclear Data ND 62 1K MCA and a Centronics 154 printer. Corrected areas for the respective photopeaks were used for the calculation of elemental concentrations. Pure elemental standards were used as primary standards, and NBS SRM 1577a and IAEA H-4 were used as comparators for quality assurance.

RESULTS

In general, the precision of our INAA analysis is within the range of ±5% at ppm (μg/g) levels and ±5–10% at ppb (ng/g) levels, based on replicate analyses. It is observed that our INAA data for various elements in biological standards NBS SRM 1577a (Bovine Liver) and IAEA H-4 (Animal Muscle) are in good agreement with the certified or literature reported values (Table 1). Typical γ-ray spectra for cancerous and normal tissues, along with that of Bovine Liver, are shown in Fig. 1. Ranges of elemental abundances and mean values, along with the standard deviation of the mean concentration obtained by INAA and RNAA procedures, are given in Table 2. Enhancement or depression is expressed as the percent change in the mean value. Some other nuclides, such as those of Tb, Eu, Cr and so on, were also identified in some cases, but could not be quantified because of nonavailability of suitable standards. All values are expressed on a dry wt basis. In most cases, these concentrations are within the normal range for human tissues (*26*). Histograms of the mean concentrations of the trace elements in paired samples obtained from 6 patients are shown in Fig. 2.

DISCUSSION

It is observed from Table 2 that concentrations of all the minor and most trace elements are enhanced for all paired samples (not given separately) of cancerous and normal breast tissue. The increase in the mean concentration of Se is maximum (by almost twofold) in cancerous compared to that in normal tissue. It is followed by Co, As, P, K, Ga, Cd, Mg, and Cu, which all show an enhancement of up to 40%. However, concentration for quite a few trace elements, viz., Fe, Cs, Sr, and I, are depressed or only marginally changed. It is apparent that the change in concentration of these elements in cancerous tissue is real and establishes a definite pattern in the carcinoma of the breast.

The essentiality of Se has been recognized in recent years, but in excess it may be deleterious (*27, 28*). Several workers have observed its enhancement in cancer of the breast (*15,18,21*) and other tissues (*12,29*). Ehmann and coworkers (*30–32*) have observed significant differences in Se level for Alzheimer's disease patients, and its role in aging. On the other hand, Vernie et al. (*33*) have not found any significant change in Se levels of whole blood and plasma from breast cancer patients. Further, Se also seems to be correlated with other trace metals such as Fe, Co, and Zn. Selenium and Zn have been found to play an important role in

Table 1
Comparison of Elemental Concentrations in Standards

Element	Bovine liver SRM 1577a		Animal muscle H-4	
	This work	Lit. value	This work	Lit. value
Na (%)	0.239 ± 0.003	0.243	0.208 ± 0.020	0.206
P (%)	1.097 ± 0.028	1.110	0.693 ± 0.013	0.683
K (%)	0.994 ± 0.018	0.996	1.56 ± 0.04	1.58
Fe (μg/g)	191 ± 2.0	194	51.1 ± 0.9	49.1
Mn (μg/g)	10.04 ± 0.15	9.9	0.49 ± 0.04	0.52
Cu (μg/g)	163 ± 4	158	3.62 ± 0.23	3.65
Zn (μg/g)	127 ± 7.8	123	80.6 ± 1.6	86.0
Cd (μg/g)	0.45 ± 0.02	0.44	4.5 ± 0.3 (ppb)	4.9 (ppb)
Cs (μg/g)	0.26	NA	–	0.12
Sr (μg/g)	–	0.138	0.325	NA
Se (μg/g)	0.76 ± 0.12	0.71	0.24 ± 0.02	0.28
Br (μg/g)	9.46 ± 0.59	9.0	4.12 ± 0.05	4.10
Ga (μg/g)	1.66 ± 0.06	–	1.80 ± 0.16	–
I (ng/g)	3.30 ± 1.5		–	14.3
Sb (ng/g)	2.4	3.0	1.33	–
As (ng/g)	43.3 ± 2.0	47.0	7.77 ± 0.33	7.0

malignant breast tissue (*18,21*). Selenium deficiency has been asserted to be the cause of muscular dystrophy, pancreatic fibrosis, hepatosis, dietetosis, dietetica, cancer, and certain disorders attributed to vitamin E deficiency. Selenium and vitamin E protect membranes from oxidative degradation, and prevent exudative diathesis. Recently, Se deficiency in soils and diets in the Keshan province of China has been stated to cause heart disease (*33*). It is well known that Se is incorporated into at least 6 sulfur-containing amino acids, and Se can replace S, making them ineffective (*34*). Therefore, it is quite likely that replacement of S by Se may accelerate the growth of tumors.

Experimentally, Zn deficiency and its supplementation have each shown both inhibitory and stimulatory responses on tumor growth (*35*). It has been shown that Se and Zn are antagonistic to each other in a number of metabolic systems (*36*).

Several workers have reported the elevation of Zn and Cu levels in breast carcinoma (*16,18,21*). The present study also confirms these results with a consistently elevated level of both the elements in all 6 paired breast cancer tissues. The exact role of Zn in carcinogenesis is unknown (*37*). However, it is known to be essential for more than a hundred different metabolic functions. Copper is absorbed as an amino acid complex in the intestinal walls, and also is deposited in the brain of the patients with obvious pathological consequences (*38*). A better understanding of the interrelationships of these trace metals is needed in order to better understand their roles in regulating tumor growth.

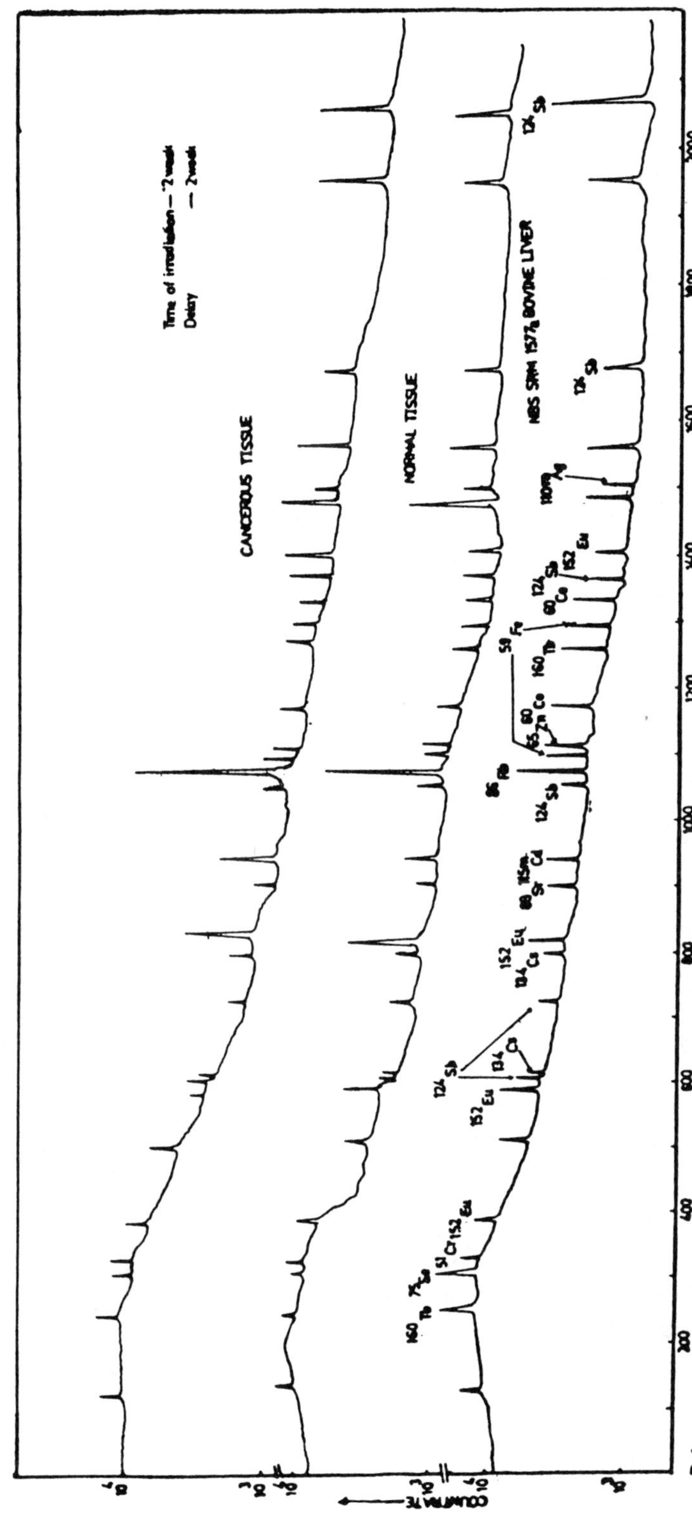

Fig. 1. Typical γ-ray spectra of cancerous and normal breast tissue, along with NBS SRM 1577a.

Table 2
Elemental Concentrations in Normal and Cancerous Human Breast Tissue

	Normal		Cancerous		
Element	Range	Mean ± SD	Range	Mean ± SD	Change, %[†]
Na (%)	0.09–0.49	0.22 ± 0.15	0.16–0.61	0.31 ± 0.14	+40.9
K (%)	0.06–0.31	0.18 ± 0.10	0.16–0.44	0.29 ± 0.11	+61.0
Mg (%)	0.04–0.17	0.11 ± 0.07	0.10–0.18	0.14 ± 0.04	+38.1
Cl (%)	0.09–0.31	0.18 ± 0.09	0.13–0.35	0.23 ± 0.08	+30.0
P (%)	0.11–0.37	0.25 ± 0.13	0.30–0.50	0.41 ± 0.08	+65.2
Fe (μg/g)	19.0–43.1	32.0 ± 8.3	20.0–45.0	25.8 ± 7.8	−19.3
Zn (μg/g)	22.0–55.0	36.6 ± 10.2	30.0–62.6	43.8 ± 10.5	+20.0
Cu (μg/g)	0.76–2.33	1.32 ± 0.55	1.11–3.62	1.80 ± 0.92	+36.4
Se (μg/g)	0.64–0.93	0.81 ± 0.08	1.20–1.84	1.58 ± 0.10	+95.0
Cd (μg/g)	0.15–0.50	0.36 ± 0.13	0.32–0.80	0.52 ± 0.18	+44.2
Mn (μg/g)	0.23–0.60	0.43 ± 0.12	0.38–0.62	0.49 ± 0.08	+13.9
Br (μg/g)	2.0–54.0	19.0 ± 21.0	6.0–70.0	25.0 ± 21.0	+31.6
Ga (μg/g)	1.4–2.3	1.9 ± 0.4	1.7–2.5	2.8 ± 0.3	+47.4
Cs (μg/g)	1.4–5.0	3.7 ± 0.6	2.0–5.5	3.2 ± 1.4	−13.5
Rb* (μg/g)	–	4.3	–	6.1	+42.8
Sr (ng/g)	10.0–105	62.3 ± 27.6	30.0–107	73.8 ± 32.7	−18.4
As (ng/g)	5.6–14.0	9.0 ± 3.0	5.0–32.0	16.6 ± 10.1	+84.4
Co (ng/g)	1.0–5.0	3.1 ± 1.1	1.3–7.0	4.9 ± 1.9	+55.4
Sb (ng/g)	0.25–0.43	0.33 ± 0.13	0.18–0.64	0.35 ± 0.19	+6.1
I (ng/g)	3.3–19.0	8.8 ± 5.6	2.5–12.0	5.2 ± 3.7	−41.1
Ag (ng/g)	25–54	43.8 ± 11.6	43–58	53 ± 5.8	+21.0

*These values are for a single case only.

[†]Expressed as ratio of change in mean value and mean concentration in normal × 100; +ve sign indicates enhancement and −ve sign indicates depression.

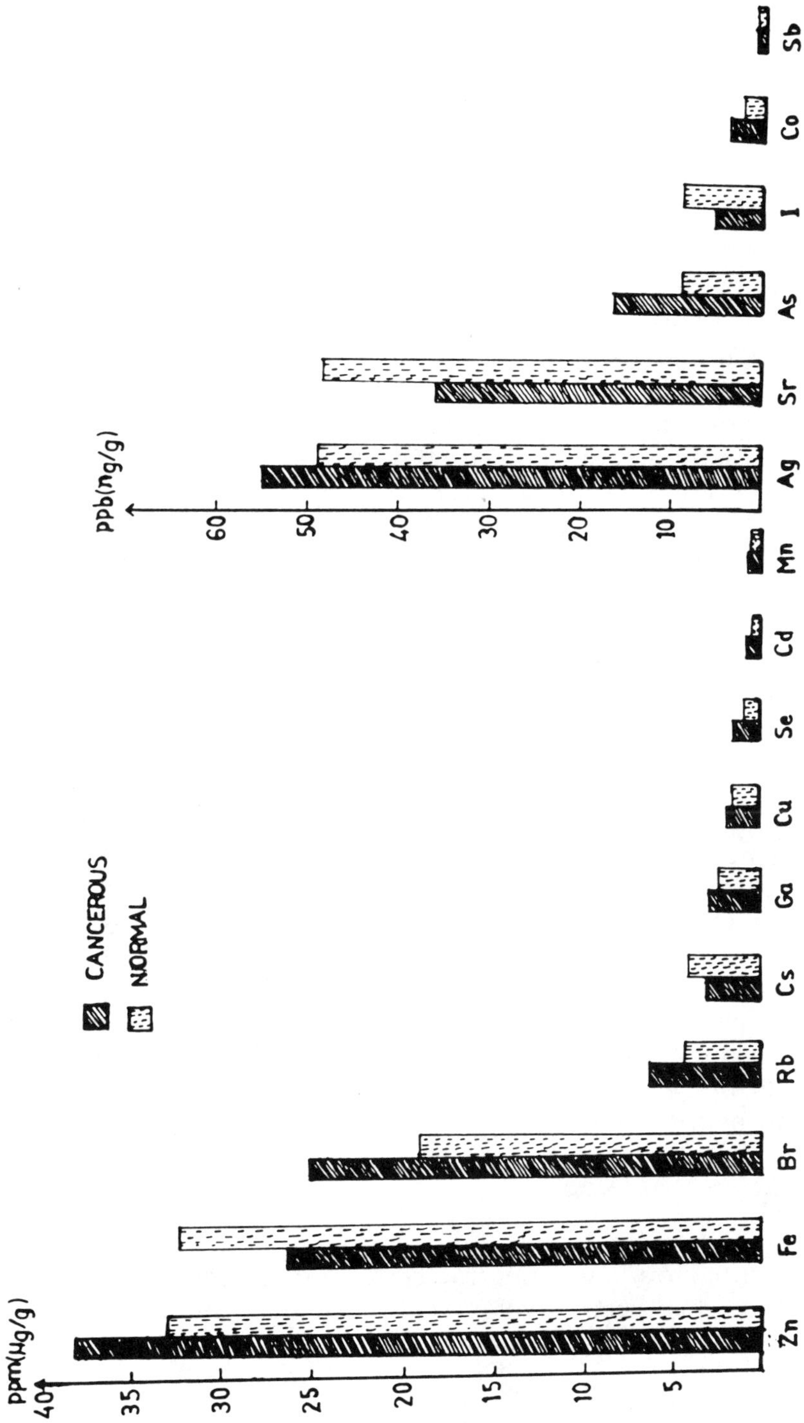

Fig. 2. Comparison of mean trace elemental contents in normal and cancerous breast tissues from 6 patients.

Rubidium has been added to the list of essential trace elements only recently, but at present it has no known physiological function. We obtained Rb data in only one case of significant enhancement, in agreement with the observaiton of Rizk and Sky-Peck (*21*). The physical properties of Rb closely resemble those of K, and may only reflect the higher concentration of K consistently found in various cancer tissues (*39*). It may simply be a reflection of the largely intracellular location of K in normal tissues, and the high cellular nature of malignancy. Our observaiton of enhancement in concentrations of minor constituents, such as K, Mg, and P, is in agreement with earlier reports (*13–16*). Seltzer at al. (*40*) have reported a twofold increase in Mg concentration in human breast cancer. Interestingly, Cs, another alkali metal, and Sr, an alkaline earth, both show depletion in cancerous tissue.

The decrease in Fe concentration of the cancerous tissue, though small, is in agreement with the observation of other workers (*12,14,15*), even though Santoliquido et al. (*16*) have found increased Fe in lymphatic leukemia. It is well known that ferritin serves as a major iron storage constituent of the tissue, but it has also been found to be associated with the blood circulation in patients with disease (*41,42*). This could possibly be the cause for depletion of Fe in cancerous breast tissue. Among the halogens, Br is enhanced, but I is depressed in cancerous tissue.

These studies only suggest that some elements are incorporated into normal cell, change its enzymatic activity, and help in reduplication of neoplastic growth. Generally, enzymatic activities change in cancerous cell, resulting in variations in trace elemental concentrations, which may act as cofactors of these enzymes.

Although present data are not conclusive, they do give guidelines for future study into the possible role and interaction of essential as well as nonessential trace elements in the carcinogenic process. The increased awareness of the role of trace elements, and their interactions in metabolism and cancer growth, suggested the need for a multielemental microanalytical method that could provide rapid analysis in the same biological specimen. Clearly NAA employing short irradiation and radiochemical procedures is most suitable for such studies. At this stage, it can only be said that further work is required to understand fully the role of trace elements in neoplastic growth.

SUMMARY

A sequential irradiation neutron activation analysis with radiochemical procedures has been developed, and permits the determination of more than 20 elements in cancerous and normal breast tissue. Concentrations of several trace elements are enhanced, whereas those of others are depressed, indicating their role in the growth of tumor.

ACKNOWLEDGMENTS

Grateful thanks are due the Department of Atomic Energy, Government of India for financial assistance. We also thank Drs. P. R. Natarajan, Satya Prakash, and S. B. Monohar for their help and cooperation in the short-irradiation work at Bombay.

REFERENCES

1. A. Furst, *Metal Binding in Medicine*, M. J. Seven and L. A. Johnson, eds. J. B. Lippincott, Philadelphia and Montreal, 1960, pp. 336–344.
2. J. Versieck, *CRC Crit. Rev. Clin. Lab. Sci.* **22,** 97 (1985).
3. T. Mega, S. Tonici, and H. Aruchi, *Nara Igaku Zasshi* **23,** 201 (1972).
4. J. W. Berg and F. Burbank, *Ann. NY Acad. Sci.* **199,** 249 (1972).
5. E. B. Wright and T. L. Dormondy, *Nature* **237,** 166 (1972).
6. M. Hrgovic, C. F. Tessmer, and B. W. Brown *Progr. Clin. Cancer* **5,** 121 (1973).
7. M. Hrgovic, C. F. Tessmer, and F. B. Thomas *Cancer,* **31,** 1337 (1973).
8. H. T. Delves, F. W. Alexander, and H. Lay *Br. J. Haematol.,* **24,** 525 (1973).
9. S. Inutsuka, S. Araki, and I. Kusaba, *Rinsho Byori* **21,** 632 (1973).
10. I. T. Tsilyunyk, M. M. Ischenko, and S. U. Rusenko, *Klin. Med.* **43,** 18 (1965).
11. M. A. Sirover and L. A. Loebb, *Science* **194,** 1434 (1976).
12. P. C. Mangal and K. B. Verma, *Indian J. Med. Res.* **69,** 290 (1979).
13. F. B. DeJorge, J. G. Sampalo, J. L. Guedes, and A. B. DeUlhoa, *Clin. Chim. Acta* **12,** 403 (1965).
14. I. L. Mulay, R. Ray, B. E. Knox, N. H. Suhr, and W. E. Dalaney, *J. Natl. Cancer Inst.* **47,** 1 (1971).
15. A. E. Schwartz, G. W. Leddicotte, R. W. Fink, and E. W. Friedman, *Surgery* **76,** 325 (1974).
16. P. M. Santoliquido, H. W. Southwick, and J. H. Oliwin, *Surg. Gynecol. Obstet.* **142,** 65 (1976).
17. I. Othman and N. M. Spyrou, *Trace Elements in Medicine and Biology,* P. Bratter and P. Schramel, eds., Walter de Gruyter Co., New York, (1980), p. 199.
18. P. C. Mangal and S. Kumar, *Indian. J. Phys.* **58A,** 355 (1984).
19. J. C. K. Lai, A. W. K. Chan, M. J. Minski, T. K. C. Leung, L. Lim, and A. N. Davison, *Metal Ions in Neurology and Psychiatry*, S. Gabay, J. Haris, and B. T. Ho, eds., Alan R. Liss, New York, 1985, p. 323.
20. W. D. Ehmann, W. R. Marksbery, E. J. Kasarskis, D. E. Vance, S. S. Khare, J. D. Hord, and C. M. Thompson, *Biol. Trace Element Research,* **13,** 19 (1987).
21. S. L. Rizk and H. H. Sky-Peck, *Cancer Res.* **44,** 5390 (1984).
22. H. K. Wankhade, D. L. Samudralwar, and A. N. Garg, *J. Radioanal, Nucl. Chem. Letters* **105,** 95 (1986).
23. R. G. Weginwar, D. L. Samudralwar, and A. N. Garg, *J. Radioanal. Nucl. Chem.* **133,** 317 (1989).
24. R. G. Weginwar and A. N. Garg, Radiochemistry and Radiation Chemistry Symposium, IGCAR, Kalpakkam, Jan 4–7, 1989, RA-01.
25. R. G. Weginwar and A. N. Garg, unpublished results.

26. G. V. Iyengar, W. E. Kollmer, and H. J. M. Bowen, eds. *Trace Elemental Composition of Human Tissues and Body Fluids*, Verlag Chemie, Heidelberg, 1978 p. 151.
27. C. D. Thomsen and M. F. Robinson, *Am. J. Clin. Nutr.* **33,** 303 (1980).
28. J. E. Spallholz, J. L. Martin, and H. E. Ganther, eds., *Selenium in Biology and Medicine,* AVI Publishing Co. Inc., Westport, USA, 1981, p. 98.
29. P. C. Mangal and J. S. Sohal, *Indian J. Exp. Biol.* **16,** 685 (1978).
30. W. D. Ehmann, W. R. Markesbery, M. Alauddin. T. I. M. Hossain, and E. H. Brubaker, *Neurotoxicology* **7,** 197 (1986).
31. D. E. Vance, W. D. Ehmann, and W. R. Markesbery, *Neurotoxicology* **9,** 197 (1988).
32. W. R. Markesbery, W. D. Ehmann, M. Alaudin, and T. I. M. Hossain, *Neurobiol. Aging* **5,** 19 (1984).
33. L. N. Vernie, M. De Vries, C. Benekhuijsen, J. J. M. Goeij, and C. Zegers, *Cancer Lett.* **18,** 283 (1983).
34. Keshan Disease Research Group of the Chinese Academy of Medical Sciences, *Chin. Med. J.* **92,** 471 (1979).
35. L. M. Cummins and J. L. Martin, *Biochemistry* **6,** 3162 (1967).
36. E. J. Underwood, *Trace Elements in Human and Animal Nutrition, 4th Ed.,* Academic Press Inc., New York, 1977.
37. A. S. Prasad, ed. *Trace Elements in Human Health and Disease, Vol. 1 Zinc and Copper,* Academic Press, New York 1976.
38. A. Lipcsey, M. Ordogh, J. Fekete, and E. Szabo, *J. Radioanal. Nucl. Chem.* **88,** 57 (1985).
39. A. Danielson and E. A. Steinnes, *J. Nucl. Med.* **11,** 250 (1970).
40. M. H. Seltzer, F. E. Rosato, and M. J. Fletcher, *J. Surg. Res.* **10,** 159 (1970).
41. P. A. E. Jones, F. Miller, M. Warwood, and A. Jacobs, *Br. J. Cancer,* **27,** 212 (1973).
42. A. Jacob and M. Warwood, *Br. J. Cancer* **34,** 162 (1976).

Instrumental Neutron Activation Analysis of Kidney Stones

S. SARMANI,* L. L. KUAN, AND M. A. A. BAKAR[1]

Department of Nuclear Science [1]Department of Biochemistry, Universiti Kebangsaan, Malaysia, 43600 Bangi, Malaysia

Received April 17, 1989; Accepted November 16, 1989

ABSTRACT

Kidney stone samples of the types calcium oxalate, uric acid, and xanthine were analyzed for their elemental contents by neutron activation analysis to study both the elemental correlation and influence of element on stone precipitation processes. Elements, such as Al, Au, Br, Ca, Cl, Co, Cr, Fe, Hg, I, K, Mg, Na, Sb, Se, Sr, and Zn, were determined quantitatively. Calcium oxalate stones contained higher concentration of all the elements analyzed compared to uric acid or xanthine stones. The concentrations of Cl, Fe, K, Na, Sr, and Zn were relatively higher than Au, Co, Cr, and Sb. A positive correlation exists between Ca and Zn, whereas a negative correlation exists between Sr and Ca. Zinc may play an important role in the formation of calcium oxalate stone.

Index Entries: Neutron activation analysis; trace elements; kidney stones; urinary calculi.

INTRODUCTION

The formation of kidney stones is the result of the precipitation of saturated oxalate, ureate, phosphate, and carbonate salt solutions in kidney. There are several suggestions offered to explain the formation pro-

*Author to whom all correspondence and reprint requests should be addressed.

cesses. Kark and Oyama (*1*) suggested there were 3 major theories of stone formation: the precipitation-crystallization theory, the matrix-nucleation theory, and the inhibitor-absence theory. The precipitation- crystallization theory implied that supersaturation of urine with crystalloids led to precipitation and crystal growth. Such a mechanism seems well supported in the formation of stones composed of uric acid and xanthine. The matrix-nucleation theory suggested that some matrix substance formed an initial nucleus upon which further stone growth occurred. The inhibitor-absence theory proposed that stone formation could occur even in patients with normal concentrations of urinary crystalloids. Generally, kidney stones can be classified into 4 types based on their chemical compositions; they are calcium oxalates, calcium oxalate mixed with apatite, magnesium ammonium phosphate, and uric acid and xanthine (*2*).

Chemical and elemental analyses of kidney stone has been carried out in order to understand its formation. However, determination of trace elements in such complex materials is often difficult at the low concentrations involved. Among the techniques capable of reliable determination of submicrogram quantities of a number of elements are neutron activation analysis and atomic absorption spectrometry. The first method has an excellent scope for multielement analysis at trace levels, both quantitative and qualitative, whereas the latter is best when applied to the determination of single elements or sequentially to a small number of them. Simultaneous multielement analyses are useful in looking at elemental correlation and the synergistic effects of trace elements on the kidney stone formation.

Applications of neutron activation to kidney stones analyses have been reported by several authors. Lewas et al. (*3*) determined Co, Cu, Mn, Mo, and Zn in kidney stone by radiochemical neutron activation analysis. Donev et al. (*4*) used instrumental neutron activation analysis to determine 12 elements in the kidney stone, which were Fe, Cr, Zn, Cd, Co, Hg, Sr, Se, Cu, Br, Na, and K. Jacimovic et al. (*5*) identified Hg, Cr, La, Sc, Co, Fe, and Sb in kidney stone sample also by instrumental neutron activation analysis. Recently Lin et al. (*6,7*) determined 19 elements in urinary stones by applying the same technique.

Neutron activation analysis is very suitable for the determination of trace elements in complex matrix, such as the kidney stone, because of the ease of sample preparation and handling. It is also capable of determining many elements simultaneously, which is very useful in looking at elemental correlation and the synergistic effect on stone formation.

In this study, instrumental neutron activation analysis was used to determine the trace element composition of kidney stones and to evaluate the relative importance of the elements in stone formation based on the statistical analysis of the data obtained. Chemical composition analysis was also carried out to determine the types of the kidney stones.

METHODS

Twenty-two kidney stone samples were obtained from patients at the General Hospital, Kuala Lumpur. The patients came from different parts of the country and comprise both the different races and social classes. The kidney stones were rinsed with distilled water to remove surface contamination and then dried at 85°C overnight in an oven. Classification of the stone types was based on the chemical composition. Analysis of the composition was carried out using Merckotest kits developed by Merck.

Each stone sample was wrapped in a polyethylene sheet, was crushed to fine powder, and then dried at 85°C to constant weight. About 100 mg of the sample was weighed accurately for the analysis. Samples and standards were then placed in a polyethylene irradiation vial. Copper wire as a flux monitor was taped to the outside wall of the vial.

An appropriate volume of the chemical standard solution was dropped onto ashless filter paper in an irradiation vial to maintain similar geometry. The vial was then dried in an oven at 65°C. Samples and standards were irradiated at 2 different irradiation conditions. Irradiations were carried at the Nuclear Energy Units TRIGA Reactor. One batch of samples and standards was irradiated for 2 min at a neutron flux of 2.2×10^{12} n cm^{-2} s^{-1}, then cooled for 10 min, and counted for 5 min for the analysis of short-lived radionuclides. For the determination of medium and long-lived radioisotopes, samples and standards were irradiated for 6 h at a neutron flux of 2.5×10^{12} n cm^{-2} s^{-1}, cooled for between 1 h and 2 wk, and counted for a period of 1–6 h depending on the activity of the samples. Activity measurements were carried out using a high resolution γ spectrometry system. The system was comprised of a hyperpure germanium detector connected to an Ortec 572 amplifier and Nuclear Data ND 6600 Multichannel Analyzer with PDP11 COMBUS Computer facility. Accuracy of the analytical technique was checked by analyzing Bowen's Kale standard reference material.

RESULTS

Table 1 shows results relating to the accuracy of the analytical technique used. By comparing the certified values with this work, it can be concluded that this analysis was precise and accurate. Data on the elemental concentration in the 3 types of kidney stones analyzed are given in Table 2. The standard deviations given in Table 2 show variations of the elemental concentrations in the samples rather than precision of the analytical methods.

Table 1
Analysis of Bowen's Kale Reference Material

Element	This work [mg/kg]	Certified value [mg/kg]
Al	79.1 ± 4.1	39.9 ± 9.9
Au	0.0214 ± 0.0042	0.0200 ± 0.0006
Br	30.6 ± 3.2	24.9 ± 2.5
Ca	41133 ± 3653	41060 ± 2220
Cl	2457 ± 288	3560 ± 43
Co	0.0506 ± 0.0140	0.0632 ± 0.0106
Cr	0.279 ± 0.012	0.3690 ± 0.0263
Fe	135.2 ± 6.3	119.3 ± 14.8
K	24246 ± 1486	24370 ± 1452
Mg	1600 ± 94	1605 ± 180
Na	2325 ± 154	2366 ± 294
Sr	76.27 ± 8.09	75.7 ± 29.3
Zn	33.88 ± 2.86	32.29 ± 2.76

DISCUSSION

From the chemical composition analysis, the kidney stone samples can be classified into calcium oxalate, uric acid, and xanthine. Seventeen elements were detected in the kidney stone samples. Ca existed as a major element in the calcium oxalate type of stones, whereas it also existed as minor element in uric acid and xanthine type stone. Al, Au, Br, Ca, Cl, Co, Cr, Fe, Hg, K, Na, Se, and Zn were found in all the samples, whereas Cs, Mn, and Sc were not detected in any of the samples. Iodine was detected in some calcium oxalate stones with an average concentration of 1825 mg/kg; however, Lin et al. (*7*) reported that as high as 5.3% iodine was found in kidney stones.

Generally, calcium oxalate stones contained higher concentrations of all the elements analyzed compared to uric acid and xanthine stones. Uric acid and xanthine stones contained lower concentrations of Zn when compared to calcium oxalate stones. This observation was also reported by Lin et al. (*7*) and Levinson et al. (*8*). Zinc may have some effects on the urolithiasis of kidney stones (*7*).

Table 2
Trace Elements in Three Types of Kidney Stones Expressed in Mg/Kg Unless Stated Otherwise

Elements	Calcium oxalate	Uric acid	Xanthine
Al	148.0 ± 73.0	64.6 ± 7.7	20.0 ± 1.4
Au	0.03 ± 0.01	0.02 ± 0.1	0.03 ± 0.01
Br	11.4 ± 8.4	34.4 ± 5.8	6.4 ± 0.7
Ca(%)	26.2 ± 3.2	1.3 ± 0.1	0.06 ± 0.01
Cl	531 ± 46	640 ± 8	208 ± 15
Co	0.15 ± 0.12	0.03 ± 0.01	0.03 ± 0.01
Cr	0.75 ± 0.57	0.42 ± 0.17	0.37 ± 0.34
Fe	97.3 ± 63.2	59.4 ± 9.5	33.7 ± 8.1
Hg	0.68 ± 0.78	1.18 ± 0.99	0.23 ± 0.01
I	1825 ± 29	41.6 ± 36.8	n.d
K	489 ± 198	274 ± 38	260 ± 2
Mg	1405 ± 1253	n.d	n.d
Na	2793 ± 1795	680 ± 3	348 ± 42
Sb	1.48 ± 1.13	n.d	n.d
Se	0.64 ± 0.31	0.25 ± 0.04	0.36 ± 0.01
Sr	101.8 ± 62	n.d	n.d
Zn	368 ± 293	24.1 ± 12.9	7.3 ± 0.6

n.d—not detected.

Sr was only detected in calcium stones that contained Mg, whereas Mg was only present in calcium oxalate stones and not in the uric acid or xanthine stones. Pearson-r correlation between Ca, Mg, and Sr was tabulated. Negative correlation between Ca and Sr was found, whereas for Ca and Mg the correlation was positive. Mg, Ca, and Sr have similar chemical properties. Since they are in the same group in the Periodic Table, they have the same ionic charge; therefore Sr will coprecipitate with Ca and Mg. The fact that the calcium concentration is 26.3% in calcium oxalate stone indicates that calcium plays an important role in the stone formation. Since magnesium is only present in calcium oxalate stones at much lower concentrations than calcium, Mg can be considered the second most significant cation in the formation of the stone.

Bromine and chlorine are present at much higher concentrations in uric acid stones compared to calcium oxalate or xanthine stones. This indicates that Br and Cl play an important role in the uric acid stone formation. Br and Cl are not essential elements, and therefore they are excreted through many ways, including their incorporation in kidney stones. Their incorporation may be through the coprecipitation of insoluble bromide and chloride salts. The importance of Br and Cl in the uric acid stone formation needs further investigation.

Gold was detected in all the samples, but this element was not reported to be present in calculi from Western patients (*9–11*). This phenomenon could be attributed to dietary habits and environmental factors. Gold is not expected to have an important role in the stone formation, because its concentrations in all types of stones analyzed are within the same magnitudes.

ACKNOWLEDGMENT

The authors would like to thank both the Nuclear Energy Unit for making their facilities available and the Department of Surgery for providing the kidney stone samples.

REFERENCES

1. R. M. Kark and J. H. Oyama, *Modern Nutrition in Health and Disease*, R. S. Goodhart and M. E. Shills, eds. Lea and Febiger, Philadelphia, 1980, pp. 1035–1038.

2. R. J. Reveillaud, M. Daudon, M. F. Protat, and G. Ayrole, *Eur. Urol. J.* **6,** 161 (1980).

3. T. Lewas, A. D. Callar, and J. Feria, *J. Radioanal. Chem.* **13,** 75 (1973).

4. I. Donev, S. Mashkarov, L. Maritchkova, and G. Gostev, *J. Radioanal. Chem.* **37,** 441 (1977).

5. L. J. Jacimovic, R. Draskovic, and B. Ostojic, *J. Radioanal. Chem.* **37,** 415 (1977).

6. S. M. Lin, C. H. Chiang, C. H. Huang, and C. L. Tseng, *Radioanal. Nucl. Chem. Lett.* **96,** 153 (1985).

7. S. M. Lin, C. L. Tseng, and M. H. Yang, *Appl. Radiat. Isot.* **38,** 635 (1987).

8. A. A. Levison, M. Nosal, M. Davidman, E. L. Prien, and R. G. Stevenson, *Invest. Urol.* **15,** 270 (1978).

9. I. Durak, A. Yasar, Z. Yurtarslani, *B. J. Urol.* **62,** 203 (1988).

10. E. Z. Hazarika and B. Rao, *Indian J. Med. Res.* **62,** 776 (1974).

11. A. Hesse, H. J. Dietze, W. Berg, and E. Hienzsch, *Eur. Urol. J.* **3,** 359 (1977).

Studies on Zinc- and Cadmium-Bound Proteins in Bovine Kidneys by Biochemical and Neutron Activation Techniques

C. K. JAYAWICKREME AND A. CHATT*

Trace Analysis Research Centre, Department of Chemistry, Dalhousie University, Halifax, NS, B3H 4J1, Canada

Received September 6, 1989; Accepted November 17, 1989

ABSTRACT

Bovine kidneys were found to contain about 78 ppm Zn and 0.78 ppm Cd. Approximately 45% of Zn and 60% of Cd were present in the cytosol fraction. More than 95% of these two metals were bound to macromolecules. Both Zn– and Cd–protein complexes were observed to be stable between pH 7 and 10.5. Separation and characterization of these proteins were carried out using several chromatographic and electrophoretic techniques in conjunction with instrumental neutron activation analysis (INAA). The results showed the presence of at least four Zn-binding proteins with mol wt > 300,000, 260,000, 89,000, and 27,000 and at least three Cd-binding proteins of mol wt > 300,000, 32,000, and 13,000.

Index Entries: Zinc; cadmium; proteins; neutron activation; chromatography; electrophoretic; pH dependence; bovine kidney.

INTRODUCTION

Zinc is one of the essential and most abundant trace elements found in living species. It is known to be associated with more than 40 enzymes

*Author to whom all correspondence and reprint requests should be addressed.

and proteins. Zinc readily forms complexes with amino acids, peptides, nucleotides, and proteins in biological media. Cadmium, on the other hand, is considered a toxic element even at low concentrations, and is associated with a fewer number of proteins. Zinc and Cd are known to occur together in many biological systems at varied ratios. Metallothioneins isolated from tissues and fluids have been reported to contain both Zn and Cd. The presence of Cd-binding proteins in mammary glands and in cytoplasm of liver and kidney has been reported (*1,2*).

Biochemical techniques coupled to INAA and/or radiotracers are being increasingly used to investigate metalloproteins. Behne et al. have investigated the utilization of ultrafiltration, gel filtration, isoelectric-focusing, and gel electrophoresis with INAA (*3*). Sabbioni et al. have studied the accumulation of Cd in rat liver Cd-binding proteins induced by intraperitoneal injections of $CdCl_2$ using radiotracers (*4*). A number of bioanalytical techniques have been used in our laboratory in conjunction with INAA to isolate and characterize protein-bound trace elements in bovine kidneys (*5–9*). The INAA technique is ideal for the determination of the metal components of metalloproteins because of its many advantages, including multielement specificity, ability to analyze milligram quantities of solids and liquids, nondestructive nature, high precision and accuracy, excellent sensitivity, freedom from reagent blanks, and either no or little interference from the matrix. The present work deals with the studies on Zn- and Cd-binding proteins in the supernatant fraction of bovine kidneys, with emphasis on the applicability of several bioanalytical techniques in conjunction with INAA.

EXPERIMENTAL

Chemicals

All chemicals used in this work were of ultrapure quality. Further purification of some chemical reagents was also done in our laboratory. Details are given elsewhere (*5*). All experiments were carried out at 4°C unless otherwise noted.

Neutron Activation Analysis

All irradiations were done in the Dalhousie University SLOWPOKE-2 Reactor (DUSR) facility at a flux of 5×10^{11} n cm^{-2} s^{-1}. Gamma-ray spectra were recorded using a Canberra coaxial Ge(Li) detector (with a FWHM of 1.90 keV at the 1332-keV photopeak of ^{60}Co, a relative efficiency of 9.4%, and a peak-to-Compton ratio of 35:1) coupled to either a Tracor Northern TN-1700 or a Nuclear Data ND-66 model 4096-channel pulse-height analyzer. The procedure for the preparation of comparator standards of Cd and Zn used for calculating the elemental levels in samples has been described previously (*5*).

Bioanalytical Methods

Fresh samples of bovine kidney of mature animals were obtained from a local supplier, and were minced and homogenized in 0.01*M* HEPES buffer (1:4 v/v) at pH 7.4. Subcellular fractionation was carried out by centrifugation at 700, 9000, 30,000, and 100,000*g*. The supernatant fraction obtained at 30,000*g* was further separated using ion exchange, gel filtration, hydroxylapatite chromatography, electrophoresis, isotachophoresis, electrofocusing, chromatofocusing, and ammonium sulfate precipitation. Each of these techniques has previously been investigated in detail for its efficiency of separating protein-bound trace elements under optimized conditions (*5*).

RESULTS AND DISCUSSION

An INAA method was developed for the determination of Zn and Cd through their long-lived nuclides, ^{65}Zn (half life = 243.8 d) and ^{115}Cd (half-life = 2.25 d). The samples were irradiated for 24 h, allowed to decay for 2–4 d and counted for 1–2 h for measuring the intensity of the 335-keV photopeak of ^{115}Cd. For the detection of the 1115-keV peak of ^{65}Zn, the samples were allowed to decay further for 15–21d and counted for 10–20 h.

To evaluate the accuracy of the INAA measurements, several certified reference materials and interlaboratory comparison standards were analyzed for Zn and Cd. A few results in ppm are given here in the following format:

- name of material (element: value obtained in this work, value certified by the issuing agency)
- IAEA mixed human diet CRM H-9 (Zn: 28.2 ± 0.4, 27.58 ± 2.62)
- NBS bovine liver SRM-1577 (Zn: 138 ± 8, 130 ± 13; Cd: <0.5, 0.27 ± 0.04)
- NRC lobster hepatopancreas RM TORT-1 (Zn: 172 ± 4, 177 ± 10; Cd: 25.2 ± 1.6, 26.3 ± 2.1)
- IAEA horse kidney CRM H-8 (Zn: 196 ± 5, 193 ± 6; Cd: 191 ± 6, 189 ± 5)
- NBS oyster tissue SRM-1566 (Zn: 835 ± 26, 852 ± 14; Cd: 3.53 ± 0.28, 3.5 ± 0.4)

The agreement between our values and certified or reported values is generally good. The precision of the measurements is around ±10%.

Ultrapure reagents must be used for studying protein-bound trace elements if the proteins are separated prior to the determination of their elemental content. All reagents employed in this work were analyzed for trace element contaminants by INAA before using them in experiments.

The levels of Zn and Cd present as contaminants in the reagents are given in Table 1. In most cases only the decision limits described by Currie (*10*) are reported because of the very low levels of Zn and Cd in the ultrapure reagents used.

Bovine kidneys were found to contain about 78 ppm Zn and 0.78 ppm Cd. Their concentrations in ppm in the subcellular fractions were as follows:

Nuclei (86.7 ± 4.8 Zn, <1.01 Cd)
Mitochondria (89.8 ± 8.2 Zn, 0.88 ± 0.43 Cd)
Lysosomes (102 ± 7 Zn, 1.04 ± 0.38 Cd)
Microsomes (107 ± 12 Zn, <1.11 Cd)
Cytosol (131 ± 11 Zn, 2.14 ± 0.19 Cd)

Nearly 45% of the total Zn was found in the cytosol fraction. Concentrations of Cd in the nuclei and microsomes could not be quantitatively measured because of the poor detection limits of this element by INAA; however, the cytosol fraction contained >55% of the total Cd in bovine kidneys. Between 0.5% (in mitochondria) and 9% (in lysosomes) of the Zn and no detectable amounts of Cd were lost on dialysis against water, indicating that much of the metals were bound to macromolecules.

The stability of Zn- and Cd-binding proteins was investigated at several pH values, namely 1.0, 3.5, 5.5, 7.0, 8.5, 10.5, and 13.0, using a number of buffer systems (*5*); the curves are illustrated in Fig. 1. The Zn–protein complexes appear to be fairly stable above pH 7.0. The stability of Cd–protein complexes slowly increases above pH 7.0 and has a maximum around pH 10.5. The Cd–protein complexes, like Zn, appear to be quite labile below pH 5.5.

The fractionation patterns of protein-bound Zn and Cd by precipitation with ammonium sulfate are shown in Fig. 2. The recoveries of total Zn and Cd were 100 and 95%, respectively. The 2*M*-precipitate (ppt) contained most of the Zn (47.6%) even though 3*M*-ppt was the richest (195.4 ppm). For Cd, the 4*M*-ppt was the most abundant (42.7%) and enriched (11.2 ppm) fraction; the 2 and 3*M* precipitates contained 17.6 and 31.9% of Cd. The results indicate the possible existence of two or three Cd- and several Zn-proteins.

A Sephadex G-150 column equilibrated with 0.2*M* ammonium acetate buffer at pH 8.8 was selected for size-exclusion chromatography. The elution profiles of Zn- and Cd-binding proteins on this column are presented in Fig. 3. For Zn, it is evident that the mixture separated into at least four different protein fractions of mol wt >300,000, 260,000, 89,000, and 27,000. More than 90% of Zn-proteins fractionated before the bed volume, and only a small amount was found at the bed volume. The Zn peak detected at the bed volume could be the same Zn found in the dialysates of the supernatant fractions. A small amount of Zn was also found in the eluants after the bed volume. This might be some of the Zn that was bound to the column material. Evans et al. (*11*) observed a fair

Table 1.
Levels of Zn and Cd in the Chemical Reagents Used

Reagent	Zn (ppm)	Cd (ppm)
Water	< 0.00008	< 0.00005
Polyethylene bags	< 0.062	< 0.0066
HEPES	< 0.020	< 0.040
Sucrose	< 0.200	< 0.030
LiOH	0.470±0.100	< 0.018
HCl	< 0.250	0.030±0.020
Ammonium Sulphate	< 0.061	< 0.038
Sephadex G-150 gel	< 0.060	< 0.080
Glacial HOAc	< 0.030	< 0.020
Conc. ammonia	0.288±0.042	< 0.004
DEAE Sephadex A-50	0.070±0.012	< 0.062
TRIS	< 0.070	< 0.030
NaCl [a]	0.010	0.100
PBE 94	< 0.080	< 0.090
PB 74	0.250±0.050	< 0.100
Imidazole buffer	0.160±0.030	< 0.018
Biogel HTP	0.540±0.40	< 0.200
Ultrodex gel	< 0.120	< 0.017
β-alanine	< 0.070	< 0.091
Phosphoric acid	2.40±0.80	< 0.170
Sulfuric acid	< 0.200	< 0.02

[a]maximum value reported by the manufacturer.

amount of binding of Zn into Sephadex G-15 and G-25 columns. However, not much binding of Zn was observed in the present work, probably as a result of the low cross-section of the Sephadex G-150 gel material compared to Sephadex G-15 or G-25. The Cd-binding proteins clearly separated into three peaks, with one at the void volume and the other

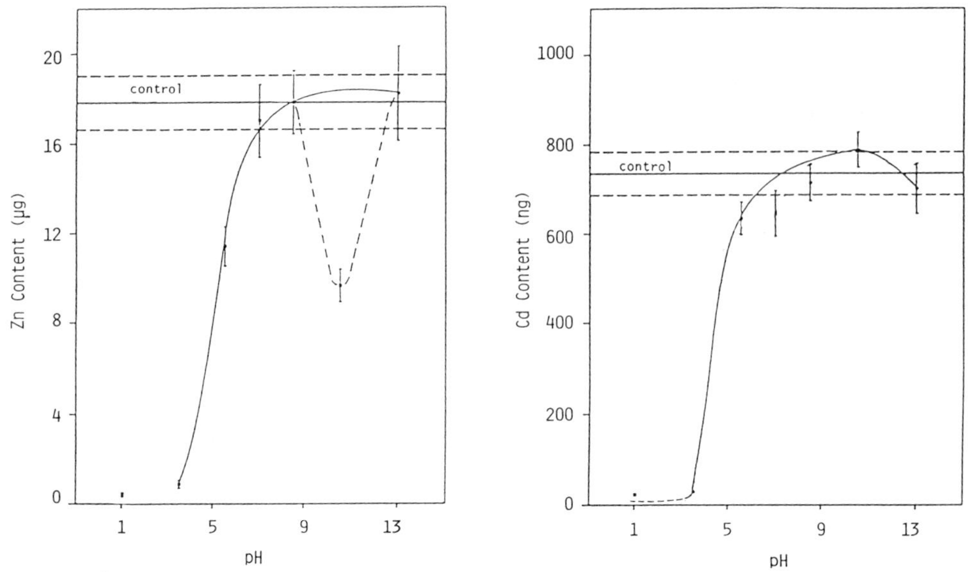

Fig. 1. Effect of pH on protein-bound Zn and Cd species: (A) Zn; (B) Cd.

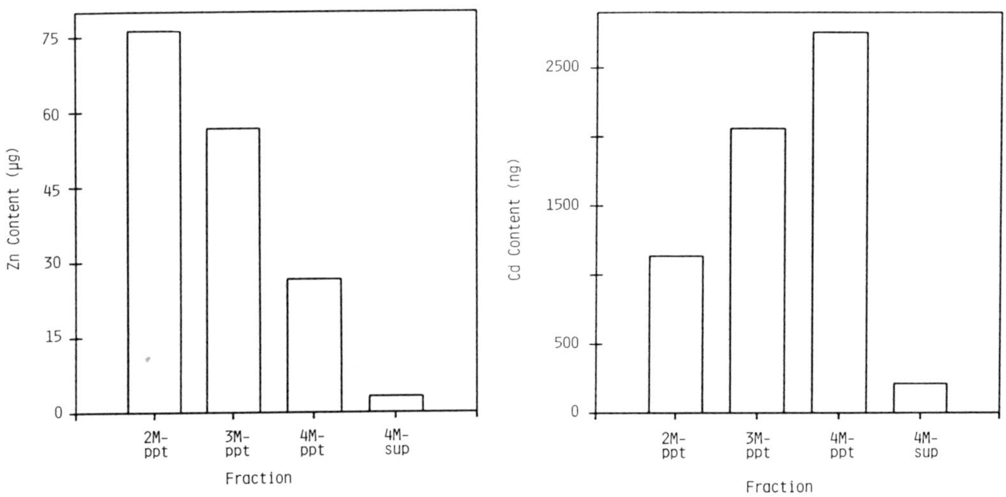

Fig. 2. Fractionation of protein-bound Zn and Cd by ammonium sulphate precipitation: (A) variation of Zn content; (B) variation of Cd content.

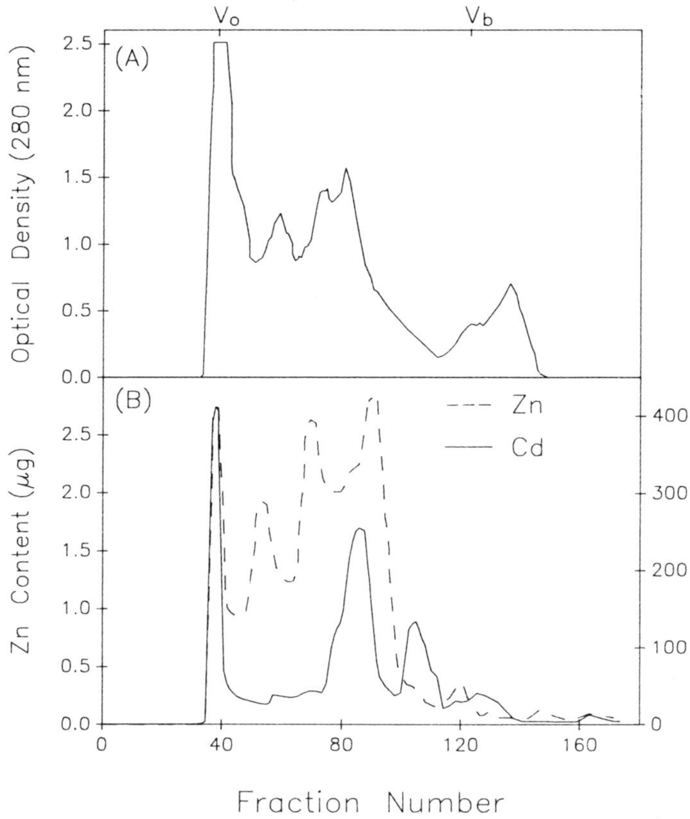

Fig. 3. Elution profile of the supernatant fraction on a Sephadex G-150 column (5 × 90 cm^2): (A) elution profile of proteins measured by optical density at 280 nm; (B) elution profile of Zn- and Cd-species monitored by INAA.

two of mol wt 32,000 and 13,000 daltons. Most of the Cd-binding proteins reported in literature are metallothioneins (*12,13*). The low-molecular-weight Cd-binding proteins detected in this study could very well be metallothioneins.

Anion-exchange chromatography was done on a DEAE Sephadex A-50 column equilibrated with 0.1*M* Tris/acetate buffer at pH 8.3. The proteins were eluted using an increasing salt gradient. The elution profile of the Zn-proteins shown in Fig. 4A, B indicates the presence of four major peaks and possible existence of additional four minor peaks.

An Ultrodex gel bed with a pH gradient of 3–9 was used in the electrofocusing experiments. The isoelectricfocusing distribution pattern for Zn-proteins in bovine kidney showed that this element was incorporated in two major proteins of isoelectric point (Ip) values of 4.95 and 5.7, and two minor proteins of 4.0 and 4.6 Ip. These four proteins could not

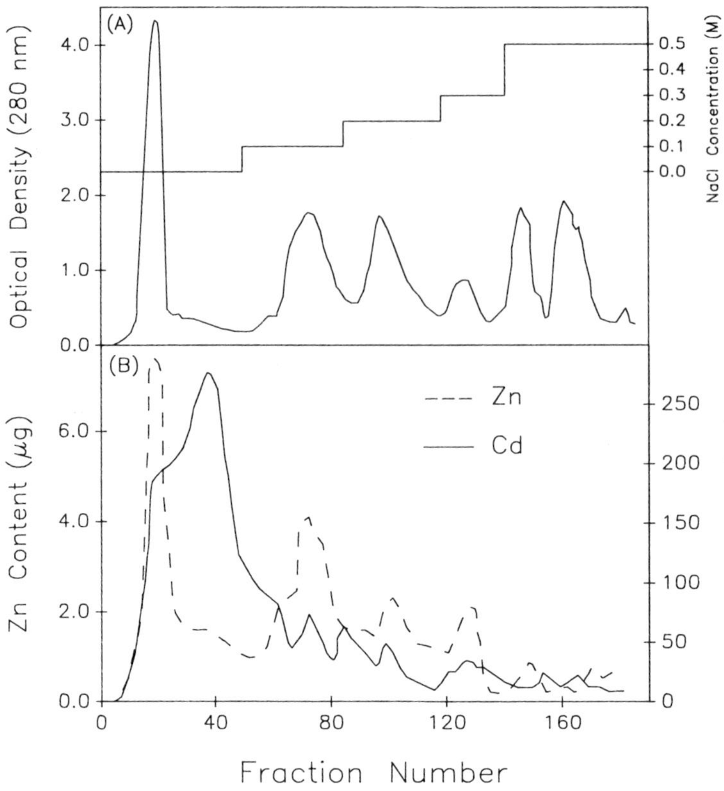

Fig. 4. Elution profile of the supernatant fraction on a DEAE-Sephadex A-50 column (5 × 20 cm^2): (A) elution profile of proteins measured by optical density at 280 nm; (B) elution profile of Zn- and Cd-species monitored by INAA.

be the same proteins observed in gel filtration (Fig. 3), because the total amount of Zn found on the gel bed was only 43%. Surprisingly, about 55% of the total Zn was detected at the anode compartment of the isoelectricfocusing experiment. This is completely opposite of what one would expect. Unlike normal anions, such as Cl^-, these Zn-species did not travel into the anode compartment, but closer to it. Therefore, one could think that this may be caused by a very acidic Zn-protein. It has also been reported that Zn can form complexes with nucleotides and, in fact, has been detected in ribonucleic acids from a variety of sources. Since nucleic acids normally possess a negative charge, this type of Zn complex may also travel towards the anode compartment. In either case, isoelectricfocusing provides an easy means of purifying this unknown Zn species, and further studies of this complex may be worthwhile.

The anion-exchange elution profile of the Cd-binding proteins on the DEAE Sephadex A-50 column is shown in Fig. 4B. Most of the Cd eluted as one major peak with two shoulders at the right and left sides. A few other small peaks were also detected. The total recovery of the experiment was about 70%. It appears that all Cd peaks observed in the gel-filtration experiment were eluted together from the anion-exchange column. The early elution of Cd-binding proteins on ion exchange suggests that their Ip values could be high and that these could be metallothioneins, which are known to have high Ip values. The fractions exhibiting high Cd levels and no optical density (Figs. 4A, B) also indicate the presence of metallothioneins. The total recovery of Cd after the isoelectricfocusing experiment was extremely poor and, therefore, did not lead to any significant conclusions. The poor recovery suggests low binding constants for these Cd–protein complexes.

Chromatofocusing experiments were done on a column packed with polybuffer exchanger 94, and the pH gradient was developed by eluting with polybuffer 74. The extremely poor yield of 20% Zn completely masked the capabilities of this technique. A majority of the Zn-proteins eluted between the pH values 5.0 and 6.0, representing about 10% of the total Zn. This resembles the Zn-proteins observed in the isoelectricfocusing experiment with Ip values of 4.95 and 5.7, which accounts for about 18% of the total Zn. It appears that all other Zn-proteins in chromatofocusing have either completely or partially denatured. The poor yield of Zn could result from the decreasing pH gradient and the formation of Zn-complexes with polybuffer 74 and the start buffer. Therefore, the application of chromatofocusing to separate Zn-proteins was of limited use in this case. The fractionation of Cd-binding proteins by chromatofocusing also did not appear to be satisfactory. The total recovery of Cd was <10%, with only 5% of Cd eluting between pH 6.5 and 7. It was obvious that the rest of the Cd-binding proteins dissociated during chromatofocusing.

Hydroxylapatite chromatography was performed on a Bio-Gel HTP column equilibrated with 0.01*M* phosphate buffer, and the proteins were eluted with increasing phosphate gradient. Isotachophoresis experiments were done on a flat Ultrodex gel bed. The existence of several Zn- and Cd-binding proteins was evident from both experiments. In hydroxylapatite chromatography, the recovery of Cd was only 26%; that of Zn was 53%. A majority of the Cd-proteins showed higher mobility than the Zn-proteins in isotachophoresis experiments. Further work needs to be done in both areas before they can be used for the purification of these proteins.

SUMMARY

Several biochemical techniques in conjunction with INAA have been used to study Zn- and Cd-bound proteins in the supernatant fraction of bovine kidneys. The presence of at least four Zn-binding proteins of mol wt > 300,000, 260,000, 89,000, and 27,000 was revealed by gel filtration. Anion-exchange chromatography also confirmed the existence of at least four different Zn-proteins. Zinc was mainly incorporated with two major proteins of Ip values 4.95 and 5.7, and two minor proteins of 4.0 and 4.6 Ip. The first two Zn-proteins could very well be the same proteins that eluted between pH 5–6 in chromatofocusing. The presence of three Cd-binding proteins with mol wt >300,000, 32,000, and 13,000 has been indicated by gel filtration. Cadmium-bound proteins eluted as a single peak from the anion-exchange column. The early elution of the Cd-proteins indicates their possible high Ip values, which resemble those of metallothioneins. Further studies on the purification and identification of both Zn- and Cd-proteins are in progress.

ACKNOWLEDGMENTS

The authors thankfully acknowledge the cooperation of the SLOWPOKE-2 Operations Group, Dalhousie University, for irradiations; J. A. Verpoorte, Department of Biochemistry, Dalhousie University, for use of his laboratory facilities and for discussions; the Natural Sciences and Engineering Research Council of Canada for a research operating grant; and the award of a Killam Postgraduate Fellowship to C. K. Jayawickreme.

REFERENCES

1. Z. A. Shaikh and O. J. Lucis, *Experientia* **27,** 1024 (1971).
2. Z. A. Shaikh and O. J. Lucis, *Arch. Environ. Health.* **24,** 418 (1972).
3. D. Behne, P. Bratter, D. Gawlik, U. Rosick, and W. Schmelzer, *Proc. Int. Symp. Nuclear Activation Techniques in the Life Sciences,* Vienna, Austria, 1979, pp. 117–127.
4. E. Sabbioni and E. Marafante, *Environ. Physiol. Biochem.* **465,** 5 (1975).
5. C. K. Jayawickreme, PhD thesis, Dalhousie University, Halifax, Canada, 1987.
6. A. Chatt and C. K. Jayawickreme, *Trans. Am. Nucl. Soc.* **49,** 162 (1985).
7. C. K. Jayawickreme and A. Chatt, *J. Radioanal. Nucl. Chem.* **110,** 583 (1987).
8. A. Chatt and C. K. Jayawickreme, *Trans. Am. Nucl. Soc.* **53,** 186 (1986).
9. C. K. Jayawickreme and A. Chatt, *J. Radioanal. Nucl. Chem.* **124,** 257 (1988).
10. L. A. Currie, *Anal. Chem.* **40,** 586 (1968).
11. G. W. Evans, P. E. Johnson, J. G. Bruchmiller, and R. W. Ames, *Anal. Chem.* **51,** 839 (1979).
12. J. H. R. Kagi and B. L. Vallee, *J. Biol. Chem.* **235,** 3460 (1960).
13. E. Sabbioni, E. Marafante, L. Amantini, L. Ubertalli, and R. Pietra, *Sci. Total Environ.* **10,** 135 (1978).

Metabolic Deposition of Selenium and Cadmium into the Hair and Other Tissues of the Guinea Pig

A. CHATT,[1] J. HOLZBECHER,[1] AND S. A. KATZ*,[2]

Trace Analysis Research Centre,
[1]Dalhousie University, Halifax, Nova Scotia, B3H 4J1, Canada;
and [2]Department of Chemistry, Rutgers University,
Camden, NJ 08102-0000

Received April 17, 1989; Accepted November 17, 1989

ABSTRACT

To better understand the significance of hair trace-element measurements and their relationships with the trace-element levels in body organs and fluids, a series of controlled animal experiments were conducted in which several trace elements were periodically measured during a 90–day chronic exposure to selenium and cadmium. Chronic selenium exposure appeared to be reflected by elevated selenium levels in the hair, kidneys, and liver. Chronic cadmium exposure, although reflected by kidney and liver elevation, appeared not to be reflected by corresponding increases in its concentration in the hair.

Index Entries: Hair; cadmium; selenium; trace elements; hair analysis.

INTRODUCTION

Much conflict and controversy surrounds the significance of hair trace-element measurements. The medical literature has been partic-

*Author to whom all correspondence and reprint requests should be addressed.

ularly critical of applying the results of hair trace-element determinations to the assessment of abnormal metabolism or dietary status (*1–3*). Some of the paramedical and psychological literature, however, has been more supportive of such applications (*4–6*). Until rather recently, some of the controversy centered around the procedures used to collect and prepare the hair samples (*7,8*). Some progress has been made in standardizing these procedures (*9,10*), establishing reference values (*11*), and developing certified reference materials (*12*). Models on trace-element compartmentalization are being studied, and intertissue relationships are being investigated (*13,14*). The metabolic deposition of selenium and cadmium into the hair and other tissues of the guinea pig is a part of these investigations.

METHODS

Forty-five 6-week-old, male, outbred Hartley guinea pigs (*Cavia porcellus*) were randomly segregated into three groups of equal size. The animals in each group were housed in individual polypropylene cages with PVC filter covers. All animals had access to the same food and water, *ad libitum*, and all animals were housed under the same environmental conditions, in strict compliance with university practice and policy.

Every Monday, Wednesday, and Friday, the 15 animals in the first group were dosed with 5.0 mL of deionized water by gastric intubation. The 15 animals in the second group were intubated with 5.0 mL of solution containing 5.0×10^{-5} mg Se/mL, and the 15 animals in the third group were intubated with 5.0 mL of solution containing 5.0×10^{-5} mg Cd/mL.

To complete the gastric intubation, the guinea pig was secured in the vertical position by an elastic band behind the upper incisor teeth, the tongue was pulled forward, and a 20-gage plastic cannula was inserted into the lumen of the esophagus. The cannula was carefully guided into the stomach, and connected to a plastic syringe containing the premeasured dose. The dose was delivered gently, and the cannula was carefully withdrawn.

At intervals of 30, 60, and 90 d, five animals from each of the three groups were shaved with an aluminum-headed clipper on the flanks. The animals were sacrificed by methoxyflurane anesthesia and partial exsanguination. The hair and blood were retained in plastic containers. The body cavities were then opened by midventral incision with a carbon steel blade. The lungs were removed with the trachea intact. The livers and kidneys were also removed. The right and left forelegs were removed, and the humeruses were scraped free of soft tissue.

The right humeruses, right lungs, and right kidneys were transferred to plastic containers, and, along with portions of the blood and hair samples, were examined by INAA (instrumental neutron activation analysis). The livers, left humeruses, left lungs, left kidneys, and the remaining portions of the blood and hair samples were prepared for selenium and cadmium determinations by AAS (atomic absorption spectrometry).

INAA was carried out at the Dalhousie University SLOWPOKE-2 Reactor. The soft tissue and hair samples were lyophilized, and the bone samples were calcined prior to irradiation of weighed specimens. The irradiations were performed at a flux of 5×10^{11} n cm^{-2} s^{-1}. The following experimental conditions were used for the various types of tissues:

1. Blood: t_i (irradiation time) = 30 s; t_d (decay time) = 10 s; t_c (counting time) = 30 s for 77m-Se.
2. Hair: t_i = 60 s, t_d = 10 s, t_c = 60 s for 77m-Se; t_i = 5 min, t_d = 60 s, t_c = 10 min for 28-Al, 80-Br, 49-Ca, 38-Cl, 66-Cu, 128-I, 27-Mg, 56-Mn, 24-Na, and 37-S; t_i = 24 h, t_d = 3d, t_c = 30 min for 76-As, 198-Au, 82-Br, 60-Co, 115-Cd, 42-K, 24-Na, 122-Sb, and 65-Zn.
3. Organs: t_i = 60 s, t_d = 15 s, t_c = 60 s for 77m-Se and 86m-Rb; t_i = 60 s t_d = 2 min, t_c = 10 min for 28-Al, 80-Br, 38-Cl, 66-Cu, 27-Mg, and 24-Na; t_i = 60 s, t_d = 1 h, t_c = 15 min for 42-K and 56-Mn; t_i = 7 h, t_d = 10 d, t_c = 2 h for 65-Zn.
4. Bone: t_i = 60 s, t_d = 15 s, t_c = 60 s for 20-F and 77m-Se; t_i = 60 s, t_d = 2 min, t_c = 10 min for 49-Ca, 38-Cl, 66-Cu, 27-Mg, 24-Na, and 87m-Sr; t_i = 60 s, t_d = 1 h, t_c = 15 min for 42-K and 56-Mn; t_i = 60 s, t_d = 60 s, t_c = 5 min for 28-Al (Al, P); t_i = 7 h, t_d = 10 d, t_c = 2 h for 65-Zn.

The livers, left humeruses, left kidneys, left lungs, and specimens of hair from the flanks were dried over phosphorus pentoxide for 5 d. The masses of the dried organs and tissues and the volumes of the blood specimens were measured, and these materials were prepared for AAS by digestion with nitric and sulfuric acids in the aluminum block heaters recommended by Environment Canada (*15*). The digested materials were transferred to 25-mL volumetric flasks and brought to volume with high-purity water.

The selenium contents of the digested organs and tissues were determined by AAS using the model 360 instrument from Perkin-Elmer. Duplicate 5.00-mL aliquots of the contents of the volumetric flasks were transferred to the reaction vessel of the Perkin-Elmer MHS-10, the selenium compounds were reduced to hydrogen selenide with sodium borohydride, and the atomic absorption of the selenium was measured in the heated quartz tube at a wavelength of 196 nm. Quantification of selenium was by reference to the absorptions of standards in 20% sulfuric acid.

The cadmium contents of the solutions obtained by digesting the guinea pig organs and tissues were also determined by AAS. Two-mL aliquots of the solutions were transferred to 5-mL volumetric flasks, treated with 2 mL of 5% diammonium monohydrogen phosphate solution, and brought to volume with 10% nitric acid solution. The absorption of replicate 20-μL injections from the 5-mL volumetric flask were measured with the Perkin-Elmer HGA 2100 in nonpyrolytic graphite tubes at a wavelength of 229 nm. Gas flow was continuous, and the drying, ashing, and atomizing times and temperatures were 30 s at 125°, 30 s at 500°, and 10 s at 1900°, respectively. The absorbances of 20-μL injections of cadmium standards in 10% sulfuric acid, 2% diammonium monohydrogen phosphate, and 20% nitric acid were used as references to quantify the cadmium in the test solutions.

RESULTS

The selenium and cadmium concentrations, as well as the concentrations of some other elements, found in guinea pig tissues by INAA are summarized in Tables 1 and 2, and the results obtained by AAS for selenium and cadmium concentrations in these tissues are presented in Table 3.

DISCUSSION

The mean INAA results listed in Table 1 for the selenium concentrations in blood, hair, and kidney of the exposed guinea pigs are similar to those determined by AAS and presented in Table 3. The corresponding control values, however, lead to divergent speculations. From the selenium results obtained by INAA, it would appear that chronic exposure to this element does not result in elevation of its concentrations in blood, hair, or kidney. The selenium results obtained by AAS also indicate that chronic selenium exposure is not reflected by elevated blood selenium levels, but that such exposure does result in elevations of the selenium concentrations in hair, kidney, and liver.

On the basis of the cadmium results obtained by AAS, chronic cadmium exposure appears to result in elevated liver and kidney cadmium concentrations. The hair and blood cadmium concentrations appear not to reflect such exposure.

The controlled chronic exposure of laboratory animals to cadmium and the resulting distributions of cadmium in various tissues has been reported by several groups of investigators. The works of Kollmer, and Kollmer and Berg, (*16,17*) and of Wesenberg et al. (*18,19*) indicate that there are relationships between hair cadmium concentrations and the concentrations of cadmium in such organs as the liver and kidney in the rat, but Machida et al. (*20*), have reported that liver and kidney cadmium

Table 1
Selenium and Cadmium Concentrations Found in Guinea Pig Tissues by INAA, mg/kg

	Duration of exposure					
	30 d		60 d		90 d	
Tissue	Se	Cd	Se	Cd	Se	Cd
Lung						
Control	1.14 ± 0.20	*	0.97 ± 0.14	*	1.00 ± 0.09	*
Exposed	1.10 ± 0.17	*	1.09 ± 0.09	*	1.09 ± 0.07	*
Kidney						
Control	2.23 ± 0.30	*	2.21 ± 0.18	*	2.26 ± 0.28	*
Exposed	2.44 ± 0.44	*	2.13 ± 0.23	*	2.24 ± 0.33	*
Hair						
Control	0.25 ± 0.14	<1	0.25 ± 0.06	<1	0.22 ± 0.02	<0.75
Exposed	0.16 ± 0.13	<0.7	0.22 ± 0.03	<0.7	0.30 ± 0.10	<0.7
Blood						
Control	*	*	0.12 ± 0.02	*	0.11 ± 0.02	*
Exposed	*	*	0.09 ± 0.02	*	0.11 ± 0.03	*
Bone						
Control						
Exposed						

* = Not determined.

Table 2
Trace Element Concentration Ranges Found in Guinea Pig Tissues by INAA

	Tissue				
Element	Lung	Kidney	Hair	Bone	Blood
Al, ppm	9.4–33.2	7.2–22.8	0.57–12.8	2.0–100	
As, ppm			0.30–1.22		
Br, ppm	16.2–31.1	7.4–16.6	9.6–22.0		
Ca, %			0.030–0.072	34.4–38.8	
Cd, ppm	0.09–0.24	0.19–0.49	0.09–0.20	0.02–0.12	0.01–0.12
Cl, ppm	8400–18940	5710–11860	2430–4440	2490–5850	
Cu, ppm	<12.5–43.2	<7.8–37.5	4.4–15.7		
F, ppm				292–990	
I, ppm			0.65–2.47		
K, ppm	6960–17620	5710–11440	870–3120	<1420–6400	
Mg, ppm	<230–1220	<220–1710	57–251		
Mn, ppm	<0.31–1.74	2.30–5.80	<0.49–1.52	0.61–8.16	
Na, %	0.73–1.54	0.42–0.99	0.037–0.27	1.02–1.45	
P, %				18.0–23.6	
Rb, ppm	6.5–21.3	7.1–20.3			
S, %			2.95–4.26		
Se, ppm	0.75–1.60	1.40–3.16	<0.02–0.45		0.06–0.16
Sr, ppm				193–470	
Zn, ppm	47–355	64–401	136–209	191–438	

concentrations increased while bone and hair cadmium concentrations remained unchanged in chronically-exposed rats. Kollmer (*21*) has also reported that the deposition of cadmium in rat hair was dependent on the growth stage of the hair. In this respect, the random growth of guinea pig hair may be a better surrogate for modeling the human situation than is the cyclic growth pattern of the rat.

Table 3
Selenium and Cadmium Concentrations Found in Guinea Pig Tissues by AAS, mg/kg

Tissue	Duration of exposure					
	30 d		60 d		90 d	
	Se	Cd	Se	Cd	Se	Cd
Liver						
Control	0.73 ± 0.09	0.40 ± 0.10	0.83 ± 0.10	0.35 ± 0.08	0.79 ± 0.11	0.30 ± 0.08
Exposed	1.10 ± 0.20	0.51 ± 0.08	1.56 ± 0.22	0.63 ± 0.10	1.70 ± 0.24	0.77 ± 0.06
Lung						
Control	0.18 ± 0.05	0.15 ± 0.03	0.22 ± 0.06	0.18 ± 0.04	0.30 ± 0.09	0.12 ± 0.02
Exposed	0.30 ± 0.08	0.12 ± 0.03	0.41 ± 0.08	0.20 ± 0.03	0.50 ± 0.07	0.20 ± 0.04
Kidney						
Control	0.72 ± 0.11	0.27 ± 0.05	0.80 ± 0.11	0.23 ± 0.04	0.81 ± 0.13	0.22 ± 0.03
Exposed	1.92 ± 0.22	0.31 ± 0.04	2.11 ± 0.18	0.37 ± 0.05	2.20 ± 0.30	0.43 ± 0.06
Hair						
Control	0.10 ± 0.02	0.12 ± 0.03	0.09 ± 0.02	0.14 ± 0.04	0.13 ± 0.03	0.15 ± 0.04
Exposed	0.17 ± 0.02	0.17 ± 0.03	0.26 ± 0.03	0.15 ± 0.03	0.26 ± 0.03	0.12 ± 0.03
Bone						
Control	0.09 ± 0.03	0.06 ± 0.03	0.10 ± 0.03	0.05 ± 0.03	0.12 ± 0.03	0.06 ± 0.03
Exposed	0.10 ± 0.03	0.06 ± 0.03	0.10 ± 0.02	0.05 ± 0.03	0.09 ± 0.03	0.06 ± 0.05
Blood						
Control	0.08 ± 0.02	0.04 ± 0.03	0.09 ± 0.03	0.06 ± 0.03	0.09 ± 0.02	0.07 ± 0.03
Exposed	0.10 ± 0.03	0.06 ± 0.04	0.10 ± 0.04	0.05 ± 0.03	0.10 ± 0.03	0.09 ± 0.03

SUMMARY

In the guinea pig, chronic selenium exposure appeared to be reflected by elevated hair, kidney, and liver concentrations of this element. Chronic cadmium exposure, although reflected by kidney and liver elevations, appeared not to be reflected by the corresponding increases of its concentration in the hair.

ACKNOWLEDGMENTS

This work was carried out with partial financial support from International Atomic Energy Agency Technical Contract No. 4374/TC as well as institutional resources from Dalhousie University and from Rutgers University.

REFERENCES

1. S. Barrett, *J. Am. Med. Assoc.*, **254,** 1041 (1985).
2. R. S. Rivlin, *Am. J. Med.*, **75,** 489 (1983).
3. K. M. Hambidge, *Am. J. Clin. Nutr.*, **36,** 943 (1982).
4. G. J. Naylor et al., *Biol. Psychol.*, **19,** 759 (1984).
5. B. Rimland and G. E. Larson, *J. Learn. Disabil.*, **16,** 1 (1983).
6. P. J. Barlow et al., *J. Ment. Defic. Res.*, **25,** 161 (1981).
7. K. M. Varier et al., *IEEE Trans. Nucl. Sci.*, **30,** 1316 (1983).
8. G. Chittleborough, *Sci. Total Environ.*, **14,** 53 (1980).
9. Y. S. Ryabukhin, *J. Radioanal. Chem.*, **60,** 7 (1980).
10. E. M. Cranton et al., *J. Holistic Med.*, **4,** 1 (1982).
11. G. V. Iyengar, *Concentrations of 15 Trace Elements in Some Selected Adult Human Tissues and Body Fluids of Clinical Interest from Several Countries: Results from a Pilot Study for the Establishment of Reference Values*, Report No. 1874, Kernforschungsanlage Jülich, Jülich, 1985.
12. S. Suzuki et al., *Bull. Atomic Energy Res. Lab. Musashi Inst. Technol.*, June, 57 (1985).
13. D. M. Mederios et al., *Nutr. Res.*, **7,** 1109 (1987).
14. A. Chatt et al., *Trans. Am. Nucl. Soc.*, **56,** 236 (1988).
15. Anonymous, *Analytical Methods Manual*, Update 1981, Trace Metals in Fish, Environment Canada, Ottawa, 1981.
16. W. E. Kollmer, and D. Berg, *J. Radioanal. Chem.*, **52,** 189 (1979).
17. W. E. Kollmer, *Sci. Total Environ.*, **25,** 41 (1982).
18. G. B. R. Wesenberg, et al., *J. Environ. Studies*, **16,** 147 (1981).
19. G. B. R. Wesenberg, *J. Environ. Studies*, **20,** 245 (1983).
20. K. Machida, et al., *Nutr. Res. Int.*, **29,** 1145 (1984).
21. W. E. Kollmer, *J. Radioanal. Chem.*, **57,** 535 (1980).

Carbon Determination in Human Teeth by Activation with He-3 Ions

M. ANWAR CHAUDHRI*,[1] AND A. C. CRAWFORD[2]

[1]Department of Medical Physics, Austin Hospital, Melbourne 3084, Australia and University of Melbourne; and [2]School of Physics, University of Melbourne, Australia

Received October 22, 1989; Accepted November 17, 1989

ABSTRACT

A nuclear analytical method, involving activation with ^{3}He ions, was developed to determine carbon content in human teeth with well-documented histories. The tooth samples were irradiated with 2.7-MeV^{3}He particles at 50 nA intensity, and the activity of ^{14}O induced through the reaction $^{12}C(^{3}He, n)^{14}O$, determined by counting the 2.31-MeV gammas. Different dental hard tissues were studied separately. A solid piece of silver steel, the carbon content of which was accurately determined by chemical means, was used as the standard. The carbon content in different teeth varied from 4–7%. The overall experimental accuracy was better than 4.5%.

Index Entries: Carbon; teeth; activation analysis; ^{3}He ions; silver steel standard; dental hard tissues (enamal, dentine, root dentine, amelodentinal junction).

INTRODUCTION

It is well established that the presence of fluorine in teeth plays an important role in preventing tooth decay. The fluorine hardens the dental enamel and makes it less prone to attack by acids. However, the role of carbon in dental health is not so well defined, although it is

*Author to whom all correspondence and reprint requests should be addressed.

suspected that the introduction of carbon into the dental matrix increases the solubility of enamel in acids and makes teeth more prone to decay (*1*). In order to try to understand the role of carbon in dental health, we initiated a project correlating the carbon contents of teeth with their health status. This fitted well into our greater program of studying the role of different elements in dental health. However, it also necessitated the development of a suitable method.

One technique for determining carbon in teeth would be to use deuteron activation, and count the positron annihilation gammas from the decay of ^{13}N from the reaction $^{12}C(d, n)\ ^{13}N$(*2*). However, this method was regarded as unsuitable for carbon determination in teeth because the tooth matrix also contains oxygen, calcium, and other elements that would give rise to positron activities too, and a number of exponentials would have to be separated in order to determine the gammas from ^{13}N. Therefore, to avoid this complex procedure, we developed a new method of carbon determination in teeth. This method involves the activation of carbon through the $^{12}C(^{3}He, n)\ ^{14}O$ reaction, which has a Q-value of −1.148 MeV. The nuclide ^{14}O, with a half-life of 70.6 s (*3*), decays by positron emission to the first excited state of ^{14}N 99.3% of the time, which then emits a prompt 2.313-MeV gamma. The advantages of using this reaction and counting the 2.313-MeV gammas are:

1. The 2.313 MeV gamma peak from ^{14}O has relatively low background and therefore, can be counted with greater accuracy.
2. At 2.7 MeV, the ^{3}He beam does not produce any interfering gammas in the 2.3-MeV region from any of the other elements present in teeth.
3. The cross-section for this reaction is reasonably large (some tens of millibarns) at ^{3}He energies greater than the threshold energy of 1.437 MeV (*4*).

By making use of this method, we determined the carbon concentrations in various dental hard tissues of a number of teeth with well-known histories.

EXPERIMENTAL AND PROCEDURE

The teeth used in this study came from 8–15-yr-old children living in South Australia, where the water is generally flouridated. The detailed histories, such as topical fluoride treatment and taking of fluoride tablets, etc., were well known, and so was the status of the health of the teeth studied. In order to eliminate the effect of external factors (such as the environment, teeth-cleaning habits, topical treatment, and so on) on the carbon content of teeth, it was decided to analyze the "inside" of the cross-sectioned teeth. For this purpose all the teeth were cross-sectioned

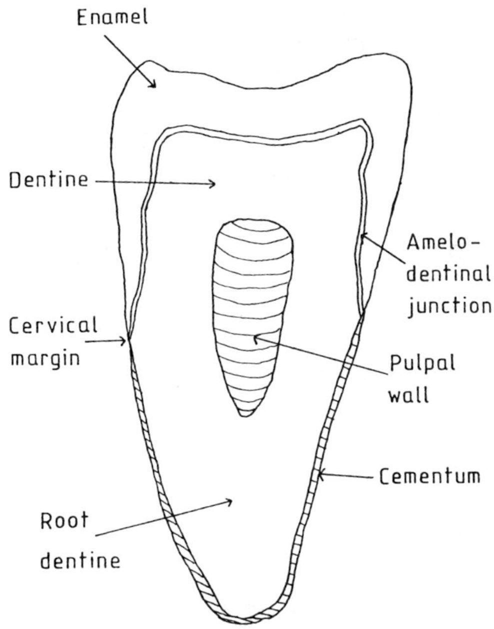

Fig. 1. A typical cross-section of a tooth, displaying various dental tissues.

through the middle with an ultra-clean saw, and the sample material prepared with the technique described by Chaudhri and Ainsworth (*5*). A typical cross-sectioned tooth is sketched in Fig. 1, with various tissues appropriately marked.

A disk of silver steel (supplied by Demco Machinery Co. of Melbourne, Australia) was used as the standard. A number of samples of this silver steel were chemically analyzed for their carbon contents, with the results showing a carbon concentration of 1.04% ± 1%, uniformly distributed throughout the steel samples (*6*).

A number of tooth samples and the standard were mounted in an A1 target ladder and placed in a scattering chamber, which was insulated from the accelerator system and acted also as the Faraday cup for charge collection and measurement. An electron suppressor ring, maintained at −600 V, was placed at the entrance of the scattering chamber. The chamber was kept at a pressure of 2×10^{-6} torr by a turbo-molecular pump. A 1-ft-long cold-trap was placed in front of the chamber to eliminate any carbon coming from the accelerator system onto the target during irradiation.

Teeth, being insulators, would become charged under bombardment with ^{3}He particle. Therefore, in order to avoid this, a 0.72-μm-thick pinhole-free nickel foil was placed about 1.5 cm in front of the target (tooth samples). The incoming ^{3}He beam would knock out a sufficient number of electrons, which would fall on the target and neutralize the buildup of positive charge during irradiation (*7*). The actual thickness of the Ni foil was determined by bombarding a thick CaF_2 target with protons and measuring the shift of the 872-keV resonance in the $^{19}F(p, \alpha\gamma)\ ^{16}O$ reaction. The average energy of the ^{3}He beam used for activation, after passing through the foil, was calculated to be 2.70 MeV. At this bombarding energy, no activity interfering with the ^{14}O activity was produced from the constituents of the tooth matrix. The ^{3}He beam was obtained from the Melbourne University Pelletron Accelerator. Its size on the target was typically 2 mm in diameter, and the intensity was 50 nA during irradiation.

The induced activity was measured with a 5 × 6 inch NaI(T1) detector placed at an angle of 125° with respect to the beam, and 11 cm away from the target. The detector was shielded by 10 cm of lead in order to avoid any background/unwanted gammas from reaching it and being counted. A stabilized pulser was also fed simultaneously into the detector and counted with a scaler in order to determine the dead-time of the detector system during operation. The induced activity in the target was measured with a PDP-Computer-Analyzer system.

Each tooth target was irradiated for 6 min with 2.70 MeV ^{3}He ions (after losing about 300 keV in the N foil) at 50 nA. This time was enough to saturate the induced activity of ^{14}O. It was made certain that the beam remained stable, with constant intensity, during this irradiation time. Afterwards the beam was switched off and the induced activity counted for 240 s, as this time period gave the optimum peak-to-background ratio for the 2.313 MeV gammas. Various teeth, as well as the steel standard, were irradiated under identical conditions, and the ^{14}O activity induced in each one counted.

An alternative method of activation, which is especially suitable for short-lived induced activities when the beam current is not stable and constant, is to measure the effective charge falling on the target during irradiation with a "Leaky integrator". The integrator consists of a parallel RC circuit with a decay constant RC set equal to $1/\lambda$, the "mean life" of the induced activity—in the present case ^{14}O. Such a "Leaky detector" was constructed and used for the current project too. The results of the two methods of carbon determination in teeth were in agreement, within experimental errors. The theory and details of this new method will be published elsewhere (*8*).

In order to ascertain that no carbon contamination was coming onto the target from the accelerator system, a number of thick aluminum targets were irradiated with ^{3}He beams, just like the tooth targets. Even after repreated irradiations, none of the targets showed any carbon buildup within the experimental capabilities of the system.

A number of experiments were also carried out in order to investigate whether the diamond saw, used to cross-section the teeth, caused any carbon contamination of the tooth surfaces. For this purpose a few

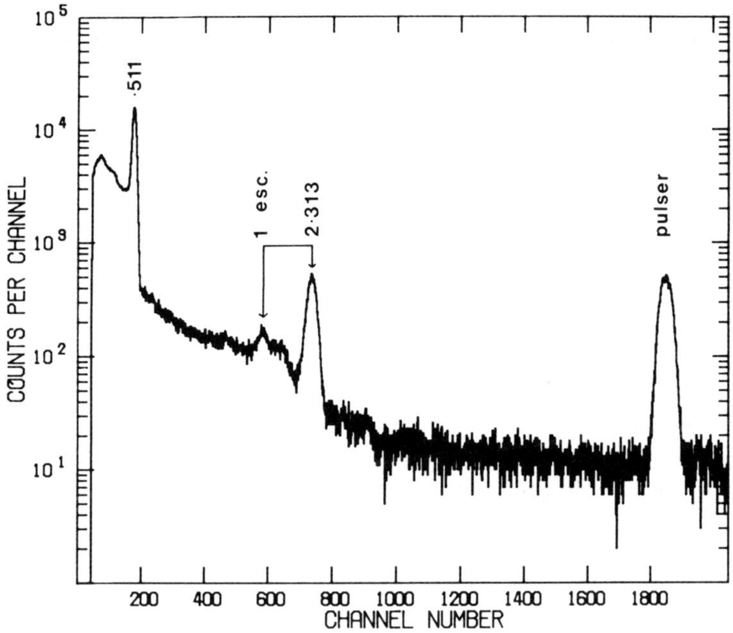

Fig. 2. A NaI(T1) gamma spectrum from the dentine region of a tooth, activated by 2.7-MeV ^{3}He particles. The 2.313-MeV gamma peak, along with its first escape peak, is well-resolved. Also shown in the spectrum is the pulser peak, as well as the peak from the 511-keV annihilation gammas.

aluminium blocks were cut with the diamond saw, using the same technique as employed for cross-sectioning the teeth. Later, the blocks were cleaned in the same way as the teeth and mounted in the scattering chamber for carbon determination. Within the limits of our experimental system, no traces of carbon were detected in either of the Al blocks. This confirmed that the diamond saw did not add any measurable carbon to the tooth surface under investigation.

RESULTS AND DISCUSSION

A typical gamma spectrum from an activated tooth is shown in Fig. 2. The 2.313-MeV gamma-ray peak from ^{14}O is very well resolved, and is appropriately marked, along with its first escape peak. Also present in the spectrum is the pulser peak and a fairly large 511-keV peak from some positron activities that were also produced in the tooth target under bombardment with 2.7-MeV ^{3}He particles. Some of this positron activity could also result from the ^{128}I activity that was being produced by the neutron capture reaction ^{127}I (n, γ) ^{128}I in the NaI detector itself—the neutrons being produced by the interaction of ^{3}He with the tooth target matrix. As stated in the experimental procedure, this spectrum was taken for 240 s after an irradiation of 6 min at ^{3}He energy of 2.7 MeV.

If the matrices of the tooth and the steel standard were identical, a comparison of the 2.313-MeV gammas from the two types of targets would give the carbon concentrations of the tooth. However, this is not the case. The tooth's matrix is that of hydroxypatite—$Ca_{10}(PO_4)_6(OH)_2$; that of the silver steel is $Fe_{14}C$. Therefore, to calculate the carbon concentration in the teeth, the following relationship was used (*8*):

$$\frac{Y}{Y_s} = \frac{F(Ei)cV}{F_s(Ei)c_sV_s} \qquad (1)$$

where

Y = number of 2.313-MeV gammas observed
c = carbon concentration in percent
V = the charge collected on the target during irradiation with ^{3}He ions

$$F(E_i) = \int_{E_t}^{E_i} \sigma(E)\left[-\frac{dE}{dx}\right]^{-1} dE \qquad (2)$$

E_i = the incident energy of the ^{3}He beam (2.7 MeV in the present case)
E_t = the threshold energy of the $^{12}C(^3He, n)\,^{14}O$ reaction, 1.437 MeV
$\sigma(E)$ = the cross-section of the reaction at energy E
$\frac{dE}{dx}$ = stopping power of the matrix for ^{3}He ions.

The same symbols with subscript "s" refer to the standard.

To solve Eqs 2 and 1, the cross-section data were taken from ref. (*4*) and that of the stopping power from ref. (*9*).

The calculated carbon concentrations in various dental hard tissues of some teeth are shown in Table 1. As can be seen from the table, the carbon concentration in various dental hard tissues of different teeth studied ranges from about 4.2 to 7%, with an overall experimental uncertainty of around 4%. A detailed analysis of the errors will be included in ref. (*8*).

SUMMARY

Activation analysis with ^{3}He ions has been developed for the determination of carbon in teeth. It has been shown to be a simple and reliable nuclear analytical method for carbon determination in complex biological matrices, such as the tooth. Furthermore, it is pointed out that this

Table 1
Carbon Content
of Dental Hard Tissues from Different Children

Tooth number	Region of tooth	% Carbon
1 (D)[a]	Amelodentinal junction	5.98 ± 0.23
1 (D)	Dentine	5.12 ± 0.20
4 (D)	Amelodentinal junction	5.65 ± 0.20
6	Dentine	5.61 ± 0.21
6	Root dentine	4.22 ± 0.17
9	Dentine	5.50 ± 0.21
10 (D)	Root dentine	7.02 ± 0.25
11	Amelodentinal junction	6.23 ± 0.25
13 (D)	Enamel	4.83 ± 0.18
16	Enamel	4.93 ± 0.19
16	Root dentine	6.46 ± 0.25

[a](D) denotes teeth that have decayed crowns. However, only the healthy part of the dental tissues were used for analysis.

technique is better than deuteron activation analysis, since it does not involve having to separate various exponentials.

REFERENCES

1. J. A. Weatherell, C. Robinson, and C. R. Hiller, *Caries Res.* **2,** 1 (1968).
2. M. L. Boettger, W. Helbig, and S. Niese, *Isotopenpraxis* **II,** 127 (1975).
3. J. C. Hardy and I. S. Towner, *Nucl. Phys.* **A254,** 221 (1975).
4. D. R. Osgood, J. R. Patterson, and E. W. Titterton, *Nucl. Phys.* **60,** 503 (1964).
5. M. A. Chaudhri and T. Ainsworth, *Nucl. Instr. Methods* **181,** 333 (1981).
6. BHP Research Laboratories, Melbourne, Australia (private communications).
7. M. A. Chaudhri and A. Crawford, *Nucl. Instr. Methods* **181,** 31 (1981).
8. M. A. Chaudhri and A. Crawford, to be published.
9. C. F. Williamson, J. P. Boujot, and J. Picard, Rapport CEA-R 3042, Saclay (1966).

Nuclear Analytical Methods and QA

Vanadium Determination at the Ultra-Trace Level in Biological Reference Materials and Serum by Radiochemical Neutron Activation Analysis

A. R. BYRNE*,[1] AND J. VERSIECK[2]

[1]J. Stefan Institute, E. Kardelj University, 61000 Ljubljana, Yugoslavia; and [2]Department of Internal Medicine, Division of Gastroenterology, University Hospital, 9000 Ghent, Belgium

Received April 17, 1989; Accepted December 11, 1989

ABSTRACT

In order to help resolve present inconsistencies of two orders of magnitude or more in reported levels of vanadium in human serum and blood, a totally postirradiation radiochemical neutron activation analysis (NAA) method was further developed and applied to some pertinent nanogram and subnanogram reference materials. In particular, the second generation human serum reference material of Versieck was found to contain a value of 0.67 ± 0.05 ng/g dry wt., corresponding to 0.061 ± 0.005 /4/ (*see* Table 1) ng/mL original fresh serum. Results are also reported for some other appropriate CRMs.

Additionally, a small-scale study in 10 normal subjects (5 m, 5 f) revealed levels similar to those in the serum reference material and in agreement with the lowest data reported in the literature. Discussion of pitfalls of vanadium determination and the use of reference materials is included.

Index Entries: Vanadium; radiochemical neutron activation analysis; ultratrace analysis; reference materials; serum.

*Author to whom all correspondence and reprint requests should be addressed.

Table 1
Results—Reference Materials

Material	Vanadium content* ng.g^{-1} dry weight	Certified value	Literature value
NBS Bovine Liver 1577a	95.0 ± 2.9 /4/	99 ± 8	
Bovine Serum: IJS Internal Reference	5.0 ± 0.17 /7/	-	
Seronorm Whole Blood, Trace Element II, Batch 905 Nycomed	20.3, 23.1 (2 separate bottles)	-	
IAEA Animal Muscle H-4	2.78 ± 0.21 /6/	3 (information values (14))	2.76 ± 0.21 /7/ (3) 3.2 ± 0.2 /4/ (15) 2.95 ± 0.10 /4/ (5) 3.09 ± 0.64 /4/ (13) 3.0 ± 0.2 /6/ (24)
NBS Milk Powder SRM 1549	0.97 ± 0.10 /6/	-	-
Human Serum Certified Reference Material Versieck (16)	0.67 ± 0.05 /4/ (0.061 ± 0.005 ng.ml^{-1} /4/)	-	-

*Mean, standard deviation, and /number of replicates/.

INTRODUCTION

The true level of vanadium in human body fluids, e.g., blood and urine, remains controversial. Discounting some very high values, the range of reported values (to make only a selection) lies from <0.3 ng/mL (*1*) in whole blood, <0.1 ng/mL in serum and erythrocytes (*2*), approx. 0.03 ng/mL in serum in females, and somewhat higher and with a wider spread in males (*3–5*); through approx. 0.7 ng/mL in serum (*6*); to several ng/mL for serum (*7–10*) and whole blood (*10,11*). This situation has been discussed in more depth in a recent book (*12*). A very recent report using AAS (*13*) gives values in broad agreement with those of Cornelis et al. (*3–5*), with 60% of serum values below 0.08 ng/mL.

A major contributor to these discrepancies of more than two orders of magnitude in results is undoubtedly the lack of biological reference materials certified for vanadium at the 1ng/g level or below. Another factor is the reluctance of many analysts to perform quality-control analyses using an internal or noncertified reference material of a suitable concentration, and the unwillingness of most workers performing studies on vanadium (and there have been many on various aspects in health and disease) to report values for such noncertified reference materials (RMs) in order to establish a consensus.

Until recently, the only nanogram-level reference material was IAEA Animal Muscle H-4, with a noncertified information value (*14*) of 3 ng/g, based on results by Cornelis et al. (*3,5*) and by Byrne et al. (*15,24*). Lately, the NIST, Washington (formerly NBS) issued a human serum reference material, SRM 909, certified for vanadium at 3.19 ± 0.55 ng/mL (or about 35 ng/g on a dry-wt basis). This unfortunately is too high to be of much use in resolving the present position. However, recently Versieck et al. (*16*) reported the certification of a second generation human serum reference material for 14 trace elements, which reproduces the values now believed to be normal for uncontaminated authentic serum. This material would therefore form an ideal CRM for vanadium, as well as providing evidence of the actual levels in serum.

In the present work, therefore, a well-validated totally postirradiation NAA method (*17*) was developed and improved, and applied to the determination of vanadium in the second generation human serum reference material referred to above, and in some other suitable materials, including NBS SRM-1549 Milk Powder. Additionally, serum from a small group of normals was analyzed. In this way it was hoped to take positive steps toward establishing a consensus of values in suitable reference materials and in human serum.

EXPERIMENTAL

Sampling and conditioning of serum samples (including Second Generation Reference Material) was done at Ghent. All the other analytical work was performed in Ljubljana.

Samples

The technique of sampling serum has been described in detail elsewhere (*16,18*). Other reference materials were weighed into carefully cleaned P.E. vials and sealed, then encapsuled in PE foil. Sample weights of approx. 1 g for the lowest levels were used.

Reagents

N-Benzoylphenylhydroxylamine (N-BPHA) was made up to 0.2% v/w in toluene. A stock solution of vanadium of 1 mg/g was made up from p.a. ammonium metavanadate (NH_4VO_3) in 1% nitric acid, and this was diluted to give a carrier solution of 200 μg/g^{-1} and an irradiation standard of 5 μg/g, both in 1% HNO_3.

Irradiation standards consisted of 0.1–0.2 mL of 5 μg/g V solution sealed in a P.E. capillary tube of the same height as the sample, encapsulated in PE foil and taped alongisde the sample. After irradiation, during counting of the sample ^{52}V fraction, about 100 μL was weighed out and diluted to 5 mL with acidified V(V) carrier in a counting vial.

Irradiations

Sample and standard were irradiated in the pneumatic transfer system (rabbit) of the TRIGA Mk II reactor of the "J. Stefan" Institute at a neutron fluence rate of $4 \times 10^{12} n/cm^2s$ for up to 12 min.

Counting

Five mL of the organic phase or standard were counted in a Canberra well-type HP Ge detector of 120 mL volume and resolution 2.2 keV (^{60}Co) to measure the 1434-keV gamma-ray peak of ^{52}V (3.75 min), generally for 7 min. A second, longer count was performed later to evaluate the chemical yield from the 1312- and 984-keV peaks of ^{48}V (16-d half-life).

Procedure

In addition to the use of a well-type HP Ge detector, other improvements over the previously published method (*17*) included increasing the volume of conc. (18*M*) H_2SO_4 used for wet-ashing to 3 mL in order to accelerate destruction and use of 0.2% N-BPHA. Extraction of V(V) from the reduced volume of acid phase (15 instead of 20 mL) was optimized by varying the volume of 10*M* HCl added until maximum extraction into the 6 mL of N-BPHA was found, for 4 mL of 10*M* HCl.

In a run, the irradiated sample was transferred to a 100-mL long-necked silica Kjeldahl flask already containing 3 mL 18*M* H_2SO_4, 200 μg V(V) carrier, and a weighed aliquot of ^{48}V tracer. After 15*M* HNO_3 was added (about 3 mL), the flask was heated over a strong gas flame in a fume hood. Additions of HNO_3 were continued until the mixture no longer charred. When a yellowish-green solution was obtained that did not darken on heating, 1 mL conc. $HClO_4$ was added and the flask heated strongly to drive off $HClO_4$ as dense white fumes. For serum samples, this step was repeated. Then the flask was plunged into cold water and about 3–4 mL H_2O cautiously added, followed by 4 mL of 10*M* HCl. The contents were poured into a 50 mL separatory funnel and the flask washed with 2–3 mL H_2O, which was also added to the funnel. Then 1 mL of 0.1*N* $KMnO_4$ was added and the funnel plugged and shaken for 15 s. N-BPHA reagent (6 mL) was added and the funnel shaken for 30 s. A strong purple color should be observed in the toluene phase. The aqueous phase was run off and the organic phase washed with 5 mL of 5*M* HCl for about 15–20 s. With a pipet, 5 mL of the organic phase was transferred to a measuring vial, placed in the well-type Ge detector, and counted.

In the case of serum samples, to obtain even better decontamination from traces of ^{24}Na, the organic phase was filtered through a 25-mm Ø circle of Whatman Phase Separating Paper held in a filter holder attached to a 20-mL injection syringe (Fig. 1). On inserting the plunger and pressing down, the filtered organic phase is rapidly delivered into the 5-mL vial.

The whole procedure takes about 10 min to perform with reasonably sized sample aliquots (0.3–0.5 g). For larger samples, of approx. 1 g serum or milk powder, an extra 2 min are required for wet ashing.

RESULTS AND DISCUSSION

Method Performance

The totally postirradiation procedure was an improved version of that reported previously (*17*), based on rapid wet-ashing in the presence of V carrier and ^{48}V tracer, followed by solvent extraction of vanadium with N-BPHA in toluene. As a result of the improvements described under Experimental, the speed, chemical yield, selectivity, and detection limit were improved.

The chemical yield of vanadium (as ^{48}V) in the organic phase was 98% and well reproducible. The effective recovery of vanadium in the 5 mL of the 6 mL organic phase measured was thus 82%; for a series of human and bovine sera, milk, bovine liver, animal muscle, and diet samples, a yield of 82.2 ± 1.5% /22/ (*see* Table 1) was found. Speed was improved by use of a larger volume of sulfuric acid (3 mL) and completion of decomposition with perchloric acid, as described later (*15,18*), rather than with 30% H_2O_2, as in the original method (*17*). This is very effective in removing traces of ^{38}Cl, which was not found in the γ-spectra.

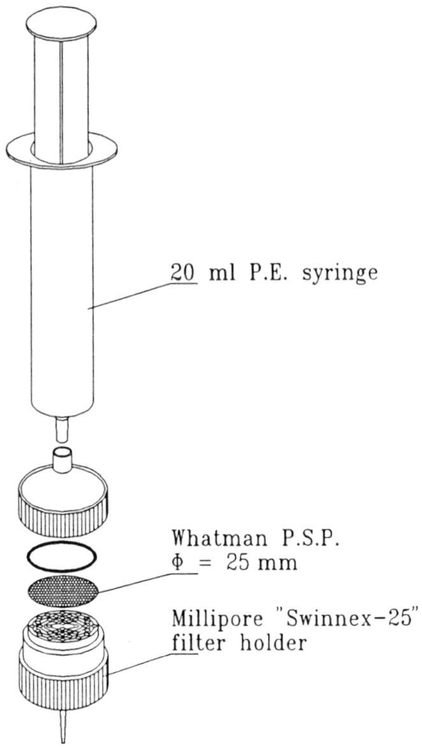

Fig. 1. Filtration device for removing traces of ^{24}Na from the vanadium-containing organic phase.

A further refinement to radiochemical purity was introduced for serum samples to improve the already excellent decontamination from ^{24}Na (freeze-dried serum is roughly 10% NaCl by weight). As the signal peak of ^{52}V was small, as a result of the modest neutron flux available, it was desirable to reduce the background signal under the peak. This was very successfully and simply accomplished by filtration of the organic phase through Whatman Phase Separating Paper (PSP), as described under Experimental. A partial γ-spectrum of a sample of Versieck's second generation human serum RM is shown in Fig. 2A; the decontamination factor for ^{24}Na is better than 10^8. With the same aim of reducing the background, the amount of ^{48}V was kept to a minimum. When ^{48}V was absent, as shown in Fig. 2B for the PSP-filtered organic phase from a human serum sample (subject 11), the background in the region of the 1434-keV ^{52}V peak is virtually zero (0.2–0.3 counts/7 min/channel).

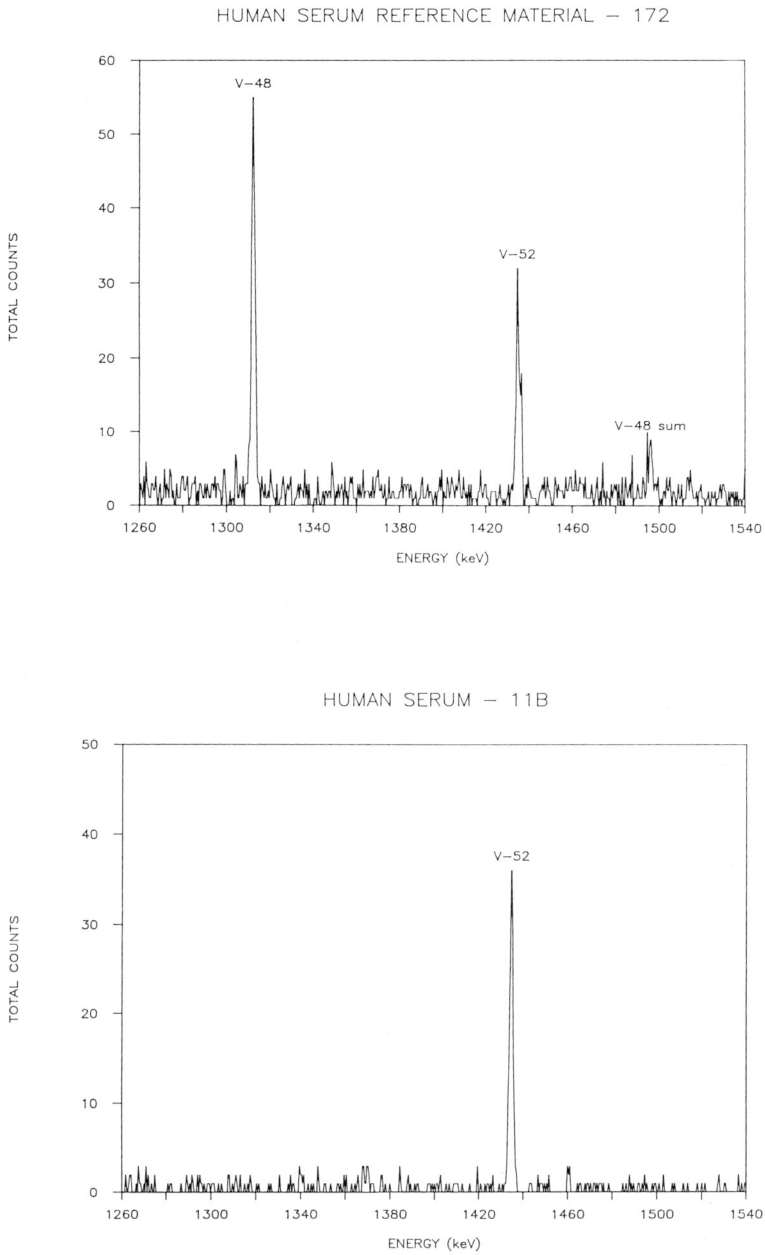

Fig. 2. Partial gamma-ray spectrum of ^{52}V separated from (A) Second Generation Human Serum Reference Material (sample 172); (B) human serum sample 11B (no ^{48}V added).

Postirradiation Radiochemical Separation of Vanadium

Although there are practical difficulties involved in this approach, in principle it is the most attractive, since it retains the advantage of NAA as a blank-free method, as well as in its authenticated chemical recovery of the analyte. It is worth pointing out that the present method and its predecessor (*15,17,19*) are virtually the only postirradiation procedures to be used in recent times for determination of vanadium in biological tissues. The statement of Simonoff et al. (*6*) that their method and that of Cornelis et al. (*3–5*) (a claim not made by these latter authors themselves) possessed blank-free characteristics is invalidated by the sample pretreatment used by Simonoff et al. (*6*) prior to irradiation, consisting of dry-ashing in Pyrex™, dissolution in acid, transfer to quartz, and redrying. Such pretreatment steps introduce the risk of contamination.

The advantages of postirradiation radiochemical NAA are not universally appreciated. It has been suggested by Stroop et al. (*20*) that rapid postirradiation separation results in poor yields, and Dick et al. (*17*) even wrote "while the values from pre-irradiation separation agree well with those from AAS, those from post irradiation are very much lower, suggesting that the very rapid separation may have a low efficiency." In fact, the advantage lies exactly the opposite way: by addition of carrier at μg or mg levels after irradiation, the chemical separation in a radiochemical procedure avoids the problems of working at ultratrace level (both loss and contamination), and the chemical yield is validated by measurement of the recovery of the carrier, which often takes the more elegant form of a radiotracer (^{48}V) measurement. In contrast, in other preseparation techniques, the recovery of the analyte is not known and is hazarded by the problems of ultratrace analysis.

Results for Reference Materials

The only currently available biological CRMs with vanadium concentrations below 100 ng/g are NBS bovine liver 1577a (99 ± 8 ng/g) and the liquid Human Serum SRM 909 (3.19 ± 0.55 ng/mL, equivalent to about 35 ng/g dry wt). At the 1-ng level, there are no CRMs. IAEA Animal Muscle H-4 has an information value (*14*) of 3 ng/g, based on work by Byrne et al. (*15,24*) and Cornelis et al. (*3,5*).

Our results are shown in Table 1; agreement with known values for SRM 1577a and H-4 is good and demonstrates the absence of systematic error at this level. NBS Milk Powder SRM 1549, with just below 1 ng/g, provides an excellent matrix for quality control purposes. The use of milk as a low-level RM for vanadium is suggested by the fact that vanadium levels in milk are known to be at the 0.1 ng/mL level from early work by Soremark (*21*), amplified by results from our laboratory for both cow's and human milk (*22–24*). For IAEA Milk Power A-11, we have previously reported (*22,23*) a very similar value of 0.91 ± 0.13 /4/ (*see* Table 1), and of 0.92 ± 0.10 /7/ (*see* Table 1) later (*24*).

More imporant, we obtained consistent results for Versieck's Second Generation Human Serum Reference Material (*16*), produced from a pool of 22 L of serum from 27 donors (hemochromatosis patients). The result of 0.67 ng/g (dry wt.) is equivalent to 0.061 ng/mL, which is in agreement with the results that Cornelis et al. (*5*) and Ishida et al. (*13*) measured in healthy persons.

In contrast, the NBS material, Serum 909, contains 3 ng/mL. Together with the fact that the Cr content is certified as 91.3 ng/mL, whereas normal levels are ca. 0.15 ng/mL, the material in all likelihood has been substantially contaminated in preparation. This is unfortunate, as the purpose of the material stated in the catalog "for evaluating the reliability of analytical methods for the determination of major, minor and trace constituents in blood serum, plasma, and similar biological fluids" cannot be met if the levels do not correspond to the much lower ones found in real samples. Besides giving a misleading impression of actual concentrations in serum, it may also promote a false feeling of confidence in the analyst in his ability to analyze real serum samples.

The question of the suitability of RMs is an important one in the context of their use in quality control. Naylor et al. (*8*), who determined vanadium in blood and urine using a preseparation evaporation step, followed by destruction in a bomb, removal of sodium on HAP, and then neutron activation and instrumental measurement, considered that "an adequate proof of this technique is shown by the analysis of NBS SRM Bovine Liver 1577". Again Simonoff et al. (*6*), who determined vanadium in serum by preashing and dissolution, followed by radiochemical NAA, wrote that "the accuracy of the analytical method described was tested by analysing 10 samples of 'standard' water obtained from the US NBS . . . vanadium content 53 ± 3 $\mu g \cdot L^{-1}$." In the first example, the level of vanadium in the SRM (about 60 ng/g) was much higher than that in the blood or urine, and of course cannot be subject to preconcentration by evaporation as can blood or urine. In the second example, the use of water instead of a complex biological matrix like serum is questionable, and the concentration of vanadium in the water was 100 times higher than they found in their samples. It is obvious that reference materials that have much higher levels of the analyte than the sample cannot provide any information about the existence of potential systematic errors, such as loss or contamination, occurring at the sample analyte level. In our opinion, it is not a coincidence that the only workers reporting values of 3 ng/g for the lowest RM available, IAEA H-4, are the same as those reporting the lowest (<100 pg/mL) values in serum (*2,5,13*).

The second generation serum RM is, in these respects, a highly suitable (in terms of concentration) and appropriate (in terms of matrix) material for those investigating ultratrace vanadium contents in biolog-

Table 2
Vanadium Concentrations Found
in 10 Normal Subjects by Duplicate Analysis

sample code	sex	age	concentration in ng/mL	
			duplicate results	mean
2	m	37	0.070 , 0.068	0.069
3	f	21	0.016 , 0.019	0.0175
4	f	37	0.031 , 0.035	0.033
5	m	42	0.229 , 0.214	0.222
6	m	30	0.096 , 0.082	0.089
7	f	40	0.071 , 0.080	0.0755
8	f	51	0.068 , 0.074	0.071
9	f	28	0.016 , 0.012	0.014
10	m	24	0.044 , 0.042	0.043
11	m	24	0.066 , 0.078	0.072

ical fluids. Of course, it is necessary for a consensus to be reached on the concentration in this CRM, and it therefore hoped and urged that those working in this field will report their values for this material. In this way, if a consensus of analytical results can be established, the extent to which present and persisting higher values for vanadium in serum can be ascribed to sampling and sample preparation, rather than to analytical differences, will become clear.

Finally in this section, the concentration we found in the bovine serum produced in the Biochemistry Department of our Institute corresponds to about 0.5 ng/mL (or close to the level reported for normal human serum by Simonoff et al. [6]). This material was *not* prepared under clean-room conditions and was collected for other purposes, its use in our study being as a test material for improving our procedure for human serum. The commercial Seronorm Whole Blood material of Nycomed was found to have a relatively high level of vanadium (Table 1). These two materials provide further evidence of the important influence of sample preparation and handling on the apparent vanadium level.

Vanadium Levels in Human Serum

A small-scale study was made on 10 normal volunteers from Ghent (5 males and 5 females) by duplicate analysis of serum. The details of the subjects and the results are shown in Table 2.

The reproducibility of the duplicate analyses is good, confirming the probable absence of systematic error in the analytical technique and the sampling procedure. The pattern of results is essentially similar to that found by Cornelis et al. (*3–5*), and similar to the values for Second Generation Human Serum CRM (derived from 27 donors). Taken together with the recent results of Ishida et al. (*13*), who found 60% of serum samples to be below 0.08 ng/mL (their limit of detection), with the remaining values ranging up to 0.24 ng/mL, there now appears to be a growing consensus on normal values in human serum.

ACKNOWLEDGMENTS

The careful and skilled laboratory assistance of Lidia Vanballenberghe and Ann Wittoek in sample collection and preparation is gratefully acknowledged. We thank I. Kregar of the IJS for the gift of freeze-dried bovine serum and our volunteers for donating blood. We are indebted to Matthias Rossbach of KFA Jülich for supplying us with ^{48}V.

The financial assistance of the Slovenian Research Council; the NIST, Washington; and the Belgian Fund for Medical Scientific Research is also acknowledged.

REFERENCES

1. A. R. Byrne and L. Kosta, *Sci. Total Environ.* **10,** 17 (1988).
2. A. R. Byrne and L. Kosta, *Sci. Total Environ.* **13,** 87 (1979).
3. R. Cornelis, L. Mees, J. Hoste, J. Ryckebusch, J. Versieck, and F. Barbier, *Proc. Int. Symp. Nucl. Act. Techn. Life Sci. 1978,* IAEA, Vienna, 1979, pp. 165–177.
4. R. Cornelis, J. Versieck, L. Mees, J. Hoste, and F. Barbier, *J. Radioanal. Chem.* **55,** 35 (1980).
5. R. Cornelis, J. Versieck, L. Mees, J. Hoste, and F. Barbier, *Biol. Trace Elem. Res.* **3,** 257 (1981).
6. M. Simonoff, Y. Llabador, A. M. Peers, and G. N. Simonoff, *Clin. Chem.* **30,** 1700 (1984).
7. D. A. Dick, G. J. Naylor, and E. G. Dick, *Psychol. Med.* **12,** 533 (1982).
8. G. J. Naylor, A. H. W. Smith, D. Bryce-Smith, and N. I. Ward, *Psychol. Med.* **14,** 767 (1984).
9. G. J. Naylor, F. M. Corrigan, A. H. W. Smith, P. Connelly, and N. I. Ward, *Br. J. Psychiatr.* **150,** 656 (1987).
10. E. Sabbioni, E. Marafante, R. Pietra, L. Goetz, F. Girardi, and E. Orvini, *Proc. Int. Symp. Nucl. Act. Techn. Life Sci. 1978,* IAEA, Vienna, 1979, pp. 179–192.
11. R. O. Allen and E. Steinnes, *Anal. Chem.* **50,** 1553 (1978).
12. J. Versieck and R. Cornelis, *Trace Elements in Human Plasma or Serum,* CRC Press Int., Boca Raton, FL, 1989.
13. O. Ishida, K. Kihira, Y. Tsukamoto, and F. Marumo, *Clin. Chem.* **35/I,** 127 (1989).

14. A. R. Byrne, C. Camara-Rica, R. Cornelis, J. J. M. de Goeij, G. V. Iyengar, G. Kirkbright, G. Knapp, R. M. Parr, and M. Stoeppler, *Fres. Z. Anal. Chem.* **326,** 723 (1987).
15. A. R. Byrne and V. Vrbič, *J. Radioanal. Chem.* **54,** 77 (1979).
16. J. Versieck, L. Vanballenberghe, A. De Kesel, J. Hoste, B. Wallaeys, J. Vandenhaute, N. Baeck, H. Steyaert, A. R. Byrne, and F. W. Sunderman, Jr., *Anal. Chim. Acta* **204,** 63 (1988).
17. A. R. Byrne and L. Kosta, *J. Radioanal. Chem.* **44,** 247 (1978).
18. J. Versieck, J. Hoste, L. Vanballenberghe, A. De Kesel, and D. van Rentergheм, *J. Radioanal. Nucl. Chem.* **113,** 299 (1987).
19. A. R. Byrne, *Radiochem. Radioanal. Lett.* **52,** 99 (1982).
20. S. D. Stroop, G. Helinek, and H. L. Greene, *Clin. Chem.* **28,** 79 (1982).
21. R. Soremark, *J. Nutr.* **92,** 183 (1967).
22. A. R. Byrne, L. Kosta, M. Dermelj, and M. Tušek-Žnidarič, *Trace Element Analytical Chemistry in Medicine and Biology*, vol. 2, P. Bratter, P. Schramel, eds., W. de Gruyter, Berlin, 1983, pp. 21–35.
23. L. Kosta, A. R. Byrne, and M. Dermelj, *Sci. Total Environ.* **29,** 261 (1983).
24. A. R. Byrne, M. Dermelj, and A. Vakselj, "Minor and Trace Elements in Breast Milk: Report of a Joint WHO/IAEA Collaborative Study," WHO, Geneva, 1989, pp. 148–151, ISBN 92 4 156121 1.

Factors Affecting the Levels Reported for Vanadium in Human Serum

K. Heydorn

Isotope Division, Risø National Laboratory, DK-4000 Roskilde, Denmark

Received April 17, 1989; Accepted December 11, 1989

ABSTRACT

Chemometric techniques may be applied to extract significant analytical information from a series of publications that present methods and results for determining trace elements in biological material. This approach was applied to the total of 28 papers published in 1971–1988 that reported determination of vanadium in normal human serum or plasma; the levels spanned four orders of magnitude.

The most important factors affecting the analytical results were found to be the choice of analytical method and the experience of the laboratory in trace-element research. Results from the most experienced laboratories with the best analytical methods were found to be correlated with the precision of the data, indicating that the correct concentration of vanadium would be <1 mg/m^3. This is in agreement with results subsequently obtained by radiochemical neutron activation analysis of eight samples of serum from Danish colleagues.

Index Entries: Chemometric techniques; Radiochemical Neutron Activation Analysis; Human serum or plasma; Vanadium concentration; Factor analysis; Literature survey; Correlation analysis.

INTRODUCTION

The discovery by Schwarz and Milne in 1971 [1] that vanadium is essential for normal growth stimulated interest in this element, and a number of animal experiments have been conducted using ^{48}V as a radioactive tracer. Vanadate has been shown to inhibit Na-K ATPase [2], and reduced membrane sodium transport has been associated with manic-depressive psychosis [3], making vanadium metabolism a possible factor in its etiology.

However, the distribution of vanadium in the human body is still poorly known, and whether actual concentrations of vanadium in human tissue and serum are sufficient to inhibit Na-K ATPase in vivo has yet to be demonstrated. Individual results for vanadium in normal human serum or plasma reported in just one year (1979) range from 0.017 [*12*] to 577 [*10*] mg/m^3, thus covering more than four orders of magnitude.

In the period 1971–1988, a total of 28 papers have presented results for vanadium in normal human serum or plasma, based on eight different analytical methods; they come from laboratories differing with respect to a number of factors that might affect the levels of vanadium observed. Their results are summarized in Table 1, together with the number of individual samples analyzed.

SIGNIFICANCE OF FACTORS

The information presented in Table 1 represents analytical data that can be processed, like any other set of data, in such a way that the most significant factors affecting any particular result can be identified. The particular result that is least affected by factors associated with the technical sources of variation [*34*] is assumed to be closest to the true concentration of vanadium.

Testing of significance on results obtained by different analytical techniques should be done with nearly assumption-free statistics, such as the rank test statistic U developed by Wilcoxon, Mann, and Whitney [*35*]. An exactly similar approach was previously used for the evaluation of results for concentrations of manganese in serum [*36*].

Analytical Method

The first criterion for accepting the results of a particular analytical method is that its limit of detection is sufficiently low to permit a determination of the level expected.

The median of the 28 results is 3.85 mg/m^3, and taking this as the expected level, we require this to be determined with less than 10% relative standard deviation. This is approximately equivalent to a requirement for detection limit $L_D \leq 1$ mg/m^3.

The following methods were found to be incapable of fulfilling this requirement:

Atomic emission spectrometry	AES
Particle-induced X-ray emission	PIXE
Flame atomic absorption spectrometry	AAS
Catalysis	CAT

Table 1
Normal Concentrations of Vanadium in Human Serum or Plasma Reported in the Period 1971–1988

Mean value mg/m³	Number of results	Analytical method	Investigator	Ref.	Year
10		Review	Schroeder	[4]	1971
56	10	CAT	Christian	[5]	1971
1.0	1	NAA	Damsgaard	[6]	1972
83	20	AES	Burger	[7]	1974
20	300	AES	Panteliadis	[8]	1975
30	3	PIXE	Vis	[9]	1977
400	9	AAS	Bello-Reuss	[10]	1979
<0.1	1	RNAA	Byrne and Costa	[11]	1979
0.034	14	RNAA	Cornelis	[12]	1979
6.6	5	NAA	Sabbioni	[13]	1979
16	10	NAA	Ward and Ryan	[14]	1979
0.10	22 ♂	RNAA	Cornelis	[15]	1980
0.03	36 ♀	RNAA	Cornelis	[16]	1981
9.8	23	ETAAS	Dick	[17]	1981
26	110	NAA	Sprague	[18]	1981
7.8	9	ETAAS	Dick	[19]	1982
3.35	10	ETAAS	Stroop	[20]	1982
98	1	ICP	Mianzhi	[21]	1984
4.1	10	NAA	Naylor	[22]	1984
<0.5	pooled	ETAAS	Pyy	[23]	1984
0.48	23	RNAA	Simonoff	[24]	1984
2.6	7	IDMS	Fassett	[25]	1985
0.6	30	ETAAS	Lewis	[26]	1985
0.67	23	RNAA	Simonoff	[27]	1986
3.38	10	NAA	Corrigan	[28]	1987
4.30	26	NAA	Naylor	[29]	1987
0.63	5	RNAA	Simonoff	[30]	1987
4.03	32	NAA	Campbell	[31]	1988
3.67	10	NAA	Lavi	[32]	1988
<0.1		Review	Iyengar	[33]	1989

and their results were excluded from further investigations.

Of the remaining 23 results, 21 are distributed among only three methods:

Electrothermal atomic absorption spectrometry	ETAAS	5
Neutron activation analysis with radiochemical separation	RNAA	7

Neutron activation analysis with preirradiation separation	NAA	9

A comparison of results from these three methods by the Mann–Whitney test is shown in Table 2. Test No. 1 indicated that NAA without radiochemical separation invariably gives high results. This method was therefore eliminated from further investigations. Test No. 2 applied to the two remaining methods showed a difference that was probably not significant.

The median of the 12 results by ETAAS and RNAA is 0.5 mg/m^3, which is now the expected level, requiring a detection limit L_D <0.2 mg/m^3. This could not be achieved by the method of ICP, which was then eliminated from further investigation.

The elimination of those results from Table 1 that are significantly affected by the choice of analytical method has thus reduced the number of potentially useful results to 13. These selected results and their coefficients of variation are shown in Table 3 in order of increasing mean value.

Experience

An important factor in a subjective judgment of analytical results is the previous record of the laboratory responsible for the published data. A more objective approach is made possible by dividing the results reported into three groups according to the experience of the laboratories, as reflected in their publications in the scientific literature. The first group consists of laboratories with previous experience in the determination of vanadium or other difficult elements at the ultratrace levels in biological materials. The second group includes laboratories with experience in the determination of elements at the trace levels in biological materials, and the third group consists of laboratories without any documented experience.

In the present case, the second and third groups were merged before a comparison was made with the laboratories having the most experience. Test No. 3 in Table 2 demonstrates a significant difference between results from experienced laboratories and from the rest in the full data set from Table 1, but not in the data set from the selected analytical methods. This shows that an important aspect of experience is to choose the best analytical method and that there is no need to exclude results from Table 3 because of insufficient experience.

The possible effect of general improvements in analytical instrumentation over the period was tested by comparing the results from 1982 and earlier with these from 1984 and later. Test No. 4 shows the absence of such influence.

Sampling

Several aspects of the sampling procedure may well influence the analytical results. The analysis of plasma rather than serum requires the

Table 2
Mann-Whitney Tests for Significant Factors

Test No.	Factors	m/n	U	P(U)
1	NAA/RNAA	9/7	0	<0.002**(*)
2	ETAAS/RNAA	5/7	5	0.05(*)
3	Experience			
	best/rest	13/14	32	<0.01**
	RNAA + ETAAS only	7/5	6	0.1
4	Vintage 1979-82/1984-87	7/6	18	>0.2
5	Plasma/Serum	3/10	9	>0.1
6	Contamination risk	6/7	13	>0.2
7	Sampling information			
	good/absent	3/5	6	>0.2
8	Quality assurance	3/10	6	0.2
9	Contamination control	8/5	16	>0.2

addition of an anticoagulant before separation from the cellular fraction of whole blood, which may introduce a significant blank value [37].

The risk of contaminating the sample during the processing and before the final measurement is very serious for a number of trace and ultratrace elements. The care exercised by the laboratories in this respect is judged by their awareness of the importance of

1. The containers and other materials with which the sample is in contact;
2. The methods used to clean these materials;
3. The environment to which the sample is exposed during the processing.

The laboratories were divided into two groups according to whether their risk of contaminating the samples was expected to be higher or lower than average.

Information presented with regard to the exact procedure for taking the blood sample from the patient also has bearing on the awareness of sampling as a source of error. In five cases no such information was given at all, and in three cases this information was presented in full detail.

All these sampling factors were tested by the Mann–Whitney Tests No. 5, 6, and 7, but no significant influence was found. The influence of sample size was also tested, but found to be without any effect.

Table 3
Selected Determinations of Vanadium in Human Serum or Plasma Reported 1979–1987

Mean value mg/m³	Coefficient of variation	Detection limit rank	Reproducibility rank absolute	relative	combined	Reference No.
0.03	32 %	5	5	11	8½	[16]
<0.1	-	7	-	-	-	[11]
0.034	42 %	5	5	10	7	[12]
0.1	73 %	5	5	7	5	[15]
<0.5	-	10	10	-	-	[23]
0.48	50 %	2	2	4	3	[24]
0.6	80 %	11	9	9	10	[26]
0.63	29 %	2	2	3	2	[30]
0.67	48 %	2	2	2	1	[27]
2.6	12 %	8	8	5	6	[25]
3.35	19 %	9	7	1	4	[20]
7.8	33 %	12	11	6	8½	[19]
9.8	62 %	-	12	8	11	[17]
Rank correlation		$r_S = 0.315$ P>0.10	$0.454^{(*)}$ P~0.05	-0.491^{*} P<0.05	$r_S = 0.11$ n.s.	

Quality Control

Although few scientific laboratories have established procedures for quality assurance of their analytical results, it is generally recognized that at least some effort should be made to ascertain the accuracy of their analytical results. Among the 13 selected results, only 3 were presented without any indication of such effort.

Control of contamination is particularly important in ultratrace analysis and may be performed quite efficiently by checking the statistical control of results from replicate samples [*38*]. Alternatively, the analytical blank may be continuously monitored. Among the selected laboratories, eight relied on the analysis of replicate samples and five did not, but no significant difference could be detected between the results reported.

Tests No. 8 and 9 show the absence of significant differences between results from laboratories with or without quality assurance, and with or without control of blank values.

Correlation Analysis

Among the factors tested by the Mann-Whitney tests in Table 2, only **analytical method** and **experience** were found to be significant, whereas contamination problems did not play any significant role. This is quite different from the previous study of concentration of manganese in human serum or plasma [*36*], in which contamination control proved to be a highly significant factor, and analytical methods were without significant influence.

The selected results in Table 3, however, span more than 2.5 orders of magnitude, and further investigations are clearly needed to find the most probable, true concentration of vanadium in serum. What has already been found indicates that technical sources of variability associated with the actual analysis are the controlling factors.

Based on reported information, the ordered results in Table 3 were ranked according to the detection limit and experimentally determined precision. Correlation with reported mean values is tested by calculating Spearman's Rank Correlation Coefficient [*35*], and probably significant correlations were found for both absolute and relative reproducibility.

This might indicate that the results with the best combined reproducibility would give the most reliable results. The last column presents this combined ranking. The value of 0.67 mg/m^3 [*27*] therefore becomes the starting point for the Synthesis of Precision [*38*] presented in Table 4.

This value proved consistent not only with other values by RNAA from the same laboratory [*24,30*], but also with one result from another laboratory [*26*] by ETAAS. The weighted mean of these four values is 0.57 ± 0.03 mg/m^3 of vanadium in human blood serum. This is the most probable result derived from the analysis of data reported in Table 1.

RADIOCHEMICAL ANALYSIS

No satisfactory explanation has been found for the much lower results reported by other investigators [*12,16*], although they used essentially the same radiochemical method [*15*] with dry-ashing before irradiation and separation by extraction of the cupferron complex after irradiation. An attempt was therefore made to carry out a few pertinent analyses by comprehensive neutron activation analysis [*39*], using a radiochemical method different from those applied previously.

In order to account for all conceivable losses during the analyses, ^{48}V was added to the serum before lyophilization, which took place in the actual container used for irradiation. Vanadium carrier was added after irradiation, and wet decomposition with nitric acid took place in a micro-

Table 4
Synthesis of Precision for Vanadium in Human Plasma or Serum

Mean value mg/m³	Method	Number of results	Standard error of the mean	Synthesis of Precision T	$P(\chi^2 \geq T)$
0.67	RNAA	23	0.067	-	
0.60	ETAAS	30	0.088	0.40	0.53
0.63	RNAA	5	0.082	0.42	0.81
0.48	RNAA	23	0.05	6.12	0.11
0.10	RNAA	22	0.016	167	***
<0.5	ETAAS	pooled		t = 2.07	*(*)

wave oven. Radiochemical separation was based on extraction with 8-hydroxyquinoline in methyl isobutylketone.

Experimental

Approximately 2.5 cm^3 of serum was weighed into a 2-dram polyethylene vial, and approximately 2 kBq of carrier-free ^{48}V (Los Alamos National Laboratory UN 2982) was added. After counting, the sample was lyophilized in vacuum.

Irradiation took place in the pneumatic tube system of the Danish Reactor DR 3 for 3 min at a neutron flux density of about $2.5 \times 10^{13}/cm^2s$.

The irradiated sample was immediately transferred to a 45-mL Teflon® cup, placed inside a Parr® 4782 Microwave Digestion Bomb. We then added 5 cm^3 of 100% fuming nitric acid, containing 0.2 mg/cm^3 of V-carrier. The sealed Parr bomb was placed in a 600-W CEM microwave oven, type MDS 81D, and heated for 25 s at 95% of full power.

The bomb was opened immediately, and the content transferred to a centrifuge tube with 5 cm^3 of methyl isobutyl ketone (MIBK). Following the reaction, we added 5 cm^3 of water and stirred the aqueous phase in a centrifuge tube with 6 cm^3 of chloroform.

The aqueous phase was neutralized by 4*M* NaOH solution, and 2 g of sodium-acetate powder was added to buffer the acidity at pH-5. Vanadium was extracted with 6 cm^3 MIBK containing 2% 8-hydroxyquinoline for 30 s, and the black MIBK phase was transferred to a polyethylene vial and counted for 5 min with a Gamma-X® HPGe detector of 36.8% relative efficiency and 1.77 keV resolution for 1332-keV ^{60}Co radiation.

After approximately 24 h, the sample was counted again to determine the chemical yield.

RESULTS AND DISCUSSION

The entire radiochemical procedure was usually completed within approximately two half-lives after pile-out. When counting begins less than 7.8 min after pile-out, the sensitivity of the analytical method exceeds 400 counts/ng of vanadium in the separated sample. This is not so different from the sensitivities available to other workers [*16,24*].

The blank correction from the ^{48}V tracer was determined by the same method, and the specific activity of the carrier-free ^{48}V at the time of delivery was found to be 1.4 ng/MBq of ^{48}V. This shows that the blank correction is negligible and does not contribute significantly to the uncertainty of the analytical result.

The accuracy of the method depends on the complete exchange between the tracer and the vanadium in the sample, and this was checked by analyzing the only reference material certified for low-level vanadium [*40*]. Results for SRM909, Human Serum, are shown in Table 5, indicating that complete exchange takes place during the microwave digestion process.

In a successful separation, the *a priori* limit of detection, assuming a 5% risk of committing errors of both the first and the second kind, is 0.7 ng of vanadium.

However, the total yield of the radiochemical separation rarely exceeds 35–40%, and the overall detection limit of the analytical method therefore becomes approximately 2 ng of vanadium. The analysis of eight samples of normal human serum obtained from the Risø Medical Centre yielded only one significantly positive result.

Samples of a second generation biological reference material of freeze-dried human serum from the University of Ghent were also analyzed, but no information was available on the true content of vanadium in this material [*41*]. Results based on four samples of approximately 200 mg dry wt are given in Table 5.

Results for both reference materials were in statistical control, confirming the absence of variability among replicate samples. Individual samples of normal human serum, however, displayed significant variability, even though they were all close to the limit of detection. Their mean value was obtained by combining all results [*42*] to represent a pooled sample of 18.4 g, and this result is presented in Table 5. It is consistent with the concentration of 0.57 mg/m^3 derived from the literature, and therefore gives experimental support to the line of reasoning that led to the exclusion of all higher results reported. However, the precision of our analytical method is insufficient to detect whether some of our samples would be much lower than the reported mean value.

Table 5
Concentrations of Vanadium in Human Serum Determined by Comprehensive Neutron Activation Analysis

Source	Number of samples	This work mg/m³ ± SD	Literature value
NIST Gaithersburg	4	3.54 ± 0.35	3.19 ± 0.55 [40]
Ghent University Hospital	4	1.00 ± 0.30	- [41]
Risø Medical Centre	8	0.72 ± 0.21	0.57 ± 0.03 [See text]

ACKNOWLEDGMENT

The author is grateful to J. Versieck of the University Hospital of Ghent for making samples of Second Generation Biological Reference Material for Trace-Element Determination available for analysis.

REFERENCES

1. K. Schwarz and D. B. Milne, *Science* **174,** 426 (1971).
2. L. C. Cantley et al., *J Biol. Chem.* **252,** 7421 (1977).
3. G. J. Naylor, *Metal Ions in Neurology and Psychiatry,* Alan R. Liss, New York, 1985, pp. 91–105.
4. H. A. Schroeder and A. P. Nason, *Clin. Chem.* **17,** 461 (1971).
5. G. D. Christian, *Anal. Lett.* **4,** 187 (1971).
6. E. Damsgaard, K. Heydorn, and B. Rietz, *Nuclear Activation Techniques in the Life Sciences,* IAEA, Vienna, 1972, pp. 119–128.
7. F. J. Burger, *Trace Element Metabolism in Animals-2,* University Park Press, Baltimore, 1974, pp. 671–74.
8. C. Panteliadis, *Spurenelemente in der Entwicklung von Mensch and Tier,* Urban & Schwarzenberg München, 1975, pp. 103–112.
9. R. D. Vis, P. M. A. Van der Kam, and H. Verheul, *Nucl. Instrum. Methods* **142,** 159 (1977).
10. Elsa N. Bello-Reuss, T. P. Grady, and D. C. Mazumdar *Ann. Intern. Med.* **91,** 743 (1979).
11. A. R. Byrne and L. Kosta, *Sci. Total Environ.* **13,** 87 (1979).
12. R. Cornelis, L. Mees, J. Hoste, J. Ryckebusch, J. Versieck, and F. Barbier, *Nuclear Activation Techniques in the Life Sciences,* IAEA, Vienna, 1979, pp. 165–177.

13. E. Sabbioni, E. Marafante, R. Pietra, L. Goetz, F. Girardi, and E. Orvini, *Nuclear Activation Techniques in the Life Sciences*, IAEA, Vienna, 1979, pp. 179–192.
14. N. I. Ward and D. E. Ryan, *Anal. Chim. Acta* **105,** 185 (1979).
15. R. Cornelis, J. Versieck, L. Mees, J. Hoste, and F. Barbier, *J. Radioanal. Chem.* **55,** 35 (1980).
16. R. Cornelis, J. Versieck, L. Mees, J. Hoste, and F. Barbier, *Biol. Trace Elem. Res.* **3,** 257 (1981).
17. D. A. T. Dick, E. G. Dick, and G. J. Naylor, *J. Physiol.* **310,** 24P (1981).
18. S. M. Sprague, A. Fregene, J. Costanino, R. W. Rosenbaun, and G. H. Mayor, *Clin. Res.* **29,** 476A (1981).
19. D. A. T. Dick, G. J. Naylor, and E. G. Dick, *Psychol. Med.* **12,** 533 (1982).
20. S. D. Stroop, G. Helinek, and H. L. Greene, *Clin. Chem.* **28,** 79 (1982).
21. Z. Mianzhi and R. M. Barnes, *Appl. Spectroscopy* **38,** 635 (1984).
22. G. J. Naylor, A. H. W. Smith, D. Bryce-Smith, and N. I. Ward, *Psychol. Med.* **14,** 767 (1984).
23. L. Pyy, E. Hakala, and L. H. J. Lajunen, *Anal. Chim. Acta* **158,** 297 (1984).
24. M. Simonoff, Y. Llabador, G. N. Simonoff, C. Beraud, P. Couzigou, C. Conri, and B. Fleury, *Trace Elem. Anal. Chem. Med. Biol.* **3,** 495 (1984).
25. J. D. Fassett and H. M. Kingston, *Anal. Chem.* **57,** 2474 (1985).
26. S. A. Lewis, T. C. O'Haver, and J. M. Harnly, *Anal. Chem.* **57,** 2 (1985).
27. M. Simonoff, C. Conri, and G. Simonoff, *Acta Pharmacol. Toxicol.* **59,** 463 (1986).
28. F. M. Corrigan, J. D. Finlayson, M. Muir, G. W. Ashcroft, and N. I. Ward, *Trace Elem. Med.* **4,** 139 (1987).
29. G. J. Naylor, F. M. Corrigan, A. H. W. Smith, P. Connelly, and N. I. Ward, *Br. J. Psychol.* **150,** 656 (1987).
30. M. Simonoff, Y. Llabador, C. Hamon, B. Berdeu, G. Simonoff, C. Conri, B. Fleury, P. Couzigou, and A. Lucena, *J. Radioanal. Nucl. Chem.* **113,** 107 (1987).
31. C. A. Campbell, M. Peet, and N. I. Ward, *Biol. Psychiatr.* **24,** 775 (1988).
32. N. Lavi and Z. B. Alfassi, J. *Radioanal. Nucl. Chem.* **126,** 361 (1988).
33. V. Iyengar and J. Woittiez, *Clin. Chem.* **34,** 474 (1988).
34. K. Heydorn, *Neutron Activation Analysis for Clinical Trace Element Research*, vol. 1, CRC, Boca Raton, FL, 1984, pp. 1–22.
35. L. Sachs, *Applied Statistics*, Springer-Verlag, New York, 1982.
36. K. Heydorn, *Proc. 2nd Int. Conf. Elements Health Disease*, Hamdard University Press, Karachi, 1988, pp. 347–359.
37. K. Heydorn, *Aspects of Precision and Accuracy in Neutron Activation Analysis*, Risø National Laboratory, Roskilde, 1980.
38. K. Heydorn, *Neutron Activation Analysis for Clinical Trace Element Research*, vol. 2, CRC, Boca Raton, FL, 1984, pp. 1–49.
39. Definitions and Terms, *J. Radioanal. Nucl. Chem.* **114,** 409 (1987).
40. Standard Reference Material 909 Human Serum, *Certificate of Analysis*, National Bureau of Standards, Gaithersburg, MD, 1988.
41. J. Versieck, Ghent, private communication, 1988.
42. K. Heydorn and B. Wanscher, *Fresenius Z. Anal. Chem.* **292,** 34 (1978).

Determination of Trace Elements in Human Serum by Inductively Coupled Plasma-Mass Spectrometry

Comparison with Nuclear Analytical Techniques

CARLO VANDECASTEELE,* HANS VANHOE, RICHARD DAMS, AND JACQUES VERSIECK

Laboratory for Analytical Chemistry, Rijksuniversiteit Gent, Institute for Nuclear Sciences, Proeftuinstraat 86, B-9000 Gent, Belgium, and Department of Internal Medicine, University Hospital, De Pintelaan 185, B-9000 Gent, Belgium

Received June 5, 1989; Accepted December 11, 1989

ABSTRACT

The determination of trace and ultratrace elements in human serum by ICP-MS is described. The accuracy of the method is tested using a "second generation" human serum reference material. Elements determined include Fe, Co, Cu, Zn, Br, Rb, Sr, Mo, and Cs. The method is compared to nuclear analytical methods (NAA, PIXE). Perspectives for the future are also discussed.

Index Entries: ICP-MS; accuracy; human serum; comparison with nuclear methods.

INTRODUCTION

For nearly 20 years, instrumental and radiochemical neutron activation analysis (NAA) has been used in this laboratory for the determination of trace and ultratrace elements in human serum (*1*). For this application,

*Author to whom all correspondence and reprint requests should be addressed.

NAA has several advantages, the most important being low detection limits and the fact that only a limited pretreatment of the sample is required. Since contamination can occur only before or during the irradiation, systematic errors from contamination are relatively easily avoided, so that accurate results may be obtained. NAA also has some drawbacks, however, the major ones being the long turn-around time and the low sample throughput. For several elements (Cr, Fe, Zn, Se, Rb, Cs, and so on) analytical results are available only after 1 mo or so, and several important elements cannot be determined or can be determined only with difficulty (Li, Al, Ni, Pb, and so on). Therefore, there is a need for an alternative method offering sufficiently low detection limits, but without the drawbacks of NAA. Proton-induced X-ray emission (PIXE) was used to determine trace elements in biological samples. However, in the case of serum it has only limited applicability, since, of the trace elements, only Fe, Cu, Zn, Se, and Br can be determined without preconcentration (*2*).

Inductively coupled plasma-mass spectrometry (ICP-MS) was therefore chosen. This method offers extremely low detection limits, at least for aqueous solutions, and, in principle, almost every element can be determined (*3*). An ICP-MS instrument was installed in this laboratory in April 1987, mainly for the determination of trace elements in biological samples, of which human serum was considered the most important from a clinical point of view. At that time, ICP-MS had not often been used for clinical samples; if it was, it was usually applied to samples with a much higher trace-element concentration than human serum. Our aim was to develop analytical methods for the determination of trace elements in real, uncontaminated serum samples. In order to achieve this goal, the instrument was installed in a dust-free room, and sample preparation was carried out in a clean bench,using high-purity water and reagents, as well as plastic pipets and measuring flasks. It was also considered of prime importance to limit sample pretreatment as much as possible.

When evaluating the performance of a new analytical technique, it is a definite asset to have available a reference material having a matrix and trace-element concentrations comparable to those occurring in real samples. A "second generation" human serum reference material prepared by Versieck et al. (*4*) was available to us in its original, liquid form. To the scientific community, it is available in freeze-dried form. The material was certified for the elements Al, Cr, Mn, Fe, Co, Cu, Zn, As, Se, Br, Rb, Mo, Cd, and Cs. For Ni, only an indicative value was given. This material is ideally suited for developing accurate methods for the determination of trace and ultratrace elements in human serum by ICP-MS and for comparing, e.g., ICP-MS with NAA and other nuclear analytical techniques.

We started our investigation by analyzing this material for the elements already certified, using five-fold dilution with 0.14*M* nitric acid as

Table 1
Polyatomic Interferences

Element	Selected nuclide	Major interferences
Al	^{27}Al	$^{12}C^{15}N$?; $^{13}C^{14}N$?
Cr	^{52}Cr	$^{35}Cl^{17}O$; $^{35}Cl^{16}OH$; $^{40}Ar^{12}C$; $^{36}S^{16}O$
Mn	^{55}Mn	$^{37}Cl^{18}O$; $^{37}Cl^{17}OH$
Fe	^{57}Fe	$^{40}Ca^{17}O$; $^{40}Ca^{16}OH$
Co	^{59}Co	$^{36}Ar^{23}Na$; $^{43}Ca^{16}O$
Ni	^{60}Ni	$^{23}Na^{37}Cl$
Cu	^{65}Cu	$^{33}S^{16}O^{16}O$; $^{32}S^{33}S$; $^{48}Ca^{17}O$
Zn	^{66}Zn	$^{32}S^{16}O^{18}O$; $^{32}S^{34}S$
As	^{75}As	$^{40}Ar^{35}Cl$
Se	^{82}Se	$^{12}C^{35}Cl^{35}Cl$?; $^{32}S^{32}S^{18}O$?; $^{32}S^{34}S^{16}O$?

the only sample pretreatment. A 0.14*M* nitric acid blank was subtracted and 100 μg/L standard solutions of the elements of interest were used. The results for several elements with mass number <82 are in general, unsatisfactory. For the elements with mass number >82 generally better results are obtained. The reason is that, because of the limited resolution of the quadrupole mass spectrometers used for ICP-MS, a large number of polyatomic interferences occur for elements with mass number <82. Some of the polyatomic ions that may occur are listed in Table 1. Several of these were described earlier (*5*). It follows that, without special sample pretreatment (in order to remove or reduce sources of polyatomic interferences) and using conventional pneumatic nebulization for sample introduction, the determination of Al, Cr, Mn, Ni, As, and Se in serum is excluded, since the apparent concentration significantly exceeds the real concentration.

Another problem with ICP-MS is the occurrence of matrix effects, mainly signal suppression. We demonstrated (*6*) that 9 g/L of sodium chloride suppresses the analyte signal by about 35%. One can accurately correct for signal suppression using a particular element that is not present in detectable concentrations in the sample, e.g., In, as an internal standard. At the same time this improves the precision of the measurements. The reason that we still dilute the serum samples 5 or 10 times is that the torch, and to a lesser extent the Meinhard nebulizer, becomes clogged after a short time when undiluted serum is analyzed.

METHODS, RESULTS, AND DISCUSSION

Determination of Fe, Co, Cu, Zn, Rb, Mo, and Cs (7)

In the first instance, we concentrated on the elements Fe, Co, Cu, Zn,Rb, Mo, and Cs. For Fe, Co, Cu, and Zn, a correction for polyatomic interferences was made using a simulated blank solution containing the same concentrations of Na, S, Cl, and Ca as found in serum. For Rb, Mo, and Cs, no polyatomic interferences could be detected for the selected nuclides.

The method developed was as follows:

- Defrost serum kept in a polyethene container in a deep-freezer, and dilute the liquid with 0.14*M* nitric acid. Add In in order to obtain a final concentration of 100 μg/L.
- Prepare a blank solution containing 0.14*M* nitric acid, 100 μg/L of In, sodium nitrate, sodium chloride, cysteine, and calcium nitrate, so that the concentrations of Na, Cl, S, and Ca are the same as found in human serum.
- In the mass-scanning mode of data acquisition, measure the samples for a total acquisition time of ca. 2 min. The following analysis sequence was applied in order to avoid systematic errors from memory effects: first the blanks were measured, then several samples, and, at the end of the sequence only, the standard.

Table 2 summarizes the results for the reference serum, expressed in μg/L of liquid serum. The results are also given in μg/g or ng/g of lyophilized material. The latter values, obtained by multiplying the results in μg/L by a factor of 0.011, are compared with the certified values. In general, the results agree, within the experimental uncertainty, with the certified values.

Determination of Br

An element that originally presented some problems was Br. In fact, the two stable isotopes of Br, ^{79}Br (50.7%) and ^{81}Br (49.3%), are situated on both sides of a large $^{40}Ar_2$ peak at mass 80. This makes peak integration and mass calibration very critical. However, when the resolution (5% definition) is increased to 0.61 u, instead of the usual 0.81 u, the peaks are clearly separated, as shown in Fig. 1. The result obtained for Br was 4299 ± 79 μg/L. Figure 2 compares the result obtained with those obtained by NAA and PIXE. The agreement is excellent.

Table 2
Comparison of the Results of the Second-Generation Biological Reference Material (Human Serum)

Element	Selected nuclide	ICP-MS result, mean ± 95% CL	Certified value
iron	^{57}Fe	2210 ± 350 μg/l	
		24.3 ± 3.9 μg/g	25.9 ± 1.5 μg/g
cobalt	^{59}Co	0.24 ± 0.05 μg/l	
		2.6 ± 0.6 ng/g	3.6 ± 0.6 ng/g
copper	^{65}Cu	1047 ± 52 μg/l	
		11.5 ± 0.6 μg/g	11.1 ± 0.4 μg/g
zinc	^{66}Zn	939 ± 21 μg/l	
		10.3 ± 0.2 μg/g	9.6 ± 0.2 μg/g
rubidium	^{85}Rb	159.8 ± 1.0 μg/l	
		1.76 ± 0.01 μg/g	1.85 ± 0.33 μg/g
molybdenum	^{98}Mo	0.67 ± 0.08 μg/l	
		7.4 ± 0.9 ng/g	7.5 ± 0.8 ng/g
cesium	^{133}Cs	0.92 ± 0.05 μg/l	
		10.1 ± 0.6 ng/g	10.0 ± 2.3 ng/g

Determination of Sr

Once the possibilities for determining the elements already certified had been studied, a logical next step was to exploit the possibilities for additional elements, again using conventional pneumatic nebulization for sample introduction. One example is Sr, which is determined using its major isotope, ^{88}Sr (82.6%). The result obtained, 25.5 ± 0.6 μg/L, agrees fairly well with the value of 32 ± 1.3 μg/L (0.35 ± 0.014 μg/g dry wt) obtained by Lin et al. (*8*) by NAA using ^{85}Sr.

Determination of Li, B, Ba, and Bi

Preliminary experiments indicated also that the elements Li, B, Ba, and Bi can be determined in fivefold diluted serum. Experiments are underway in order to ensure the accuracy of the results.

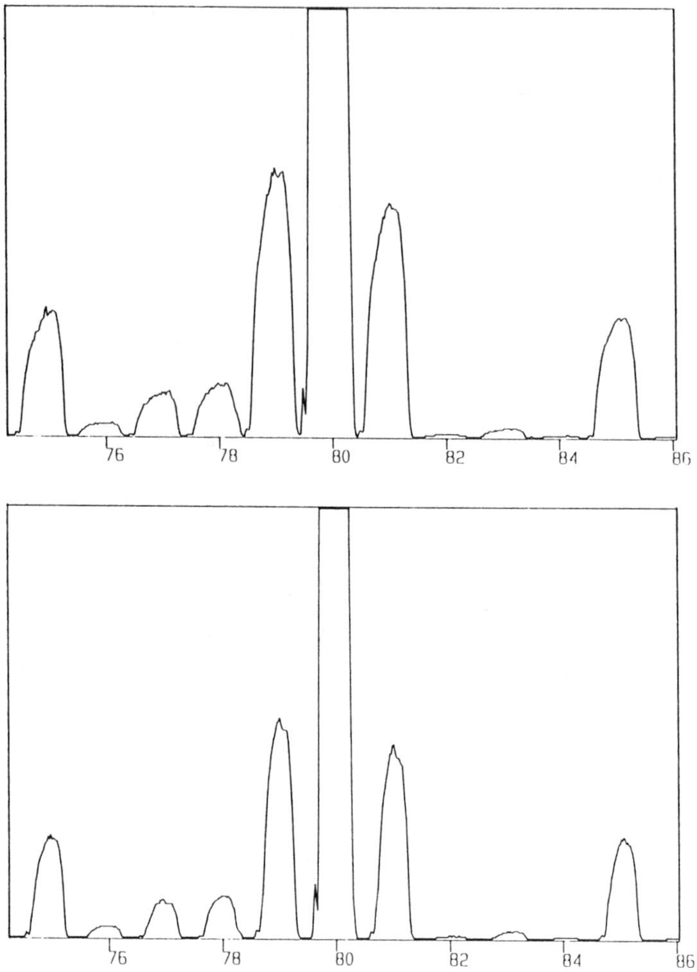

Fig. 1. Mass spectrum for serum measured with a resolution of 0.81 u (top) and 0.61 u (bottom).

CONCLUSIONS

It is clear that ICP-MS is an accurate method for the determination of trace elements in serum, if one corrects for interferences from polyatomic ions and suppression effects from the matrix. It must be emphasized that, for all the results presented, the only sample pretreatment consisted of dilution and addition of an internal standard. Thus far, no ashing of the sample (wet or dry) to remove the protein matrix or the chloride ions, nor any chemical separation or preconcentration, was applied.

Nevertheless, the number of trace elements that can be determined reliably in human serum without chemical separation is already compa-

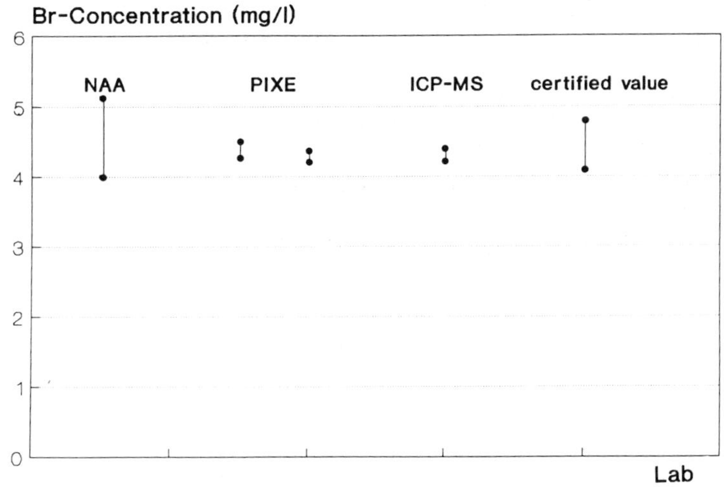

Fig. 2. Comparison of the Br results for the human serum reference material with those from other analytical techniques.

rable to what can be achieved by NAA using a low-neutron-flux (10^{12}) nuclear reactor. We are convinced that sample pretreatment, possibly combined with, e.g., flow injection for sample introduction, will increase the number of detectable elements. In addition, sample introduction by electrothermal volatilization may, for several elements, present some advantages. The use of a drying step before the analysis leads to a greatly reduced population of polyatomic ions deduced from the plasma (ArO) or from major matrix constituents (CaO and the like). Use of an ashing step, possibly combined with application of a matrix modifier, may allow one to eliminate other potential interferences (NaCl). In this way, peaks from ClO, NaCl, ArCl, and so on can be significantly reduced.

ACKNOWLEDGMENTS

Grateful acknowledgment is made to J. Hoste for his interest in this work, to L. Vanballenberghe for preparing the samples, and to the IIKW and NFWO for financial support.

REFERENCES

1. J. Versieck, A. Speecke, J. Hoste, and F. Barbier, *Z. Klin. Chem. Klin. Biochem.* **11,** 143 (1973).
2. W. Maenhaut, *Nucl. Instr. Meth. Phys. Res.* **B35**, 388 (1988).
3. G. M. Hieftje and G. H. Vickers, *Anal. Chim. Acta* **216,** 1 (1989).

4. J. Versieck, L. Vanballenberghe, A. De Kesel, J. Hoste, B. Wallaeys, J. Vandenhaute, N. Baeck, and F. W. Sunderman, Jr., *Anal. Chim. Acta.* **204**, 63 (1988).
5. S. H. Tan and G. Horlick, *Appl. Spectroscopy* **40**, 445 (1986).
6. C. Vandecasteele, M. Nagels, H. Vanhoe, and R. Dams, *Anal. Chim. Acta* **211**, 91 (1988).
7. H. Vanhoe, C. Vandecasteele, J. Versieck, and R. Dams, *Anal. Chem.* **61**, 1851 (1989).
8. Lin Xilei, D. Van Renterghem, R. Cornelis, and L. Mees, *Anal. Chim. Acta* **211**, 231 (1988).

Aspects of Accuracy and Precision in the Determination of As and Sb in Biological Materials by Neutron Activation Analysis

JOOST R. W. WOITTIEZ* AND MARCO GEUSEBROEK

Netherlands Energy Research Foundation E C N, Division of Chemistry and Material Sciences, PO Box 1, 1755 ZG Petten, The Netherlands

Received April 17, 1989; Accepted December 11, 1989

ABSTRACT

Aspects of accuracy and precision of the determination of As and Sb in biological materials, using neutron activation with hydride generation, are discussed. It proves necessary to perform a yield determination for each sample. Both radiotracers and reactivation have been applied, and their practical use for yield correction is discussed. Results for several RMs are presented.

Index Entries: Arsenic; antimony; hydride generation; separation; biological materials; reference materials; neutron activation analysis; accuracy; precision.

INTRODUCTION

The importance of the elements As and Sb in life sciences studies is common knowledge now (*1*). From a recent review on trace-element contents in human clinical specimens, it appears, however, that As is

*Author to whom all correspondence and reprint requests should be addressed.

poorly documented (*2*), and data for Sb are extremely scarce (*3*). Probably this is mainly a consequence of analytical problems in the assay of these elements. Below a level of about 30 μg/kg, hydride generation atomic absorption spectrometry starts to suffer seriously from interferences for both elements (*4*). Facing the need for accurate and precise results at this low level, the possibilities of neutron activation analysis with postirradiation separation (RNAA) provide an option. The literature mentions several techniques of separating As and Sb from the radioactive matrix (*5–16*). It was decided to focus the analytical effort on improvement of an existing postirradiation hydride generation technique (*15*). This paper presents the results of these efforts.

EXPERIMENTAL

Materials

Laboratory Equipment

The vials and other laboratory equipment used before irradiation were made of TFE or high-density polyethylene. The device in which the separation was performed was identical for As and Sb (*21*) (Fig. 1).

Calibration Materials

As calibration materials, dilutions of Merck atomic absorption solutions were used. Standards were analyzed instrumentally (*17*).

Reference Materials

Several NBS SRMs and BCR CRMs have been analyzed, together with real samples, to check the validity of the method.

Methods

Sample Preparation

All preirradiation manipulations were carried out in a dust-free laboratory. Samples of about 300 mg of biological material were weighed into 0.8-mL polyethylene vials. The vials, containing either standard, blank, sample, or RM, were packed in an irradiation capsule and provided with an iron flux monitor. Standards, blanks, SRMs, and unknown samples were irradiated simultaneously.

Irradiation

Samples were irradiated for 4 h in a neutron flux of $5 \times 10^{12} \cdot cm^{-2} \cdot s^{-1}$ in the Poolside Rotating Facility of the Euratom High Flux Reactor in Petten. The containers were returned to the laboratory after 1 d of decay.

Digestion

Samples were unpacked and transferred to 250-mL glass beakers. Each sample to be digested was provided with either (a) ^{74}As and ^{125}Sb

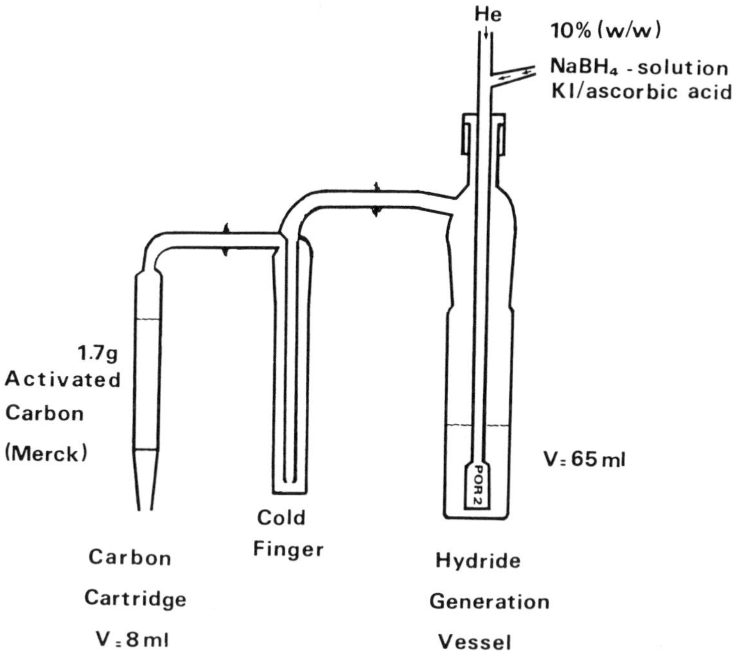

Fig. 1. Hydride generation device for the simultaneous postirradiation determination of As and Sb.

tracer solution with 5 μg As(V) and Sb(III) carrier solution (yield determination with radiotracers) or (b) 50 μg As(V) and 50 μg Sb(III) (yield determination using reactivation). To each beaker, 10 mL of a saturated $Mg(NO_3)_2$ solution was added, after which the beakers were heated on a hot plate until dry. This was repeated once. The digestion procedure was completed by a 1-h heating period in a muffle oven at 450°C. After cooling, the white MgO fusion cake was redissolved in 100 mL of 8M HCl.

Postirradiation Separation

The 8M HCl solutions were transported to a hydride generation vessel [*21*]. The vessel was flushed with He while the sample was introduced. As soon as possible, an active carbon cartridge was connected and the system closed. Hydride generation then took place using 2 × 2.5 mL 10%(w/w) $NaBH_4$ solution. As(V) has to first be reduced to As(III) by using 1 mL 0.2*M* KI/10%(w/w) ascorbic acid solution. The hydrides formed were trapped on active carbon cartridges. The carbon was poured into test tubes that were vigorously vortexed and counted.

Counting

Samples, standards, and blanks were counted for 2000 s on a Ge detector (21% efficiency, FWHM = 1.78 keV at 1332.5 keV). The gamma-

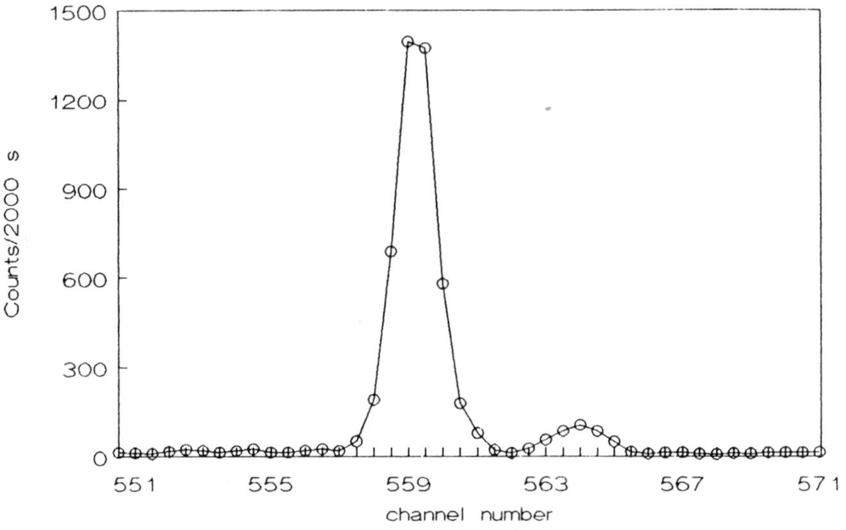

Fig. 2. Spectrum from 551 to 571 keV for NBS SRM 1577A bovine liver.

ray peaks of ^{76}As (E_γ = 559.1 keV, T1/2 = 26.4 h) and ^{122}Sb (E_γ = 564.1 keV, T1/2 = 2.70 d) were used for evaluation.

Determination of Chemical Yield

The efficiency of the hydride generation procedure was determined for each individual sample. This was done by using the radiotracers ^{74}As (E_γ = 596 keV, T1/2 = 17.77 d) and ^{125}Sb (E_γ = 428 keV, T1/2 = 2.77 y). Also, reactivation was applied for recovery determination. Here, the active carbon cartridges were irradiated for 2 h in a flux of $3 \times 10^{11} \cdot cm^{-2} \cdot s^{-1}$ and counted after a one-d decay for 2000 s on a Ge detector.

RESULTS AND DISCUSSION

Blanks

Blank values for an 0.8-cm^3 polyethylene sample vessel were <0.5 ng for both As and Sb. Since no cross-contamination of ^{76}As or ^{122}Sb in the postirradiation separation was found, it was decided not to take any blank value into account.

Spectral Interferences

As Fig. 2 shows, the spectrum of NBS SRM 1577A, Bovine Liver, from 551 to 571 keV is free from interference from ^{82}Br activity at 554 keV. Byrne (7) mentions the occurrence of an ^{76}As peak at 562.8 keV, with a gamma abundance of 3.6% compared to the 559.1 line. The detector used is not capable of separating the 562.8- and 564.1-keV peaks. Conse-

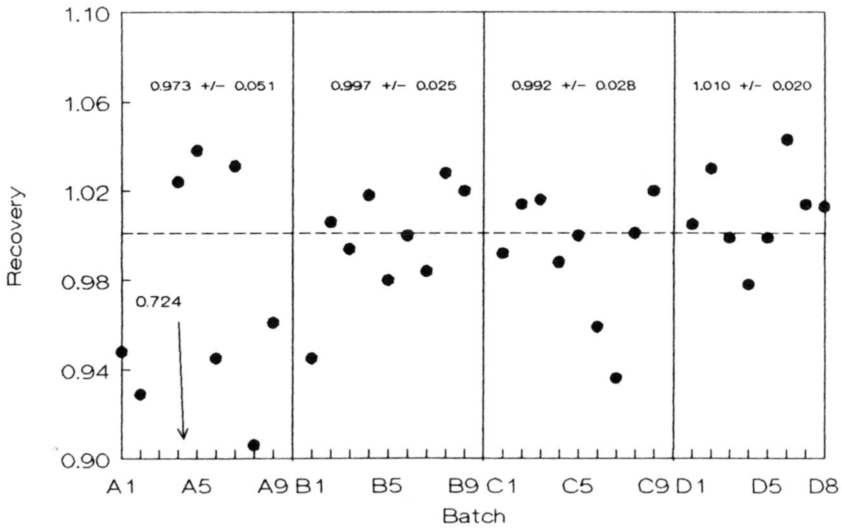

Fig. 3. Results for the recovery of As using reactivation.

quently, for samples with a high As-to-Sb ratio, the peak area of the 564.1-keV peak was corrected with 0.036 × the number of counts of the 559.1-keV peak.

Sensitivity

The specific count rates for ^{76}As and ^{122}Sb were 107 and 56 $ng^{-1}\cdot(2000\ s)^{-1}$ immediately after a 4-h irradiation in a thermal-neutron flux of $5 \times 10^{12}\cdot cm^{-2}\cdot s^{-1}$. Detection limits (*18–19*) for NBS SRM 1577A are 0.8 μg/kg As and 1.5 μg/kg Sb in the case of yield determination by reactivation, and approx. 5 μg/kg using radiotracers. Longer irradiation and counting times may improve these figures if necessary.

Determination of the Chemical Yield

Arsenic

The yield of the radiochemical procedure is typically over 98%. There are, however, large differences between individual determinations as shown in Fig. 3. The beneficial effect of yield correction for each single determination using ^{74}As is illustrated in the determination of As in mussel tissue, as the mean value ($N = 6$) changes from 5.87 to 6.19 mg/kg, and the relative standard deviation from 4.3 to 1.9%. Results for the analysis of several reference materials and candidate reference materials using yield correction with ^{74}As are given in the next section. In the case of reactivation, one has to accept an additional statistical and systematic error in the mere assay of the carrier. Both these errors must be small to result in any beneficial correction effect. The significance of this demand becomes manifest because the hydrides are not homogeneously distributed in the active carbon cartridge. The radiotracer method also corrects for actual unknown geometry differences, but the reactivation method refers, in fact, to an entirely different sample. The statement is illustrated in Table 1, which shows that yield correction by reactivation does not result in an improvement in the determination of arsenic in NBS SRM 1577A.

Despite this disadvantage, one cannot expect high and constant yields only. About 2% of our reactivation results show extremely low yields, down to 70%. In these cases, there is a definite relation between a poor analytical result and a low yield. For one batch, it was possible to actually compare results corrected with both radiotracer and reactivation. Results, as shown in Table 2, emphasize the advantage of the radiotracer technique for As.

Antimony

The discussion for As, with obvious adjustments, in general also holds for Sb. Typical recoveries are around 95%. A drawback of the use of radiotracer is illustrated by the background enhancement caused by the 636-keV gamma of ^{125}Sb at the 564.1-keV peak of ^{122}Sb and, thus, the induction of an undesired increasing limit of detection. This effect is demonstrated in Table 3.

Application

The technique for As has been applied to several NBS SRMs and BCR CRMs. Table 3 shows the results. There is a good agreement with the certified values.

Antimony has been determined in NBS SRM 1577A to be 2.3 ± 1.0 μg/kg ($N = 11$). The literature mentions 3.81 ± 0.46 (*7*) and 3.3 ± 0.2 (*5*), and the NBS information value is set at 3 μg/kg.

As a demonstration of the maintenance of precision of the technique, even in the 1–15 μg/kg range, Table 4 gives the results of the analysis of agricultural candidate reference materials. Details on the project are given by Ihnat (*20*).

CONCLUSIONS

From this work, it is concluded that application of postirradiation separation of As and Sb by hydride generation provides a sensitive analytical tool. Correction of the yield must be done for each separate sample, by either method.

ACKNOWLEDGMENT

The authors gratefully acknowledge the comments from J. G. van Raaphorst on the manuscript.

Table 1
The Effect of Yield Correction Using Reactivation on the Determination of As in NBS SRM 1577A

	Uncorrected result (μg/kg)	yield	corrected result (μg/kg)
	41.3	0.906	45.6
	51.9	0.961	53.9
	50.9	1.000	50.9
	44.2	0.984	45.0
	43.4	0.959	45.2
	54.5	0.936	58.2
	37.2	1.001	37.2
	39.0	1.020	38.3
	48.5	1.043	46.4
	48.3	1.014	47.7
	44.0	1.013	43.5
average	45.7	0.985	46.5
st. dev.	5.5	0.041	6.2
rsd.	12.0 %	4.1 %	13.3 %

Table 2
Results for the Determination of As and Sb after Yield Determination by Both Reactivation and Radiotracers

Sample	Sb (μg/kg)		As (μg/kg)	
	Re-activation	Radio-tracer	Re-activation	Radio-tracer
Gluten 149	5.0	5.9	15.6	15.3
Gluten 149	5.2	<4.9	16.9	12.8
Gluten 500	4.0	<4.1	14.1	14.1
Gluten 925	5.5	<4.9	11.5	15.4
GLuten 1083	3.9	<4.4	12.7	13.6
average	4.7	<5	14.2	14.2
standard dev.	0.7	–	2.2	1.1
rel. st. dev.	15.4 %	–	15.3 %	7.9 %
NBS SRM 1577A	3.2	<5.3	53.9	45.6
NBS SRM 1577A	2.1	<4.2	45.2	46.3
NBS SRM 1577A	4.1	<3.9	58.3	45.5
NBS SRM 1577A	3.3	<4.1	46.5	44.5
average	3.2	<4	51.0	45.5
standard dev.	0.8	–	6.2	0.7
rel. st. dev.	26 %	–	12.2 %	1.6 %

Table 3
Application of the Analysis for As of NBS SRMs and BCR CRMs[a]

Sample	Individual results	average (st.dev.)	ref. value (st.dev.)
NBS SRM 1572 Citrus leaves	3180; 3180; 3320; 3120 (*)	3200 (85)	3100 (300)
NBS SRM 1577 Bovine Liver	58.1; 52.1; 59.5; 60.0; 56.8 (*)	57.3 (3.2)	55 (5)
NBS SRM 1577A Bovine Liver	45.6; 53.9; 50.9; 45.0; 58.2; 37.2; 38.3; 46.4; 47.7; 43.5 (**)	46.7 (6.5)	47 (6)
NBS SRM 1577A Bovine Liver	45.6; 46.3; 45.5; 44.5 (*)	45.5 (0.7)	47 (6)
BCR CRM 184 Bovine Muscle	30.2; 27.0; 26.3; 28.7; 27.7; 28.9 (*)	28.1 1.4	not certfied
BCR CRM 185 Bovine Liver	23.8; 25.4; 26.8; 25.6; 24.6; 25.0 (*)	25.2 (1.0)	24 (3)
BCR CRM 186 Pig Kidney	72.6; 71.4; 74.4; 71.2; 70.5; 73.6 (*)	72.3 (1.5)	63 (9)
BCR grass sample	49.7; 52.3; 57.1; 51.9; 51.0; 53.7 (*)	52.6 2.5	not known

(*) = Yield determination using ^{74}As.
(**) = Yield determination by reactivation.

[a]Results are in μg/kg. Ref. values are NBS or BCR certified values. Values in parenthesis are standard deviations. All analyses of BCR materials were done prior to certification.

Table 4
Application of the Analysis for As and Sb of Canadian Agricultural Candidate Reference Materials[a]

Sample	As (μg/kg) average	As standard deviation	As N	Sb (μg/kg) average	Sb standard deviation	Sb N
Wheat Gluten	14.2	2.1	6	5.5	0.8	6
Corn Bran	2.4	0.9	6	4.0	0.6	6
Bovine muscle powder	9.5	1.1	6	2.9	0.2	6
Whole egg powder	9.6	0.9	6	2.5	1.3	6

[a]All yield determination by reactivation.

REFERENCES

1. L. Friberg et al., *Handbook on the Toxicity of Metals,* 2nd ed., Elsevier, Amsterdam, 1986.
2. G. V. Iyengar and J. R. W. Woittiez, *Clin. Chem.* **43/3,** 474–481 (1988).
3. J. R. W. Woittiez, PhD thesis, University of Amsterdam (1983).
4. J. G. van Raaphorst, personal communication.
5. A. Fajgelj et al., *J. Radioanal. Nucl. Chem. Lett.,* **128/2,** 93–102 (1988).
6. A. R. Byrne, *Anal. Chim. Acta* **59,** 91 (1972).
7. A. R. Byrne, *Fresenius Z. Anal. Chem.* 326, 733–735 (1987).
8. K. Heydorn, *RISO Report R-419,* (1980)
9. R. R. Greenberg, *Anal. Chem.* **58 (12),** 2511–2516 (1986).
10. W. C. Cunningham, *J. Radioanal. Nucl. Chem.* **113 (2),** 423–430 (1987).
11. J. R. W. Woittiez and G. V. Iyengar, *Fresenius Z. Anal. Chem.,* 332, 657–661 (1988).
12. K. Samsahl et al., *Anal. Chem.* **40,** (1), 181–187 (1968).
13. P. S. Tjioe, *J. Radioanal. Chem.* **16,** 153 (1973).
14. C. A. Weers, PhD thesis, University of Amsterdam (1980).
15. D. Hoede et al., *ECN Report 203,* (1987).
16. C. Y. Wu et al., *J. Radioanal. Nucl. Chem.* **112 (1),** 133–141 (1987).
17. A. R. Byrne, *J. Radioanal. Nucl. Chem. Lett.* **93 (4),** 245–252 (1985).
18. L. A. Curry, *Anal. Chem.* Vol **40 (3),** (1968)
19. V. C. Rogers, *Anal. Chem.* Vol **42 (7),** (1970) 807–808.
20. M. Ihnat, *Fresenius Z. Anal. Chem.* **332,** 539–545 (1988).
21. J. R. W. Woittiez and M. Geusebroek, ECN report 89, in press.

Homogeneity and Evaluation of the New NIST Leaf Certified Reference Materials

DONALD A. BECKER

Center for Analytical Chemistry, National Institute of Standards and Technology, Gaithersburg, MD 20899

Received June 16, 1989; Accepted December 11, 1989

ABSTRACT

The NIST has produced and is in the process of certifying two new leaf CRMs, SRM1515 Apple Leaves and SRM 1547 Peach Leaves, as replacements for the no longer available NBS Orchard Leaves and the almost depleted Citrus Leaves. These two new materials have been processed and are being thoroughly evaluated and should provide the most advanced natural matrix botanical trace-element reference materials available. Caution should be used in determining a basis weight (drying) for these CRMs because of their very fine particle size. Homogeneity has been established by instrumental neutron activation analysis on both leaf materials for five elements, to date, to better than 1.5% (1 *s*) for 100-mg sample sizes.

Index Entries: Certified reference materials; standard reference materials; trace-element analysis; neutron activation analysis; homogeneity; botanical materials.

INTRODUCTION

The first natural matrix, widely distributed reference material was Bowen's Kale, produced and evaluated by Humphrey J.M. Bowen of Reading University, UK (*1*). This material demonstrated to the scientific community the usefulness of such reference materials, and shortly afterwards the National Bureau of Standards (NBS) brought forth the first such material to be certified for trace-elemental content (*2*). Eventually, SRM 1571, Orchard Leaves, was certified by NBS for 25 elements, with information values and/or compiled literature values for 42 additional elements (3), for a total of 67 elements. The Orchard Leaves SRM has been out of stock for a number of years and was replaced by SRM 1572, Citrus Leaves. The Citrus Leaves SRM was certified by NBS for 22 elements, with information values and/or compiled literature values for 35 additional elements, for a total of 57 elements.

The National Institute of Standards and Technology (NIST; formerly NBS) has obtained and prepared a new orchard leaves reference material, to be certified for elemental content. Actually, the previous orchard leaves are being replaced by two new leaf materials: Apple Leaves, SRM 1515; and Peach Leaves, SRM 1547. These two materials have been obtained and prepared for certification work, and this paper discusses some of the work on these materials to date.

Since the intent was to improve these new leaf SRMs in relation to both the old leaf SRMs and other currently available botanical reference materials, a number of new or revised procedures were developed to better evaluate the SRM preparation procedures. These included: hand-picking the apple leaves followed, prior to grinding, by washing, to help remove dirt and siliceous material (a significant problem with Orchard Leaves; washing was not done with the peach leaves, to determine differences, if any); jet-milling and air-classification to obtain very small particle size, thus allowing evaluation of homogeneity at smaller sample sizes; a detailed moisture-content and moisture-uptake evaluation, along with weight-loss measurements during drying; and a dissolution study both to help establish the "insolubles" content of the leaves, if any, and to evaluate the ability of different dissolution techniques to totally solubilize the leaf materials.

PROCESSING

The apple leaves were collected and initially processed by Cyril B. Smith of the Department of Plant Nutrition, Pennsylvania State University. Leaves were hand-picked from the midshoot portions of trees that were free from trace-element sprays and pesticides. Leaves were placed into approximately 650 paper grocery bags and refrigerated until they could be washed and ground. They were washed in a dilute solution of a nonionic, phosphorus-free detergent, dipped in tap water, and then rinsed three times in distilled water. They were dried, in cardboard trays with clean paper liners, in an oven at 65°C. They were then ground in a stainless steel Wiley mill to pass a 40-mesh screen. The processing was in batches, with emission spectroscopic analysis of each batch showing no detectable differences between batches.

The peach leaves were obtained through R. A. Isaac at the Soil Testing and Plant Analysis Laboratory of the College of Agriculture, University of Georgia. They provided approximately 600 pounds of dried leaves, including preliminary grinding and sieving.

At NIST, the leaves were further ground against themselves in a jet-mill and air-classified to pass 200 mesh. They were thoroughly blended in a large V-blender, and then radiation-sterilized in bulk through exposure to cobalt-60 gamma radiation. The apple and peach leaves were each bottled in approximately 5000 clean amber glass bottles, with 50 g of leaves/bottle.

DRYING STUDY

The finely ground leaf materials were evaluated for weight loss during drying and moisture pickup by J. Moody and T. Vetter of the NIST Inorganic Analytical Research Division. The drying techniques evaluated were: convection oven at 85°C, vacuum oven at room temperature with liquid nitrogen cold trap, desiccator drying using anhydrous magnesium perchlorate desiccant, and freeze-drying. The space available here is not sufficient to present these data; however, only desiccator-drying and true freeze-drying (with the sample frozen prior to application of vacuum and kept at −5°C during the entire process) provided acceptable weight-loss results. The conventional oven and vacuum oven techniques were not acceptable. Under the two recommended drying conditions, weight losses were approximately 2.0% for the apple leaves and 2.3% for the peach leaves.

Samples of both apple leaves and peach leaves from bottles used for the homogeneity analyses (15 bottles for each leaf SRM) were again checked for weight losses, using the approved freeze-drying technique. These data provided confirmatory values of 1.68 ± 0.19% (1 *s*) weight loss for the apple leaves, and 2.08 ± 0.11% (1 *s*) weight loss for the peach leaves.

It is believed that these materials are very sensitive to the drying technique because of their extremely fine particle size. Even very gentle drying techniques that had been used successfully for other conventionally ground botanical materials gave substantially higher weight losses for these jet-milled materials, due most likely because of release and loss of the natural volatile oils in the leaves. Thus, for actual analyses of these materials, it may be preferable that undried samples be used, with weight-loss corrections determined on separate samples and applied to the final calculations.

A moisture-uptake study was also made on these materials, prior to the homogeneity study, to help evaluate appropriate sample-handling protocols. In our laboratory atmosphere (21°C), the undried apple leaves picked up 0.15% by weight in 4.0 min, whereas the dried apple leaves picked up 1.12% in 4.0 min. For the peach leaves, the weight gain was 0.23% for the undried material and 0.88% for the dried leaf material. These data indicate that analytical uncertainties can be significantly reduced by working with the undried leaf material, and thus, for the homogeneity study, undried samples were taken for analysis and the appropriate correction for weight loss upon drying made in the final calculations. In addition, sample weighing was completed as quickly as possible after opening the sample bottles.

DISSOLUTION STUDY

In order to better understand and eliminate some of the dissolution problems that had been found for Orchard Leaves, dissolution studies are being performed on both leaf materials. In these studies, scientists from a number of different analytical groups at NIST are applying their chosen dissolution methods to leaf samples that have been irradiated in the NBS Reactor. After dissolution is complete, each solution is filtered, and the filters and dissolution containers are counted on a germanium detector and gamma-ray analysis system to measure any remaining or insoluble residue. This dissolution evaluation is being coordinated by R. Greenberg, and the results will be reported in detail elsewhere (*4*). However, it can be stated that to date many of the dissolution techniques showed, on the filter paper, one or more elements present above the percent level (relative to the total amount contained in the material).

EXPERIMENTAL

Since these leaf materials were ground to extremely fine particle sizes through use of the jet-mill, it was expected that a very high degree of homogeneity would be obtained. The homogeneity study was thus designed to establish whether a sample size smaller than the usual 250 mg could be specified, while still maintaining appropriate certification uncertainties. The homogeneity study was made by INAA with a sample size of 100 mg.

Experimental conditions were as follows: Short-half-life isotopes were measured after a 60-s irradiation at 2×10^{13} $n \cdot cm^{-2}s^{-1}$ fluence rate, with four counts, usually at 2, 9, 22, and 60 min decay. Counting geometries were always 10 cm for these counts, and data were corrected for differences in irradiation time (using copper-foil fluence monitors), decay time before counting, live time of count, decay during counting period, pulse pileup, and weight losses during drying. For these preliminary results, the elements determined under these conditions were Al, Ca, Mn, and K.

Activation products with half-lives of 1 d or more were determined after a second irradiation on these same samples, this time for 4 h at a fluence rate of 6×10^{13} $n \cdot cm^{-2}s^{-1}$. Samples were counted twice, after approximately 6 and 30 d. These 4-h irradiations were actually two irradiations of 2 h each. After the first irradiation, each rabbit was immediately flipped end-for-end and irradiated for an additional 2 h. With the specific configuration of the NBSR pneumatic tubes, this exactly corrects for the linear fluence dropoff in these tubes, providing all samples with exactly the same neutron exposure (within measurement uncertainties of ±0.25%). For these preliminary results, the element lanthanum was measured under these conditions.

In addition to those corrections specified above, conditions were monitored as closely as possible to eliminate or minimize all sources of error in these analyses.

RESULTS

This study is still in process, and the elements evaluated for homogeneity to date are Al, Mn, K, Ca, and La for both materials. The results for SRM 1515, Apple Leaves, are contained in Table 1, and results for SRM 1547 are contained in Table 2. Evaluation of the data shows that for Apple Leaves SRM, three of the five elements measured have observed variations that fall *within* the 5–95% range of expected variances for a normal population described by the given counting statistics (σ). For Peach Leaves SRM, four of the five elements have observed experimental uncertainties that fall *outside* of that range, suggesting that washing of Apple Leaves may have slightly improved the observed sample variability. However, the measured homogeneity of both these materials is still very good, with "Additional Variance" requirements of, at most, not much over 1%. It should be noted that the "Additional Variance" value includes all experimental uncertainties in the entire measurement process (preparation, irradiation, counting, and data reduction), except for the counting statistics (σ).

In order to evaluate the effect of the copper-foil fluence monitor correction on the short half-life measurements, potassium values shown in Table 1 for Apple Leaves were recalculated without using the fluence monitor correction factor. The copper-foil data had an experimentally observed standard deviation of 1.014% (1 *s*) for 42 irradiations ($n = 42$) over a 56-d period, whereas the counting statistics for these data (σ) averaged 0.11%. Without the copper-foil fluence correction factor, the Apple Leaf potassium values had SD = 1.07%, σ = 0.56% (since counting statistics are the same), and an "Additional Variance" of 1.00%. Since use of the fluence normalization factors reduced the experimentally observed variability (*s*) from 1.07 to 0.55%, this demonstrates the need for such fluence monitoring when high-precision data are required and when irradiations are performed over a relatively long period of time. Further, the above data were calculated with only 13 data points for the non-fluence-corrected potassium values, because one irradiation was known to be about 12% longer than the others as a result of operator error. However, when the fluence monitor correction was applied, this value was indistinguishable from the other 13 values and, thus, was included in the fluence-corrected data in Table 1.

Table 1
Homogeneity Results for SRM 1515, Apple Leaves

Element	n	s^a	σ^b	Excess Variance[c] Observed?	Additional[d] Variance
Potassium	14[e]	0.55%	0.56%	No	0.0%
Manganese	14	0.63%	0.16%	Yes	0.62%
Aluminum	15	0.72%	0.61%	No	0.42%
Calcium	13[f]	0.92%	0.91%	No	0.30%
Lanthanum	15	1.07%	0.26%	Yes	1.07%

[a]s about unweighted mean (in % relative).
[b]*A priori* uncertainty based only on counting statistics (in % relative).
[c]Determination of Excess Variance based upon agreement between s and σ (i.e., whether s fell between 5–95% probability for a normal population).
[d]Additional variance required to bring "probability of exceeding" to 50 ± 2% for a normal population.
[e]One homogeneity sample was lost for all determinations except aluminum.
[f]A second sample was counted at a different geometry, thus could not be directly compared to the others.

Table 2
Homogeneity Results for SRM 1547, Peach Leaves

Element	n	s^a	σ^b	Excess Variance[c] Observed?	Additional[d] Variance
Potassium	15	1.03%	0.45%	Yes	0.95%
Manganese	15	0.74%	0.10%	Yes	0.76%
Aluminum	15	1.44%	0.61%	Yes	1.35%
Calcium	15	0.89%	0.91%	No	0.0%
Lanthanum	15	0.79%	0.17%	Yes	0.74%

[a]s about unweighted mean (in % relative).
[b]*A priori* uncertainty based only on counting statistics (% relative).
[c]Determination of Excess Variance based upon agreement between s and σ (i.e., whether s fell between 5–95% probability for a normal population).
[d]Additional variance required to bring "probability of exceeding" to 50 ± 2% for a normal population.

In conclusion, the data in Tables 1 and 2 indicate that these two new SRMs are homogeneous at the 100-mg sample size to better than ±1.5% at the 1 *s* confidence level for the five elements evaluated. Further homogeneity analyses on seven additional elements in these materials are reported elsewhere (*5*).

ACKNOWLEDGMENT

The considerable efforts of R. Alvarez of the NIST Office of Standard Reference Materials in the planning and contracting of these materials is gladly acknowledged.

REFERENCES

1. H. J. M. Bowen and P. A. Cawse, *Activation Analysis: Principles and Applications,* J. M. Lenihan and S. J. Thompson, eds., Academic (1965).
2. G. W. Smith, and D. A. Becker, *Nuclear Activation Techniques in the Life Sciences,* IAEA, Vienna (1967).
3. E. S. Gladney, et al., *Compilation of Elemental Concentration Data for NBS Clinical, Biological, Geological and Environmental SRMs,* NBS Tech. Note 260–111, US Department of Commerce (1987).
4. R. R. Greenberg, et al., *Fres. Z. Anal. Chemie* (in press).
5. D. A. Becker, *NIST Reactor: Summary of Activities July 1988 through June 1989,* NIST Technical Note 1272, US Dept. of Commerce (December, 1989), pp. 238–243.

Instrumental Neutron Activation Analysis of Standard Reference Material 1941, Organics in Marine Sediment

Element Content and Homogeneity

SUSAN F. STONE,* BARBARA J. KOSTER, AND ROLF ZEISLER

Center for Analytical Chemistry, National Institute of Standards and Technology, Gaithersburg, MD 20899

Received June 8, 1989; Accepted December 11, 1989

ABSTRACT

The National Institute of Standards and Technology has developed a new Standard Reference Material 1941, "Organics in Marine Sediment." In addition to the organic constituents, over 30 elements have been determined by instrumental neutron activation analysis and prompt-gamma activation analysis. The homogeneity of the material was investigated and relative standard deviations of single-element concentrations in 250-mg samples were found to be 1% or less with regard to major inorganic constituents and rare earth elements. A slightly higher relative SD was found for elements that may stem from biological or anthropogenic input. The element concentrations determined in this work are discussed in comparison to concentrations in other similar reference materials. Concentrations for 31 elements will be included for information on the certificate.

Index Entries: Elemental analysis; environmental monitoring; gamma spectrometry; instrumental neutron activation analysis (INAA); prompt gamma activation analysis (PGAA); sediment; standard reference material.

*Author to whom all correspondence and reprint requests should be addressed.

Biological Trace Element Research Editor: G. N. Schrauzer © 1990 by The Humana Press Inc.

INTRODUCTION

Certified reference materials (CRMs) are widely used to validate analytical methods and to assess the quality of data generated in programs in which many laboratories and individuals are determining the analytes of interest. For example, the National Oceanic and Atmospheric Administration's (NOAA) National Status and Trends (NS&T) program collects data and analytical results from more than 200 sites in US coastal and estuarine waters (*1*). These data are generated by 10 participating laboratories, and hence need a common base for comparison. Natural matrix certified reference materials (CRMs) have been available since 1971 (*2*), preceded only by such intercomparison materials as Bowen's Kale (*3*). However, these CRMs had been characterized for inorganic trace constituents only. During the past decade, the advent of reliable analytical techniques for the determination of organic trace constituents, particularly those of environmental concern, has led to programs in the CRM-issuing agencies to develop and certify a number of natural materials for organic trace constituents. In the area of marine sediments, the International Atomic Energy Agency (IAEA), Bureau of Reference Communities (BCR), and the National Research Council of Canada (NRCC) have issued several materials. However, for each material, only a limited number of analytes or groups of analytes are normally determined. No single sediment material has been prepared and characterized for polychlorinated biphenyls (PCBs), polycyclic aromatic hydrocarbons (PAHs), chlorinated pesticides, and inorganic constituents.

To meet the need for a single CRM containing all analytes of interest to marine monitoring programs, the National Institute of Standards and Technology (NIST, formerly the National Bureau of Standards) has developed Standard Reference Material (SRM) 1941, Organics in Marine Sediment (*4*). The certificate will provide certified concentrations for 11 PAHs, and non-certified concentrations for 24 PAHs, 15 PCB congeners, and seven chlorinated pesticides. Since trace elements may be correlated with organic pollutant levels and vice versa, it may be useful to know the inorganic constituents in this material. Uncharacterized constituents may also influence results of analytical procedures, depending on their concentration in the sample. In addition, certification-quality inorganic data on this SRM can provide the user with another benchmark sample for monitoring inorganic constituents in a sediment matrix. Also, the homogeneity of the material can be assessed easily with an analytical technique that requires only small samples and virtually no sample preparation. Hence, an NAA approach was used to determine the concentrations of 30 major and trace elements, combining the methods of neutron-capture PGAA and INAA to determine as many elements as possible in a single aliquot.

MATERIALS AND METHODS

SRM 1941 Sample Collection and Preparation

The sediment used for the preparation of SRM 1941 was collected in the Chesapeake Bay, at the mouth of Baltimore (MD) Harbor. The details of the collection and preparation are described by Schantz et al. (*4*). The SRM is provided as a dried, homogenized material. In addition to the SRM 1941 samples, various control materials were included in the different analyses for quality assessment. These included SRM 1645, River Sediment; SRM 1646, Estuarine Sediment; SRM 2704, Buffalo River Sediment; and Certified Reference Material SD—N-115-1/2, Sediment, from the IAEA. Aliquots of 200–300 mg from two bottles of the reference material were used, as received, for the analyses. Samples were pelletized in a KBr pellet press and packaged in Teflon™ film for PGAA or in linear polyethylene film for INAA.

Prompt-Gamma Activation Analysis

The PGAA was performed at the National Bureau of Standards Reactor facility (*5*). The neutron fluence rate at 20 MW reactor power is $3.5 \cdot 10^{12} n \cdot m^{-2} \cdot s^{-1}$. The pelletized samples were positioned so that they were at 45° with reference to the neutron beam. Spectral data were collected with a gamma spectrometer (relative efficiency of 25%; resolution of 2.0 keV full width at half maximum [FWHM] at 1332 keV) combined with a split annulus NaI(Tl) detector. Counting times ranged from 8 to 16 h/sample. Spectral data were collected in the Compton suppressed mode with electronics as described previously (*5*) and were transferred through a Nuclear Data (ND) micro multichannel analyzer (μMCA) interfaced to a VAX 730. Data were stored on disk, and peak searches were done on each spectrum using ND software (*6*), with additional fits using a channel summing program implemented on the VAX.

For quantitation of elemental concentrations, monitor activation analysis was used, with titanium foils counted between samples as flux monitors. Sensitivities for pertinent energy lines of elements (counts $\cdot s^{-1} \cdot mg^{-1}$) were tabulated in a spreadsheet, along with corrections for interferences, background, and enhancement effects (*7*). Peak areas, sample mass, counting time, and appropriate titanium-foil count rates were entered into the spreadsheet, and the concentrations for the listed elements were calculated.

Instrumental Neutron Activation Analysis

The INAA procedure consisted of two irradiations followed by gamma counting after appropriate decay times. The first irradiation was 30 s at a fluence rate of $2 \cdot 10^{17} n \cdot m^{-2} \cdot s^{-1}$ in the RT-4 pneumatic irradiation facility at 15 MW reactor power (*8*). Each irradiation was carried out on a single pellet of SRM 1941, together with one of the standards. All irradiations were completed within a 6-h period with the reactor in thermal equilibrium and no major changes in control settings. Thus, possible fluence rate changes were expected to be much smaller than 1%. Comparison of specific activities in the standards irradiated during this run confirmed the expected stability. This was followed by a 16-h irradiation in the same irradiation facility. Samples, control materials, and standards were irradiated in two sets, packaged in the center of each irradiation container. To compensate for the linear neutron flux gradient along the axis of the container, the irradiation was performed in two equal time intervals of 8 h each, with an 180° inversion of the rabbit at the midpoint of irradiation.

The gamma spectrometry was based on high-resolution germanium spectrometers linked to ND micro multichannel analyzers (μMCAs)/VAX 730 or ND6620 MCAs. Performance criteria of the spectrometers were selected to match the requirements of the complex gamma spectra of the sediments. For the assay of short-lived nuclides, an intrinsic germanium detector (25% relative efficiency and 1.72 FWHM at 1332 keV) was used coupled to an Ortec 572 amplifier and the ND581/ND599 ADC and loss-free counting module, utilizing Westphal's principle of loss correction for a decaying source (*9*). An Ortec 644 fast busy delay aided in the timing for the ND599 module. Count rates up to 70,000 s^{-1} can be handled by the system with a typical detector resolution of 1.95 keV FWHM for 1332 keV. Sample assays were started after the count of the standards, approximately 500 s decay. Counting time for standards and samples was 300 s. The samples were counted twice after the long irradiation: the first count after 5–8 d, and the second count after 7–8 wk. The spectra from the first counts were collected with an intrinsic germanium detector (relative efficiency 28.5% and 1.8 keV FWHM resolution). A Tennelec 244 amplifier with built-in pile-up rejector made the system capable of handling 20000 cts/s without degradation in resolution. This allowed for a high sample throughput using a 4-h counting time for each sample with the help of a sample changer. A Gamma-X spectrometer with a relative efficiency of 7% and 1.65 keV FWHM resolution connected to a ND 6600 system was used for the second set of counts. Additional counts during the first counting cycle were also done using this higher resolution spectrometer to better resolve the complex spectra and to provide control for the quantitation of the shorter lived nuclides.

Quantitative evaluation was accomplished using the comparator method. Comparator standards were prepared from primary solutions of high-purity metals or compounds in ultrapure acids and water. These standards were combined into multielement standards, which were then pipeted onto Whatman filter papers. These were air-dried and then formed into pellets. The VAX computer or ND6600 (as appropriate) was used for data storage and processing, with ND software for peak searches, nuclide identification, decay corrections (using data from ref. *10*), and concentration calculations (*7,11*). Spectral interference corrections were calculated directly for the specific detectors used in counting, whenever possible. Recent work by James (*12*) was used to confirm the best selection of energies for elemental determination.

RESULTS AND DISCUSSION

For the INAA assay of the sediment samples, the optimum performance of the gamma spectrometers was essential. The gamma spectrometers had to have high count-rate capabilities, while maintaining excellent resolution. As an example, the resolution obtained with the Gamma-X spectrometer in the second count of the ^{152}Eu, ^{65}Zn, ^{46}Sc triplet at 1112, 1115.4, and 1120 keV is illustrated in Fig. 1. High count rates caused by major activities from certain nuclides had to be tolerated in each of the counts in order to quantitatively determine elements with lower intensity nuclides of similar or shorter half-lives. An additional advantage of utilizing the high count-rate capabilities of the described system was a considerable time saving in counting, allowing a rapid throughput on the sample changer. This was particularly advantageous for the determination of As, since the 26.3-h half-life of ^{76}As resulted in a significant loss of activity during the first series of counts after the long irradiation. Also, since a second count of the samples for short-lived nuclides, after the short irradiation, would not have yielded significantly improved results for Mn and K, it was worthwhile to handle the high count rate with the loss-free system. However, some additional uncertainty is introduced by the loss-free mode, which has been included in Table 1.

The recommended element concentrations in SRM 1941 were determined by this NAA procedure and are listed in Table 1. Uncertainties represent observed standard deviations of the calculated means. Included are the results for one control material, IAEA-SD-N-1/2, the certified values and confidence intervals of which have been published in the literature (*13*). No significant deviations have been observed for this material, except for Ce, Ag, and Th, which are 5–10% lower, and Mn, which is about 10% higher than the upper limit of the reported confidence interval of the certificate. The result for Se is 15% lower than the lower limit of the reported noncertified confidence interval; however, no such bias was apparent in the other reference materials that were analyzed.

The homogeneity of this material, as determined for inorganic elements, was very good; observed SD of 1% were found for the major

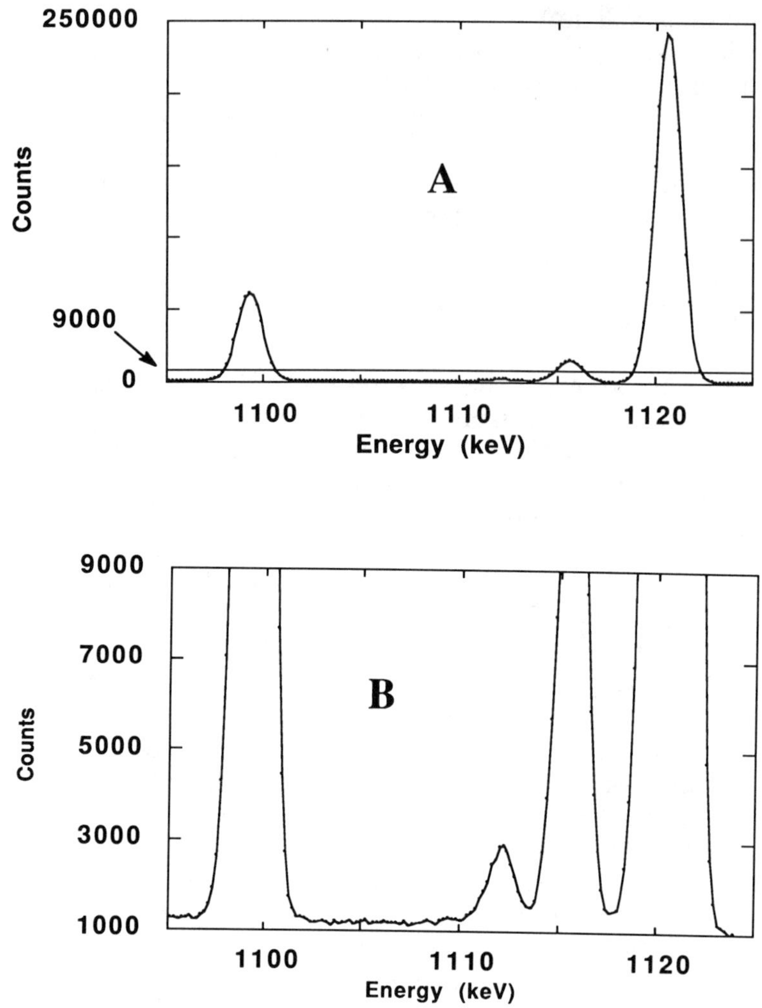

Fig. 1. A: Full scale (linear) view of a spectrum region from a SRM 1941 sample showing the Eu-Zn-Sc triplet at 1112-1120 keV.
B: Expanded view of the same region showing the resolution between the three peaks.

constituents, such as Fe and Cr, for Sc, and for most of the rare earth elements (REE), in aliquots of 200–250 mg. For example, the range of values for Sc was 32.56–33.57 mg/kg, with an observed SD of 0.37 (1.1%) and a reduced chi-squared (X^2) of 0.928. This indicates that most of the variation was caused by the counting error. The ranges for As and Sb were slightly larger. With similar counting statistics, the reduced X^2 for As and Sb were 9.4 and 12.0, respectively. This indicates probable inhomogeneity. These elements and others, such as Se and Ag, which show larger uncertainties, are characteristic of biological and anthro-

Table 1
Instrumental Neutron Activation Analysis Results.
Element Concentrations in mg/kg, Dry Weight Basis

Element	Gamma Energy Used	SRM 1941 Mean	SRM 1941 Uncertainty[2]	IAEA-SD-N-1/2 Value	IAEA-SD-N-1/2 Uncertainty[3]
B[1]	477	75.5	1.7	87.7	1.5
Na	1368, 2754	12910	290	10740	120
Al	1778	64800	2400	37290	650
Si[1]	3539, 4933	222000	8000	272000	11000
S[1]	841	16410	840	3000	1000
Cl	1642, 2167	16400	400	9460	120
K[1]	770	15800	120	15200	580
Sc	889	34.34	0.42	6.913	0.060
Ti[1]	320	17220	310	2812	93
V	1434	812	31	79.0	1.3
Cr	320	634.5	9.7	150.8	2.4
Mn	846	787.8	9.5	856.1	5.7
Fe	1099,1292	105400	1000	36230	380
Co	1173,1332	27.48	0.14	11.455	0.074
Zn	1115	1012	43[4]	437.6	6.5
As	559	75.4	4.0	56.30	0.44
Se	264	10.11	0.50	1.76	0.27
Rb	1076	92.1	1.3	69.6	2.2
Ag	657	1.24	0.47	1.54	0.25
Sb	1691	15.20	0.42	3.55	0.061
Cd[1]	558	2.32	0.29	12.30	0.52
Cs	795	4.78	0.13	4.36	0.10
La	1596	359	12	29.11	0.29
Ce	145	271.8	3.6	54.77	0.77
Sm[1]	333	25.69	0.40	5.52	0.20
Eu	778	2.184	0.060	1.065	0.042
Tb	298	2.15	0.14	0.669	0.0010
Gd[1]	182	15.16	0.37	5.43	0.35
Hf	482	22.43	0.29	8.48	0.12
Ta	1221	16.43	0.45	0.990	0.12
Th	311	25.63	0.25	6.73	0.10
U	(1596)	(22)		≤6.6	

[1]PGAA.
[2]$ts/\sqrt{n}$ for $P = 0.05$.
[3]1s counting statistics.
[4]Includes an allowance for uncertainties in peak integration.

pogenic sources. Such elements as As and Se may occur with the organic compounds in the sediment, hence the observed variability may be an indication of the homogeneity of the organic compounds.

The composition of this material appears unusual in many respects, and is discussed in detail elsewhere (*4*). The material contains high levels of S, Fe, Ti, Sc, and many of the REE, compared to other sediment reference materials; e.g., many of the REE had concentrations higher than other similar reference materials by a factor of 10 or more. This unusual composition in addition to the normal complexity of a sediment spectra meant that great care was necessary both during the analysis and when considering interferences, especially those involving the REE.

Some elements, such as Na, Cl, K, Ti, Cd, and Sm, could be obtained by both INAA and PGAA. The results for Na, Cl, Ti, and Sm were comparable, but again, because of the complex matrix, one technique generally obtained more precise results than the other. The more precise results were reported in Table 1. For K, Cd, and Sm, the PGAA results were used.

The characterization of inorganic constituents in SRM 1941, Organics in Marine Sediment, has proven to be useful, both for elemental analyses, and for added information that has proven to be helpful for the organic analyses. For example, knowing about the high sulfur content of SRM 1941 resulted in the development of an additional clean-up procedure to accomplish the organic analyses. The material will also be a useful control material in the inorganic analyses of samples from environmental monitoring programs, providing a probable upper boundary for such samples.

ACKNOWLEDGMENTS

The authors acknowledge the assistance of D. L. Anderson from the Food and Drug Administration for his advice in the analysis of the PGAA data. Certain commercial equipment, instruments, or materials are identified in this paper in order to specify the experimental procedure adequately. Such identification does not imply recommendation or endorsement by the National Institute of Standards and Technology, nor does it imply that the materials or equipment identified are necessarily the best available for the purpose.

REFERENCES

1. G. G. Lauenstein and J. A. Calder, *Progress in Environmental Specimen Banking*, S. A. Wise, R. Zeisler, and G. M. Goldstein, eds. NBS Spec. Publ. 740, US Government Printing Office, Washington, DC 1988, 19–30.
2. SRM 1571, Orchard Leaves, *Certificate*, National Bureau of Standards, Gaithersburg, MD 20899 (1981).
3. H. J. M. Bowen, *Proceedings of the Society, Analytical Chemistry Conference, Nottingham*, P. W. Shallis, ed; W. Heffer, Cambridge, UK 1965, 25.
4. M. M. Schantz, B. A. Benner, Jr., S. N. Chesler, B. J. Koster, S. F. Stone, R. Zeisler, and S. A. Wise, submitted to *Fresenius Z. Anal. Chem.*
5. D. L. Anderson, M. P. Failey, W. H. Zoller, W. B. Walters, G. E. Gordon, and R. M. Lindstrom, *J. Radioanal. Nucl. Chem.* **63**, 97 (1981).
6. VAX/VMS NAA package, Nuclear Data, Jan. 1985 with updates through Aug. 1988.
7. D. L. Anderson, Y. Sun, M. Failey, and W. H. Zoller, *Geostandards Newsletter* **9**, 219 (1985).
8. D. A. Becker, *J. Radioanal. Nucl. Chem.* **110**, 393 (1987).
9. G. P. Westpahl, *J. Radioanal. Chem.* **70**, 387 (1982); Austrian Patent 368,291; US Patent 4,476,384.

10. From Nuclear Properties spreadsheet by C. A. Stone, using data extracted from the NUDAT database at the National Data Center at Brookhaven National Laboratory (Feb. 1988).
11. T. Meyers, *Neutron Activation Analysis Package Operational Instruction Documentation,* Nuclear Data, 1978.
12. D. J. James and P. N. Boothe, *J. Radioanal. Nucl. Chem.* **123,** 295 (1988).
13. *Certificate for IAEA/Sediment SD-N-1/2,* International Atomic Energy Agency, International Laboratory of Marine Radioactivity, Musee Oceanographique, 98000, Monaco, 1985.

Applicability of Microwave Acid Digestion to Sample Preparation of Biological Materials for Analysis by Particle-Induced X-Ray Emission (PIXE)

T. PINHEIRO,[1] H. DUFLOU, AND W. MAENHAUT*

Instituut voor Nucleaire Wetenschappen, Rijksuniversiteit Gent, Proeftuinstraat 86, B-9000 Gent, Belgium; and [1]IAEA Fellow, Permanent affiliation—LNETI, Physics Department, E.N. 10, P-2685 Sacavem, Portugal

Received April 17, 1989; Accepted December 11, 1989

ABSTRACT

A microwave acid digestion method for the preparation of biological samples for PIXE analysis is presented. The precision and accuracy of the entire PIXE analytical procedure, including the microwave digestion step, were evaluated by analyzing eight certified reference materials. For elements heavier than K, and for concentration levels from 2 μg/g upward, the total random error of a single analysis is in the range of 2–5%. The accuracy is better than 5%. The detection limits are down to 0.3 μg/g.

Index Entries: Particle-induced X-ray emission; PIXE; sample preparation; microwave digestion; acid digestion; trace elements; reference materials; precision; accuracy; detection limits.

INTRODUCTION

Particle-induced X-ray emission (PIXE) is a powerful analytical technique that allows for simultaneous determination of sub-μg/g levels of

*Author to whom all correspondence and reprint requests should be addressed.

several elements in small biological samples (*1*). Consequently, numerous papers have been devoted to the use of PIXE for analyzing biological and biomedical materials (*2*). It became clear that the sample and target preparations are critical steps in the analysis of such sample types. Therefore, considerable efforts have been made to develop and improve appropriate sample and target preparation procedures. They involve, e.g., making thick target pellets from the dry biological material, powdering the freeze-dried sample by means of the brittle-fracture technique, low- and high-temperature ashing, and wet digestion in an open or closed Teflon™ vessel (*3–10*). However, most published methods are rather time-consuming, and in several of them there is serious risk of contamination.

Recently, acid digestion in a microwave oven has been introduced as an elegant and fast procedure for decomposing and dissolving various materials for trace-element analysis using atomic absorption and emission techniques (*11*). In the present work, we investigated the applicability of microwave digestion to the PIXE analysis of animal and plant tissues. In order to assess the precision and accuracy of the method, eight certified reference materials were analyzed, and the PIXE results were compared with the certified values. The detection limits obtainable when using the microwave digestion method were also determined.

EXPERIMENTAL

Microwave Digestion Procedure

The digestion experiments were carried out with an Amana microwave oven normally used for domestic purposes (Model RS 560A, Amana Refrigeration, Amana, IA). The oven is equipped with a rotating antenna which guarantees a homogeneous distribution of the microwave radiation. The oven power (max. 700 W) and time can be selected with 10% and 1 s increments, respectively. All digestions were done in closed Teflon™ vessels of 60 mL capacity, each provided with a threaded cover (Model 561R2, Savillex Corporation, Minnetonka, MI). In all cases, 100–200 mg subsamples were used that had been dried overnight at 75°C. As a digesting solvent, nitric acid was exclusively employed. Furthermore, the weight fraction of the dry biological material in the digestion mixture was always kept equal to 10%. Concentrated digestion solutions are required if one wants to minimize the amount of solution to be pipeted during target preparation and still have enough biological material on the target.

Several parameters were examined in order to obtain a general digestion procedure for different types and amounts of biological materials and for a varying number of vessels in the oven. These parameters were the oven power, the digestion time, and the acid concentration. In the procedure finally established, 1 mL of high-purity nitric acid (14*M*) was added per 100 mg of dry biological sample in the Teflon™ vessel. The nitric acid contained Y (9 μg/mL) as internal standard in the PIXE analysis. The threaded cover was screwed onto the vessel using a constant-torque wrench (torque of 15 Nm). Up to four vessels were placed inside the oven, together with a "safety" vessel filled with 7.5 mL of vegetable oil/100 mg of sample in the oven. The purpose of the oil is to absorb the excess microwave power and to prevent damage to the antenna. The oven was then operated at 50% full power for a period linearly related to the amount of biological material in the oven, i.e., 5 min/100 mg of the total amount of biological sample.

Sample and Target Preparation

The microwave digestion procedure was applied to eight certified reference materials. Six certified reference materials came from the National Institute of Standards and Technology (NIST, formerly NBS), i.e., Spinach (SRM 1570), Orchard Leaves (SRM 1571), Citrus Leaves (SRM 1572), Tomato Leaves (SRM 1573), Pine Needles (SRM 1575), and Bovine Liver (SRM 1577a). The two other reference materials were from (a) the European Community Bureau of Reference (BCR), i.e., Milk Powder (BCR 63); and (b) the International Atomic Energy Agency (IAEA), i.e., Animal Muscle (H4). From each material, four to eight subsamples were taken and subjected to the digestion procedure.

PIXE targets were prepared by pipeting 10 μL of the digestion solution onto a 1.5-μm thick Kimfol™ polycarbonate film, mounted on a 25-mm diameter polyethylene target frame. The film had been treated with nitric acid (14*M*) and a 0.05% polyvinylpyrrolydon solution to make it hydrophylic. The targets were dried in a vacuum dessicator. Typically, four targets were prepared and analyzed from each digested sample. The sample mass on the target film corresponded to about 50% of the mass of the original biological material and was virtually the same for Bovine Liver as for the plant materials. This mass-reduction factor of 2 in our microwave digestion procedure can be compared to the factor of about 4–5 obtained for the same materials in Teflon™ bomb digestion in a regular oven (*10,12*).

PIXE Analysis

All analyses were carried out with the experimental PIXE set-up of the University of Ghent (*4,13*). The bombardments were done in vacuo using a proton beam of 2.4 MeV with a 0.5-cm^2 area at the target. A 660-μm thick Mylar™ absorber was placed in front of the Si(Li) detector.

The beam current was typically 150 nA, and the X-ray count rate was always lower than 1500 cps. The preset charge for each target bombardment was 90 μC. The PIXE spectra were analyzed by means of the nonlinear least-squares fitting program AXIL-84 (*14*). Corrections for proton energy degradation and X-ray attenuation in the target (matrix effects) were carried out as described by Maenhaut et al. (*9*).

RESULTS AND DISCUSSION

Detection limits for the entire PIXE analytical procedure, including the microwave acid digestion step, were calculated from PIXE spectra of actual sample targets. The detection limits were based on the 3(background)$^{1/2}$ criterion, where the background was summed over a spectral region of 3 FWHM at the principal X-ray line of each element. The detection limits for NIST Bovine Liver are given in Table 1. They are expressed in μg/g original dried material and apply to a preset charge of 100 μC. A comparison is made with the corresponding detection limits obtained after Teflon™ bomb acid digestion in a regular oven (*10,12*). Because of the mass-reduction factor in our microwave digestion procedure being smaller than in the Teflon™ bomb method (*see above*), somewhat higher detection limits are observed after microwave digestion. For the energy region of the PIXE spectrum up to about 5 keV, where the secondary electron bremsstrahlung background is particularly important, the difference between the detection limits in both methods is in the range of 20–40%. For the spectral region corresponding to the K X-ray lines of Cu to Y, a difference of only about 10% is observed. When considering all elements listed in Table 1, the detection limits after microwave digestion are on the average 18% higher than after Teflon™ bomb digestion.

Up to 18 elements were observed above the detection limit (and could thus be quantified) in the various certified reference materials. These elements were K, Ca, Ti, V, Cr, Mn, Fe, Ni, Cu, Zn, As, Se, Br, Rb, Sr, Mo, Ba, and Pb. However, the PIXE results for Br were only between 8 and 26% of either the reference values or the data given by Gladney et al. (*15*). This indicates that this element is lost to a considerable and varying extent while opening the Teflon™ vessels after the acid digestion or during the preparation and drying of the targets. Therefore, bromine will not be discussed further. The concentration data for the other elements were first averaged over the four different targets of the same subsample, and the means were then averaged over the four to eight subsamples. Standard deviations were calculated for the first and second averages, and they are indicated by s_t and s_p, respectively. These standard deviations were compared with those expected from counting statistics (s_c). Tables 2 and 3 present the PIXE concentration data and the various standard deviations for two of the eight reference materials

Table 1
Detection Limits[a] in PIXE Analysis of NIST Bovine Liver[b]

Element	Microwave Oven (closed vessel)	Regular Oven (Teflon bomb)
K	73	52
Ca	12	9.5
Ti	2.0	1.6
V	1.1	0.89
Cr	0.71	0.58
Mn	0.52	0.43
Fe	0.42	0.34
Ni	0.31	0.27
Cu	0.31	0.27
Zn	0.29	0.26
As	0.32	0.29
Se	0.31	0.29
Br	0.32	0.30
Rb	0.38	0.36
Sr	0.44	0.41
Y	0.53	0.48
Mo	1.0	0.84
Ba (L)	7.9	6.1
Pb (L)	0.88	0.81

[a]On a μg/g freeze-dried material basis.

[b]Comparison of values obtained after microwave acid digestion in a closed Teflon™ vessel with those obtained after Teflon™ bomb acid digestion in a regular oven. (PIXE experimental conditions: proton beam energy 2.4 MeV, 660 μm Mylar™ absorber in front of Si(Li) detector, preset charge 100 μC.)

analyzed, i.e., for NIST Orchard Leaves (SRM 1571) and NIST Bovine Liver (SRM 1577a).

For all materials and elements, s_t was generally of the same order as s_c, even for the lowest s_c values (which were about 1%). This indicates that the digestion solution is homogeneous, and that the sample deposits on the targets have a similar mass thickness (important for the matrix

Table 2
Concentration Data Obtained by PIXE After Acid Digestion of the Sample in a Closed Vessel in a Microwave Oven[a]

	Overall PIXE Average				
	$\bar{x} \pm s_p$	$\%s_p$	$\%s_t$	$\%s_c$	Ds_ps_t
Element	N = 5		Nn = 20		
K	14200 ± 200	1.4	1.7	2.0	1.1
Ca	21300 ± 300	1.4	1.8	1.8	1.1
Ti	22 ± 4	18.2	15.4	5.1	16.5
Cr	2.5 ± 1.0	40.0	15.5	11.5	39.2
Mn	91 ± 2	2.2	1.6	1.9	2.1
Fe	274 ± 14	5.1	4.1	1.8	4.7
Ni	1.40 ± 0.18	12.9	9.9	8.9	11.9
Cu	13.2 ± 0.3	2.3	3.0	2.5	1.7
Zn	25.0 ± 0.5	2.0	2.7	2.3	1.4
As	11.9 ± 0.2	1.7	6.2	5.0	
Rb	11.5 ± 0.3	2.6	6.2	4.7	
Sr	36.2 ± 0.7	1.9	3.5	3.0	0.8
Ba	38 ± 1	2.6	7.6	7.4	
Pb	47 ± 1	2.1	3.4	3.2	1.2

For the explanation of s_p, s_t and s_c, *see* text.
N: Number of subsamples.
Nn: Number of targets analyzed, i.e., $N \times 4$.
$Ds_ps_t = 100 \times (1/\bar{x}) \times [s_p^2 - s_t^2/n]^{1/2}$, where $\bar{x}$ is the mean elemental concentration and $n = 4$, the number of targets analyzed.
[a]Results for NIST Orchard Leaves (SRM 1571) in µg/g. Overview of the precision evaluation.

correction). The standard deviation s_p, on the other hand, tended to be systematically higher than the expected value $s_c/\sqrt{n}$, with $n = 4$. This may indicate that there is a contribution to s_p from the inhomogeneity of the reference material on a 100–200-mg scale, and from the random error in the concentration of the internal standard (from weighing error, pipeting error, and irreproducible retention of such volatiles as water after drying at 75°C). By averaging over all elements with a percentage s_c ($\%s_c$) better than 5% and over the eight reference materials, an overall mean value for the residual standard deviation $[s_p^2 - s_c^2/n]^{1/2}$ of 2.0% was calculated (av-

Table 3
Concentration Data Obtained by PIXE After Acid Digestion of the Sample in a Closed Vessel in a Microwave Oven[a]

Element	Overall PIXE Average $\bar{x} \pm s_p$ N = 5	$\%s_p$	$\%s_t$ (Nn = 20)	$\%s_c$ (Nn = 20)	Ds_ps_t
K	10600 ± 300	2.8	1.7	2.0	2.7
Ca	137 ± 9	6.6	6.5	8.5	5.7
Mn	10.7 ± 0.1	0.9	2.3	2.9	
Fe	202 ± 3	1.5	1.4	1.8	1.3
Cu	166.7 ± 1.3	0.8	1.3	1.8	0.4
Zn	130.0 ± 0.8	0.6	1.8	1.9	
Se	0.68 ± 0.05	7.4	18.3	17.6	
Rb	13.1 ± 0.4	3.1	3.8	4.4	2.4
Mo	3.9 ± 0.2	5.1	19.1	14.4	

For the explanation of s_p, s_t, s_c, N, Nn, and Ds_ps_t, *see* footnotes of Table 2.
[a]Results for NIST Bovine Liver (SRM 1577a) in μg/g. Overview of the precision evaluation.

erage of 59 values; outlying values for Ti in SRMs 1570, 1571, 1573, and 1575 were excluded). It can thus be concluded that the total random error of the results from a single target analysis is in the range of 2–5%, when the error from counting statistics is < 5% (this corresponds roughly to concentrations from 2 μg/g up).

The accuracy of the entire PIXE analysis procedure (comprising calibration, beam-handling, sample/target preparation, specimen homogeneity, contamination control, spectral data processing, and correction for matrix effects) was evaluated by comparing the overall means with the certified values. Only data with good ($\%s_c < 5\%$) or acceptable ($\%s_c <$ 15%) error from counting statistics were retained in this comparison. For each element the absolute values of the percentage difference, relative to the certified value, were averaged over the eight certified reference materials. The results are presented in Table 4. The mean absolute percentage difference is generally better than 7%, except for Cr and Ni. However, the PIXE data for those two elements exhibited only acceptable precision from counting statistics ($\%s_c$: 5–15%). In addition, at levels of a few micrograms/gram, Cr PIXE data could suffer from systematic error as

Table 4
Comparison Between Certified Values and PIXE Data[b]
for 16 Elements in 8 Reference Materials

Element	Number of certified values	Average absolute % difference, relative to certified value
K	7	4.3 (7)[a]
Ca	7	7.2 (7)
Cr	5	[27 (4)]
Mn	7	3.1 (6)
Fe	8	6.9 (8)
Ni	2	[20 (2)]
Cu	8	4.3 (8)
Zn	6	3.1 (6)
As	6	8.2 (2)
Se	3	[5.0 (1)]
Br	1	
Rb	7	4.0 (7)
Sr	6	2.1
Mo	3	[12 (1)]
Ba	1	
Pb	7	1.9 (4)

[a]The numbers in parentheses designate on how many values the averages are based; brackets are used to indicate that all PIXE data for the certified element had $\%s_c$ values in the range of 5–15%.

[b]With $\%s_c < 15\%$.

a result of the spectrum analysis (*16*). The certified values for Ni had a very high uncertainty, especially in Citrus Leaves (SRM 1572). If both elements are excluded, the overall mean absolute difference is 4.6%. The analytical procedure presented here can thus provide data accurate to better than 5%.

The results clearly indicate that microwave acid digestion is a valuable method for the preparation of biological materials for subsequent PIXE analysis. Compared to Teflon™ bomb digestion in a regular oven, microwave digestion is much faster, whereas it provides essentially the same precision and accuracy, and only slightly higher detection limits. It still has to be investigated whether the procedure can also be applied to the digestion of milligram amounts of biological samples for subsequent analysis by PIXE. At present, experiments to examine this are in progress.

ACKNOWLEDGMENTS

T. P. gratefully acknowledges the International Atomic Energy Agency for granting her a fellowship. H. D. and W. M. thank the Belgian Nationaal Fonds voor Wetenschappelijk Onderzoek and the Interuniversitair Instituut voor Kernwetenschappen for research support. J. Cafmeyer is thanked for technical assistance.

REFERENCES

1. S. A. E. Johansson and J. L. Campbell, *PIXE: A Novel Technique for Elemental Analysis*, Wiley, Chichester (1988).
2. W. Maenhaut, *Nucl. Instr. Methods* **B35,** 388 (1988).
3. N. F. Mangelson, M. W. Hill, K. K. Nielson, and J. F. Ryder, *Nucl. Instr. Methods* **142,** 133 (1977).
4. W. Maenhaut, L. De Reu, H. A. Van Rinsvelt, J. Cafmeyer, and P. Van Espen, *Nucl. Instr. Methods* **168,** 557 (1980).
5. J. Pallon and K. G. Malmqvist, *Nucl. Instr. Methods* **181,** 71 (1981).
6. W. Maenhaut, L. De Reu, and J. Vandenhaute, *Nucl. Instr. Methods* **B3,** 135 (1984).
7. J. Galuszka, L. Jarczyk, E. Rokita, A. Strzalkowski, and M. Sych, *Nucl. Instr. Methods* **B3,** 141 (1984).
8. E. Clayton, *Nucl. Instr. Methods* **B22,** 145 (1987).
9. W. Maenhaut, J. Vandenhaute, and H. Duflou, *Fresenius Z. Anal. Chem.* **326,** 736 (1987).
10. H. Duflou, W. Maenhaut, and J. De Reuck, *Biol. Trace Elem. Res.* **13,** 1 (1987).
11. H. M. Kingston and L. B. Jassie, *Introduction to Microwave Sample Preparation*, ACS Professional Reference Book, Washington (1988).
12. H. Duflou, PhD thesis, University of Ghent, 1988.
13. W. Maenhaut and H. Raemdonck, *Nucl. Instr. Methods* **B1,** 123 (1984).
14. W. Maenhaut and J. Vandenhaute, *Bull. Soc. Chim. Belg.* **95,** 407 (1986).
15. E. S. Gladney, C. E. Burns, D. R. Perrin, I. Roelandts, and T. E. Gills, *1982 Compilation of Elemental Concentration Data for NBS Biological, Geological, and Environmental Standard Reference Materials*, NBS (US) Spec. Publ. 260–88 (1984).
16. J. L. Campbell, W. Maenhaut, E. Bombelka, E. Clayton, K. Malmqvist, J. A. Maxwell, J. Pallon, and J. Vandenhaute, *Nucl. Instr. Methods* **B14,** 204 (1986).

Trace Element Composition of Biological Materials

New Essential Trace Elements for the Life Sciences

FORREST H. NIELSEN

US Department of Agriculture, Agricultural Research Service, Grand Forks Human Nutrition Research Center, Grand Forks, ND 58202

Received April 17, 1989; Accepted December 11, 1989

ABSTRACT

The possible importance of some new essential trace elements in nutrition is discussed. Most likely, insufficient intake of a specific trace element becomes obvious only when the body is stressed in some way that enhances the need for that element; this has been supported by recent findings with selenium. The trace elements boron and copper may be of nutritional significance in a manner similar to selenium. When the diets of animals and humans are manipulated to cause possible changes in cellular integrity or in hormone responsiveness, a large number of responses to dietary boron occur. The findings indicate that boron is important for optimal calcium and, thus, bone metabolism. High dietary cystine and fructose exacerbate the signs of copper deficiency in rats; this indicates that the response to copper deficiency by humans would vary with the amino acid and carbohydrate composition of the diet. There is some evidence that chromium, molybdenum, nickel, arsenic, and vanadium may also be of nutritional significance under stress conditions. In other words, an increasing number of studies have been performed that have examined the importance of trace element nutriture in various forms of nutritional, metabolic, hormonal, or physiologic stress in animals and humans. These studies indicate that situations will be found in which a trace element is of nutritional significance. It is likely that some of the trace elements are more important in human nutrition than is now generally acknowledged.

Index Entries: Trace elements; boron, copper, selenium, chromium, molybdenun, nickel, arsenic, vanadium.

INTRODUCTION

At present, only seven trace elements have defined essential functions in humans. These elements are cobalt, copper, iodine, iron, molybdenum, selenium, and zinc. Essential functions have been identified for manganese in animals, but not in humans. Signs of chromium deficiency have been described for humans, but a specific biochemical role for chromium has not been demonstrated conclusively. A number of other elements in addition to the aforementioned nine elements have been suggested to be essential nutrients, including arsenic, boron, bromine, cadmium, fluorine, lead, lithium, nickel, silicon, tin, and vanadium. Deficiencies of only four elements—cobalt as vitamin B-12, iodine, iron, and zinc—occur with known sufficient frequency in humans to be of concern to health professionals. Nonetheless, the trace elements are often suspected of being the missing link in some of the unexplained human diseases, such as atherosclerosis, osteoporosis, osteoarthritis, hypertension, and ischemic heart disease. Efforts to demonstrate that trace element deficiencies are the missing links generally have been unsuccessful. In the following, it is suggested that perhaps some of the failures have occurred because the experimental approach has not been correct in the past. Recent studies examining the need for various trace elements by animals under some form of nutritional, metabolic, hormonal, or physiologic stress have indicated that these are situations in which some of the trace elements may be of nutritional significance.

Factors Affecting Trace Element Requirements

Although trace elements play key roles in a variety of processes necessary for life, the occurrence of overt simple or uncomplicated deficiencies of any of the trace elements is probably relatively uncommon because of powerful homeostatic mechanisms that the human body possesses. However, there are situations that may make a trace element nutritionally significant. These include:

1. Inborn error of metabolism that affects absorption, retention, or excretion
2. Alterations in metabolism and/or biochemistry as a secondary consequence to malnutrition, disease, injury, or stress
3. Marginal deficiencies (slight deviation from an optimal intake of an essential nutrient) induced by various dietary manipulations or by direct or indirect interaction with another nutrient or drug and
4. The enhanced requirement for a trace element caused by a sudden or severe change in the system requiring that element.

The preceding probably can be summarized by the statement that the insufficient intake of a specific trace element becomes obvious only when the body is stressed in some way that enhances the need, or interferes with the utilization, of that element.

Recently, Tapp and Natelson (*1*) presented the formula:

$$\text{Pathological Effects} = \text{Stress} \times \text{Organic Vulnerability}$$

This formula seems quite applicable to trace element nutrition. In other words, pathological effects are not likely to be seen if a trace element deficiency (organic vulnerability) is not multiplied by some significant stress. Likewise, pathological effects are not likely to be large if stress is not accompanied by an organic vulnerability or an inappropriate lack of a trace element. However, the multiplication of a suboptimal intake of a specific trace element times the presence of some nutritional, metabolic, hormonal, or physiologic stress affected by that element most likely would lead to serious pathological consequences. The preceding concept is supported by findings with selenium.

Selenium was first shown to be nutritionally important by using vitamin E-deficient animals (*2*). Close examination of published data indicates that a very limited number of deficiency signs are caused exclusively by selenium deficiency; most signs appear when vitamin E or antioxidant metabolism is suboptimal (*3,4*). Human diseases involving selenium apparently are not simple selenium deficiencies (*5*). For example, it has been suggested that Keshan disease, which responds to selenium supplementation, also involves another factor. Suggested possibilities include various toxins, hypoxia, or infectious agents, particularly viruses (*6*).

Examining the possibility that other trace elements are important in human nutrition in a manner similar to selenium has revealed several candidates of potential nutritional concern, including boron, chromium, copper, molybdenum, nickel, arsenic, silicon, and vanadium. This is especially true for boron and copper, which will be discussed most extensively here.

NUTRITIONAL SIGNIFICANCE OF BORON

Animal Studies

Seventy years after boron was first suggested to be essential for plants (*7*), an experiment was conducted indicating that boron might be essential for chicks. Hunt and Nielsen (*8*) reported that boron deprivation depressed growth and elevated plasma alkaline phosphatase activity in chicks fed inadequate cholecalciferol. Subsequent experiments suggested that cholecalciferol deficiency enhanced the need for boron and that boron might interact with cholecalciferol metabolism, which in turn affected calcium, phosphorus, or magnesium metabolism (*9*). Since those experiments, it has been found that the response to changes in dietary boron is markedly influenced by dietary methionine, potassium, magnesium, cholecalciferol, aluminum, and calcium (*10–14*). For example, in weanling rats fed a casein-based diet not supplemented with methionine, but containing luxuriant amounts of arginine (probably methionine-deficient) and marginal potassium, the interaction between magnesium and boron was different from that seen in rats fed optimal amounts of methionine, arginine, and potassium. In the former instance, the interaction was characterized by the deprivation of either boron or magnesium making the deficiency signs (e.g., depressed growth) of the other more marked (*10*). In the latter case, magnesium deprivation similarly affected boron-deprived and boron-supplemented rats (*12*).

Generally, the preceding studies have shown that, when the diet was manipulated to cause possibly changes in cellular membrane integrity (potassium or magnesium deficiency) or in hormone responsiveness (magnesium or cholecalciferol deficiency, aluminum toxicity), a large number of responses to dietary boron occurred. On the other hand, when the animal was fed a diet apparently optimal in all respects, the response to dietary boron was not very marked. These findings suggested that the need for boron was not crucial, or was quite low, when the animal was not under any nutritional or metabolic stress, but that there was an enhanced need for boron when the animal needed to respond to a stressful situation that adversely altered hormonal or cellular membrane status.

Human Studies

Studies with humans also indicate that boron may be of nutritional concern under certain metabolic or nutritional stressful situations, for example, with a low dietary intake of magnesium, or when hormonal changes occur (menopause), which causes an increased loss of calcium from bone. A study done with 12 postmenopausal women housed in a metabolic unit showed that dietary boron markedly affected several indices of mineral metabolism (*15*). After the women had consumed a conventional diet supplying about 0.25 mg boron/d for 119 d, they were given a boron supplement of 3 mg/d for 48 d. As shown in Table 1, boron supplementation markedly reduced the urinary excretion of calcium; the depression seemed more marked when dietary magnesium was low. Boron supplementation also markedly elevated the serum concentrations of estradiol-17β and, as shown in Table 1, ionized calcium; the elevation seemed more marked in the magnesium-low women.

Recently, another study with five men over the age of 45, five postmenopausal women on estrogen therapy, four postmenopausal women not on estrogen therapy, and one premenopausal woman fed a magnesium-low diet (115 mg/2000 kcal) gave findings indicating that boron affects calcium metabolism (*16*). After the subjects had consumed a

Table 1
Effect in Postmenopausal Women of Boron and Aluminum on Plasma Ionized and Urinary Calcium[a]

Dietary Treatment		Low-Mg Diet		Adequate-Mg Diet	
B	Al	Plasma iCa	Urinary Ca	Plasma iCa	Urinary Ca
mg/day	mg/day	%	g/24 hr	%	g/24 hr
0.25	0	49.7	0.117	49.5	0.132
0.25	1000	49.1	0.124	49.1	0.128
3.25	0	52.1	0.065	51.0	0.104
3.25	1000	51.6	0.073	51.5	0.113
		Analysis of Variance - p Values			
Boron		0.001	0.0001	0.002	0.02
Aluminum		NS	NS	NS	NS
Boron x Aluminum		NS	NS	NS	NS

[a]Experimental conditions described in reference (*15*).

conventional diet supplying about 0.23 mg boron/2000 kcal for 63 d, they were given a boron supplement of 3 mg/d for 49 d. When the last 42 d of depletion and the last 35 d of repletion were compared, several indicators of calcium status were found to be significantly different ($p < 0.05$). As shown in Table 2, plasma ionized calcium was increased and serum osteocalcin was decreased by the boron supplementation. When all 15 individuals were used in the comparisons, serum 25-hydroxycholecalciferol (29.1 vs 32.3 ng/mL) was lower, and serum calcitonin (74.1 vs 59.0 pg/mL) and glucose (93 vs 88 mg/dL) were higher during boron depletion than during boron repletion.

Table 2 shows that the subjects who were receiving estrogen therapy to prevent calcium loss from bone had the highest plasma ionized calcium and lowest serum osteocalcin. Boron supplementation after 63 d of boron depletion tended to make the values of the other two groups more similar to those of the women receiving estrogen therapy. These types of changes were the same for serum 25-hydroxycholecalciferol, calcitonin, and glucose. Thus, the boron supplementation was construed as being beneficial to calcium metabolism. In other words, boron probably has an important role in the maintenance of normal bones. Moreover, this role becomes more apparent under conditions in which increased calcium loss from the bone is quite likely.

Table 2
Effect of Boron on Plasma Ionized Calcium in Individuals Fed a Low Magnesium Diet[a]

Dietary Boron	Plasma iCa		Serum Osteocalcin
mg/day	mg/dL	%	ng/mL
Men over age 45 (n = 5)			
0.25	4.86	52.1	3.7
3.25	4.92	53.2	3.6
p values	0.06	0.004	0.74
Postmenopausal women (n = 4)			
0.25	4.90	52.4	3.8
3.25	4.97	53.6	3.5
p values	0.008	0.06	0.58
Postmenopausal women on estrogen therapy (n = 5)			
0.25	4.95	53.2	2.8
3.25	5.04	54.3	1.8
p values	0.02	0.02	0.08
Above combined plus one premenopausal woman (n = 15)			
0.25	4.91	52.6	3.3
3.25	4.98	53.8	2.8
p values	0.0001	0.0001	0.06

[a]Brief description of experimental conditions in text and in reference (*16*).

NUTRITIONAL SIGNIFICANCE OF COPPER

Animal Studies

Copper deficiency in experimental animals causes many sequelae (*17,18*). A number of abnormalities seem to have implications for human health and include impairment of glucose metabolism, abnormal lipid metabolism and hypercholesteremia, cardiac necrosis, dissecting aneurysms, osteopenia and bone fracture, impaired myelination, anemia and leukopenia, and congenital anomalies. When rats die from copper deficiency, the most common cause is sudden death through structural failure of the heart or arteries.

Recently, a number of dietary factors were found to affect the nature and severity of the signs of copper deficiency. These include sex and genetic makeup of the animal (*19–23*), and dietary carbohydrate and sulfur amino acid content (*22,23*).

Fields and coworkers (*19,20*) have shown that the use of fructose instead of starch as the dietary source of carbohydrate markedly increased the severity and extent of copper deficiency signs in rats. They also reported (*19,20*) that male Sprague-Dawley rats fed a copper-deficient high-fructose diet displayed anemia, hypertrophic hearts, and mortality; female rats fed the same diet did not.

Examples of how genetic makeup, sex, and dietary sulfur amino acids affect copper deficiency are shown in Table 3. With males, regardless of copper status, a supplement of 6 or 12 mg of methionine/g diet elevated plasma cholesterol. In these groups, copper deficiency did not cause a noteworthy additional increase in plasma cholesterol in the male Sprague-Dawley rats, but did cause an apparent increase in the male Long-Evans rats. In all other groups of males fed supplemental sulfur amino acids, copper deficiency elevated plasma cholesterol; the only exception was the groups of male Long-Evans rats fed the 6 mg cystine/g diet. With the females, adding sulfur amino acids to the diet did not increase plasma cholesterol concentrations. Copper deficiency elevated plasma cholesterol concentrations in all groups of females.

The effect of sex and genetic factors on the hematocrit response to copper deficiency was especially noticeable when the rats were fed the 6 mg cystine/g diet. With this diet, the copper-deficient females and Long-Evans males showed only a mild depression in hematocrit. The copper-deficient Sprague-Dawley males exhibited a more severe depression in hematocrit. The effect of sex on the hematocrit response was also quite apparent when the rats were fed the 16 mg arginine/g diet. The females did not show a depression in hematocrit when deprived of copper, whereas the males did.

Human Studies

Epidemiologic studies and observations on experimental animals indicating that a low intake of dietary copper can adversely affect cardiovascular health, and affect the metabolism of some predictors of heart disease including plasma cholesterol (*17,18*) became of interest to nutritionists when it was found that many diets contained substantially less copper than was previously believed (*24*). Apparently less than 25% of diets in the US contain the 2 mg of copper thought to be required daily. However, attempts to produce signs of copper deficiency in adult humans have not yielded consistent changes in examined copper status indicators, e.g., plasma cholesterol. Milne and coworkers (25) fed eight men and eight women diets low in copper for periods ranging from 42–120 d in four separate studies. Only one man showed definite signs of copper depletion (*25,26*) with significantly depressed plasma copper and erythrocyte superoxide dismutase, and increased plasma cholesterol. Only two of the other men exhibited similar trends; however, the changes were not significant and were within the normal range. Two men also showed impaired glucose tolerance (*27*). No changes in serum ceruloplasmin were found in the men.

Table 3
Effect of Genetic Makeup, Sex, and Dietary Sulfur Amino Acids on Hematocrits and Plasma Cholesterol in Rats[a]

Treatment[b]		Hematocrit			Plasma cholesterol		
Cu	Amino acid	Female SD[c]	Male SD	Male LE[d]	Female SD	Male SD	Male LE
μg/g	mg/g	%			mg/100 mL		
0	16 Arginine	41.4	28.8	27.4	96	99	109
6	16 Arginine	42.3	41.8	38.8	85	90	103
0	2 Methionine	32.3	19.9	–	120	128	–
6	2 Methionine	40.5	41.0	–	89	90	–
0	6 Methionine	31.9	21.3	21.8	124	114	160
6	6 Methionine	41.3	40.5	38.5	82	111	125
0	12 Methionine	24.5	22.6	20.8	120	119	161
6	12 Methionine	41.7	40.5	38.7	78	115	137
0	6 Cystine	38.1	19.0	34.9	94	133	114
6	6 Cystine	41.0	40.5	38.7	81	87	111
0	12 Cystine	26.3	15.2	29.2	116	130	123
6	12 Cystine	41.7	39.9	38.3	78	97	107
0	6 Met + 6 Cys	28.9	18.9	17.7	128	126	160
6	6 Met + 6 Cys	42.2	40.2	38.3	78	112	108
		Analysis of variance—p values					
Copper		0.0001	0.0001	0.0001	0.0001	0.0001	0.001
Amino acid (AA)		0.0001	0.0001	0.0001	0.03	0.001	0.0008
Cu × AA		0.0001	0.0001	0.0001	0.006	0.0004	NS
Error mean square		5.1	6.6	7.8	187	171	750

[a]Experimental conditions described in references (22,23).
[b]Amounts of copper (copper sulfate) and specific amino acids supplemented to the diet. Six rats were assigned to each treatment group.
[c]Sprague-Dawley.
[d]Long-Evans.

In the one study with women, plasma copper and erythrocyte superoxide dismutase were unaffected by the dietary copper manipulations. However, copper depletion depressed enzymatic ceruloplasmin and cytochrome-c oxidase in platelets and mononucleated white cells (25).

The described animal studies suggest that the inability to obtain consistent signs of copper deficiency in adult humans may be caused by factors of sex, genetic makeup, dietary sulfur amino acids, and dietary carbohydrate. This suggestion is supported by a recent study conducted at the Grand Forks Human Nutrition Research Center (unpublished observations). In this study, ten healthy men were fed a mixed Western diet that supplied about 0.6 mg copper/2500 kcal. The diet contained a high amount of dairy and casein-based products; thus, it contained 2.4 g methionine and 0.7 g cystine/2500 kcal. After an equilibration period of 15 d, the men were placed on four dietary periods of 36 d each when the following treatments were presented:

1. Basal diet
2. Basal diet supplemented with 2 mg copper/d
3. Basal diet supplemented with 2 g cystine/d and
4. Basal diet supplemented with 2 mg copper and 2 g cystine/d.

When comparisons were made that contained all ten subjects, there were very few statistically significant changes in the variables examined. However, the subjects seemed to separate into two groups; the groups were those whose serum cholesterol did not increase during copper depletion with the high cystine diet (dubbed the nonresponders) and those whose serum cholesterol increased at least 15 mg/dL during copper depletion with the high cystine diet (dubbed the responders). Performing statistical comparisons on the two groups separately yielded a number of significant differences. An example is shown in Table 4.

Table 4 shows that, if genetic variability is accepted as being present (responders and nonresponders), serum cholesterol is affected by an interaction between dietary copper and cystine. When the nonresponders were fed high cystine, copper deficiency decreased serum cholesterol, but when they were fed low cystine, copper deficiency tended to increase serum cholesterol. When the responders were fed high cystine, copper deficiency increased serum cholesterol, but when they were fed low cystine, copper deficiency had no effect.

The preceding discussion indicates that copper is another trace element whose nutritional need is markedly influenced by several nutritional, metabolic, and physiologic factors. Future research is needed to identify clearly the stressors in humans that enhance the need for copper.

Table 4
Effect in Men of Copper, Cystine, and Their Interaction on Serum Cholesterol

	Total Cholesterol, mg/dL		
Dietary Period[a]	All Subjects	Non-responders[b]	Responders[c]
LoCu LoCys	220	223	214
LoCu HiCys	220	214	229
HiCu LoCys	215	214	216
HiCu HiCys	225	239	204
	Analysis of Variance - p Values		
Copper	NS	NS	0.05
Cystine	NS	NS	NS
Cu x Cys	NS	0.01	0.03

[a]LoCu = 0.6 mg copper/d; HiCu = 2.6 mg copper/d; LoCys = 0.7 g cystine/d; HiCys = 2.7 g cystine/d.

[b]Nonresponders were subjects whose serum cholesterol did not increase during copper depletion with the high cystine diet.

[c]Responders were subjects whose serum cholesterol increased at least 15 mg/dL during copper depletion with the high cystine diet.

NUTRITIONAL SIGNIFICANCE OF OTHER TRACE ELEMENTS

Chromium

Chromium is another trace element whose need by humans apparently is influenced by nutritional or physiologic stress. The dietary need for chromium seems to change when normal insulin-dependent metabolism of carbohydrate, protein, and fat is upset. Stress, including trauma, infection, surgery, and intense heat or cold, elevates the secretion of hormones, which alters glucose metabolism and apparently affects chromium metabolism. In experimental animals, the stress of a low protein diet, controlled exercise, acute blood loss, or infection aggravated the signs of depressed growth and survival caused by chromium-deficient diets (*28*). In humans, severe trauma and exercise elevated the excretion of chromium in urine (*29*).

Molybdenum

Epidemiologic findings have implicated molybdenum deficiency in the incidence of esophageal cancer in Africa, China, and Russia (*30*). Cancer is often caused by xenobiotic compounds. The molybdoenzymes xanthine oxidase, aldehyde oxidase, and sulfite oxidase may be involved in the detoxification of xenobiotic compounds (*31*). Possibly, animals and humans stressed by an exposure to certain xenobiotics have an enhanced need for molybdenum.

Nickel

Vitamin B-12 status was found to affect the response of methionine/methyl group-depleted rats to nickel deprivation (*32*). An interaction between nickel and vitamin B-12 affected growth, kidney weight/body weight ratio, plasma concentrations of copper, iron and molybdenum, liver concentrations of calcium, copper, and molybdenum, and kidney concentrations of copper, manganese, and nickel. With almost all the variables affected by dietary nickel, the effects were influenced by vitamin B-12 status. With many of the variables, vitamin B-12 deprivation made, or tended to make, the nickel-supplemented rats essentially the same as the nickel-deprived rats. As a result, it was suggested that vitamin B-12 is necessary for the optimal expression of the biological role of nickel. Moreover, the findings indicate that animals or humans with suboptimal vitamin B-12 or methionine metabolism may have an enhanced need for dietary nickel.

Arsenic

Studies with rats, chicks, and hamsters have revealed that the nature and severity of the signs of arsenic deprivation are affected by several dietary manipulations, including variations in the concentrations of zinc, arginine, choline, methionine, taurine, and guanidoacetic acid, all of which can affect methyl-group metabolism (*33,34*) and, thus, polyamine synthesis in which arsenic apparently has a role (*35*). It seems possible that situations might occur that would enhance the dietary need for arsenic.

Vanadium

Recently it was found that some haloperoxidases from red and brown algae (*36,37*) and from lichens (*38*) require vanadium to be active.

In rats, it was found that, as dietary iodine increased from 0 to 0.33 μg/g diet, thyroid peroxidase activity decreased, with the decrease more marked in vanadium-supplemented (38.1 to 12.3 μ/mg protein) than vanadium-deprived (18.7 to 10.3 μ/mg protein) rats (*39*). Perhaps there is an enhanced need for vanadium in animals and humans with subnormal thyroid status.

CONCLUDING STATEMENT

An increasing number of studies have been performed examining the importance of trace element nutriture in various forms of nutritional, metabolic, hormonal, or physiologic stress in animals and humans. These studies indicate that situations will be found in which a trace element is of nutritional significance. Thus, it is likely that some of the trace elements are more important in human nutrition than is now generally accepted.

REFERENCES

1. W. N. Tapp and B. H. Natelson, *FASEB J.* **2,** 2268 (1988).
2. K. Schwarz and C. M. Foltz, *J. Amer. Chem. Soc.* **79,** 3292 (1957).
3. R. J. Shamberger, *Biochemistry of the Essential Ultratrace Elements,* E. Frieden, ed., Plenum, NY, pp. 201–237, 1984.
4. O. A. Levander, *Trace Elements in Human and Animal Nutrition,* 5th ed., W. Mertz, ed., Academic, Orlando, FL, pp. 209–279, 1986.
5. O. A. Levander, *Annu. Rev. Nutr.* **7,** 727 (1987).
6. G. Yang, J. Chen, Z. Wen, K. Ge, L. Zhu, X. Chen, and X. Chen, *Adv. Nutr. Res.* **6,** 203 (1984).
7. H. Agulhon, *Ann. Inst. Pasteur* **34,** 321 (1910).
8. C. D. Hunt and F. H. Nielsen, *Trace Element Metabolism in Man and Animals,* vol. 4, J. McC Howell, J. M. Gawthorne, and C. L. White, eds., Australian Academy of Science, Canberra, Australia, pp. 597–600, 1981.
9. C. D. Hunt, T. R. Shuler, and F. H. Nielsen, *4: Spurenelement Symposium,* M. Anke, W. Baumann, H. E. Braunlich, and C. Bruckner, eds., Karl-Marx Univ. and Friedrich-Schiller Univ., Liepzig and Jena, DDR, pp. 149–155, 1983.
10. F. H. Nielsen, T. R. Shuler, T. J. Zimmerman, and E. O. Uthus, *Biol. Trace Element Res.* **17,** 91 (1988).
11. F. H. Nielsen, T. R. Shuler, T. J. Zimmerman, and E. O. Uthus, *Magnesium,* **7,** 133 (1988).
12. F. H. Nielsen, T. J. Zimmerman, and T. R. Shuler, *Proc. N.D. Acad. Sci.* **42,** 61 (1988).
13. C. D. Hunt, and F. H. Nielsen, *Proc. N.D. Acad. Sci.* **41,** 50 (1987).
14. T. R. Shuler and F. H. Nielsen, *Proc. N.D. Acad. Sci.* **41,** 49 (1987).
15. F. H. Nielsen, C. D. Hunt, L. M. Mullen, and J. R. Hunt, *FASEB J.* **1,** 394 (1987).
16. F. H. Nielsen, *FASEB J.* **3,** A760 (1989).
17. L. M. Klevay, *Biol. Trace Element Res.* **5,** 245 (1983).

18. L. M. Klevay, *Med. Hypotheses* **24,** 111 (1987).
19. M. Fields, C. Lewis, D. J. Scholfield, A. S. Powell, A. J. Rose, S. Reiser, and J. C. Smith, *Proc. Soc. Exp. Biol. Med.* **183,** 145 (1986).
20. M. Fields, C. G. Lewis, T. Beal, D. Scholfield, K. Patterson, J. C. Smith, and S. Reiser, *Proc. Soc. Exp. Biol. Med.* **186,** 183 (1987).
21. H. Nederbragt and L. F. M. Van Zutphen, *Biol. Trace Element Res.* **14,** 217 (1987).
22. F. H. Nielsen, *J. Trace Elements Exp. Med.* **1,** 157 (1988).
23. F. H. Nielsen, *FASEB J.* **2,** A1422 (1988).
24. L. Klevay, *Inflammatory Diseases and Copper,* J. R. J. Sorenson, ed., Humana, Clifton, NJ, pp. 123–136, 1982.
25. D. B. Milne, L. M. Klevay, and J. R. Hunt, *Trace Elements in Man and Animals,* vol. 6, L. S. Hurley, C. L. Keen, B. Lönnerdal, and R. B. Rucker, eds., Plenum, NY, pp. 451–452, 1988.
26. L. M. Klevay, L. Inman, L. K. Johnson, M. Lawler, J. R. Mahalko, D. B. Milne, H. C. Lukaski, W. Bolonchuk, and H. H. Sandstead, *Metabolism* **33,** 1112 (1984).
27. L. M. Klevay, W. K. Canfield, S. K. Gallagher, L. K. Henriksen, H. C. Lukaski, W. Bolonchuk, L. K. Johnson, D. B. Milne, and H. H. Sandstead, *Nutr. Rpts. Int.* **33,** 371 (1986).
28. W. Mertz and E. E. Roginski, *J. Nutr.* **97,** 531 (1969).
29. J. S. Borel and R. A. Anderson, *Biochemistry of the Essential Ultratrace Elements,* E. Frieden, ed., Plenum, NY, pp. 175–199, 1984.
30. X. M. Luo, H. J. Wei, and S. P. Yang, *J. Nat. Can. Inst.* **71,** 75 (1983).
31. K. V. Rajagopalan, *Enzymatic Basis for Detoxification,* vol. 1, W. B. Jacoby, ed., Academic, NY, pp. 295–309, 1980.
32. F. H. Nielsen, T. J. Zimmerman, T. R. Shuler, B. Brossart, and E. O. Uthus, *J. Trace Elements Exp. Med.* **2,** 21 (1989).
33. F. H. Nielsen and E. O. Uthus, *Biochemistry of the Essential Ultratrace Elements,* E. Frieden, ed., Plenum, NY, pp. 319–340, 1984.
34. F. H. Nielsen, *Essential and Toxic Elements in Human Health and Disease,* A. Prasad, ed., Alan R. Liss, NY, pp. 277–292, 1988.
35. E. O. Uthus, R. Poellot, B. Brossart, and F. H. Nielsen, *FASEB J.* **3,** A1072 (1989).
36. H. Vilter, *Phytochem.* **23,** 1387 (1984).
37. B. E. Krenn, H. Plat, and R. Wever, *Biochem. Biophys. Acta.* **912,** 287 (1987).
38. H. Plat, B. E. Krenn, and R. Wever, *Biochem. J.* **248,** 277 (1987).
39. E. O. Uthus and F. H. Nielsen, *FASEB J.* **2,** A841 (1988).

Neutron Capture Prompt-γ Activation Analysis of Foods

D. L. ANDERSON,*[,1] W. C. CUNNINGHAM,[1] AND E. A. MACKEY[1,2]

[1]*Food and Drug Administration, Division of Contaminants Chemistry, Washington, DC, USA; and* [2]*Department of Chemistry and Biochemistry, University of Maryland, College Park, MD*

Received June 1, 1989; Accepted December 11, 1989

ABSTRACT

The suitability of neutron capture prompt-γ activation analysis (PGAA) for multielement analysis of foods was investigated. A total of 22 elements was observed in 40 food and mineral supplements. Hydrogen, B, C, N, Na, S, Cl, and K concentrations were determined in NIST RM 8431a Mixed Diet and in a wet diet composite made from FDA Total Diet Study collections. Because the neutron flux is so low for PGAA, the method is nondestructive and reanalysis of analytical portions is possible. Both diet materials were analyzed before and after freeze-drying to check for element losses during this process. No losses were found for RM 8431a, but significant losses of B and Na were observed for the wet composite. The measured loss of hydrogen for the wet composite was not consistent with the assumption that the lost mass was water only.

Index Entries: Neutron capture prompt-γ activation analysis (PGAA); food and mineral supplements; food analysis; freeze-drying losses; diet composites.

*Author to whom all correspondence and reprint requests should be addressed.

INTRODUCTION

Neutron capture prompt-γ activation analysis (PGAA) has been shown to be a useful in-beam technique for determination of up to 21 major, minor, and trace elements in geochemical materials (*1*) and up to 13 elements in biological reference materials (*2*). This study was conducted to explore the capabilities of PGAA for determination of the elemental composition of foods. Two general categories were analyzed: food supplements and human diet composites. These provided such a diverse range of food matrices that special calibration techniques were required to overcome a matrix effect resulting from hydrogen. Standard Reference Materials (SRM) from the National Institute of Standards and Technology (NIST, formerly the National Bureau of Standards) were analyzed as control materials. Because the technique is nondestructive, analytical portions could be reanalyzed. Some of the diet composites were analyzed both before and after freeze-drying to check for element losses during this process.

EXPERIMENTAL

All analytical portions were heat-sealed in Teflon™ bags. Forty food and mineral supplements in tablet or capsule form were randomly selected to include a variety of matrices (*3*). Ten analytical portions (~750 mg each from 3 bottles) of NIST Reference Material (RM) 8431a and two portions (one 1-g and one 2-g portion) of a wet diet composite were analyzed along with SRM 1571 (Orchard Leaves) and SRM 1577 (Bovine Liver) as controls. The wet diet portions were from a separate composite of prepared foods from market basket #K74 of FDA's Total Diet Study (*4*). The two portions were taken from the outer surface of a frozen block of the composite and may not have been representative of the composite as a whole. In particular, a slight excess of liquid is thought to have been selected. Ten portions of RM 8431a and four portions each of SRM 1571 and SRM 1577 were pressed to form pellets and analyzed before freeze-drying. For each of two bottles of RM 8431a, two portions were analyzed (in pellet form) before and after freeze-drying to check for element losses. One additional portion from each bottle was left as powder to monitor the drying efficiency for the pellets. The powder portions were analyzed only after drying.

For freeze-drying, the Teflon™ bags holding the analytical portions were slit and weighed prior to being placed in an evacuable desiccator. The desiccator was placed in a freezer and the portions frozen to a temperature of −30°C. A 26-millitorr vacuum was slowly applied, and then the freezer was shut off. The portions were vacuum-dried for an additional 24 h before venting. Immediately afterwards, the analytical portions (still in their bags) were weighed, resealed, weighed a second time, and then stored over P_2O_5.

ANALYTICAL PROCEDURES

Irradiation and counting took place at the NIST 20-MW reactor PGAA facility. This facility and analytical procedures have been discussed in detail elsewhere (*2,5,6*). The Teflon™-bagged portions were suspended in a thermal neutron beam (fluence rate = $3.3 \times 10^8/cm^2 \cdot s$) and γ-counted for 4–5 h. Some counts were longer (10–15 h) to improve counting statistics for certain elements (e.g., Cd). The γ-ray counting equipment included a 27% efficient Ge detector with a Nuclear Data (ND) 16,384 channel μMCA interfaced to a DEC VAX 11/730. Spectral reduction was performed using ND "PEAK" software and a user-interactive program "SUM" encoded by R. M. Lindstrom of NIST's Center for Analytical Chemistry.

For in-beam PGAA of thick (~5 mm) test portions, element sensitivities (counts/s·mg for a given γ-ray line) are enhanced primarily by H in the matrix. The enhancement is 1–2%/wt% H, and special standards must be prepared for proper calibration (*7*). Most elements' sensitivities were measured in at least two matrices of significantly different H concentrations to establish the degree of enhancement. The percent enhancements for H and B, for example, were characterized in detail with two sets of prepared standards. A set of liquid standards consisted of $B(OH)_3/H_2O/D_2O$ solutions ranging from ~0 to 11% H by weight. Homogeneous pellets of graphite/mannitol (G/M) mixtures, with H varying from ~0 to 8% by weight, were used as solid standards for H. Similar G/M mixtures with B_4C powder added were used as solid standards for B. The G/M pellets used were those prepared and run previously by Kitto (*7*). Figure 1 shows that, for these experiments, there was 1.00% enhancement/wt% H for H itself ($r^2 = 0.92$) and 1.85%/wt% H for B ($r^2 = 0.96$). Enhancement effects for Mg, Al, Si, Ti, Mn, Fe, Cu, Zn, and Sr have not yet been characterized, and no correction was applied for these elements. However, they were observed only in relatively low-H matrices, so that matrix effects were probably small. Stoichiometric compounds or metal foils were used for standards. The total analytical uncertainty expressed below includes uncertainties arising from this enhancement effect, counting statistics, flux and geometry variations, portion mass variations during irradiation (which are very small), and nuclide interferences and blanks, where appropriate.

RESULTS AND DISCUSSION

All concentrations presented for diet materials and SRM are on a freeze-dried weight basis unless noted, whereas concentrations for the supplements are on an as-received basis. SRM 1571 and SRM 1577 were run as controls. Findings are presented in Table 1.

In the food and mineral supplements, 22 elements were observed by PGAA (Table 2). Hydrogen, B, and Cl concentrations were measured in

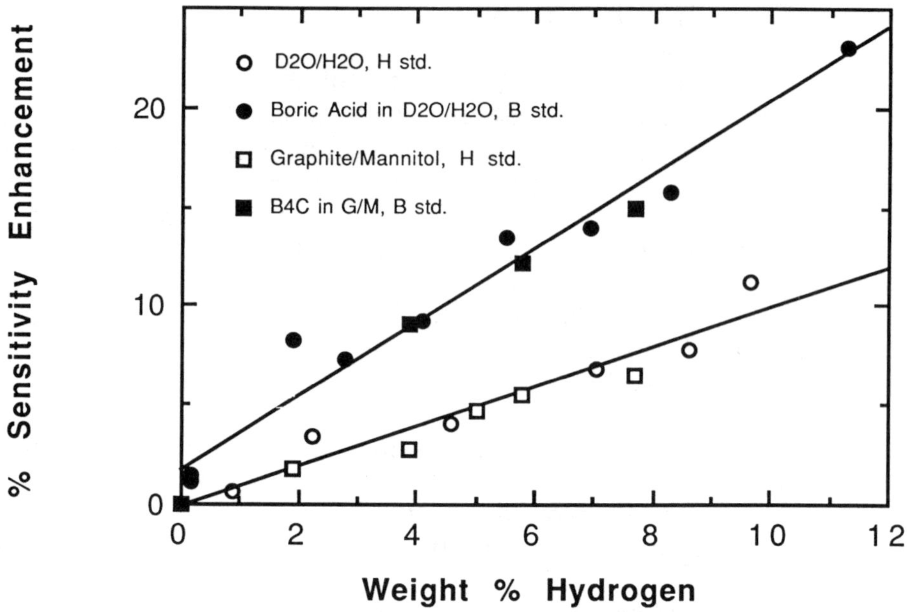

Fig. 1. Plot of % sensitivity enhancement for H and B vs wt% H in prepared standards.

all 40 supplements; however, in contrast Cu could be measured in only one supplement. The potentially toxic trace element Cd was observed only at levels considered to be nontoxic in the supplements (*3*). Large amounts of Ti indicated the presence of a TiO_2 coating on some of the tablets. The presence of boron above ~2 μg/g indicated an herbal component in the supplement (according to the product name and/or labeling). In other work being conducted on the supplements, discrepancies were observed between atomic spectrometric methods and instrumental neutron activation analysis (INAA). The PGAA findings supported explanations for the observed discrepancies. PGAA showed that, in cases where the atomic methods' measured concentrations were lower than those for INAA, the matrix in question had relatively high Si content. This suggests that nondigestible Si residue had tied up other elements. A complete multitechnique evaluation of the food and mineral supplements is discussed by Sinex, et al. (*3*).

Concentrations for H, B, C, N, Na, S, Cl, and K were determined in RM 8431a (Table 3). One bottle (#78) showed a significantly larger B concentration than did the other two, whereas the other elements showed good bottle-to-bottle agreement. The observed freeze-drying losses for RM 8431a (Table 4) indicated that concentrations for the eight elements were not affected by the drying procedure. For the wet composite material, however, significant losses were measured for B and Na (Table 5). Apparently, much of the B was present in solution as boric acid; this has been shown (*10,11*) to volatilize during freeze-drying unless

Table 1
PGAA Analysis of SRM 1571 Orchard Leaves and SRM 1577 Bovine Liver

	SRM 1571 Orchard Leaves			SRM 1577 Bovine Liver		
	This work[a]	NIST	Compilation (*8*)	This work[a]	NIST	Compilation(*8*)
Element	$\bar{x} \pm s$	Certificate	mean	$\bar{x} \pm s$	Certificate	mean
H %	5.87±0.06	-	5.84±0.26	6.83±0.08	-	6.97±0.16
B μg/g	31.6±0.4	33±3	33±3	0.92±0.06	-	2.8±0.8
C %	46.2±2.6	-	46.1±0.5	51.2±2.0	-	50.6±1.1
N %	2.76±0.18	2.76±0.05	2.72±0.04	11.10±0.30	10.6±0.6	10.5±0.2
Na μg/g	-	82±6	89±15	2590±130	2430±130	2395±200
P %	-	0.21±0.01	0.2±0.018	1.2±0.2	1.1	1.13±0.12
S μg/g	2350±260	1900	2040±240	7160±210	-	7900±1000
Cl μg/g	790±20	690	730±40	2850±50	2700	2680±140
K μg/g	1.45±0.01	1.47±0.03	1.44±0.07	0.99±0.03	0.97±0.06	0.98±0.06
Ca %	2.22±0.12	2.09±0.03	2.04±0.12	-	0.0124±0.0006	0.0122±0.0014
Cd μg/g	-	0.11±0.01	0.12±0.02	0.32±0.03	0.27±0.04	0.28±0.02

[a]Mean, $\bar{x}$, and standard deviation, s, for four portions.

Table 2
Observed Concentration Ranges and Determination Limits for Food and Mineral Supplements by PGAA

Element	n[a]	Range	L_Q[b]
H (%)	40	0.83-9.82	0.02
B (µg/g)	40	0.69-312	0.6
C (%)	35	22-81	20
N (%)	31	1.2-8.9	1.5
Na (%)	28	0.02-6.27	0.14
Mg (%)	12	0.4-13.2	1.7
Al (%)	5	0.13-0.90	0.1
Si (%)	23	0.5-14.4	1.2
P (%)	7	0.8-11.6	1
S (%)	28	0.04-3.75	0.1
Cl (%)	40	0.004-2.24	0.012
K (%)	32	0.02-3.19	0.11
Ca (%)	26	0.1-31.1	0.6
Ti (µg/g)	7	150-8400	300
Mn (%)	4	0.04-2.00	0.07
Fe (%)	10	0.12-4.05	0.4
Cu (%)	1	0.47	0.5
Zn (%)	4	0.4-13.0	0.5
Sr (µg/g)	1	800	400
Cd (µg/g)	12	0.12-1.8	0.8
Sm (µg/g)	10	0.16-0.65	0.2
Gd (µg/g)	7	0.43-0.69	0.5

[a]Number of observations.
[b]Determination limit, L_Q, from Currie (9). (Limit of detection, L_D, $\sim 0.3 \times L_Q$.)

stabilized by an appropriate base. There is no explanation for the apparent loss of Na, and INAA experiments are under way that will clarify these preliminary results. Furthermore, the H mass loss, as determined by PGAA and compared to the weighed total mass loss, was smaller than that necessary to be accounted for by H_2O only. The material lost during freeze-drying was 9.5–10% H by weight, instead of 11.19% (pure H_2O).

Table 3
PGAA Analysis of RM 8431a Mixed Diet (Dry Wt Concentrations)[a]

Element	BOTTLE # 78	98	169	All Portions	Total	
	(n=3)	(n=3)	(n=4)	(n=10)	Analytical	
	$\bar{x} \pm s$	$\bar{x} \pm s$	$\bar{x} \pm s$	$\bar{x} \pm s$	uncertainty[b]	L_Q
H (%)	6.82±0.06	6.75±0.06	6.49±0.10	6.67±0.10	0.11	0.02
B (µg/g)	4.22±0.04	3.78±0.06	3.69±0.11	3.87±0.25	0.15	0.5
C (%)	47.7±2.0	49.7±3.8	45.6±2.3	47.9±3.7	4.5	15
N (%)	3.28±0.03	3.35±0.15	3.19±0.14	3.26±0.13	0.12	1.4
Na (µg/g)	3440±200	3580±30	3320±200	3420±180	140	1100
S (µg/g)	2300±560	2120±220	2010±140	2130±320	120	1800
Cl (µg/g)	5070±30	5030±20	4870±70	4980±100	60	110
K (µg/g)	7780±50	7690±10	7530±130	7650±130	110	860

[a]Mean, $\bar{x}$, and standard deviation, s, for n portions.
[b]*See* text.

Table 4
PGAA Analysis of RM 8431a Mixed Diet: Drying Study of Bottles 78 & 98

	Bottle 78 (n=2)[a]		Bottle 98 (n=2)[a]		All Portions (n=4)[b]	
Element	before drying	after drying	before drying	after drying	before drying	after drying
H (%)	6.84±0.19	6.73±0.16	6.78±0.18	6.78±0.17	6.81±0.05	6.75±0.16
B (μg/g)	4.20±0.21	3.93±0.22	3.80±0.17	4.00±0.26	4.00±0.20	3.96±0.08
C (%)	46.8±4.4	47.0±3.0	47.8±4.2	47.8±4.6	47.3±1.0	47.4±1.8
N (%)	3.29±0.55	3.48±1.00	3.29±0.50	3.48±1.00	3.29±0.08	3.48±0.10
Na (μg/g)	3410±180	3660±290	3590±160	3930±260	3500±160	3800±170
S (μg/g)	2010±190	2530±560	2160±260	2240±400	2100±240	2400±400
Cl (μg/g)	5070±100	5070±70	5020±80	5090±80	5050±40	5080±20
K (μg/g)	7810±220	7920±250	7680±180	7650±290	7740±60	7780±170

[a]Error expressed is propagated total analytical uncertainty for two portions.
[b]Error expressed is the standard deviation, s, for the four portions.

Table 5
Freeze-Drying Losses[a] in FDA Total Diet Study Wet Composite

	Portion #1			Portion #2		
	Total µg/portion			Total µg/portion		
Element	wet	dried	% lost	wet	dried	% lost
H[b]	96810	7880	91.9±2.3	200500	16270	91.9±2.6
B	1.1	0.74	34±11	2.07	1.21	42±11
Na	1481	1295	12.6±6.4	3120	2583	17.2±7.7
Cl	1607	1600	-	3335	3337	-
K	1185	1104	6.9±8.5	2377	2305	3.0±8.6

[a]Error shown for % lost is total analytical uncertainty as defined in text.
[b]Measured hydrogen loss corresponds to 10.0 ± 0.2% of weighed mass loss for portion #1 and 9.5 ± 0.2% of weighed mass loss for portion #2.

Although the wet portions might have contained a small excess of liquid and these findings may be slightly exaggerated, significant losses would still be expected in homogeneous material.

SUMMARY AND CONCLUSION

PGAA provides unique advantages for food analysis. It is a multielement, instrumental technique that can be used for virtually any matrix. In spite of the fact that significant element sensitivity enhancements were encountered primarily because of the high H content of the biological materials, the relationships could be accurately characterized. PGAA requires only minimal test sample preparation, has a short analytical turnaround time, and determines many significant elements. It can determine B in a wide range of materials and can determine Cd down to the sub-µg/g range. Relative to conventional NAA, the neutron flux is ~5 orders of magnitude lower. Accordingly, very little delayed radioactivity is produced, and damage does not appear to occur from γ-heating. The technique is, therefore, ideally suited for element-loss measurements, because a test portion can be analyzed many times. Using PGAA, predried diet composite materials were shown to be stable in terms of element losses during freeze-drying, and therefore suitable for a reference material. However, wet composites suffered significant losses, especially of B, which indicates that nutritional information should not be derived from predried material. Experiments are under way using homogenized composites to more accurately assess the problem.

Like conventional NAA, PGAA is fundamentally based on nuclear properties and can serve as a confirmatory technique for other analytical techniques. In addition, however, because prompt γ-rays result from transitions or isotopes not used in conventional NAA, it can be considered an independent technique. In this study, PGAA provided a distinct advantage over INAA for H, B, C, N, S, Cd, and Gd, and was able to confirm Na, Cl, K, and Ca.

ACKNOWLEDGMENTS

Some of this work is from a dissertation to be submitted to the Graduate School, University of Maryland, by Elizabeth A. Mackey in partial fulfillment of the requirements for the Ph. D. degree in chemistry.

REFERENCES

1. D. L. Anderson, Y. Sun, M. P. Failey, and W. H. Zoller, *Geostandards Newsletter* **9,** 219 (1985).
2. M. P. Failey, D. L. Anderson, W. H. Zoller, G. E. Gordon, and R. M. Lindstrom, *Anal. Chem.* **51,** 2209 (1979).
3. S. A. Sinex, S. C. Hight, W. C. Cunningham, D. L. Anderson, and S. G. Capar, *J. Agric. Food Chem.* (in preparation).
4. P. Lombardo, *Environmental Epidemiology,* F. C. Kopfler and G. F. Craun, eds., 141, 1986.
5. D. L. Anderson, M. P. Failey, W. H. Zoller, W. B. Walters, G. E. Gordon, and R. M. Lindstrom, *J. Radioanal. Chem.* **63,** 97, 1981.
6. M. P. Failey, Ph. D. Thesis, University of Maryland, College Park, MD (1979).
7. M. E. Kitto, Ph. D. Thesis, University of Maryland, College Park, MD (1987).
8. E. S. Gladney, B. T. O'Malley, I. Roelandts, and T. E. Gills, *NBS Special Publication,* 260–111, (1987).
9. L. A. Currie, *Anal. Chem.* **40,** 586 (1968).
10. L. McCarthy, Masters Thesis, University of Maryland, College Park, MD (1988).
11. W. B. Clarke, M. Koekebakker, R. D. Barr, R. G. Downing, and R. F. Fleming, *Appl. Radiat. Isot.* **38,** 735 (1987).

Activation Analysis of Human Diet Samples with Epithermal Neutrons

L. TOBLER, V. FURRER, AND A. WYTTENBACH

Paul Scherrer Institute, CH-5232 Villigen PSI, Switzerland

Received April 17, 1989; Accepted December 11, 1989

ABSTRACT

Activation of diet samples by epithermal reactor neutrons is a very sensitive method for the determination of iodine. The method was developed and optimized for its determination. However, it turned out that with the same measurement valid results were also obtained for Br, Cl, K, Na, and P. The procedure involves the irradiation of the sample in a boron nitride (BN) container for 1 min, followed by γ-counting for 20 min after a decay time of 7–15 min. The necessary sample weight is 0.5 g (dry), and no chemical operations have to be performed. The capacity is 15 samples per working day, and results are available immediately.

Index Entries: Diet samples, Br, Cl, K, I, Na, P concentrations in; epithermal neutron activation analysis; boron nitride, filter of; accuracy; reproducibility; determination limits.

INTRODUCTION

Epithermal activation analysis of diet samples was primarily developed and optimized for the determination of iodine (*1,2*). Because the half-life of ^{128}I is only 25 min, irradiation and measuring times can be kept short. This is a prerequisite for dealing with many samples per day and for keeping the unit cost low.

It has since turned out that the same procedure also allows the determination of bromine (via ^{80}Br, $t_{1/2}$ = 17.5 min), chlorine (via ^{38}Cl, $t_{1/2}$ = 37.2 min), potassium (^{42}K, $t_{1/2}$ = 12.4 h), sodium (^{24}Na, $t_{1/2}$ = 15 h), and phosphorus (via ^{28}Al, $t_{1/2}$ = 2.24 min). This information is obtained without additional handling, and no chemical operations are necessary. The method is sensitive, rapid, and reliable. The capacity is 15 samples/d, and results are available immediately.

METHODS

Samples

Duplicates of freeze-dried and homogenized total diet samples (0.4–0.5 g) are weighed in polyethylene containers (vol 0.85 cm^3). These samples have been obtained by the IAEA coordinated research program on human daily dietary intakes of nutritionally important trace elements. Heating separate samples to 105°C for 1 h reduced their weights by 5–12%. All quoted concentrations in this work refer to dry sample weight.

As a multielement working standard the reference material BCR-151 (milk powder: 220 mg) was used, containing the following concentrations:

I	5.36 μg/g	Na	4591 μg/g
Cl	10,150 μg/g	K	18,100 μg/g
Br	24.3 μg/g	P	10,995 μg/g

All concentrations were determined in our laboratory.

BN-Filter

The BN-containers, with a wall thickness of 4 mm (330 mg B/cm^2), were made in our institute from 2.5-cm diameter boron nitride rods. The outer dimensions of the cylindrical container are: height 28 mm and diameter 21 mm. BN attenuates the initial thermal neutron flux (2.4×10^{13} n/cm^2 s) by a factor of 300 and, thus, accentuates the activation of nuclides with large resonance integrals. Neutron fluxes inside the BN-shield are 1.5×10^{11} n/cm^2 s epithermal and 8×10^{10} n/cm^2 s thermal. Reducing the ^{24}Na and ^{38}Cl activities, which are most abundant in the thermal activation of diet samples, is absolutely necessary for the determination of iodine by instrumental neutron activation analysis. However, the reduction of these activities does not exclude their quantitative determination in the samples.

Boron nitride is a more appropriate absorber material than Cd when working with short-lived activation products, because BN does not give rise to high levels of short-lived activities, which would hinder the handling of samples after irradiation. The following long-lived activities of isotopes were found in irradiated BN: ^{24}Na, $^{46,47}Sc$, ^{151}Sm, and $^{175,181}Hf$. However, activity levels for these isotopes were low even after several irradiations of the containers.

A disadvantage of B is its heat production by the reaction $^{10}B\ (n, \alpha)\ ^{7}Li$, which limits the irradiation time to 1 min. Beyond this time,

Table 1
Utilized and Interfering Nuclear Reactions

Element	Utilized reaction	Energy, keV	Half-life, min	Interfering reactions
I	$^{127}I(n, \gamma)^{128}I$	442.9	25.0	None
Br	$^{79}Br(n, \gamma)^{80}Br$	616.3	17.4	None
Cl	$^{37}Cl(n, \gamma)^{38}Cl$	1642.4	37.2	$^{41}K(n, \alpha)^{38}Cl$
K	$^{41}K(n, \gamma)^{42}K$	1524.7	744	$^{42}Ca(n, p)^{42}K$
Na	$^{23}Na(n, \gamma)^{24}Na$	1368.6	900	$^{24}Mg(n, p)^{24}Na$ $^{27}Al(n, \alpha)^{24}Na$
P	$^{31}P(n, \alpha)^{28}Al$	1779.0	2.25	$^{27}Al(n, \gamma)^{28}Al$ $^{28}Si(n, p)^{28}Al$

the temperature would rise above 100°C inside the container and, thus, soften the polyethylene sample containers too much. As an additional safeguard, the samples are frozen for 15 min in liquid nitrogen before irradiation, which reduces the rise in temperature.

Irradiation

An assembly of polyethylene sample container, a boron nitride (BN) shield, and associated polyethylene cushions are transferred to the reactor "SAPHIR" in a transport rabbit. Transport is done by a pneumatic transfer system (transfer time = 10 s). The irradiation time is 1 min. Every sample is irradiated separately, and every fifth irradiation is reserved for a working standard. The activities obtained for the working standards show that the flux variation within one working day is smaller than 1%.

Counting

The irradiated samples are removed from the BN-shield after irradiation. γ-ray spectra are collected for 20 min, following a decay time of 7–15 min (dependent on the dead time of the samples) with a well-type intrinsic Ge-detector (PGT, FWHM: 2.8 keV, absolute efficiency: 4.3% at 1332 keV) and a Nuclear Data 9900 μVAX-based multi-user data acquisition and processing system. A decay time shorter than 7 min is precluded by the reaction ^{24}Na (n, p) ^{23}Ne, because the γ-line of ^{23}Ne (440 keV, $t_{1/2}$ = 37 s) would interfere with the line of ^{128}I (443 keV).

Utilized and Interfering Nuclear Reactions

Utilized nuclear reactions for the elements determined in diet samples are listed in Table 1. The actually observed half-life of ^{80}Br is greater than 17.4 min because of the concurrent reaction $^{79}Br(n,\gamma)^{80m}Br$ (4.4 h) $\rightarrow$ ^{80}Br. During the first hour after irradiation, a half-life of 20 min has been found appropriate for the decay correction of ^{80}Br.

Table 2
Sensitivity Reproducibility (DF = 24) and Determination Limit in Diet Samples

Element	Typical concn., μg/g	Registered counts	Energy, keV	Expt. reprod., %	Determ. limit, μg/g
I	0.27	5000	443	5	0.10
Cl	13000	84000	1642	2	500
Br	9	37000	616	2	1
K	6400	4000	1525	5	3000
Na	8000	120000	1368	2	300
P	3900	12000	1779	3	1500

Table 3
Test for Accuracy[a]

	NBS SRM 1549, nonfat milk powder			
	This work mean (n = 12)	COV%	Literature[b]	Factor[c]
I	3.50	2.4	3.38	1.04
Br	12.0	2.4	11.85	1.01
Cl	10898	1.7	10900	1.00
K	16518	5.6	16900	0.98
Na	4946	2.5	4970	1.00
P	10662	3.0	10500	1.01
	NBS RM 8431, mixed diet			
	This work mean (n = 3)	COV%	Literature[d]	Factor[c]
I	0.82	1.7	–	–
Br	10.5	1.9	–	–
Cl	5148	1.9	–	–
K	7758	4.8	7900	0.98
Na	3196	1.7	3120	1.02
P	3596	2.2	3320	1.08
	NBS SRM 1572, citrus leaves			
	This work mean (n = 4)	COV%	Literature[e]	Factor[c]
I	1.89	2.4	1.84	1.03
Br	8.08	1.7	8.36	0.97
K	17913	2.4	18200	0.98

[a]All concentrations are given in μg/g for dry samples. Weight loss was determined in this work at 105°C, 1 h on separate samples.

[b]Certified values, except Br (concensus value).

[c]Factor = this work/literature.

[d]N. Miller-Ihli and W. Wolf, *Anal. Chem.* **58**, 3225 (1986).

[e]Certified values, except Br (concensus value).

Nuclear interferences arise from reactions of epithermal and thermal neutrons with elements other than the ones being determined (Table 1). Interferences are negligible in the determination of Br, Cl, K, and I, but can be substantial with Na and P. A concentration ratio of Al/P = 0.011 or Si/P = 0.027 increases the observed ^{28}Al activity (and thus the deduced P concentration) by 10% with the experimental conditions in this work. Literature values for meat and milk (*3,4*) show that the Al/P and Si/P concentration ratios are substantially lower than the above values, but those of plants (*3*) are usually greater. Therefore, the experimentally determined P results for diet samples can be too high by an amount that depends on the individual foodstuff that makes up the actual sample. A reliable correction of the results is only possible if the Al and Si contents of the samples are known, which is not usually the case.

RESULTS

The sensitivity of the method, the experimental reproducibility, and the determination limits are given in Table 2. The determination limits are based on the spectral data from this work and are calculated according to the equation given by Currie (*5*). The sensitivity is adequate for all six elements in diet samples. The experimental reproducibility was evaluated by the analysis of duplicates (DF = 24). Counting statistics is a major contribution to reproducibility.

Accuracy was determined by the analysis of 3 NBS—Standard Reference Materials. Table 3 gives the obtained results. The differences to certified or literature values (*3,6*) are within 4%. The positive difference (+8%) for P in RM 8431 (mixed diet) may be explained by interfering reactions from Al and Si. Na and P cannot be determined in SRM 1572 (citrus leaves) because of the nuclear interferences mentioned above.

REFERENCES

1. W. B. Stroube, W. C. Cunningham, and G. J. Lutz, *J. Radioanal. Nucl. Chem. Articles,* **112,** 341–346 (1987).
2. A. Wyttenbach, L. Tobler, and V. Furrer, *Instrumentelle Multielementanalyse,* B. Sansoni, ed., VHC, Weinheim FRG, pp. 137–140, 1985.
3. R. W. Seward, *NBS Special Publication* 260, 47–49 (1988).
4. G. V. Iyengar, W. E. Kollmer, and J. M. Bowen, *The Elemental Composition of Human Tissues and Body Fluids,* Verlag Chemie, Weinheim, 1978.
5. L. A. Currie, *Anal. Chem.* **40,** 586–593 (1968).
6. N. Miller-Ihli and W. Wolf, *Anal. Chem.* **58,** 3225 (1986).

Determination of Selenium in Duplicate Diets of Residents of Pinhel, Portugal, by Neutron Activation

M. F. REIS,[1,2] J. HOLZBECHER,[1] E. MARTINHO,[2] AND A. CHATT[1,*]

[1]*Trace Analysis Research Centre, Department of Chemistry, Dalhousie University, Halifax, N.S., B3H 4J1, Canada; and* [2]*Laboratorio Nacional de Engenharia e Tecnologia Industrial, 2685, Sacavem, Portugal*

Received April 17, 1989; Revised August 30, 1989; Accepted December 11, 1989

ABSTRACT

A rapid cyclic instrumental neutron activation analysis (CINAA) method has been used to determine the selenium content of 27 duplicate diet samples from each of the 27 districts surrounding Pinhel, Portugal. The accuracy and precision of the CINAA method have been evaluated by analyzing certified reference materials and observed to be within ±5–10% for samples containing at least 40 ppb of selenium. The detection limit has been found to vary between 26–42 ppb selenium depending on the sample composition. The average daily dietary intake has been calculated as 37 μg of selenium per day.

Index Entries: Selenium; food; neutron activation analysis.

INTRODUCTION

Selenium is considered an essential trace element. This was first recognized in 1957, when it was demonstrated that selenium could prevent necrotic liver degeneration in rats (*1*). In 1972, selenium was

*Author to whom all correspondence and reprint requests should be addressed.

shown to be a functional component of the enzyme glutathione peroxidase (2) that assures the maintenance of the structural integrity of liver and other cell membranes; selenium has also been considered to be required for the maintenance of fertility, functioning of the eye, the heart, and the immune system. Recently, it has been reported as a possible cancer-protecting agent (3), and for this function alone, it could be regarded as one of the most important trace elements. Selenium can be, however, highly toxic in many of its chemical forms at high concentrations, whereas its deficiency can be responsible for several diseases (4).

Dietary intake of selenium has been identified as the major source of this element for humans (5). As a result, there is an increasing interest in the determination of selenium in food samples. Some of the commonly used methods based on food balance sheets, food records, food frequency, diet history, or short interviews do not provide accurate information on the average daily dietary intake of elements. This, however, can be reliably obtained by analyzing duplicate diets and individual food items (6,7).

Analytical methods of high precision, accuracy, and sensitivity are required for the measurement of the levels of selenium that are sometimes very low in diet and food samples. Instrumental neutron activation analysis (INAA) method offers high sensitivity in combination with very little or no sample handling and treatment. The determination of selenium by INAA through its long-lived nuclide, ^{75}Se (118.5 d), requires, however, lengthy irradiation, decay, and counting periods leading up to several weeks of total experimental time. Alternatively, the short-lived nuclide ^{77m}Se (17.4 s) can be conveniently utilized. The combination of high isotopic abundance, relativity large cross-section, and short time to reach saturation activity indicates that this short-lived nuclide can be produced in high yields. To improve precision and detection limits, a rapid cyclic instrumental neutron activation analysis (CINAA) method can be used. This method involves cycling of the sample between the irradiation and counting positions and the accumulation of the γ-ray spectrum obtained in each cycle. The signal-to-noise ratio for the short-lived nuclides is significantly enhanced in this manner. In addition, the method is capable of simultaneous measurement of short- and medium-lived nuclides in the same experiment. Dalhousie University SLOWPOKE-2 Reactor is an ideal tool for this type of activation analysis. Fast transfer times and the presence of the analyzer systems on site make it possible to work with very short-lived nuclides. Its characteristics have been previously described (8,9).

Except for some big cities, the diet in Portugal is generally the diet of farmers and fishermen. Basically simple, yet nourishing, it takes advantage of what happens to be locally grown and available. The diet changes according to seasons, feast days, and other important events in local life. Pork is the most popular meat. Salt-dried cod is one of the staples, and it is available in all but the most remote mountain regions. Portuguese diet can be divided into seafood and inland cooking categories. Seafood dishes predominate along the coast; in the inland areas, thick soups, stews, and heavy dishes are generally prepared. This paper presents results of the selenium determination in Portuguese diet samples collected in winter from Pinhel—an inland area.

EXPERIMENTAL

Sample Collection and Preparation

In each of the 27 districts surrounding the city of Pinhel located near the mountains of northern Portugal, one participant from one household was selected to collect daily samples of duplicate diets over a 7-d period. Each participant was asked to keep a written record of the origin and frequency of foods and beverages consumed daily. The duplicate diets were blended, and a portion of the mixture was stored in a freezer. After collection, the diets were allowed to thaw overnight. On the next day, two equal portions were taken from each of the individual daily diets. One portion was stored in a large freezer. The seven second portions were combined. The resulting mixture was homogenized, and a sample was taken from it. These samples were weighed and stored in a freezer until they were freeze-dried. No detectable loss of selenium during freeze-drying was observed. Seven portions of approximately 0.8 g of each freeze-dried diet samples were weighed into small polyethylene vials (1.2 mL), heat-sealed, and placed in medium-sized irradiation vials (7 mL), and then heat-sealed again for CINAA.

Standards

Selenium and multielemental comparator standards were prepared by depositing microlitre portions, containing microgram quantities, of ultrapure (SPEX) solutions on a sucrose support, wetting them with ultrapure water to achieve homogeneity and evaporating them to dryness at low heat under an infrared lamp. The standards were heat-sealed in small polyethylene vials.

Irradiation and Counting Conditions

All samples and standards were irradiated in Dalhousie University SLOWPOKE-2 Reactor (DUSR) at a flux of $5 \times 10^{11}\ n\ cm^{-2}\ s^{-1}$. Following preliminary experiments, the number of cycles was selected as three, and the irradiation and counting times as 20 s. For certain food samples containing large amounts of Na, Cl, and Al, the number of cycles had to be reduced to one or two because the dead-time values for the second or third cycle exceeded 10%. γ-ray spectra of the irradiated samples and standards were recorded using a PGT Ge(Li) detector with a full-width at half-maximum (FWHM) of 2.4 keV at the 1332-keV photopeak of ^{60}Co coupled to a Tracor Northern TN-1700 multichannel analyzer (MCA) for CINAA and an Aptec hyperpure Ge detector with an FWHM of 2.08 keV at the 1332 keV photopeak of ^{60}Co in conjunction with a high throughput pulse processor connected to a Nuclear Data ND66 MCA for one-shot irradiation.

RESULTS AND DISCUSSION

To evaluate the accuracy of the CINAA and INAA methods, a number of standard reference materials (SRM) from the US National Bureau of Standards (NBS) and of certified reference materials (CRM) and intercomparison samples from the International Atomic Energy Agency (IAEA) were analyzed for selenium. The results are shown in Table 1. The agreement with certified values is excellent for IAEA Horse Kidney, NBS Bovine Liver, and IAEA Mixed Human Diet. The other results are also comparable with those reported in literature. The reproducibility of the methods varied from ±5–10% in the diet samples containing 40 ppb or more Se.

The detection limits for Se, defined by $2\sqrt{\mu B}$ where μB is the number of background counts under the photopeak (162 keV), were calculated. In the presence of large amounts of Na, Cl, and Al, the detection limits for selenium could be high. Because of the considerable quantities of salt in the Portuguese diet (for example, 1.6% Na, 2.4% Cl, and 11.5 ppm Al as measured in this work), the detection limits for selenium were within the range of 26–42 ppb.

The means of seven replicates of each of the 27 Portuguese duplicate diet samples have been calculated and shown in Fig. 1. The arithmetic and geometric means are 99 and 92 ppb, respectively. The agreement between the median and arithmetic mean, after rejections using the Chauvenet's criterion (*10*), suggests that the frequency distribution approaches Gaussian. The range, calculated as an interval within which 95% of the values fall, extends from 36–145 ppb Se.

The average daily dietary intake of selenium has been calculated for this population group. The frequency distribution is shown in Fig. 2. The values have been found to range from 7–67 μg/d with arithmetic and geometric means of 42 and 38 μg/d of selenium (after rejection of values, 37 and 35 μg/d), respectively. The median is 36 μg/d. The mean daily intake is comparable to the value of 30 μg reported from Finland, is lower than 50–60 μg generally found in Western Europe, and is significantly lower than 90–100 μg reported for Canada and the USA (*11*). The value observed in this work is outside the safe and adequate intake range of 50–200 μg/d recommended for healthy adults by the US Food & Nutrition Board (*12*).

Table 1
Selenium Content of Selected Biological Reference Materials by CINAA[a]

Reference Material	Selenium Content (ppm)	
	This Work	Reported Value
IAEA H-8, Horse Kidney	4.65 ± 0.20	4.67 ± 0.30
NIES-5, Human Hair	1.61 ± 0.07	1.4
IAEA MA-A2, Fish Flesh	1.31 ± 0.15	1.7 ± 0.3
NBS SRM 1567, Wheat Flour	1.00 ± 0.06	1.1 ± 0.2
NBS SRM 1577a, Bovine Liver	0.72 ± 0.02	0.71 ± 0.07
NBS SRM 1568, Rice Flour	0.29 ± 0.03	0.4 ± 0.13
NBS SRM 1575, Pine Needles	0.04 ± 0.01	-
IAEA H-9, Mixed Human Diet	0.10 ± 0.01	0.11 ± 0.01

[a]The results are mean values of seven replicate samples except for the Mixed Human Diet ($n = 3$).

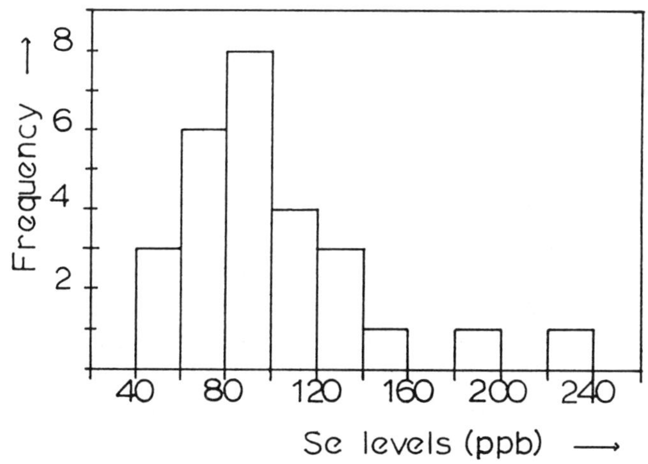

Fig. 1. Frequency distribution of Se levels (ppb) in 27 Portuguese diet samples.

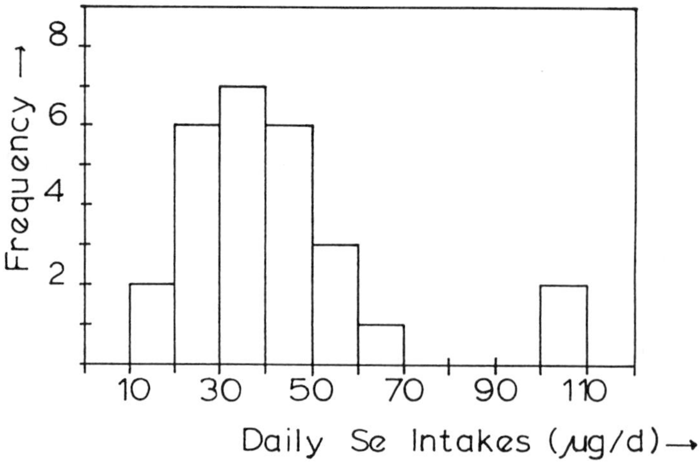

Fig. 2. Frequency distribution of average daily dietary intakes of Se (μg/d).

Several studies have identified fish and meat as good sources of selenium (*11,13,14*). No reliable data for selenium in food items, whether imported or produced in Portugal, are available. Linear regressions between selenium levels in daily individual diets and consumption frequency of fish, meat, or fish and meat together, have been therefore calculated. No significant correlations are evident perhaps because the contribution from these sources was modest because of the relatively limited consumption of fish and meat. There is little information on the Se content of Portuguese soil and vegetables. The amount of Se in vegetables has been reported in literature (*13*) to vary with the area in which it is grown (*13*). Further interpretation of the results obtained in this work will require the determination of selenium in Portuguese soil samples and in individual food items.

ACKNOWLEDGMENTS

The authors thankfully acknowledge the cooperation of Dr. M. Aguas and Dr. N. Petisca of Laboratorio Nacional de Investigacao Veterinaria, Portugal, for assistance in selecting the sampling region; of the technical and administrative staff, in particular Eng. S. Gomes, Eng. J. Gamboa, Sr. M. Caetano, and Dr. P. Martinho, of COVICOA Cooperative–Pinhel, for assistance in collecting the diets; of all the participants in the project; and the SLOWPOKE-2 Operations Group, Dalhousie University, for their assistance in irradiations. The research was financially supported by grants from LNETI (Portugal) and NSERC (Canada).

REFERENCES

1. K. Schwarz and C. M. Foltz, *J. Am. Chem. Soc.* **79,** 3292 (1957).
2. J. T. Rotruck, A. L. Pope, H. E. Ganther, A. S. Swanson, D. G. Hafeman, and W. G. Hoekstra, *Science* **179,** 588 (1973).
3. G. N. Schrauzer, *Advances in Nutritional Research,* vol. 2, H. Draper, ed., Plenum, NY, pp. 219–244, 1979.
4. R. J. Shamberger, *Biochemistry of the Essential Ultratrace Elements,* vol. 3, E. Frieden, ed., Plenum, NY, pp. 213–228, 1984.
5. H. Sakuri and K. Tsuchiya, *Environ. Physiol. Biochem.* **5,** 107–118 (1975).
6. A. Chatt, H. S. Dang, B. B. Fong, C. K. Jayawickreme, L. S. McDowell, and D. L. Pegg, *Proc. Int. Conf. on Methods and Applications of Radioanalytical Chemistry* Kona, Hawaii, 1987.
7. S. Bingham, *The Diet Factor in Epidemiological Research,* Joseph G. A. J. Hautvast and W. Klaver, eds., Wageningen, The Netherlands, pp. 106–122, 1982.
8. D. E. Ryan, D. C. Stuart, and A. Chattopadhyay, *Anal. Chim. Acta* **100,** 87 (1978).
9. D. E. Ryan, A. Chatt, and J. Holzbecher, *Anal. Chim. Acta* **200,** 89 (1987).
10. J. Holzbecher and D. E. Ryan, *Clinical Biochemistry* **15,** 80–82 (1982).
11. L. S. McDowell, P. R. Giffen, and A. Chatt, *J. Radio. Nuc. Chem.* **110,** 519–529 (1987).
12. *U.S. Food and Nutrition Board,* Recommended Dietary Allowances 9th rev. ed., National Academy of Sciences, Washington, DC (1980).
13. J. Thorn, J. Robertson, D. H. Buss, and N. G. Bunton, *Br. J. Nutr.* **39,** 391 (1978).
14. J. N. Thompson, P. Erdody, and D. C. Smith, *J. Nutr.* **105,** 274 (1975).

Multielement Determination in Rice, Wheat, and Barley by Instrumental Neutron Activation Analysis

S. M. AL-JOBORI, K. M. SHIHAB, M. JALIL, A. SAAD, AND A. MOHSIN

Nuclear Research Center, Tuwaitha, Baghdad, Iraq

Received April 17, 1989; Accepted December 11, 1989

ABSTRACT

INAA has been used for the determination of Na, Mg, Al, Cl, K, Sc, Cr, Mn, Fe, Co, Cu, Zn, As, Se, Br, Rb, Sr, Mo, and W in grains of rice, wheat, and barley, which were collected from different plant fields in Iraq. Samples and standards were irradiated in the IRT-5000 reactor, at neutron fluxes of 2×10^{13} $cm^{-2} \cdot s^{-1}$ and 3.2×10^{11} $cm^{-2} \cdot s^{-1}$. Interferences of photopeaks with each other were considered, and reaction interferences were calculated and determined experimentally. Accuracy of our method was assessed by the analysis of IAEA standards Wheat Flour and Bovine liver. A good agreement has been achieved between the present results and recommended values. The precision and detection limit were determined for all elements in all types of grain.

Index Entries: INAA; biological trace analysis; foodstuffs.

INTRODUCTION

The elemental composition of plant material reflects the composition of the soil or other nutrient medium. Trace elements are known to have an important effect on physiological processes that occur in all living systems. A number of them exist in living bodies at very low concentrations. It has been suggested that their action might be catalytic, as shown in the syntheses of some hormones, activity of some enzymes, and in the setting of acid-base equilibria. The trace elements are incorporated with animals, especially humans, through the foods. Therefore, the first step in the study of their behavior is the knowledge of their concentration in foods, their distribution in different organs, and their toxicity, if any. A relationship between the trace elements levels in the diet and the occurrence of certain diseases has been reported by several investigators (*1* and references therein).

Several trace elements are considered as essential, nonessential, or toxic for animal life. However, all elements in high concentrations may become detrimental to organisms (*2*). The nutritional role and the essentiality of trace elements, as well as their biochemical pathological significance to humans and animals, have been recently reviewed by several investigators (*3,4*). It is therefore important to measure the trace element content of foodstuffs in an attempt to increase our knowledge of the distribution of elements both geographically and throughout the food chain (*5*). Since rice, wheat, and barley are the most common foodstuffs of the people of Iraq, their trace elements content is of considerable interest. The present work is part of a comprehensive program planned to study trace elements in foodstuffs (water, diets, crops, plants, vegetables, fruits, and milk).

Nuclear methods, especially NAA, are playing an increasingly important role in the determination of elemental concentrations in biological samples. They have been widely used as reliable techniques for multielement analysis in biological samples (*6–10*). Rice also has been analyzed by employing this technique (*11,12*). This is because of the sensitivity, accuracy, and specificity of the technique.

EXPERIMENTAL

Different cultivars of Iraqi and foreign grains of rice, wheat, and barley were collected from different plant fields in clean polyethelene bags. The samples were washed using IAEA method (*13*) and dried in an oven at 60°C. The dried samples were then crushed to fine powder with an agate mortar.

About 400 mg of each sample and (SRM 1571) ORCHARD LEAVES standard were irradiated for 6 h in a vertical channel of IRT-5000 reactor at a flux of 3.2×10^{13} $cm^{-2} \cdot s^{-1}$ and for 5 min with the pneumatic system at a flux of 2×10^{11} $cm^{-2} \cdot s^{-1}$.

γ-rays of samples and standard were measured using 30 cc Ge(Li) and 80 cc well-type HPGe detectors, each coupled to an HP-1000 computer through a 4096 channel ADC. The energy resolutions for the two detectors were 2.5 and 2 kev (FWHM), respectively, for the 1332 keV γ-ray of ^{60}Co.

Table 1
Data for Irradiation, Cooling, and Counting Times for Elements Measured by INAA

Irradiation time	Decay time	Counting time	Elements measured
5 min	5 min	1000 s	Mg, Al, Cl, Mn
6 h	4 d	2000 s	Na, K, Ca, Cu, As, Br, Mo, W
6 h	30 d	3000 s	Sc, Cr, Fe, Co, Zn, Se, Rb, Sr

The computer was programmed to identify each peak in the spectrum, check its shape and full-width-half-maximum, and to calculate the area under the peak by a least-squares fit. Using the computer program BABIL (*14*), it was possible to resolve doublet peaks.

γ-ray spectra of samples were taken several times after irradiation. Table 1 shows the data for irradiation, cooling, and counting times for the elements measured by INAA.

Foodstuffs and other biological samples usually contain much Na, K, and Br. The radionuclides ^{24}Na, ^{42}K, and ^{82}Br are produced in such samples by NAA, and therefore dominate the γ spectra to such an extent that the determination of some elements, such as Hg and As, becomes more complicated.

INTERFERENCES

Two types of interferences are involved in NAA, namely, interferences from other radionuclides and interferences from two or more different nuclear reactions producing the same radionuclide. The first type of interference can be eliminated by specific radiochemical separations in which the interfering radionuclides are removed from the system of interest; otherwise, the contribution of the other radionuclides can be subtracted from the photopeak of interest, using another γ-line of the interfering radionuclide as a reference line. Such correction was done for Cu because of the presence of Zn and Na. All other interferences are of the second type and are determined experimentally by irradiation of pure elements with weights many times that expected in the study samples. The results are always of negligible influence, because of the fact that the ratio of thermal neutrons to fast neutrons is about 100 at the position of irradiation. In the case of Al, the value was calculated and subtracted from the original value. The β spectra of ^{24}Na, ^{32}P, and ^{59}Fe produced usually in biological samples have been eliminated by using Plexiglas™ measuring sample holder.

Table 2
Range and Mean Concentrations of Elements Determined in Rice Samples (ppm)

Element	Iraq			Japan	Iran
	Range	Mean SD	Detection limit		
Na	8.76–25.76	15.56 ± 5.3	0.08	15.61 ± 0.65	11.78 ± 0.41
Mg	48.3–128.8	83.52 ± 24.31	8.38	–	–
Al	1.55–2.62	2.09 ± 0.303	0.24	–	–
Cl	75.1–318.5	167.9 ± 66	1.53	168.8 ± 11.3	32.67 ± 5.33
K%	0.053–0.097	0.074 ± 0.01	0.0006	0.193 ± 0.0093	0.0978 ± 0.00920
Sc	0.48–0.95	0.76 ± 0.172	0.35	–	–
Cr	0.05–0.18	0.11 ± 0.06	0.05	0.73 ± 0.35	1.18 ± 0.03
Mn	2.39–6.6	3.96 ± 1.2	0.09	17.48 ± 0.74	20.5 ± 0.7
Fe	2.0–5.44	4.5 ± 1.1	1.5	–	–
Co	0.016–0.10	0.037 ± 0.02	0.01	–	–
Cu	1.55–3.17	2.340 ± 0.50	0.08	4.8 ± 0.31	3.59 ± 0.69
Zn	7.12–14.65	10.6 ± 2.4	0.12	19.91 ± 1.2	20.22 ± 3.83
As	0.07–0.13	0.105 ± 0.02	0.01	0.034 ± 0.005	0.03 ± 0.005
Se	0.099–0.137	0.11 ± 0.02	–	–	–
Br	0.26–0.42	0.31 ± 0.04	0.02	1.24 ± 0.69	1.55 ± 0.13
Rb	0.20–0.50	0.33 ± 0.08	0.08	–	–
Mo	0.205–0.396	0.293 ± 0.050	0.05	–	–
W	0.07–0.46	0.246 ± 0.197	0.05	–	–

Range and Mean Concentrations of Elements Determined in Rice Samples (ppm) (*Con't*)

	Italy	Thailand Siam Panta	Holland Surlnm Panta	Belgium Super Patna	USA Kresto
Na	10.8 ± 0.6	–	–	–	–
Mg	–	–	–	–	–
Al	–	–	–	–	–
Cl	–	–	–	–	–
K%	0.131 ± 0.007	–	–	–	–
Sc	–	–	–	–	–
Cr	0.35 ± 0.04	1.875 ± 0.84	1.349 0 .56	0.655 ± 0.405	1.143 ± 0.71
Mn	6.5 ± 0.4	–	–	–	–
Fe	41.7 ± 3.4	24.3 ± 3.1	64.9 ± 18.6	139.8 ± 6.1	130.0 ± 10.4
Co	0.026 ± 0.008	0.026 ± 0.014	0.044 ± 0.021	0.031 ± 0.016	0.055 ± 0.019
Cu	1.13 ± 0.06	–	–	–	–
Zn	11.4 ± 0.7	–	–	–	–
As	–	–	–	–	–
Se	0.023 ± 0.003	0.029 ± 0.003	0.032 ± 0.012	0.253 ± 0.03	0.265 ± 0.036
Br	0.37 ± 0.03	–	–	–	–
Rb	1.604 ± 0.38	2.446 ± 0.617	2.939 ± 0.977	3.2 ± 0.404	6.4 ± 1.04
Mo	–	–	–	–	–
W	–	–	–	–	–

Table 3
Concentration of Elements Determined in Eight Cultivars of Wheat

Element	American wheat	Canadian wheat	Abu-Garaib	Maxipak	Nouri 70	Jirardo 574	Kokoret C-70	Saber Bag	D.L.[a]
Na[b]	16.68	16.86	78.80	58.60	24.94	47.00	108.9	11.60	0.4
Mg%	0.153	0.150	0.116	0.126	0.132	0.103	0.140	0.144	0.005
Al	19.50	21.20	19.80	13.90	14.60	12.30	15.30	14.30	1.200
Cl	654	569	1052	945	823	906	709	630	8.720
K%	0.304	0.298	0.280	0.350	0.370	0.375	0.550	0.490	0.004
Sc	2.0	2.27	1.320	3.240	1.470	0.830	1.610	2.400	0.260
Cr	0.110	–	–	–	–	–	–	0.230	0.080
Mn	42.10	42.60	46.20	49.70	43.40	29.40	37.40	52.40	0.500
Fe	37.50	48.74	34.00	47.50	34.10	38.77	39.80	44.70	1.600
Co	0.037	0.028	0.051	0.047	0.040	0.028	0.031	0.110	0.002
Cu	4.51	4.520	6.590	6.790	5.060	6.460	8.400	5.570	0.500
Zn	25.1	25.00	16.83	27.67	20.58	22.06	20.80	32.20	0.110
As	–	–	0.055	0.007	–	0.060	–	–	0.020
Se	0.660	1.360	0.240	–	0.490	0.240	0.880	–	0.120
Br	10.22	5.710	3.570	3.220	3.300	1.820	1.460	3.050	0.030
Rb	1.78	1.880	0.610	0.870	0.450	0.680	0.720	2.010	0.340
Sr	–	–	11.70	–	12.00	–	–	–	2.900
Mo	0.14	0.250	–	0.440	0.260	0.340	0.260	–	0.110

[a]Detection limits.
[b]All concentrations in ppm, except for Mg and K in percent.

Table 4
Concentration of Elements Determined
in Three Cultivars of Barley

Element	Cleapper	Numar	Arivat	D.L.[a]
Na[b]	78.90	158.8	88.46	0.42
Mg%	0.099	0.135	0.116	0.005
Al	35.8	45.1	34.3	1.7
Cl	1222	1486	1242	8.8
K%	0.620	0.820	0.450	0.004
Sc	4.60	7.56	–	0.3
Cr	0.090	0.370	0.160	0.08
Mn	12.60	23.01	16.51	0.47
Fe	46.00	74.10	54.90	1.5
Co	0.029	0.04	0.030	0.01
Cu	6.80	14.7	8.31	0.6
Zn	23.90	30.30	23.20	0.15
As	0.066	–	–	0.05
Se	0.44	0.65	–	0.19
Br	10.70	6.42	7.63	0.17
Rb	0.60	0.67	0.67	0.11
Sr	–	9.8	3.3	2.9
Mo	0.59	0.31	–	0.1

[a]Detection limits.
[b]All concentrations in ppm, except for Mg and K in percent.

Table 5
Accuracy of NAA Method for Wheat Flour
and Bovine Liver Reference Material in This Work

	Wheat flour		Bovine liver	
Elements	X ± SD[a]	Recommended value	X ± SD[a]	Recommended value
Na[b]	15.3 ± 0.76	–	2417 ± 207.7	2430 ± 130
K%	0.32 ± 0.02	–	1.08 ± 0.05	0.97 ± 0.06
Fe[c]	37.9 ± 1.14	84.3	262.7 ± 40.2	270 ± 20
Co	0.041 ± 0.002	0.037	0.210 ± 0.022	0.18
Cu	5.6 ± 0.77	2.9	182 ± 13.04	193 ± 10
Zn	35.9 ± 2.15	33.2	138.5 ± 14.2	130 ± 10
Br	15.5 ± 1.5	–	10.6 ± 0.9	–
Rb	2.11 ± 0.13	–	18.0 ± 1.8	18.3 ± 1

[a]X: Mean concentration. SD: standard deviation.
[b]All concentrations in ppm except for K in percent.
[c]Ref. (*7*) gives the same result.

RESULTS AND DISCUSSION

The range and mean concentrations of 18 elements in 12 types of rice grains are listed in Table 2. In this table, each value represents the average of two or more independent determinations. The values obtained in the present investigation were compared with the results of other countries (*11,12*). This comparison can be considered as a guide for the number and approximate concentrations of the trace elements. Also, there is a good agreement between the values obtained and the values quoted for rice from several countries, except for that of Fe, Mn, and Rb, which were found to be higher in the samples from Japan, Italy, Iran, Thailand, and America. This also applies for Se, which was found to be richer in the samples from Belgium and America. These variations in elemental concentrations may be the result of the variation in the climatic conditions as well as other affecting factors. It has been found that the concentrations of toxic elements (Hg and Sb) are lower than the detection limits. The concentrations of 18 elements in eight types of wheat and three types of barley grains are listed in Table 3 and Table 4, respectively.

The accuracy of our method has been assessed by the analysis of IAEA standard wheat flour and Bovine liver. A good agreement has been achieved between the present results and the recommended values for the standard as shown in Table 5, except for Fe in wheat flour, which is in agreement with reference (*7*). The concentrations of the elements Na, K, Br, and Rb determined in IAEA standard "wheat flour" were not certified. Precision is an evaluation of random errors in the experiment. It can be caused by sampling, weighing, random contamination, instrumental fluctuations, and counting statistics. These propagative errors are cumulative and characterized by relative SD values, so that for precision estimation, the rice sample "IR-232" and wheat sample "Saber Beg" have been analyzed by six runs. The relative SD values for determined elements in rice and wheat were $<10\%$.

The detection limits for measured compounds were calculated using the formula introduced by Currie (*15*). The results are given in Table 2 for rice grains and in Tables 3 and 4 for wheat and barley, respectively.

REFERENCES

1. J. M. A. Lenihan and S. J. Thomson, eds., *Activation Analysis, Principles And Application*, Academic, London, 1965.
2. K. Schwarz, *Life Sciences, Proceed. of Symp. Bled.*, IAEA, Vienna, 3 (1977).
3. H. J. M. Bowen, *Life Sciences, Proced. of Symp., Bled*, IAEA, Vienna, 393 (1972).
4. E. J. Underwood, *Human Nutrition*, 3rd Ed., Academic, New York, 1971.
5. D. K. Teherani, R. Rehwoldt, J. Washvttl, and I. Kiss, *Acta Alimentaria*, **4**, 405 (1975).

6. H. A. Vander Stoot, G. D. Wals, C. A. Weers, C. A. Das, and H. A. Das, *Anal. Chem.* **52,** 112 (1980).
7. P. S. Tjioe, J. J. M., Degoeij, and J. P. W., Houtman, *J. Radioanal. Chem.* **37,** 511 (1977).
8. R. M. Awadallah, M. K. Sherif, A. H. Amrallah, and F. Grass, *J. Radioanal. and Nucl. Chem. Articles* **98,** 325 (1986).
9. W. D. Wei-Zhi Tain, Ehmann, *J. Radioanal. and Nucl. Chem. Articles* **89,** 109 (1985).
10. R., Demiralp, *1st Balkan Conf. on Activation Analysis,* Varna, Bulgaria, May 6–8, 1985.
11. Z. Abedin Zadah, A. Owlya, and H. Rafii, *Radiochem. Radioanal. Letters* **32,** 315 (1978).
12. D. K. Teherani, *J. Radioanal. and Nucl. Chem. Letters,* **117,** 113 (1987).
13. Yu S. Ryabukhin *Report of the advisory group on the applications of nuclear methods in environmental research.* IAEA, Vienna, May 1976.
14. A. Hussain, M. Sc. Thesis, University of Baghdad, 1977.
15. L. A. Currie, *Anal. Chem.* **40,** 586 (1968).

Nuclear and Atomic Analytical Techniques in Environmental Studies in South America

A. S. PASCHOA

Pontifícia Universidade Católica do Rio de Janeiro (PUC/RJ), Depto. de Física, C. P. 38071, Rio de Janeiro, RJ 22453

Received April 17, 1989; Accepted December 11, 1989

ABSTRACT

The use of nuclear analytical techniques for environmental studies in South America is selectively reviewed since the time of earlier works of Lattes with cosmic rays until the recent applications of the PIXE (particle-induced X-ray emission) technique to study air pollution problems in large cities, such as São Paulo and Rio de Janeiro. The studies on natural radioactivity and fallout from nuclear weapons in South America are briefly examined.

Index Entries: Environmental studies; radioactivity measurements; accelerator/reactor-based analytical techniques; PIXE.

INTRODUCTION

The use of nuclear analytical techniques in South America dates back to the end of World War II. At that time, Cesar Lattes, Ugo Cammerini, and Gleb Wataguin had built a cloud chamber at the University of São Paulo. Lattes contacted Giuseppe Occhialini and Cecil Powell, who were at the University of Bristol, to ask to work with the Ilford emulsion that was being used by them to obtain photomicrophotographs of protons and α particles (*1*). This contact marked the beginning of a fruitful cooperation that led eventually to the detection of π mesons, first at the

Pic du Midi, in the Pyrinees, and later at the meteorological station at Chacaltaya Mountain in Bolivia. After spending a little over one year at the University of California, Berkeley, Lattes returned to Brazil in 1949, and established with other young Brazilian scientists the Centro Brasileiro de Pesquisas Físicas (CBPF), in Rio de Janeiro. Apparently, the work of Lattes with cosmic rays in Chacaltaya was the first application of a nuclear technique to environmental studies in South America.

This article reviews selectively and briefly the reports of the use of nuclear and atomic analytical techniques in South America for environmental studies since the mid-1950s. Unfortunately, a comprehensive review cannot be presented at this opportunity, because the responses to most inquiries have not yet been received. The number and diversity of applications of nuclear and atomic analytical techniques in South America have been surprisingly large since the pioneer work of Lattes.

FALLOUT STUDIES

A report by the National Academy of Sciences (NAS), published in 1956, that discussed the use of radioactivity in atmospheric studies, among other topics, recognized two important sources of naturally occurring radioactivity in the atmosphere: cosmic-ray interactions in the stratosphere, and the rock and soil of the earth's crust (*2*). This same NAS report also addressed some meteorological aspects of the effects of atomic radiation, including the debris from nuclear weapons tests and the possibility of an accidental release to the atmosphere of large amounts of fission products, in consequence of damage or destruction of a nuclear power reactor or fuel reprocessing facility (*2*). The 1956 NAS report was published almost simultaneously with the first Brazilian studies on the fallout from the nuclear weapons tests. These studies from Brazil started being reported in 1956 (*3,4*), and further reports continued to be published at least until 1971 (*5–19*). Studies on the fallout from nuclear weapons tests in the atmosphere occurred also in other South American countries (*20–32*). These environmental studies were instrumental in the development of a local competence in nuclear analytical techniques, such as γ and α spectrometry, and associated radiochemistry.

Figure 1 is a histogram of the number of kilotons released per year from nuclear explosions in the earth's atmosphere. The dark bars represent the number of kilotons released per year in the Southern Hemisphere, which in some years constituted a significant fraction of the total. Most of the fallout in this hemisphere occurred as a consequence of such releases. The years between 1956 and 1962, with a short moratorium in 1959 alone, nuclear bombs totaling 108,000 kilotons were exploded. Data on the 1963 explosions, if any, are not readily available.

An unusually large radioactive particle was detected in Rio de Janeiro on June 25, 1957 (*11*). The origin of this particle was traced back, based on its radioactive decay properties, to a nuclear explosion that

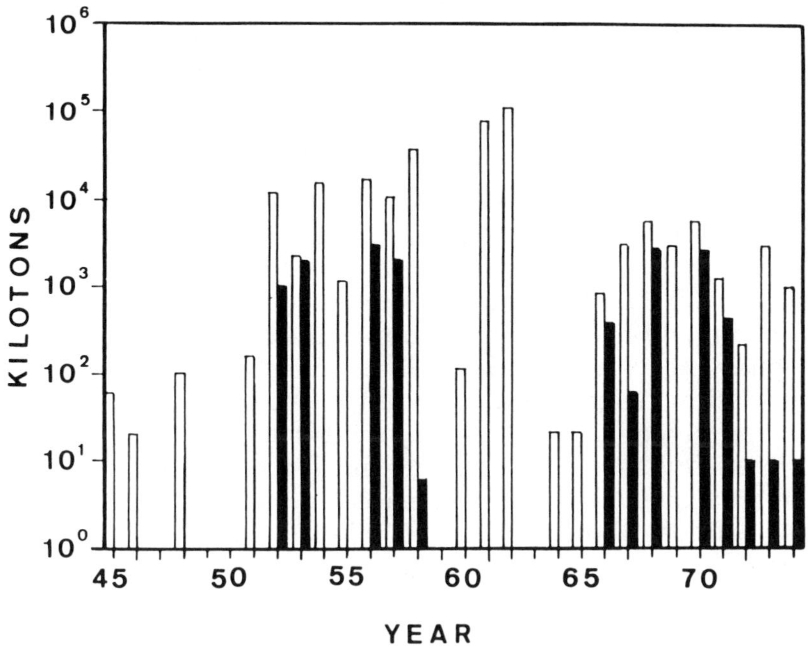

Fig. 1. Number of kilotons released per year from nuclear explosions: white and black bars indicate releases in the Northern and Southern Hemispheres, respectively. (Data taken from refs. *75–77,80,82*.)

occurred in the Christmas Islands on May 15, 1957. This was among the first experimental evidence that the "stratospheric fallout" originated in the other hemisphere, though these islands are located just a few degrees into the north latitude.

The second nuclear moratorium was signed by the UK, USA, and USSR in 1963. However, the ^{90}Sr activity already injected into the atmosphere by the nuclear weapons tests had reached 7.1×10^{15} Bq (19 mCi). The variation of the ^{90}Sr activity during the years between 1957 and 1965 in the air of both hemispheres was used by the United Nations Scientific Committee on the Effects of Atomic Radiation (UNSCEAR) to estimate the deposition of other radionuclides in each of the earth's hemispheres (*78*).

The United States Naval Research Laboratory (USNRL) established, between 1957 and 1962, in cooperation with the governments of Bolivia, Chile, Equador, and Peru, an observational network to determine the ^{90}Sr concentration in the air near the 80°W meridian (*79*). After the 1963 moratorium, the Health and Safety Laboratory (HASL) expanded the observation network around the 80°W meridian to include 49 stations between the latitudes 90°S (South Pole) all the way to 81° 40 N (Thule, Greenland) (*80*). Local efforts, mainly by scientists in Argentina and Brazil, made possible the development of nuclear analytical techniques needed to examine independently the nuclear weapons fallout in their geographical areas. Data obtained locally and through the USNRL indicated that the surface deposition of ^{95}Zr in intermediate latitudes, for the period between 1966 and 1973, was larger in the Southern than in the Northern Hemisphere. This is illustrated in Fig. 2.

Pu-238 was injected into the atmosphere as a result of nuclear explosions in an amount estimated to be about 3.7×10^{13} Bq (10 kCi), based on the ^{90}Sr measurements. However, an activity of about 6.3×10^{13} Bq (17 kCi) ^{238}Pu was introduced into the stratosphere as a consequence of the accident during the re-entry of the SNAP-9A device in the atmosphere on April 24, 1964 (*78,81,82*). The accident occurred at about 50 km above the Indian Ocean. and ^{238}Pu contamination of the stratosphere was detected by balloons sent to an altitude of 33 km at 34°S on August, 1964 (*78*).

The window of opportunity to launch a spaceship this year towards the direction of Venus and then to Jupiter makes the use of ^{238}PU as a long-term energy source a possibility much too real to be ignored. The potential consequences of spreading accidentally a large amount of ^{238}Pu at a low latitude, though the probability of such occurrence can be considered minute, cannot be disregarded during the launching preparations.

STUDIES ON NATURAL RADIOACTIVITY

The problem of the biological effects of chronic exposure to low levels of ionizing radiation provided the incentive for a cooperative venture that involved investigators from Brazil and the United States to study the Brazilian areas of high natural radioactivity. Such studies were helpful to establish nuclear analytical techniques in several Brazilian universities and research institutes. As a consequence, a number of scientists were trained in the use of these techniques. A large number of articles and reports were published in connection with these studies (*33–56*).

The natural radioactivity studies in Brazil evolved from radioecological investigations associated with the areas of high natural radioactive background (such as Araxa, Guarapari, and Pocos de Caldas), to the whole body counting of some inhabitants of Guarapari (*33–56, 83–85*). A population of approx 6000 lived on a monazite-bearing sand bar in Guarapari in the early 1960s. A survey of external radiation dose was carried out by using lithium fluoride dosimeters. The average annual dose distribution resulting from the survey is presented in Fig. 3 (*44,85*). The high external doses associated with ^{224}Ra body burdens up to approx 10 Bq (0.27 nCi) inferred from thoron in breath measurements prompted an effort to measure directly these burdens by using a whole body counter. First an effort was made to measure body burdens using a whole body counter with the shielding made of sugar (*85*). The high humidity of Rio de Janeiro in the summer of 1967 made the shield start dissolving. A low-level radioactivity room was then built with 18 metric tons of steel as shield. The ceiling and floor are 0.20 m (8 in) in thickness

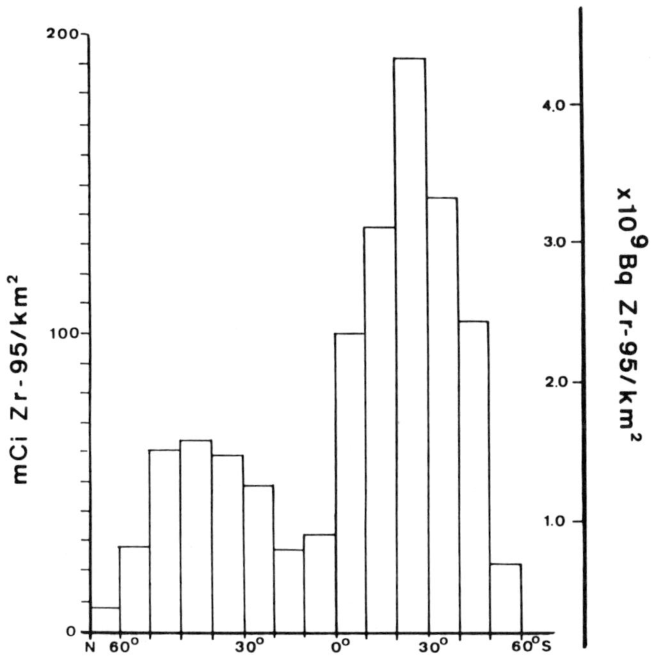

Fig. 2. Zr-95 deposition in low and intermediate latitudes of the Northern and Southern Hemispheres between the years 1966 and 1973. (Data taken from ref. *77*.)

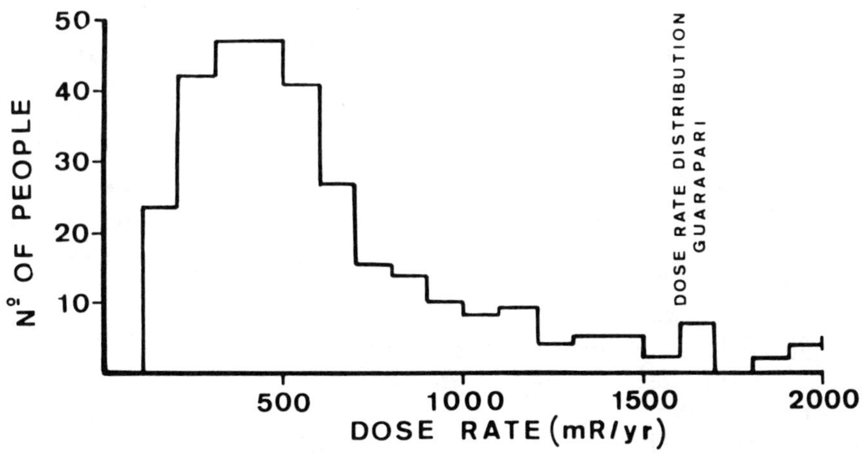

Fig. 3. Distribution of the average annual external radiation dose to the population of Guarapari for 1965/1966. (Data taken from refs. *44,85*.)

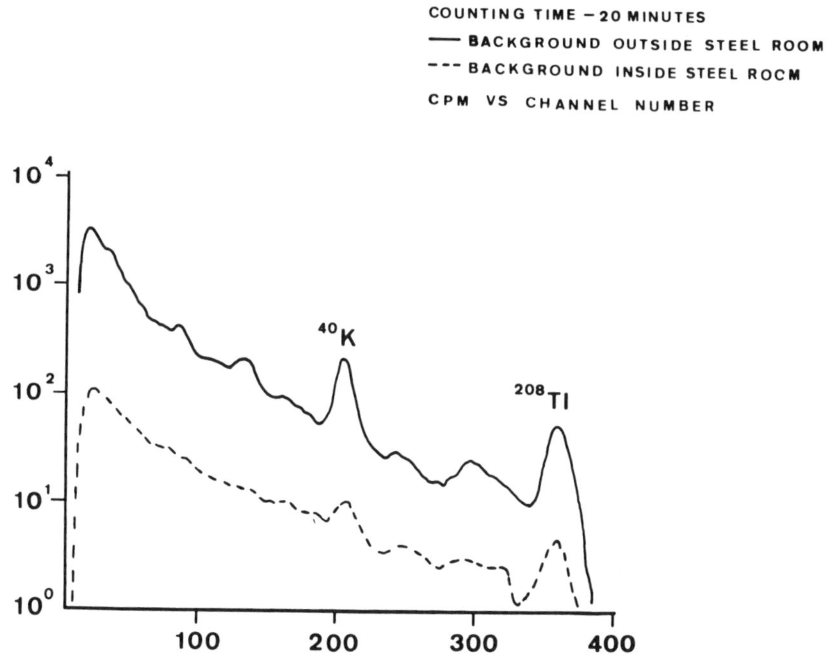

Fig. 4. Background spectra, outside and inside the low-radioactivity steel room, measured with a NaI(Tl) detector. (From ref. *47*.)

and the side walls 0.10 m (4 in). The interior volume dimensions are 2 × 1 × 1.5 m. Figure 4 shows background spectra outside and inside the low-radioactivity steel room, as measured with a 0.15 m (6 in) × 0.10 m (4 in) NaI(Tl) detector for 20 min in each case. Figure 5 illustrates the chair and phantom geometry used to calibrate the steel whole body counter, mostly to detect lung contamination by thoron inhalation. Table 1 presents the minimum detectable activity (MDA) of the chair geometry for several γ-ray energies, based on a 1-h counting time. Unfortunately, the MDAs were not low enough to detect the body burdens of the Guarapari inhabitants who were brought to the laboratory in Rio de Janeiro to be measured. It was concluded that the body burdens were mostly the result of thoron inhalation. As a consequence, the activity decayed with the 10.6-h half-life of ^{212}Pb, making the ^{208}Tl activity lower than the 63 Bq (1.7 pCi) MDA.

The whole body counter was instrumental, however, in determining temporary ^{212}Pb body burdens of two scientists and several workers in a gas mantle factory, because of thoron inhalation. The fabrication process of the gas mantle used ThO_2, maintained in a tank with poor air circulation and/or renovation. As a consequence, the thoron concentration in the air near the ^{232}Th liquid sources (as nitrate) reached levels as high as 222 Bq · L^{-1} (6 nCi · L^{-1}). Figure 6 illustrates the ^{212}Pb body burdens of workers of the gas mantle factory as a function of the day of the week, based on the measurement of the ^{208}Tl photopeak. The scientists and the

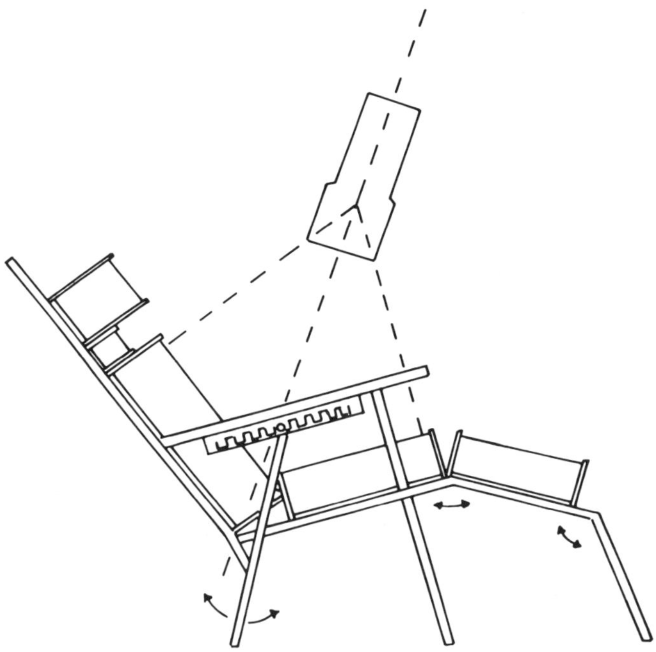

Fig. 5. Chair and plastic phantom counting geometry to calibrate the whole body counter. (From ref. *47*.)

Table 1
Minimum Detectable Activity as a Function of γ-Ray Energy

Nuclide	Energy, MeV	Minimum detectable activity Bq	nCi
^{137}Cs	0.662	333	9.0
^{54}Mn	0.835	307	8.3
^{60}Co	1.173	89	2.4
^{22}Na	1.271	159	4.3
^{40}K	1.460	189	5.1
^{214}Bi	1.760	118	3.2
^{88}Y	1.840	104	2.8
^{208}Tl	2.615	63	1.7

workers were exposed voluntarily and involuntarily, respectively, to $Th(NO_3)_4$ fumes in the gas mantle factory. In all cases, the body burdens decayed with the ^{212}Pb half-life, as expected from the earlier information available from the previous work with the Guarapari inhabitants.

The experience accumulated with the Brazilian studies on natural radioactivity, and with those on radioactive fallout from the nuclear

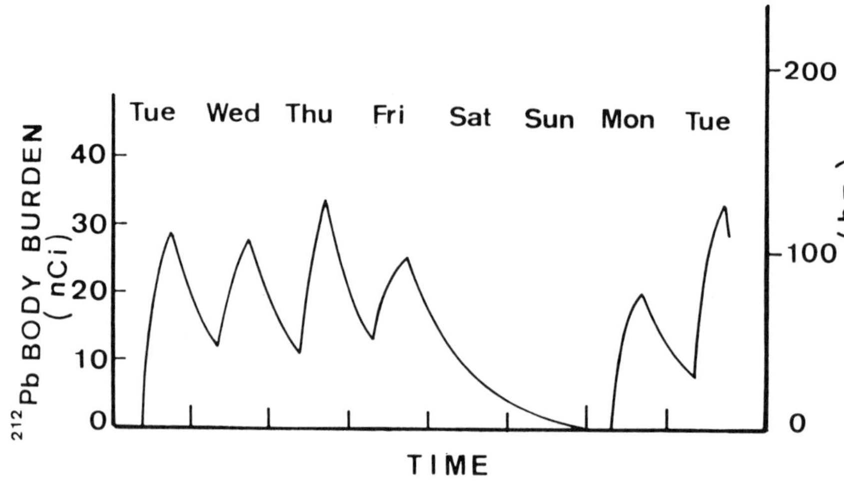

Fig. 6. Weekday variation of the ^{212}Pb body burdens of scientists and workers exposed to thoron inhalation. (Based on data taken from refs. *49,50*.)

weapons tests, proved to be very helpful to assess the consequences of the accidental release of about 3.7×10^{13} Bq (1 kCi) of ^{137}Cs from a deactivated teletherapy capsule in Goiania in 1987 (*86*).

ACCELERATOR-BASED TECHNIQUES

Accelerator-based atomic analytical techniques, such as PIXE (particle-induced X-ray emission), started being used simultaneously at PUC/RJ in Rio de Janeiro and at the Universidade de São Paulo (USP), in São Paulo in the second part of the 1970s (*57,58*). The investigators from Rio de Janeiro used PIXE to study a number of environmental and biological samples (*58,59–62*), whereas the group from São Paulo concentrated their PIXE efforts to study air pollution problems (*57,63–67*). An overview of the activities of other South American groups that are using PIXE techniques for air pollution studies is about to be published (*68*). As a consequence of lack of materials to make appropriate targets for irradiation, it is a common practice for laboratories of the region to establish cooperative research efforts with European and American groups that use similar accelerator-based techniques to study environmental and biological problems.

Photofission has recently been used at the linear accelerator at USP as an analytical technique to measure uranium in environmental samples. The Comissao Nacional de Energia Nuclear (CNEN) of Brazil operates a cyclotron that also has capabilities for nuclear analytical techniques, such as charged-particle activation analysis, to be used in environmental studies. This cyclotron is one of the tools used now in joint environmental studies involving CNEN, PUC/RJ, and the Universidade Federal do Rio de Janeiro (UFRJ).

REACTOR-BASED TECHNIQUES

Nuclear analytical techniques based on research reactors have been used in South America for a long time now. However, the application of these techniques to environmental studies has been meager (*69–72*). The South American activities in reactor-based neutron activation analysis (NAA) until 1984 were reviewed by Guinn (*73*). Argentina, Brazil, Chile, and Venezuela are the only South American countries with active programs in reactor-based NAA to apply the technique to environmental studies. The IEA-R1 research reactor at the Instituto de Pesquisas Energeticas e Nucleares (IPEN/CNEN), in São Paulo, is the source of neutrons for NAA results being reported at this Conference (*87*).

THE AMAZON BASIN

Since 1986, the International Atomic Energy Agency and the Government of Brazil (IAEA) have been negotiating an agreement to study the Amazon Basin, concerning (1) the routing of atmospheric water vapor, and (2) the water budgets and water use of continuous forest, isolated forest, and cleared areas. The Amazon project will take advantage of a number of nuclear and atomic analytical techniques to be implemented mostly by the Centro de Energia Nuclear na Agricultura (CENA) at the Universidade de São Paulo, Piracicaba. The Instituto Nacional de Pesquisas Amazonicas (INPA), the Instituto Nacional de Pesquisas Espaciais (INPE), and the World Wildlife Fund are cooperating to make this project feasible. Other South American countries with direct interest in the Amazon Basin are expected to join the IAEA-supported Amazon project. The general aspects of this project were mentioned by the President of Brazil in a recent speech, but the detailed description of the project is not of public knowledge as yet.

Nuclear and atomic analytical techniques will play important roles in the Amazon project, in accordance with internal reports from the government of Brazil just made available for potential participating scientists from Brazil and abroad (*74*). Competent help will be needed to implement the application of nuclear and atomic analytical techniques to the complex environment of the Amazon Basin. Material resources will be made available to the extent that internal and external public opinion exert pressure. However, well-trained human resources are scarce in South America, and cannot be created fast enough for the needs of the Amazon project. As a consequence, well-trained scientists concerned with the future of the Amazon Basin should start thinking about cooperating with the gigantic effort necessary to protect the still largely untamed tropical rain forest.

CONCLUDING REMARKS

1. Lattes started using nuclear techniques in South America to study cosmic rays.
2. Fallout from nuclear weapons tests in the atmosphere motivated investigations in South America that took advantage of the then-existing nuclear analytical techniques.
3. Studies on natural radioactivity in Brazil and elsewhere in South America were helpful to train local scientists in the use of nuclear analytical techniques for environmental studies.
4. Accelerator-based and research reactor-based atomic and nuclear analytical techniques are currently being used in several universities and laboratories of South American countries.
5. The challenge of protecting the environment of the Amazon Basin constitutes the ultimate test for the use of nuclear and atomic analytical techniques in an international framework of scientific cooperation.

REFERENCES

1. C. M. G. Lattes, *Topics on Cosmic Rays,* Editora da UNICAMP, Campinas, 1984.
2. National Academy of Sciences-National Research Council, *The Biological Effects of Atomic Radiation, Summary Reports,* Washington, **1** (1956).
3. A. Aron, *An. Acad. Brasil. Ciênc.,* **28,** 423 (1956).
4. G. Kegel, *An. Acad. Brasil. Ciênc.,* **28,** 447 (1956).
5. E. Penna Franca, A. Pinto Coelho, and C. Chagas, *An. Acad. Brasil. Ciênc.,* **29,** 553 (1957).
6. L. Marquez, N. L. Costa, and I. G. Almeida, *Radioactive Products in the Soil and in the Atmosphere,* Comissão Nacional de Energia Nuclear, Instituto Brasileiro de Bibliografia e Documentação, Rio de Janeiro, p. 55, 1958.
7. M. Campos and E. Peralles Pinto, *Radioactive Products in the Soil and in the Atmosphere,* Comissão Nacional de Energia Nuclear, Instituto Brasileiro de Bibliografia e Documentaçaõ, Rio de Janeiro, p. 65, 1958.
8. E. Penna Franca and A. Pinto Coelho, *Radioactive Products in the Soil and in the Atmosphere,* Comissão Nacional de Energia Nuclear, Instituto Brasileiro de Bibliografia e Documentação, Rio de Janeiro, p. 99, 1958.
9. G. Kegel, *Radioactive Products in the Soil and in the Atmosphere,* Comissaõ Nacional de Energia Nuclear, Instituto Brasileiro de Bibliografia e Documentação, Rio de Janeiro, p. 103, 1958.
10. E. Penna Franca and A. Pinto Coelho, *Proceedings of the Second International Conference on the Peaceful Uses of Atomic Energy,* vol. XVIII, United Nations, Geneva, p. 512, 1958.

11. B. Gross, A. Aron, E. Meyer, and F. X. Roser, *Proceedings of the Second United Nations International Conference on the Peaceful Uses of Atomic Energy,* vol. XVIII, United Nations, Geneva, p. 575, 1958.
12. B. Gross, A. Aron, E. Meyer, F. X. Roser, S. Costa Ribeiro, and A. C. Olinto, *Acta Physica Austriaca* **2,** 187–195 (1958).
13. B. Gross and A. Aron, *Proceedings of the Second United Nations International Conference on the Peaceful Uses of Atomic Energy,* vol. XVIII, United Nations, Geneva, p. 580, 1958.
14. E. Penna Franca and A. Pinto Coelho, *An. Acad. Brasil Cieñc.,* **31,** 49 (1959).
15. B. Gross, E. Meyer, and A. Aron, *Radioatividade do "fallout" no Rio de Janeiro,* Comissão Nacional de Energia Nuclear, Instituto Brasileiro de Bibliografia e Documentação, DTC 3, p. 1, 1959.
16. E. Penna Franca, *Segundo Simposio Interamericano sobre la Aplicacioń de la Energia Nuclear para Fines pacíficos,* C.I.E.N., Organización de Estados Americanos, Washington, p. 63, 1960.
17. E. Penna Franca, P. L. dos Santos, and A. S. Paschoa, *An. Acad. Brasil. Ciênc.,* **41,** 583 (1969).
18. E. Meyer, *An. Acad. Brasil. Ciênc.,* **43,** 331 (1971).
19. A. H. Mendonça and E. Penna Franca, *An. Acad. Brasil. Ciênc.,* **43,** 813 (1971).
20. D. Beninson and E. Ramos, *Contaminación por I-131 debida a "fallout,"* Comisión Nacional de Energia Atómica, Informe No. 75 (1962).
21. D. Beninson, A. Migliori, H. Mugliaroli, and E. van der Elst, *4th Interamerican Symposium on the Peaceful Applications of Nuclear Energy,* vol. II, Pan American Union, Washington, pp. 85–92, 1962.
22. C. Benech and C. M. Franchi, *Arch. Soc. Biol. Montevideo,* **26,** 81–86 (1963).
23. C. Benech and C. M. Franchi, *Sr-90 in dientes de niños,* Grupo de estudio sobre Higiene Radiofísica, International Atomic Energy Agency and Pan American Health Organization, Buenos Aires, Nov. 1966.
24. C. Benech, *Radioiodo debido a "fallout" en ganado,* Ministerio de Ganadería y Agricultura, Depto. de Prensa y Biblioteca, Boletín Informativo No. 1194, Montevideo, 1967.
25. C. Benech and C. M. Franchi, Arch. *Soc. Biol. Montevideo,* **27,** 26–29 (1969).
26. D. Beninson, A. Migliori de Beninson, and C. Menossi, *Fallout radioactivo debido a las explosiones en el pacífico Sur en el período 1966–1977,* Comisión Nacional de Enewrgia Atómica, Dependiente de la Presidencia de la Nación, Gerencia de Protección Radiológica y Seguridad, Inf. RS 28/49, 1971.
27. Comisión Nacional de Energia Atómica, *Dependiente de la Presidencia de la Nación,* Gerencia de Pretección Radiológica y Seguridad, División Estudios Ambientales, Compilación de los resultados del monitoraje de Sr-90 y Cs-137 debido al fallout en la Republica Argentina, Inf. RS 34/66, 1972.
28. C. Benech and S. Cubas de Porta, *Arch. Pediatría del Uruguay,* **45,** 232–239 (1974).
29. S. Cubas de Porta and C. Benech, *Arch. Pediatría del Uruguay,* **46,** 149–152 (1975).
30. C. Benech and S. Cubas de Porta, *Sintaxis,* **2,** 65–76 (1976).
31. C. Benech. A. Kun, and S. Cubas de Porta, *An. Fac. Odontologia,* **16,** 55–60 (1978).
32. C. Benech and S. Cubas de Porta, *Revista de Medicina y Biología Nuclear (ALASBYMN)* **12,** 13–14 (1980).
33. F. X. Roser and T. L. Cullen, *Radioactive Products in the Soil and Atmosphere,* Instituto Brasileiro de Bibliografia e Documentacão, pp. 5–27, 1958.

34. T. L. Cullen, *An. Acad. Brasil. Ciênc.*, **29,** 545–552 (1958).
35. F. X. Roser and T. L. Cullen, *An. Acad. Brasil. Ciênc.*, **34,** 23–35 (1962).
36. F. X. Roser and T. L. Cullen, *Environmental Radioactivity in High Background Areas of Brazil,* Institute of Physics, Pontifical Catholic University, Rio de Janeiro, pp. 1–105, 1962.
38. F. X. Roser and T. L. Cullen, *Science,* **138,** 145–146 (1962).
39. F. X. Roser and T. L. Cullen, *The Natural Radiation Environment I,* J. A. S. Adams and W. M. Lowder, eds., Univ. Chicago Press, Chicago, pp. 825–872, 1964.
40. M. Eisenbud, H. Petrow, R. T. Drew, F. X. Roser, G. Kegel, and T. L. Cullen, *The Natural Radiation Environmental I,* J. A. S. Adams and W. M. Lowder, eds., 837–854, Univ. Chicago Press, Chicago, pp. 837–854, 1964.
41. F. X. Roser, G. Kegel, and T. L. Cullen, *The Natural Radiation Environment I,* J. A. S. Adams and W. M. Lowder, eds., Univ. Chicago Press, Chicago, pp. 855–872, 1964.
42. F. X. Roser, G. H. Kegel, L. Hainberger, T. L. Cullen, and R. T. Drew, *A Study of Artificial and Natural Radioactive Contamination in Brazil,* Instituto de Fisica, Pontificia Universidade Católica do Rio de Janeiro, USAEC-AT (30-1)-2577, pp. 1–83, 1964.
43. E. Penna Franca, J. C. C. Almeida, J. Becker, M. Emmerich, F. X. Roser, G. Kegel, L. Hainberger, T. L. Cullen, H. Petrow, R. T. Drew, and M. Eisenbud, *Health Phys.*, **11,** 699 (1965).
44. T. L. Cullen, *Health Phys.*, **12,** 970–971 (1966).
45. E. Penna Franca, M. Fiszman, N. Lobão, C. Costa Ribeiro, H. Trindade, P. L. dos Santos, and D. Batista, *Health Phys.*, **14,** 95–99 (1968).
46. T. L. Cullen, I. M. Antunes, A. S. Paschoa, H. Quintella, L. Hainberger, H. A. S. Andrade, and S. V. Duarte, *A Study of Natural Radioactivity in Brazil,* Pontificia Universidade Católica, Rio de Janeiro, NYO-2577-10, pp. 1–43, 1968.
47. T. L. Cullen, I. M. Antunes, A. S. Paschoa, and H. Quintella, *A Study of Natural Radioactivity in Brazil,* Pontificia Universidade Católica do Rio de Janeiro, NYO-2577-11, pp. 1–25, 1969.
48. E. Penna Franca, M. Fiszman, N. Lobão, H. Trindade, C. Costa Ribeiro, and P. L. dos Santos, *Health Phys.*, **19,** 657–662 (1970).
49. T. L. Cullen and A. S. Paschoa, *A Study of Natural Radioactivity in Brazil,* Pontifícia Universidade Católica do Rio de Janeiro, pp. 1–55, 1972.
50. E. Penna Franca, C. Costa Ribeiro, D. E. Gonzales, T. I. Cullen, and M. Barcinski, *The Natural Radiation Environment II,* J. A. S. Adams, W. M. Lowder, and T. F. Gesell, eds., CONF. 720805, 1975.
51. L. Hainberger, I. R. Oliveira Paiva, H. Andrade, G. Zundel, and T. L. Cullen, *Radiological Data and Reports* **15,** 483 (1974).
52. E. Penna Franca, *Areas of High Natural Radioactivity,* T. L. Cullen and E. Penna Franca, eds., Acad. Brasil. Ciênc., pp. 29–48, 1975.
53. T. L. Cullen, *Areas of High Natural Radioactivity,* T. L. Cullen and E. Penna Franca, eds., pp. 49–64, 1975.
54. T. L. Cullen and A. S. Paschoa, *Natural Radioactivity in Man's Environment,* Health Physics Society, p. 301, 1976.
55. T. L. Cullen, A. S. Paschoa, E. Penna Franca, C. Costa Ribeiro, M. Barcinski, and M. Eisenbud, in the *5th Int'l. Congress of the Int'l. Radiation Protection Association,* vol. III, pp. 335–338, 1980.
56. M. Eisenbud, K. Krauskopf, E. Penna Franca, W. Lei, R. Ballad, P. Linsallata, and K. Fujimore, *Environ. Geol. Water Sci.* **6,** 1–9 (1984).

57. A. S. Paschoa, E. C. Montenegro, W. Margulis, A. G. de Pinho, C. V. Barros Leite, and G. B. Baptista, *Nuclear Methods in Environmental and Energy Research,* James R. Vogt, ed., CONF-771072, University of Missouri, Columbia, pp. 21–48, 1977.
58. C. M. Q. Orsini and L. C. S. Bouéres, *Nucl. Instr. and Meth.,* **142,** 27 (1977).
59. E. C. Montenegro, G. B. Baptista, L. V. de Castro Faria, and A. S. Paschoa, *Nucl. Instr. and Meth.,* **168,** 471–484 (1980).
60. A. S. Paschoa, G. B. Baptista, E. C. Montenegro, C. V. Barros Leite, and T. L. Cullen, *Nuclear Methods in Environmental and Energy Research,* James R. Vogt, ed., CONF-800433, University of Missouri, Columbia, pp. 147–158 (1980).
61. G. B. Baptista, E. C. Montenegro, A. S. Paschoa, and C. V. Barros Leite, *Nucl. Instr. and Meth.,* **181,** 263–267 (1981).
62. A. S. Paschoa, G. B. Baptista, G. M. Mauricio, C. V. Barros Leite, Y. B. Lerner, and P. F. Issler, *Nucl. Instr. and Meth. in Phys. Research* **B3,** 352–356 (1984).
63. C. M. Q. Orsini and L. C. S. Bouéres, *Rev. Bras. Fis.,* **9,** 747 (1979).
64. L. C. S. Bouéres and C. M. Q. Orsini, *Nucl. Instr. and Meth.* **181,** 417–424 (1981).
65. P. Artaxo and C. Orsini, *Nucl. Instr. and Meth. in Phys. Research,* **B22,** 259–263 (1987).
66. M. H. Tabacniks, C. Orsini, and P. Artaxo, *Nucl. Instr. and Meth. in Phys. Research,* **B22,** 315–318 (1987).
67. P. Artaxo, H. Storms, F. Bruynseels, R. Van Grieken, and M. Maehaut, *J. Geophys. Research,* **93,** 1605–1615 (1988).
68. II Seminário Latinoamericano de Analisis por Técnicas de Rayos-X, SARX-87, Bahia Blanca, Argentina.
69. L. P. Geraldo, O. Y. Mafra, and E. M. Tanaka, *J. Radioanal. Chem.,* **49,** 115 (1979).
70. A. S. Paschoa, C. A. N. Oliveira, and O. Y. Mafra, *Trans. Amer. Nucl. Soc.,* **39,** 359–360 (1981).
71. A. S. Paschoa, A. L. Ferro, and O. Y. Mafra, *Trans. Amer. Nucl. Soc.,* **45,** 60–61 (1983).
72. A. S. Paschoa, O. Y. Mafra, D. O. Cardoso, and A. C. S. Rocha, *Nuclear Tracks and Radiation Measurements,* **8,** 469–472 (1984).
73. V. P. Guinn, *Nuclear Methods in Environmental and Energy Research,* CONF-840408 (DOE), p. 23.
74. R. N. Alves, Comissão Nacional de Energia Nuclear (CNEN), personal communication (1989).
75. M. W. Carter and H. H. Moghissi, *Health Phys.,* **33,** 55 (1977).
76. W. J. Gibbs, J. R. Moroney, D. J. Stevens, and E. W. Titerton, *Fallout over Australia from Nuclear Weapons Tested by France in Polynesia during June and July 1972,* Publishing Service, Canberra, 1972.
77. A. S. Paschoa and C. Benech, *An. Acad. Brasil. Cienc.* To be published (1989).
78. United Nations Scientific Committee on the Effects of Atomic Radiation. Report to the General Assembly. Official Records of the Twenty-First Session, Supplement No. 14 (A/6314), United Nations, New York (1968).
79. United States Naval Research Laboratory, *Radiological Health Data Repts.,* **2,** 477 (1961).
80. H. W. Feely, L. E. Toonkel, and M. Schoenberg, *Health and Safety Laboratory,* Environmental Quarterly, Appendix B, HASL-328, 1977.
81. United Nations Scientific Committee on the Effects of Atomic Radiation,

Report to the General Assembly, vol. I, Annex A, United Nations, NY, E.72.IX.17, p. 19, 1972.
82. United Nations Scientific Committee on the Effects of Atomic Radiation. *Report to the General Assembly*, Annex C, United Nations, NY, E.77.IX.1, p. 115, 1977.
83. F. X. Roser, T. L. Cullen, L. Hainberger, N. L. Costa, H. Gomes, I. M. Antunes, and A. S. Paschoa, *A Study of Natural Radioactivity in Brazil*, Pontificia Universidade Catolica, Rio de Janeiro, NYO-2577-7, pp. 1–90, 1966.
84. A. S. Paschoa, *Environmental Transfer to Man of Radionuclides Released from Nuclear Installations*, IAEA/CEC, Luxemburg, vol. II, V/7400/84-EN, p. 827, 1984.
85. T. L. Cullen, I. M. Antunes, A. S. Paschoa, J. Pohl-Ruling, L. Hainberger, G. Zundel, H. A. S. Andrade, and I. R. O. Paiva, *A Study of Natural Radioactivity in Brazil*, Pontifícia Universidade Católica, Rio de Janeiro, NYO-2577-8, pp. 1–68, 1967.
86. International Atomic Energy Agency, *Report of the Goíania Accident*, IAEA, Vienna, 1988.
87. M. Salki, M. B. A. Vasconcellos, and J. A. A. Sertie, *This Conference* (1989).

Use of Nuclear Physics Methods in Life Sciences in the USSR

A. A. Kist

Institute of Nuclear Physics, Academy of Sciences of the Uzbek SSR, Tashkent, USSR

Received April 17, 1989; Accepted December 11, 1989

ABSTRACT

Analytical nuclear physics methods, as well as radioanalytical, X-ray, and radiometric techniques are widely used in the Soviet Union in life sciences studies. Main approaches to the study of natural materials by activation analysis, including bulk analyses, localized analyses, speciation, and in vivo experiments are described. Examples of applications in agriculture to increase protein content of wheat, to improve crop yield, and to enhance resistivity of cotton plants to diseases are given. Activation analysis is used in medicine for prognosis and study of processes of pathogenesis. Detailed information on trace elements is valuable in environmental studies, and general regularities in biogeochemistry may be clarified with activation analysis.

Index Entries: Nuclear analytical methods; activation analysis; trace elements; agriculture; medicine; environment; biogeochemistry; biosphere; hydrosphere; forensic science; glaciers; mineral prospecting; toxic elements; industrial emission.

INTRODUCTION

In the Soviet Union, the use of nuclear physics methods in the life sciences has been increasing because of the relative simplicity, high sensitivity, efficiency, and multielement nature of the methods. These methods are closely related to radioanalytical, X-ray, and radiometric techniques, but this chapter is restricted to consideration of only activation analysis.

METHODS

Activation analysis is usually performed using research or analytical reactors designed for that purpose, but one can also use nuclear power reactors. The most widely used technique is multielement instrumental neutron activation analysis (INAA) based on high resolution γ-ray spectrometry. Neutron generators are used predominantly to detect N, P, and K in soils and plant and, occasionally, to determine O, F, and some heavy metals. Linear electron accelerators, betatrons, and microtrons are used for multielement instrumental photon activation analysis (IPAA). Cyclotrons are sometimes used in life sciences, both for multielement instrumental analysis and for proton-induced X-ray emission (PIXE). Nuclidic neutron sources (^{252}Cf, ^{124}Sb-Be, ^{238}Pu-Be, and so on) are useful for analysis of large soil and plant samples, as well as for in vivo experiments.

Automatic devices for instrumental analysis allow one to study several hundred samples per day. A variety of radiochemical methods are used, often involving group separations for removal of interfering nuclides. Thermochromatography is a promising technique, since it allows simultaneous destruction of activated samples, evaporation of newly formed nuclides, and their separation according to the boiling temperature in thermochromatographic columns. Samples are preconcentrated using coprecipitation, selective, and nonselective sorption, ion exchange, and extraction. These methods are used mainly for study of surface water or glacier waters containing background levels of elements. An example is the concentration of Au by ion exchange by overloading of ionic sorption capacity. The procedure involves the use of a microcolumn (about 100 mg of anionite) for Au sorption from 0.5–3 L of water followed by INAA of the polyethylene column with ionite. The method has been used to analyze several thousand samples in a hydrogeochemical search for Au (*1*). Coprecipitation with hydroxide was used to determine Au concentrations in waters of the Pacific and Atlantic Oceans (*2*).

Studies of speciation of elements have become increasingly important. For example, in the case of soil analyses, samples are treated by buffer solutions to determine the amount of the element accessible to plants. In plants, tissues, and animal fluids, activation analysis is widely used to study subcellular structures, such as proteins and nucleic acids. For example, gel filtration is used to determine the distribution of trace elements among different protein fractions. Figure 1 shows the distribution curves for some of the elements in rat-liver cytosol (*3*).

The combination of filtration, extraction, thin-layer chromatogra-

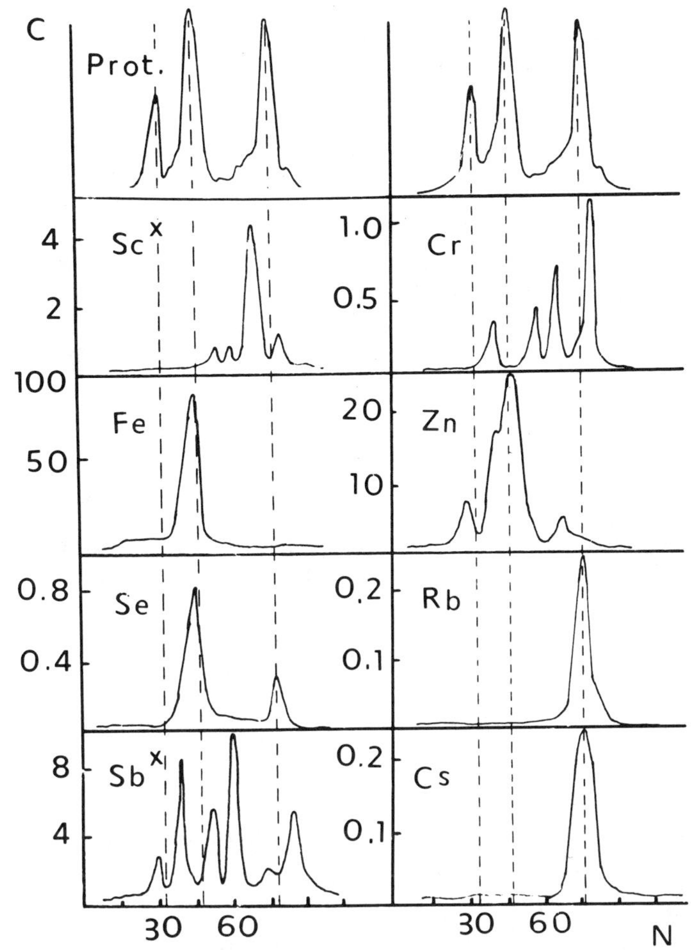

Fig. 1. Elements in rat liver cytosol (protein-relative units, Sc and Sb—ppb, Cr, Fe, Zn, Se, Rb, and Cs—ppm).

phy, freezing, sorption, ion exchange, electrodialysis, and gel filtration with subsequent activation analysis allows the detection of ionic, dissolved, organic and pesticide, complex, and colloid elements' speciation in waters (Fig. 2). Similarly, filters with various pore size, cascade impactors, freezing, bubbling, and sorption allow one to study elements' behavior in the air.

Localized analyses, such as fractional analysis, scanning, and autoradiography (including individual trace detection) are especially important in life sciences. For example, they have been used to show the difference in the levels of native iodine and injected ^{131}I in microscopic studies of thyroid glands (4). An autoradiographic study of the distribution of injected ^{6}Li was used to develop the procedure of injection of Li

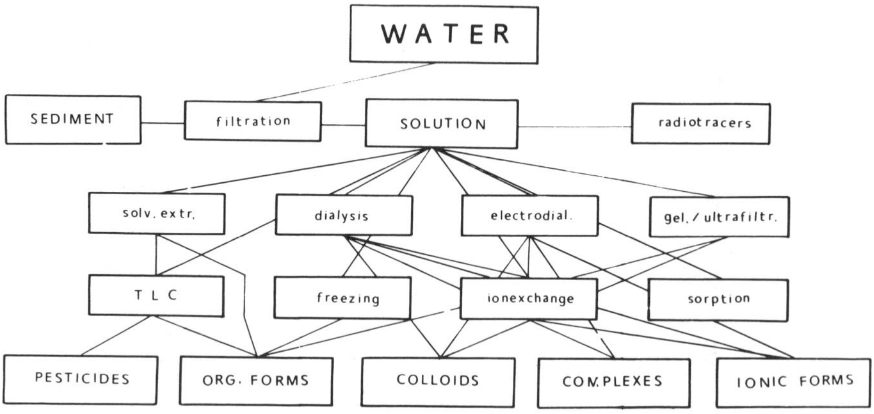

Fig. 2. The scheme of water analysis.

accessible to tumors for neutron-capture therapy (4). Techniques for local autoradiographic determination of U, B, and other elements have also been developed.

Activation analysis in vivo involves the use of nuclidic neutron sources (^{238}Pu-Be or ^{252}Cf). It is used, for example, to estimate concentrations of Na, Cl and K in human limbs for differential diagnosis of tumors in limb bones, for diagnosis of nonmalignant and systemic skeletal diseases, and for the control of the locomotor system, in particular, in the case of Ca metabolism breakdown (5).

In developing these methods, the investigators had to solve many problems, e.g., loss of elements during heating of the biological material in air or inert gases. Large fractions of many elements are lost even if ashing is carefully controlled. It has been shown that the dynamics of the element loss during heating can yield information on the valence state of an element in the material analyzed. The possible errors caused by surface contamination of samples, radiation processes during irradiation, and so on, were also studied. Soviet analysts have participated in the certification of various standard reference materials. Procedures for making artificial standards that simulate compositions of various natural materials for activation analysis have been proposed (6).

Periodic All-Union conferences on activation analysis and its applications in environmental monitoring and medicine are held in the Soviet Union. Several thousand Soviet papers on activation analysis of natural materials have been published, of which only a few will be described.

Agriculture

Many studies of relationships between the elemental compositions of soil, plants, and tissues and crop yield, disease rate, or the quality of agricultural crops have been made. The practical significance of such works is shown by the following samples.

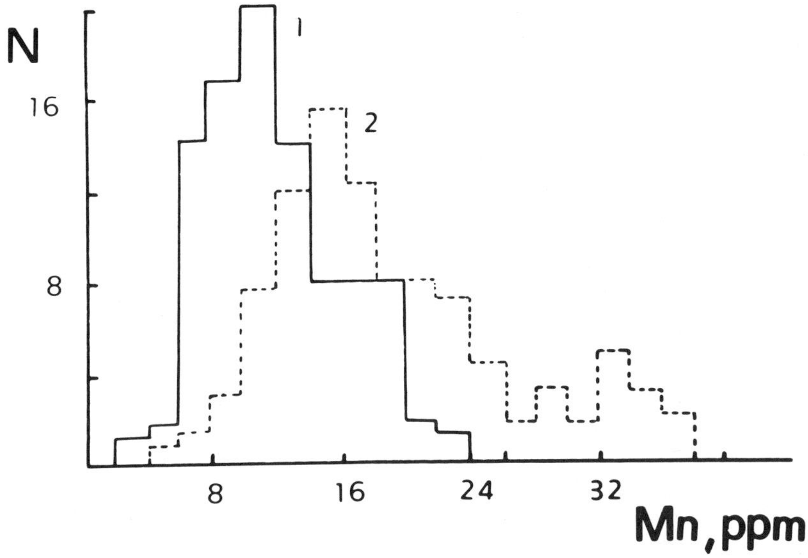

Fig. 3. Frequency distributors of manganese concentrations in cotton seeds (1—sowing materials, 2—first harvest).

Manganese is a trace element, but in soil and plants, its concentration correlates with those of the nutrition elements, especially P. A device has been designed that allows for 60–100 analyses/h. Strong correlations are observed between the results of bulk analyses and those of aqueous extracts. As a result of 30,000 analyses, large-scale maps of Mn concentration in soils in Uzbekistan were generated. They show that the cotton yield is correlated with Mn concentration in soils. On the basis of these maps, $MnSO_4$ was applied to more than 5000 ha, and the average gain in crop yield was 500 kg/ha. In addition, the magnitude of the effect of wilt (a disease causing considerable crop losses) on cotton plants is affected by the amount of Mn in plants, in particular, in seeds. As an example, Fig. 3 shows histograms of the Mn distribution in individual seeds and in the seeds of the first harvest. The second peak in the bimodal distribution pattern appears to be connected with the affliction of plants by wilt. This was confirmed by analysis of >9000 seeds, which were sorted into three groups according to the Mn content and sowing procedure. The three groups were characterized by different degrees of affliction by wilt, and the degree rose with increasing Mn concentration (Fig. 4). This example suggests that seeds should be sorted prior to sowing to select the cotton plants more resistant to wilt. Similar approaches can, perhaps, be used for other agricultural plants, even if the mechanism of pathological plant processes has not been studied.

Activation analysis has been used to determine N in plants to estimate their protein content. The reactions $^{14}N(d,p)^{15}N$, and $^{14}N(d,\alpha)^{12}C$ were used (*7*). Computer control and automatic devices made possible the analysis of 5000 wheat grains/d. A difference in protein content in

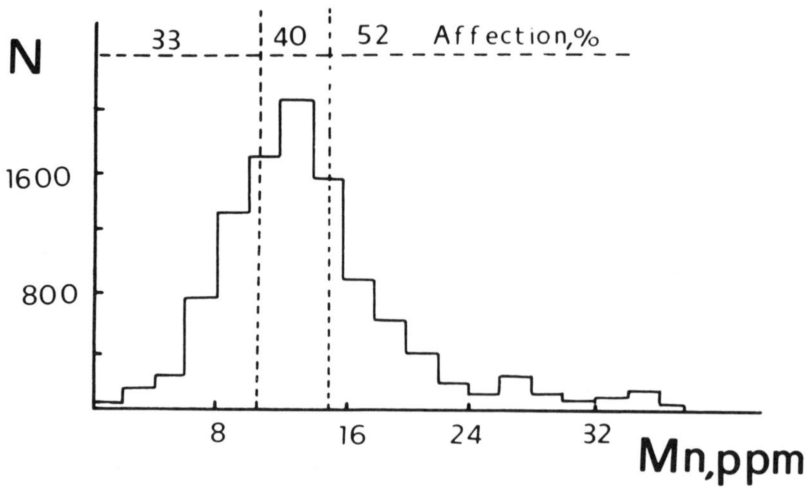

Fig. 4. Cotton seed sorting according to manganese concentration.

different parts of the grains was found. Shielding of the germs during activation makes it possible to sort and sow the examined grains. Thus, several hundred thousand grains were sorted and the six-generation study produced lines with a 2% increase of protein relative to the base value, whereas crop yield remained the same.

Medicine

Changes in elemental compositions of various organs and tissues of humans and animals in various physiological states have been studied. These investigations covered a great variety of diseases, and convincing regularities in the distribution of elements (including some elements whose roles are unknown, e.g., Sc in diabetes) have been found. For examples, high concentrations of I were detected in some eye tissues, suggesting the participation of this element in photoreception. The correlation between Mn content and sugar levels in blood and urine in diabetes was observed, and the possibility of prolonging the insulin effect by simultaneous injection of Mn has been shown. Although many interesting results have been obtained, their explanation and practical application are still far from complete. Activation methods based on stable nuclides are used to study the dynamics of drugs, e.g., of some antitumor preparations. Enriched stable nuclides are used, for example, to detect extracellular fluids in the human body.

Activation analysis may be combined with separations to determine an element's speciation, for example, to determine I levels while studying thyroid gland pathogenesis. Much work has been on elemental compositions of nuclei, mitochondria, cytoplasm, and histone in healthy tissues and tumors experimentally and clinically. The results suggest a possible specific role played by Zn, Mn, and some other elements in the process of malignation (*8*).

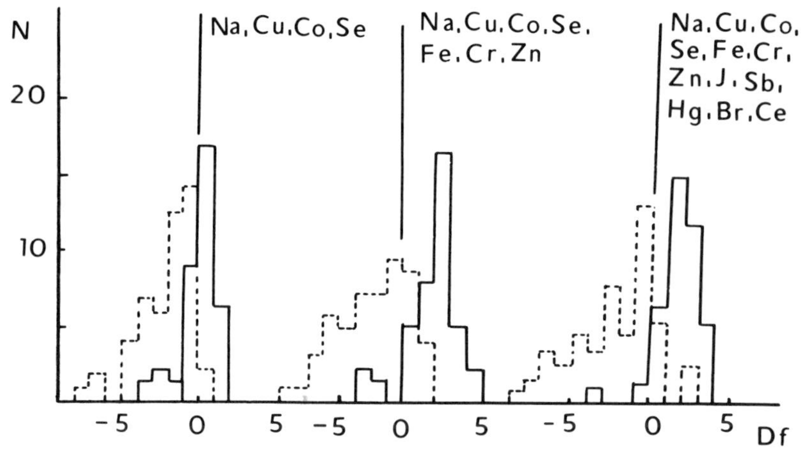

Fig. 5. Discriminant analysis results for diabetes (—— pathology, —— nonpathology).

Recently there has been much interest in the composition of human hair. Factor analysis has shown marked differences in the elemental composition in healthy subjects and those affected by various diseases. As an example, Fig. 5 shows results of discriminant analyses for diabetes. Differences in elemental hair composition of twins, their parents, and their children were studied, as well as the relationship between the elemental composition of newborns' hair and the amniotic fluids, the dependence of hair composition on sex, age, blood group (where a series of unexpected correlations has been found), occupation, and so on. The method offers great possibilities for study of occupational diseases, the impact of environment and working conditions, and so forth.

From data presented by various authors and those from WHO on disease rates, one can possibly make global maps based on the elemental hair composition data. Even the incomplete data available suggests close correlations between hair composition and death rates from several diseases.

Environmental Studies

Despite the disadvantages of long turnaround time and the inability to determine concentrations of the major contaminant, lead, activation analysis is widely used in environmental investigations. In addition to the advantages described above, it allows determination of elements' speciations, including possible measurement of individual pesticides or the total content of the compounds containing that element. A major advantage of the method is its ability to determine many elements over wide concentration ranges in various media, i.e., water, fallout, aerosols, soil, plants, and natural fuels and their combustion products.

Activation analysis is also widely used to determine background concentration levels of various elements at background stations and

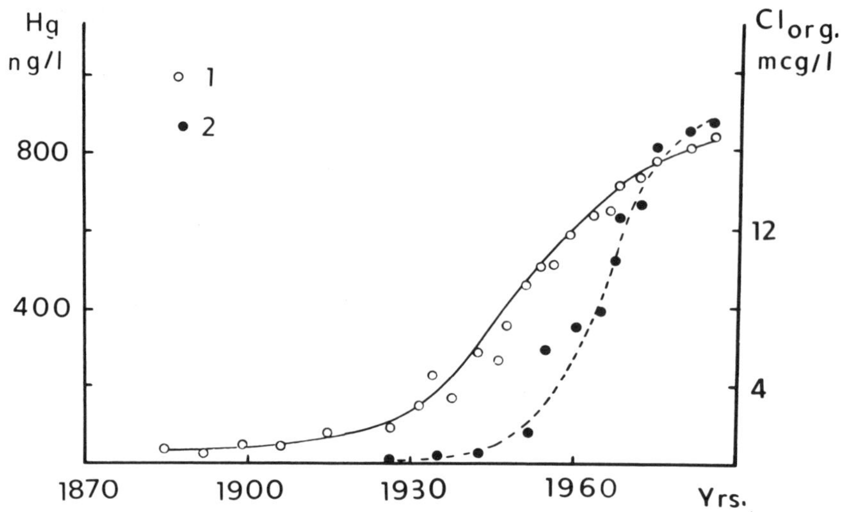

Fig. 6. Mercury (*1*) and organic forms of chlorine (2) in glacial water (Abramovsglacier, Pamir mountains).

nature reserves. Laboratories at background stations apply the isotopic dilution technique and prepare samples for further treatment by activation analysis. In this way, the cores of glaciers of Arctic and Antarctic regions and central Asian mountains were analyzed. For example, Fig. 6 shows the curve if dissolved Hg concentrations in one of the Pamir glaciers during the last century. Although the Hg content in large particles dispersed in the glacier waters was nearly constant during this period and it increased only slightly in suspended fine particles, the concentration of dissolved Hg grew considerably. The location of the background station allows us to assume that the curve characterizes global processes. There is also a marked increase in the sum of organic forms of Cl in the same samples (Fig. 6).

In regional monitoring, activation analysis is used not only to estimate average levels, but also to identify possible sources of contamination. For impact monitoring, activation analysis is used to study compositions of water and air near industrial enterprises, as well as in their workshops. Contours of the concentrations of toxic elements in snow near the contamination sources are quite interesting. Such studies show that the discharge of some elements into the atmosphere sometimes accounts for much of the total loss of the elements during processing. A very important observation is the considerable role of coal burning in contamination of the biosphere by Hg. Earlier data on Hg content in coals obtained by emission spectral analysis of ash seem to be very understated. Many studies are being conducted to evaluate the state of the environment based on the elemental composition of hair, saliva, and teeth. Activation methods are used to study salt transfer processes in the Aral sea, which is drying out.

A method of collection of submicrometer particles and gas-phase species from air on organic thin-film sorbents has been developed. The method has much higher sensitivity and lower blank levels than the traditional technique for trapping of gas-phase species. A surprising result of this method is the observation of large variations in gas-phase Au concentrations. The results of older studies of Au content in various natural materials may have been overstated because of contamination by Au from the gas phase (*9*).

Forensic Science

To some extent, forensic studies are related to life sciences. For example, examination of the saliva elemental composition on cigaret stubs revealed that a high Au concentration may indicate the presence of gold crowns or fillings. Modern studies of elemental compositions are used for identification, as well as for other specific purposes in forensic toxicology. Multielemental analysis is applied for identification of possible source areas of drugs of plant origin.

Biogeochemistry

Activation analysis is widely used in various biogeochemical studies. This method is often the most advantageous for evaluation of the average elemental abundances in the hydrosphere and biosphere.

The composition of sphagnum moss near the location of the Tunghus meteorite fall is very interesting. Abnormal contents of Fe, Co, Zn, Br, Rb, La, and Hg in the layer corresponding to the year 1908 and some concentration ratios (e.g., Ce/La) are not characteristic of the known types of meteorites (*10*).

Tens of thousands of Au determinations are performed in the search for deposits by biogeochemical methods. Activation analysis is quite useful in the deserts of the Soviet Union, where thick layers of sand and sediments make detection of geochemical anomalies difficult and the dry climate slows migration of elements towards the surface. Thus, as the root system of desert plants may be tens of meters long, the plants were found to "pump out" Au and other elements from a considerable depth and accumulate it. This was how some prospective metal deposits were found (*11*).

Enhanced knowledge of elemental compositions of biospheric materials has made it possible to develop phenomenological models that simplify the numerical representations of elemental compositions of biospheric materials and make it easy to compare them. Distinct regularities in manifestations of toxicity and deficiencies of elements in living organisms have been found. Also, certain regularities were observed in variations of elemental compositions in phylo- and onthogenesis. The models developed allow one to check the reproducibility of any analytical method and the sample-preparation procedure. In addition, the possibility exists to predict the state of the biosphere as a result of its contamination by chemical elements (*12*).

SUMMARY

This brief review deals with just a few examples taken from numerous works of Soviet scientists dealing with activation analysis techniques and their application in life sciences. In view of the importance of these investigations, it is hoped that they will be carried out on a larger scale. Furthermore, the results of the application of nuclear methods in life sciences may give some hope to those concerned with environmental protection.

REFERENCES

1. G. S. Nikanorov and A. A. Kist, *Zhurnal Anal. Khimii.* **22,** 2438 (1967).
2. A. I. Ryabinin, A. S. Romanov, Sh. Khatamov, and R. Khamidova, *Geokhimia* **N1,** 158 (1974).
3. T. A. Sherbakova, V., A. Kovalyova, I. N. Mikholskaya, A. A. Kist, and L. I, Zhuk, *Proc. of the USSR-Japan Joint Symposium on Analytical Chemistry, Tashkent,* p. 250, 1988.
4. V. E. Zaichik and A. V. Tkachev, *Bull. Exp. Biol. i Meditsiny* **N10,** 118 (1970).
5. V. E. Zaichik, V. M. Kalashnikov, and V. A. Bizer, *Yaderno-Fizicheskie Medoty Elementnogo Analiza v Meditsine, Obninsk,* pp. 58–74, 1980.
6. M. A. Kolommiytsev, V. Yu. Dundua, N. V. Pachulia, and N. V. Bagdavadze, *J. Radioanal. Chem.* **22,** 95 (1974).
7. A. E. Melenevskyi, O. F. Nemets, B. A. Tichyi, D. M. Grodzinskyi, G. S. Ponomaryev, B. I. Bernshtein, V. V. Morgun, and V. F. Logvinenko, *Vtoraya Vsesoyuznaya Konferentsiya po Selskokhozaystvennoy Radiologii, Tez. Dokl. Obninsk,* t. 2, pp. 14–15, 1984.
8. E. L. Andronikashvili, L. M. Mosulishvili, and A. I. Byelokobylskyi, *Primenenie Aktivatsionnogo Analiza v Meditsine, Tbilisis, Metsnierba,* pp. 7–15, 1977.
9. A. A. Kist and L. I. Zhuk, *Geokhimia* **N6,** 892 (1987).
10. S. P. Golenetskyi and V. V. Stepanok, *Vzaimodeystvie Meteoritnogo Veshestva s Zemley, Novosibirsk, Nauka,* pp. 102–115, 1980.
11. Sh. Khatomov, E. M. Lobanov, and A. A. Kist, *Aktivatsionnyi Analiz Biologichiskih Objektov, Tashkent, Fan,* pp. 3–7, 1967.
12. A. A. Kist, *Fenomenologiya Biogeokhimii i Bioneorganicheskoi Khimii, Tashkent, Fan,* 1987.

Applications of Nuclear Analytical Methods in the Life Sciences as Exemplified by Recent Research Programs of the IAEA

ROBERT M. PARR* AND EDUARDO CORTES-TORO

International Atomic Energy Agency,
PO Box 100, A-1400 Vienna, Austria

Received April 17, 1989; Accepted December 11, 1989

ABSTRACT

More than 300 research nuclear reactors are presently in operation in 56 countries, most of which are suitable for neutron activation analysis (NAA). Typical applications in the life sciences are exemplified by recent research programs of the IAEA involving the participation of institutes in 47 countries. Topics of interest include applications of NAA (together with XRF and PIXE) in occupational health, nutrition, and environmental studies, including analyses of solid wastes and aerosols. The IAEA supports such programs not only by providing funding, but also through quality assurance and information services.

Index Entries: Aerosols; analytical quality assurance; environment; foodstuffs; hair; neutron activation analysis; nutrition; occupational health; PIXE; solid wastes; XRF.

INTRODUCTION

Facilities for undertaking analyses by nuclear analytical methods are widely distributed throughout the world. According to the IAEA's own

*Author to whom all correspondence and reprint requests should be addressed.

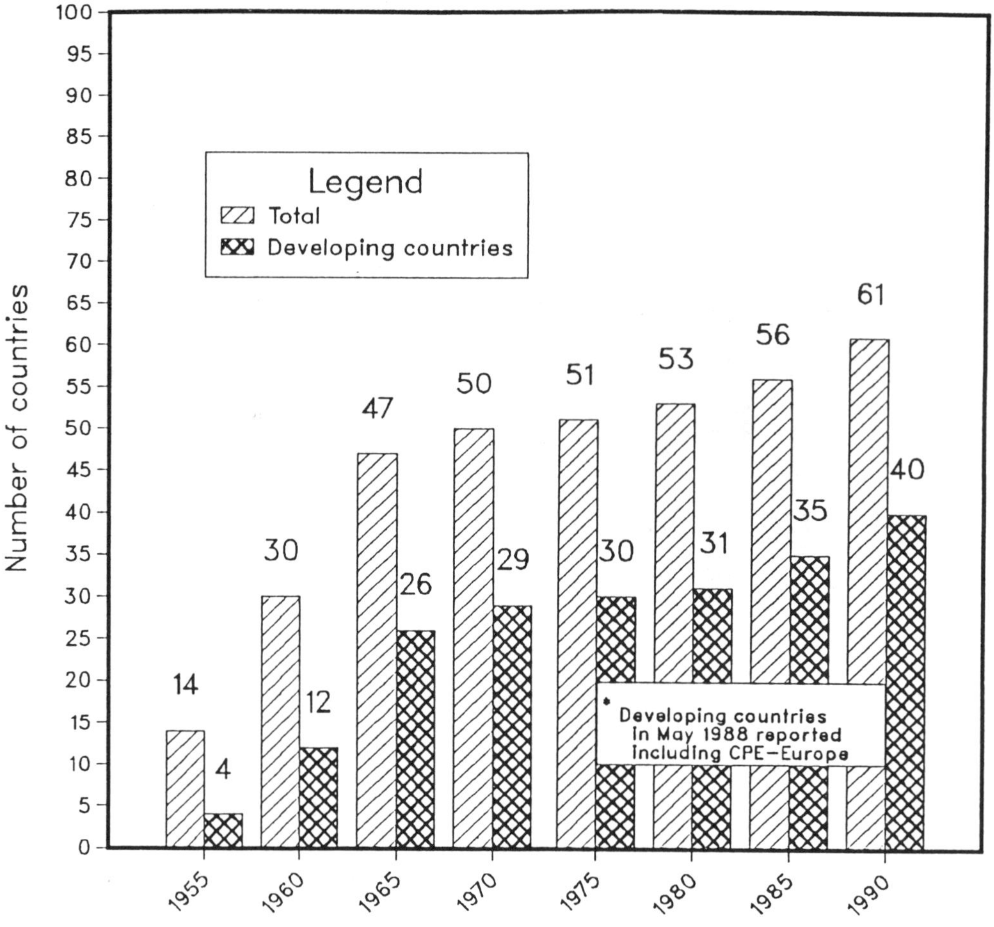

Fig. 1. Research nuclear reactors in IAEA Member States.

statistics (*1*), there are at present more than 300 research nuclear reactors in operation in 56 countries (including more than 70 in approx 35 developing countries), and these numbers are expected to increase further during the next few years (Fig. 1). Although approx one-third of existing research nuclear reactors are low-power critical assemblies or other reactor types unsuitable for neutron activation analysis (NAA), this still leaves approx 200 operational reactors with which NAA studies can presently be carried out. Comparable data are lacking for other nuclear analytical methods, but the IAEA, in its own programs, has already provided support for about 50 energy-dispersive X-ray fluorescence (ED-XRF) systems, and 20 neutron generators, and has also provided components for several particle-induced X-ray emission (PIXE) systems.

Typical applications of these research facilities in the life sciences are exemplified by several recent and current research programs, which are summarized in the following section. Emphasis is placed in these programs not so much on the development of new techniques, but rather on the application of existing technology to solve specific problems in nutritional and health-related environmental research.

THE IAEA RESEARCH CONTRACT PROGRAM

General Aspects

Among UN organizations, the IAEA is almost unique in promoting technology transfer to developing countries by means of a wide-ranging research contract program. The present program (1988 statistics) provides funding of just over US $4 million, which is used to support approx 120 coordinated research programs (CRPs) in many different subject areas, and about 1300 individual research contracts and research agreements. The money provided by the IAEA is not intended to cover the full costs of the research; very often it serves as "seed money," which attracts considerably more funding from national sources. Individual CRPs typically have 10–15 participants from both developed and developing countries. They have a total duration of 3–5 y, though individual contracts are awarded for only 1 yr at a time. As a rule, participants in developing countries are eligible for research contracts paying $5000–8000/y. Participants in developed countries are normally expected to contribute their knowledge and expertise under so-called research agreements, which involve no costs to the IAEA apart from those incurred by participation in research coordination meetings. These are usually held at intervals of around 1½ y.

Applications of nuclear analytical techniques in the life sciences form only a small section of the IAEA's total research contract program, but nevertheless encompass six recent and ongoing CRPs and annual funding of approx $320,000. Some examples are given below.

Coordinated Research Programs on Nuclear Analytical Techniques in the Life Sciences

CRPs of recent or current interest in the life sciences area deal with applications of nuclear analytical methods (mainly NAA) in the following fields:

1. Applications of nuclear-related techniques in occupational health studies
2. Trace elements in human hair (studies of possible correlations with internal body burdens)
3. Dietary intakes of nutritionally important minor and trace elements
4. Toxic trace elements in foodstuffs

Table 1
IAEA Member States That Have Participated in Relevant CRPs During the Period 1986–89

Argentina	2	Ghana	1	Peru	1
Australia	3	Greece	1	Poland	1
Bangladesh	3	Guatemala	1	Spain	1
Belgium	1	Hungary	2	Sudan	1
Bolivia	1	India	2	Sweden	2
Brazil	5	Indonesia	2	Thailand	4
Bulgaria	1	Iran	1	Turkey	2
Canada	4	Italy	2	UK	1
Chile	1	Jamaica	1	Uruguay	1
China	5	Japan	4	USA	3
Czechoslovakia	2	Korea, Rep.	1	USSR	1
Ecuador	1	Malaysia	2	Venezuela	1
Egypt	3	Netherlands	2	Viet Nam	1
Finland	1	Norway	1	Yugoslavia	5
Germany (GDR)	1	Pakistan	2	Zaire	1
Germany (FRG)	1	Paraguay	1		

5. Nuclear techniques for the study of solid wastes (including coal fly-ash, leaching studies, and aerosol analyses)
6. Trace-element analysis in agro-industrial products.

Altogether, 47 Member States have participated in these CRPs during the period 1986–89 (Table 1).

The CRP on occupational health (1983–87) has demonstrated that a variety of nuclear-related analytical techniques can be usefully applied in occupational health studies. Examples include NAA and PIXE analyses of hair, blood, and other tissues from occupationally exposed workers; PIXE analysis of air-particulate samples collected in factories; radioisotope XRF analysis (in vivo) of lead in bone of exposed workers; and studies in experimental animals for the purpose of elucidating the metallo-biochemical behavior of toxic elements.

The CRP on trace elements in human diets (1984–90) provides a good example of the applicability of NAA in nutrition research. In this program, representative total diets are being collected in 14 countries (2). These are analyzed in selected reference analytical laboratories according to a protocol designed to provide adequate analytical quality assurance. The elements of interest are listed in Table 2, which also identifies the analytical techniques being used as the primary source of information (referred to as the "reference method"), together with those methods that are being used for backup or quality assessment (QA) analyses. In all cases, as an additional component of QA, the reference analyses and backup/QA analyses are being carried out in a different laboratory by a different analyst. As illustrated in Table 1, NAA is being used as the primary analytical method for 14 of these elements (As, Cl, Cr, Fe, Hg, I,

Table 2
Applicability of Neutron Activation Analysis and Other Analytical Methods[a] in Research on Dietary Intakes of Minor and Trace Elements (*See Text*)

Element	Reference method	Backup and QA methods	Certification of diet RM[b]
Al	AAS	NAA ICP	*
As	NAA	NAA	*
Ca	ICP		
Cd	AAS	NAA	
Cl	NAA	NAA	*
Cr	NAA	AAS NAA	*
Cu	ICP	NAA	
F	?	NAA ISE	*
Fe	NAA	ICP AAS	*
Hg	NAA	NAA AAS	*
I	NAA	NAA	*
K	NAA		*
Mg	ICP		
Mn	NAA	AAS ICP	*
Mo	NAA	AAS NAA	*
Na	NAA		*
Ni	AAS	NAA	
P	ICP		
Pb	AAS	AAS	
Se	NAA	AAS	*
Sn	NAA	NAA	NC
V	NAA	NAA	NC
Zn	NAA	AAS ICP	*

[a]AAS: atomic absorption spectrometry; ICP: inductively coupled plasma atomic emission spectrometry; ISE: ion specific electrode; NAA: neutron activation analysis.

[b]Elements for which NAA was an important source of data used in calculating the provisional certified concentrations in IAEA mixed human diet, H-9, are indicated by an asterisk (*); NC = not certified.

K, Mn, Mo, Na, Se, Sn, V, and Zn), and is providing quality control data for five additional elements (Al, Cd, Cu, F, and Ni).

A new mixed human diet reference material (H-9) has been prepared for use as a quality control material in this CRP, as well as for general availability under the IAEA's analytical Quality Control Services Program. For 14 of the 23 elements of primary interest listed in Table 1, NAA was an important source of the data used for calculating the provisional certified concentrations (*3*).

This CRP builds on the experience gained in the course of an earlier research program, conducted jointly with the World Health Organization

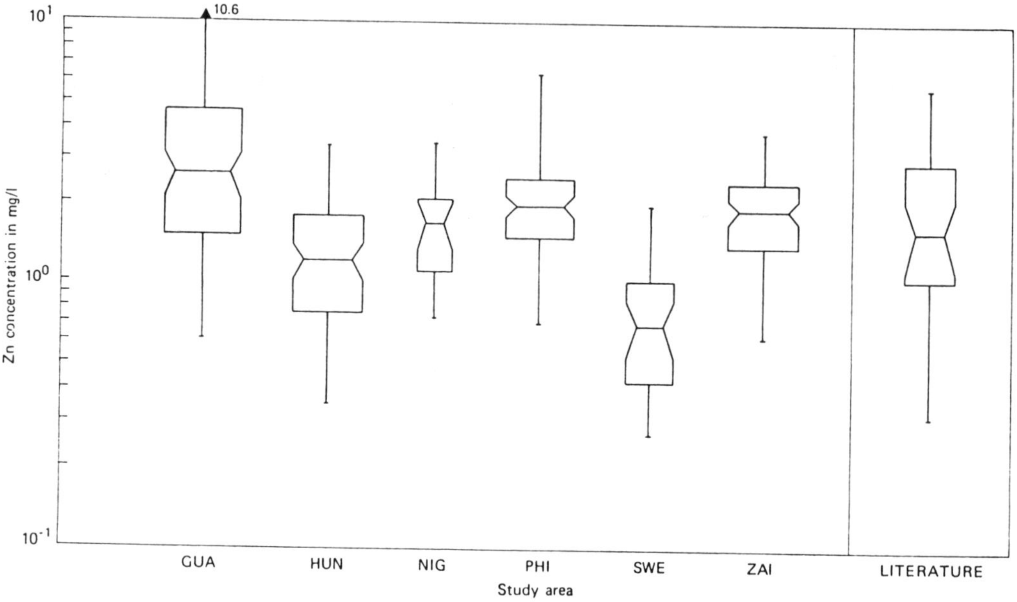

Fig. 2. Box plots showing amounts of zinc in human breast milk specimens from six study areas (Guatemala, Hungary, Nigeria, Philippines, Sweden, Zaire); comparison with data from the scientific literature. All values in mg/L of original milk (*4*).

(WHO) on minor and trace elements in breast milk. The results of that study have just become available (*4*). Figure 2 illustrates the data obtained by NAA for zinc in breast milk samples collected in six different countries. These data, together with the results of the current CRP on dietary intakes of minor and trace elements, are expected to have an important bearing on the forthcoming reassessment of dietary requirements of trace elements, which is due to be made by an Expert Committee of WHO, FAO, and IAEA at a meeting in June 1990.

The CRP on hair analysis (1984–88) is concerned with the question of whether trace-element levels in hair can be used to measure human body burdens, in particular whether there are meaningful correlations between trace-element levels in hair and in internal body organs. The elements of primary interest are four toxic elements As, Cd, Pb, and Hg, together with three essential elements with which these sometimes interact: Cu, Se, and Zn. Many of the measurements, e.g., of trace elements in hair and autopsy samples, are being made by NAA; metabolic studies in laboratory animals are also included in this CRP. Evidence obtained so far (*5*) indicates that hair is a useful indicator of the internal body-burden of mercury; for other toxic elements, the levels in hair do not appear to correlate well with those in internal organs. A full report on this CRP is expected to be available by the end of 1990.

This CRP is expected to lead to a new one, starting in 1990, focused more narrowly on the problem of environmental exposure to mercury, particularly in population groups that may have an excessive intake resulting from the consumption of contaminated fish. This project will be carried out in collaboration with the WHO, and will involve the determination of total mercury and methyl mercury in various samples, including human hair.

The CRP on toxic elements in food and drinking water (1986–90) has as its purpose the monitoring of compliance with national and international regulations and guidelines (e.g., those of the Code Alimentarius Commission [*6*]) pertaining to maximum permissible levels of toxic elements in individual foodstuffs and drinking water. NAA and other nuclear-based analytical techniques are used mainly to provide validation support for national food monitoring programs, and to carry out small-scale pilot studies.

The CRP on heavy metals in solid waste products (1987–91) is concerned with applications of NAA and other nuclear-based techniques in research of the following kinds: environmental impact assessment of individual pollution sources (e.g., coal-fired power stations); problems of disposal of coal fly-ash, sewage sludge, and so on (e.g., leaching of toxic elements into ground water); and both short- and long-range atmospheric transport of toxic heavy metals.

This is a relatively new CRP that has not yet had much time to produce results. Nevertheless, it is already evident that nuclear methods are particularly appropriate for multielement studies of solid waste products and aerosol samples. A followup study on aerosol composition in remote and polluted environments is now under active consideration for possible implementation starting in 1991. This may be carried out jointly with the Background Monitoring Network (BAPMoN) program of the World Meteorological Organization (WMO). As evidenced by the overviews presented in Tables 3 and 4, which are based on the conclusions of a recent IAEA Advisory Group, nuclear techniques are highly appropriate for the analysis of aerosols collected with a variety of different sampling devices on several different kinds of filters.

Quality Assurance (QA)

One of the main problems in coordinating such CRPs is that of ensuring the comparability of the data produced in different laboratories using a variety of different techniques. To this end, the IAEA has developed recommendations on analytical QA, and assists the participants in its CRPs to validate their methods by providing appropriate reference materials and by monitoring laboratory performance with the aid of "blind" QA samples. Figure 3 illustrates the kinds of results obtained. These QA assurance exercises draw upon the wide range of intercomparison samples and reference materials that are available under the IAEA's own Analytical Quality Control Services (AQCS) program (*7,8*).

Table 3
Overview of the Appropriateness of Widely Used Aerosol Samplers with Respect to the Analytical Techniques Used

Sampler	No. of Stages	INAA	XRF	PIXE
Hi-Vol	1	YES	NO	NO
Stacked Filter Units	2	YES	YES	YES
Dichotomous Sampler	2	YES	YES	NO
Cascade Impactors (CIs):				
Anderson (8-13 Lpm)	8-10	YES	NO	NO
Batelle (1 Lpm)	6-10	NO	NO	YES
Sierra (Hi-Vol CI)	8-10	YES	NO	NO

Table 4
Overview of the Appropriateness of Different Kinds of Filters with Respect to the Analytical Techniques Used

Filter	INAA	XRF	PIXE
Cellulose ester (Millipore)	+++	++	++
Polycarbonate (Nuclepore)	+++	+++	+++
PTFE (Teflon)	+++	+++	+
Cellulose (Whatman 41)	+++	+	+

+ Highly recommended.
+ + Less recommended.
+ + + Can be used, but are not recommended.

OTHER SUPPORT MECHANISMS

In addition to the research contract program described above, the IAEA has several other ways in which it can support applications of nuclear methods in the life sciences. For scientists in developing countries, the most important of these include technical cooperation, fellowships, and training courses. Space does not permit the authors to provide further details here. The interested reader is invited instead to apply to the IAEA, or to the authors directly, for more information.

Of more general interest to scientists in all countries (not only developing ones) is the Agency's publications program. Some recent publica-

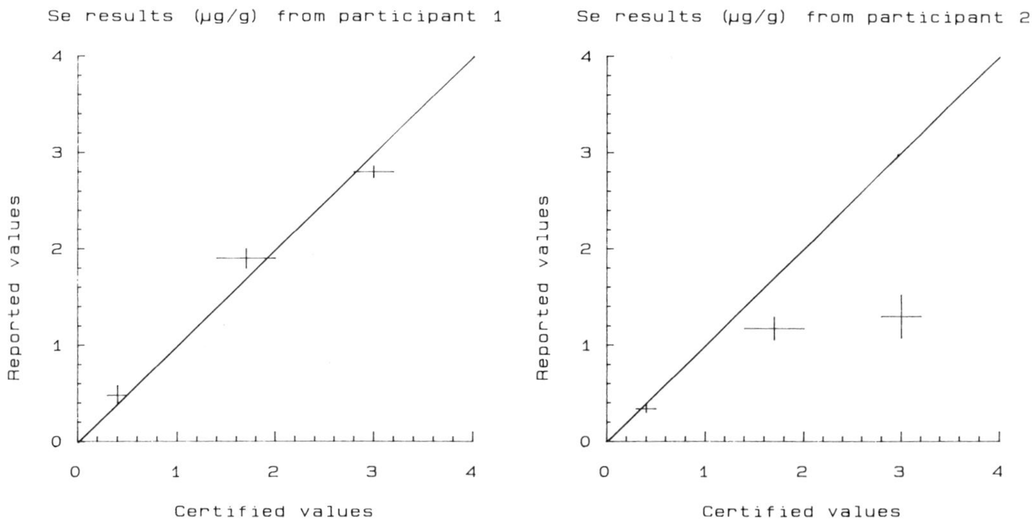

Fig. 3. Validation of analytical methods for selenium in two laboratories using certified biological reference materials (BRMs). Results from participant 1 indicate satisfactory validation. Results from participant 2 demonstrate that unreliable conclusions might have been drawn had only one BRM (the one with the lowest values) been used.

tions relevant to the subject of this paper are listed in references (*9–15*). All of these are available cost-free on request.

The IAEA also has a major publications effort relating to INIS, the International Nuclear Information System, which is a computerized database and abstracting service operated by the IAEA in collaboration with its Member States. It covers publications in the scientific literature (both conventional and nonconventional publications) on topics relevant to the Agency's field of interest. A recent development that will greatly facilitate the more widespread dissemination of information from INIS is the availability of the database, starting in 1989, in the form of an optical compact disk (CD-ROM).

Information on the numbers of documents in INIS also provides an interesting overview of the status of the different nuclear-related techniques as applied in the life sciences, as well as in other subject areas. Table 3 presents some relevant statistics showing that, although the applications of nuclear techniques to biological materials are less numerous than is the case for inorganic matrices, such as rocks and ores, they nevertheless represent a significant fraction of the total workload of these methods in the Agency's Member States.

Table 5
Numbers of Documents in INIS (Up to July 1988) on NAA, XRF, and PIXE Combined with Selected Elements and Matrices

	NAA	XRF	PIXE
Antimony, Sb	904	124	16
Arsenic, As	551	98	35
Bromine, Br	918	241	150
Cadmium, Cd	774	343	50
Chromium, Cr	1131	299	104
Copper, Cu	1067	454	207
Iodine, I	358	133	12
Lanthanum, La	896	155	34
Lead, Pb	605	447	152
Mercury, Hg	639	76	25
Selenium, Se	880	126	75
Vanadium, V	491	152	33
Zinc, Zn	1219	419	179
Aerosols, dust	432	187	164
Alloys	529	267	47
Biological materials	831	126	133
Blood	304	78	90
Bone	182	44	21
Coal, fly-ash	390	102	24
Fish, seafoods	90	3	2
Foods, foodstuffs	357	39	19
Hair	195	26	64
Heart	35	12	3
Liver	279	30	39
Meteorites, chondrites, tektites	231	12	14
Ores	835	516	21
Rice, flour	58	7	4
Rocks	962	301	33
Steels	281	127	23

CONCLUSIONS

Applications of nuclear analytical methods in the environmental and life sciences continue to be an important research activity in many of the IAEA's member states, even though there has been an enormous growth, in recent years, in the use of other, competing, nonnuclear analytical methods for the determination of trace elements (methods such as AAS and ICP-AES). NAA still seems to be the method of choice in many investigations, particularly for multielement studies and for analyses of difficult matrices, i.e., those in which the trace elements are present only at extremely low concentrations or where high concentrations of interfering elements lead to systematic errors. PIXE and XRF, on the other hand, appear to find their main applications in studies of air particulates.

ACKNOWLEDGMENTS

Some of the information presented in this report is drawn partially from the programs of other present and former colleagues in the IAEA The authors would like to acknowledge in particular their collaboration with Y. Muramatsu, R. Schelenz, H. Vera Ruiz, and R. Rosenberg.

REFERENCES

1. Nuclear Research Reactors in the World, IAEA Vienna, Reference Data Series No. 3 (1988).
2. R. M. Parr, *Trace Element Analytical Chemistry in Medicine and Biology,* de Gruyter & Co., Berlin 4, pp. 157–164, 1987.
3. R. M. Parr, *J. Radioanal. Nucl. Chemistry, Articles,* **123/1,** 259–271 (1988).
4. Minor and Trace Elements in Breast Milk, Report of a Joint WHO/IAEA Collaborative Study, WHO, Geneva (1989).
5. Y. Muramatsu and R. M. Parr, *Sci. Total Environ.* **76,** 29–40 (1988).
6. Contaminants, Joint FAO/WHO Food Standards Programme, FAO, Rome, Codex Alimentarius Commission, CAC/VOL. XVII–Ed. 1, (1984).
7. Analytical Quality Services Programme 1990, IAEA. Vienna.
8. R. M. Parr, R. Schelenz, and S. Ballestra *Fresenius Z. Anal. Chem.* **332,** 518–523 (1988).
9. Quality assurance in biomedical neutron activation analysis, IAEA, Vienna, IAEA-TECDOC-323 (1984).
10. Muramatsu Y. and Parr R. M., Survey of currently available reference materials for use in connection with the determination of trace elements in biological and environmental materials, IAEA. Vienna, IAEA/RL/128 (1985).
11. Health-related monitoring of trace element pollutants, using nuclear techniques (1985), IAEA, Vienna, IAEA-TECDOC-330 (1985).
12. Nuclear-based techniques for the in vivo study of human body composition (1986), IAEA, Vienna, IAEA/RL/131 (1986).
13. Nuclear techniques for analysis of environmental samples (1986), IAEA, Vienna, IAEA/RL/135 (1986).
14. Comparison of nuclear analytical methods with competitive methods (1987), IAEA, Vienna, IAEA-TECDOC-435 (1987).
15. Application of short-lived activation products in neutron activation analysis of bio-environmental specimens (1987), IAEA, Vienna, IAEA/RL/141 (1987).

Determination of Mercury in Human Serum and Packed Blood Cells by Neutron Activation Analysis

JACQUES VERSIECK,*,[1] LIDIA VANBALLENBERGHE,[1] ANN WITTOEK,[1] GERDA VERMEIR,[2] AND CARLO VANDECASTEELE[2]

[1]Department of Internal Medicine, Division of Gastroenterology, University Hospital, De Pintelaan 185, B-9000 Ghent, Belgium; and [2]Laboratory for Analytical Chemistry, Institute for Nuclear Sciences, University of Ghent, Proeftuinstraat 86, B-9000 Ghent, Belgium

Received April 17, 1989; Accepted December 11, 1989

ABSTRACT

A method is described for the determination of mercury in human blood serum and packed blood cells employing neutron activation analysis. Great attention was devoted to the collection and manipulation of the samples. The accuracy and precision of the method were tested by analyzing biological reference materials and by comparing the concentrations measured in a number of serum samples to those obtained by another, independent technique (cold vapor atomic absorption spectrometry) in the same samples. The article reports the levels measured in blood serum and packed blood cells samples from 15 adult volunteers, as well as the figures determined in a "second-generation" biological reference material (freeze-dried human serum), prepared and conditioned at the University of Ghent.

Index Entries: Radiochemical neutron activation analysis for mercury; "second-generation" biological reference material; mercury in human blood serum; mercury in human packed blood cells; reference values for mercury.

*Author to whom all correspondence and reprint requests should be addressed.

INTRODUCTION

Mercury is an element of great toxicologic importance that caused several health problems in humans. Symptoms of poisoning were first observed in relation to its use as a therapeutic, in the earliest instances in the form of ointments for inunction. In the Middle Ages, clinical symptoms began to make their appearance with the mercurial treatment of syphilis. Cases of industrial poisoning were first described in the 16th and 18th centuries. At times, mercury poisoning was widely spread among goldsmiths and mirror workers. The expression "mad as a hatter" comes from the signs shown by workers in the treatment of furs with mercuric nitrate. Recently, outbreaks of methylmercury intoxication were observed in Iraq, as a consequence of the consumption of bread made from seed grain treated with mercurial fungicides, and in Japan (Minamata disease), following the dumping of mercury-containing industrial wastes in the Minamata bay and the consumption of fish caught in the bay.

It will be evident that, especially in the case of a toxic element, such as mercury, reliable quantitative figures are badly needed both for issuing regulations about permissible levels, as well as for monitoring individuals at risk. Although in recent years considerable progress has been made, reference values for mercury levels in human blood plasma or serum and packed blood cells are not yet generally accepted. Table 1 catalogs some of the published figures for plasma or serum mercury: they vary from a mean ± SD of 0.551 ± 0.256 ng/mL (*1*) to 20.5 ± 8.2 ng/mL (*13*). With the intention to contribute to a definitive solution of the problem, it was decided to develop an additional method for the determination of the element employing radiochemical neutron activation analysis. An extra, important incentive was, in a second phase, to certify the mercury level in the "second-generation" biological reference material (freeze-dried human serum) made available to the scientific community in 1988 (*14*).

METHODS

Blood samples were collected and processed according to the rigorous, previously *in extenso* described, protocol developed at the University of Ghent (*15*). In the case of mercury, an additional, very important point is that, before analysis, samples must be sealed in quartz containers or (when in polyethylene) stored in hermetically closed and impermeable boxes, so that they cannot take up the element from the air. This risk should not be neglected, even when working in a clean laboratory: indeed, part of the mercury in the air is not in a particulate, but in a

Table 1
Plasma or Serum Mercury Levels Previously Reported by a Number of Researchers

Investigators	Analytical technique	Mean	S.D.	Range
		(all values in ng/mL)		
Vermeir et al., 1988 (1)	AAS	0.551	0.256	0.236 - 0.863
Negretti de Brätter et al., 1987 (2)	AAS	1.03	0.51	0.05 - 4.1
Komarek et al., 1986 (3)	AAS			0.6 - 2.3
Corrigan et al., 1987 (4)	NAA	1.65	0.62	0.69 - 2.57
Wester, 1973 (5)	NAA	1.8	0.6	
Magos and Clarkson, 1972 (6)	AAS	2.2	0.9	1.0 - 4.3
Tejning, 1967 (7)	NAA	2.33[a]	1.11[a]	
Birke et al., 1972 (8)	NAA	3.28	0.51	
Newton et al., 1983 (9)	AAS			5
Kubasik et al., 1972 (10)	AAS	11		7 - 18
Ward and Pim, 1984 (11)	NAA	12	3	
Tejning, 1967 (12)	NAA	13.74[b]	2.98[b]	
Kostadinov and Djingova, 1981 (13)	NAA	20.5	8.2	

[a]Normal Swedish population.
[b]Heavy fish eaters from the Lake Vänern region.

gaseous form, which is not retained by the HEPA-filters controlling the air supply.

Samples were irradiated at a flux of $1.8 \cdot 10^{12}$ $n \cdot cm^{-2} \cdot s^{-1}$ for approx 7 h. After irradiation they were mineralized, along with 1 mg of Hg^{2+} and 1 mg of Cu^{2+} carriers, on a hot plate with magnetic stirrer using 14–20 mL (depending on the sample weight, which varies from approx 500 mg to approx 750 mg dry wt) of a 3:1:1 mixture of nitric, perchloric, and sulfuric acid in a modified Bethge apparatus (*16*), equipped with two, superimposed water coolers to avoid losses of mercury. The mineralization procedure takes approx 4 h. The temperature has to be increased progressively, especially in the beginning. After an initial decomposition with open tap (condensate returning to the digestion flask), contents were heated with closed tap, so that nitric acid distilled and condensed in the reservoir. After some of the perchloric acid also distilled, the condensate was cautiously admitted again to the digestion flask. In general, this operation was repeated three to five times to assure a perfect mineralization of the organic material.

After rinsing the water coolers and the reservoir with some 12 mL of demineralized water and after transfer of the samples to a separatory funnel with an additional 14 mL of demineralized water, the pH was slowly adjusted to 1.5 with 14*M* and 7*M* ammonia. After addition of ammoniumpersulfate, $(NH_4)_2S_2O_8$ (two times the sample weight) as oxidant (mixing time: 15 min) and after washing the aqueous solution with two successive 20-mL vol of chloroform (both discarded after 4 min of shaking), mercury was extracted with two successive 20-mL vol of 0.005*M* nickel diethyl-dithiocarbamate in chloroform (collected together in an Erlenmeyer with a volume of 150 mL after 4 min of shaking), according to a modification of the method described by Greenberg (*17*).

After separation, the organic phase was washed two times for 2 min with 20 mL of an aqueous solution of 5% Na_2EDTA and 1% NaOH, containing 10 mg Br^- (as NaBr) per mL, with the purpose of decreasing the ^{82}Br ($T_{1/2}$ of 35.30 h) activity in the organic, mercury-containing phase. Finally, the ^{197}Hg ($T_{1/2}$ of 64.1 h, γ-line of 77.3 kEv) activity was counted (counting time: 42 h) on a HPGe detector coupled to a 4000-channel analyzer. In some of the serum samples and in nearly all of the packed blood cells samples, ^{203}Hg ($T_{1/2}$ of 46.612 d, γ-line of 279.2 kEv) was also detected.

Yield determinations for mercury were done using tracer experiments. The following figures were obtained—means ± SD: 98.4 ± 1.3%, range: 96.5–100.0%.

In the experimental conditions of this study, roughly 1% of the ^{82}Br activity was ultimately found in the mercury phase (exceptionally 2% or more). Upon measurement of a real sample—especially in the case of serum—even this small fraction gives rise to a considerable activity. The activities of ^{24}Na and ^{64}Cu, also formed during irradiation, entail no serious problems.

RESULTS AND DISCUSSION

The accuracy and precision of the method were exhaustively tested. Analyses of reference materials gave the following values—NBS oyster tissue, SRM 1566: 39.0 ± 6.6 ng/g (certified value: 57 ± 15 ng/g); NBS bovine liver, SRM 1577: 16.3 ± 1.6 ng/g (certified value: 16 ± 2 ng/g); and IAEA milk powder, A-11: 3.64 ± 0.35 ng/g (consensus value: 3.2 ± 0.6 ng/g). It is evident that, with the exception of NBS oyster tissue, the agreement between the obtained and the certified (or consensus) values is very satisfactory. The reason for the discrepancy in the case of NBS oyster tissue remains to be elucidated. For NBS nonfat milk powder, SRM 1549, only upper limits could be established, which ranged from < 0.50 ng/g–< 0.80 ng/g; when using a longer irradiation time and a higher neutron flux, the following values were obtained in three samples: 0.16 ng/g, 0.20 ng/g, and 0.32 ng/g (certified value: 0.3 ± 0.2 ng/g).

A comparison of the results obtained by neutron activation analysis (current method) and cold vapor atomic absorption spectrometry (*1*) in seven serum samples yielded a correlation coefficient (r) of 0.958 (t = 7.666, $P < 0.001$). It must be added, however, that neutron activation analysis systematically gave a somewhat higher value, the equation of the regression line being $y = 0.793x - 0.104$ (y = results obtained by cold vapor atomic absorption spectrometry; x = results obtained by neutron activation analysis).

To test the reproducibility of the method, six analyses were performed of one serum sample: they gave a mean ± SD of 1.461 ± 0.157 ng/mL (coefficient of variation, or CV, of 10.75%). Besides, all determinations been done in duplicate, the reproducibility was also estimated from the difference between the two results according to the formula

$$s = \sqrt{\Sigma d^2/2N}$$

d being the difference between the two results of a pair and N the number of pairs. Calculations gave the following figures—for serum: s = 0.196 ng/mL and for packed blood cells: s = 0.911 ng/g wet wt.

Determinations of mercury in our "second-generation" biological reference material (freeze-dried human serum) gave the following values: 7.035 ± 0.753 ng/g (means ± SD) and 6.080–8.240 ng/g dry wt (range). Based on these results and those of other researchers, namely 6.24 ± 0.99 ng/g and 5.19–7.81 ng/g dry wt (*1*) and 6.5 ± 0.7 ng/g and 5.3–7.0 ng/g dry wt (*18*), the level can now be certified to be 6.6 ng/g dry wt (95% confidence limits: 6.2–7.0 ng/g dry wt).

In 15 adult volunteers (six patients with ophthalmologic disorders, but without other important diseases; seven university employees, namely two medical doctors, two analytical chemists, three laboratory technicians; and two medical students), a serum mercury concentration was measured (mean ± SD) of 1.156 ± 0.554 ng/mL with a range of 0.404–2.456 ng/mL and a packed blood cells mercury concentration was measured (mean ± SD) of 5.320 ± 3.048 ng/g with a range of 1.446–12.512 ng/g wet wt. The higher serum levels obtained in the current study when compared to those reported in an earlier publication (*1*) may be explained, first, by the fact that neutron activation analysis systematically gave a somewhat higher value (vide supra) and, second, by the characteristics of the subjects. The calculation of the correlation factor between the mercury level in serum, on the one hand, and in packed blood cells, on the other, showed $r = 0.773$ ($t = 4.389$, $P < 0.001$).

As already mentioned under Methods, in the case of packed blood cells (with the exception of the lowest values), mercury could be determined both via its ^{197}Hg and ^{203}Hg isotopes—be it with poor counting statistics for the latter. The agreement turned out to be very satisfactory: $r = 0.901$ ($t = 12.110$, $P < 0.001$) and $y = 0.819x + 1.135$ (y = results obtained via the ^{203}Hg, x results obtained via the ^{197}Hg isotope). Assuming an hematocrit of 45%, the mean whole blood mercury level in the adult volunteers may be estimated to be situated somewhere in the vicinity of 3.20 ng/mL.

SUMMARY

The newly developed radiochemical method permits the determination of mercury in human blood serum and packed blood cells of healthy adult persons. A comparison of the figures obtained in this study with those reported in a number of previous studies (Table 1) strongly suggests that the mean level in serum of environmentally or occupationally unexposed people must have been definitely overestimated (roughly by a factor of 10 or 20) in the past. This is particularly true for the data reported by Kubasik et al. (*10*), Ward and Pim (*11*), and Kostadinov and Djingova (*13*). In the case of the data published by the latter two groups of researchers, this impression is strengthened by the poor quality of the analytical results for several other trace elements determined in their laboratories.

ACKNOWLEDGMENTS

The authors thank Professor Fabrice Barbier and Professor Richard Dams for their interest and advice. The financial support of the Fund for Medical Scientific Research (Grants 9.0005.86 and 3.0050.87) and the Inter-University Institute for Nuclear Sciences in Belgium is also gratefully acknowledged.

REFERENCES

1. G. Vermeir, C. Vandecasteele, E. Timmerman, R. Dams, and J. Versieck, *Mikrochim. Acta*, III, 305 (1988).

2. V. E. Negretti de Brätter, P. Brätter, J. Müller, P. Weigert, S. Götze, and G. Karkut, *Trace Element Analytical Chemistry in Medicine and Biology*, vol. 4, P. Brätter and P. Schramel, eds., de Gruyter, Berlin, pp. 151–156, 1987.
3. J. Komárek, V. Otruba, and M. Černiková, *Chem. Listy*, **80,** 206 (1986).
4. F. M. Corrigan, J. D. Finalayson, G. Stevenson, G. W. Ashcroft, and N. I. Ward, *Trace Elements Med.*, **4,** 117 (1987)
5. P. O. Wester, *Acta Med. Scand.* 194, 505 (1973).
6. L. Magos and T. W. Clarkson, *J. Assoc. Off. Anal. Chem.* **55,** 966 (1972).
7. S. Tejning, *Report No. 67 02 06, Dept. of Occupational Medicine*, University Hospital, Lund, 1967, pp. 3. Quoted by H. Ackefors, *Proc. Roy. Soc. Lond. B.* **177,** 365 (1971).
8. G. Birke, A. G. Johnels, L. -O. Plantin, B. Sjöstrand, S. Skerfving, and T. Westermark, *Arch. Environ. Health*, **25**, 77 (1972).
9. J. A. Newton, I. M. House, G. N. Volans, and F. J. Goodwin, *Human Toxicol.* **3,** 535 (1983).
10. N. P. Kubasik, H. E. Sine, and M. T. Volosin, *Clin. Chem.* **18,** 1326 (1972).
11. N. I. Ward and B. Pim, *Biol. Trace Element Res.* **6,** 469 (1984).
12. S. Tejning, *Report No. 67 08 31, Dept. of Occupational Medicine*, University Hospital, Lund, 1967, pp. 30. Quoted by H. Ackefors, *Proc. Roy. Soc. Lond. B.* **177,** 365 (1971)
13. L. Kostadinov and R. Djingova, *J. Radioanal. Chem.* **63,** 5 (1981).
14. J. Versieck, L. Vanballenberghe, A. De Kesel, J. Hoste, B. Wallaeys, J. Vandenhaute, N. Baeck, H. Steyaert, A. R. Byrne, and F. W. Sunderman, Jr., *Anal. Chim. Acta*, **204,** 63 (1988).
15. J. Versieck and R. Cornelis, *Trace Elements in Human Plasma or Serum*, CRC Press, Boca Raton, FL, pp. 48–53, 1989.
16. P. O. Bethge, *Anal. Chim. Acta*, **10,** 317 (1954).
17. R. R. Greenberg, *Anal. Chem.* **58,** 2511 (1986).
18. A. R. Byrne, Personal communication (1986).

Determination of Main and Trace Element Contents in Human Brain by NAA and ICP-AES Methods

E. ANDRÁSI,[1,*] J. NÁDASDI,[1] ZS. MOLNAR,[1,2]
L. BEZUR,
AND L. ERNYEI

[1]*Institute of Inorganic and Analytical Chemistry, L. Eötvös University, Budapest, Muzeum krt. 4/B. 1088;* [2]*Reactor Laboratory, Technical University, Budapest Müegyetem rkp. 9. 1111;* [3]*Institute of General and Analytical Chemistry, Technical University, Budapest, Gellért Tér 4. 1111*

ABSTRACT

A study was undertaken to determine the average values for elements in normal human brain (11 individuals, age group 65–75). Twelve brain parts were selected from both hemispheres. Determinations were carried out by NAA and ICP-AES. The main elements (Na, K, Mg, Ca, Fe, P, S) and trace elements (Al, B, Co, Cr, Cu, Mn, Ni, Pb, Zn) were investigated. Quality control was ensured by using NBS Bovine Liver SRM. The results obtained with independent methods were compared, and the data show a good correlation. On the basis of these investigations, the regional distribution of elements can be given.

Index Entries: Main and trace elements in human brain; wet digestion; NAA; ICP-AES; NBS Bovine Liver SRM.

*Author to whom all correspondence and reprint orders should be addressed.

Biological Trace Element Research Editor: G. N. Schrauzer © 1990 by The Humana Press Inc.

INTRODUCTION

The increasing concern about the health risk from exposure of humans to environmental pollutants is often directed to trace elements, which can cause acute and chronic diseases. In consequence of this, knowledge of the elemental composition of human tissues and body fluids is of great importance in biomedical research on trace metals.

MATERIALS AND METHODS

Sample Collection and Preparation

The brains originated from individuals with no known neurological disorders. The anatomic examination did not show damage. The tissue samples were dissected from the brains about 26 h after death. The stainless-steel knife was soaked for a week in 1:1 nitric acid, and it was then rinsed with double-distilled water prior to use. The soaking solution was analyzed by ICP-AES and the quantity of the unbound impurities was found to be of no importance (Table 1). Talc-free gloves were used, and the leptomeninges were removed. The samples were pressed between sheets of ashless filter paper (MN 600 W) to remove blood contamination. The samples were conserved in formalin. The formalin blank and the problem of leaching were investigated (Table 2). All sample containers were polyethylene, and were washed in bidistilled nitric acid and deionized water prior to use. After determining the wet sample mass, the conservated tissue samples were dried to a constant mass in cleaned quartz vials (36 h, 105°C). The dried to wet wt ratio (D/W) was determined for each sample (Table 3). To remove organic substances, a wet digestion method was applied. For wet digestion, bidistilled Carlo Erba 65% nitric acid was used in a Parr-bomb (150°C, 2 h, sample mass: 100 mg), and the resultant solution was diluted to 5 mL with double redistilled water for analysis. The efficiency of decomposition was investigated (*1*). Multielement calibration standards were prepared by dissolving ultrapure salts of different elements in 1:1 diluted nitric acid.

Instrumental Neutron Activation Analysis (INAA)

The samples, dried until constant mass was achieved, were analyzed directly by NAA in the nuclear research reactor of the Technical University Budapest. The samples (ca. 100 mg) and the comparator (ruthenium powder) were encapsulated in polyethylene vials (*2*). The brain samples were irradiated in two steps by thermal neutrons with a flux of 2.10^{12} neutron cm^{-2} s^{-1}. After 300 s irradiation, 90 s cooling period, and 10 min counting, the Na, K, Mg, Cu, and Mn contents were determined. The period of the second irradiation was 8 h, the cooling period lasted for 1 w to 1–5 m, 8 h counting was set, and Co, Cr, Fe, and Zn concentrations

Table 1[a]
Analytical Results for the Soaking Solution
of the Dissection Knife, Measured by ICP-AES (μ/mL)

Al	B	Co	Cr	Cu	Mn	Ni	Pb	Zn
0.07 ±0.01	0.04 ±0.005	[b]	[b]	[b]	0.05 ±0.005	[b]	[b]	0.02 ±0.005

[a]These values are based on nine measurements.
[b]These element concentrations were below the detection limit

Table 2[a]
Ion Concentrations in Formalin, Measured by ICP-AES (μg/mL)

	Formalin blank, (before conservation)	Overflowed formalin blank, (after a few months of conservation)
Al	0.10 ± 0.02	0.09 ± 0.02
B	0.40 ± 0.04	0.40 ± 0.04
Co	[b]	[b]
Cr	[b]	[b]
Cu	0.10 ± 0.01	0.09 ± 0.01
Mn	0.03 ± 0.005	0.04 ± 0.005
Ni	[b]	[b]
Pb	[b]	[b]
Zn	0.60 ± 0.04	0.67 ± 0.03
Ca	8.0 ± 0.8	11.0 ± 1.2
Na	3.0 ± 0.3	39 ± 4
K	1.0 ± 0.2	65 ± 15
Mg	2.0 ± 0.4	4.2 ± 0.8
Fe	0.40 ± 0.06	0.40 ± 0.04
P	0.6 ± 0.1	40 ± 6
S	3.0 ± 0.3	4.8 ± 0.4

[a]These values are based on nine measurements.
[b]These element concentrations were below the detection limit.

Table 3
Dried to Wet Wt Ratios for a Few Brain Parts

Brain Parts	D/W ratios[a]
Corpus callosum	0.276 ± 0.008
Cortex cerebri	0.205 ± 0.006
Putamen left side	0.203 ± 0.006
right side	0.204 ± 0.006
Thalamus left side	0.205 ± 0.006
right side	0.202 ± 0.006
Hippocampus left side	0.1733 ± 0.007
right side	0.1708 ± 0.007
Pallidum Caudatum	0.230 ± 0.006
(mean of left and right sides)	0.165 ± 0.006

[a]The ratios are based on results of nine measurements.

were determined. The γ-spectra were detected by a Ge (Li) detector and evaluated by a Canberra-80 analyzer connected to a TPA computer. Corrected for background, peak areas were calculated to determine the concentration of elements.

Inductively Coupled Plasma Atomic Emission Spectrometry (ICP-AES)

Simultaneous multielement analysis of the digested samples was carried out on a Labtest Plasmalab ICP spectrometer for 16 elements in an argon–argon plasma (incident power 1,3 kW; 27,12 mHz; observation height 16 mm). Sample introduction was based on a flow-injection of 130 μL volume sample using a pneumatically operated injection valve. All the important parameters of the valve operation (rinse time, delay time, integration time) can be set from the computer program. As a result of very accurate timing and operation, it was possible to use only the middle section of the peaks, resulting in an increase of signal-to-background ratio and in an increased speed of operation (3). Background correction and an element-specific interference correction (Fe) was implemented in the analytical program. An off-line drift correction was applied using a special program handling all the measured data, resulting in good reproducibility.

RESULTS AND DISCUSSION

It can be seen that the leaching of trace elements is insignificant, because these elements are bound to the proteins and the formalin fixed their structure. Taking the concentration of main elements into account, their leaching is also negligible. Digestion carried out in a Parr-bomb is advantageous because of the small risk of contamination and loss, as well as the complete decomposition. This digestion has only one disadvantage, i.e., the inhomogeneity problem is relatively high because of the small sample mass (100 mg) that can be digested in a Parr-bomb.

NBS Bovine Liver SRM was analyzed to prove the accuracy of these methods. The composition of this Standard Reference Material and the human brain is similar. These figures are in a good agreement with certified values (Table 4). The concentration variations were not greater than 17% relative SD of the mean. A typical value of 7% was observed at levels above the detection limits of the techniques. Brain parts, of which sufficient quantities were possessed, were analyzed by the above-mentioned independent methods. The main and trace element concentrations determined by these authors in different laboratories are in a good agreement with one another, if the possible inhomogeneity within the brain parts is taken into account (Table 5).

From these investigations, it can be concluded that INAA has many advantages. The INAA method is unique because it eliminates the usual

Table 4
Analytical Results for NBS Bovine Liver SRM 1755;
Concentration Mean ± SD (μg/g Dry Wt)[a]

Element	Certified values	Present results	
		NAA	ICP-AES
Na	2340 ± 130	2300 ± 100	2200 ± 40
K	9700 ± 600	9500 ± 400	9500 ± 570
Mg	604 ± 9	605 ± 100	603 ± 80
Ca	124 ± 6	–	116 ± 10
Fe	268 ± 8	315 ± 30	284 ± 10
Cu	193 ± 10	185 ± 10	205 ± 10
Mn	10.3 ± 1.0	9.8 ± 0.7	11.5 ± 0.8
Zn	130 ± 13	149 ± 10	147 ± 5

[a]The concentrations given in this table are based on results of three measurements.

problems of trace elemental analysis (loss by adsorption and contamination of reagents), since there is no sample treatment. A small sample quantity (50–100 mg) is enough. The INAA method has good detection limits for many elements. The relative SD is approx 5–20%. Only poor precision was available for Co and Cr, because of their small quantity in brain samples. It is disadvantageous that the instrumentation is available only in few laboratories. The analysis time depends on the radioactive-decay half-life.

The ICP-AES method is currently one of the most powerful analytical methods for determination of components of dissolved samples. ICP source, combined with a polychromator, can provide a simultaneous multielement technique. The advantages of the flow-injection method are the reduction of sample consumption and the high speed of analysis (100–150 samples/h). For one analysis, a 600 μL sample volume is sufficient.

The ICP-AES technique is almost insensitive to disturbing effects, resulting in a high detection power and a wide linear range. However, in the biological samples, Fe, Ca, Mg, and Al are present, sometimes at relatively high concentration. These elements can be critical from the point of view of spectral interferences. This problem was solved by the combination of background-correction and blank measurement. In most cases, the effect of the correction decreased the error down to the range of the detection limit (*3*).

The ICP-AES method is a high-precision method, the relative SD being approx 1–5%. Co, Cr, Ni, and Pb concentrations of human brain part solutions are below the detection limit of the ICP-AES technique, so these elements can only be measured with poor precision, if at all. The greatest disadvantage of this technique is the dilution (about 50 times during the sample preparation).

The present results show a nonhomogeneous distribution of the elements in the human brain. It was found that the elements had gener-

Table 5
Analytical Results for a Few Human Brain Parts Determined by NAA and ICP-AES; Concentration Mean ± SD (μg/g Dry Wt)[a]

Brain parts	Element	NAA	ICP-AES
Thalamus left side	Na	6300 ± 200	6700 ± 200
	K	9500 ± 200	9300 ± 300
	Mg	840 ± 50	670 ± 30
	Fe	210 ± 30	230 ± 10
	Cu	20.2 ± 2	17.9 ± 1
	Mn	2.5 ± 0.2	2.2 ± 0.2
	Zn	58.8 ± 4	66.1 ± 3
Thalamus right side	Fe	230 ± 20	280 ± 20
	Cu	18.0 ± 1.5	21.0 ± 1
	Mn	1.9 ± 0.1	2.2 ± 0.1
	Zn	56.2 ± 2	61.4 ± 4
Corpus callosum	Fe	160 ± 10	150 ± 8
	Cu	15.0 ± 1	13.3 ± 1
	Mn	1.4 ± 0.1	1.5 ± 0.1
	Zn	49.3 ± 2	46.5 ± 2
	P	–	13,700 ± 400
	S	–	5800 ± 200
Putamen left side	Na	8900 ± 250	9000 ± 300
	K	10,200 ± 200	12,600 ± 200
	Mg	690 ± 20	790 ± 20
	Fe	840 ± 100	760 ± 40
	Cu	28.1 ± 3	29.2 ± 3
	Mn	3.3 ± 0.2	3.5 ± 0.2
	Zn	91.7 ± 10	78.1 ± 4
Cortex cerebri	Fe	260 ± 30	320 ± 20
	Cu	24.3 ± 2	27.5 ± 3
	Mn	2.0 ± 0.1	2.4 ± 0.1
	Zn	69.1 ± 5	78.8 ± 6
	P	–	13,000 ± 500
	S	–	6700 ± 700

[a]The concentrations are based on results of three measurements.

ally higher levels in gray than in white matter regions. Regions corresponding to each other in both hemispheres show almost identical concentration values for the various elements. The distribution of Mg, P, and S is relative homogeneous. An enrichment of Fe and Mn was observed in pallidum, putamen, and caudatum. The lowest Fe concentration can be seen in corpus callosum.

A summary of these data on a few normal brain parts and comparisons with literature values are given in Table 6. There is, generally, a good conformity with the literature (*4–7*). Exceptions are Mn in cortex cerebri, and Zn in corpus callosum.

Table 6
Elemental Data for a Few Normal Human Brain Parts and Recent Literature Comparisons (μg/g Dry wt)

Brain parts		This work[a]	Means for other determinations		
Cortex cerebri	Fe	277 ± 40	280[4]	240[7]	
	K	13,100 ± 1700	12,230[4]	17,200[7]	
	Mn	1.89 ± 0.4	1.43[4]	1.25[7]	
	Na	7090 ± 1390	7250[4]		
	P	13,800 ± 1000	15,120[4]		
	Zn	70 ± 10	59.7[4]	71[7]	
Corpus callosum	Fe	150 ± 20	145[4]	123[6]	134[7]
	K	8520 ± 1000	8740[4]	10,700[7]	
	Mn	1.4 ± 0.2	1.24[4]		
	Na	5520 ± 1120	4915[4]		
	Zn	48 ± 6	27.3[4]	24.3[6]	28[7]
Hippocampus	Fe	240 ± 20	222[5]	238[6]	
	K	12,500 ± 1500	13,500[5]		
	Zn	86 ± 5	83.2[5]	96.3[6]	
Putamen	Mn	2.65 ± 0.5	2.67[7]		
Pallidum	Mn	2.30 ± 0.3	2.18[7]		
Caudatum	Mn	2.54 ± 0.5	2.48[7]		

[a]These values are the means of element concentrations from both hemispheres of 11 persons measured by NAA and ICP-AES.

SUMMARY

In the first part of this research project, 16 elements in brain parts were measured, which originated from 12 different regions of both hemispheres of 11 normal human brains (mean age: 70). The brain samples were analyzed by both INAA and ICP-AES methods, which are multielement techniques allowing the measurement of μ/g levels with good precision in small biological samples. To minimize systematic errors, the NBS Bovine Liver SRM was used.

REFERENCES

1. E. Andrási, M. Mohos, L. Bezur, and Zs. Molnár, *Trace Element Analytical Chemistry in Medicine and Biology*, vol. 5., Walter de Gruyter, Berlin, New York, p. 261, 1988.
2. T. Bereznai, D. Bódizs, and G. Keömley, *J. Radioanal. Chem.* **36** (1977).
3. L. Bezur, L. Ernyei, and E. Pungor, Abstracts, Winter Conference on Plasma and Laser Spectrochemistry, Lyon, p. 241 (1987).
4. W. D. Ehmann, W. R. Markesbery, M. Alauddin, T. I. M. Hossain, and E. H. Brubaker, *Neurotoxicology* **7/1** 197–206 (1986).

5. C. M. Thompson, W. R. Markesbery, W. D. Ehmann, Y-X. Mao, and D. E. Vance, *Neurotoxicology* **9/1** 1–8 (1988).
6. A. Höck, U. Demmel, H. Schicha, K. Kasperek, and L. E. Feinendegen, *Brain* 98, 49–64 (1975).
7. H. Duflou, W. Maenhaut, and J. De Reuck, *Trace Element Analytical Chemistry in Medicine and Biology*, vol. 5., Walter de Gruyter, Berlin, New York, p. 483, 1988.

Levels of Some Trace Elements in Selected Autopsy Organs, and in Hair and Blood Samples from Adult Subjects of the Italian Population

G. INGRAO,[1,*] P. BELLONI,[1] S. DI PIETRO,[1] AND G. P. SANTARONI[2]

[1]*ENEA Casaccia PAS-SCAMB C.P. 2400, 00100 Roma, Italy*
[2]*Istituto Nazionale della Nutrizione via Ardeatina 546, 00179 Roma, Italy*

Received April 17, 1989; Accepted December 11, 1989

ABSTRACT

This study, which is part of a research program for the determination of trace element reference levels in various human tissues for the Italian population, presents the concentrations of Se, Hg, Cr, Cs, Sc, Rb, Zn, Fe, Co, and Sb in lung, liver, spleen, and kidney autopsy samples taken from 14 adult subjects of the Italian population who died from accidental causes. Concentrations of the same trace elements are given also for blood and hair samples taken from subjects of the general Italian population and of a population group living in a small coastal town that has an average annual fish consumption well above the national average. The analytical method used was Instrumental Neutron Activation Analysis. The levels and distribution of the trace elements in the various human organs examined are analyzed and discussed.

Index Entries: Trace elements; human tissues; neutron activation analysis.

*Author to whom all correspondence and reprint requests should be addressed.

INTRODUCTION

A great number of research studies carried out worldwide during the past few decades have demonstrated the important role of many trace elements in various biochemical processes (*1,2*). It has been shown that the normal growth of living organisms depends on the continuous and adequate supply of certain trace elements. The beneficial biological effect of an element depends on its concentration in the organism, and with increasing concentrations even essential trace elements may give rise to detrimental effects. Knowledge of the normal levels in human tissues and body fluids is indispensable to detect any alteration resulting from inadequate intake of trace elements through the diet, disease conditions, or environmental pollution.

A research program to establish the reference range of trace element concentrations in various human tissues for the Italian general population, and to study the transfer pathways of trace elements from the environment to humans, with particular attention to the food chain, was started in our laboratory in collaboration with the Institute of Legal Medicine of Rome University and the National Institute of Nutrition.

This article presents the levels of Se, Hg, Cr, Cs, Sc, Rb, Zn, Fe, Co, and Sb in lung, liver, spleen, and kidney autopsy samples taken from 14 adult subjects of the Italian population who died from accidental causes. Concentrations of the same trace elements are given also for blood and hair samples taken from subjects of the general Italian population and of a population group living in a small coastal town that has an average annual fish consumption of 27.1 kg, well above the value of 12.5 kg for the Italian population (*3*).

MATERIALS AND METHODS

The human autopsy samples were provided by the Institute of Legal Medicine of Rome University. A standard procedure was followed to remove a representative sample of about 200–300 g from each organ. The samples, after being squeezed and drained to remove most of the blood present in the organs, were enclosed in clean glass containers and kept in a freezer at $-40°C$ until they were freeze-dried. After homogenization, the samples were stored in polyethylene containers. Blood samples were collected with plastic syringes and immediately after enclosed in pure quartz vials. The hair samples were washed, following the IAEA procedure (*4*), with distilled water and acetone.

The analytical method used for the determination of the trace element concentrations was Instrumental Neutron Activation Analysis. The samples and standard reference materials (IAEA and NBS), enclosed in pure quartz vials, were irradiated in the 1 mW TRIGA reactor located in the Casaccia research center for about 14 h in a thermal-neutron flux of approx $2.6 \cdot 10^{12} n \cdot cm^{-2} \cdot sec^{-1}$. The continuous rotation of the irradiation facility ensured a uniform neutron flux for all the samples. After a 2–4 wk cooling time, the samples were measured by γ spectrometry, using a high-purity germanium detector with a relative efficiency of about 20% and a resolution (FWHM) of 1.9 kEv at the 1332 kEv peak. The Camberra computer program Spectran F was used for the analysis of the γ spectra.

RESULTS AND DISCUSSION

Human Autopsy Samples

The samples analyzed in this study were taken from 14 adult subjects, 10 men and four women of the Italian general population, who died from accidental causes. Except for two young subjects, 20 and 25 years old, the ages ranged from 51–79 yr.

Tables 1, 2, 3, and 4 summarize the data of the trace element concentrations in lung, liver, spleen, and kidney, respectively. These organs were selected because, in general, they act as accumulation sites for most of the trace elements. The range of values reported in the literature, when available, is also shown in the same Table for each trace element.

Concise comments on the levels of trace elements measured in the four organs examined are now reported for each element:

Selenium

The intersubject variability is about a factor of 3–4 for all the four organs. This element appears to accumulate more in the kidney, in agreement with the data reported by Iyengar et al. (*5*). The lung presents the lowest concentration. The ratio between the mean concentrations in the kidney and lung is approx 2.5.

Mercury

The intersubject variability is very high: it is about a factor of 75 for the kidney and 218 for the lung. The analysis of the parameters that generally are considered important in determining the mercury body burden (occupation, dietary habits, age, sex) does not explain the high variability observed. This is probably because of the small number of subjects examined. Also for this element the kidney is the principal organ of accumulation, as reported by other authors (*6*). The spleen shows the lowest concentrations, the mean value in this organ being a factor 9 less than in the kidney.

Chromium

Chromium concentration is in most samples below the detection limit of 100 ng/g. The high values observed in the lungs of two subjects could be the result of exposure to environmental pollution.

Table 1
Trace Element Concentrations in Lung (Dry Wt)

Element	N[a]	Mean ± SD	Range	Reported range, *5, 7, 9, 10*
Co (ng/g)	12	116 ± 53	25–212	57–3600
Cs (μg/g)	12	0.5 ± 0.4	0.11–1.8	0.12
Fe (mg/g)	12	1.7 ± 1.4	0.4–5.6	0.5–1.3
Hg (ng/g)	12	373 ± 996	16–3532	<300–8800
Rb (μg/g)	12	50 ± 20	16.2–86.6	28.6
Sb (ng/g)	11	82.4 ± 59.6	<2–233	33–749
Sc (ng/g)	12	7.1 ± 5.6	1.2–17	
Se (μg/g)	12	0.7 ± 0.3	0.40–1.22	0.38–1.1
Zn (μg/g)	12	74.2 ± 75.6	38.0–311.6	19–70.5

[a]Number of samples above detection limit.

Table 2
Trace Element Concentrations in Liver (Dry Wt)

Element	N[a]	Mean ± SD	Range	Reported range, *5, 7, 9, 10*
Co (ng/g)	13	129 ± 52	32–215	59–2790
Cs (μg/g)	13	0.6 ± 0.6	0.06–2.3	0.05
Fe (mg/g)	13	1.2 ± 1.0	0.3–3.8	0.03–3.7
Hg (ng/g)	13	319 ± 294	5.9–951	14–1843
Rb (μg/g)	13	55.8 ± 20	12.9–92.1	37.6
Sb (ng/g)	9	32.5 ± 48	<2–155	10–408
Sc (ng/g)	10	2.0 ± 1.2	<1–6.4	
Se (μg/g)	13	1.1 ± 0.3	0.4–1.8	0.3–2.6
Zn (μg/g)	13	174 ± 146	55.8–608	49–770

[a]Number of samples above detection limit.

Table 3
Trace Element Concentrations in Spleen (Dry Wt)

Element	N[a]	Mean ± SD	Range	Reported range, *5, 7, 9, 10*
Co (ng/g)	13	73.8 ± 68.4	14.6–287	
Cs (μg/g)	13	0.5 ± 0.38	0.06–1.6	0.06
Fe (mg/g)	13	1.8 ± 1.5	0.2–6.3	1.4
Hg (ng/g)	13	116 ± 135	3.6–490	
Rb (μg/g)	13	56.8 ± 18.1	23.5–79.6	30.1
Sb (ng/g)	11	52.9 ± 63.3	<2–232	45–73
Sc (ng/g)	10	2.8 ± 5.6	<1–6.4	
Se (μg/g)	13	0.9 ± 0.4	0.60–2.12	0.17–1.2
Zn (μg/g)	13	76.7 ± 38.5	47.3–199	77.5

[a]Number of samples above detection limit.

Table 4
Trace Element Concentrations in Kidney (Dry Wt)

Element	N[a]	Mean ± SD	Range	Reported range, *5, 7, 9, 10*
Co (ng/g)	12	63.1 ± 34.1	25.2–145	26–811
Cs (μg/g)	12	0.5 ± 0.3	0.07–1.1	0.03
Fe (mg/g)	12	0.3 ± 0.1	0.13–0.43	0.24
Hg (ng/g)	12	1062 ± 1453	69–5218	
Rb (μg/g)	12	36.2 ± 18.0	12.9–75.5	19
Sb (ng/g)	8	25.1 ± 28.6	<2–89.6	19–166
Sc (ng/g)	10	1.3 ± 0.7	<1–2.5	
Se (μg/g)	12	1.9 ± 0.6	0.95–3.28	0.3–3.8
Zn (μg/g)	12	108.6 ± 40.4	43.1–180	152

[a]Number of samples above detection limit.

Cesium

Although the intersubject variability is high, going from a factor of 14 in the kidney to a factor of 40 in the liver, the mean concentrations in the four organs are very similar. This result can be explained by the chemical similarities of this element and potassium, which is distributed rather uniformly in the human body. The high variability is probably the result of the fact that, in general, the amount of cesium present in the diet, and consequently absorbed in the body, depends on the geological characteristics of the area where the subject lives.

Scandium

The concentration of this element in all the examined organs is very low, being in the ppb region. The intersubject variability ranges from a factor of 5 in the liver to a factor of 14 in the lung, which appears to be the main accumulation organ for this element. The mean concentration in the lung is in fact a factor of 6 higher than in the kidney, the organ with the lowest mean concentration. As for chromium, the levels measured in the lung should be correlated to the lifetime exposure to air pollution.

Rubidium

The intersubject variability is contained within a factor of 7. This element, like cesium, although it does not have a definite biological role in the human metabolism, resembles potassium in its metabolic behavior and, as might be expected, its distribution is relatively uniform in the four organs. The body burden generally depends more on the geological characteristics of the area of residence than on dietary habits.

Zinc

The intersubject variability ranges from a factor of 4 in the kidney and spleen to a factor of 8 in the lung. The liver is the main organ of accumulation, in agreement with the data reported by lyengar et al. (5). The lung and spleen have mean concentrations that are a factor of 2 lower.

Iron

Although intersubject variability is limited to a factor of 3 in the kidney, it is more than a factor of 10 in the other organs. The spleen and lung, as found in other studies (2), are the organs that present the highest concentration of Fe. The concentration in the kidney is generally a factor of 5 lower. The high levels measured in the lung can be attributed to the substantial contribution from the relatively high amount of residual blood in this organ.

Cobalt

The intersubject variability ranges from a factor of 6 in the kidney to a factor of 20 in the spleen. The observed variability could depend on particular dietary habits of the subjects (like beer consumption). The liver is the main organ of accumulation. A similar mean concentration of Co is present in the lung. The Co levels in the other two organs are only a factor of 2 smaller.

Antimony

The intersubject variability is rather high, ranging from a factor of 7 in the lung to a factor of 90 in the liver. Higher concentrations of this element are observed in the lung. The lowest values are present in the kidney, with a mean concentration value a factor of 3 smaller than in the lung.

In conclusion, for most elements the intersubject variability is rather high, in some cases more than a factor of 10. Hg, Sb, Cr, and Cs are the elements that show the highest intersubject variability, whereas Se is the element with the lowest variability. Nevertheless, the concentration levels of the trace elements in lung, liver, spleen, and kidney found in this work are in general of the same order of magnitude as the values reported in the literature. One exception is cesium, for which higher values were found in this study.

In spite of the intersubject variability observed in the measured concentration levels, it is possible to observe a certain trend in the distribution of some trace elements in the four organs examined. Hg, Sc, and Fe are the elements showing the highest interorgan variability. On the contrary, Cs and Rb are the elements more uniformly distributed in the four organs. The highest concentrations of Cr, Sb, and Sc are observed in the lung, as expected for elements present mostly in air pollution. The liver presents higher concentrations of Zn and Co. Se and Hg accumulate more in the kidney, whereas Fe is more concentrated in the spleen.

Table 5
Trace Element Concentrations in Blood Samples

Element	N[a]	Mean ± SD	Range	Male Subjects Italian population, 12
Co (ng/g)	6	3.3 ± 2.5	1.1–7.8	1.5–4.4
Cs (μg/g)	6	3.4 ± 1.0	1.9–4.4	2.5–56
Fe (mg/g)	6	563 ± 38	506–610	234–616
Hg (ng/g)	6	24 ± 10	15–43	<5–8.6
Rb (μg/g)	6	3.8 ± 0.5	3.0–4.5	1.0–7.0
Sb (ng/g)	3	1.4 ± 1.4	<0.5–3.0	<0.5–1.6
Se (μg/g)	6	178 ± 50	125–267	66–119
Zn (μg/g)	6	6.7 ± 1.6	4.5–8.5	3.9–8.6

[a]Number of samples above detection limit.

Blood and Hair Samples

To verify if dietary habits could explain the high variability of the mercury levels in the autopsy samples, a comparison was made of the concentrations of this and other trace elements in blood and hair samples from subjects of a population group more exposed to the risk of mercury contamination, with the values reported for subjects of the general Italian population. Hair and blood samples were collected from subjects living in a small coastal town where the average annual fish consumption is a factor of 2 higher than the national average.

The mean concentrations and ranges of some trace elements measured in the blood samples are reported in Table 5. The values observed in blood samples collected from other Italian population groups (*12*) are also shown for comparison.

It is clear from the data reported in this Table that the Hg blood levels for the population group examined in this study are much higher than the values found for other Italian population groups. The values measured in this study are of the same order of magnitude as the concentration levels measured in a group of miners exposed occasionally to dust of mercury ore (*8*). Also, the Se concentration is significantly higher than the values measured in other Italian population groups. No significant differences were observed for other elements.

Table 6 shows the mean concentration and range of some trace elements measured in hair samples collected from 90 subjects (57 males and 33 females), with ages ranging from 11–71 yr. The data measured for other Italian population groups are shown for comparison.

The data relative to the subjects for whom both hair and blood samples were collected were analyzed to check if correlations were pre-

Table 6
Trace Element Concentrations in Hair Samples (All Subjects)

Element	N[a]	Mean ± SD	Median	Range	Italian mean values, *(11)*	Reported range, *5, 7, 9, 10*
Co (ng/g)	87	296 ± 320	136.2	<10–1337	130 ± 180	7–2600
Cr (μg/g)	90	2.5 ± 3.2	1.1	0.06–13.6	4.0 ± 12.5	0.02–65
Fe (μg/g)	84	98.2 ± 94	67.3	<5–378	203 ± 530	6–2400
Hg (μg/g)	90	10.4 ± 6.6	8.8	1.4–30	2.4 ± 4.5	0.1–58
Sb (ng/g)	74	155 ± 135	88.1	<10–628	260 ± 680	90–8800
Sc (ng/g)	60	5.4 ± 6.6	1.43	<0.5–28.4	2 ± 1	2–22
Se (μg/g)	86	0.8 ± 0.4	0.75	<0.1–1.83	1.4 ± 3.1	0.64–2.53
Zn (μg/g)	90	185 ± 59	181.3	40.5–376	208 ± 137	26–915

[a]Number of samples above detection limit.

sent between the elemental concentrations in hair and blood. Statistically significant correlations were observed only for Se and Hg, with correlation coefficients $r = 0.91$ and $r = 0.89$, respectively. This result confirms that the subjects studied were exposed to these elements constantly and not just occasionally.

It is evident from an examination of the data reported in Table 6 that the mean Hg hair concentration in the population group having a high fish consumption is much higher than the levels observed in other Italian population groups. Also, for the hair samples, the mercury levels measured in this study are of the same order of magnitude as the values found in the hair of a group of miners occasionally exposed to dust of mercury ore (*8*). The concentrations of the remaining trace elements do not differ significantly from the values measured in other Italian population groups. This result shows that, in subjects not professionally exposed, the hair mercury concentration is strongly correlated to the mercury intake through the diet.

Examining the distribution of the trace element concentrations in all the subjects, it was found that Zn and Se have approximately a normal distribution, the skewness coefficient being 0.6, as expected for essential elements. Hg, Fe, Co, and to a larger extent Cr, Sb, and Sc, present an asymmetrical distribution that is approximately log-normal.

In order to examine if the trace element concentrations vary according to sex, the data have been analyzed separately for male and female subjects. It appears that males have in general higher mean concentrations than women. The difference is statistically significant ($P < 0.05$) for Hg and Cr, although it is not significant for the other elements.

No significant correlations were found between trace elements concentrations and age. The only significant correlation found between elemental pairs was between Cr and Co, with a correlation coefficient $r = 0.85$.

REFERENCES

1. L. E. Feinendegen and K. Kasperek, *Trace Element Analytical Chemistry in Medicine and Biology*, P. Bratter and P. Schramel, eds., Proc. 1st international Workshop Neuherberg, FRG, pp. 1–19, 1980.
2. E. J. Underwood, *Trace Elements in Human and Animal Nutrition*, (4th ed., Academic, 1977.)
3. C. Nauen, G. Tomassi, G. P. Santaroni, and H. Josupeit, *UNEP Workshop on Pollution of the Mediterranean*, Proc. VI international Workshop, Cannes, France, pp. 571–583, 1982.
4. *Activation Analysis of Hair as an Indicator of Contamination of Man by Environmental Trace Element Pollutants*, Report IAEA/R1/50, 1978.
5. G. V. Iyengar, W. E. Kollmer, and H. J. M. Bowen, *The Elemental Composition of Human Tissues and Body Fluids: A Compilation of Values for Adults.* Verlag Chemie, NY, 1978.
6. G. Kazantis, *Metals in the Environment: Mercury*, H. A. Waldron ed., Academic, NY, 1980.

7. G. V. Iyengar, *Concentration of 15 Trace Elements in Some Selected Adult Human Tissues and Body Fluids of Clinical Interest from Several Countries: Results from a Pilot Study for the Establishment of Reference Values*. KFA ISSN 0366-0885, Julich, 1985.
8. L. Cigna Rossi, G. F. Clemente, and G. P. Santaroni, *Arch. Environ. Health* **31, N.3,** pp. 160–165 (1986).
9. T. G. Aalbers, J. P. W. Houtman, and B. Makkink, *Clin. Chem.* **33, No, 11,** 2057–2064 (1987).
10. M. Yukawa, M. S. Yasumoto, K. Amano, and M. Terai, *Arch. Environ. Health* **35, No. 1,** 36–44, (1980).
11. G. F. Clemente, L. Cigna Rossi, and G. P. Santaroni, Proc. IAEA Conference "Nuclear Activation Techniques in the Life Sciences", *IAEA-SM-227/13*, 527–543, 1978.
12. L. Cigna Rossi, G. F. Clemente, and G. P. Santaroni, *Rev. Environ. Health* **111, No 1,** 19–42, (1978).

Trace Element Distribution in Human Eye

D. TLUČHOŘ, B. ŠTVERÁK, AND L. CIGÁNEK

Institute for Research, Production, and Application of Radioisotopes, Radiová 1, 102 27 Prague, Czechoslovakia

Received April 17, 1989; Accepted December 11, 1989

ABSTRACT

A hundred samples of four types of eye tissues and hair from a set of 10 donors from the city of Prague were analyzed using INAA. The hair samples of the set of donors were taken as the indicator of the environmental conditions. Within the framework of the set of results obtained for 30 chosen elements, the differences between various types of tissues and sex were tested. The lowest trace element deposition was found for the lens tissue. The iris differs predominantly by the higher contents, particularly of metal elements. Slightly higher contents of practically all trace elements under study were found in all eye tissues taken from male donors.

Index Entries: Trace element; neutron activation analysis; eye tissues; mineralization; γ-spectrometry; significance test.

INTRODUCTION

The increasing number of eye diseases attracts attention to the investigation of the mechanisms of the origin of these diseases and to the effects of ecological parameters contributing to their formation. Eye cataracts belong to the most frequent eye diseases. At present, it is assumed that the formation of a cataract is a complex process, with metabolic changes and disturbances being the most relevant causes. These changes

*Author to whom all correspondence and reprint requests should be addressed.

can be affected by various trace elements acting as catalysts in enzymatic processes.

The trace element effects depend on a number of factors, such as age, sex, lactation, illnesses, and so on. The decisive factor need not be the extent of the exposure to certain elements, but rather the changes of this exposure.

Because of these facts, a large number of studies using predominantly neutron activation analysis, X-ray fluorescence analysis, and atomic absorption spectroscopy were devoted to the problem of determination of trace element contents depending on environmental factors. Lakoma (*1*) applied a combination of NAA with atomic absorption. A correlation between the contents of some trace elements, mostly in the eye lens, and such factors as age, sex, or lens turbidity were investigated in various studies (*1–6*).

The statistical evaluation of the analytical results obtained from different sets of experiments were often laden with large errors. Various authors draw different conclusions from the concentrations of certain elements or from their variations. From the analytical point of view, it is therefore necessary to determine the weight of the correlation between the trace element contents and a number of life factors, at least with such factors as sex, age, profession, health conditions, and ecological situation in the respective locality of the origin of the samples.

This study was motivated by the attempt (1) to minimize the effect of the parameters mentioned above on the deposition of the trace elements in the lens, and (2) to enhance the difference between the normal transparent lens and the turbid one. This article is thus directly linked with previous results (*7*). The results obtained can be used as a basis for the investigation of the trace element dynamics in various tissues of a human eye.

EXPERIMENTAL

A set of 100 samples was chosen, composed of tissues of 10 donors that

1. Did not suffer eye diseases before their death
2. Lived for the last 10 years in the same locality (i.e., in the city of Prague) and
3. Equally represented both sexes.

Unfortunately, it was not possible to achieve the span of donors' ages less than 10 y. For each donor, four tissues were taken from both eyes: cornea, iris, vitreous humor, and lens. Two hair samples from each donor were added to these eight samples. Both the sampling and the tissue preparation were made by specialists using titanium instruments.

INAA was used for the analysis of samples. Following the procedure recommended by IAEA, the surface impurities were removed from the hair samples by fivefold washing: in acetone, three times in water, and again in acetone. The samples were lyophilized and weighed.

Before irradiation, the samples were sealed in ampules of fused silica. Sets with 10 samples in each were activated for 20 h in a nuclear reactor using a thermal neutron flux of (2–5) · $10^{13}s^{-1}cm^{-2}$. Reference materials (mostly "Horse kidney" H-8 IAEA) and standard solutions were activated together with the samples. After activation, both samples and reference materials were mineralized by HNO_3 in autoclaves with PTFE cover (150°C, 10 MPa, 2 h). Perfect and quantitative transfer of samples from the irradiation ampules into the measuring cuvets was the principal aim of the mineralization, as well as the establishment of identical geometrical conditions for the measurement of different types of tissues.

The samples were measured in two series using the γ-spectrometric equipment (detector ICG 121 cm^3, FWHM = 1.75 kEv (1332.5 kEv), supplied by PGT; MCA Canberra S-85, computer PDP 11/23). First, the radionuclides with half-lives of 10–50 h were measured after a 8–12 h decay time; a second measurement for the detection of long-lived nuclides was made about 1 mo later. The concentrations of elements were calculated using the relative method. The spectra evaluation and their statistical treatment were made on the PDP 11/23 computer using our own software.

RESULTS AND DISCUSSION

Differences among the trace element contents in various tissues were determined. Moreover, the differences according to the donor sex were also investigated. For the evaluation of other factors (age, health conditions, profession), the number of cases cannot be considered as fully representative. All errors, reliability intervals, and significance tests were evaluated using the reliability level of 95%.

In Table 1, the average concentrations of 30 elements are listed for the different types of tissues. The table also contains the standard error, the range obtained, and the limits of detection calculated according to ref. (*8*) using the relation

$$LC = 2.23 \times \sqrt{B}$$

where B is the background under the peak. The detection limit, given in Table 1, is the average value from all 20 measurements of the given type of tissue. The other data listed in Table 1, i.e., the average, the standard error, and the range (min. and max.) were obtained only for the actually detected contents of the respective element in the tissue. The number of such results is given in the Sample size column. Therefore, it may happen that the average detection limit, calculated using also the least favorable results, can exceed some results for element contents obtained under the optimum conditions (higher sample weight, longer time of

Table 1
Average Contents of Trace Elements in Various Types of Eye Tissues and Hair (the Values are Given in μg/g)

Element	Tissue	Sample Size	Average	Standard error	Detection limit	Minimum	Maximum
Ag	Cornea	20	0.38	0.15	0.01	0.01	2.35
	Iris	20	0.174	0.025	0.047	0.053	0.569
	Vit. humor	18	0.069	0.018	0.044	0.011	0.329
	Lens	16	0.0087	0.0011	0.0030	0.0028	0.0180
	Hair	20	2.17	0.76	0.19	0.36	12.30
As	Cornea				0.73		
	Iris	1	2.89		1.18		
	Vit. humor				1.99		
	Lens	1	0.20		0.14		
	Hair	6	3.28	0.74	0.81	1.10	9.57
Au	Cornea	10	0.0213	0.0083	0.0071	0.0003	0.0850
	Iris	14	0.061	0.023	0.016	0.005	0.311
	Vit. humor	8	0.045	0.028	0.015	0.002	0.207
	Lens	8	0.0070	0.0029	0.0015	0.0004	0.0200
	Hair	17	0.115	0.064	0.014	0.007	1.120
Ba	Cornea	4	1.78	0.85	2.62	0.64	4.31
	Iris	12	47	16	11	4	178
	Vit. humor	9	3.28	0.94	3.15	1.07	9.24
	Lens	4	1.31	0.88	0.50	0.14	3.92
	Hair	3	28	12	44	10	52
Br	Cornea	17	4.4	1.5	1.0	1.5	28.1
	Iris	15	40	17	2	2	199
	Vit. humor	17	17.7	1.4	2.8	8.1	32.2
	Lens	19	1.24	0.09	0.21	0.49	1.78
	Hair	15	6.6	2.0	1.4	1.5	34.3

Ca	Cornea	7	2031	1597	9905	129	11600
	Iris	1	496		22735		
	Vit. humor	6	1855	382	24067	628	2840
	Lens	6	418	223	2462	30	1420
	Hair	5	11986	4775	17657	181	35200
Cd	Cornea	2	20	18	12	3	38
	Iris	1	25		21		
	Vit. humor				25		
	Lens				2.3		
	Hair	1	13		14		
Ce	Cornea	10	0.175	0.077	0.036	0.014	0.830
	Iris	13	1.3	1.1	0.1	0.1	15.0
	Vit. humor	7	1.02	0.93	0.04	0.04	6.62
	Lens	6	0.112	0.058	0.011	0.014	0.0375
	Hair	12	3.5	1.7	0.4	0.1	16.9
Co	Cornea	20	0.0289	0.0029	0.0021	0.0157	0.0603
	Iris	20	0.0914	0.0073	0.0087	0.0484	0.1650
	Vit. humor	19	0.0337	0.0076	0.0019	0.0080	0.1480
	Lens	19	0.0059	0.0012	0.0003	0.0014	0.0222
	Hair	20	0.78	0.18	0.03	0.07	3.26
Cr	Cornea	20	1.38	0.48	0.05	0.15	7.62
	Iris	20	1.61	0.27	0.22	0.74	5.16
	Vit. humor	19	1.21	0.52	0.06	0.17	10.30
	Lens	19	0.47	0.19	0.01	0.02	3.27
	Hair	20	60	12	1	6	163

Table 1
Average Contents of Trace Elements in Various Types of Eye Tissues and Hair (the Values are Given in μg/g) (*Con't*)

Element	Tissue	Sample Size	Average	Standard error	Detection limit	Minimum	Maximum
Cs	Cornea	20	0.14	0.12	0.01	0.01	2.40
	Iris	20	0.113	0.011	0.012	0.045	0.209
	Vit. humor	19	0.066	0.010	0.003	0.005	0.174
	Lens	19	0.020	0.003	0.001	0.008	0.044
	Hair	20	0.277	0.060	0.036	0.011	1.090
Eu	Cornea	20	0.0127	0.0029	0.0008	0.0018	0.0426
	Iris	20	0.0465	0.0091	0.0042	0.0042	0.1320
	Vit. humor	19	0.0141	0.0045	0.0008	0.0008	0.0690
	Lens	18	0.0019	0.0004	0.0001	0.0002	0.0047
	Hair	20	0.167	0.046	0.012	0.012	0.743
Fe	Cornea	20	43.1	5.0	2.0	18.0	100.0
	Iris	20	192	16	8	83	295
	Vit. humor	19	29.6	4.8	1.8	6.6	82.6
	Lens	19	9.4	2.8	0.4	1.2	49.5
	Hair	20	407	70	29	39	1233
Hf	Cornea	9	0.0069	0.0033	0.0026	0.0013	0.0327
	Iris	6	0.0254	0.0096	0.0102	0.0017	0.0554
	Vit. humor	5	0.0106	0.0039	0.0027	0.0021	0.0248
	Lens	7	0.0100	0.0073	0.0007	0.0008	0.0535
	Hair	1	0.0220		0.0356		
Hg	Cornea	20	0.113	0.035	0.012	0.014	0.570
	Iris	20	2.3	1.1	0.1	0.2	19.5
	Vit. humor	18	0.31	0.20	0.02	0.02	3.74
	Lens	14	0.055	0.022	0.004	0.007	0.273
	Hair	19	2.41	0.44	0.06	0.14	6.29

In	Cornea	3	0.117	0.010	0.181	0.107	0.137
	Iris				0.271		
	Vit. humor				0.089		
	Lens	1	0.027		0.021		
	Hair	3	0.93	0.26	0.79	0.77	1.15
Ir	Cornea	6	0.00020	0.00009	0.00008	0.00005	0.00059
	Iris	5	0.0029	0.0025	0.0003	0.0003	0.0127
	Vit. humor	4	0.00025	0.00016	0.00007	0.00005	0.00071
	Lens	7	0.00021	0.00007	0.00002	0.00002	0.00052
	Hair	7	0.00091	0.00012	0.00098	0.00048	0.00143
K	Cornea	11	2336	573	1293	560	7560
	Iris	14	3669	555	2032	302	7320
	Vit. humor	11	15602	3811	9092	3640	45300
	Lens	19	4803	226	383	3250	6420
	Hair	8	926	314	4965	75	2380
La	Cornea	3	0.24	0.17	0.23	0.03	0.57
	Iris	7	1.73	0.85	0.36	0.09	6.76
	Vit. humor	3	8.2	3.6	0.6	1.2	13.4
	Lens	4	0.097	0.023	0.105	0.050	0.159
	Hair	10	1.17	0.40	0.25	0.14	4.19
Mo	Cornea				11.2		
	Iris	11	35.0	8.9	20.9	11.1	95.9
	Vit. humor				19.2		
	Lens				2.2		
	Hair	2	17.9	8.8	16.3	9.0	26.7

Table 1
Average Contents of Trace Elements in Various Types of Eye Tissues and Hair (the Values are Given in μg/g) (*Con't*)

Element	Tissue	Sample Size	Average	Standard error	Detection limit	Minimum	Maximum
Na	Cornea	20	17460	936	13	11200	25700
	Iris	20	12420	1051	16	1520	21700
	Vit. humor	19	176289	20054	88	88200	409000
	Lens	19	4064	360	3	1860	8400
	Hair	20	10992	3845	16	124	75100
Ni	Cornea	11	1.82	0.50	0.82	0.19	5.21
	Iris	4	1.05	0.16	3.67	0.62	1.41
	Vit. humor	9	1.34	0.78	1.02	0.07	7.53
	Lens	7	1.51	0.84	0.22	0.12	6.44
	Hair	15	53	16	13	4	225
Rb	Cornea	19	1.13	0.15	0.19	0.30	2.53
	Iris	20	5.5	1.0	0.8	1.0	17.3
	Vit. humor	19	8.1	1.5	0.2	2.2	23.9
	Lens	19	3.00	0.60	0.05	0.74	9.56
	Hair	11	4.1	1.5	2.6	0.6	16.4
Sb	Cornea	20	0.178	0.088	0.004	0.013	1.670
	Iris	20	4.10	0.57	0.02	0.34	8.83
	Vit. humor	19	0.207	0.091	0.004	0.012	1.640
	Lens	19	0.020	0.007	0.001	0.004	0.138
	Hair	20	2.04	0.78	0.07	0.13	16.00
Sc	Cornea	20	0.00173	0.00028	0.00020	0.00044	0.00456
	Iris	19	0.0098	0.0016	0.0005	0.0035	0.0292
	Vit. humor	19	0.00267	0.00052	0.00018	0.00051	0.00818
	Lens	17	0.00030	0.00005	0.00005	0.00014	0.00083
	Hair	20	0.354	0.0080	0.0032	0.0024	0.1280

Se	Cornea	20	0.290	0.025	0.031	0.014	0.570
	Iris	20	3.73	0.36	0.13	0.15	19.50
	Vit. humor	19	0.60	0.14	0.04	0.02	3.74
	Lens	19	0.497	0.031	0.010	0.007	0.273
	Hair	13	1.00	0.14	0.37	0.14	6.29
Sn	Cornea	3	0.97	0.45	1.03	0.30	1.83
	Iris	3	2.81	0.15	4.17	2.53	3.06
	Vit. humor	3	0.71	0.28	1.07	0.42	1.27
	Lens	3	1.62	0.99	0.26	0.18	3.52
	Hair	5	14.5	2.1	11.8	9.8	21.1
Ta	Cornea	3	0.00187	0.00050	0.00172	0.00088	0.00246
	Iris	2	0.0076	0.0011	0.0069	0.0065	0.0086
	Vit. humor	4	0.00238	0.00080	0.00114	0.00133	0.00473
	Lens	7	0.00047	0.00011	0.00026	0.00018	0.00091
	Hair	3	0.05	0.10	0.02	0.01	0.07
W	Cornea	3	3.1	1.2	1.2	0.8	4.6
	Iris				1.7		
	Vit. humor	1	8.2		3.4		
	Lens				0.21		
	Hair	3	0.98	0.39	1.39	0.46	1.32
Zn	Cornea	20	22.7	3.9	0.1	7.5	50.2
	Iris	20	116	13	1	37	246
	Vit. humor	19	13.9	2.2	0.1	3.3	45.1
	Lens	19	18.7	1.6	0.0	9.9	33.0
	Hair	20	204	40	1	23	749

measurement). Generally, it can be stated that the highest contents of trace elements have been found in hair samples. This fact could be expected, since the trace element deposition in the tissue depends on their deposition dynamics in the tissue, which is only a one-way process in hair. In the case of eye tissues, the highest and the lowest contents were found in iris and lens, respectively. Transparent lenses were measured in this work, and they exhibited a lower contamination with some elements than turbid lenses (*7*). This comparison will be discussed elsewhere. Alkali metals, i.e., Na, K, Rb, and Cs, have a rather unique position. The highest contents of these elements were found in vitreous humor (except for Cs); on the other hand, a higher amount of Cs was measured in the cornea. The ratio Na/K in the lens (0.89) is lower in comparison with other eye tissues, where this ratio amounts to about 7.4. This ratio is still lower in the turbid lenses (*7*). The cornea differed from all other tissues only by a higher content of Ag. As mentioned above, iris contains, in most cases, the highest concentration of trace elements. However, it is remarkable that it also contains elements like La or Mo that are rarely detected in other tissues.

The effect of sex on the trace element deposition in eye tissues was tested for each type of the tissue. However, the number of degrees of freedom evidently decreased. Generally, it can be stated that the concentrations of all elements measured in the tissues under study are slightly higher for males than females. However, for various chosen aspects (elements, tissue, sex), no significant differences were observed using statistical tests. This fact is already obvious from the global survey of the entire set of results and the fact that it could not be corroborated by the tests because of the low number of degrees of freedom in the chosen sets. More significant differences between the sets of males and females were found for Fe and Zn in the cornea, for Zn in the iris, for Na and Ba in the vitreous humor, and for Cr, Fe, and Hg in hair.

CONCLUSION

From the results obtained, it can be deduced that the deposition of trace elements in the eye lens is practically the lowest in comparison with other eye tissues. This can be explained by the fact that the lens is the only eye tissue that is not directly nurtured by blood.

The different ratios of the Na/K concentrations in different tissues indicate a different function of the Na/K pump in the alimentation of cells of these tissues by means of the ATP-aza enzyme. This ratio also indicates the degree of the energy transfer in the tissue cells. If this ratio decreases substantially under the value of 1, it indicates a pathological state of the tissue. The high contents of trace elements, predominantly of metals, in the iris demonstrate their participation in the formation of the iris color.

REFERENCES

1. E. L. Lakomaa, and P. Eklund, *Int. Symp. Nuclear Activation Techniques in the Life Sci.* Vienna, IAEA/SM-227/1, 1979.
2. N. A. Kryzenkova, L. I. Zuk, and A. A. Kist, *Int Symp. Met. Nucl. Act.*, Ljubljana, IAEA/SM-157/74, 1974.
3. M. Ördögh, and P. Rácz, *Proc. Int. Conf. Mod. Trends Activ. Anal.*, München, 1967.
4. T. Yamaguchi, M. Bando, A. Nakajima, M. Tera, and M. Suzuki-Yasumoto, *J. Radioanal. Chem.* **57,** 169 (1980).
5. B. Štverák, L. Cigánek, and Z. Dubská, *Radioisotopy* **21,** 303 (1980).
6. T. Kouris, C. Papadopoulou, G. D. Kanias, F. Bairaktari-Kouri, and G. P. Theodossiadis, *Proc. Int. Conf. Mod. Trends Activ. Anal.*, Copenhagen, 1986.
7. B. Štverák, D. Tlučhoř, and L. Cigánek, *J. Radioanal. Nucl. Chem. Articles* **120(2),** 345 (1988).
8. L. A. Curie, *Anal. Chem.* **40,** 586 (1968).

Pigmentation and Temporal Effects on Trace Elements in Hair

S. AUFREITER* AND R. G. V. HANCOCK

SLOWPOKE Reactor Facility and Department of Chemical Engineering and Applied Chemistry, University of Toronto, Toronto, Canada, M5S 1A4

Received April 17, 1989; Accepted December 11, 1989

ABSTRACT

Variations in trace element concentration in the head and facial hair of five individuals with ever-increasing amounts of white hair were examined. Hair was collected from the scalp, cheeks, and chin from one donor on a regular basis since 1984. Samples were separated into white and pigmented fractions, and analyzed at SLOWPOKE-Toronto by INAA for the short-lived, isotope-producing elements Br, Ca, Cl, Cu, I, Mg, Mn, Na, S, and Zn. Temporal concentration variations of these elements over time, and variations of the elemental concentrations in pigmented and white hair were established.

Index Entries: Age; calcium; INAA; pigmentation; trace elements; white compared to pigmented hair.

INTRODUCTION

Many studies have applied the results of hair elemental analysis to questions of nutritional differences (*1,2*), forensic identification (*3*), toxicology (*4*), occupational exposure (*5*), population studies (*6*), and medicine (*7,8*). Most studies examined whether the levels of elements in hair could be used to identify differences in people's physiology, genes, or environment. This is justifiable, since the sources of trace elements in

*Author to whom all correspondence and reprint requests should be addressed.

hair are either endogenous or exogenous. However, the hair of individuals within groups is quite variable, and comparisons of data appear to be best made using large groups of hair donors (*9*). To further complicate matters, even hair taken from different areas on the same head may vary considerably. Nevertheless, many studies have reported finding differences in the elemental levels in hair from different groups of people.

Differences in elemental concentrations with age and sex of the donors have been found (*1,2,9–11*), so that age-matching of control and experimental groups, or separation of groups by age and sex, has been recommended. A complicating factor is pigmentation of the hair: Anke and Schneider (*12*) found that black hair has lower levels of manganese than blonde hair, and that manganese is significantly decreased in white relative to pigmented hair, whereas all other elements tested showed no differences. Loss of pigmentation may be genetically determined, stress-related, and independent of age. When hair of adults is of interest, differences in element concentrations that relate to hair pigmentation may confuse the interpretation of results.

This study focused on the differences between the elemental concentrations of white and pigmented hair of five individuals, with one individual followed over a period of about 4 yr. All the donors were white males, above 35 years of age, and graying. Four were in excellent health, considering their advancing years, whereas one, though physically active, had extensive arterial disease. Three of the donors also contributed beard samples.

METHODS

Hair was obtained from haircuts and beard trimmings. Although specific sampling sites varied from individual to individual, samples were always taken from the same areas of the head of any one person; sampling sites were chosen based on the availability of both pigmented and white hair. Thus, subject 1 gave hair from above the forehead (20 samples), the sideburns (seven samples) and from the chin (18 samples); subject 2 donated hair on two occasions from five areas of the scalp; subjects 3 and 4 gave hair from the nape of the neck, and subject 5 supplied hair from the back of his head and from his beard.

All samples were separated by hand into white and pigmented portions, using stainless-steel tweezers. Each sample was washed according to the IAEA recommended procedure (*13*) using washes of acetone and deionized water, and stored in clean polyethylene vials.

Samples were analyzed at the SLOWPOKE Reactor Facility of the University of Toronto, by sequential irradiation for 5 min at a neutron flux of 1.0×10^{12} $n \cdot cm^{-2} \cdot sec^{-1}$. After a 1-min delay, during which time the irradiated hair was transferred into a fresh polyethylene vial, each

Table 1
The Effect of Washing on Dark and White Hair Samples from a Single Individual[a]

		Subject 1, 10/84			
		Scalp hair		Beard hair	
Element		Unwashed	Washed	Unwashed	Washed
S(%)	D	5.3 ± 0.3	5.3 ± 0.2	4.1 ± 0.3	4.6 ± 0.3
	W	5.4 ± 0.4	5.5 ± 0.3	5.2 ± 0.4	5.1 ± 0.5
Na	D	170 ± 10	12 ± 2	580 ± 10	64 ± 6
	W	120 ± 10	27 ± 2	510 ± 20	57 ± 3
Cl	D	1070 ± 20	530 ± 10	2590 ± 30	1350 ± 20
	W	930 ± 20	750 ± 20	2600 ± 30	1470 ± 30
Br	D	3.0 ± 0.3	2.1 ± 0.2	6.5 ± 0.5	4.3 ± 0.3
	W	2.8 ± 0.4	3.7 ± 0.4	4.9 ± 0.5	3.5 ± 0.5
I	D	0.6 ± 0.1	0.8 ± 0.1	0.6 ± 0.2	0.8 ± 0.1
	W	0.7 ± 0.1	0.8 ± 0.1	≤0.66	0.6 ± 0.1
Al	D	13.5 ± 0.5	7.4 ± 0.2	24.3 ± 0.7	19.0 ± 0.4
	W	10.1 ± 0.5	9.2 ± 0.4	21.3 ± 0.9	14.4 ± 0.6
Ca	D	780 ± 70	270 ± 30	570 ± 70	250 ± 40
	W	500 ± 60	260 ± 40	330 ± 80	350 ± 80
Cu	D	8 ± 1	14 ± 1	15 ± 3	13 ± 1
	W	16 ± 2	16 ± 1	15 ± 3	10 ± 2
Mg	D	40 ± 10	21 ± 4	60 ± 10	30 ± 7
	W	≤41	31 ± 7	≤48	≤51
Mn	D	0.73 ± 0.07	0.78 ± 0.02	0.9 ± 0.1	0.73 ± 0.06
	W	1.02 ± 0.08	0.65 ± 0.03	0.7 ± 0.1	0.41 ± 0.04

[a]All concentrations in ppm unless listed.

sample was assayed using a 5 min counting interval, and was recounted for 10 min after a decay period of about 1.5 h. The levels of the 11 elements measured compared well with levels found by others (*11*).

RESULTS AND DISCUSSION

White and pigmented hair showed some differences in element levels even before washing. Analysis of scalp and beard hair samples from subject 1 before washing showed differences in sodium, aluminum, calcium, and manganese levels (Table 1). This indicates that white and pigmented hair retains exogeneous material to different degrees. After washing, no great differences were observed between the pigmented and white hair of these samples from subject 1, but there was a factor of two in the sodium, chlorine, and aluminum concentrations between hair and beard. Examination of subsequent washed samples showed a number of elemental concentration differences between white and pigmented hair.

Table 2
Means of Element Concentrations Determined in Samples Donated by Subject 1 Between September, 1984 and June, 1988[a]

Element		Scalp $n \leq 20$	Beard $n \leq 18$	Sideburns $n \leq 7$
S%	D	5.5 ± 0.2	5.1 ± 0.4	5.2 ± 0.2
	W	5.3 ± 0.4	4.9 ± 0.3	5.3 ± 0.3
Na	D	30 ± 20	40 ± 30	30 ± 20
	W	30 ± 20	40 ± 30	50 ± 40
Cl	D	600 ± 200	1000 ± 400	1200 ± 400
	W	600 ± 300	1100 ± 300	1500 ± 300
Br	D	3 ± 1	3 ± 2	6 ± 3
	W	2 ± 1	4 ± 2	6 ± 3
I	D	0.6 ± 0.2	0.5 ± 0.2	–
	W	0.4 ± 0.2	0.5 ± 0.1	–
Al	D	9 ± 4	15 ± 5	11 ± 5
	W	8 ± 3	13 ± 5	11 ± 5
Ca	D	700 ± 200	400 ± 100	360
	W	500 ± 200	200 ± 80	170
Cu	D	14 ± 3	13 ± 4	9 ± 2
	W	13 ± 2	12 ± 3	9 ± 1
Mg	D	53 ± 2	50 ± 10	45 ± 2
	W	40 ± 20	36 ± 7	37 ± 4
Mn	D	0.5 ± 0.3	0.4 ± 0.2	0.3 ± 0.2
	W	0.4 ± 0.2	0.3 ± 0.2	0.2 ± 0.1
Zn	D	180 ± 40	200 ± 60	–
	W	200 ± 80	190 ± 50	–

[a]Values in ppm unless noted.

Almost 50 hair samples of different types were obtained from subject 1, to be divided into white and dark fractions. Table 2 shows the means of the results obtained from analysis of the pigmented hair samples and of the white hair samples from subject 1 collected between September, 1984 and June, 1988.

Consistent differences can be seen in the mean levels of calcium, magnesium, and manganese, with higher levels occurring in the pigmented hair. The remaining elements did not show reproducible differences, except between scalp hair relative to the beard. Ranges of the concentrations found in scalp hair and beard are shown in Table 3. The results of analysis of hair from subjects 3, 4, and 5 are shown in Table 4.

Two samples obtained from subject 3 show pigmented to white hair variations in levels of iodine, calcium, magnesium, and manganese. The white and pigmented hair collected from subject 4 showed differences in the levels of iodine and zinc. Differences in calcium and magnesium levels could not be assessed because of the elevated detection limits resulting from the high Cl content. The results of analysis of subject 5's hair showed reverse differences in levels of sodium, chlorine, iodine, and

Table 3
Ranges of Concentration of Elements in Hair of Subject 1[a]

Element	Scalp		Beard	
	Dark	White	Dark	White
S(%)	5.0 – 5.7	4.5 – 5.9	4.4 – 5.8	4.2 – 5.5
Na	10 – 77	9 – 94	10 – 100	13 – 86
Cl	270 – 1060	220 – 1370	410 – 1950	500 – 1490
Br	1.4 – 6.1	0.8 – 5.5	1.3 – 8.2	1.0 – 9.7
I	0.26 – 1.04	0.19 – 0.87	0.27 – 0.80	0.34 – 0.69
Al	4 – 23	4 – 17.2	8 – 31	7 – 28
Ca	270 – 1040	240 – 820	250 – 640	150 – 490
Cu	9 – 19	10 – 16	9 – 23	9 – 17
Mg	21 – 90	19 – 69	35 – 81	24 – 47
Mn	0.14 – 1.06	0.10 – 0.86	0.13 – 0.73	0.09 – 0.71
Zn	113 – 255	107 – 435	80 – 318	104 – 265

[a]In ppm unless noted.

Table 4
Element Concentrations in the Scalp Hair of Three Donors, Between 55 and 67 Years of Age[a]

Element		Subject 3		Subject 4,	Subject 5,
		5/86	10/86	6/86	11/84
S(%)	D	4.8 ± 0.2	5.3 ± 0.3	4.8 ± 0.3	6.2 ± 0.4
	W	5.0 ± 1.0	4.8 ± 0.2	4.5 ± 0.2	6.1 ± 0.5
Na	D	42 ± 0.3	22 ± 2	35 ± 2	44 ± 4
	W	38 ± 4	31 ± 2	26 ± 1	75 ± 6
Cl	D	2030 ± 20	890 ± 20	2600 ± 20	1210 ± 20
	W	2340 ± 20	880 ± 10	2800 ± 20	1510 ± 30
Br	D	13.6 ± 0.4	6.2 ± 0.4	4.7 ± 0.3	7.8 ± 0.4
	W	13.5 ± 0.3	5.2 ± 0.3	3.9 ± 0.3	8.9 ± 0.6
I	D	1.3 ± 0.1	5.1 ± 0.2	0.6 ± 0.1	1.0 ± 0.2
	W	0.7 ± 0.1	2.7 ± 0.2	≤0.36	1.5 ± 0.3
Al	D	14 ± 1	12.2 ± 0.4	10.6 ± 0.4	18.7 ± 0.6
	W	12 ± 2	10.5 ± 0.3	16.5 ± 0.3	48 ± 1
Ca	D	320 ± 50	540 ± 50	≤140	190 ± 50
	W	160 ± 40	240 ± 30	140 ± 20	≤230
Cu	D	15 ± 2	28 ± 2	18 ± 2	25 ± 2
	W	14 ± 7	22 ± 1	15 ± 2	19 ± 3
Mg	D	130 ± 10	37 ± 9	≤34	30 ± 3
	W	76 ± 8	≤21	≤31	≤45
Mn	D	0.48 ± 0.04	0.14 ± 0.02	0.40 ± 0.05	0.41 ± 0.05
	W	0.25 ± 0.04	0.10 ± 0.02	0.36 ± 0.03	0.39 ± 0.07
Zn	D	132 ± 3	137 ± 6	147 ± 4	≤310
	W	131 ± 4	103 ± 4	129 ± 5	≤510

[a]Values are in ppm unless noted.

aluminum, with white hair concentrations higher than pigmented hair. Differences in calcium and magnesium were not clear because of high detection limits; however, a beard sample from subject 5 showed lower levels of calcium, magnesium, and manganese in the white hair fraction (Ca 1210 ± 60 ppm pigmented, 900 ± 50 ppm white; Mg 154 ± 8 ppm pigmented, 140 ± 7 ppm white; Mn 0.90 ± 0.05 ppm pigmented, 0.54 ± 0.06 ppm white). Copper, chlorine, bromine, and iodine were approximately twice as concentrated in pigmented beard relative to white beard samples. The data obtained from subjects 1 and 2 are summarized in a different form in Table 5. This Table presents the means of the ratios of elemental concentrations in pigmented relative to white hair in each sample. The associated ranges are found in Table 6.

Relative to the stability exhibited by sulfur, these data show calcium to be an element of particular interest in the hair of subject 1. Levels of calcium, magnesium, and manganese are elevated in the pigmented hair of this healthy man of about 40 yr. Data from subject 2 also showed indications of elevated levels of calcium in pigmented hair, though not of magnesium and manganese. Dark hair from subject 2 also had higher bromine, and considerably lower sodium and iodine, than his white hair.

Data from subject 1 were also examined with respect to temporal changes. Some elements (sulfur, zinc, and copper) varied little over time, whereas there was considerable variation in the rest of the elements, with possibilities of cyclical, seasonal changes in the concentrations of chlorine, iodine, manganese, and magnesium. Sodium and aluminum showed a decreasing trend over the 4-yr sampling period.

The higher levels of calcium in pigmented hair than white hair was very consistent over time. An increase in the calcium concentrations of scalp hair with time was not reflected in the values found for the beard hair.

SUMMARY

Of the 11 elements determined in hair in this study, sulfur, chlorine, and copper showed no apparent differences between white and dark fractions of the same hair sample. Calcium levels were found to be higher in pigmented hair of 4 out of 5 subjects; magnesium and manganese were higher in most of the dark hair fractions, relative to white hair, collected from subject 1. The other experimental subjects sometimes confirmed these trends. Subjects 2, 3, and 4 showed marked differences in iodine levels; subjects 2, 4, and 5 in sodium levels. Bromine, aluminum, and zinc differed in only one or two subjects.

Our experimental population may be too small to allow extrapolations to be made to large groups of people in terms of element variability. However, the significance of the variability observed here should not be

Table 5
The Means of Ratios of Element Concentrations in Pigmented to White Hair of Two Donors

Element	Subject 1			Subject 2, scalp
	Scalp $n \leq 20$	Beard $n \leq 18$	Sideburns $n \leq 7$	$n \leq 9$
S	1.04 ± 0.07	1.04 ± 0.09	0.97 ± 0.06	1.00 ± 0.05
Na	1.1 ± 0.5	1.1 ± 0.4	0.8 ± 0.4	0.43 ± 0.4
Cl	1.2 ± 0.4	0.9 ± 0.2	0.8 ± 0.3	1.1 ± 0.2
Br	1.2 ± 0.3	0.9 ± 0.2	1.1 ± 0.2	1.4 ± 0.4
I	1.1 ± 0.3	1.1 ± 0.4	(0.9)	0.38 ± 0.3
Al	1.0 ± 0.2	1.2 ± 0.3	1.1 ± 0.1	1.3 ± 0.8
Ca	1.5 ± 0.6	1.7 ± 0.5	2.2 ± 0.6	1.3 ± 0.2
Cu	1.1 ± 0.1	1.1 ± 0.2	1.0 ± 0.2	1.05 ± 0.1
Mg	1.4 ± 0.6	1.5 ± 0.3	1.3 ± 0.5	1.1 ± 0.3
Mn	1.3 ± 0.4	1.5 ± 0.3	1.3 ± 0.2	1.1 ± 0.4
Zn	1.0 ± 0.3	1.1 ± 0.3	1.0 ± 0.4	1.4 ± 0.8

Table 6
Ranges of the Means of Ratios of Concentrations in Pigmented Hair to Those in White Hair of Subjects 1 and 2

Element	Subject 1		Subject 2,
	Scalp	Beard	Scalp
S	0.94 – 1.18	0.90 – 1.23	0.94 – 1.10
Na	0.40 – 4.7	0.54 – 1.83	0.12 – 1.10
Cl	0.71 – 2.61	0.72 – 1.59	0.67 – 1.42
Br	0.57 – 2.10	0.51 – 1.23	0.86 – 1.43
I	0.38 – 1.53	0.38 – 1.53	0.028 – 0.86
Al	0.80 – 1.24	0.73 – 2.15	0.29 – 3.54
Ca	1.04 – 3.67	0.71 – 3.33	1.06 – 1.59
Cu	0.89 – 1.27	0.76 – 1.36	0.73 – 1.23
Mg	0.68 – 2.77	0.95 – 2.06	0.75 – 1.68
Mn	0.54 – 2.15	0.30 – 2.87	0.57 – 1.60
Zn	0.47 – 1.66	0.64 – 1.58	1.09 – 2.49

ignored, especially important being the "loss" of calcium in white hair, which may obscure physiological effects resulting from disorders found in older populations.

ACKNOWLEDGMENTS

This work was supported by an NSERC infrastructure grant to the University of Toronto SLOWPOKE Reactor Facility.

REFERENCES

1. L. M. Klevay, *Am. J. Clin. Nutr.* **23,** 284 (1970).
2. L. M. Klevay, *Am. J. Clin. Nutr.* **23,** 1194 (1970).
3. A. K. Perkons, *J. Forensic Sci.* **11,** 50 (1966).
4. R. E. Jervis, *Nuclear Activation Techniques in the Life Sciences*, IAEA-SM 227/67, p. 627 (1979).
5. R. E. Jervis, *J. Radioanal. Chem.* **37,** 751 (1977).
6. T. Takeuchi, T. Hiashi, J. Takada, Y. Hayashi, M. Koyama, H. Kozuka, H. Tsuyi, Y. Kusaka, S. Ohmi, M. Shinogi, A. Aoki, K. Katayama, and T. Tomiyama, *J. Radioanal. Chem.* **70,** 29 (1982).
7. H.-J. Schneider and M. Anke, *Zschr. inn. Med.* **21**, 802 (1966).
8. S. P. Moo, *J. Radioanal. Chem.* **77,** 141 (1983).
9. G. J. Evans and R. E. Jervis, *J. Radioanal. and Nuclear Chem. Articles* **110,** 613 (1987).
10. H. G. Petering, *Arch. Environ. Health* **23,** 202 (1971).
11. J. G. Evans, *The Use of Hair as an Indicator of Occupational Exposure*, M.A.Sc. Thesis, Department of Chemical Engineering and Applied Chemistry, University of Toronto, p. 124 (1984).
12. M. Anke and H.-J. Schneider, *Zschr, inn. Med.* **21,** 794 (1966).
13. Y. Ryabukhin, *J. Radioanal. Chem.* **60,** 7 (1980).

Preliminary Study of the Distribution of the Toxic Elements As, Cd, and Hg in Human Hair and Tissues by RNAA

G. S. ZHUANG, Y. S. WANG, M. G. TAN, M. ZHI, W. Q. PAN, AND Y. D. CHENG

Shanghai Institute of Nuclear Research, Academia Sinica, P. O. Box 8204, Shanghai, 201849, People's Republic of China

Received April 17, 1989; Accepted December 11, 1989

ABSTRACT

In order to study the relationships between trace element concentrations of hair and internal body burdens, a radiochemical NAA technique has been used for determination of the elements As, Cd, and Hg in autopsy samples of liver, kidney-cortex, lung, and hair from 24 male persons who died by accident. High significant positive correlations were observed between the As concentration in hair and in kidney-cortex, and between Cd and Zn concentrations in kidney-cortex. The contents of Cd, both for lung and kidney-cortex, were related to the smoking habits of the subjects.

Index Entries: Neutron activation analysis; toxic element; As; Cd; Hg; correlation; liver; kidney-cortex; lung; human hair.

INTRODUCTION

In recent years, trace elements in Chinese human hair, foodstuffs, and dietary nutritional status have been determined and evaluated (*1–3*). Human hair has a number of characteristics as a practical indicator for evaluating environmental pollutants in human health. However, serious uncertainty still exists for the interpretation of hair analytical data in environmental health studies, because of the lack of knowledge about whether there is a quantitative relationship between mineral concentrations in hair and internal body burdens.

The International Atomic Energy Agency (IAEA) started a Coordinated Research Program in 1984 on the significance of hair mineral analysis as a means of assessing internal body burdens of environmental mineral pollutants. As a part of this program, the main aims of the present study are to assess the toxic elements As, Cd, and Hg in contents of hair and internal organs from autopsies, in order to

1. Investigate whether these element contents in hair can be used as an indicator of body burdens.
2. Try to clarify whether there are any correlations between elements in the same tissues, and
3. Try to find the relationship between trace element body burden and smoking habits of the subjects.

Thus, autopsy samples of scalp hair, kidney-cortex, liver, and lung were collected and analyzed by RNAA.

MATERIALS AND METHODS

Collection and Pretreatment of Sample

Twenty-four male autopsies between 35–60 yr of age were obtained from the Institute of Forensic Medicine in China. All autopsy samples were identified clearly by a pathologist, and tissue samples of liver, lung, kidney-cortex, and scalp hair were collected within 24–48 h after accidental death of persons who were proven to have been in good health before the accident. Information concerning the residence, profession, length of service, and smoking habits of each subject was obtained. To remove external contamination, the surface of the soft tissues was cut off using a titanium knife, diced into 1–2 cm cubes on a clean bench, and stored in a freezer at a temperature below −30°C until taking out for analysis.

HOMOGENIZATION

The brittle-fracture technique (BFT) was used to homogenize all the samples in our laboratory. Details of this method were discussed by Iyengar (4). Soft samples were placed in the Teflon™ vessel, along with the Teflon™ ball, and the lid was closed tightly. The vessel was then cooled in liquid nitrogen for a few minutes, and vibrated for 5 min at 3000 cycles/min with a commercially available Mikro-Dismembrator. Hair samples taken only less than 5 cm from the proximal end were cut into 2-mm pieces and washed according to the IAEA recommended procedure (5) with acetone, water, water, water, and acetone again. After air-drying overnight at room temperature, hair samples were homogenized by the brittle-fracture technique as described above.

RADIOCHEMICAL NEUTRON ACTIVATION ANALYSIS

A simple group separation procedure based on extraction for the determination of the elements As, Cd, Hg, Cu, and Zn in biological samples was developed for this program (*6*). Zinc-diethyldithiocarbamate $Zn(DDC)_2$ and methylisobutyl ketone iodide were chosen as reagents. The elements could be extracted successively and quantitatively from strong mineral acids without adjusting the pH of the solution, and separated into two groups suitable for γ-ray spectrometry. Samples of 100–200 mg dry wt were double-sealed into polyethylene bags and irradiated in a swimming pool-type reactor at a thermal-neutron flux density of 1×10^{13} n cm^{-2} s^{-1} for 44 h. After a cooling time of 1–3 d, the samples were digested with microgram quantities of carrier in concentrated nitric acid and sulfuric acid at 150°C for 3.5 h in Teflon™ bomb, and then extracted by the separation scheme. The procedure of RNAA was described in detail elsewhere (*6*).

RESULTS AND DISCUSSION

The analytical results for 24 autopsy subjects are given in Tables 1 and 2 and Figures. 1–3. The values are reported both on dry- and wet-wt bases, in mg/kg. In Table 1, the arithmetic mean (AM), standard deviation (SD), and geometric mean (GM) are reported for each element in each tissue. The range (minimum [MIN] and maximum [MAX] values) and the median (MED) are also presented. In the following discussion, dry-wt values are used.

Hg (Mercury)

Hg is one of the important toxic elements in this CRP. Chronic exposure to mercury vapor may affect other tissues besides the nervous system. From the results shown in Table 1, the content of Hg shows a large variability both for hair and internal tissues. The level of Hg content is higher in hair (0.09–3.22 mg/kg) than in kidney-cortex (0.09–1.64 mg/kg), liver (0.09–0.93 mg/kg), and lung (0.01–0.02 mg/kg). It seems to be in accordance with the results for the order in concentration of Hg from 36 subjects in Sweden presented by Muramatsu (*7*). The contents of Hg in hair are usually highest in populations consuming methylmercury from fish. The mean value of hair Hg in this work (1.18 mg/kg) is comparable to that of people residing in Italy (1.5 mg/kg), New Zealand (1.3 mg/kg) (*8*), France (1.33 mg/kg), and Iraq (1.4 mg/kg) (*9*). However, no signifi-

Table 1
Analytical Results of Toxic Elements for 24 Male Chinese Autopsies (mg/kg)

	Hair, dry	Kidney cortex Dry	Kidney cortex Wet	Liver Dry	Liver Wet	Lung Dry	Lung Wet
Hg							
AM	1.18	0.62	0.15	0.37	0.11	0.10	0.02
SD	0.65	0.46		0.26		0.04	
MIN	0.41	0.09	0.018	0.09	0.027	0.01	0.003
MED	1.00	0.47		0.34		0.09	
MAX	3.22	1.64	0.032	0.93	0.22	0.20	0.04
Cd							
AM	0.20	116.2	24.5	8.6	2.62	1.90	0.36
SD	0.14	74.1		5.32		1.30	
MIN	0.04	21.1	4.43	1.79	0.51	0.15	0.03
MED	0.17	90.7		6.58		1.89	
MAX	0.59	320.3	65.6	21.0	6.13	5.68	1.28
As							
AM	0.40	0.11	0.056	0.16	0.047	0.10	0.020
SD	0.22	0.06		0.11		0.07	
MIN	0.08	0.02	0.004	0.06	0.017	0.02	0.005
MED	0.34	0.11		0.12		0.07	
MAX	0.97	0.30	0.065	0.58	0.16	0.27	0.05

cant relationships are found between Hg in hair and different tissues, even though a positive correlation between Hg in hair and kidney-cortex has been reported by Y. Muramatsu (*7*).

As (Arsenic)

The contents of As show a large individual variability, and there is a difference of more than 10-fold in each sort of tissue. The contents of As in tissues decrease in the order of hair, liver, kidney-cortex, and lung. It is interesting to note the significant positive correlation between As concentration in hair and kidney-cortex (*see* Fig. 1) from a statistical analysis for 24 autopsy subjects (correlation coefficient r : 0.751). The ratio of hair As to kidney As is 4.10 in this work. However, no relationships are found between As in hair and liver or lung. In comparison with the literature, the mean value of hair As (0.40 mg/kg) is slightly higher than the results presented by other countries (*8,9*). It may be related to the pollution and food structure in the Shanghai area. The As daily intakes (52.2 μg) in Shanghai shown by Sun Laiyan (*3*) are higher than in other areas of China and some western countries (*10*).

Cd (Cadmium)

The results from Table 1 show an individual difference of more than 10-fold for Cd mean values of each sort of tissue, and the contents of Cd decreased in the order of kidney-cortex, liver, lung, and hair. A differ-

Table 2
Correlation Between Hair and Internal Tissues for Trace Elements of 24 Male Autopsies

Elements	Tissues[a]	*r* value[b]	*p* value[c]
As	Kidney cortex	0.751	<0.01
	Liver	0.161	<0.05
	Lung	0.130	>0.05
Cd	Kidney cortex	0.418	<0.05
	Liver	0.05	>0.05
	Lung	−0.017	>0.05
Hg	Kidney cortex	0.175	>0.05
	Liver	0.028	>0.05
	Lung	−0.082	>0.05

[a]*n*–number of pairs hair; tissues are 24 for every element.
[b]*r*–value is degree of correlation in linear regression analysis.
[c]*p*–value is probability of correlation.

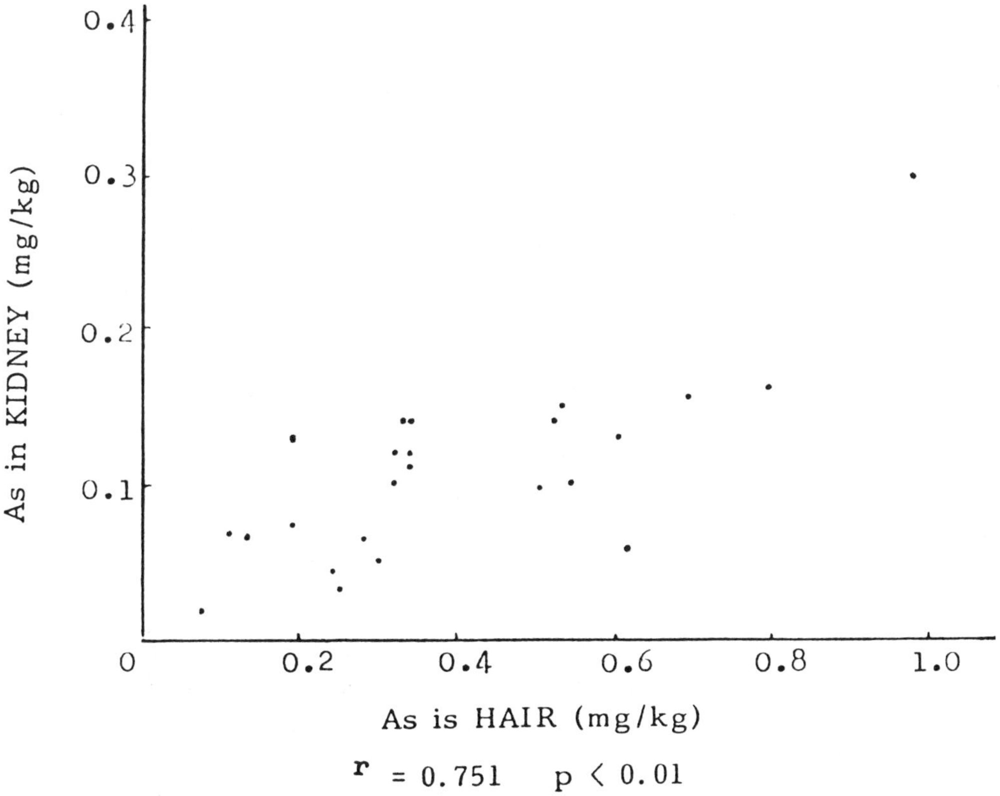

Fig. 1. Scatter plot of As contents of hair and kidney-cortex in 24 individuals.

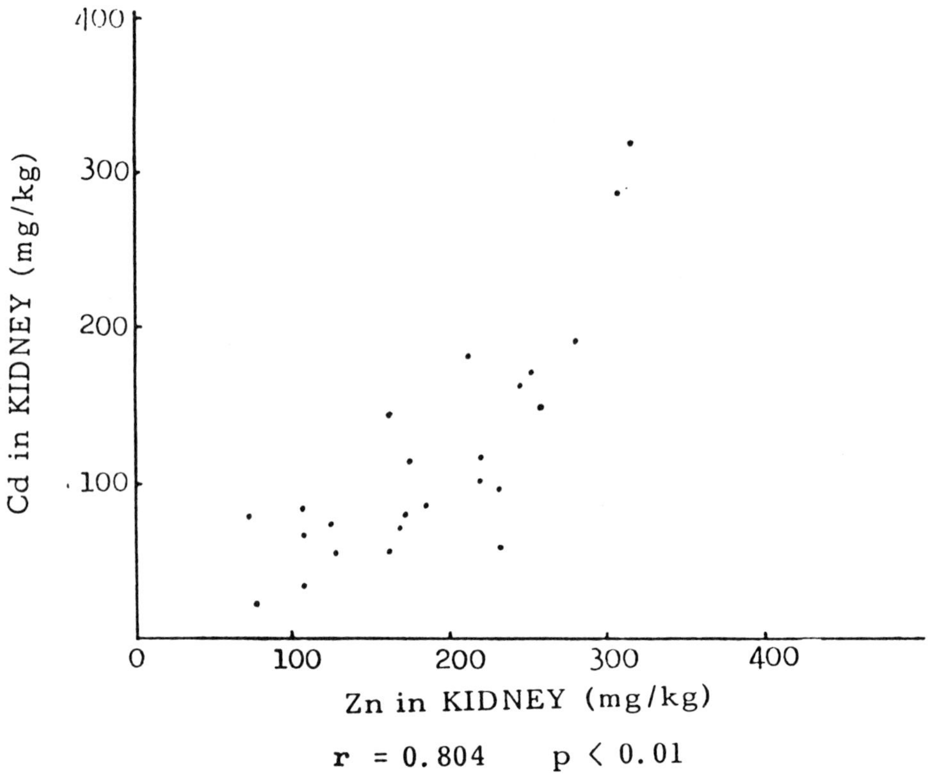

Fig. 2. Scatter plot of Cd and Zn contents in kidney-cortex for 24 individuals.

ence of more than 500-fold of Cd mean values is observed between kidney-cortex and hair, and a middle correlation is shown between these two tissues. However, no significant relationships are observed between hair and liver or lung. The concentration levels in this work are comparable to the results presented by other researchers for hair (*11*) and tissues (*12*).

Cigaret smoking is an important source of cadmium. If the 24 subjects are divided into three subgroups according to their smoking habits (group A—nonsmoker, group B—5–20 cigarets/d, and group C—more than 20 cigarets/d), it can be seen that the concentrations of Cd are related to the smoking habits both for kidney-cortex and lung tissues (*see* Fig. 3). The concentrations of Cd in kidney-cortex increase with the smoking habits. There is a significant difference between group A and group C. The mean value of Cd for group C is twice that for group A.

A comparison of the Cd concentration in lungs for nonsmokers and smokers shows a significant difference. However, there are no significant differences between group B and group C for Cd in kidney-cortex and lung, and no significant effect on tissues of smoking for the trace elements As and Hg could be found. The kidney damage may be the result of the fact that cadmium competes with zinc for binding sites in various

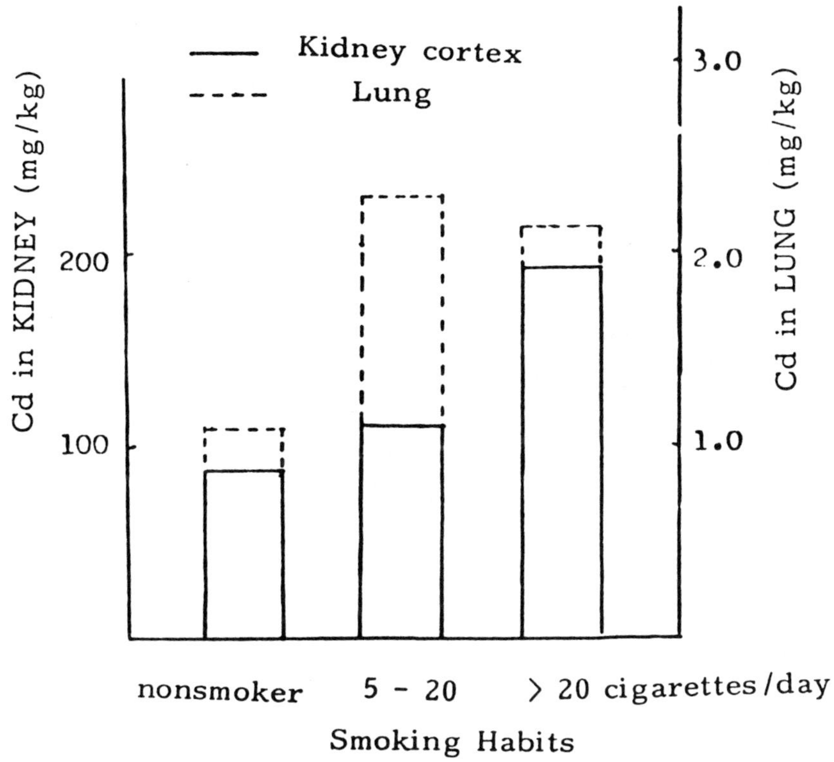

Fig. 3. Influence of smoking habits on Cd concentration in kidney-cortex and lung samples.

enzymes and other proteins. It is found that there is a significant relationship between the concentrations of Cd and Zn in kidney-cortex for normal persons (Fig. 2). The mean value of the Cd/Zn ratio in kidney-cortex is 0.60 ± 0.23. However, the exact mechanism in these detoxification reactions is still unknown, and no relationships between Cd and Zn in other tissues have been found.

ACKNOWLEDGMENTS

The collection and pathological evaluation of the material for the investigation was made by Drs. Y. J. Li and W. L. Zhang. We are grateful for their help. The authors also wish to thank Dr. Y. Q. Li for his assistance in the analytical work.

REFERENCES

1. Q. Junfa et al., *Nuclear Techniques* **10(3)**, 50–52 (1986).
2. S. F. Zheng, *The Journal of Cancer in China* **4(3)**, 174 (1982).
3. S. Laiyan, *Determination and Evaluation of Toxic Elements in Chinese Foodstuffs.* A Presentation for the IAEA Second Research Co-ordination Meeting on

Nuclear Techniques for Toxic Elements in Foodstuffs, 19–22 Beijing, China, pp. 19–22, April, 1988.
4. G. V. Iyengar and K. Kasperek, *J. Radioanal. Chem.* **39,** 301–316 (1977).
5. IAEA: Activation analysis of hair as an indicator of contamination of man by environmental trace element pollutants, IAEA/RL/50, Oct. 1978.
6. G. S. Zhuang and Y. S. Wang, *J. Radioanal. Chem.* **129(2),** 59–64 (1989).
7. Y. Muramatsu and R. M. Parr, *The Sci. Total Environ.* **76,** 29–40 (1988).
8. D. Airey, *The Sci. Total Environ.* **31(2),** 157–80 (1983).
9. Al-Sharistani et al., Nuclear Activation Techniques in the Life Sciences 1978 IAEA, Vienna, 1979.
10. J. W. Copins Peerboom, *The Sci. Total Environ.* **42,** 1–27 (1985).
11. N. B. Kim et al., Health-related monitoring of trace element pollutants using nuclear techniques, IAEA-TECDOC-330, 169–174, 1985.
12. J. Bacso, et al. Preliminary results on trace element concentrations in human hair, brain, lung, kidney, liver and bone. A research report on the Second Research Co-ordination Meeting, Neuherberg, FRG 2–5, Sept. 1985.

Study of Correlation of Se Content in Human Hair and Internal Organs by INAA

Y. D. CHENG, G. S. ZHUANG, M. G. TAN, M. ZHI, AND W. ZHOU

Shanghai Institute of Nuclear Research, Shanghai, People's Republic of China

Received April 17, 1989; Accepted December 11, 1989

ABSTRACT

The correlations of essential element Se between human hair and kidney-cortex, liver, and lung from the same subjects were investigated by instrumental neutron activation analysis, using the reaction $^{76}Se\ (n, \gamma)\ ^{77m}Se$, for 24 Chinese autopsies. The concentration of Se is higher in kidney-cortex (2.04–5.36 mg/kg) than in liver (0.73–2.29), lung (0.50–1.85), and hair (0.37–1.43). It is important to know that there are significant relationships of Se concentration between hair and the other three internal organs. The correlation coefficient by linear regression analysis are 0.639, 0.570, and 0.635 for liver, lung, and kidney-cortex, respectively; and the P values are all less than 0.01 for the three tissues.

Index Entries: Selenium; instrumental neutron activation analysis; trace element research in humans.

INTRODUCTION

Many advances have been made recently in the field of trace element analysis, providing a large number of data of interest in biological and

Biological Trace Element Research Editor: G. N. Schrauzer

Table 1
Analytical Results of Se Content in Human Hair, Liver, Lung, and Kideny-Cortex for 24 Male Chinese Autopsies (mg/kg)

		Kidney-cortex		Liver		Lung	
	Hair, Dry	Dry	Wet	Dry	Wet	Dry	Wet
AM	0.62	3.65	0.75	1.45	0.43	0.93	0.19
SD	0.24	0.82		0.38		0.32	
MED	0.63	3.67		1.40		0.85	
MIN	0.37	2.04	0.43	0.73	0.23	0.50	0.10
MAX	1.43	5.36	1.03	2.29	0.65	1.55	0.33

environmental studies. The main technique now used is instrumental neutron activation analysis. In the present study, the Se determinations were made with the $^{76}Se(n, \gamma)\ ^{77m}Se$ reaction (*1*).

Most researchers (*2,3*) have concentrated on measuring the trace element contents of patient or normal human hair. It is difficult to interpret the cause of diseases without knowledge of the correlations of trace elements between human scalp hair and internal organs. Serious uncertainty still exists for interpretation of hair analytical data in environmental health studies because of the lack of knowledge about whether there is a quantitative relationship of trace element concentrations between hair and internal body burdens.

For this purpose, analyses and comparisons have been conducted of Se in human hair and kidney-cortex, liver, and lung from the same subjects. A moderate relationship has been found for Se in hair with that in liver (*4*). Since the number of samples analyzed in this study (*4*) is rather small, it is difficult to draw a firm conclusion. This article is further concerned with the relationships between Se concentration in hair and the other three organs from 24 Chinese autopsies, by instrumental neutron activation analysis.

MATERIALS AND METHODS

Specimens from 24 male (35–60 yr) subjects were analyzed, the specimens being supplied by the Institute of Forensic Medicine in China. Specimens of liver, kidney-cortex, lung, and scalp hair were collected within 48 h after death (subjects who died of accidental death, but were otherwise free from disease). Using a titanium knife, the samples were cut into small cubes. Homogenization of each sample was carried out in liquid nitrogen, using the brittle-fracture technique, and then lyophilized and ground into fine powder, and store at a temperature below −10°C. Hair samples were washed with water and acetone, and then dried.

The Se standard was prepared by dropping a small volume of an appropriate Se solution onto clean filter paper. The 100 mg of lyophilized samples and the standards were wrapped and sealed in foils of poly-

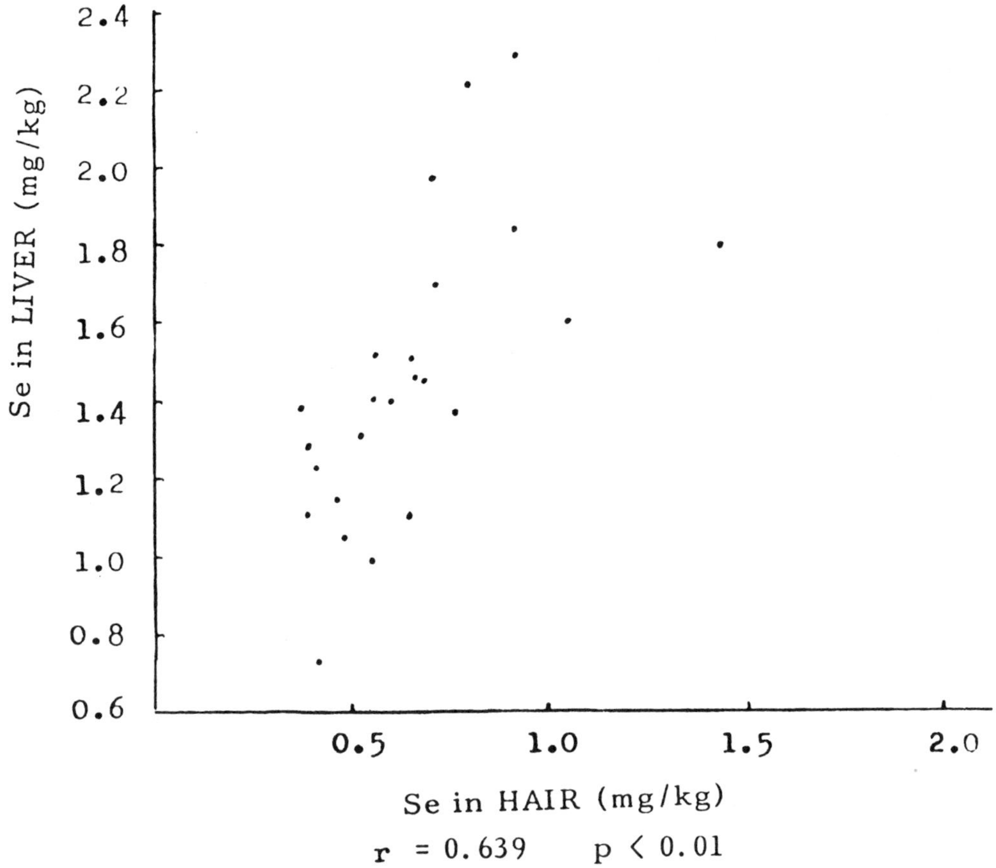

Fig. 1. Scatter plot of Se contents of hair and liver in 24 individuals.

ethylene, and then irradiated in the Slowpoke reactor, which produces a thermal-neutron flux of $8 \times 10^{11} n/cm^2/s$. It is equipped with a pneumatic transfer system, which can rapidly deliver the activated sample to a counting position. The γ-rays of ^{77m}Se (161.9 kEv) were detected with a germanium detector with 2.0 KeV resolution at 1332 KeV. For all the measurements, the irradiation time was 60 s, the decay time 5 s, and the counting time 60 s. The spectra were collected with a 4096-channel analyzer and stored in a disk.

RESULTS AND DISCUSSION

Selenium is one of the essential elements needed by humans to maintain health. The only known function of Se is the selenoenzyme glutathione peroxidase (GSHPX), which protects biomembranes against oxidative destruction. Selenium is considered to be of vital importance in human nutrition. The results of the measured Se contents in human hair,

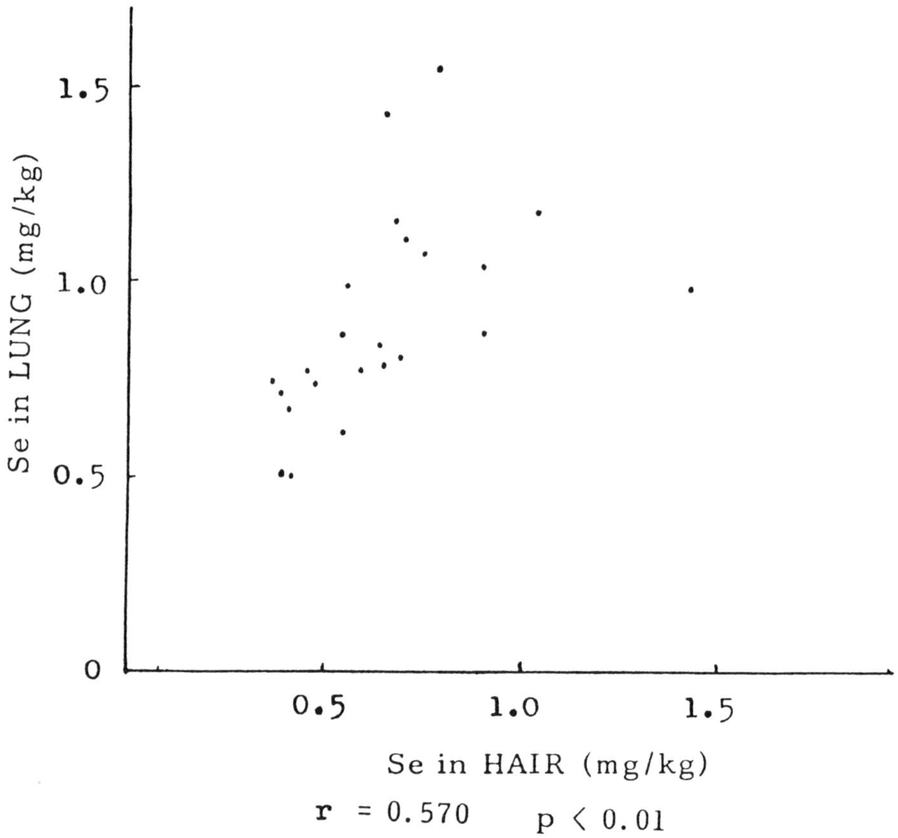

Fig. 2. Scatter plot of Se contents of hair and lung in 24 individuals.

Table 2
Correlation Between Hair and Internal Tissues for the Trace Element Se of 24 Male Autopsies

	Hair		Kidney-cortex		Liver	
Tissues	*r*-value	*p*-value	*r*-value	*p*-value	*r*-value	*p*-value
Kidney-cortex	0.635	0.01				
Liver	0.639	0.01	0.421	0.05		
Lung	0.570	0.01	0.498	0.01	0.549	0.01

The experimental results show that hair analysis may be a useful tool for assessing the internal body burdens of Se.

liver, lung, and kidney-cortex for 24 subjects are summarized in Table 1. They are expressed both dry and wet-wt bases for soft tissues. It can be seen that the concentrations of Se are higher in kidney-cortex (2.04–5.36 mg/kg) than in liver (0.73–2.29), lung (0.50–1.85), and hair (0.37–1.43). The mean value of Se in human hair (0.62 mg/kg) is in good agreement with the value presented by Ohta (5) for 22 subjects living in Japan. It

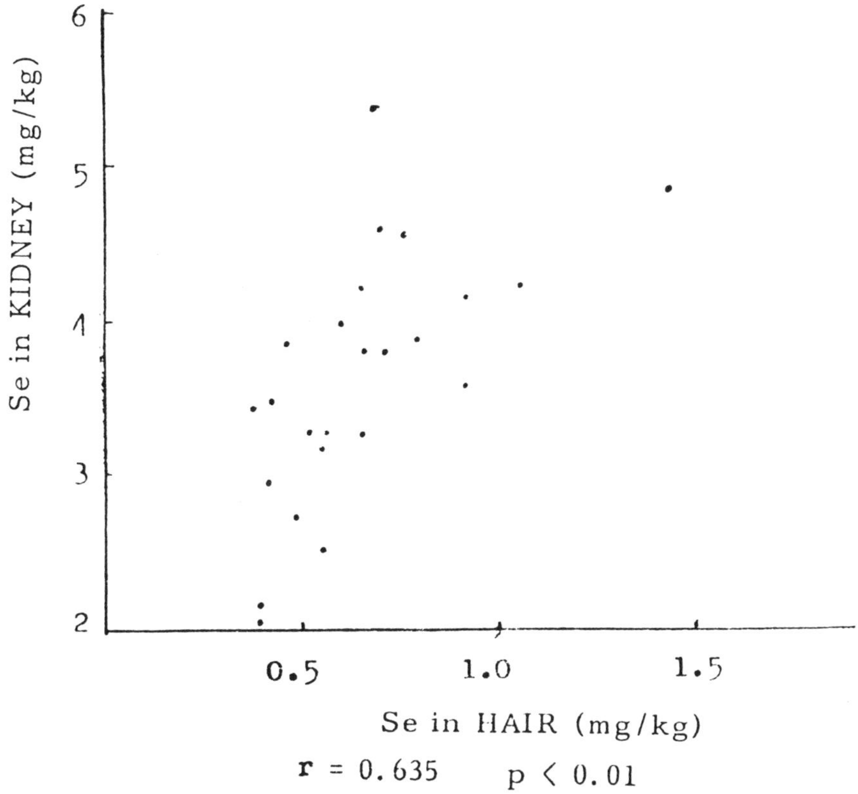

Fig. 3. Scatter plot of Se contents of hair and kidney-cortex in 24 individuals.

may be related to the similar dietary habits between Japan and Shanghai, China. Muramatsu and Parr (*6*) reported that a weaker correlation was found between Se concentration in hair and kidney-cortex. It is interesting to find that there are significant positive correlations for concentrations of Se between hair and liver, lung, and kidney-cortex (Figures 1–3). The correlation coefficients obtained in linear regression analysis, are 0.639, 0.570, and 0.635, respectively. The *P*-values are all less than 0.01 for the three tissues (Table 2).

REFERENCES

1. A. P. Nanmov, I. O. Abugassa, and V. V. Turbin, *J. Radioanal. Nucl. Chem. Letters* **105/3,** 141–150 (1986).
2. X. P. Wang, Y. D. Cheng, et al., *J. Trace Element* **3,** (1985).
3. U. Tomza and W. Maenhant, *J. Radioanal. Nucl. Chem. Letters* **86/4,** 209–220 (1984).
4. Y. D. Cheng, G. S. Zhuang, et. al., *J. Trace Elements in Experimental Medicine* **1,** 19–21 (1988).
5. Y. Ohta, *J. Radioanal. Nucl. Chem.* **114,** 75–82 (1987).
6. Y. Muramatsu and R. M. Parr, *Sci. Total Environm.* **76,** 29–40 (1988).

Determination of Inorganic Components in Brazilian Medicinal Plants by Neutron Activation Analysis

M. SAIKI,[1,*] M. B. A. VASCONCELLOS,[1]
AND J. A. A. SERTIÉ[2]

Instituto de Pesquisas Energéticas e Nucleares, IPEN—CNEN/SP, Radiochemistry Division, CEP 05508, São Paulo-SP, Brasil; and Universidade de São Paulo, Instituto de Ciências Biomédicas, Department of Pharmacology, CEP 05508, São Paulo, SP Brasil

Received April 17, 1989; Accepted December 11, 1989

ABSTRACT

Instrumental neutron activation analysis (INAA) has been applied to multielemental determinations of medicinal extracts obtained from the plants. *Cordia Verbenacea DC, Folidago Microglossa DC,* and Petiveria Alliacea.

Concentrations of the elements Al, Br, Ca, Cl, Co, Cs, Fe, K, La, Mg, Mn, Na, Rb, Sb, and Zn have been determined in dried extracts of these herbs by short and long irradiations under a thermal neutron flux of 10^{11}–10^{13} n/cm^{2}s in the IEA-R1 nuclear reactor. The NBS Tea Leaves (1572) and NIES Pepperbush (1) reference materials were analyzed simultaneously with the plant extracts.

The results obtained in these analyses have shown a good accuracy and reproducibility of the method. The relative errors and the relative standard deviations were less than 10% for most of the elements analyzed.

Index Entries: Neutron activation analysis; medicinal herbs; analysis of plant extracts; *Cordia Verbenacea; Folidago Microglossa; Petiveria Alliacea.*

*Author to whom all correspondence and reprint requests should be addressed.

INTRODUCTION

The use of medicinal herbs is widespread in Brazil. In the last years there has been an increasing interest in using natural medicines, in order to avoid the risks of side effects and the rising cost of medicines obtained from synthetic chemical products.

Although the efficiency of some natural drugs for curing diseases has been proven, the specialists in this subject have warned against the indiscriminate use of these products because they can cause serious intoxications.

The belief that the medicinal herbs do not present side effects has led people to use them without any medical orientation. In addition, people have usually used medicinal plants for a long period to treat chronic diseases; many times, people start using them just to follow a trend.

It is very important to determine the elemental composition of medicinal plants or their extracts, for pharmacological assays as well as to study the effect of the environment on changes in their composition. The analysis of inorganic components in these kinds of materials is essential for the evaluation of their usefulness in relation to their medicinal value. The chemical elements may take part in the active component of the medicinal plant as a healing agent. Besides this, some elements can be closely related to human health or disease when their excess or deficiency induce physiological and metabolic changes.

On the other hand, the concentration of elements in plants depends on many factors, such as the oxidation state of the element, type and pH of the soil, species and age of the plant, and also on the fertilizer used. Consequently, the efficiency of a medicinal plant may also depend on the region where it was cultivated.

Many papers have been published about the determination of elements in several types of plants (*1–4*) and few about herb extracts (*5*).

The purpose of this work was to apply the method of instrumental neutron activation analysis (INAA) to determine inorganic components in herb extracts for further contribution to studies of their toxicity, medicinal action, or active component.

MATERIALS AND METHOD

Extract Samples Obtained from Medicinal Herbs

Medicinal extracts obtained from three species of herbs were analyzed by INAA in this work. *Cordia Verbenacea DC*, from the family of Boraginaceae, is a perennial bush plant used for many centuries by Brazilian Indians to treat infections and rheumatisms. The bioflavonoids produced by their leaves show antiinflammatory activity and can be used as a medicine without causing side effects (*6*) such as stomach lesions, hormonal changes, and liquid retention in the body. Besides this, the leaves of this herb produce essential oils that contain substances responsible for their peculiar smell. Certain insects and beetles that destroy the orchard leaves are attracted by this smell, so this herb is planted among the orange and lemon trees to be used as a "trap plant." *Folidago Microglossa DC* (Compositae) is known in Brazil by the name of Brazilian Arnica and the aerial parts of this plant are used as an antiinflammatory and cicatrizing drug. *Petiveria Alliacea* (Phytolaccaceae), known in Brazil by the name "Guiné" is a plant widely used in popular medicine "to calm nerves" and as an anesthetic, diuretic, analgesic, and abortive drug too.

The extracts of these plants were provided by the Faculty of Pharmaceutical Sciences of the University of the State of São Paulo. The fresh material of these herbs was extracted using a 70% alcoholic solution at room temperature, as described in ref (*6*). After the filtration, the dark green solution was concentrated under reduced pressure at 50°C to obtain an aqueous viscous material that was lyophilized afterwards. The dried extracts are stored in a dessicator.

Instrumental Neutron Activation Analysis of the Samples

Dried extracts of the herbs (50–100mg) were weighed and heat-sealed in clean polyethylene bags or in polyethylene capsules imported from the Free University of Amsterdam for irradiation together with synthetic standards prepared in our laboratory using standard solutions of the elements to be determined.

These standard solutions were prepared by dissolving high-purity metal, oxide, or salt of elements with adequate reagents. An appropriate volume of these standard stock solutions was pipeted onto a small sheet of Whatman No. 42 filter paper and dried in a dessicator at room temperature. The amounts of elements in these standards were about 10 times higher than those found in the samples. The sheets of filter paper were also put into polyethylene bags or capsules.

Irradiations were performed in the IEA-R1 nuclear research reactor. The irradiation conditions, decay times, and nuclear data of the nuclides used in this study are shown in Table 1.

The occurence of the interfering reactions ^{27}Al (n,p) ^{27}Mg, ^{31}P (n,α) ^{28}Al and ^{28}Si (n,p) ^{28}Al in the determinations of Mg and Al was also examined by irradiation of the samples with epithermal neutrons using 1.0-mm-thick cadmium containers.

The irradiated samples and standards were counted using an ENERTEC hyperpure Ge detector coupled to an EG & G ORTEC 4096 channel plus height analyzer connected to a Monydata PC 200 Plus microcomputer. The counting system had a resolution (FWHM) of 2.5 keV for the 1332-keV gamma-ray of ^{60}Co and 1.4 keV for the 122-keV gamma-ray of

Table 1
Experimental Conditions and Nuclear Data of the Nuclides Used in This Study

Irradiation Time	Thermal Neutron Flux $n.cm^{-2}.s^{-1}$	Decay Time	Radioisotopes (Gamma ray energies measured in keV)
1 min	3.7×10^{11}	2 min	^{28}Al (1778) ; ^{27}Mg (1014) ; ^{80}Br (617);
		60 min	^{38}Cl (1642) ; ^{56}Mn (1810)
30 min	1.7×10^{12}	15 h	^{42}K (1525) ; ^{56}Mn (846 , 1810) ; ^{82}Br (776); ^{24}Na (1368)
8 h	10^{13}	5 - 20 d	^{82}Br (776) ; ^{47}Ca (1296) ; ^{60}Co (1173) ; ^{134}Cs (796) ; ^{59}Fe (1099) ; ^{140}La (1596) ; ^{24}Na (1368) ; ^{86}Rb (1076) ; ^{122}Sb (564) ; ^{65}Zn (1115)

^{57}Co. Samples and standards were measured in the polyethylene capsules or bags used for their irradiations. The presence of impurities in these involucres was previously examined; and it was found to be negligible in comparison with the amounts of elements present in the sample. The gamma-ray spectra were processed by using the modified version of the FALA (*7*) program, written in Pascal language. This program locates peak positions and calculates gamma-ray energies and net areas.

Two certified reference materials, Citrus Leaves (1572) provided by the National Bureau of Standards (NBS) USA and Pepperbush No. 1 from the National Institute for Environmental Studies (NIES), Japan, were analyzed in this work for the evaluation of the accuracy of the method. These reference materials were dried as recommended in their respective certificates (*8,9*). With the drying process there was a weight loss of 6.47% for Citrus Leaves and of 7.97% for Pepperbush. These values were used to calculate the weights of materials analyzed on a dried basis.

RESULTS AND DISCUSSION

Table 2 shows the results of elemental concentrations obtained for *Cordia Verbenacea* and *Folidago Microglossa* extracts, and Table 3 presents the results for extracts obtained from the roots and aerial parts of *Petiveria Alliacea* herb. These reported values are the averages and standard deviations from at least four determinations. Acceptable precisions, with relative standard deviations <10%, were obtained for most of the elements analyzed. Less precise results were obtained for Co, Cs, Sb, and Mg. The induced activity of ^{60}Co, ^{134}Cs and ^{122}Sb was not high because of the small concentrations of these elements in the samples (<1 ppm). Magnesium is present in relatively high concentration; however, the 1014-keV gamma-ray photopeak of ^{27}Mg is not very pronounced, resulting in poor counting statistics.

Concerning the elements analyzed, Al, Br, Cl, Cs, Fe, K, Mg, Mn, Na, Rb, and Zn were found in all the extracts. Calcium was not detected in the extracts obtained from the roots and aerial parts of *Petiveria Alliacea*. Table 2 shows little difference in the concentrations of Al, Br, Ca, Cl, Co, Cs, Fe, K, Mg, Mn, Rb, and Zn found in *Cordia Verbenacea* and *Folidago Microglossa* extract.

Also, the results obtained for *Petiveria Alliacea* extract (Table 3) show that the concentrations found for Al, Br, Cl, Co, Fe, K, Mg, Mn, Na, Rb, and Zn in the extracts from roots are practically the same as those found in the extracts from aerial parts. Especially for Cl, this was not expected, since Cl is generally found in higher concentration in the green parts (leaves) of the plant than in the roots.

Table 2
Concentrations of Elements in Extracts Obtained from Leaves of *Cordia Verbenacea* and *Folidago Microglossa* by INAA

Element	Cordia Verbenacea Extract	Folidago Microglossa Extract
Al (ppm)	240 ± 24 (10.0)[a]	192 ± 21 (10.9)
Br (ppm)	56 ± 4 (7.1)	34 ± 3 (8.8)
Ca (%)	0.84 ± 0.08 (9.5)	0.31 ± 0.02 (6.4)
Cl (%)	3.19 ± 0.08 (2.5)	1.06 ± 0.04 (3.8)
Co (ppb)	135 ± 35 (25.9)	112 ± 11 (9.4)
Cs (ppb)	940 ± 180 (19.1)	964 ± 124 (12.8)
Fe (ppm)	175 ± 16 (0.9)	140 ± 9 (6.4)
K (%)	4.6 ± 0.2 (4.3)	2.95 ± 0.18 (6.2)
La (ppb)	-	130 ± 14 (10.9)
Mg (ppm)	4107 ± 1230 (30.9)	2746 ± 576 (20.9)
Mn (ppm)	19.5 ± 1.4 (7.1)	58 ± 3 (5.2)
Na (%)	1.4 ± 0.1 (7.1)	0.020 ± 0.002 (10.0)
Rb (ppm)	109 ± 7 (6.4)	135 ± 5 (4.0)
Sb (ppb)	-	740 ± 100 (13.5)
Zn (ppm)	63 ± 4 (6.0)	112 ± 5 (4.5)

[a] Values in parentheses are relative standard deviations.

It is important to note that the medicinal extracts analyzed here have shown a high concentration of K, varying from 3–10%, and that potassium salts participate actively in the cardiac rhythm and in constipation. Also, Mg present in these medicinal extracts can neutralize the acidity and avoid side effects of stomach lesions when these drugs are taken orally by patients.

From the pharmacological and toxicological point of view, the concentration of the elements present in the three species studied are too low to cause any kind of effect.

The accuracy of the method was checked by analyzing Citrus Leaves and Pepperbush reference materials. Table 4 shows the results obtained

Table 3
Concentrations of Elements in Extracts from *Petiveria Alliacea* (Guiné) Obtained by INAA

Element	Extract from Roots	Extract from Aerial Parts
Al (ppm)	424 ± 28 (6.6)[a]	118 ± 15 (13.0)
Br (ppm)	325 ± 33 (10.1)	160 ± 8 (5.0)
Cl (%)	1.38 ± 0.16 (11.6)	1.60 ± 0.09 (5.6)
Co (ppb)	756 ± 102 (13.5)	840 ± 43 (5.1)
Cs (ppb)	552 ± 65 (11.8)	721 ± 46 (6.4)
Fe (ppm)	69 ± 2 (2.9)	14.8 ± 0.9 (6.1)
K (%)	10.4 ± 1.5 (1.4)	9.2 ± 0.5 (5.4)
Mg (ppm)	1558 ± 564 (36.2)	1525 ± 525 (34.4)
Mn (ppm)	270 ± 36 (13.3)	334 ± 28 (8.4)
Na (%)	0.61 ± 0.04 (6.5)	0.19 ± 0.01 (5.3)
Rb (ppm)	320 ± 20 (6.2)	373 ± 11 (2.9)
Sb (ppb)	241 ± 49 (20.3)	1109 ± 121 (10.9)
Zn (ppm)	134 ± 7 (5.2)	451 ± 19 (4.2)

[a]Values in parentheses are relative standard deviations.

in these analyses, together with certified values presented by NBS and NIES. We have determined 15 elements; most of them are in reasonably good agreement with these certified values. For the elements Ca, Cs, K, Mn, Na, Rb, and Zn, relative errors <10% were obtained. Also in these samples, the least precise results were obtained for Co and Cs.

ACKNOWLEDGMENTS

This work was partially supported by the Brazilian National Research Council (CNPq) and Financial Agency for Studies and Projects (FINEP).

Table 4
Concentrations of Elements in NBS Citrus Leaves (1572) and in NIES Pepperbush (No. 1) Reference Materials Obtained by INAA

Element	Citrus Leaves		Pepperbush	
	This work	Ref (8)	This work	Ref (9)
Al (ppm)	137 ± 16 (11.6)[a]	92 ± 15	705 ± 17 (2.4)	
Br (ppm)	8.0 ± 0.5 (6.2)	8.2 (b)	1.4 ± 0.2 (12.5)	
Ca (%)	3.30 ± 0.27 (8.2)	3.15 ± 0.10	1.41 ± 0.04 (2.9)	1.38 ± 0.07
Cl (ppm)	660 ± 102 (15.4)	414 (b)	-	
Co (ppb)	33 ± 6 (18.7)	20 (b)	172 ± 49 (28.5)	230 ± 30
Cs (ppb)	107 ± 16 (15.4)	98 (b)	1274 ± 88 (6.9)	1200
Fe (ppm)	111 ± 11 (9.9)	90 ± 10	220 ± 12 (5.3)	205 ± 17
K (%)	1.84 ± 0.09 (4.9)	1.82 ± 0.06	1.53 ± 0.07 (4.7)	1.51 ± 0.06
La (ppb)	228 ± 36 (15.7)	190 (b)	252 ± 30 (11.9)	
Mg (%)	0.46 ± 0.06 (13)	0.58 ± 0.03	0.393 ± 0.04 (10.4)	0.408 ± 0.02
Mn (ppm)	24 ± 3 (12.5)	23 ± 2	2228 ± 110 (4.9)	2030 ± 170
Na (ppm)	170 ± 11 (6.7)	160 ± 20	109 ± 7 (6.3)	106 ± 13
Rb (ppm)	4.94 ± 0.17 (3.5)	4.84 ± 0.06	75 ± 3 (4.6)	75 ± 4
Sb (ppb)	45 ± 4 (8.9)	40 (b)	-	
Zn (ppm)	30 ± 3 (10.0)	29 ± 2	342 ± 24 (7.0)	340 ± 20

[a]Values in parentheses are relative standard deviations.
[b]Information values from Ref (*8*).

REFERENCES

1. Ch. L. Ndiokwere, *J. Radioanal. Nucl. Chem.*, **85,** 325 (1984).
2. S. Mukhammedov, Kh. Tillaeva, and N. B. Badalov. *Sov. At. Energy* **61(6),** 1043 (1987).
3. M. Koyama, M. Shirakawa, J. Takada, Y. Katayama, and T. Matsubara, *J. Radioanal. Nucl. Chem.*, **112(2)** 489 (1987).
4. S. B. Aidid, *J. Radioanal. Nucl. Chem.*, **120(2),** 335 (1988).
5. L. Marichkova and O. Kostarova, *Proceedings of Balkan Conference on Activation Analysis*, Varna, Bulgaria May 6–8, 1985, in INIS-mf-11126, 121–3, 1985.
6. J. A. A. Sertié, A. C. Basile, S. Panizza, A. K. Matida, and R. Zelnik, *Planta Medica*, **1,** 7 (1988).
7. F. W. Lima and L. T. Atalla, *J. Radioanal. Chem.*, **20,** 769 (1974).
8. NBS Certificate of Analysis for SRM 1572 Citrus Leaves, Office of Standard Reference Materials, US Dept. of Commerce, Washington DC 1982.
9. NIES Certificate of Analysis for NIES Pepperbush No. 1 Reference Material, National Institute for Environmental Studies, Japan, 1980.

PIXE Determination of Essential Trace Elements in Some Traditional Chinese Medicines

LUN XIAO[1,*] AND JUN-FA QIN[2]

[1]Institute of Atomic Energy, PO Box 275 Beijing; and [2]Institute of Nuclear Research, PO Box 8204, Shanghai, People's Republic of China

Received April 17, 1989; Accepted December 11, 1989

ABSTRACT

The essential trace elements in 30 traditional Chinese medicines, (24 tonics and 6 nontonics) were determined by proton-induced X-ray emission. The authors' previous suggestion that traditional Chinese medicines may be classified by the order of magnitude of their essential trace elements, thus indicating their pharmacological effects, is not justified. The pharmacological effect of a trace element or its essentiality may be dependent on some ligand that can be chelated with it. A nonlinear mapping algorithm, however, shows that the 30 traditional Chinese medicines are nearly separated into two groups, indicating their tonic or nontonic pharmacological effects.

Index Entries: PIXE; traditional Chinese medicines; tonic; nonlinear mapping algorithm.

INTRODUCTION

Essential trace elements (*1*) are required and indispensable for maintenance of life. Their absence results in the death of the organ. Severe deficiency of an element that results in death is difficult to produce, particularly if the element is only required in very low concentration. A broader characterization widely accepted for essential elements is that a

*Author to whom all correspondence and reprint requests should be addressed.

Biological Trace Element Research Editor: G. N. Schrauzer © 1990 by The Humana Press Inc.

deficient intake consistently results in impairment of a function for optimal to suboptimal and supplementation with physiological levels of that element, but not with others, prevents or cures the impairment. If deficiency of a certain essential element really causes disease, instead of the disease causing the deficiency, then it is natural to treat the disease with drugs or pharmaceuticals abundant in that element.

Traditional Chinese medicines, after thousands of years of practice, really do cure diseases with remarkable effectiveness. However, the pharmacokinetics of traditional Chinese medicines is not yet understood in terms of modern pharmacological concepts. The lack of scientific proof of the effects of traditional Chinese medicines has often made scientists doubt their efficacy. However, we cannot underestimate their effects, because the long history of Asian people's experience has proven that human functions have really been restored to normal levels through their use. In order to understand the pharmacokinetics of the traditional Chinese medicines, their constituents must be determined and synthesized, as was first done in the case of ephedrin. However, essential elements in few of these traditional Chinese medicines have been determined. We started to determine the essential elements in these medicines, hoping to find their effective essential elements so they may be used to treat patients deficient in those elements. Then we may investigate their roles at molecular levels, e.g., in enzymes, hormones, or the like.

The essential trace elements in some traditional Chinese medicines, two tonics and three antifebriles, were determined by instrumental neutron activation analysis (2). The similar patterns for the tonics and antifebriles, respectively, suggested that traditional Chinese medicines may be classified or subclassified by the patterns or orders of magnitude of their essential trace elements, thus indicating their pharmacological effects. We now present some analytical data of essential trace elements in 30 traditional Chinese medicines, 24 tonics, and 6 nontonics, by proton-induced X-ray emission.

METHODS

The samples (3) were washed clean, dried, and ashed. The ash residue was dissolved in 6*N* HNO_3, and 100 ppm of yttrium was added as an internal standard.

The proton beam passed through an aluminum scattering foil and a 0.6-cm tantalum collimator. The proton energy decreased from 3.5 to 3.1 MeV. The beam current was in the range of 10–20 nA. Each of the samples was bombarded by 3.1-MeV protons for about 10 min, for a total charge of 5–10 μC. The automatic PIXE analysis system (4) of Shanghai Institute of Nuclear Research consists of an automatic target, proton

Table 1
Essential Trace-Element Patterns
of Some Traditional Chinese Medicines

Group	Pattern	Medicine Nos.
1	Fe>Mn>Zn>Cu	1,2,3,4,5,6,
2	Fe>Zn>Mn>Cu	7,8,9,10,11,12,13,14,15,16
3	Fe>Mn>Cu>Zn	17(Korean ginseng) 18,19,20
4	Fe>Zn>Cu>Mn	21,22,23,24,25
5	Fe>Cu>Zn>Mn	26,27,28
6	Fe>Cu>Mn>Zn	---
7	Mn>Zn>Fe>Cu	29
8	Mn>Fe>Zn>Cu	30

beam current measuring system, and an automatic X-ray spectra acquisition and analysis system.

The content C_x of element X in the sample was calculated by

$$C_x = (N_x/N_s) \cdot (K_{xs}/R_{xs})\ Cs$$

where N_x is the total count of the characteristic X-ray peak of element X in the sample, N_s is the same for the yttrium standard K_{xs} is the normalized correction factor of the matrix absorption effect of X relative to the standard, and R_{xs} is the relative sensitivity of X to S.

RESULTS AND DISCUSSION

Essential trace elements Fe, Mn, Zn, Cu, Cr, Ni, and Sr (possibly essential) of 30 traditional Chinese medicines were analyzed by PIXE. The data are arranged in order of magnitude or pattern of the contents of the abundant essential trace elements, Fe, Mn, Zn, and Cu, in Table 1, where the medicines are numbered for convenience. The pharmaceutical names or Chinese names of the medicines corresponding to the numbers are listed in the Appendix.

Six of the medicines listed in Table 1, (Nos. 4, 15, 20, 25, 29, and 30) are not considered tonics by traditional Chinese doctors, but the other 24 medicines are. Antifebriles were not investigated.

We see from Table 1

1. That the most abundant essential trace element in tonics is Fe;
2. That the most abundant essential trace element in groups 7 and 8 is Mn; and

Appendix:
Pharmaceutical or Chinese Names

Nos.	Medicine names
1	Semen Cuscutae
2	Shengdi
3	Radix Acanthopanacis Senticosi
4	Radix Gentianae
5	Haiersheng
6	Radix Morindae Officinalis
7	Radix Asparagi
8	Rhizoma Polygonati
9	Fructus Ligustri Lucidi
10	Rhizoma Polygonati Odorati
11	Fructus Tribuli
12	Radix Ophiopogonis
13	Radix Astragali
14	Radix Aconiti Lateralis Preparata
15	Radix Curcumae
16	Fructus Cnidii
17	Korean Radix Ginseng
18	Fructus Lycii
19	Cortex Eucommiae
20	Herba Houttuyniae
21	Rhizoma Atractylodis Macrocephalae
22	Radix Glycyrrhizae
23	Radix Polygoni Multiflori
24	Rhizoma Anemarrhenue
25	Fructus Aurantii Immaturus
26	Rhizoma Dioscoreae
27	Herba Cistanches
28	Shasheng
29	Fructus Bruceae
30	Rhizoma Alismatis

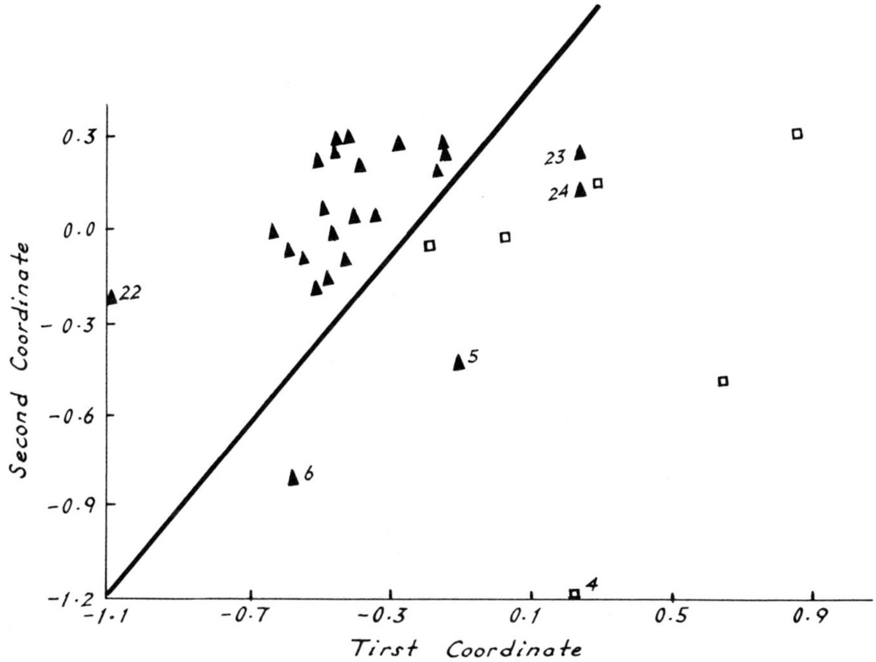

Fig. 1. Nonlinear mapping algorithm: ▲, tonic; □, nontonic.

3. That the tonics may be subclassified into six groups by the patterns or orders of magnitude of the essential trace elements Fe, Mn, Zn, and Cu.

However, according to traditional Chinese doctors, there are different kinds of tonics, and tonic medicines in each of groups 1, 2, 3, 4, and 5 do not have a similar pharmacological effect.

We suspected that medicines Nos, 4, 15, 20, and 25 possess a tonic action, usually neglected by traditional Chinese doctors, just like a similar traditional Chinese medicine, Radix Puerariae, which was usually not considered a tonic, but was suspected and then found (*5*) to possess tonic action in *Materia Medica*, the Chinese "medical bible."

We now find that medicines abundant in Fe are not necessarily tonic. A traditional Chinese medicine named "the herb geese won't eat," containing as much as 5000 ppm Fe, is nontonic. Similarly, Nos. 4, 15, 20, and 25 really may be nontonic, although they are abundant in Fe. In these cases, the role of ligands may be supported. To be pharmacologically effective or essential, the trace element may need to be combined or chelated with some ligand (porphyrin, for example), in order to be physiologically absorbed to prevent or cure impairment caused by deficiency of that element.

The most abundant essential trace element in Nos. 29 and 30, which are both antiphlogistic, is Mn, but we are not sure that Mn is the most abundant trace element in other antiphlogistic medicines.

Nonlinear mapping algorithms (*6*) of Fe, Mn, Zn, Cu, Cr, Ni, and Sr in the 30 traditional Chinese medicines shown in Fig. 1, shows that nontonics scatter around a region with tonics Nos. 5, 6, 23, and 24, and that no nontonics fall into the tonic region, a cluster of 19 tonics, with tonic No. 22 far away.

REFERENCES

1. L. Xiao, Y. H. Zhang, Y. G. Liu, and J. F. Qin, "INAA of Essential Trace Elements in Some Traditional Chinese Medicines," orally presented at the 2nd International Conference on Nuclear and Radiochemistry, Brighton, UK, July 1988, 11–15.
2. W. Mertz, *Science* **213,** 1322 (1981).
3. M. C. Li et al., Nucl. Instr. *Methods* **181,** 309 (1981).
4. M. C. Li et al., Ibid., **181,** 37 (1981).
5. Y. L. Xu, L. Tao, J. F. Qin, et al., *The Nonlinear Mapping Algorithm and Its Application*, Annual Report, Shanghai Institute of Nuclear Research, Shanghai, 1984, pp. 43–45.
6. X. F. Zhang et al., "Nonlinear Mapping Algorithm of PIXE Data in Some Traditional Chinese Medicines," to be presented at the MTAA-8 Conference, Beijing, May 28–June 1, 1990.

Inductively Coupled Atomic Emission Spectrometry and Neutron Activation Analysis for the Determination of Element Reference Values in Human Lung Tissue

E. SABBIONI,*[,1] G. R. NICOLAOU,[1] R. PIETRA,[1] E. BECCALONI,[2] E. CONI,[2] A. ALIMONTI,[2] AND S. CAROLI[2]

Commission of the European Communities, J.R.C.-Ispra, Radiochemistry Division, I-21020 Ispra (Va), Italy, and Istituto Superiore di Sanitá, Viale Regina Elena 299, 00161 Rome, Italy

Received June 21, 1989; Accepted December 11, 1989

ABSTRACT

An investigation was undertaken in order to assess the performance of neutron activation analysis and inductively coupled plasma atomic emission spectrometry techniques for determining reference values for minor and trace elements in human lungs of urban subjects. Results show that in both instances experimental conditions must be carefully optimized to guarantee reliability of experimental data. Strict criteria for tissue sampling and pretreatment also had to be set. Provisional reference values for ca. 50 elements could thus be established.

Index Entries: Inductively coupled plasma atomic emission spectrometry; neutron activation analysis; trace elements in human lungs, analysis; minor elements in human lungs, analysis; reference values for elements in human lungs.

*Author to whom all correspondence and reprint requests should be addressed.

INTRODUCTION

Following epidemiological studies, there is undisputed evidence that lungs are a critical target organ of air pollutants, including trace elements. The knowledge of trace-element content in human lungs is essential (i) for establishing "normal criteria" for levels of trace elements in pulmonary tissues of selected population groups; (ii) in the diagnosis of trace-element-related diseases; and (iii) for setting recommendations for airborne-trace-element exposure. Unfortunately, a survey of the data on human beings published so far shows that lung trace-element content varies by several orders of magnitude and that the literature data are of questionable analytical quality (*1*). Since legal decisions in limiting the exposure to elements also depend on such data, their reliability is mandatory.

The considerable disagreement among literature values is not only caused by the natural biological variability. Another reason is that little, if any, attention was given to contamination or loss during sampling and handling of the samples before the actual analysis, thus invalidating in practice even determinations of excellent analytical quality. There is a great need for laboratories with different expertise to produce data of similar quality, so that the "normal levels" of trace elements may be determined for individuals and populations residing in different areas.

The work described here is part of the research activity on trace elements in human carried out in the context of the Environmental Protection Program at the JRC-Ispra (Italy) in collaboration with the Istituto Superiore di Sanitá (Rome) and the University of Surrey (Guildford, UK).

OBJECTIVES

The main goal of the research was to establish accurate and reliable reference values of trace elements for subsequent use in toxicological studies by narrowing the range of the reported literature levels and minimizing the analytical sources of errors. In particular, the following aspects are considered:

1. Quality-assurance studies of trace element analysis by combining such reference techniques as neutron activation analysis (NAA), with such routine techniques as inductively coupled plasma atomic emission spectroscopy (ICP-AES). Studies are carried out on parameters affecting the analytical determinations, including those related to collection, sample handling, removal of organic matrix, and preconcentration methods. In this context, animal tissues containing radioisotopically labeled trace elements in "metabolized form," as generated from administration of radiotracers to laboratory animals, are used.

2. Systematic determinations of trace-element levels in the lungs of unexposed subjects living in the various EEC countries, to be compared with the pulmonary tissues of selected groups of workers and clinical cases. Since the anatomical complexity of the lungs, which contributes to the architecture of various zones, is well documented, 13 sampling points were considered in order to establish possible differences in the element distribution.

An overview of the research is presented, showing:

1. The study of some factors affecting the determination of trace elements in human lungs (with hilar lymph nodes). Radiotraces have been administered to rats, and the radioisotopically-labeled lung tissues were used to check the losses during different analytical steps, such as freeze-drying and ashing procedures.
2. The techniques used, including NAA and ICP-AES. In particular, radiochemical separations of the induced radioisotopes in neutron-activated samplings have been developed in order to reach the maximum sensitivity.
3. Reference values for trace elements in the lungs of Italian subjects.

MATERIALS AND METHODS

Preparation of Radiolabeled Rat Tissues

The different radiotracers have been supplied by the Radiochemical Centre, Amersham, UK, or produced by NAA at the Triga Mark II (University of Pavia) and HFR (JRC-Petten) reactors as well as by proton activation of metallic targets at the MC 40 Scanditronix cyclotron of the JRC-Ispra.

Groups of three rats (200–250 g) had been injected with the radioisotopically labeled elements at concentrations ranging from 10 to 500 ng element/rat. Lungs were removed, washed with physiological saline solution, and used to check the losses of the elements during sample handling for the preparation of the tissues to be analyzed by NAA and ICP-AES.

Sample Collection

Both lungs and part of the trachea were used for analysis of the elemental content. The lungs were taken during autopsy from nonsmoking subjects, aged more than 50 yr, deceased from nonpathological causes, who during their lifetime had not been occupationally exposed to contaminating elements. All the subjects were from the same urban area

(Rome), which minimized inconsistencies arising from differences in diet and urban air composition.

Upon removal, the lungs were laid on a Teflon™ surface in a clean, dust-free environment and specimens of about 2 g in weight were collected from 13 sampling points; one subpleura, one lymph node, one bronchial, and three internal bulk sites of each lung, and one sample from the tracheal bifurcation. The samples were collected by the Institute of Pathological Anatomy of the Catholic University, Rome, not later than 15 h after death, so as to minimize any redistribution of elements in the organ caused by any movement of fluids. The dissection of the lung into different segments was carried out by special titanium nitride coated surgical instruments that minimized the risk of metal contamination. Titanium nitride coated instruments were employed in the sample collection in order to avoid any contamination during this step (*2*). Upon collection, samples were placed in previously decontaminated polyethylene vials and were securely sealed and stored at −20°C. Samples were always taken in duplicate, so that two equivalent sets of specimens were available to be analyzed by the two techniques mentioned above.

Analytical Techniques

Figure 1 illustrates the overall procedures employed in each method. In the case of NAA, the samples were immediately frozen at −80°C for 24 h and then freeze-dried for 48 h. They were then analyzed by Instrumental and Radiochemical NAA (INAA and RNAA, respectively).

INAA measurements involved short irradiations and counting periods (170 s) and were carried out with the Triga reactor available at the University of Pavia (Italy).

RNAA was employed in order to remove Na, present in the samples at high levels, and hence to improve the detection limits for other elements (*3*). The samples were irradiated for about 8 h at a reactor flux of 10^{13} neutrons /cm^2 sec with the Triga reactor. The radiochemical separation began 1 d after irradiation. Mineralization of each sample was performed in a Teflon™ bomb containing 3 mL of HNO_3 and 1 ml of multicarrier solution for 12 h at 165°C. The acid and ^{82}Br present were then evaporated at 60°C, and each sample was eluted through a set of chromatographic columns filled with CHELEX 100 and TDO resins. The columns and eluate were then counted on a high resolution Ge(Li) detector. The working scheme for RNAA, including the elements determined, is illustrated in Fig. 2.

ICP-AES requires samples in liquid form. Therefore, a preanalytical treatment was necessary to convert the solid samples into solutions. This was achieved by using a relatively moderate temperature (400 ± 10°C) for the ashing step, which allows high efficiency of combustion and low risk of contamination or loss (*4*).

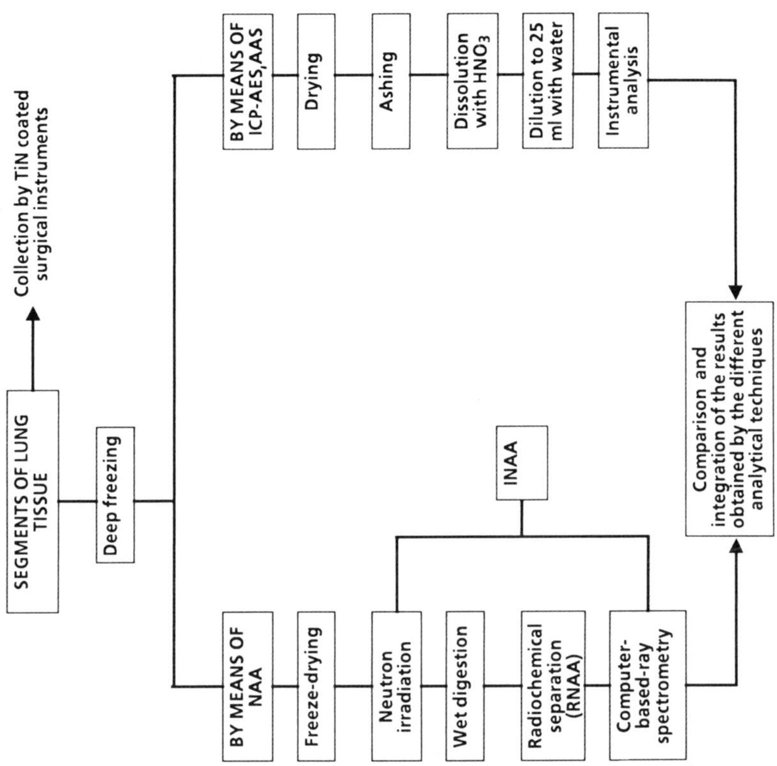
SEGMENTS OF LUNG TISSUE
Collection by TiN coated surgical instruments
Deep freezing
BY MEANS OF NAA
Freeze-drying
Neutron irradiation
Wet digestion
Radiochemical separation (RNAA)
Computer-based-ray spectrometry
INAA
BY MEANS OF ICP-AES,AAS
Drying
Ashing
Dissolution with HNO_3
Dilution to 25 ml with water
Instrumental analysis
Comparison and integration of the results obtained by the different analytical techniques

Figure 1

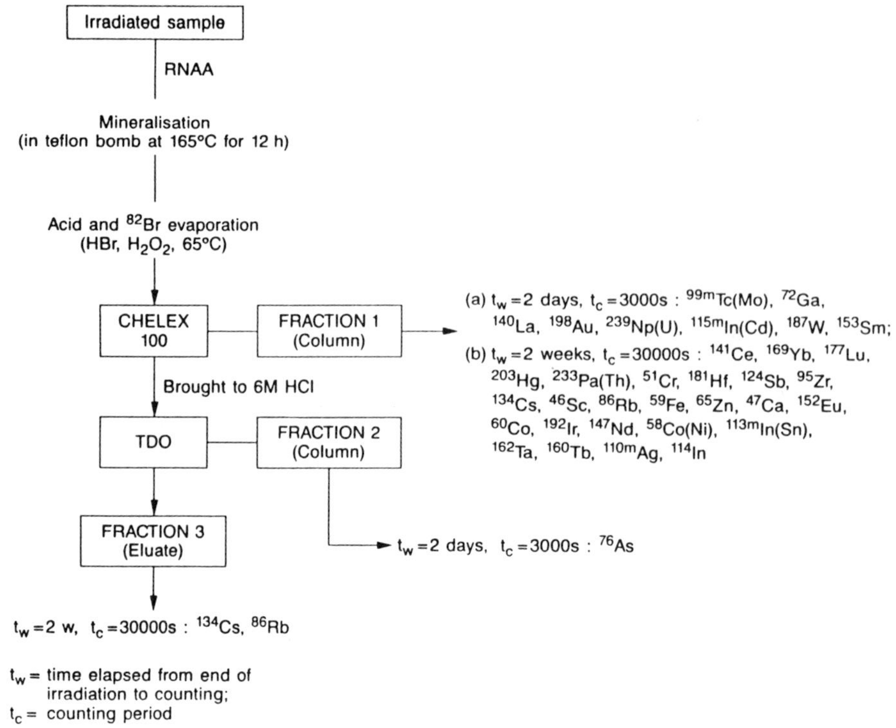

Figure 2

RESULTS AND DISCUSSION

The accuracy of the NAA and ICP-AES procedures was tested by carrying out a complete analysis cycle on NBS Orchard Leaves (SRM 1571) and NBS Bovine Liver (SRM 1577 a), respectively. A good agreement ($p = 0.001$) between certified and measured values was obtained, as shown in Table 1. The apparent discrepancies that occur for some elements indicate that not all the possible systematic errors plaguing the two types of analyses have been eliminated. The determined concentrations are, in general, lower than the certified ones, indicating a possible loss of material at some point in the determination process. However, the fact that some values are simply qualified, not certified, for the reference materials employed may well have contributed to these discrepancies.

Table 2 shows the chemical yield of the elements during freeze-drying and dry ashing as determined by "metabolized radiotracers." The recoveries can be considered satisfactory for practical purposes. The only exceptions seem to be As, Hg, and Sb, for which losses $>10\%$ have been reported during the ashing procedure.

Table 1
Analysis of NBS Bovine Liver and NBS Orchard Leaves by ICP-AES and NAA, Respectively

	Element Concentration (μg/g)			
	NBS Bovine Liver SRM 1577a		NBS Orchard Leaves SRM 1571	
	Certified	Found	Certified	Found
Al	2.0	2.5	310	285
As			11.5	9.2
Au			0.0018	0.0014
Cd	0.44	0.43	0.122	0.112
Co	0.21	0.15	0.17	0.14
Cs			0.032	0.035
Cu	158	155		
Fe			280	270
La			1.2	0.9
Mg	600	572	6070	5960
Mn	9.9	8.7		
Sb			3.0	2.5
Se			0.08	0.075
Sr	0.138	0.132	37	32
V	0.098	0.081	0.60	0.55
W			0.020	0.014
Zn	123	123	25.0	23.0
Zr			2.0	1.2

The importance of avoiding contamination of the samples during collection is fundamental. Thus, the TiN-coated tools used in this step were checked for release of elements by dipping them for 36 h in a 10% HNO_3 solution, repeating the cycle three times. It was found that a noticeable release of Cr, Fe, Mn, Ni, and V during the first cycle had been reduced, by the third cycle, to a release resulting in concentrations of these elements two and three orders of magnitude lower than their content in lung tissues. This points out that such releases can be minimized by conditioning the instruments in diluted HNO_3.

The elemental concentrations determined in the lungs of nine subjects are shown in Table 3. These data are the mean values and concentration ranges pertaining to the pulmonary pair. The corresponding results found in the literature are included for comparison. The use of the two analytical techniques has resulted in the determination of a total of 48 elements.

Table 2
Loss of Trace Elements During Freeze-Drying[a] and Dry-Ashing[b] of Lung Tissues of Rats Administered Radiotracers

Metabolized radiotracer	Freeze-drying	Dry-ashing
^{110}Ag (I)	<1	<1
^{74}As (III)	7	15
^{133}Ba (II)	<1	<1
^{106}Bi (III)	2	7
^{109}Cd (II)	<1	<1
^{58}Co (II)	<1	<1
^{51}Cr (III)	1.5	2
^{203}Hg (II)	3.8	18
^{54}Mn (II)	0.7	1
^{63}Ni (II)	<1	<1
^{203}Pb (II)	0.5	1
^{86}Rb (I)	<1	<1
^{124}Sb (III)	1.2	13
^{75}Se (IV)	<1	<1
^{113}Sn (IV)	<0.5	2
^{44}Ti (IV)	<1	<1
^{201}Tl (I)	0.9	3
^{48}V (IV)	<1	1.4
^{187}W (VI)	<1	1.2
^{65}Zn (II)	<1	<1

[a]Operating time, 24–48 h; operating pressure, 0.1 mbar; starting temperature, −80°C; final temperature, 25°C.
[b]6 h at 100°C + 6 h at 200°C + 6 h at 400°C.

The elements Ag, As, Au, B, Ba, Be, Bi, Cd, Ce, Co, Cr, Cs, Eu, Ga, Hf, Hg, Ir, La, Li, Lu, Mn, Mo, Ni, Pb, Rh, Sb, Sc, Se, Sm, Sr, Ta, Tb, Th, U, V, W, Yb, and Zr occur at the ng/g level (wet wt). The concentrations of Al, Cu, Rb, Sn, Ti, and Zn are at the μg/g level, whereas Ca and Fe concentrations are at the mg/g level. A comparison of the literature data with the values obtained in this study reveals, for some elements, a much wider range of concentrations in the former. This may well result from analytical inconsistencies; differences in age, sex, smoking habits, and diet; biological variations between individuals; and variation in urban and occupational environment.

Table 3
Elemental Concentrations and Ranges
in Human Lung Tissues (ng/g wet wt)

	Literature[a]	This study	
		Mean	Range
Ag	1 - 10	2.4	0.2 - 10
Al§*	1 - 50	11	3 - 22
As	1 - 100	7.6	0.4 - 35
Au	(0.1 - 10)	3.95	0.17 - 15
B	b	410	-
Ba	(30 - 1100)	170	-
Be	(0.1 - 30)	5	-
Bi	(2 - 200)	650	-
Ca*	50 - 300	100	82 - 160
Cd§	10 - 1000	100	30 - 370
Ce	(1 - 400)	30.5	3 - 106
Co§	2 - 100	55	6 - 82
Cr§	2 - 500	107	16 - 200
Cs	3 - 300	38.7	6 - 120
Cu§	1000 - 5000	1700	530 - 6800
Eu	(0.1 - 4)	2.2	0.2 - 13
F	b	-	-
Fe*	40 - 500	105	18 - 250
Ga	(7 - 120)	6.8	0.6 - 54
Hf	(0.5 - 50)	6.5	0.7 - 60
Hg	(49 - 127)	11.8	2.3 - 27
In	b	200	-
Ir	(0.03 - 0.3)	0.08	0.004 - 0.2
La	(1 - 200)	14	2.3 - 100
Li	b	90	b
Lu	(0.1 - 10)	0.4	0.005 - 2.7
Mg§*	20 - 250	95.7	42 - 240
Mn§	(10 - 3000)	137	65 - 240
Mo	1 - 1000	51	1 - 98
Nd	(6 - 90)	180	-
Ni	(10 - 1000)	47.7	10 - 87
Pb	20 - 500	240	-
Rb	500 - 1000	6992	1300 - 27000
Rh	b		-
Sb	(2 - 100)	28.1	0.9 - 181
Sc	(0.1 - 7)	2.2	0.15 - 19
Se	50 - 500	82.8	11 - 260
Sm	(0.92 - 6.6)	1.3	0.12 - 8
Sn	(100 - 4000)	935	60 - 3000
Sr	(10 - 1000)	0.270	-
Ta	(4 - 50)	3.9	0.1 - 18
Tb	(0.4 - 3)	4.2	2.8 - 6
Th	(1 - 20)	2.3	0.12 - 8.1
Ti	(100 - 20000)	3307	1100 - 8000
U	(1 - 400)	2.5	0.3 - 6.8
V§	(0.5 - 500)	47.8	18 - 270
W	(70 - 600)	1.3	0.2 - 3.9
Yb	(0.2 - 6)	4.3	1.2 - 11
Zn§	1000 - 30000	9585	1500 - 22000
Zr	(50 - 200)	426	75 - 2800

[a]Reference 5.
[b]Figures within parentheses are literature speculative values.
*μg/g wet wt.
§Mean between the NAA and ICP-AES values from this study.

Table 4
Elemental Concentration in Bifurcation, Bronchus, Subpleura (ng/g^{-1} wet wt)

	Bifurcation (n = 9)	Bronchus (n = 18)	Subpleura (n = 18)
Ag	0.7 ± 0.07	0.8 ± 0.1	1.5 ± 0.3
Al*§	4.2 ± 0.6	4.6 ± 0.9	7.4 ± 0.7
As	7.6 ± 1.4	9.1 ± 1.7	7.8 ± 1.5
Au	0.6 ± 0.2	1.1 ± 0.3	3 ± 0.6
B	570	570	470
Ba	150	150	140
Be	4	4	5
Bi	740	740	540
Cd§	130 ± 30	173 ± 19	88 ± 15
Ce	10.5 ± 2.7	9.2 ± 1.4	16.5 ± 2.4
Co§	70 ± 16	37 ± 6	29 ± 4
Cr§	82 ± 20	73 ± 10	60 ± 6
Cs	65 ± 12	56 ± 9	49 ± 12
Cu§	1300 ± 380	1438 ± 170	1856 ± 180
Eu	3.3 ± 0.8	3.2 ± 0.4	2.1 ± 0.3
Fe	62 ± 12	51 ± 6	92 ± 9
Ga	2.5 ± 1.2	4.4 ± 1.7	8.4 ± 3
Hf	1.9 ± 0.4	5.7 ± 1.7	1.7 ± 0.2
Hg	14.6 ± 1.9	8.2 ± 0.6	9.3 ± 0.8
In	110	150	200
Ir	0.046 ± 0.016	0.092 ± 0.007	0.18 ± 0.09
La	30.3 ± 6.2	22.3 ± 5.6	11.9 ± 5.7
Li	40	30	30
Lu	0.32 ± 0.12	0.68 ± 0.2	0.6 ± 0.2
Mg*§	118 ± 20	160 ± 15	132 ± 28
Mn§	120 ± 22	118 ± 7	120 ± 8
Mo	912 ± 23	56 ± 4	57 ± 5
Nd	100	150	160
Ni	150	72 ± 9	144 ± 15
Pb	350	170	13
Rb	4400 ± 400	5650 ± 620	7800 ± 1000
Sb	50.8 ± 21	50.6 ± 13	31 ± 12
Sc	3.9 ± 2.3	2.1 ± 0.5	11 ± 0.3
Se	96.4 ± 17	76 ± 10	84 ± 14
Sm	1.4 ± 0.7	1.8 ± 0.4	1.5 ± 0.8
Sn	130 ± 66	650 ± 180	466 ± 129
Sr	670	120	130
Ta	2.1 ± 0.8	1 ± 0.1	4.5 ± 1
Tb	0.7 ± 0.5	1.1 ± 0.2	4.4 ± 0.3
Th	3.9 ± 0.9	4.3 ± 0.6	2.8 ± 0.5
Ti	3528 ± 700	4738 ± 720	3750 ± 370

(*Cont.*)

Table 4 (*Cont.*)

	Bifurcation (n = 9)	Bronchus (n = 18)	Subpleura (n = 18)
U	2.1 ± 0.4	4.3 ± 1.3	2.2 ± 0.3
V§	38 ± 4.7	60 ± 11	58 ± 12
W	3.2 ± 1.1	1.3 ± 0.4	1.8 ± 0.4
Yb	0.3 ± 0.2	0.8 ± 0.3	1.2 ± 0.2
Zn§	9800 ± 1500	10200 ± 2200	9780 ± 1700
Zr	75.6 ± 14.1	239 ± 14	684 ± 9.5

The values are expressed as mean ± SE.
*: μg/g wet wt.
§: Mean between the NAA and ICP-AES values.
n: Total number of samples analyzed.

Trachea, bronchus, and subpleura were analyzed as well as lung tissues. These tissues exhibit, in general, an equal or lower content of the elements than do lung tissues, as is shown in Table 4. The investigation of the trace-element distribution over the different lung segments revealed a higher content of Au, Co, Cr, Cs, Sb, Sc, and V in upper lobes, a result also established by other investigators (*5,6*). For the remaining elements, no particular pattern of distribution was obtained, although clear inhomogeneity of the elements throughout the lungs is evident. This inhomogeneous distribution of the elements emphasizes the need for consistency in the choice of sampling points in such studies, especially for epidemiological purposes.

Although this investigation was performed on a relatively small number of subjects, it has resulted in adequate information concerning the elemental composition and distribution of pulmonary tissues. However, more extensive investigations are in progress, in order to strengthen the validity of these preliminary results and to allow for correlation of these findings with such factors as age, sex, diet, and urban environment.

REFERENCES

1. C. Vanoeteren, R. Cornelis, and E. Sabbioni, *Critical evaluation of normal levels of major and trace elements in human lung tissues*, Report EUR 10440 EN, Commission of the European Communities, Luxembourg, 1985.
2. F. Brossa, B. Looman, R. Pietra, and E. Sabbioni, *The use of titania and alumina as protective coatings on metallic prosthesis to prevent metal release*, 6th CIMTEC World Congress on High Technology of Ceramics, Milan, June 24–28, 1986.
3. R. Pietra, E. Sabbioni, M. Gallorini and E. Orvini, *J. Radioanal. Nucl. Chem.* **102,** 69 (1986).

4. S. Caroli, E. Coni, A. Alimonti, E. Beccaloni, E. Sabbioni and R. Pietra, *Analysis* **16** Suppl. 9, 10, 75–80 (1988).
5. P. Bartsch, A. Collignon, G. Weber, G. Robaye, J. M. Delbrouck, I. Roelandts, and J. Yujie, *Arch. Environ. Health* **37,** 111 (1982).
6. C. Vanoeteren, R. Cornelis, and J. Versieck, *Sci. Total Environ.* **54,** 217 (1986).

Author Index

Subject Index